Handbook of
Neurochemistry
SECOND EDITION

Volume 8
NEUROCHEMICAL SYSTEMS

Handbook of Neurochemistry

SECOND EDITION

Edited by Abel Lajtha
Center for Neurochemistry, Wards Island, New York

Volume 1 • CHEMICAL AND CELLULAR ARCHITECTURE
Volume 2 • EXPERIMENTAL NEUROCHEMISTRY
Volume 3 • METABOLISM IN THE NERVOUS SYSTEM
Volume 4 • ENZYMES IN THE NERVOUS SYSTEM
Volume 5 • METABOLIC TURNOVER IN THE NERVOUS SYSTEM
Volume 6 • RECEPTORS IN THE NERVOUS SYSTEM
Volume 7 • STRUCTURAL ELEMENTS OF THE NERVOUS SYSTEM
Volume 8 • NEUROCHEMICAL SYSTEMS
Volume 9 • ALTERATIONS OF METABOLITES IN THE NERVOUS SYSTEM
Volume 10 • PATHOLOGICAL NEUROCHEMISTRY

Handbook of
Neurochemistry

SECOND EDITION

Volume 8
NEUROCHEMICAL SYSTEMS

Edited by
Abel Lajtha
Center for Neurochemistry
Wards Island, New York

PLENUM PRESS • NEW YORK AND LONDON

Library of Congress Cataloging in Publication Data

Main entry under title:

Handbook of neurochemistry.

Includes bibliographical references and indexes.
Contents: v. 1. Chemical and cellular architecture—v. 2. Experimental neurochemistry—[etc.]—v. 8. Neurochemical systems—[etc.]
1. Neurochemistry—Handbooks, manuals, etc.—Collected works. 2. Neurochemistry. I. Lajtha, Abel. [DNLM: 1. Neurochemistry. WL 104 H434]
QP356.3.H36 1982   612'.814   82-493

ISBN 978-1-4684-7020-8   ISBN 978-1-4684-7018-5 (eBook)
DOI 10.1007/978-1-4684-7018-5

©1985 Plenum Press, New York
Softcover reprint of the hardcover 2nd edition 1985
A Division of Plenum Publishing Corporation
233 Spring Street, New York, N.Y. 10013

All rights reserved

No part of this book may be reproduced, stored in a retrieval system, or transmitted in any form or by any means, electronic, mechanical, photocopying, microfilming, recording, or otherwise, without written permission from the Publisher

Contributors

Bernard W. Agranoff, Neuroscience Laboratory Building and Mental Health Research Institute, University of Michigan, Ann Arbor, Michigan 48109

S. Arch, Biological Laboratories, Reed College, Portland, Oregon 97202

Nicolas G. Bazan, Lions Eye Research Laboratories, LSU Eye Center, Louisiana State University Medical Center School of Medicine, New Orleans, Louisiana 70112

M. Billingsley, Section on Biochemical Pharmacology, National Heart, Lung, and Blood Institute, National Institutes of Health, Bethesda, Maryland 20205

Laszlo Z. Bito, College of Physicians and Surgeons, Columbia University, New York, New York 10032

Carla Perrone Capano, Institute of General Physiology, Faculty of Sciences, University of Naples, and International Institute of Genetics and Biophysics, Naples, Italy

Priscilla S. Dannies, Department of Pharmacology, Yale University School of Medicine, New Haven, Connecticut 06510

Philippa M. Edwards, Division of Molecular Neurobiology, Institute of Molecular Biology, Laboratory of Physiological Chemistry and Rudolf Magnus Institute for Pharmacology, State University of Utrecht, 3508 TB Utrecht, The Netherlands

H. Gainer, Laboratory of Developmental Neurobiology, National Institutes of Health, Bethesda, Maryland 20205

Candace J. Gibson, Department of Pathology, University of Western Ontario, London, Ontario N6A 5C1, Canada

Willem Hendrik Gispen, Division of Molecular Neurobiology, Rudolf Magnus Institute for Pharmacology, and Institute of Molecular Biology, State University of Utrecht, 3584 CH Utrecht, The Netherlands

Antonio Giuditta, Institute of General Physiology, Faculty of Sciences, University of Naples, and International Institute of Genetics and Biophysics, Naples, Italy

László Gráf, The Institute for Drug Research, 1325 Budapest, Hungary

Roger Guillemin, Laboratories for Neuroendocrinology, The Salk Institute, La Jolla, California 92037

I. Hanbauer, Section on Biochemical Pharmacology, National Heart, Lung, and Blood Institute, National Institutes of Health, Bethesda, Maryland 20205

L. Hertz, Department of Pharmacology, University of Saskatchewan, Saskatoon, Saskatchewan S7N 0W0, Canada

François B. Jolicoeur, Department of Psychiatry, Faculty of Medicine, University of Sherbrooke, Sherbrooke, Quebec J1H 5N4, Canada

B.H.J. Juurlink, Department of Anatomy, University of Saskatchewan, Saskatoon, Saskatchewan S7N 0W0, Canada

E. Ronald de Kloet, Rudolf Magnus Institute for Pharmacology, Medical Faculty, University of Utrecht, 3521 GD Utrecht, The Netherlands

D. Kuhn, Section on Biochemical Pharmacology, National Heart, Lung, and Blood Institute, National Institutes of Health, Bethesda, Maryland 20205

Allen S. Levine, Neuroendocrine Research Laboratory, Minneapolis Veterans Administration Medical Center, Minneapolis, Minnesota 55417; and Departments of Medicine and Food Science and Nutrition, University of Minnesota, Minneapolis–St. Paul, Minnesota 55455

Bruce S. McEwen, The Rockefeller University, New York, New York 10021

Henry McIlwain, Department of Biochemistry, St. Thomas's Hospital Medical School, London SE1 7EH, England

John E. Morley, Neuroendocrine Research Laboratory, Minneapolis Veterans Administration Medical Center, Minneapolis, Minnesota 55417; and Departments of Medicine and Food Science and Nutrition, University of Minnesota, Minneapolis–St. Paul, Minnesota 55455

Jacques Nunez, Unité de Recherche 35 de l'INSERM, and Equipe de Recherche sur la Biochimie de la Régulation Hormonale, CNRS, Hospital H. Mondor, 94 Creteil, France

Chandan Prasad, Departments of Medicine (Section of Endocrinology) and Biochemistry, Louisiana State University Medical Center, New Orleans, Louisiana 70112

Thomas C. Rainbow, Department of Pharmacology, University of Pennsylvania, Philadelphia, Pennsylvania 19104

T. Sanjeeva Reddy, Lions Eye Research Laboratories, LSU Eye Center, Louisiana State University Medical Center School of Medicine, New Orleans, Louisiana 70112

Francis Rioux, Department of Physiology and Pharmacology, Faculty of Medicine, University of Sherbrooke, Sherbrooke, Quebec J1H 5N4, Canada

Serge St.-Pierre, Department of Physiology and Pharmacology, Faculty of Medicine, University of Sherbrooke, Sherbrooke, Quebec J1H 5N4, Canada

Peter Schotman, Division of Molecular Neurobiology, Institute of Molecular Biology, Laboratory of Physiological Chemistry and Rudolf Magnus Institute for Pharmacology, State University of Utrecht, 3508 TB Utrecht, The Netherlands

Loes H. Schrama, Division of Molecular Neurobiology, Institute of Molecular Biology, Laboratory of Physiological Chemistry and Rudolf Magnus Institute for Pharmacology, State University of Utrecht, 3508 TB Utrecht, The Netherlands

Hitoshi Shichi, Institute of Biological Sciences, Oakland University, Rochester, Michigan 48063

S. Szuchet, Department of Neurology, University of Chicago, Chicago, Illinois 60637

Gyula Telegdy, Institute of Pathophysiology, University Medical School, Szeged, Hungary

Yasuzo Tsukada, Department of Physiology, Keio University School of Medicine, Tokyo, Japan

H. Dick Veldhuis, Rudolf Magnus Institute for Pharmacology, Medical Faculty, University of Utrecht, 3521 GD Utrecht, The Netherlands

Tony L. Yaksh, Department of Neurosurgical Research, Mayo Clinic, Rochester, Minnesota 55905

Gigliola Grassi Zucconi, Institute of Cell Biology, Faculty of Sciences, University of Perugia, Perugia, Italy

Henk Zwiers, Division of Molecular Neurobiology, Rudolf Magnus Institute for Pharmacology, and Institute of Molecular Biology, State University of Utrecht, 3584 CH Utrecht, The Netherlands

Preface

The content of Volume 8 of the *Handbook of Neurochemistry* is a perfect example and sample of what occupies neurochemists in the late 1980s. What occupies them are questions, concepts, and technology that either did not start with the nervous system, or rapidly moved out of its exclusivity (see, for instance, chapters on neurotensin, beta-lipotropin, behavioral and neurochemical effects of ACTH, cholecystokinin, etc.). Thus, the neurochemist is more and more seen as a biochemist occupied by questions, concepts, and technology that are not unique to the nervous system, even though the ultimate substrate of these questions, as well as the ultimate functions so studied and occasionally explained, are of the nervous system.

Look at the case of the hypothalamic hypophysiotropic peptides, also called hypothalamic releasing factors, or hypothalamic releasing hormones. These are all small-to-medium-size polypeptides originally characterized in extracts of the hypothalamus on the basis of bioassays directed at studying their effects on one or another of the secretions of the adenohypophysis. We know now that TRFs (See Chapter 8), the thyrotropin and prolactin releasing factor, somatostatin, the hypothalamic inhibitor of the secretion of growth hormone, as well as LRF, the hypothalamic decapeptide stimulating the secretion of pituitary gonadotropins, are to be found in parts of the brain other than the hypothalamus, where their function is obviously not hypophysiotropic. Such is also the case for CRF—the corticotropin and beta-endorphin releasing factor. Our original statement about the uniquely hypothalamic location of the growth hormone releasing factor—GRF, somatocrinin, has recently been challenged by colleagues using polyclonal antisera different from those used by us, and they may well be right.

Furthermore, we know that these peptides, first of the hypothalamus, then of the brain, are also found in the spinal cord, peripheral nerves, and in various tissues of the gastrointestinal tract including the pancreas, and, in some forms with relatively minor structural variation, the skin of amphibians and even in various anatomical structures of invertebrates.

Similarly, all the biologically active peptides originally characterized in the gastrointestinal tract, the lungs, the kidney, and the teguments (of amphibians) have been recognized as such or with minor structural variations in the brain, the spinal cord, and the peripheral nervous system of mammalians.

There is excellent evidence that in all locations we are dealing with the same polypeptide sequence or molecule.

In one location, these peptides will be called hormones, and will be called neuromediators or neuromodulators in other locations. I have already alluded to the fact that the hypothalamic hypophysiotropic peptides are also called hypothalamic hormones. We obviously have a problem of terminology. Should we ignore it?

The word hormone was originally introduced into the literature by Ernest Henry Starling on June 20, 1905, in the first of his Croonian Lectures delivered before the Royal College of Physicians in London on "The Chemical Correlation of the Functions of the Body". Starling frames the thinking within which the definition of hormones will appear: "The chemical adaptation or adaptations of the body, like those which are carried out through the intermediation of the central nervous system, can be divided into two main classes: 1) those which are involved in consequence of changes impressed upon the organism as a whole from without; and 2) those which, acting entirely within the body, serve to correlate the activities, in the widest sense of the term, of the different parts and organs of the body." Then, after some discussion of the first category, he moves on to the second and says the following: "These chemical messengers, however, or hormones (from *ormao*, I excite, or arouse) as we might call them, have to be carried from the organ where they are produced to the organ which they affect, by means of the bloodstream, and the continually recurring physiological needs of the organism must determine their repeated production and circulation through the body".

Most of the substances which we call hormones to this day do meet these criteria; this is the case for the secretory products of all the endocrine glands. These circulate far and wide from their organ of production to the receptors of their target organs, where they somehow excite, stimulate, or cause some positive effect in the cells of that tissue. Physiologically meaningful levels of these hormones can now be measured by all sorts of exquisitely sensitive methods in the peripheral blood with the added feature that there is always a demonstrable arterial/venous difference in the concentration of these substances when measured in the inflow or outflow blood to and from the organ known to be the source of the hormone in question.

In the case of the hypothalamic peptides involved in the control of pituitary functions, there is reliable evidence that they can be demonstrated by bioassay or radioimmunoassay in the effluent blood from the hypothalamus when tapped in the portal vessels along the pituitary stalk. There is also no doubt that there is a difference in the concentrations found in the hypothalamic portal blood when compared to peripheral blood. This, however, is where the problems begin in calling these substances hormones. Reliable methodology shows that the peripheral levels of circulating thyrotropin releasing factor (TRF) or gonadotropin releasing factor (LRF) are so low as to be of no physiological significance. In the case of somatostatin, the matter is even more complex. First, somatostatin is universally an inhibitor (of one secretion or another), and thus can hardly be called a "hormone", a name that etymologically implies "stimulation, excitation". But perhaps more important, it is now well recognized that somatostatin has a ubiquitous distribution (though not random) ranging from the central nervous system to multiple locations of the gastrointestinal tract and the pancreas as we discussed above. Immunoreactive and bioactive somatostatin, in fact forms of somatostatin of various molecular sizes, can be demonstrated to circulate in peripheral blood (jugular vein in laboratory animals, antecubital vein in man), but in concentrations that appear to be far below what can be calculated to be its binding or affinity constants. There is, however, excellent evidence that much larger concentrations of immunoreactive and bioactive somatostatin can be shown in more localized circulation such as in the effluent vein of the pancreas, where we know that somatostatin is present in the delta cells. There is also good evidence that these local plasma concentrations of local somatostatin can vary considerably as a function of physiological or experimental situations (absorption of meals, injection of various peptides, such as cholecystokinin, or endorphins, or drugs such as opiates, or arginine). If TRF, LRF, and somatostatin are to be called hormones and considered as such, then it must be said that they do not qualify as the classical hormones.

But things are even more complex. We know now that immunoreactive, as well as bioactive, TRF, LRF, somatostatin, the "gut hormones" cholecystokinin (see Chapter 5), VIP (vasoactive intestinal peptide), gastrin, etc., are found within discrete neurons, either in the cell body or in peripheral endings from which they certainly have to be released for physiological purposes which are not well understood at the moment. In such circumstances and setups, neither TRF, LRF, nor somatostatin or any of the other "gut hormones" behaves as, or meets the criteria of, hormones. They seem to be involved in localized controls. It is probably also the case when trying to understand how pancreatic somatostatin could modify the secretion of insulin and glucagon by the nearby cells of the islets.

Because of their neuronal locations, TRF, LRF, somatostatin (and perhaps the other biologically active peptides such as neurotensin, endorphins, enkephalins, VIP, etc.) have been proposed as neurotransmitters, as are catecholamines or acetylcholine. But somatostatin is not a neurotransmitter when released by the delta cells of the endocrine pancreas to affect the glucagon secretion by a nearby alpha cell, reaching either through gap junctions or extracellular space. Noradrenalin in and out of neurons is the neurotransmitter, while adrenalin is the hormonal form in and from the adrenal medulla, with the small amounts of noradrenalin found in the adrenal medulla leaving us in a quandary.

It is thus obvious that the current terminology is wanting. Either we have to redefine what it is that we mean by hormone or some additional terminology has to be proposed.

The question is what to do with these ubiquitous molecules which have local effects that can range from angstroms to microns (gap junctions, extracellular spaces) in and from cells which include neurons, to centimeters, when dealing with local either splanchnic or pituitary locations. Such substances do not fit in well with the definition of a hormone or of a neurotransmitter. We have the word *paracrine*, as originally proposed by Feyrter to describe precisely the suspected secretory activities of what we now know to be the peptide-secreting cells of the gut. The etymology of the word is obvious and implies a local or nearby use or function for what is being secreted. I personally think that the word paracrine is excellent and should be used often in relation to the problem we are discussing here. But paracrine is an adjective, and to my knowledge Feyrter, in his difficult German, used it exclusively as such, referring to paracrine secretory cells and paracrine secretion. We could perhaps coin the word "parahormone" or "parhormone", but neither is euphonic or easy to pronounce in either French or English, or German for that matter. Several years ago I proposed the word *cybernin*, from the Greek "kurbenetes", meaning "pilot" or "rudder" of a boat, implying the local nature of the command or information involved. This is also the root of the well-known word cybernetics, even though I could never ascertain whether Norbert Wiener implied any localization (of information) when he decided to use the word (according to Littré, the word "cybernétique" was coined by Ampère to define "la partie politique qui s'occupe des moyens de gouverner"). I never pushed very hard for the implantation of the word cybernin, feeling that it was another word, another root, without a clearly defined mission. The word, however, is being used by more and more people, thus appearing to fulfill a role. So, how should we use it? First of all, what is a cybernin? A cybernin is a polypeptide biosynthesized, processed, and released by a cell or group of cells, that represents information that will affect the function of another cell or group of cells in the vicinity of the first cell or group of cells. Such a simple definition excludes all steroids, prostaglandins, or molecules such as cyclic-AMP. Would beta-endorphin, ACTH, which certainly are hormones when secreted by the pituitary, be considered as cybernins when relating to their presence in hypothalamic neurons and when released either at nerve endings from these neurons (a statement which is only a proposal at the moment, albeit a logical one) or when released in the portal vessels of the pituitary of the median eminence and as measured in the down-flow blood along the pituitary stalk? The answer is yes. I would say that, in these circumstances, beta-endorphin and ACTH are seen, act, and should be considered, as cybernins. When we describe their action it will be referred to as a cybernin action rather than a hormonal action. Somatostatin will act as a cybernin when it modifies the secretion of insulin and glucagon in nearby pancreatic cells and originates from pancreatic delta cells; it will also be acting as a cybernin when proceeding to the adenohypophysis, inhibiting the secretion of growth hormone, when originating from the hypothalamus. I would prefer to consider somatostatin as a cybernin rather than as a neurotransmitter if it can be shown that it is actually released at some axonal or dendritic ending and that it modifies the response of another neuron to any one of the classical neurotransmitters. In fact, the word cybernin may turn out to be the optimal noun for the adjective paracrine. The polypeptidic growth factors recently proposed by Sporn and Todaro as "autocrine" secretions in the ultimate of paracrine function could also be considered as cybernins.

The success of any nomenclature is based on need. The need for the use of the word cybernin is probably not compelling. However, since the words hormone, neuromediator, mediator, modulator, are either too restrictive or too vague in their definition or implied use, maybe there will be some feeling of comfort in the use of the word cybernin.

I have discussed above the ubiquity of these biologically active peptides. There is also evidence, though not so widely established, that the "receptor" molecules for these peptides in their ubiquitous locations are identical or share extensive common binding properties for related ligands. Thus what is specific (for the nervous system) if are neither the ligands, nor the receptors? There is evidence that not even the post- receptor-binding *type* of response is. What *may be* specific is the final result of the activation of a highly specialized neuron. What *is* specific finally is the functional response as seen in a system of neurons. And while (neuro)chemistry is necessary to understand the function of each cellular unit of that system, (neuro)chemistry does not tell us what will be the function of that system or how that function will take place. This is where the power, the immense power of reductionism as we know it currently comes to a halt, a pause, possibly a standstill—an "impossibility to proceed owing to exhaustion", says the Oxford dictionary. Rather than accepting this latest meaning of the word I'd rather use it as implying standing still, as in wait of some new paradigm, a true revolution in science, according to the terminology of Thomas Kühne. How else could we engage in the neurochemical study of sleep while at the same time attempting to characterize sleep peptides?

Perhaps one of those new ways will be found in the emerging studies of what I will simply call nonlinear dynamics to refer to the mode of thinking or the concept that complexity when it reaches such a height that it is seen as chaos or apparent randomness can still generate its own periodic order. Physicists, chemists, and mathematicians like Haken, Prigogine, and Thom have all proposed mathematics to support this vastly novel way of thinking for the biologist and have already shown the powerful predictive value of such mathematics in "explaining" the emergence of complex biological structures in embryogenesis, complex periodic functions in multicompartmental biological systems, etc. Psychiatrists like Mandell have already proposed that such thinking could lead to new approaches to understand normal and abnormal patterns of behaviour, in some ways defining normal brain functions and behaviour from the absence of the characteristics of mental disorders.

And this is where so much of the subject matter of this volume comes back in focus: Many of the molecules discussed here have already been shown to affect the amplitude or frequency of one or another of these nonlinear dynamics events as observed in the central nervous system. Perhaps here will be found a new generation of still reductionist questions, unless some genial holist takes us away from modern obscurantism in its glorious achievements.

Roger Guillemin

Contents

Chapter 1

Thyroid Hormones
Jacques Nunez

1. Introduction 1
2. Thyroid Function and the Development of Neuroendocrine Control 2
   2.1. Thyroid Hormone Synthesis 3
   2.2. Thyroid Hormone Secretion 4
   2.3. Hormonal Control of Thyroid Function 5
   2.4. Ontogenesis of the Hypothalamic–Pituitary–Thyroid Axis 6
3. Thyroid Hormone Effects on Cell Acquisition and Neuronal Differentiation 8
   3.1. Cell Division and Cell Acquisition 9
   3.2. Cell Migration and the Development of the Neuropil 10
   3.3. Synaptogenesis and Development of Nerve Cell Transmitters 12
   3.4. Myelination 12
4. Uptake, Metabolism, and Mechanism of Action of Thyroid Hormones 14
   4.1. Transport and Uptake of Thyroid Hormones by the CNS 15
   4.2. The Concept of "Active Form" and he Peripheral Conversion of Thyroxine to Triiodothyronine 16
   4.3. Thyroid Hormone Binding Proteins and Receptors 17
   4.4. Thyroid Hormones and Protein Synthesis 19
   4.5. The Role of the Cytoskeleton during Brain Differentiation 19
5. Neonatal Thyroid Screening and Replacement Therapy 22
6. Conclusions 22
   References 23

Chapter 2

Mechanisms of Gonadal Steroid Actions on Behavior
Thomas C. Rainbow and Bruce S. McEwen

1. Introduction 29

2. Activational Effects of Gonadal Steroids on Behavior 30
   2.1. Genomic Mechanisms of Steroid-Activated Behaviors 32
   2.2. Target Cells for the Activation of Feminine Sexual
        Behavior 34
   2.3. Identification of Proteins 35
3. Organizational Actions of Gonadal Steroids on Behavior 40
4. Organizational and Activational Effects of Gonadal Steroids and Their
   Relevance for the Study of Memory Formation 43
References 44

Chapter 3

Adrenocortical Hormone Action
  E. Ronald de Kloet and H. Dick Veldhuis

1. Introduction 47
2. Function of Adrenocortical Hormones 48
3. Steroid–Cell Interaction 49
   3.1. Cellular Uptake 49
   3.2. Intracellular Biotransformation of Steroids 51
   3.3. Activation of the Receptor and Transformation to Its DNA-Binding State 51
   3.4. Translocation of the Steroid–Receptor Complex and Initiation of
        Genomic Effect 52
   3.5. Steroid–Nerve Cell Interaction 53
4. Adrenal Steroid Receptors in Brain 54
   4.1. Cellular Localization of Corticosterone 54
   4.2. Cellular Localization of Mineralocorticoids and
        Dexamethasone 54
   4.3. Competition for Cell Nuclear Localization of
        Corticosterone 58
   4.4. Cell Nuclear Uptake *in Vitro* in Tissue Slices 58
   4.5. Binding of Adrenal Steroids to Soluble Receptors 59
   4.6. Heterogeneity in Adrenal Steroid Receptor Systems 60
   4.7. Fractionation of Adrenal Steroid Receptor Sites 62
   4.8. Corticoid Receptors and CBG-like Proteins 63
5. Regulation of the Adrenal Steroid Receptor System 64
   5.1. Neurotropic Substances 64
   5.2. Compensatory Changes in Number of Steroid Receptor
        Sites 66
   5.3. Ontogeny and Aging 67
6. Effects of Adrenal Steroids on Brain Chemistry 68
   6.1. Neurotransmitters 69
   6.2. Neuropeptides 73
   6.3. Proteins 73
   6.4. Enzymes 74

6.5. Morphology	74
6.6. Other Actions	75
7. Control of Brain Function by Adrenal Steroids	75
7.1. Neuroendocrinology	76
7.2. Behavior	78
8. Concluding Remarks	81
References	82

Chapter 4

Neurotensin
François B. Jolicoeur, Francis Rioux, and Serge St.-Pierre

1. Introduction	93
2. Distribution of Neurotensin in the Central Nervous System	93
3. Metabolism of Neurotensin in the Central Nervous System	95
4. Electrophysiological Studies	96
5. Neurobehavioral Effects of Central Administration of Neurotensin	96
5.1. Body Temperature	97
5.2. Muscular Tone	98
5.3. Motor Activity	98
5.4. Reactivity to Noxious Stimuli	99
5.5. Other Central Effects	101
6. Structure–Activity Studies	102
7. Interactions of Neurotensin with Dopamine and Other Neurotransmitters	104
7.1. Dopamine	105
7.2. Neuroleptic Properties of Neurotensin	106
7.3. Interactions with Other Neurotransmitters	108
8. Summary	109
References	110

Chapter 5

Cholecystokinin
John E. Morley and Allen S. Levine

1. Introduction	115
2. Central Nervous System Distribution	116
3. Cholecystokinin-Converting Enzymes	118
4. Cholecystokinin Receptors in the Brain	119
5. Direct Effect of Cholecystokinin on Neurons	120
6. Neuropharmacology of Cholecystokinin	120
6.1. Hyperglycemia	121
6.2. Hypothermia	121

6.3. Analgesia 121
6.4. Rotational Syndrome 122
6.5. Central Nervous System Depression and Exploratory Behaviors 122
7. Effects of Cholecystokinin on Pituitary Hormone Release 123
8. Cholecystokinin and Satiety 124
9. Monoamines and Cholecystokinin 127
10. Conclusion 129
References 130

Chapter 6

β-Lipotropin
László Gráf

1. Introduction 137
2. Structure 138
   2.1. Amino Acid Sequence 138
   2.2. Conformation 142
3. Biological Properties 142
4. Biosynthesis in the Pituitary 145
   4.1. Pulse–Chase Studies: The Pathway of the Biosynthesis 145
   4.2. Nucleotide Sequencing of Cloned cDNA for the Precursor and Protein Sequencing of the Biosynthetic Intermediates 147
   4.3. Proteinases Involved in the Processing of the Precursor and in the Release of the Final Products 150
5. Occurrence in the Pituitary, Periphery, and Brain 151
6. Physiological Significance 153
References 154

Chapter 7

Prolactin
Priscilla S. Dannies

1. Introduction 159
2. Synthesis and Processing of Prolactin 159
3. Regulation of Prolactin Production 161
   3.1. Agents That Regulate Prolactin Production by Acting Directly at the Pituitary Gland 161
   3.2. Regulation at Levels Other Than the Pituitary Gland 165
4. Physiological Patterns of Prolactin Secretion 165
   4.1. Prolactin Secretion in Man 165
   4.2. Prolactin Secretion in Rats 166
   4.3. Factors That Regulate Physiological Changes in Prolactin Secretion 167

5. Intracellular Mechanisms by Which Prolactin Production Is
   Controlled 168
   5.1. Release 168
   5.2. Synthesis 169
6. Effects of Prolactin 169
   6.1. Prolactin Receptors 170
   6.2. Effects of Prolactin 171
7. Pathology 172
8. Conclusion 172
   References 172

Chapter 8

Thyrotropin-Releasing Hormone
  Chandan Prasad

1. Introduction 175
2. Distribution and Modulation of Endogenous TRH
   Concentration 175
   2.1. In Central Nervous System 175
   2.2. In Gastrointestinal Tract and Pancreas 177
   2.3. In Body Fluids and Other Tissues 178
3. Metabolism of TRH 179
   3.1. Biochemical Pathways 179
   3.2. Properties of the Enzymes 181
   3.3. Regulation of TRH Metabolism 182
   3.4. Possible Physiological Significance of TRH Metabolism . 182
4. Multiple Biological Effects of TRH 184
   4.1. Behavioral Effects 184
   4.2. Endocrine Effects 185
   4.3. Cardiovascular Effects 187
   4.4. Gastrointestinal Effects 187
   4.5. Respiratory and Other Effects 188
5. Thyrotropin-Releasing Hormone as a Possible Neurotransmitter or
   Neuromodulator 188
   5.1. Uneven Distribution in Mammalian Brain and Spinal Cord . 188
   5.2. Synaptosomal Localization 190
   5.3. Uptake and Release 190
   5.4. The TRH Receptor 191
   5.5. Action of TRH on Neurons 192
   5.6. Effect on Putative Neurotransmitter Metabolism 193
6. Discussion 193
   References 194

Chapter 9

Role of Calmodulin in the Regulation of Neuronal Function
M. Billingsley, I. Hanbauer, and D. Kuhn

1. Introduction	201
2. Neurotransmitter Synthetic Enzymes	201
2.1. Tryptophan Hydroxylase	202
2.2. Tyrosine Hydroxylase	205
3. Calcium- and Calmodulin-Stimulated Phosphorylation of Membrane Proteins in Nervous Tissue	206
3.1. Presynaptic Protein Phosphorylation	207
3.2. Calmodulin-Stimulated Phosphorylation in Postsynaptic Densities	208
3.3. Phosphorylation of Myelin	209
3.4. Drug Effects on Ca^{2+}·CaM-Dependent Phosphorylation of Membrane Proteins	209
4. Postsynaptic Receptor Regulation	210
4.1. Regulatory Proteins	210
4.2. Desensitized and Supersensitive Dopamine Receptors	211
References	213

Chapter 10

Effects of Gastrointestinal Peptides on the Nervous System
Gyula Telegdy

1. Introduction	217
2. Secretin	218
2.1. Distribution	220
2.2. Effect on Brain Function	220
2.3. Effect on Endocrine System	220
2.4. Conclusion	220
3. Vasoactive Intestinal Polypeptide	220
3.1. Distribution	221
3.2. Peripheral Action	221
3.3. Central Effects	222
3.4. Effect on Endocrine System	222
3.5. Conclusion	223
4. Gastrin	223
4.1. Peripheral Effects	225
4.2. Central Effects	225
4.3. Endocrine Effects	226
4.4. Conclusion	226
5. Motilin	226
5.1. Distribution in the Nervous System	226
5.2. Peripheral Effects	227
5.3. Central Action	227

5.4. Endocrine Action	227
5.5. Conclusion	227
6. Bombesin	227
6.1. Distribution	228
6.2. Peripheral Effects	228
6.3. Central Effects	228
6.4. Endocrine Effects	228
6.5. Conclusion	229
7. Angiotensin	229
7.1. Distribution	230
7.2. Peripheral Action	231
7.3. Central Action	231
7.4. Conclusion	232
8. Pancreatic Polypeptide	232
8.1. Distribution	232
8.2. Peripheral Effect	234
8.3. Central Effects	234
8.4. Conclusion	235
9. Insulin	235
9.1. Distribution	235
9.2. Conclusion	236
References	236

Chapter 11

Peptidergic Systems
Peter Schotman, Loes H. Schrama, and Philippa M. Edwards

1. Introduction	243
2. Anatomic Considerations Relevant to the Role of Peptides as Neurohormones	245
3. Localization of Peptidergic Systems	247
3.1. General Considerations	247
3.2. Peptides of Pituitary Origin	247
3.3. Peptides of Hypothalamic Origin	249
3.4. Peptides Originating in Extrahypothalamic Brain and of Uncertain Origin	255
3.5. Peptides in Spinal Cord and Peripheral Nerves	257
4. Peptides as Neuroregulators	258
5. Interactions between Peptidergic and Aminergic Systems	260
5.1. Dopaminergic Systems	260
5.2. Noradrenergic Systems	262
5.3. Serotonergic Systems	263
6. Interaction between Peptidergic and Nonmonoaminergic Systems	263
6.1. Cholinergic Systems	263

6.2. Amino Acid Systems 264
6.3. Peptide–Peptide Interactions 264
7. Coexistence of Peptides and Other Agents 265
8. Biochemical Effects 266
   8.1. Receptors and Second Messengers 266
   8.2. Trophic Functions of Peptides 270
   8.3. RNA and Protein Metabolism 271
9. Final Comments 274
   References 275

Chapter 12

Neurosecretion
   S. Arch and H. Gainer

1. Introduction 281
2. Morphology 283
3. Electrophysiology 286
4. Synthesis of Neurosecretory Products 288
5. Axonal Transport 297
6. Secretion 298
7. Concluding Remarks 302
   References 303

Chapter 13

Control of Monoamine Synthesis by Precursor Availability
   Candace J. Gibson

1. Introduction 309
2. Plasma Amino Acid Composition and Brain Precursor
   Concentration 309
   2.1. Sources of Circulating Tryptophan and Tyrosine 309
   2.2. Brain Uptake of Tryptophan and Tyrosine 312
   2.3. Hormonal Influences on Plasma Amino Acid Pattern 313
3. Monoamine Biosynthetic Pathways 314
   3.1. Tryptophan Availability and Serotonin Synthesis 315
   3.2. Tyrosine Availability and Catecholamine Synthesis 316
   3.3. Requirements for Precursor Control of Monoamine
        Synthesis 319
4. Consequences of Precursor Control of Monoamine Synthesis .. 319
5. Future Research 321
   References 322

Chapter 14

Cerebral Subsystems and Isolated Tissues
  Henry McIlwain

1. Introduction 325
2. Preparative Methods 326
   2.1. Obtaining the Isolate Minimally Altered 327
   2.2. Choice and Criteria of Preparative Methods 327
   2.3. Alternative Methods 328
   2.4. Incubation and Superfusion Techniques 329
3. Types of Observations Feasible 330
   3.1. Metabolic Responses to Applied Agents 330
   3.2. Neurotransmitter Synthesis, Metabolism, and Output 330
   3.3. Electrical Responses to Stimulating Agents 331
   3.4. Histological Examination of Isolates 331
4. Neocortex 332
5. Piriform Lobe: Olfactory Cortex and Lateral Olfactory Tract .. 332
6. Hippocampus 333
   6.1. Dentate Gyrus 334
   6.2. The CA1 Region 334
   6.3. The CA3 Region 336
   6.4. Long-Lasting Potentiation of Synaptic Transmission in Hippocampal Preparations 336
7. Optic Tract, Lateral Geniculate, and Superior Colliculus 337
8. Other Systems 338
9. Outlook 339
   References 339

Chapter 15

Memory
  Bernard W. Agranoff

1. Introduction 343
   1.1. The Role of Macromolecular Synthesis in Behavior—Interventive and Correlative Approaches 343
   1.2. Further Progress in the Role of Macromolecular Synthesis in Mediation of LTM Formation 344
2. Recent Progress in Biochemical Approaches to Behavior 345
   2.1. A Phyletic Sampling 345
   2.2. Reduced Systems 346
   2.3. Imprinting 348
   2.4. Kindling 348
   2.5. Interventive Agents 349
   2.6. Correlative Changes 350
   2.7. Neurochemical Studies on Memory in Humans 351
   References 353

Chapter 16

Neurochemical Correlates of Learning Impairment
Yasuzo Tsukada

1. Introduction 357
2. Strategies for Neurochemical Studies 358
   2.1. Correlative Studies 358
   2.2. Changes of Learned Behavior Caused by Administration of Metabolic Inhibitors or Experimental Brain Damage 359
   2.3. Isolation and Identification of Memory Substances 363
   2.4. Neurochemical Correlates of Learning Impairment in Animals with Maldeveloped Brain 364
3. Conclusion 370
   References 371

Chapter 17

Behavioral and Neurochemical Effects of ACTH
Willem Hendrick Gispen and Henk Zwiers

1. Introduction 375
2. ACTH 375
   2.1. ACTH: Peptide Hormone and Neuropeptide 375
   2.2. ACTH as a Messenger 376
3. ACTH and Behavior in Experimental Animals and Man 378
   3.1. Conditioned Avoidance Behavior 378
   3.2. The Stretching and Yawning Syndrome 381
   3.3. Excessive Grooming 381
   3.4. Sexual Behavior 388
   3.5. Social Behavior 389
   3.6. Behavioral and Clinical Studies in the Human 390
   3.7. Aging 391
4. Neurochemical Mechanism of Action of ACTH 392
   4.1. Neurotransmitters 393
   4.2. Cyclic AMP 394
   4.3. Protein Phosphorylation 395
   4.4. Lipid Phosphorylation 399
   4.5. Synaptic Plasma Membrane Fluidity 401
5. Concluding Remarks 402
   References 403

Chapter 18

Pain Transmission
Tony L. Yaksh

1. Pain-Transmitting System 413
   1.1. Peripheral Afferents 413

1.2. Ascending Pathways by Which High-Intensity Stimulation Gains Access to Brain Centers	423
1.3. Supraspinal Systems in Pain Transmission	425
1.4. Summary of the Rostral Projection System	430
2. Modulatory Systems That Control the Processing of Sensory-Evoked Activity	432
2.1. Spinal Cord	432
2.2. Descending Modulation	433
References	436

Chapter 19

The Neurochemical Study of Sleep
   Antonio Giuditta, Carla Perrone Capano, and Gigliola Grassi Zucconi

1. Introduction	443
2. A Brief Outline of the Biology of Sleep	444
3. Hypotheses on the Functions of Sleep	447
4. Possible Approaches in the Neurochemical Study of Sleep	448
4.1. Deprivation Studies	448
4.2. Studies during Sleep	449
5. Sleep Neurochemistry	450
5.1. Energy and Intermediary Metabolism	450
5.2. Amino Acids and Proteins	458
5.3. Nucleic Acids	464
6. Sleep-Inducing Factors	467
7. Conclusion	470
References	473

Chapter 20

Composition of Intraocular Fluids and the Microenvironment of the Retina
   Laszlo Z. Bito

1. Introduction	477
2. Relevant Literature and the Scope of This Chapter	478
3. The Intraocular Fluid Compartments	478
4. Definitions and Methodology	479
5. Concentration Gradients within the Intraocular Fluid Compartments	481
6. The Transport Function of the Blood–Ocular Barrier Systems	482
7. Species Differences and the Reliability of Information on Human Intraocular Fluids	483
8. The Concentration of Major Solutes in Mammalian Intraocular Fluids and in the Retinal Microenvironment	483
8.1. Sodium, Chloride, and Bicarbonate	483
8.2. Potassium	486

8.3. Calcium and Magnesium	486
8.4. Inorganic Phosphate	489
8.5. Glucose and Lactic Acid	489
8.6. Ascorbic Acid	492
8.7. Amino Acids	494
8.8. Osmolality	496
8.9. Soluble Proteins and Enzymes	497
9. The Homeostasis of the Microenvironment of the Mammalian Retina	499
10. The Microenvironment of the Avian Retina and Its Homeostasis	499
11. Recommendations for the Composition of Artificial Physiological Solutions for Retinal Research	500
12. Conclusions	502
References	503

Chapter 21

Retina
Nicolas G. Bazan and T. Sanjeeva Reddy

1. Introduction	507
2. Cellular Organization	509
3. Photoreceptor Cells	512
4. Chemical Composition and Metabolism	513
4.1. Nucleic Acids	514
4.2. Proteins	514
4.3. Carbohydrates	516
4.4. Lipids	517
5. Neurotransmitter Metabolism	532
5.1. Acetylcholine	535
5.2. GABA	537
5.3. Other Amino Acid Neurotransmitters	538
5.4. Dopamine	538
5.5. Serotonin	539
5.6. Neuropeptides	539
6. Cyclic Nucleotide Metabolism	541
6.1. Cyclic Nucleotides	541
6.2. Guanylate and Adenylate Cyclases	541
6.3. Dopamine-Sensitive Adenylate Cyclase	541
6.4. Phosphodiesterase	543
7. Retinoid Binding Proteins in Retina	543
8. Renewal of Visual Cells	545
9. Effect of Light on Retina	546
10. Axoplasmic Transport	547
11. Nutritional Studies	548
11.1. Essential Fatty Acid Deficiency	548

11.2.	Vitamin E Deficiency	548
11.3.	Vitamin A Deficiency	549
11.4.	Taurine Deficiency	550
11.5.	Zinc Deficiency	550
12.	Effect of Ischemia	550
13.	Diabetic Retinopathy	551
14.	Retinal Degeneration	551
14.1.	Carbohydrate Metabolism	552
14.2.	Protein and Nucleic Acid Metabolism	552
14.3.	Lipid Peroxides	553
14.4.	Cyclic Nucleotide Metabolism	554
14.5.	Gyrate Atrophy	554
15.	Retinal Regeneration	555
16.	Conclusions and Perspectives	556
	References	559

Chapter 22

Vision
Hitoshi Shichi

1.	Introduction: Anatomy and Physiology of the Retina	577
2.	Photoreceptors	579
2.1.	Assembly and Breakdown	579
2.2.	Structure of Disk Membranes	581
2.3.	Properties of Disk Membranes	582
3.	Rhodopsin	584
3.1.	Chemical Properties of the Opsin Protein	584
3.2.	Physical Properties of the Opsin Protein	586
3.3.	Chromophore and Photochemistry	587
4.	Visual Transduction	592
5.	Neurotransmitters in the Retina	596
	References	598

Chapter 23

Cell Cultures
L. Hertz, B. H. J. Juurlink, and S. Szuchet

1.	Introduction	603
2.	Principles Used to Obtain Monotypic Cultures	604
2.1.	Cell Lines	604
2.2.	Primary Cultures	606
3.	Astrocytes	613
3.1.	Primary Cultures	613
3.2.	Cell Lines	621

4. Neurons 624
   4.1. Neuronal–Glial Cocultures 624
   4.2. Monotypic Primary Cultures of Neurons 624
   4.3. Cell Identity and Cell Markers 628
   4.4. Neurochemical Use of Primary Cultures of Neurons 631
   4.5. Neuronal Cell Lines 633
5. Oligodendrocytes 634
   5.1. Strategies to Obtain Monotypic Cultures of
        Oligodendrocytes 634
   5.2. Procedures for Cell Isolation 635
   5.3. Establishment of Oligodendrocyte Cultures from Cells Isolated
        by Gradient Centrifugation 639
   5.4. Properties of Cultured Oligodendrocytes 645
   5.5. Cell Lines 650
6. Concluding Remarks 651
   References 652

Index 663

1

Thyroid Hormones

Jacques Nunez

1. INTRODUCTION

The bulk of information accumulated during the past 30 years has confirmed that thyroid hormones exert direct effects on brain organization and function. Most of the brain abnormalities observed as a consequence of early hypothyroidism both in humans and experimental animals are permanent: they are reversed only if early replacement therapy with adequate amounts of thyroid hormones is performed. Thyroid hormones appear to be a major epigenic factor, which allows the innate developmental program to be accomplished.

In a review published in 1960, Eayrs[1] described the major features of thyroid hormone action in the immature brain: (1) retardation in the time of appearance of the "innately organizational developmental landmarks," (2) changes in adaptive behavior with "a close relationship between the performance of the hypothyroid rat and the age at which the thyroid has been destroyed," (3) evidence that the excitable properties of the nervous system are related to the activity of the thyroid at birth, (4) changes in brain weight, cortical histology, etc., and (5) reversibility of the effects by thyroid hormone administration in relation to the age at which therapy is begun.

Thyroid hormone excess also leads to physiological, biochemical, and behavioral defects that are also permanent.[2] Eayrs[1] also pointed out that "probably the most significant of the several abnormalities observed is a hypoplastic neuropil" and that "the reduced probabilities of interaction between neurons associated with a hypoplasia of the neuropil may be invoked to explain the changes which occur both in the electrical activity of the brain and the behavior of the cretin."

Twenty years later there is little to add to Eayrs's analysis and conclusions. Additional structure targets, metabolic processes, and biochemical reactions have been studied and found to be impaired as a consequence of neonatal hypothyroidism. All of the major maturation processes contributing to the normal maturation of the brain seem to be under thyroid hormone control: cell

---

Jacques Nunez • Unité de Recherche 35 de l'INSERM, and Equipe de Recherche sur la Biochimie de la Régulation Hormonale, CNRS, Hopital H. Mondor, 94 Creteil, France.

acquisition, cell migration, proliferation of neuronal processes, synaptogenesis, and myelin formation. The timing, rate, and overall development of all of these parameters seem to be impaired both in hypothyroidism and hyperthyroidism. This suggests that a precise hormone concentration is required to synchronize the different phases of brain maturation.

These observations raise several important questions:

1. Do thyroid hormones have a direct and specific effect on each of the developmental processes, cell division, migration, and different steps of neuronal differentiation?
2. What is the precise thyroid hormone concentration required to fulfill brain maturation? Is it constant whatever the stage of brain maturation? In other words, are the ontogenesis of thyroid function and the timing of brain development synchronized in the whole animal?
3. Do thyroid hormones induce in the developing brain a set of specific proteins for each given developmental process to occur, and by what mechanism? How does thyroid hormone interfere with innately organized development, i.e., with the genetic program that dictates the organization of the neuronal network?
4. Are the data obtained in experimental hypothyroidism established postnatally in the rat of some interest in understanding, screening, and treating neonatal hypothyroidism, in the infant?

We try in this chapter to analyze these questions with the aid of the available data and according to the following subdivisions:

1. Thyroid function: ontogenesis, regulation of thyroid hormone synthesis and secretion, ontogenesis of neuroendocrine control.
2. Thyroid hormone effects on the developing brain: cell acquisition, cell migration and differentiation, synaptogenesis, myelin formation.
3. Mechanism of action of thyroid hormones: general models, conversion of T_4 and T_3, thyroid hormone receptors, effects on different enzymatic activities and on protein and RNA synthesis; the role of the cytoskeleton in cell division and nerve outgrowth.
4. Neonatal thyroid screening, replacement therapy.

No attempt has been made to refer to all articles published on these topics. The reader should refer to the detailed reviews that have been quoted in the different parts of this chapter.

2. THYROID FUNCTION AND THE DEVELOPMENT OF NEUROENDOCRINE CONTROL

The differentiation occurring during the critical period of brain development is strictly dependent on proper concentrations of circulating thyroid hormones. This parameter varies during development and depends on the level of thyroid hormone production and secretion and the rate and extent of the peripheral metabolism of thyroid hormones.

Thyroid hormone production and secretion are stimulated by TSH, a polypeptide hormone produced by the pituitary. The production and secretion of TSH are in turn controlled by TRH, a small peptide hormone produced in the hypothalamus. Finally, thyroid hormones exert a negative feedback control on TSH secretion. These major parameters develop more or less sequentially during embryogenesis and/or postnatal life in different species.

Peripheral thyroid hormone metabolism also contributes to the adjustment of the circulating hormone concentration, first by producing inactive derivatives and second by transforming thyroxine, the major iodothyronine produced by the thyroid gland, to 3,5,3'-triiodothyronine, which is three to five times more potent than thyroxine and is considered the "active form' of the hormone.

The different aspects of these problems have been reviewed recently in detail: the ontogenesis of the hypothalamopituitary function by Fisher et al.,[3] the control of thyroid function and growth by Dumont et al.,[4,5] and the mechanisms of thyroid hormone synthesis by Nunez[6] and Nunez and Pommier[7]; the problems related to the metabolism of thyroid hormones and the conversion of T_4 and T_3 (which have been reviewed in detail by Chopra et al.[8]) are analyzed elsewhere in this chapter. We therefore summarize here only the major aspects of these problems, whenever possible, in relation to the effects of thyroid hormones during the development of the CNS.

2.1. Thyroid Hormone Synthesis

Thyroid hormone synthesis requires inorganic iodide, a glycoprotein, thyroglobulin, and two enzymes, thyroid peroxidase and a H_2O_2-generating system.[6,7] A thyroid peroxidase–H_2O_2 species catalyzes the first step of thyroid hormone synthesis, the iodination of thyroglobulin. Approximately 30 tyrosine residues (out of ~140) of thyroglobulin can be iodinated by thyroid peroxidase, yielding ~12 monoiodotyrosine and ~15 diiodotyrosine residues. Actually, depending on the availability of inorganic iodide and other factors, the iodine content of thyroglobulin may vary in vivo from zero to 40 iodine atoms/molecule of protein.

Only one fraction of the iodotyrosine residues (~8) present in thyroglobulin undergo a second enzymatic reaction, which is catalyzed by the same thyroid peroxidase (but another enzyme–H_2O_2 species) and produces thyroid hormone residues (coupling reaction). In vivo and in vitro, maximally iodinated thyroglobulin contains about three residues of thyroxine (T_4) and less than one residue of 3,5,3'-triiodothyronine (T_3). Two other iodothyronines, 3,3',5'-triiodothyronine (reverse T_3 or rT_3) and 3,3'-diiodothyronine (3,3'-T_2), are also present in the thyroid gland[8] (Fig. 1). When the iodination level of thyroglobulin is low, the T_3 content may exceed the T_4 content.

Figure 2 summarizes the steps of thyroid hormone synthesis and shows the cellular sites of the thyroid follicle where each of these steps occurs.

The concentration of iodide in the thyroid gland, and, therefore, the production of thyroid hormones, greatly depends on (1) the dietary intake of inorganic iodide, (2) eventually, the presence in the food of natural inhibitors of

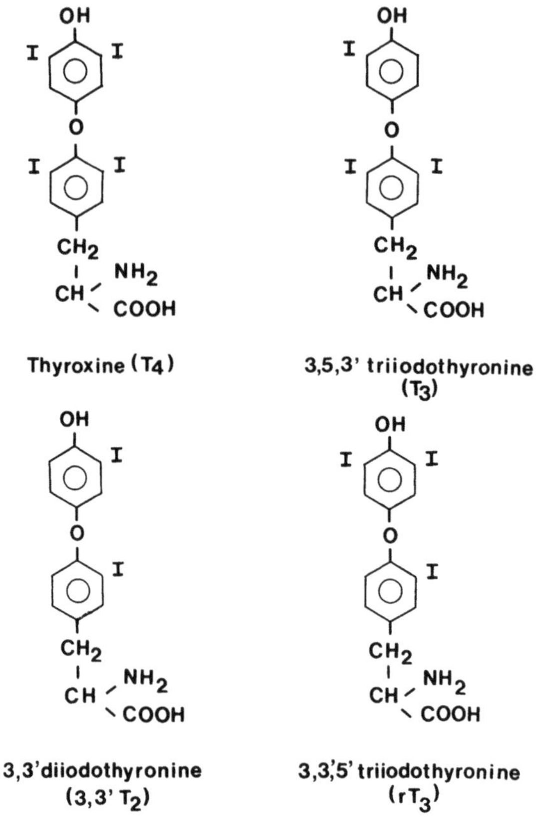

Fig. 1. Major iodothyronines.

Structures shown: Thyroxine (T$_4$); 3,5,3' triiodothyronine (T$_3$); 3,3'diiodothyronine (3,3' T$_2$); 3,3',5' triiodothyronine (rT$_3$).

thyroid peroxidase, and (3) some therapeutic treatments, for instance, by antithyroid drugs or iodinated drugs utilized for x-ray studies.

The antithyroid drugs that are used to treat hyperthyroid patients cross the placenta barrier and may induce fetal hypothyroidism. Iodide deficiency is responsible for endemic goiters and for neonatal cretinism in infants, principally in developing countries. Another major type of neonatal cretinism results from sporadic hypothyroidism. This disease, which is the origin of most cases of cretinism in industrialized countries, seems to have a variety of causes: agenesis or ectopy of the thyroid gland, defective peroxidase or absence of this enzyme, absence or decrease in concentration of thyroglobulin, etc.

2.2 Thyroid Hormone Secretion

Once the iodination and thyroid hormone synthesis are terminated, thyroglobulin is stored in the colloid vesicle. This store of iodide and of hormones is mobilized whenever necessary to maintain the blood hormone concentration. The mobilization process implies several steps, which are summarized in Fig. 3: phagocytosis of the thyroglobulin present in the colloid vesicle with the

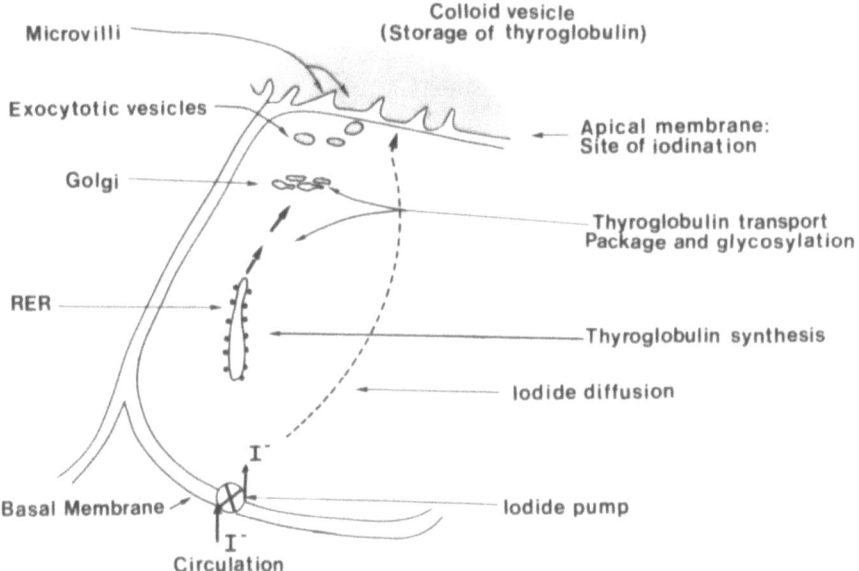

Fig. 2. Noniodinated thyroglobulin, synthesized at the level of the RER, is then glycosylated and "packed" in the Golgi apparatus. The exocytotic vesicles thus formed fuse with the apical membrane of the cell; at this level the iodination reaction and the coupling reaction occur. The iodinated thyroglobulin is then stored in the colloid vesicle.

formation of internal colloid droplets; fusion of these endocytotic vesicles with phagolysosomes; proteolytic digestion and release of free hormones and free iodotyrosines. The free hormones are secreted into the blood; free iodotyrosines are deiodinated by a specific enzyme, with the released iodide being used in a second cycle of hormone synthesis (second iodide pool).

2.3. Hormonal Control of Thyroid Function

Almost all steps of thyroid hormone synthesis and secretion are regulated by thyrotropin (TSH). Most of the effects of TSH are mediated by cyclic AMP, i.e., by cyclic-AMP-dependent phosphorylation reactions; TSH also stimulates phosphatidylinositol turnover and other pathways independently of cyclic AMP (for reviews see refs. 4,5).

Other signals operate in the thyroid gland: cholinergic and adrenergic agonists, prostaglandins, etc.; Ca^{2+} seems to be involved as a second or third messenger; cyclic GMP has also been proposed (see refs. 4,5) as a second messenger.

Iodide, in addition to being a substrate of thyroid peroxidase, has several regulatory effects. The formation and the rate of secretion of thyroid hormone therefore seem to be controlled by a number of endogenous and exogenous (metabolic, hormonal, and nervous) factors.

The secretion of TSH, the major regulatory signal of thyroid function, is itself under double control: TRH, a small peptide produced in the hypothalamus

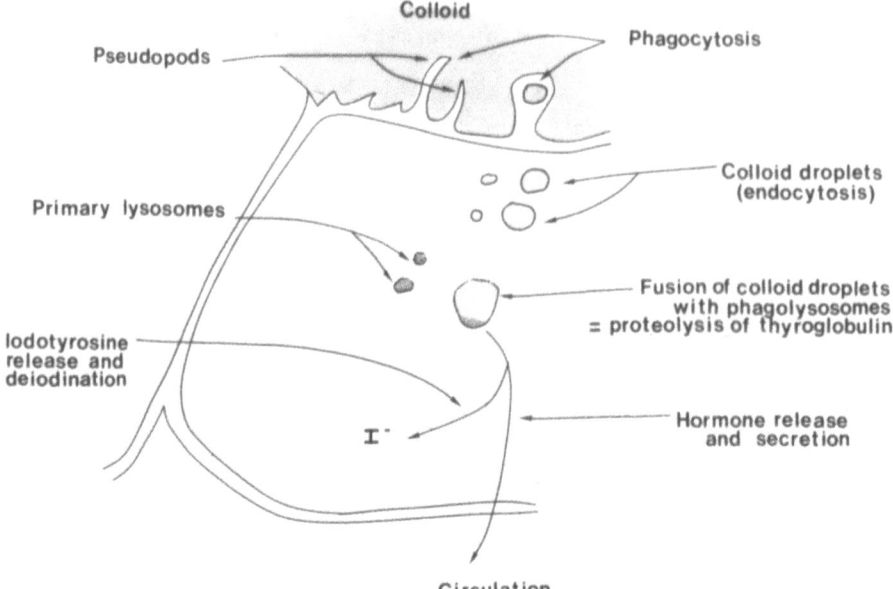

Fig. 3. Thyroid hormone secretion includes several steps. The iodinated thyroglobulin stored in the colloid vesicle is taken up by the cell (by a phagocytotic and pynocytotic process). The colloid droplets (endocytotic vesicles) thus formed fuse with primary phagolysosomes; the iodinated thyroglobulin is digested by the lysosomal protease with the release both of the hormones (T_4 and T_3) and the nonhormonogenic iodotyrosines. The hormones are secreted, whereas the iodotyrosines are deiodinated by a "microsomal" enzyme.

(and in other brain regions), stimulates, whereas thyroid hormone inhibits (negative feedback), TSH secretion.

The existence of a series of regulatory systems intrinsic to the thyroid gland, combined with a cascade of positive and negative exogenous hormonal and nervous effects, brings the secretion of thyroid hormones under tight control. The ontogenesis of such a hypothalamic–pituitary–thyroid axis has been studied in detail in the past few years, since brain development greatly depends on proper levels of circulating thyroid hormones.

2.4. Ontogenesis of the Hypothalamic–Pituitary–Thyroid Axis

It seems well established that the placenta is practically impermeable to thyroid hormones. The fetus will therefore depend for its brain development only on the hormone that has been secreted by its own thyroid gland.

Fisher et al.[3] have reviewed the literature (Fig. 4) in the field, and we briefly summarize their main conclusions here:

1. In man and rat, the thyroid gland is almost completely differentiated during fetal life. Thyroid hormones are produced at midgestation in the human fetus (74th day) and at the end of gestation in the rat (20 days).
2. Thyrotropin increases in the pituitary and serum before birth in man (16–22 gestational weeks). Fetal hypothalamic TRH and hypothalamus

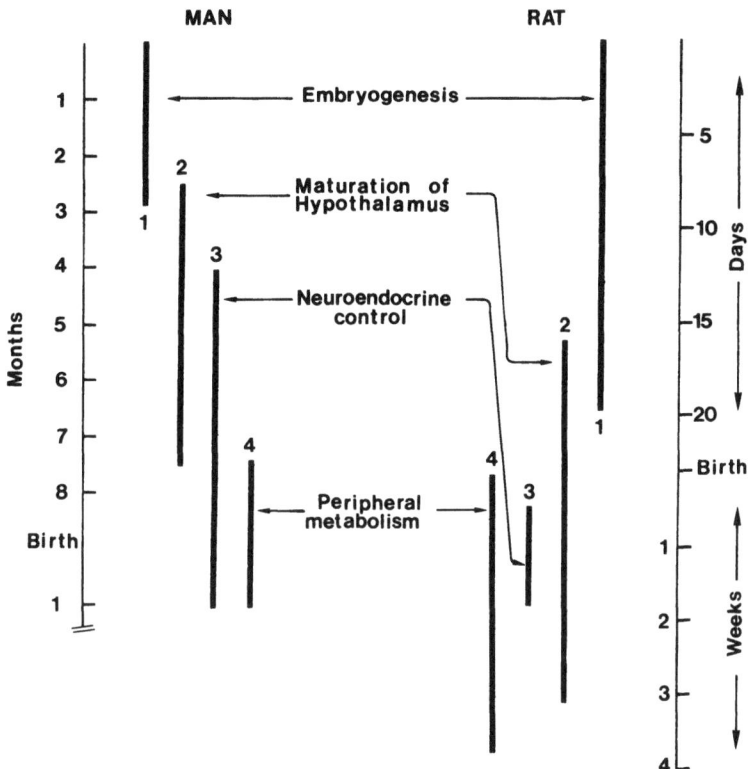

Fig. 4. Ontogenesis of thyroid function in man (left) and rat (right). The vertical bars show the approximate duration of the four major steps of this process: thyroid embryogenesis, maturation of hypothalamus (TRH production), neuroendocrine control (TSH secretion, T_4-dependent negative feedback), and peripheral metabolism (conversion of T_4 to T_3 or rT_3). (Recalculated from Fisher et al.[3])

maturation occur a little earlier, suggesting that "histological maturation of the hypothalamus and/or maturation of the pituitary portal system are the critical factors conditioning increasing function of the pituitary–thyroid axis."[3] In the rat, serum and pituitary TSH and maturation of the pituitary response to TRH occur postnatally (4 to 18 days after birth), i.e., after hypothalamic maturation.

3. The negative feedback (inhibition of TSH secretion by thyroid hormones) is one of the latest maturation events in both species.
4. Thyroid hormone serum levels reach adult values in the human fetus at ~40 weeks of gestation, whereas in rat, the hormone content is very low at birth and maximal ~10 days later.

In both species the distribution of serum hormonal iodine among T_4, T_3, and rT_3 depends not only on the maturation of the axis but also on the onset of the peripheral conversion of T_4 to either rT_3 or T_3. The rT_3 is high at birth in the rat and until the end of the first postnatal month in man. The decrease in rT_3 levels and the increase in T_3 levels depend on the late maturation of specific enzymes (5'-deiodinase or 3-deiodinase) that catalyze, respectively, these two monodeiodination reactions (see Chopra et al.[8] and Section 4 of this chapter).

Finally, several reports suggest that the ontogenesis of the hypothalamic–pituitary–thyroid axis is affected by early hypo- or hyperthyroidism[10,11] (P. Walker and J. H. Dussault, personal communication); rats treated soon after birth with antithyroid drugs exhibit decreased TSH secretion at an adult stage.

3. THYROID HORMONE EFFECTS ON CELL ACQUISITION AND NEURONAL DIFFERENTIATION

In the rat, neuronal differentiation in the cortex begins after birth and is maximal ~10 days later (critical period). In the cerebellum, the differentiation process begins later, approximately during the 10- to 25-day period, and takes place after a phase of intensive cell proliferation. In the cortex, 50% of the adult cell population is already present at birth (and 100% of the neurons; postnatal cell proliferation in the cortex involves mostly the glial cells). Only ~3% of the final cell population is already present in the cerebellum at birth, and most of the interneurons are formed in the interval from 1 to 15 days.[12]

Since thyroid hormone is present in very small quantities in the newborn rat,[13] and since the rat placenta is practically impermeable to thyroid hormones (see ref. 3), it is possible that only the postnatal stages of brain development are under thyroid hormone control in this species. However, it is not known whether identical and maximal amounts of hormones are required at each maturation step: it might be that prenatal and early postnatal developmental events are sensitive to very low doses of hormone. In addition, cell acquisition continues for much longer periods after birth in the hippocampus and in the olfactory bulbs of the rat.

In other animal species, similar sequential events occur in the different parts of the brain. Moreover, the critical period of neuronal differentiation always begins approximately at the end of the ontogenesis of the thyroid function, i.e., when the thyroid hormone concentration begins to increase in the blood. This is true whatever the species and whatever the period, prenatal, perinatal, or postnatal, at which neuronal differentiation occurs.[14]

Three major cellular events occur sequentially during brain maturation: (1) the undifferentiated cells actively divide into a few proliferative zones (the subependymal zone for the cortex, the external granular layer in the cerebellum, the proliferative zone in the hippocampus) and (2) the new cells begin then to migrate and (3) to differentiate. The differentiation process includes several cellular and functional changes: neurite outgrowth, synaptogenesis, neurotransmitter specification, myelination, etc. Since cell division, cell migration, and differentiation apparently occur postnatally and sequentially in the cerebellum of the rat, most studies have been performed in this structure in order to establish whether or not thyroid hormones exert their effects on one or more than one of these processes.

3.1. Cell Division and Cell Acquisition

The weights of the brain and cerebellum are both decreased when hypothyroidism is established in late pregnancy. This suggests that thyroid hormones might have some effect on the number of cells in the CNS. Several mechanisms might explain such an effect of hypothyroidism: (1) direct effects on the mechanisms of cell division, (2) acceleration of migration and differentiation, which would prematurely decrease the pool of immature dividing cells present in the proliferative zones, (3) increase in cell death, or (4) changes in the size and weight of each cell (for instance by modifying the DNA/protein content ratio in each cell). The example of the cerebellum shows that it is very difficult to decide among these possibilities.

In the euthyroid rat, the Purkinje cells are already present at birth. The basket cells, granule cells, Golgi cells, and stellate cells are formed postnatally and more or less sequentially in the external granular layer (EGL).[15] The process of cell division begins at birth, is maximal ~10 days later, and disappears at approximately day 21.[16] Legrand and Rebiere have also shown that the external granular layer disappears almost completely at this age in the euthyroid rat, whereas it is still present in hypothyroid animals.[17-19] Hamburgh *et al.* also observed prolonged [^{3}H]thymidine incorporation in the proliferative zone of the hypothyroid rat cerebellum.[20]

Nicholson and Altman[16] have also performed systematic short-term labeling experiments on the cells present in the EGL by injecting [^{3}H]thymidine into young rats at different stages of postnatal development and measuring the number of labeled cells in this layer 30 min later. In control animals, labeling was maximal at day ~10 and terminated at day 21. In contrast, in the EGL of the hypothyroid rats, cell proliferation was lower at day 10 but higher at day 15. Injecting excess thyroid hormones resulted in a decrease in the total number of labeled cells found in this zone at day ~15 and ~21, whereas at day ~5, more labeled cells seemed to be present than in the controls. However, the number of labeled cells also seemed to be higher at day 5 in the EGL of the hypothyroid animals. Despite these discrepancies, it was concluded that hypothyroidism slows, and hyperthyroidism accelerates, the process of cell division in the cerebellum. This conclusion seemed to be in agreement with the observed changes in thymidine kinase[21] and in the DNA content.[22]

To answer the same type of question, Balasz *et al.*[23] and Patel *et al.*[28] have calculated the amount of newly formed DNA at different periods of postnatal brain development. Cell acquisition appears to be unchanged by hypothyroidism in the forebrain but decreased 30% below control levels in the cerebellum during the second week after birth. This reduction in the number of cells is not the result of a change in either the length of the cell cycle or the duration of the DNA synthetic phase.[26] However, the proliferative activity lasts longer in the EGL,[27,29] thus explaining why the final number of cells present in the cerebellum of the adult hypothyroid rat is almost normal. These conclusions are consistent with the early findings of Legrand,[17,18] Rebiere and Legrand,[19] and Nicholson and Altman.[16]

In contrast, the final number of cells is permanently decreased in the cerebellum of the adult hyperthyroid rat. In the hippocampus, hypothyroidism leads not only to a decrease in the number of cells at early stages of development but also to a permanent deficit[30] in the adult animal.

Thus, although these data clearly show that thyroid hormones affect the number of replicating cells, the timing of proliferation, and also (depending on the brain region and/or on the hormonal conditions) the final number of cells, they do not permit one to say whether they directly regulate the mechanisms of cell division. Several other mechanisms might account for these observations: for instance, changes in the output rates from the proliferative zones might be involved in the retardation or acceleration of cell acquisition depending on the thyroid state. Finally, it seems that the main effect of hyperthyroidism is to markedly reduce the final number of basket cells and granule cells; this might depend on the early termination of cell proliferation in the EGL.[31] Less clear results were obtained in hypothyroidism, although Clos and Legrand[32] have reported a decrease in the number of basket cells.

3.2. Cell Migration and the Development of the Neuropil

The changes in the number of interneurons in the internal granular layer and the increase in astroglial cells might be secondary to the severe hypoplasia of the neuropil that clearly occurs in the cerebellum of the hypothyroid rat. A permanent and dramatic reduction in the arborization of the dendritic tree of the Purkinje cell has been observed by Legrand[18]; The length of the primary dendritic trunk is increased, and a deficit in the number, density, and branching of the dendritic spines was noticed. In contrast, hyperthyroidism accelerates development of spines.[19,23] Similar findings have been reported in the cortex by Eayrs et al.[34–36]: thyroid deficiency reduces the length and the branching of the pyramidal neurons, the density of the axonal terminals, and the number of spines.

All of these events therefore suggest that the rate of growth of the neurite is under thyroid hormone control both in the cortex and in the cerebellum. Cell migration might also be considered highly dependent on neurite outgrowth. The newly formed cells that cease to divide in the EGL begin to migrate through the molecular layer until they reach the internal granular layer (IGL), where they establish their contacts with the Purkinje cells. The most abundant synaptic contacts are made in this area between the axons of the granule cells and the tertiary spines of the Purkinje cells.

A permanent deficit in the number of branches and of spines of the Purkinje cell, together with a retardation in the timing of granule cell migration, might be sufficient to explain a reduction in the number of available presynaptic and postsynaptic sites and therefore an increase in cell death.[37] Reduced migration rates of the basket cells together with a decrease in available sites for the parallel fibers of the granule cell might also increase the number of cell deaths. Finally, the changes in the number of astrocytes (Bergman cells) and in the amount of glial cytoplasm[16,32] in the molecular layer might be secondary to the increased cell death observed in the neuropil.

Cell migration might be considered the first differentiation event occurring when cell proliferation is terminated. Nicholson and Altman[16] and Lauder[38] sought to determine whether thyroid hormones influence the time of onset of granule cell differentiation in the EGL, i.e., the termination of cell proliferation. They performed long-term labeling experiments by injecting [^{3}H]thymidine on a given day of development to hypothyroid and hyperthyroid rats and allowing them to survive up to 2 months. After killing the rats, they calculated the number of labeled cells still present in the EGL and the time required for the labeled cell to migrate from the EGL to the molecular layer and the internal granular layer (transit time). They found that hyperthyroidism markedly decreases, and hypothyroidism increases, the transit time. The same authors[39,40] have measured the rate of migration of the granule cell and the rate of growth of the parallel fiber in euthyroid and hyperthyroid animals. Although both rates were markedly increased in hyperthyroidism, their ratio was identical to that found in euthyroid animals. In hypothyroidism both rates were so low that ratios could not be measured. The same authors[16] also observed that in hypothyroidism the granule cells that have begun to migrate towards the internal granular layer "pile up" in the molecular layer. This defect does not seem to imply that cell mobility is suppressed by thyroid hormone deficiency, since these cells have been able to reach the molecular layer; they simply seem to be unable to move further, perhaps because the average length of the parallel fiber is shorter, as shown by Lauder[39] and Lauder and Bohn[40] and as deduced by Crepel[41,42] from the results of detailed electrophysiological experiments. Thus, the transit time and the rate of migration might depend on the rate of growth of the parallel fiber.

At the end of this analysis we are left with two possible interpretations. The first assumes that thyroid hormones essentially regulate only one of the cellular processes occurring during brain maturation, i.e., some mechanism related to nerve growth. Such a major or unique primary effect might explain the changes in proliferation, in the rates of migration, and in the number of cell deaths, the reduction in synaptogenesis, and the increase in the glial population seen in the developing cerebellum of the hypothyroid rat. Excess of thyroid hormone might accelerate neurite outgrowth, thus desynchronizing the developmental pattern.

The second possibility is that thyroid hormones exert multiple and distinct, i.e., pleiotropic, effects on each (or most) of the parameters of brain maturation. This would imply that (1) the response to thyroid hormones by a given cell type would be different depending on its stage of maturation and (2) different cell types would respond differently to thyroid hormones. Such a pleiotropic mechanism of action must be considered seriously for this class of hormones not only for the differentiating brain but also for other peripheral immature and mature target cells (see Section 4).

Both mechanisms imply that the effect(s) of thyroid hormone is direct, as proposed by Eayrs in 1960,[1] and not secondary to undernutrition or to other hormones and growth factors; undernutrition and thyroid hormone deficiency[1,43-45] actually produce very different effects. Recent data have shown, however, that excess of thyroid hormone increases the NGF concentration in the CNS.[46,47] However, since hypothyroidism does not alter the NGF concentration, the significance of this finding remains unclear. On the other hand, different groups have been able to grow primary cultures of brain cells in defined media and to show that they respond to thyroid hormones.[48–51] Romijn et al.[51] have, for instance, observed that the addition to such cultures of low concentrations of T_3 markedly increases the growth of nerve fibers. Since primary cell cultures contain different cell populations in different proportions, with variations depending on the culture medium and on the survival period, further proof is required for a direct effect of T_3 on neuronal differentiation.

3.3. Synaptogenesis and Development of Nerve Cell Transmitters

It might be expected that decreased rates of cell migration and of neurite outgrowth as well as changes in cell population would reduce the probability of interaction between neurons. The use of histological techniques has shown a reduction of the neuropil and in the number of synapses in the sensorimotor cortex,[34–36] in the visual cortex,[52] and in the cerebellum.[19,43,53–55] The detailed studies of Nicholson and Altman[53,54] also showed that the number of varicosities per parallel fiber in the cerebellum is markedly modified by the thyroid status; the number of synapses of the parallel fiber is also probably decreased as a consequence of the diminution of the number of basket cells.[32] Other parameters are modified by the thyroid state, and this may reflect changes in synaptogenesis: cholinesterase and acetylcholinesterase activity,[55,56] synaptosomal protein concentration,[43] and circadian changes in serotonin IV acetyltransferase.[57] GABA-transaminases,[58] etc. As a general rule, early thyroid hormone deficiency seems to reduce these parameters, whereas early hyperthyroidism has a transient but marked accelerating effect.

The recent availability of sensitive radioreceptor assays has shown that excess thyroid hormone accelerates the acquisition of muscarinic receptors in the cerebral cortex of the mouse but not in the cerebellum; 30 days after birth the number of receptors is, in contrast, reduced in the cortex, cerebellum, caudate–putamen, and hippocampus.[59]

The concentration of several neurotransmitters has also been evaluated in different discrete brain nuclei by radioimmunologic and radioenzymatic methods.[60] Thyroid hormone deficiency increased the concentration of substance P in 19 nuclei out of the 32 tested. The distribution of catecholamines (or opioids) was found to be unchanged with the exception of dopamine in three regions (out of 32). Finally, TRH concentration was drastically reduced in the median eminence and in three other extrahypothalamic zones. All of these effects were reversed after administration of thyroxine.

Smith et al.[61] have found that the developmental increase in β-adrenergic receptor binding is permanently depressed in thyroid deficiency in both the forebrain and, to a greater extent, the cerebellum (total number reduced by 50%). In contrast, hyperthyroidism had not significant effect. Similar studies have been performed by Patel et al.[62] on the muscarinic and GABA receptors.

3.4. Myelination

Myelin is a complex structure, probably an expansion of the oligodendrocyte membrane in the CNS and of the Schwann cell in the PNS. Its formation is also under thyroid hormone control. The myelination process begins in the young rat during the first postnatal week, increases in intensity during the next 2 weeks, and continues at a much lower rate during a prolonged period of time. Thyroid hormone deficiency both decreases the myelin content and retards its formation; hyperthyroidism accelerates myelination.[20,63-68] Such changes in the amount of myelin and in the duration of its formation are likely to alter the activity of the CNS (and PNS) since, according to the conventional point of view, myelin sheets play an essential role in the conduction of the nervous influx.

These changes in myelination might be secondary to some direct effect of thyroid hormones on the maturation of the oligodendrocyte[69] or related to the area of the axons, as discussed by Clos and Legrand.[63,66] These authors have found that a linear relationship exists between the circumference of an axon and the number of laminae in its myelin sheath. It is obvious that myelin will be deposited only if the corresponding normal nerve fibers are formed.[35]

Thus, the effect of thyroid hormones on the myelination process might be at least partly indirect. In addition, and since some myelin is formed in the CNS before the onset of thyroid function, it does not seem that thyroid hormones are absolutely necessary for myelin to be formed; rather, they exert a stimulatory and synchronizing effect.

A better insight into the problem might be acquired by studying the components of myelin, simple and complex lipids, structural proteins, associated enzymatic activities, etc. Balasz et al.,[65] by analyzing some of these parameters, found no significant changes in the composition of the myelin from hypothyroid rats.

Some complex lipids, such as the cerebrosides and sulfatides, are very abundant in myelin and might be considered, to a certain extent, to be good markers of this structure[65,67]; several authors[67,69,70] have found that hypothyroidism leads to a 30 to 40% decrease in their level in the brain.

Some enzymatic activities more or less related to the synthesis of these complex lipids have also been measured. Wisocki and Segal[71] have shown, for instance, that thyroid hormone deficiency decreases the specific activity of the UDP-galactose:sphingosine galactosyltransferase and of 2',3'-nucleotide 3'-phosphohydrolase; after T_3 injection into the hypothyroid animals, the recovery in enzyme activity varied depending on whether animals were measured in the brain or in the spinal chord and on the day at which such replacement therapy was performed (the first 8 days being critical). Similar findings have been reported for the enzymes of galactosyldiacylglycerol metabolism.[72]

In vitro studies have also been performed. The first experiments of this type[20] showed that adding thyroid hormones to surviving brain explants increases *in vitro* myelin synthesis. More recently, Bhat et al.,[73,74] using a primary culture system enriched in oligodendrocytes prepared from fetal mouse brain, found a clear stimulatory effect of nanomolar concentrations of T_3 on the incorporation of sulfate into sulfolipids. The enzyme that catalyzes the sulfation reaction, glycolipid PAPs phosphotransferase, was also clearly stimulated by T_3, as was 2',3'-cyclic nucleotide 3'-phosphotransferase, another enzymatic marker of myelin; T_3 did not affect either the uptake of sulfate or the synthesis of adenoside 3'-phosphatephosphosulfate (PAPs), which is the substrate of the sulfating enzyme. These results suggest a direct effect of T_3 on the synthesis of sulfatides by the oligodendrocytes.

To study the effects of thyroid hormones, some proteins associated with myelin [proteolipid, basic protein, and a glycoprotein (MAG)] might be used as markers of this structure.[76-78] Walters and Morell[79] have recently reevaluated this problem, reaching the conclusion that the synthesis of these proteins is retarded by thyroid deficiency; the composition of the proteins of myelin prepared from hypothyroid rats resembles that of younger euthyroid animals. They also confirmed the findings of Matthieu et al.,[76] which showed thyroid-hormone-dependent changes in the molecular weight of MAG.

Altogether, these data suggest that thyroid hormones influence the synthesis of several components of myelin by accelerating the maturation process. Is this effect direct? Even the experiments performed with primary cultures do not provide a complete answer to this question, since these systems contain heterogeneous cell populations. Moreover, how can the reduction and retardation in myelin formation seen in the whole hypothyroid brain be reconciled with the increase in Bergman glia observed in the cerebellum[16,32] and in S-100 protein?[80] Is this latter effect of hypothyroidism caused by the increased pycnosis and decreased synaptogenesis seen in the molecular layer? We are left again with several assumptions to explain the effect of thyroid hormones on myelin formation.

4. UPTAKE, METABOLISM, AND MECHANISM OF ACTION OF THYROID HORMONES

The dominant view during the last two decades is that thyroid hormones stimulate the synthesis of specific proteins in almost all tissues of vertebrates. The general characteristics of this model (Fig. 5) can be summarized as follows:

1. Thyroxine and triiodothyronine are secreted by the thyroid gland (see Section 2) and transported in the blood by different serum proteins (for a review see Robbins et al.[81]). The free hormone fraction is then taken up by the target cells.
2. The active form of the hormone is triiodothyronine, which is formed both from thyroxine by peripheral deiodination (for a general review see Chopra et al.[8]) and by the thyroid gland (see Section 2).
3. Cellular binding proteins are present in almost all cell types. A nuclear binding protein is considered to be the true hormonal receptor. The hormone bound to the receptor would then interact in an unknown manner with the DNA to induce (or repress) the synthesis of specific messenger RNAs, which in turn are translated at the ribosomal level to specific proteins (for a general review see Baxter and Eberhardt[82]).

Other models are possible, including specific action at the membrane,[83] mitochondrial,[84,85] or translational[86,87] level.

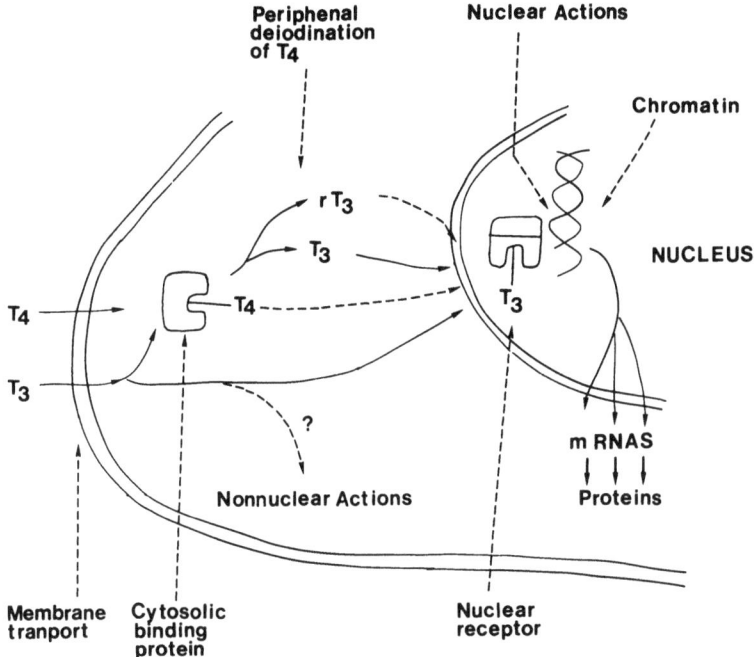

Fig. 5. The different steps of thyroid hormone action are represented schematically. Both T_4 and T_3 are transported from the blood into the cell. T_4 is converted to either T_3 or rT_3 by specific monodeiodinating enzymes. Both T_4 and T_3 are probably bound to a cystosolic binding protein and transferred to nucleus. The "nuclear model" considers that at this level T_3 is the active hormone that binds to the nuclear receptor, triggering some unknown event at the level of the chromatin. The result of this interaction is the derepression (or repression) of a limited domain of genes.

4.1. Transport and Uptake of Thyroid Hormones by the CNS

Most of the studies in this field have been performed by injecting labeled triiodothyronine into rats and measuring the radioactivity present in the brain or in different regions of the brain. Such a methodology suffers from a number of difficulties, since the amount of label found depends not only on the uptake value but also on the amount of endogenous unlabeled hormone present at the time of injection, which obviously changes with the iodide intake and with the period of brain maturation (see Section 2). Another parameter that should be taken into account is the degradation rate, which can also be very high and variable depending on the age, sex, etc. After the early publications by Courrier et al.[88] and Jensen and Clark,[89] who showed that thyroid hormones accumulate in the CNS, Ford et al.[90-92] have shown that there is a preferential accumulation of triiodothyronine in the neurons of the cerebellum (Purkinje cells and granule cells), cortex (granular layer), diencephalon, brainstem (nuclei), and spinal cord. More recently, Dratman et al.[93] have confirmed that nerve cells and neuropil in different brain regions concentrate and retain administered T_3. Clear-cut differences in the distribution of the label were seen in different brain regions. For example, the Purkinje cells were more strongly labeled than the basket cells; the intensity in neuropil was different in different regions of the cortex and much stronger than in white matter.

Thyroid hormone uptake also varies with the species and sex, and T_3 is taken up faster than T_4.[94] Thyroid deficiency seems to decrease the uptake, and hormone excess to increase this process. However, the authors were aware of the fact that these differences might depend on a variety of factors (changes in blood flow, rates of glucose utilization, and hormone degradation, etc).

A few groups[95,96] have also shown that thyroid hormone uptake increases with development in the different brain regions.

4.2. The Concept of "Active Form" and the Peripheral Conversion of Thyroxine to Triiodothyronine

The physiological responses to thyroxine *in vivo* (for instance, an increase in oxygen consumption) are in general manifest after a long time lag. This lag has been considered to be the time required to transform thyroxine to "an active form." The discovery of 3,5,3'-triiodothyronine (both in the thyroid gland and in blood), an iodothyronine having a potency three to five times higher than that of thyroxine, has suggested that 5'-monodeiodination in the peripheral tissues might lead to the formation of T_3, the active form of T_4. This conclusion has been extensively documented since the early findings of Pitt-Rivers *et al.*[97] and of Braverman *et al.*[98] (for a review see ref. 8).

Thyroxine is also actively deiodinated to 3,3'5'-triiodothyronine (T_3) or reverse T_3 (rT_3) in most tissues by a distinct enzyme, 3-monodeiodinase; rT_3 can be further rapidly deiodinated to 3,3'-T_2. Three points might be of importance (see ref. 8) in regard to T_3 and rT_3 formation during the maturation of the CNS: (1) in man the serum concentration of rT_3 is much higher in the newborn than in the adult; (2) the peripheral conversion of T_4 to T_3 is decreased in the human and sheep fetus; and (3) several hormonal treatments, fasting, various diseases, etc. markedly change the T_3 and rT_3 serum contents.

In the brain, for instance, the actual concentration of T_4, T_3, and rT_3 will depend, at each developmental stage, on a number of factors, including (1) the rate of production and secretion of T_4 and T_3 by the thyroid gland, (2) the conversion of T_4 to T_3 or rT_3 by the peripheral tissues (for instance, the liver and the kidney), (3) the output of T_3 and rT_3 from these tissues to the circulation and their uptake by the brain, (4) the conversion of T_4 to T_3 and rT_3 in the different regions of the CNS, and (5) a number of other physiological, pathological, and therapeutic factors.

Crantz and Larsen[99] have shown that most of the endogenous T_3 present in the cerebral cortex and the cerebellum of the adult rat is derived from T_4 by 5'-deiodination; T_3 is also further deiodinated to 3,3'-T_2 and then to 3-T_1[100]. A 5'-deiodinase activity is present in all regions of the CNS,[101,102] but the maximal mean rates have been observed in the cerebellar and cerebral cortex and the minimal in the spinal chord. Hypothyroidism markedly increases this enzymatic activity.[101,102]

A recent report[103] suggests the presence of two 5'-deiodinase activities in microsomal fractions of cortex homogenates: a propylthiouracile (PTU)-sensitive and a PTU-insensitive pathway; T_4 would be converted to T_3 only via the latter pathway. In contrast, both systems catalyze the deiodination of rT_3 to $3,3'$-T_2.

The conversion of T_4 to rT_3 also varies depending on the brain region: the activity is higher in the upper part of the neuroaxis, cortex, striatum, midbrain, and hypothalamus, whereas it is reduced in the cerebellum, brainstem, and spinal cord. No 5'- or 3'-deiodinase has been found in the PNS.[101] Finally, the level of 3-deiodinase activity adapts to the thyroid state.[100]

In the brain these deiodinating activities (T_4 to rT_3 and T_3 to $3,3'$-T_2) have mainly been found in the synaptosomal fraction; the particulate fraction prepared from neonatal or fetal rat brains had higher activity than that from adult brain.[104] These data are in agreement with those of Dratman et al.,[105] who have found that T_3 is present in nerve endings.

Other degradative pathways have been described in the CNS: metabolism of the alanine side chain of T_4 and T_3 with the production of acetic and propionic derivatives[106]; formation of iodotyrosines resulting from the breakdown of the phenoxy ether bridge.[95] However, stepwise deiodination seems to be the dominant pathway, yielding T_3 or rT_3 and their less iodinated metabolites. The rates, extent, and preferential deiodinating pathways in the CNS seem to depend greatly on the stage of postnatal development, the CNS region, the thyroid state, and other physiological conditions. The actual concentration of the active hormone in each region of the CNS and at each stage of brain maturation will therefore depend on a number of factors, some of which are intrinsic to the brain itself.

4.3. Thyroid Hormone Binding Proteins and Receptors

Several proteins with binding properties for thyroid hormones have been detected in a variety of cell types and with different subcellular localizations.

4.3.1. Plasma Membrane Binding Sites

High-affinity sites (3×10^8 and 4×10^7 or 4×10^6 M^{-1}) for thyroid hormones have been found in the hepatocyte plasmic membrane.[107,108] These sites might be related to the transport, metabolism, and (eventually) action of thyroid hormones. An effect of low concentrations of thyroid hormones (10^{-9}–10^{-10} M) that seems to be independent of protein synthesis has been described in the fetal cardiac cell[109] and in the thymocyte.[110] Binding sites for thyroid hormones with similar affinities have been detected in the sarcoplasmic reticulum.[108]

4.3.2. Cytosolic Binding Sites

Their affinity for thyroxine ($\sim 10^8$–10^7 M^{-1}) is higher than that for T_3. The binding activity is present in most tissues.[111–116] It is a well-defined protein with a molecular weight of $\sim 70,000$ and is present in large amounts ($\sim 1\%$ of total soluble protein) in the brain of the newborn rat.[117] The number of sites decreases with age both in the forebrain and in the cerebellum; in this latter region almost no sites are found at day 30. Hypothyroidism seems to increase the number of sites postnatally (Lennon, A. M., Osty, J., and Nunez, J., unpublished results). The physiological function of this soluble binding protein is unknown. It might play a role in the transport, storage, metabolism (T_4 to T_3 conversion), or action of thyroid hormones. The decrease in the number of binding sites during development suggests that, whatever its function, this binding protein plays an important role during brain maturation. It is also interesting to note that, in contrast, the number of sites increases with development in the liver.[115,116]

4.3.3. Mitochondrial Binding Sites

Mitochondrial binding sites with high affinity for T_3 (1×10^{11} M^{-1}) have been reported by Sterling et al.,[85] but their existence remains controversial.[118]

4.3.4. Nuclear Binding Sites

Nuclear high-affinity binding sites (1×10^{11} M^{-1}) have been detected in a variety of cell types and are considered to represent the major (if not the unique) receptor of thyroid hormones.[119,120] They have been partially purified after solubilization by high salt concentrations or DNAse treatment and seem to belong to a class of nonhistone proteins belonging to the chromatin.[121-125]

Their involvement in the mechanism of thyroid hormone action is implied by several observations:

1. There exists, in general, a good correlation between the relative affinities of several analogues of thyroid hormones and their biological potencies[126,127] and conformations.[128]
2. A good correlation has also been found among the intranuclear concentration of T_3, the nuclear receptor occupancy, and the effects of T_3 on the synthesis of growth hormone,[129] TSH release by the pituitary,[130] α-glycerophosphate dehydrogenase and malic enzyme activities,[131] and activity of different lipogenic enzymes,[132], etc.
3. T_3 binding to the nuclear receptor is also correlated with an increase in total mRNA synthesis[133] and that of specific mRNAs; growth hormone,[129] α_2U-globulin,[134] malic enzyme.[135]

The nuclear receptor activity seems to be modulated by proteins extracted from chromatin[124,125] or by histones.[136] However, the mechanism by which the interaction of the hormone–receptor complex modulates gene expression and the synthesis of specific mRNA remains unknown.

The nuclear receptor is also present in the brain nuclei[137-139] with properties very similar to those of the activity found in other tissues. Eberhardt et al.[139] have reported that in the adult rat the anterior hypophysis contains two to three times more binding sites than the cerebral hemispheres and ten times more than the liver. The number of nuclear binding sites also changes with postnatal development: it decreases twofold during the first 2 postnatal weeks and then remains constant until 6 months.

Neonatal hypothyroidism increases the number of binding sites by 40%.[140] Recent data also suggest that a chromatin-associated fraction modifies the activity of brain nuclear receptors.[141]

4.4. Thyroid Hormones and Protein Synthesis

During the decade 1960–1970, several authors examined the effects of thyroid hormones on the incorporation of labeled precursors into proteins and nucleic acids in the developing brain. The impression one has after reading an extensive literature is that the rates of protein and RNA synthesis are depressed in the hypothyroid developing brain (see refs. 95,142–145 for reviews), but the mechanism of such effects remains unclear.

Although experimental evidence is still lacking, the conventional view is that the primary action of thyroid hormones involves changes in the genetic expression of the different cell types that participate in the construction of the neuronal network. This implies that thyroid hormones selectively stimulate (or repress) a limited number of genes, thus increasing (or diminishing) the synthesis of a few specific mRNAs coding for a limited number of proteins. In addition, depending on the cell type (or the stage of maturation of a given cell type), different genetic domains[146] would be under thyroid hormone control during development.[147] According to such a model, and if the "domain" is restricted to a small number of proteins, one would not expect to see significant changes in total protein and RNA synthesis even if the effect of thyroid is "pleiotropic" (i.e., several mRNAs and proteins are under the control of thyroid hormones in a given cell type and at a given stage of development).

Little is known in this respect as far as brain development is concerned: the heterogeneity in brain cell populations and the fast changes in cell numbers and their differentiation, probably preclude detailed *in vivo* studies. Most of the results in this field have actually been obtained with well-defined cell lines that respond *in vitro* to thyroid hormones by producing new mRNA and protein species. Ivarie *et al.*[146] have shown, for instance, that thyroid hormone and corticosteroids influence a limited and highly specific domain in each cell line, "approximately 1% or less of the detected gene products." Corticosteroid and thyroid hormone domains may overlap by corepressing or coinducing the same gene product; induction by one hormone and repression by the other have also been noted. It might be that the multiple changes in enzymatic activities seen in the developing brain of the euthyroid, hypothyroid, or hyperthyroid rat will be explained by such a model, and progress in this field probably depends on the establishment of homogeneous cell culture systems responding *in vitro* to thyroid hormones.

4.5. The Role of the Cytoskeleton during Brain Differentiation

The three major cellular events occurring during brain differentiation, cell proliferation, cell migration, and neurite outgrowth, are probably highly dependent on the properties of different systems which form the cytoskeleton.

The mitotic spindle is composed of microtubules, and the movement of chromosomes during mitosis probably depends on the activity of contractile fibers. Similarly, neurite outgrowth in cell culture is efficiently inhibited[148-150] by the same antimitotic drugs that block cell division (colchicine, vinblastine, maytansine, etc.). In addition, microtubules represent the main linear structure present in axons and dendrites.[148,151,152]

A large body of evidence also shows that the same pool of tubulin molecules is used either to form the spindle in the dividing cell or to form the microtubule network that covers the cytoplasm underneath the plasma membrane in the differentiated cell (see ref. 153). Other data, including those obtained in the developing brain of normal or hypothyroid rats,[154,155] confirm that the concentration of tubulin does not seem to be the rate-limiting factor for microtubule assembly.

This suggests that specific and different signals are operating to channel the preformed tubulin pool towards the formation either of the spindle microtubules or of the cytoskeletal microtubules.

Cell mobility and cell migration also depend on the properties of the cytoskeleton. The contractile elements that are responsible for the amoeboid movement of the nerve cells and the forward progression of the neurites probably depend on the various sorts of filaments that are very abundant in the growth cone (including nonmuscle actin filaments). Poisoning these filaments with drugs (such as cytochalasin) stops neurite outgrowth in nerve cell cultures. Similarly, treating the cultures with colchicine, a drug that inhibits microtubule assembly, results in neurite regression.[148-150]

These data provided the basis of a model that has been developed in our laboratory in the past few years. This model suggests that (1) tubulin assembly is promoted by polymerizing factors, (2) different polymerizing factors are utilized to form the microtubules, which are present either in the spindle or in the neurites, by using the same pool of tubulin molecules, (3) microtubules are good markers of neurites and of neurite outgrowth, (4) thyroid hormones accelerate cell migration and neurite outgrowth by timing the changes in the polymerizing factors during brain development.

A clear answer to the first question came from analysis of the *in vitro* assembled microtubules that can be prepared from crude brain supernatants in appropriate media[156] and purified by one or several cycles of assembly-disassembly.[157] Polyacrylamide gel electrophoresis (PAGE) of such recycled microtubules revealed that several minor components (15 to 20% of the total microtubule protein) are present beside tubulin, the major microtubule component. At least two of these minor microtubule-associated proteins (MAPs) are able to promote the polymerization of tubulin, which when completely purified is unable to form microtubules at reasonable concentrations. Microtubules prepared from adult brain contain two major MAPs: MAP_2, with a mol. wt. of 300,000,[158,159] and τ, with a mol. wt. ~70,000.[160] The τ factor is composed of four closely spaced bands when analyzed by one-dimensional PAGE, whereas more than 12 isoproteins are detected after two-dimensional analysis.

The first assumption we made at this stage was that the rate of neurite outgrowth increases with brain development because the rate of tubulin assembly increases as a result of changes in the concentration or activity of MAPs. A good correlation was found among these three parameters by studying *in vitro* microtubule assembly in rat brain preparations[161–165]:

1. In the fetus (15–18 days of gestation) and soon after birth, the rate and overall *in vitro* assembly were very low when compared to the adult.
2. The rates of assembly increased with age and could be restored by adding adult MAPs.
3. The rate of assembly was lowered at day 15 when hypothyroidism was established 5 days before birth; addition of MAPs restored normal rates.
4. Thyroid hormone administration at birth restored normal rates of assembly at day 15. The percentage of reversibility was related to the age at which therapy was begun.
5. Spontaneous restoration of the rates was observed when hypothyroidism produced during late fetal life was maintained until adulthood.
6. Animal species such as the guinea pig, which are considered as having a mature brain at birth, also showed reduced rates of assembly, but only with fetal preparations.

An answer to the second question raised above, i.e., whether the factors present at early stages are different from those responsible for microtubule assembly at an adult stage, was provided by comparing the composition of MAPs prepared from *in vitro* assembled microtubules at different stages of brain development: PAGE analysis showed that at day 3 the τ fraction contains only two bands (τ_{slow} and τ_{fast}) with neither of them migrating as the four adult closely spaced bands (τ 1 to 4).[166] There were similar findings for the fetal guinea pig τ.[167] In addition, young τ_{slow} and τ_{fast} were purified and shown (1) to be both active but less so than adult τ 1 to 4 in promoting pure tubulin assembly, and (2) to have a different peptide mapping, thus suggesting that different genes are coding for the τ entities at young and adult stages.[167]

In contrast, the four adult τs have very similar peptide mappings.[160] They might be generated either by posttranslational processing or coded by redundant genes of slightly different size.

The MAPs prepared from microtubules of 15-day-old hypothyroid rats were also analyzed: one of the MAPs, τ_3, was missing, whereas young τ_F was still present at concentrations higher than in the euthyroid controls of the same age (unpublished results).

More questions are raised by these data than have been solved. It is not clear, for instance, whether the young τs are specific for the dividing cell, whereas adult τs are required for neurite outgrowth. Recent data also suggest that the mechanism of microtubule assembly might be more complicated than suspected. For instance, Matus *et al.*[168] have recently reported that MAP_2 is present only in dendrites and not in axons, as shown by immunofluorescent techniques. Recent unpublished experiments also suggest that the different MAPs in addition to promoting tubulin assembly may also regulate the size of the microtubule.

A number of mechanisms might therefore modulate the number and the size of the microtubules by changing qualitatively or quantitatively one or more

MAPs during brain development. These changes might occur with a different timing depending on the brain region, the cell types, the receptivity of each cell type to thyroid hormones, and the domain of gene products controlled by these hormones.

5. NEONATAL THYROID SCREENING AND REPLACEMENT THERAPY

Considerable interest has been shown in the past few years in neonatal thyroid screening. Screening programs have been developed in several industrialized countries in order to detect those infants (1/4000 to 1/5000) who have low T_4 production just after birth. It is outside the scope of this chapter to review in detail the large literature published on this subject. The reader should therefore refer to the excellent surveys edited by Fisher and Burrow[169] and Burrow and Dussault.[170] We only summarize here the main lines of interest of these studies.

1. As indicated above, neonatal hypothyroidism may result from a number of factors operating at the level of the thyroid gland: severe or mild iodine deficiency, metabolic abnormalities, thyroid agenesis or ectopy, treatment of the mothers during pregnancy with antithyroid or other drugs. In addition, hypothyroidism may result from defective hypothalamic–pituitary activity, delayed onset of TSH elevation, defective transport of thyroid hormone in the blood, reduced conversion of T_4 to T_3 (low-T_3 syndrome), etc.
2. Detection of these various cases of postnatal hypothyroidism can be accomplished by both T_4 and TSH measurements, which are usually performed during the first week with infant blood or by using cord blood at birth.
3. The presence of high TSH (>80 U/ml) or very low T_4 (<4 µg/ml) is indicative of defective thyroid function and justifies priority medical evaluation and possibly replacement therapy.
4. Replacement therapy is performed as soon as possible. Oral T_4 administration is the preferred approach. The amount of hormone delivered must be adjusted on the basis of all possible clinical signs, and periodic evaluations are recommended. Excessive or inadequate therapy may be detrimental to neurological development and function. Since, as indicated above, thyroid function begins before birth in man, and since it is impossible by definition to know the potential mental ability of each individual, the degree of success of such replacement therapy is and will remain open to discussion.

6. CONCLUSIONS

Several general statements might be proposed at the end of this survey.
The onset of thyroid function, the production of thyroid hormones, the maturation of neuroendocrine control, and the regulation of the concentration

of circulating hormones seem to be synchronized with the critical period of brain development.

Thyroid hormones seem to be required to synchronize the various differentiation events occurring during this period. Two extreme possibilities have been discussed on the basis of the available data. According to the first one, thyroid hormones would essentially synchronize neurite outgrowth in the different neuronal cell types. Thyroid hormone deficiency or excess would result in desynchronization of the various developmental events with a series of consequences (reduced synaptogenesis, cell death, impaired myelinogenesis, etc). The second possibility, which does not contradict the first, would be that thyroid hormones have direct and specific effects on each of the events occurring during the critical period of brain differentiation.

The primary mechanism of action of thyroid hormones in the brain remains unknown. By analogy with the data obtained in other systems, it might be proposed that thyroid hormones also act by modifying gene expression. Several lines of evidence are in favor of such an assumption: (1) thyroxine is converted in the brain to 3,5,3'-triiodothyronine, as in other tissues; (2) nuclear receptors and other cellular binding proteins are present in the brain; and (3) the activity of several enzymes and probably their synthesis seem to be under thyroid hormone control.

As emphasized in the Introduction and as demonstrated by Eayrs[34] in the cortex and by Legrand[18] in the cerebellum, the most dramatic effect of hypothyroidism that is seen during brain development is a hypoplastic neuropil. Tight control of two major parameters, distance and direction, is probably essential to establish the adult neuronal network. The distance between two interconnected neurons is obviously directly related to the length of the dendrites and axons. The establishment of contact between two interconnected neurons will therefore essentially depend on the distance and rate of migration of each neuron, the rates of neurite growth, and the length and orientation of their axons and dendrites. It might be, therefore, that each cell type expresses, according to the timing of its genetic program, a reduced set of gene products directly responsible for the definition of these parameters. Epigenic factors such as thyroid hormones are probably required to synchronize the expression of these gene products in the different nerve cells that will establish contact in the adult mature brain.

REFERENCES

1. Eayrs, J. T., 1960, *Br. Med. Bull.* **16**:122–127.
2. Schapiro, S., 1977 *Gen. Comp. Endocrinol.* **10**:214–228.
3. Fisher, D. A., Dussault, J. H., Sack, J., and Chopra, I. J., 1977, *Recent Prog. Horm. Res.* **33**:59–116.
4. Dumont, J., and Lamy, F., 1980, *Thyroid Gland* (M. De Visscher, ed.), Raven Press, New York, pp. 153–168.
5. Dumont, J., Boeynaems, J. M., De Coster, C., Erneux, C., Lamy, F., Lecocq, R., Mockel, J., Unger, J., and Van Sande, J., 1978, *Adv. Cyclic Nucleotide Res.* **9**:723–734.
6. Nunez, J., 1980, *The Thyroid Gland* (M. De Visscher, ed.), Raven Press, New York, pp. 39–60.

7. Nunez, J., and Pommier, J., 1982, *Vitam. Horm.* **39**:175–229.
8. Chopra, I. J., Solomon, D. M., Chopra, U., Wu, S. Y., Fisher, D. A., and Nakamura, H., 1978, *Recent Prog. Horm. Res.* **34**:521–567.
9. Roche, J., Michel, R., Wolf, W., and Nunez, J., 1965, *Biochim. Biophys. Acta.* **19**:308–317.
10. Bakke, J. L., Lawrence, N., and Wilber, J. F., 1974, *Endocrinology* **25**:406–411.
11. Azizi, F., Vaganakis, A. G., Bollinger, J., Reichlin, S., Braverman, L. E., and Ingbar, S. H., 1974, *Endocrinology* **94**:1681–1688.
12. Patel, A. J., Rabié, A., Lewis, P. D., and Balász, R., 1976, *Brain Res.* **104**:33–48.
13. Dussault, J. H., and Labrie, E., 1975, *Endocrinology* **97**:1321–1324.
14. Querido, A., and Swaab, D. F., 1975, *Brain Development and Thyroid Deficiency*, North Holland, Amsterdam, pp. 27–32.
15. Altman, J., 1969, *Handbook of Neurochemistry*, Volume 2 (A. Lajtha, ed.), Plenum Press, New York, pp. 137–182.
16. Nicholson, J. L., and Altman, J., 1972, *Brain Res.* **44**:13–23.
17. Legrand, J., 1965, *C.R. Acad. Sci.* [D] (*Paris*) **261**:544–547.
18. Legrand, J., 1967, *Arch. Anat. Microsc. Morphol. Exp.* **56**:205–244.
19. Rebiere, A., and Legrand, J., 1972, *Arch. Anat. Microsc.* **61**:105–126.
20. Hamburgh, M., Mendoza, L. A., Burkart, J. F., and Weil, F., 1971, *Cellular Aspects of Neuronal Growth and Differentiation* (D. C. Please, ed.), University of California Press, Berkeley, pp. 321–328.
21. Weichsel, M. E., Jr., 1974, *Brain Res.* **78**:455–465.
22. Gourdon, J., Clos, J., Coste, C., Dainat, J., and Legrand, J., 1973, *J. Neurochem.* **21**:861–871.
23. Balász, R., Kovacs, S., Teichgräber, P., Cocks, W., and Eayrs, J. T., 1968, *J. Neurochem.* **15**:1335–1379.
24. Patel, A. J., and Balász, R., 1975, *Metabolic Compartmentation and Neurotransmission* (S. Berl, D. D. Clark, and D. Schneider, eds.), Plenum Press, New York, pp. 363–383.
25. Patel, A. J., Rabie, A., Lewis, P. D., and Balász, R., 1976, *Brain Res.* **104**:33–48.
26. Balász, R., 1977, *Thyroid Hormones and Brain Development* (G. D. Grave, ed.), Raven Press, New York, pp. 287–298.
27. Lewis, P. D., Patel, A. J., Johnson, A. L., and Balász, R., 1976, *Brain Res.* **104**:49–62.
28. Patel, A. J., Lewis, P. D., Balász, R., Bailey, P., and Lai, M., 1979, *Brain Res.* **172**:57–72.
29. Lauder, J. M., 1978, *Brain Res.* **142**:25–39.
30. Rabie, A., Patel, A. J., Clalel, M. C., and Legrand, J., 1979, *Dev. Neurosci.* **2**:183–194.
31. Lauder, J. M., 1977, *Brain Res.* **126**:31–51.
32. Clos, J., and Legrand, J., 1973, *Brain Res.* **63**:450–455.
33. Geloso, J. P., Hemon, P., Legrand, J., Legrand, C., and Jost, A., 1968, *Gen. Comp. Endocrinol.* **10**:19n197.
34. Eayrs, J. R., and Taylor, S. M., 1951, *J. Anat.* (*Lond.*) **83**:350–358.
35. Eayrs, J. R., and Horn, G., 1955, *Anat. Rec.* **121**:53–61.
36. Eayrs, J. T., 1961, *Growth* **25**:175–189.
37. Rabie, A., Clavel, M. C., and Legrand, J., 1980, *Brain Res.* **190**:409–414.
38. Lauder, J. M., 1979, *Dev. Biol.* **70**:105–115.
39. Lauder, J. M., 1977, *Thyroid Hormones and Brain Development* (G. D. Grave, ed.), Raven Press, New York, pp. 235–252.
40. Lauder, J. M., and Bohn, M. C., 1980, *Progress in Psychoneuroendocrinology* (F. Brambilla, G. Racagni, and D. De Wied, eds.), Elsevier-North Holland Biomedical Press, Amsterdam, pp. 603–620.
41. Crepel, F., 1974, *Exp. Brain Res.* **20**:403–420.
42. Crepel, F., 1975, *Brain Res.* **85**:157–160.
43. Rabié, A., and Legrand, J., 1973, *Brain Res.* **61**:267–278.
44. Patel, A. J., Balász, R., and Johnson, A. L., 1973, *J. Neurochem.* **20**:1151–1165.
45. Lewis, P. D., Balász, R., Patel, A. J., and Johnson, A. L., 1975, *Brain Res.* **83**:235–247.
46. Walker, P., Weichsel, M. E., Jr., Fisher, D. A., Guo, S. M., and Fisher, D. A., 1979, *Science* **204**:427–429.
47. Walker, P., Weil, M. L., Weichsel, M. E., and Fisher, D. A., 1981, *Life Sci.* **28**:1777–1788.

48. Snyder, E. Y., and Kim, S. Y., 1979, *Neurosci. Lett.* 13:225–230.
49. Honnegger, P., Lenoir, D., and Favrod, P., 1979, *Nature* 282:305–308.
50. Honnegger, P., and Lenoir, D., 1980, *Brain Res.* 199:425–434.
51. Romijn, H. J., Mc Habets, A. M., Mud, M. T., and Wolters, P. S., 1982, *Dev. Brain Res.* 2:583–589.
52. Cragg, B. G., 1970, *Brain Res.* 18:297–307.
53. Nicholson, J. L., and Altman, J., 1972, *Brain Res.* 44:25–36.
54. Nicholson, J. L., and Altman, J., 1972, *Science* 176:530–532.
55. Lefranc, G., George, Y., and Tusques, J., 1968, *C.R. Soc. Biol. (Paris)* 162:219–224.
56. Geel, S. E., and Timiras, P. S., 1967, *Endocrinology* 80:1069–1074.
57. Yuwiler, A., and Brammer, G. L., 1981, *J. Neurochem.* 37:985–991.
58. Krawiec, L., Garcia Argiz, C. A., Gomez, C. J., and Pasquini, J. M., 1969, *Brain Res.* 15:209–218.
59. Ben Baruch, G., Egozy, Y., Kloog, Y., Mashiach, S., and Sokoloky, M., 1981, *Endocrinology* 109:235–239.
60. Dupont, A., Dussault, J. H., Rouleau, D., Di Paolo, T., Coulombe, P., Gagné, B., Merand, Y., Moore, S., and Barden, N., 1981, *Endocrinology* 106:2039–2045.
61. Smith, R. M., Patel, A. J., Kingsbury, A. E., Hunt, A., and Balśz, R., 1980, *Brain Res.* 198:375–387.
62. Patel, A. J., Smith, R. M., Kingsbury, Hunt, A., and Balśz, R., 1980, *Brain Res.* 198:389–402.
63. Clos, J., and Legrand, J., 1969, *Arch. Anat. Microsc.* 58:339–354.
64. Walravens, P., and Chase, H. P., 1969, *J. Neurochem.* 16:1477–1484.
65. Balśz, R., Brooksbank, B. W. L., Davison, A. N., Eayrs, J. T., and Wilson, D. A., 1969, *Brain Res.* 15:219–232.
66. Clos, J., and Legrand, J., 1970, *Brain Res.* 22:285–297.
67. Dalal, J., Valcana, T., Timiras, P. S., and Einstein, E. R., 1971, *Neurobiology* 1:211–224.
68. Clos, J., Rebiere, A., and Legrand, J., 1973, *Brain Res.* 63:445–449.
69. Bass, N. H., and Young, E., 1973, *J. Neurol. Sci.* 18:155–173.
70. Rosman, N. P., and Malone, M. J., 1977, *Thyroid Hormones and Brain Development* (G. D. Grave, ed.), Raven Press, New York, pp. 169–194.
71. Wysocki, S. W., and Segal, W., 1972, *Eur. J. Biochem.* 28:183–189.
72. Flynn, T. J., Deshmuhk, D. S., and Piekinger, R. A., 1977, *J. Biol. Chem.* 252:5864–5870.
73. Bhat, N. R., Sarlieve, L., Rao, S. G., and Pieringer, R. A., 1979, *J. Biol. Chem.* 254:9342–9344.
74. Bhat, N. R., Rao, G. S., and Pieringer, R. A., 1981, *J. Biol. Chem.* 256:1167–1171.
75. Malone, M. J., Rosman, N. P., Szoke, M., and Davis, D., 1975, *J. Neurol. Sci.* 26:1–11.
76. Matthieu, J. M., Reier, P. J., and Sawchak, J. A., 1975, *Brain Res.* 84:443–451.
77. Rosman, N. P., Malone, M. J., and Szoke, M., 1975, *J. Neurol. Sci.* 26:159–166.
78. Valcana, T., Einstein, E. R., Sejtey, J., Dalal, K. H., and Timiras, P. S., 1975, *J. Neurol. Sci.* 25:19–27.
79. Walters, S. N., and Morell, P., 1981, *J. Neurochem.* 36:1792–1801.
80. Rende, M., Zucco, M., Cocchia, D., and Michetti, F., 1982, *Dev. Brain Res.* 2:590–595.
81. Robbins, J., Cheng, S. Y., Gershengorn, M. C., Glinnoer, D., Cahnmann, H., and Edelhoch, H., 1978, *Recent Rev. Horm. Res.* 34:477–519.
82. Baxter, J. D., and Eberhardt, N. L., 1979, *Recent Prog. Horm. Res.* 35:97–153.
83. Maxfield, F. R., Willingham, M. C., Pastan, I., Dragsten, P., and Cheng, S. Y., 1981, *Science* 211:63–65.
84. Sokoloff, L., 1977, *Thyroid Hormones and Brain Development* (G. Grave, ed.), Raven Press, New York, pp. 73–89.
85. Sterling, K., Milch, P. O., Lazarus, J. H., Sakurada, T., and Brenner, M. A., 1978, *Science* 201:1126–1129.
86. Tata, J. R., Ernster, L., Lindberg, O., Arrhenius, E., Pedersen, S., and Hedman, R., 1963, *Biochem. J.* 86:408–428.
87. Correze, C., and Nunez, J., 1969, *Bull. Soc. Chim. Biol.* 51:909–917.
88. Courrier, R., Moreau, A., Marois, M., and Morel, F., 1949, *C. R. Soc. Biol. (Paris)* 143:935–938.

89. Jensen, I. M., and Clark, D. E., 1951, *J. Lab. Clin. Med.* **38**:663–670.
90. Ford, D. H., and Gross, J., 1958, *Endocrinology* **62**:416–437.
91. Ford, D. H., 1961, *Gen. Comp. Endocrinol.* **1**:59–69.
92. Ford, D. H., and Rhines, R., 1967, *Brain Res.* **6**:481–488.
93. Dratman, M., Futaesaku, Y., Crutchfield, F. L., Berman, N., Payne, B., Sar, M., and Stumpf, W. E., 1982, *Science* **215**:309–312.
94. Ford, D. H., Fishman, S. K., and Rhines, R., 1962, *Gen. Comp. Endocrinol.* **2**:480–489.
95. Ford, D. H., and Cramer, E. B., 1977, *Thyroid Hormones and Brain Development* (G. D. Grave, ed.), Plenum Press, New York, pp. 1–14.
96. Eberhardt, N. L., Valcana, T., and Timiras, P. S., 1976, *Psychoneuroendocrinology* **1**:399–409.
97. Pitt-Rivers, R., Stanbury, J. B., and Rapp, B., 1971, *J. Clin. Endocrinol. Metab.* **15**:616–620.
98. Braverman, L. E., Ingbar, S. H., and Sterling, K., 1970, *J. Clin. Invest.* **49**:855–864.
99. Grantz, F. R., and Larsen, P. R., 1980, *J. Clin. Invest.* **65**:935–938.
100. Kaplan, M. M., and Yaskoski, K. A., 1980, *J. Clin. Invest.* **66**:551–562.
101. Kaplan, M. A., Mc Cann, U. D., Yaskoski, K. A., Larsen, P. R., and Leonard, J. L., 1981, *Endocrinology* **109**:397–403.
102. Cheron, R. G., Kaplan, M. M., and Larsen, P. R., 1980, *Endocrinology* **106**:1405–1408.
103. Visser, T. J., Leonard, J. L., Kaplan, M. M., Reed, P. R., and Larsen, P., 1981, *Biochem. Biophys, Res. Commun.* **101**:1297–1303.
104. Tanaka, K., Inada, M., Ishi, H., Naito, H., Nishikawa, M., Mashio, Y., and Imura, M., 1981, *Endocrinology* **109**:1619–1624.
105. Dratman, M. B., Crutchfield, F. L., Axelrod, J., Colburn, R. W., and Thoa, N., 1976, *Proc. Natl. Acad. Sci. USA* **73**:941–944.
106. Naidoo, S., and Timiras, P. S., 1979, *Dev. Neurosci.* **2**:213–224.
107. Pliam, N. B., and Goldfine, I. O., 1977, *Biochem. Biophys Res. Commun.* **79**:166–172.
108. Gharbi, J., and Torresani, J., 1979, *Biochem. Biophys. Res. Commun.* **88**:170–177.
109. Segal, J., and Gordon, A., 1977, *Endocrinology* **101**:150–156.
110. Segal, J., and Ingbar, S. H., 1979, *J. Clin. Invest.* **63**:507–517.
111. Hamada, S., Torizuka, K., Mikyake, T., and Fusaze, M., 1970, *Biochim. Biophys. Acta* **201**:479–492.
112. Davis, P. J., Handwerberg, B. S., and Glaser, F., 1974, *J. Biol. Chem.* **249**:6208–6217.
113. Dillman, W., Surks, M. I., and Oppeinheimer, J. H., 1974, *Endocrinology* **95**:492–498.
114. Yoshida, K., and Davis, P. J., 1977, *Biochem. Biophys. Res. Commun.* **78**:697–705.
115. Geel, S. E., 1977, Nature **269**:428–430.
116. Dozin-Van Roye, B., and De Nayer, P., 1978, *FEBS Lett.* **96**:152–154.
117. Lennon, A. M., Osty, J., and Nunez, J., 1980, *Mol. Cell. Endocrinol.* **18**:201–214.
118. Greif, R. L., and Sloane, G., 1978, *Endocrinology* **103**:1899–1902.
119. Oppeinheimer, J. H., Koerner, D., Schwartz, H. L., and Surks, M. I., 1972, *J. Clin. Endocrinol. Metab.* **35**:330–333.
120. De Groot, L. J., and Strausser, J. L., 1974, *Endocrinology* **95**:74–83.
121. Surks, M. I., Koerner, D. H., and Oppeinheimer, J. H., 1975, *J. Clin. Invest.* **55**:50–60.
122. Torresani, J., and De Groot, L. J., 1975, *Endocrinology* **96**:1201–1209.
123. Gruol, D. J., 1980, *Endocrinology* **107**:994–999.
124. Jump, D. B., Seelig, S., Schwartz, H. L., and Oppeinheimer, J. H., 1981, *Biochemistry* **20**:6781–6789.
125. Samuels, H. H., Stanley, F., Casanova, J., and Shao, T., 1980, *J. Biol. Chem.* **255**:2499–2508.
126. Bolger, M. B., and Jorgensen, E. C., 1980, *J. Biol. Chem.* **255**:10217–10278.
127. Koerner, D., Shwartz, H. L., Surks, M. I., Oppeinheimer, J. H., and Jorgensen, E. C., 1975, *J. Biol- Chem.* **250**:6417–6423.
128. Cody, V., 1978, *Recent Prog. Hormone Res.* **34**:437–474.
129. Samuels, M. H., and Shapiro, L. E., 1976, *Proc. Natl. Acad. Sci. U.S.A.* **73**:3369–3373.
130. Silva, J. E., and Larsen, P. R., 1976, *J. Clin. Invest.* **61**:1247–1259.

131. Oppenheimer, J. H., Coulombe, P., Schwartz, H. L., and Gutfeld, N. W., 1978, *J. Clin. Invest.* **61**:987–997.
132. Mariash, C. N., Kaiser, F. E., and Oppenheimer, J. H. N., 1980, *Endocrinology*, **106**:22–27.
133. Dillman, W. H., Mendecki, J., Koerner, D., Schwartz, H. L., and Oppenheimer, J. H., 1978, *Endocrinology* **102**:568–575.
134. Kurtz, D. T., Sippel, A. E., and Feigelson, P., 1976, *Biochemistry* **15**:1031–1036.
135. Towle, H. C., Mariash, C. N., Schwartz, H. L., and Oppenheimer, J. H., 1981, *Biochemistry* **20**:3486–3492.
136. Eberhardt, N. L., Ring, J. C., Latham, K. R., and Baxter, J. D., 1979, *J. Biol. Chem.* **254**:8538–8539.
137. Oppenheimer, J. H., Schwartz, H. L., and Surks, M. I., 1974, *Endocrinology* **95**:897–903.
138. Schwartz, H. L., and Oppenheimer, J. H., 1978, *Endocrinology* **103**:943–948.
139. Eberhardt, N. L., Valcana, T., and Timiras, P. S., 1978, *Endocrinology* **102**:556–561.
140. Valcana, T., and Timiras, P. S., 1978, *Mol. Cell. Endocrinol.* **11**:31–41.
141. De Nayer, P., and Dozin-Van Hoye, B., 1981, *Biochem. Biophys. Res. Commun.* **98**:1–6.
142. Oklund, S., and Timiras, P. S., 1977, *Thyroid Hormone and Brain Development* (G. D. Grave, ed.), Raven Press, New York, pp. 33–44.
143. Sokoloff, L., 1977, *Thyroid Hormone and Brain Development* (G. D. Grave, ed.), Raven Press, New York, pp. 73–91.
144. Valcana, T., and Eberhardt, L., 1977, *Thyroid Hormone and Brain Development* (G. D. Grave, ed.), Raven Press, New York, pp. 271–286.
145. Krawiec, L., Montalbano, C. A., Duvilanski, B. H., De Guglielmone, A. E. R., and Gomez, C., 1977, *Thyroid Hormone and Brain Development* (G. D. Grave, ed.), Raven Press, New York, pp. 315–334.
146. Tomkins, G. M., 1975, *Science* **189**:760–763.
147. Ivarie, R. D., Morris, J. A., and Eberhardt, N. L., 1980, *Recent Prog. Horm. Res.* **39**:195–235.
148. Yamada, K. M., Spooner, B. S., and Wessels, M. K., 1970, *Proc. Natl. Acad. Sci.* **66**:1206–1212.
149. Seeds, N. W., Gilman, A. G., Amano, T., and Niremberg, M. W., 1970, *Proc. Natl. Acad. Sci. U.S.A.* **66**:160–167.
150. Daniels, M. P., 1972, *J. Cell Biol.* **53**:164–176.
151. Peters, A., and Vaughn, J. E., 1967, *J. Cell Biol.* **32**:113–119.
152. Lyser, K. M., 1968, *Dev. Biol.* **17**:117–152.
153. Weber, K., and Osborn, M., 1979, *Microtubules* (K. Roberts and J. F. Hyams, eds.), Academic Press, New York, pp. 279–313.
154. Nunez, J., Lennon, A. M., Mareck, A., Francon, J., and Fellous, A., 1980, *Radioimmunoassay of Hormones* (E. Albertini, ed.), Excerpta Medica, Amsterdam, pp. 127–135.
155. Fellous, A., Francon, J., Virion, A., and Nunez, J., 1975, *FEBS Lett.* **57**:5–8.
156. Weisenberg, R. G., 1972, *Science* **177**:1104–1105.
157. Shelanski, M. L., Gaskin, F., and Cantor, R. C., 1973, *Proc. Natl. Acad. Sci. U.S.A.* **70**:765–768.
158. Sloboda, R. D., Dentler, W. L., and Rosenbaum, J. L., 1976, *Biochemistry* **15**:4497–4505.
159. Murphy, D. B., Johnson, K. A., and Borisy, G. G., 1977, *J. Mol. Biol.* **117**:33–52.
160. Cleveland, D. W., Hwo, S. Y., and Kirschner, M. W., 1977, *J. Mol. Biol.* **116**:207–225.
161. Fellous, A., Francon, J., Lennon, A. M., and Nunez, J., 1976, *FEBS Lett.* **64**:400–403.
162. Francon, J., Fellous, A., Lennon, A. M., and Nunez, J., 1978, *Eur. J. Biochem.* **85**:43–53.
163. Francon, J., Fellous, A., Lennon, A. M., and Nunez, J., 1977, *Nature* **266**:188–190.
164. Fellous, A., Lennon, A. M., Francon, J., and Nunez, J., 1979, *Eur. J. Biochem.* **101**:365–376.
165. Lennon, A. M., Francon, J., Fellous, A., and Nunez, J., 1980, *J. Neurochem.* **35**:804–813.
166. Mareck, A., Fellous, A., Francon, J., and Nunez, J., 1980, *Nature* **284**:353–355.
167. Francon, J., Lennon, A. M., Fellows, A., Mareck, A., Pierre, M., and Nunez, J., 1982, *Europ. J. Biochem.* **129**:465–471.

168. Matus, A., Bernhardt, A., and Hugh-Jones, T., 1981, *Proc. Natl. Acad. Sci. U.S.A.* **78**:3010–3014.
169. Fisher, D. A., and Burrow, A. (eds.), 1975, *Kroc Foundation Symposia Series*, Volume 3, Raven Press, New York.
170. Burrow, G. N., and Dussault, J. M. (eds.), 1980, *Neonatal Thyroid Screening*, Raven Press, New York.

2

Mechanisms of Gonadal Steroid Actions on Behavior

Thomas C. Rainbow and Bruce S. McEwen

1. INTRODUCTION

In all known species of vertebrates, gonadal steroid hormones are powerful regulators of reproductive behavior. In laboratory studies, there is substantial evidence that the behavioral effects of gonadal steroids occur exclusively through actions on the brain. The ablation of particular regions of the rat hypothalamus will eliminate the facilitation of feminine sexual behavior by estrogen (E_2) and progesterone (P), whereas the direct application of picogram quantities of E_2 and P to the same regions is sufficient to stimulate sexual behavior.

We review here what is known about the molecular and cellular events that mediate the behavioral actions of gonadal steroids. Although the behavioral phenomenology of gonadal steroids has been studied for over 30 years,[6,7] it has only been recently that sufficiently sensitive neurochemical methods have been developed to allow the determination of cellular mechanisms. The advent of these techniques has in turn stimulated new behavioral studies designed to provide information relevant to the neurochemist, such as the precise anatomic locus of steroid hormone action and a detailed time course of the appearance of behavior after physiological doses of steroids. We feel that the determination of the cellular and molecular events responsible for steroid action is one of the most powerful approaches for elucidating the functions of the mammalian brain. The identification of the cells responsible for steroid effects and the mapping of their connections will allow for the first time the determination of circuits that mediate mammalian behaviors. The determination of the molecular events that change synaptic function in these pathways to alter behavior might provide insights into more complicated forms of synaptic plasticity, such as that which mediates associative learning.

Thomas C. Rainbow • Department of Pharmacology, University of Pennsylvania, Philadelphia, Pennsylvania 19104.   *Bruce S. McEwen* • The Rockefeller University, New York, New York 10021.

Our interest in the neurochemistry of steroid action resulted in part from our previous work on the biochemistry of memory[8,9] and the need to find analogues of long-term behavioral modification more amenable to localization and cellular analysis. Both the formation of memories and the activation of sexual behavior by gonadal steroids probably involve the alteration of synaptic function in specific brain cells. In both cases, it is necessary to know the anatomic location of the relevant cells before the biochemical mechanisms are studied. As we shall discuss, much of the progress that has been made in determining the cellular mechanisms of steroid effects can be attributed to the identification of the relevant cells. By contrast, the neuronal pathways responsible for learning and memory in the rat are almost as undefined today as they were 16 years ago, when one of us (B.M.) studied the biochemistry of memory. We are encouraged by the progress that has been made in determining the cellular mechanisms of how gonadal steroids influence behavior and feel that both the solution to this problem and the approach to solve it are relevant to clarifying the neural basis of memory.

2. ACTIVATIONAL EFFECTS OF GONADAL STEROIDS ON BEHAVIOR

Both sexual and nonsexual behaviors in adult rodents are influenced by gonadal steroids. The nonsexual behaviors include aspects of feeding, drinking, aggression, and activity.[10] Whereas sexual behavior is completely dependent on gonadal steroids, nonsexual behaviors are only modulated by gonadal hormones, making them less suitable for studies on biochemical mechanisms. The behavior in rodents that has attracted the attention of many neurochemists and physiologists is feminine sexual behavior, the expression of which is totally dependent on estrogen and progesterone. Although masculine sexual behavior is similarly dependent on testosterone, the disappearance of the behavior after castration is very slow, as is its restoration by testosterone.[11] By contrast, the onset of feminine sexual behavior is relatively rapid, occurring within hours or minutes after exposure to E_2 or P.

An attractive feature of feminine sexual behavior for mechanistic studies is that it is easy to quantify. Feminine sexual behavior in rodents can be divided into two stereotyped components, receptivity and proceptivity, which are affected differently by E_2 and P. Sexual receptivity, or lordosis, is the posture the female assumes in response to a mount from a male. The lordosis posture (Fig. 1) is a stereotyped dorsiflexion of the spine and elevation of the rump and head. The purpose of the reflex is to facilitate intromission by the male. A common way to quantify this behavior is to express as a ratio the number of lordotic postures in response to an arbitrary number of mounts from a male. This ratio is known as the lordosis quotient. It is also possible to quantify the quality of the lordosis posture, ranking the degree of spine dorsiflexion on a three-point scale.[12]

The lordosis reflex is totally dependent on exposure to estrogen. Ovariectomized rats, deprived of circulating estrogens, have LQs close to zero when

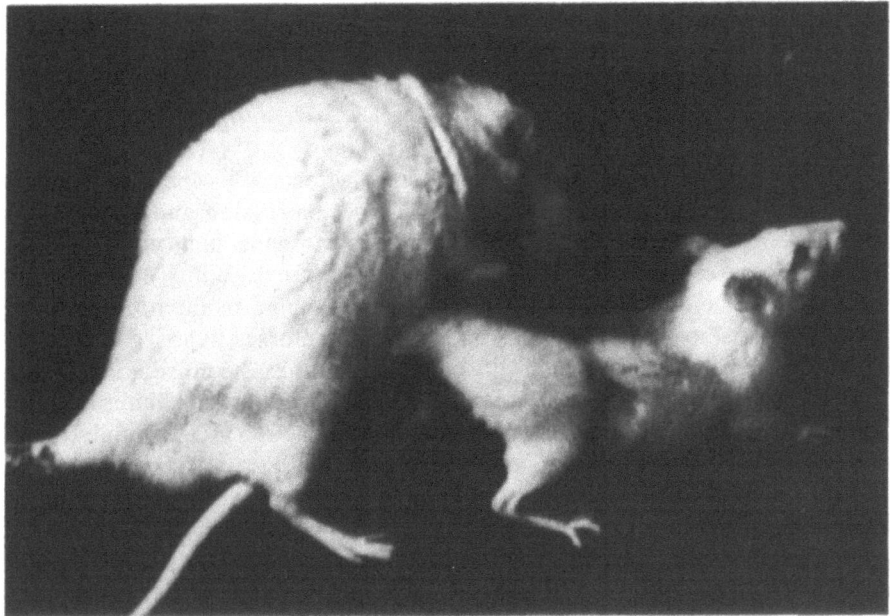

Fig. 1. Female rat in lordosis posture (photograph courtesy of D. Pfaff).

tested with stud males. Although chronic exposure to extrogen alone is sufficient to activate sexual receptivity[13] (B. Parsons, T. C. Rainbow, and B. S. McEwen, unpublished observations), the facilitation of lordosis behavior with physiological exposures of E_2 requires the additional administration of progesterone. The administration of P to a rat that has been primed at least 18–24 h before with E_2 will result in a elevation of LQ, an improvement in the lordosis quality score, and the appearance of of entirely new behaviors, known collectively as proceptive behaviors. The purpose of proceptive behaviors is to entice the males to intromit. There are three main proceptive behaviors—hopping, which is a short leap terminating in a receptive crouching posture, darting, a short run, also abruptly ending in a crouching posture, and ear wiggling, a rapid vibration of the ears. These behaviors are typically quantified by measuring their rate of occurrence with an event recorder.

The activation of proceptive behaviors is a unique action of progesterone. Although high doses of E_2 given either acutely or chronically will stimulate proceptive behaviors[14] (B. Parsons, T. C. Rainbow, and B. S. McEwen, unpublished observations), this is probably mediated through occupation of brain progestin receptors, for which E_2 has a weak affinity.[15,16] Under the conditions of the rat estrus cycle, where blood levels of E_2 are elevated for 48 h before a rat comes into heat, the only behavioral action of E_2 would be to prime the brain for the facilitation of receptive behaviors by progesterone.[13] With longer durations of exposure to a proestrus level of E_2, sexual receptivity will be activated by estrogen alone. Thus, estrogen and progesterone synergize to activate feminine sexual behavior, with E_2 having the effect of priming the brain for the actions of P as well as selectively activating receptive behavior, and P facilitating receptive behavior and selectively activating proceptive behavior.

2.1. Genomic Mechanisms of Steroid-Activated Behaviors

There is much evidence that the activational effects of E_2 and P on behavior are mediated through changes in neuronal gene expression and protein synthesis. As in well-characterized peripheral target organs such as the uterus, the brain possesses specific intracellular receptors for E_2 and P, which translocate into the cell nucleus after steroid hormones bind to them.[2,16,17] In the uterus, the translocation of cytoplasmic steroid receptors activates gene expression and protein synthesis, which subsequently modifies the physiology of uterine cells. In the brain, the occupation of nuclear receptors correlates with the appearance and disappearance of sexual behavior,[16,18,19,20] and inhibitors of gene expression block the behavioral effects of E_2 and P.[21-23] This is essentially the same evidence that supports the genomic model of steroid action in peripheral target organs.

Although the genomic model of steroid action has been successfully applied to peripheral targets for over a decade, it is only recently that evidence has been supplied to apply this explanation to the behavioral actions of E_2 and P. The delay in accepting the genomic model had mostly to do with the behavioral actions of progesterone. Although intracellular estrogen receptors were identified relatively early in rat brain,[26,27] it was not until the introduction of the synthetic progestin R5020 that a ligand existed that could measure the low level of progestin receptor in brain tissue.[15] Up to that point, the inability to find progestin receptors in brain tissue, along with the fact that P can act very quickly to activate behavior (15 min), led to the suggestion that progesterone might facilitate sexual behavior by acting on membrane receptors.[28] However, the identification of neural progestin receptors with [^{3}H]R5020 resulted in studies showing that cytoplasmic progestin receptors translocate to the nucleus[16,29] and that the extent and time course of the nuclear translocation correlate with the activation of behavior by P[15,18] (Fig. 2).

Additional support for the genomic model came from studies in which the behavioral effects of P were blocked by inhibition of cerebral protein synthesis.[21,22,30,31] Again, there was previous evidence that inhibitors of RNA and protein synthesis (actinomycin D and cycloheximide) would block the priming and receptive actions of E_2.[24,25] It was difficult to apply the same inhibitors to the rapid behavioral actions of P because their high toxicity in rats precluded using doses that would rapidly inhibit macromolecular synthesis. However, we found that rats could tolerate large doses of the protein synthesis inhibitor anisomycin, which has been extensively used in studies on the biochemistry of memory.[32] The subcutaneous administration of anisomycin inhibited cerebral protein synthesis by 90% and rapidly blocked the activation of receptivity and proceptivity by P (Fig. 3). The rapid inhibition of sexual behavior by anisomycin raised the possibility that the behavioral deficit might simply reflect a generalized impairment of movement. We controlled for this in part by showing that under conditions of high E_2 priming, anisomycin would selectively

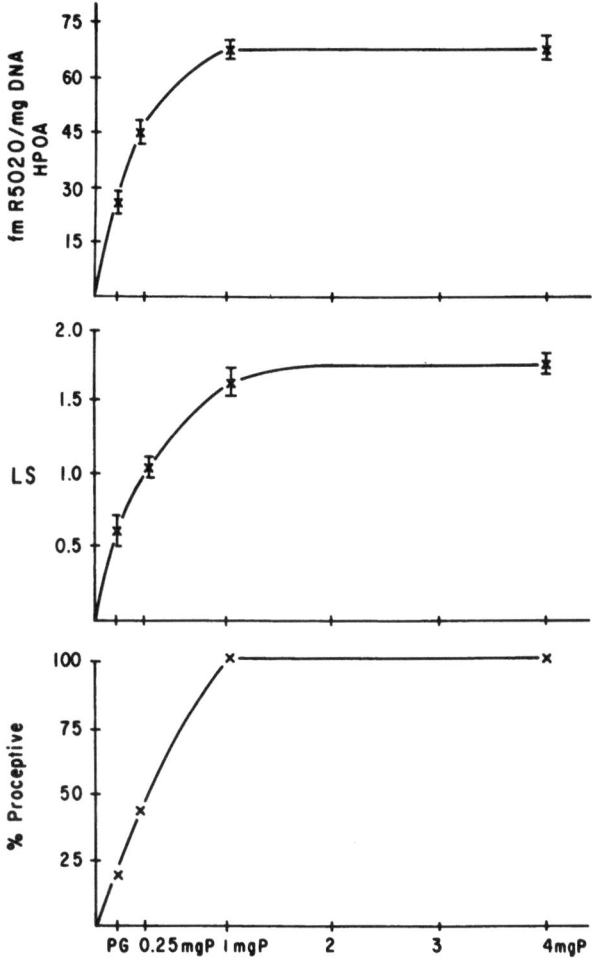

Fig. 2. Correlation between nuclear progestin receptor occupation and proceptive sexual behavior. Ovariectomized rats were given subcutaneous injections of 20 µg EB once a day for 3 days. They were then given a subcutaneous injection of 0.25, 1, or 4 mg P in 0.1 ml propylene glycol (PG) vehicle or PG vehicle alone. Nuclear progestin receptor measurements were made 1 h after injection of P. Sexual behavior was determined 2 h after administration of P. Top: Amount of progestin receptor in MBH–POA nuclei 1 h after injection of 0.25, 1, or 4 mg P or PG vehicle alone. Values are the means ± S.E. for six or seven determinations. $F = 6.65$, $P < 0.01$, one-way ANOVA: $P < 0.05$ for 4 mg P and 1 mg P vs. 0.25 mg P and PG, and for 0.25 P or PG (Duncan's t). Middle: Lordosis quality score 2 h after injection of the same doses of P or PG vehicle: $n = 5$, for all groups $H = 13.7$. Kruskal–Wallis, $P < 0.01$; $P < 0.05$ for 4 mg P and 1 mg vs. 0.25 mg P and PG and for 0.25 vs. PG (Mann–Whitney U). Bottom: Percentage of rats showing proceptive behaviors (hop–darting or ear wiggling) 2 h after injection of P or PG vehicle: $n = 5$ for all groups; $P < 0.01$, χ^2 PG vs. 1 mg or 4 mg P. (Reprinted with permission from Rainbow et al.[16])

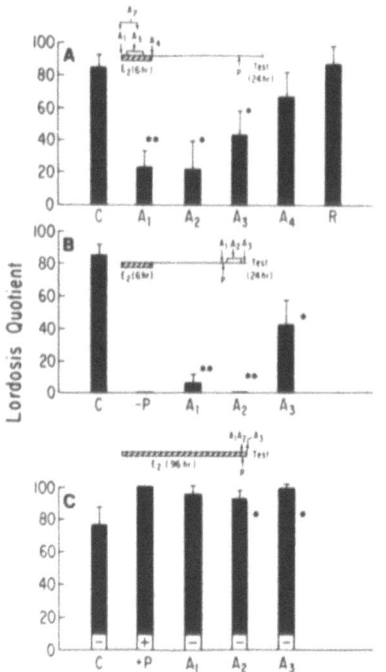

Fig. 3. Effects of anisomycin (Ani) on activation of sexual behavior by E_2 and P. Anisomycin (100 mg/kg) in saline or saline alone was given during exposure to gonadal steroids. Ovariectomized rats received E_2 in Silastic capsules inserted under the skin. Silastic capsules were removed after 6 h (A, B) or left in place for 96 h (C). Tests of sexual behavior were made either 24 h (A, B) or 96 h (C) after E_2 was administered. P (1 mg in 0.1 ml sesame oil) was given 5 h before the tests. Results are means ± S.E.M. A: Inhibition of E_2 activation of behavior. C: Controls ($n = 11$), ovariectomized rats given Silastic E_2 and P before testing. All rats received injections of saline with hormone treatment. A_1 ($n = 9$): Ani given 15 min before insertion of Silastic E_2 capsules. A_2 ($n = 5$): Ani given 15 min before placement of E_2 capsules and again 2.5 h later. A_3 ($n = 6$): rats received injections of Ani between 30 min and 4 h after insertion of E_2 capsules. A_4 ($n = 6$): Ani given immediately after removal of Silastic E_2 capsules. R ($n = 9$): reversal rats treated with Ani were given E_2 and P 1 week later, as before, and retested for sexual behavior. B: Inhibition by Ani of P facilitation of sexual behavior. C: Controls ($n = 11$), ovariectomized rats given injections of saline with hormone treatment. $-P$ ($n = 6$): rats given E_2 and injections of sesame oil 5 h before testing. A_1 ($n = 6$): Ani given 15 min before injection of P. A_2 ($n = 5$): Ani given 1–4 h after P. A_3 ($n = 6$): Ani given 15 min before testing. C: Effects of Ani on receptivity after 96 h E_2 and proceptivity induced by P. +, presence of proceptive behaviors (hopping, darting, and ear wiggling); −, no proceptivity. C ($n = 8$): Controls, ovariectomized rats given Silastic E_2 capsules for 96 h. No proceptive behavior. +P ($n = 8$): P given 5 h before testing. Seven animals showed proceptive behavior. All rats also received injections of saline. A_1 ($n = 4$): Ani given 5 h before testing. A_2 ($n = 6$): Ani given 15 min before injection of P. No proceptivity. A_3 ($n = 4$): Ani given 1 h before testing. No proceptive behavior. Numbers of rats showing proceptivity in groups A_2 and A_3 were significantly fewer than in the +P treatment group ($P < 0.02$, Mann–Whitney U-test, two-tailed test). **Significantly different from control groups, $P < 0.002$, Mann–Whitney U-test, two-tailed test. *$P < 0.02$, Mann–Whitney U-test. (Reprinted with permission from Rainbow et al.[21])

block the proceptive actions of P but would not impair sexual receptivity established by chronic exposure to E_2 (Fig. 3). Similarly, anisomycin blocks the proceptive actions of P in hamsters, where the effect of P is to make the animal less active.[31] It is likely that the rapid inhibition of sexual behavior by anisomycin results from a loss of rapidly turning over proteins important for the activation of behavior by P.

2.2. Target Cells for the Activation of Feminine Sexual Behavior

Although the evidence is relatively strong that E_2 and P activate behavior by affecting the synthesis of specific proteins, conclusive evidence can only be obtained by identifying the proteins induced by steroids, As it is likely that these proteins would be produced in a relatively small number of neurons, the isolation of these proteins would require the localization in rat brain of the target cells for the activation of feminine sexual behavior. A great deal of progress has been made over the past 10 years in mapping the neural circuits responsible for feminine sexual behavior. This is largely a result of the use of steroid autoradiography to localize brain receptors for E_2 and P.[33,34] The intracellular receptors for E_2 and P are highly concentrated in particular hypothalamic nuclei and limbic nuclei and are essentially absent from most other brain regions. There is now good evidence that one of the steroid-concentrating nuclei, the ventromedial nucleus of the hypothalamus (VMN), and specifically its lateral portion, contains the target cells for the activation of feminine sexual behavior. The ablation of the VMN will disrupt the activation of receptive and proceptive behaviors,[1,35] whereas the direct application to the VMN of E_2 and P in picogram amounts is sufficient to facilitate feminine sexual behavior.[3,4] There is no impairment of sexual behavior when other steroid-concentrating nuclei are lesions.[1] Similarly, picogram amounts of E_2 or P will only facilitate sexual behavior when placed in the VMN.[4,5]

The neuronal connections of the VMN have been identified by anterograde and retrograde tracing techniques.[1] A major projection area of the VMN is the central gray. The severing of this specific projection will disrupt lordosis behavior, as will the ablation of the midbrain central gray.[36,37] The central gray in turn projects to the medullary reticular formation. The elimination of this projection will also impair lordosis behavior.[1] The medullary reticular formation projects to motoneurons in the spinal cord, which control the back muscles that make the lordosis posture.[1] The pathways that control proceptive behavior have not been so extensively mapped, but knife cuts that sever the axonal connections between the central gray and the VMN will also impair sexual proceptivity.[35] Thus, both sexual receptivity and proceptivity are activated through target cells in the VMN that connect directly or indirectly to the central gray.

2.3. Identification of Proteins

It is likely that E_2 and P modify the activity of VMN target cells by the induction of specific proteins, because the direct application of anisomycin to the VMN will block the activation of sexual behavior (Figs. 4,5). Given that sexual receptivity and proceptivity are activated by the induction of specific proteins in the VMN, what progress has been made in identifying these proteins? Although numerous macromolecular changes are known to occur in the whole hypothalamus, relatively few of these changes are known to occur in the VMN (Table I). For example, treatment of rats with E_2 will increase the number of 5-HT$_1$ receptors in membranes made from whole hypothalamus but does not affect 5-HT$_1$ receptors in the VMN.[38,39] There are only four known macromolecular changes that occur in the VMN in response to gonadal steroids: a 30% decrease in monoamine oxidase (MAO) activity after estrogen treatment that is reversed by the administration of P,[40] a 30% decrease in glutamic acid decarboxylase (GAD) activity after estrogen treatment, which is also produced by P in the absence of E_2 priming,[41] a 20–40% increase in mus-

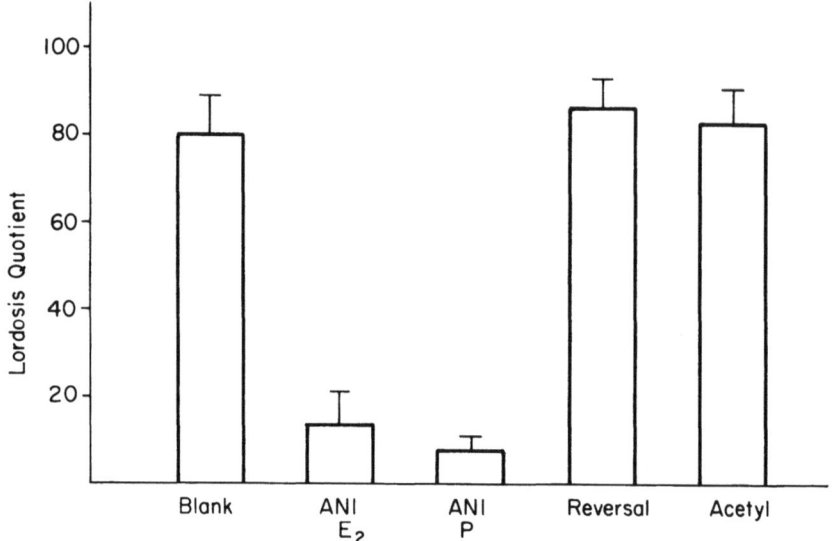

Fig. 4. Inhibition of sexual behavior by the application of anisomycin (ANI) to the ventromedial nucleus (VMN): OVX rats were given estradiol (E_2) from a Silastic capsule left under the skin for 6 h. Sexual behavior was measured 24 h after exposure to E_2 and 4–6 h after s.c. injection of 1 mg progesterone (P) in 0.1 ml propylene glycol. The lordosis quotient (LQ: percentage of lordotic postures in response to ten mounts from a stud male) was determined for each animal. At 10 min before insertion of Silastic capsules or injection of P, crystalline anisomycin was introduced into the VMN through bilateral cannulas as described in the text. Values are means + S.E.M. Blank ($n = 8$): Empty cannulas were inserted into the VMN before administration of E_2 or P. ANI E_2 ($n = 8$): ANI given 10 min before implantation of E_2 Silastic capsules ($P = 0.002$, Kruskal–Wallis; $P < 0.002$, Mann–Whitney U, compared to blanks), ANI P ($n = 8$): ANI inserted into the VMN before administration of P ($P < 0.002$, Mann–Whitney U). Reversals ($n = 11$): Rats from the ANI E_2 and ANI P groups were reprimed with gonadal steroids 1 week after inhibitor treatment. AcetylANI ($n = 6$): At 10 min before administration of E_2 or P, rats received implants of acetylanisomycin, a derivative of ANI that does not inhibit protein synthesis. (Reprinted with permission from Rainbow et al.[22])

carinic cholinergic receptor binding after E_2 exposure, which is not affected by P,[42] and a 400% increase in [^{3}H]R5020 binding to progestin receptors after estrogen treatment.[43]

In order for any macromolecular change in the VMN to be considered necessary or sufficient in the activation of feminine sexual behavior, it must meet certain requirements. First, it must appear at the same time as or before the earliest appearance of feminine sexual behavior. This would be 18–24 h after steroid administration for the priming action of E_2 on progestin-dependent behavior,[44] 15–20 min for the facilitation of receptive and proceptive behaviors by P in an E_2-primed rat,[18] and 48–72 h for the activation of receptive behavior by E_2 alone.[13] Next, it should be blocked by agents that block the activation of sexual behavior (anisomycin, antiestrogens) and not occur in the VMN of male rats, which show little or no feminine sexual behavior. It should also appear in the VMN after exposure to steroid doses that are minimally sufficient to elicit behavior.[19,20] Finally, sexual behavior should be impaired by treatments

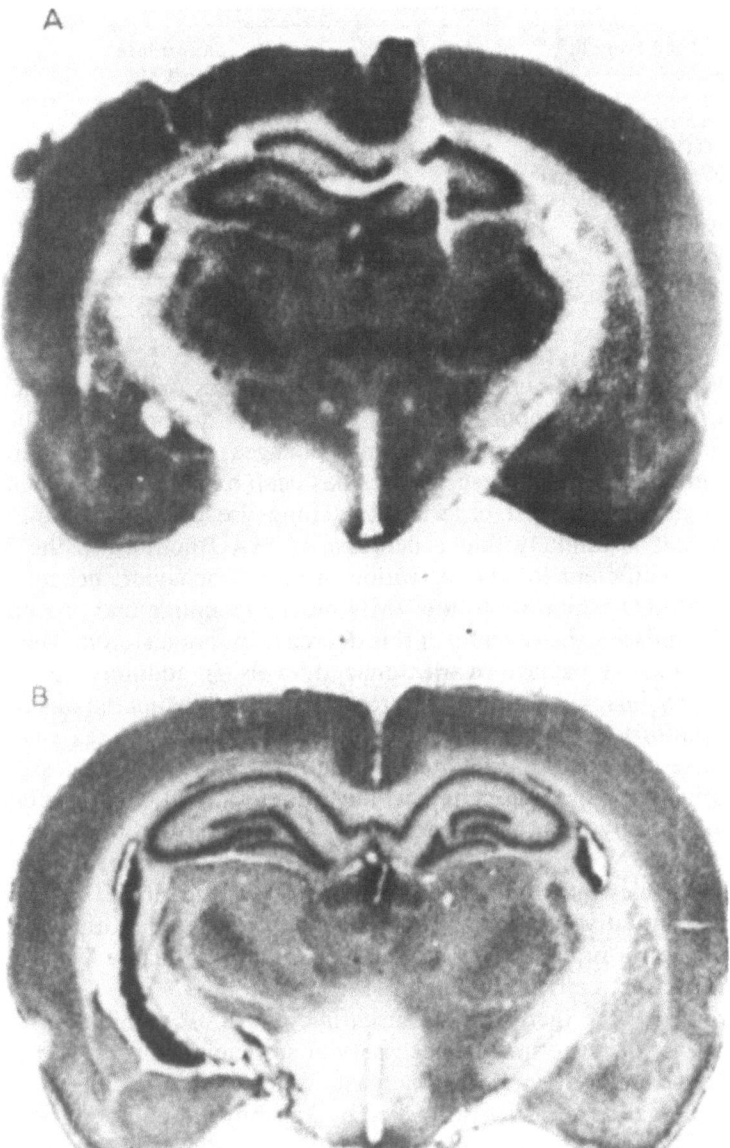

Fig. 5. Localization of protein synthesis inhibition by amino acid autoradiography. Rats received i.p. injections of 500 µCi [^{35}S]methionine 30 min before sacrifice. Autoradiograms were made by placing 32-µm brain sections against X-ray film. A: Inhibition of protein synthesis at 2 h after application of ANI to the left side of the VMN. At 2 h after application of ANI, the area of protein synthesis inhibition was largely confined to vicinity of the cannula tip. There was some diffusion of the inhibitor laterally into the medial forebrain bundle. The white area lateral to the zone of inhibition is the optic tract ($n = 3$). B: Inhibition of protein synthesis 4 h after insertion of ANI on the left side of the VMN. By 4 h after exposure to ANI, the area of protein synthesis inhibition had spread to the right side of the VMN. There was also diffusion of the inhibitor dorsally into the zona incerta ($n = 3$). (Reprinted with permission from Rainbow et al.[22])

Table I
Estrogen-Regulated Neurochemical Events

Estrogen-regulated event in hypothalamus	Reference	Occurrence in VM	Reference
Increased β-adrenergic receptors	78	?	—
Increased 5-HT$_1$ receptors	38	No	39
Increased muscarinic receptors	42	Yes	42[a]
Increased progestin receptors	15,19,44	Yes	43
Decreased type A MAO	79	Yes	40
Decreased GAD		Yes	41

[a] T. C. Rainbow, L. S. Snyder, D. Berck, and B. S. McEwen (unpublished data).

that would eliminate the effect of the macromolecules on neuronal activity in the VMN, such as the application of enzyme inhibitors or receptor blockers.

The steroid-induced macromolecular changes in the VMN were found fairly recently, so none of these changes has been examined closely enough to say that is satisfies all these criteria. Still, some already appear more relevant than others. It is unlikely that a decrease in MAO activity in the VMN is necessary or sufficient for the activation of sexual behavior, because the application of MAO inhibitors to the VMN blocks receptive and proceptive behaviors.[45] Similarly, the reversal of this decrease by progesterone[40] might only bring MAO activity back to ovariectomized levels. In addition, this change is not blocked by anisomycin.[40] However, it is still possible that the MAO changes are contributatory but not sufficient events in the activation of sexual behavior. The decrease in GAD activity after E_2 treatment might be a necessary event for the activation of sexual behavior, but it is not a sufficient one because it is also produced by P in the absence of E_2, a treatment that does not facilitate sexual behavior. No further studies have been reported on this macromolecular change, so its relevance to feminine sexual behavior remains to be determined. GABA is quantitatively a major neurotransmitter in the hypothalamus and in the rest of the rat brain,[46,47] so it would not be unreasonable for E_2 and P to affect VMN target cells by regulating GABA synthesis.

By contrast, the increases in muscarinic and R5020 binding in the VMN satisfy many of the criteria to be necessary or sufficient events in the activation of feminine sexual behavior. The increase in muscarinic binding occurs at 24 h after a minimal dose of E_2, is blocked by anisomycin, and does not occur in the VMN of males (T. C. Rainbow, L. S. Snyder, D. Berck, and B. S. McEwen, unpublished data). Similarly, the application of muscarinic blockers to the hypothalamus, though not exclusively to the VMN, will inhibit sexual behavior, whereas muscarinic agonists will facilitate sexual behavior.[48,49] The increase in VMN progestin receptor levels is also absent in males after E_2 treatment.[50] Although similar information is lacking on VMN progestin receptors, the induction of whole hypothalamic progestin receptors closely parallels the 18 to 24-h time course for the priming of progestin-dependent sexual behavior. The induction of hypothalamic progestin receptors will also occur after exposure to a minimally sufficient dose of E_2, and it is also blocked by treatment with anisomycin.[19,21] There is no *a priori* reason to think that VMN progestin receptors would behave differently. Thus, it is plausible that the increases in muscarinic and progestin receptors in the VMN are necessary events for the activation of behavior by E_2.

The increases in muscarinic and progestin receptors would only be relevant towards the behavioral actions of E_2; there are currently no candidates for molecular events that might mediate the activation of proceptive behavior by P. There is also no evidence that the induction of muscarinic or progestin receptors is the exclusive event in the VMN that mediates the behavioral actions of estrogen. This could only be determined by enumerating all of the molecular effects of E_2 on the VMN and subjecting each to the criteria for being a necessary or sufficient event in the activation of behavior.

What are the prospects for finding additional steroid-induced proteins in the VMN? The strategy so far has been to examine particular proteins that *a priori* might be important in synaptic function, such as neurotransmitter enzymes or receptors. This approach would be greatly aided by the identification of the neurotransmitters used by the VMN target cells. This is a somewhat complicated problem because only 20–30% of the cells in the lateral portion of the VMN possess estrogen or estrogen-inducible progestin receptors (J. M. Morrell and D. W. Pfaff, personal communication). These cells are morphologically indistinguishable from non-steroid-target cells (D. W. Pfaff, personal communication), so the identification of their neurotransmitters requires combining steroid autoradiography with immunocytochemistry. It has been shown with the combination method that some of the oxytocin- or vasopressin-containing cells in the paraventricular nucleus also concentrate [^{3}H]estrogen.[51] The application of the combination method to the VMN should provide information for biochemical studies about which neurotransmitter receptors or enzymes might be effected by gonadal steroids.

Given that the neurotransmitters used by VMN target cells might be currently unknown molecules, it is useful to consider other strategies for identifying steroid-induced proteins. Instead of selecting particular enzymes or receptors for examination, the technique of two-dimensional gel electrophoresis could be used to look for effects of E_2 and P on many proteins at once. Two-dimensional gel electrophoresis has been used to identify steroid-modulated proteins in peripheral target cells but has not yet been applied to the brain. In hepatoma cells, glucocorticoids alter the synthesis of seven proteins out of the 1000 or so proteins that can be resolved on a 2-D gel.[52] Thus, the main advantage of 2-D gel electrophoresis is that it can simultaneously measure changes in the rates of synthesis of 1000 separate proteins—perhaps 10–20% of the total number of proteins in a mammalian cell.[53] Its main disadvantage is that the proteins are only characterized by their molecular weight and electrical charge, so their functions are unknown. However, previous studies have managed to clarify the functional significance of unknown neuronal protein bands or spots.[54-56] This was done in part by making antibodies to the proteins and localizing them to various components or regions of the brain.[54,55] It is likely that similar approaches could clarify the functions of any steroid-modulated proteins found in the VMN.

It would also be helpful to clarify the contribution of axonal transport of VMN proteins to the behavioral actions of E_2 and P. If transport of proteins to the central gray is required to activate behavior, it may be that the central gray is a more appropriate site to look for steroid-induced proteins than is the VMN. By combining steroid autoradiography with a fluorescent retrograde tracer method, it has been shown that 20–30% of the estrogen-labeled cells in the VMN project to the central gray.[57] Some evidence that axonal transport is required comes from a recent study in which the application of colchicine to the VMN impaired the activation of receptive behavior by chronic exposure to E_2 alone.[58] In this study, receptive behavior first occurred 48 h after continuous exposure to E_2 from a silastic capsule. The application of colchicine to the VMN 24 h before the insertion of the silastic capsule impaired receptive behavior at 48–130 h after E_2 exposure. The colchicine-treated rats showed normal levels of receptive behavior by 6 days after E_2 administration. Similarly, the application of colchicine after lordosis was established by chronic exposure to E_2 caused a decline in receptive behavior beginning 4 h after administration of colchicine.

This evidence is consistent with the possibility that axonal transport of VMN proteins to the central gray is involved in the activation of receptive behavior by chronic exposure to E_2. It is also possible that the impairment in lordosis behavior resulted in part or solely from the disruption of axonal transport in neurons that project only within the VMN.[58] There are still other explanations for the impairment of behavior by colchicine,[58] and additional studies are required to clarify the role of axonal transport in the activation of receptive behavior. It would seem prudent, however, to also examine the VMN region for local neurochemical events that might mediate the activation of receptive behavior by chronic exposure to E_2 as well as by acute P administration. Since the activation of behavior by P can occur within 20 min,[18] it is unlikely that even the fast rate of axonal transport from the VMN to the central gray is sufficient to mediate the behavioral actions of P.[58,59]

3. ORGANIZATIONAL ACTIONS OF GONADAL STEROIDS ON BEHAVIOR

We have seen that much progress has been made in determining the cellular mechanisms that activate feminine sexual behavior. It is more or less established that receptive and proceptive behaviors occur through the induction of specific proteins in neurons of the ventromedial hypothalamic nucleus. It is possible that two of these proteins are the muscarinic cholinergic receptor and the progestin receptor, both of which appear to be relevant to the activation of sexual behavior by E_2.

In addition to activating behavior in adult rats, gonadal steroids influence the neonatal nervous system to determine whether at maturity the rat shows masculine or feminine patterns of sexual behavior. These actions of gonadal steroids are termed organizational effects, in part to emphasize their permanence and their possible involvement with changes in neuronal structure. In rats, organizational effects of gonadal steroids occur only during a 10-day interval lasting from immediately before to a week after birth. This interval is known as the critical period. The castration of a male rat pup during the critical period will cause it in adulthood to show low levels of masculine sexual behavior (mounting, intromission) and high levels of feminine sexual behavior in response to gonadal steroids. Similarly, treatment of a female rat pup with testosterone will cause it to display high levels of masculine sexual behavior and low levels of feminine sexual behavior when gonadal steroids are given as an adult.

By comparison with activational effects, less is known about the cellular mechanisms that mediate the organizational actions of gonadal steroids. This is to some extent because the behavioral phenomenology of organizational effects is more complicated than it is for activation of sexual behavior. Sexual differentiation of behavior in the rat can be divided into two separate components—masculinization and defeminization. As in peripheral organs, the normal inclination of the rat brain is to differentiate according to a feminine plan unless testicular secretions impose a masculine phenotype. A normal male rat shows low levels of feminine sexual behavior because testosterone has defeminized its neonatal brain. The same rat shows high levels of masculine sexual behavior because testosterone has masculinized the brain during the critical period. It is possible to interrupt defeminization without affecting masculinization, resulting in male rats that show high levels of both masculine and feminine sexual behavior.[60] Similarly, an adult female rat shows high levels of feminine sexual behavior and low levels of masculine behavior because its brain is neither masculinized nor defeminized during the critical period.

The process of defeminization appears to be mediated by the aromatization of testosterone into estrogen,[61,62] whereas masculinization may result from a combination of the actions of aromatizable and nonaromatizable androgens.[63] Thus, masculinization and defeminization appear to result from separable neuronal events, perhaps under the control of different hormones. It is also possible that the different components of masculine and feminine sexual behavior, such as receptivity and proceptivity, undergo separate masculinizations or defeminizations.

The current view of brain sexual differentiation is that testosterone, through its pathways of metabolism, promotes morphological differences in cell number, size, and patterns of synaptic connections within the preoptic area and hypothalamus.[61] This may occur through the stimulation of neurite growth[64] as well as through postulated hormone effects on neuronal proliferation or cell survival.[65] In addition, testosterone may influence other neuronal features such as neurotransmitter phenotype and the "program" for regulation of enzymes or receptors by gonadal steroids in adult life[66,67] (T. C. Rainbow, L. S. Snyder, D. Berck, and B. S. McEwen, unpublished data).

By analogy with the activational effects of gonadal steroids, it would seem likely that these organizational actions also occur through gene activation and protein synthesis. The evidence at this point so far is consistent with the genomic hypothesis, but it is much more limited than the evidence for activational effects. The brains of neonatal rats possess specific cytoplasmic receptors for estrogens and androgens.[68-70] As in adult rats, these receptors translocate into the nucleus in response to gonadal steroids.[69,71] The administration of estrogen or androgen receptor anatagonists to neonates will also block defeminization and masculinization.[61,63] However, in contrast to activational effects, it has not yet been shown that the degree and duration of nuclear receptor occupation correlate with the extent of masculinization or defeminization. Similarly, it is not known if inhibitors of RNA and protein synthesis would block masculinization or defeminization, although genomic inhibitors will disrupt anovulatory sterility, a related organizational effect of testosterone.[72]

If we make the reasonable assumption that gonadal steroids produce organizational effects via gene activation and protein synthesis, there are two main obstacles in identifying these proteins: it is not known when these proteins are produced after exposure to steroids, and it is not known where they are synthesized within the neonatal rat brain. The time course of protein production could simply be determined by inhibiting cerebral protein synthesis at specific times after steroid exposure—the same approach that was used for activational effects. However, it is not so easy to identify the target cells for organizational effects. The brain of a neonatal rat is roughly 90% smaller than an adult rat brain. This makes it correspondingly more difficult to locally implant steroids into particular hypothalamic nuclei. It also appears that implanted steroids diffuse more readily in neonatal brain. When $[^3H]E_2$ was implanted directly into the VMN of neonates, there was significant spread of E_2 to all other brain regions by 72 h after insertion (M. Y. McGinnis and B. S. McEwen, personal communication). There is much less diffusion of $[^3H]E_2$ when similar-sized implants are made into adult brain.[3]

It would be helpful for future implant studies to define the minimum duration of exposure to E_2 or testosterone that was sufficient to produce organizational effects. By analogy with activational effects,[19,20] it is possible that only a few hours of steroid exposure[73] are necessary to complete the final events required to organize behavior—an interval over which sufficient diffusion might not occur. It might also be possible to use local implants of protein synthesis inhibitors to identify the target cells for organizational effects if the relevant proteins are synthesized rapidly after steroid exposure, as they are for activational effects.[21]

In spite of the lack of direct evidence, it is a reasonable assumption that the target cells for organizational effects are located among the steroid-concentrating nuclei of the hypothalamus and preoptic area. In adult rats, there are the cells that either modulate or, in the case of the VMN, are sufficient to activate sexual behavior, so they would be prime candidates as organizational target cells. In support of this, it is known that in adult brains there are biochemical and anatomic sex differences in these regions. A portion of the medial preoptic area has more cells in males than in females,[74] and sex differences occur in dendritic processes in the medial preoptic and arcuate nuclei.[75-77] There are lower levels of estrogen receptors in the medial preoptic nucleus of males than in females and lower levels of estrogen-inducible progestin receptors in the periventricular preoptic area and the ventromedial nucleus in males than in females.[50] Male rats have fewer 5-HT_1 receptors in the medial preoptic nucleus than do females and, in the same nucleus, show a greater induction of 5-HT$_1$ receptors than females in response to estrogen treatment.[67] By contrast, males show less induction of 5-HT$_1$ receptors in the arcuate nucleus after E$_2$ treatment.[67] Males, unlike females, also show no increase in acetylcholinesterase activity in the bed nucleus of the stria terminalis in response to estrogen[66] and show no increase in muscarinic receptors in the VMN after estrogen treatment (T. C. Rainbow, L. S. Snyder, D. Berck, and B. S. McEwen, unpublished data).

Some of these sex differences are reversed by neonatal castration of males or by neonatal treatment of females with testosterone,[76–78] making it conceivable that they might mediate behavioral musculinization or defeminization. It is difficult to test this experimentally until the target cells for organizational effects are more specifically identified. Even without knowing the target cells for organizational effects on particular behaviors, it might be possible to exclude some sex differences from further consideration because they are not produced by minimally sufficient conditions of steroid exposure.

4. ORGANIZATIONAL AND ACTIVATIONAL EFFECTS OF GONADAL STEROIDS AND THEIR RELEVANCE FOR THE STUDY OF MEMORY FORMATION

As we have discussed, gonadal steroids modify the synaptic properties of their target cells to activate or organize sexual behavior. The time course of activational effects ranges from minutes for the activation of proceptive behavior by progesterone to days for the activation of masculine sexual behavior by testosterone. The organizational effects of testosterone or estrogen on sexual behavior occur during the 10-day-long critical period after birth and persist for the life of the rat. A considerable amount of progress has been made in identifying the cellular events that mediate the activation of feminine sexual behavior, although less progress has been made in elucidating the cellular basis of organizational effects.

Although previous workers in the field of memory and learning have not commented on the similarity, we feel that there are some important parallels between the phenomenology of memory formation and the behavioral actions of steroid hormones. Operationally, memory may be defined as a specific behavioral change in an animal resulting from a specific environmental input. Like the behavioral actions of steroids, memories can endure over intervals ranging from minutes to entire lifetimes. It is also assumed that specific memories result from the modification of the synaptic properties of specific cells. This represents another parallel with the behavioral actions of steroids. Given these similarities, it is entirely possible that memories and steriods might modify synapses by similar mechanisms involving genomic activation and increased production of specific gene products.

The advantage of studying how steroid hormones modify synapses is that the target cells are largely known or can be discovered. The identification of these target cells makes it possible to study the neurochemical events that mediate steroid actions. By contrast, the cellular location of any kind of memory formation in the rat is largely unknown. Thus, until the anatomic location of memory formation is better defined, it is likely that the behavioral actions of gonadal steroid hormones will be one of the better model systems for studying the kinds of cellular changes that may be involved in learning and memory.

REFERENCES

1. Pfaff, D. W., 1980, *Estrogens and Brain Function*, Springer-Verlag, New York.
2. McEwen, B. S., Davis, P. G., Parsons, B., and Pfaff, D. W., 1979, *Annu. Rev. Neurosci.* **2**:65–112.
3. Davis, P. G., McEwen, B. S., and Pfaff, D. W., 1979, *Endocrinology* **104**:898–903.
4. Rubin, B. S., and Barfield, R. J., 1980, *Endocrinology* **106**:504–509.
5. Rubin, B. S., and Barfield, R. J., 1983, *Endocrinology* **113**:797–805.
6. Beach, F. A., 1948, *Hormones and Behavior: A Study of Interrelationships between Endocrine Secretions and Patterns of Overt Response*, Paul B. Hoeber, New York.
7. Phoenix, C. H., Goy, R. W., Gerall, A. A., and Young, W. C., 1959, *Endocrinology* **65**:369–382.
8. McEwen, B. S., 1968, *Modern Perspective in Psychiatry* (J. Howells, ed.), Oliver & Boyd, Edinburgh, pp. 87–107.
9. Rainbow, T. C., 1979, *Neurochem. Res.* **4**:297–312.
10. Beatty, W. W., 1979, *Horm. Behav.* **12**:112–163.
11. Davidson, J. M., and Bloch, G. J., 1969, *Biol. Reprod. (Suppl.)* **1**:67–92.
12. Hardy, D. F., and DeBold, J. F., 1971, *Horm. Behav.* **2**:287–297.
13. McGinnis, M. Y., Krey, L. C., MacLusky, N. J., and McEwen, B. S., 1981, *Neuroendocrinology* **33**:158–161.
14. Kow, L.-M., and Pfaff, D. W., 1975, *Horm. Behav.* **6**:259–276.
15. MacLusky, N. J., and McEwen, B. S., 1980, *Endocrinology* **106**:192–202.
16. Rainbow, T. C., McGinnis, M. Y., Krey, L. C., and McEwen, B. S., 1982, *Neuroendocrinology* **34**:426–432.
17. Roy, E. J., and McEwen, B. S., 1977 *Steroids* **30**:657–669.
18. McGinnis, M. Y., Parsons, B., Rainbow, T. C., Krey, L. C., and McEwen, B. S., 1981, *Brain Res.* **218**:365–371.
19. Parsons, B., Rainbow, T. C., Pfaff, D. W., and McEwen, B. S., 1981, *Nature* **292**:58–59.
20. Parsons, B., McEwen, B. S., and Pfaff, D. W., 1982, *Endocrinology* **110**:613–619.
21. Rainbow, T. C., Davis, P. G., and McEwen, B. S., 1980, *Brain Res.* **194**:548–555.
22. Rainbow, T. C., McGinnis, M. Y., Davis, P. G., and McEwen, B. S., 1982, *Brain Res.* **233**:417–423.
23. Parsons, B., Rainbow, T. C., Pfaff, D. W., and McEwen, B. S., 1982, *Endocrinology* **110**:620–624.
24. Quadagno, D. M., Shryne, J., and Gorski, R. A., 1971, *Horm. Behav.* **2**:1–10.
25. Quadagno, D. M., and Ho, G. K. W., 1975, *Horm. Beh.* **6**:19–26.
26. Eisenfeld, A. L., and Axelrod, J., 1965, *J. Pharmacol. Exp. Ther.* **150**:469–475.
27. McEwen, B. S., Pfaff, D. W., and Zigmond, R. E., 1970, *Brain Res* **21**:17–28.
28. Feder, H. H., and Marrone, B. L., 1977, *Ann. N.Y. Acad. Sci.* **286**:331–352.
29. Blaustein, J. D., and Feder, H. H., 1980, *Endocrinology* **106**:1061–1069.
30. Glaser, J. H., Cislo, S. C., and Barfield, R. J., 1982, *Soc. Neurosci. Abstr.* **8**:69.
31. Glaser, J. H., and Barfield, R. J., 1984, *Neuroendocrinology* (in press).
32. Flood, J. F., Rosensweig, M. R., Bennett, E. L., and Orme, A. E., 1973, *Physiol. Behav.* **10**:555–562.
33. Pfaff, D. W., and Keiner, M., 1973, *J. Comp. Neurol.* **151**:121–158.
34. Warembourg, M., 1978, *Neurosci. Lett.* **9**:329–332.
35. Clark. A. S., Pfeifle, J. K., and Edwards, D., 1981, *Physiol. Behav.* **27**:597–602.
36. Sakuma, Y., and Pfaff, D. W., 1979, *Am. J. Physiol.* **237**:R285–R290.

37. Manogue, K. R., Kow, L. M., and Pfaff, D. W., 1980, *Horm. Behav.* **14**:277–302.
38. Biegon, A., and McEwen, B. S., 1982, *J. Neurosci.* **2**:199–205.
39. Biegon, A., Fischette, C. T., Rainbow, T. C., and McEwen, B. S., 1982, *Neuroendocrinology* **35**:287–291.
40. Luine, V. N., and Rhodes, J., 1982, *Soc. Neurosci. Abstr.* **8**:931.
41. Wallis, C. J., and Luttge, W. G., 1980, *J. Neurochem.* **34**:609–613.
42. Rainbow, T. C., DeGroff, V., Luine, V. N., and McEwen, B. S., 1980, *Brain Res.* **198**:239–243.
43. Parsons, B., Rainbow, T. C., MacLusky, N. J., and McEwen, B. S., 1982, *J. Neurosci.* **2**:1446–1452.
44. Parsons, B., MacLusky, N. J., Krey, L. C., Pfaff, D. W., and McEwen, B. S., 1980, *Endocrinology* **107**:774–779.
45. Luine, V. N., and Fischette, C. T., 1982, *Neuroendocrinology* **34**:237–247.
46. Iverson, L. L., and Bloom, F. E., 1972, *Brain Res.* **41**:131–143.
47. Snyder, S. H., Young, A. B., Bennett, J. D., and Mulder, A. H., 1973, *Fed. Proc.* **32**:2039–2047.
48. Clemens, L. G., and Dohanich, G. P., 1980, *Pharm. Biochem. Behav.* **13**:89–95.
49. Clemens, L. G., Humphrys, R. R., and Dohanich, G. P., 1980, *Pharm. Biochem. Behav.* **13**:81–88.
50. Rainbow, T. C., Parsons, B., and McEwen, B. S., 1982, *Nature* **300**:648–649.
51. Rhodes, C. H., Morrell, J. I., and Pfaff, D. W., 1981, *Neuroendocrinology* **33**:18–27.
52. Ivarie, R. D., and O'Farrel, P. H., 1978, *Cell* **13**:41–55.
53. O'Farrell, P. H., 1975, *J. Biol. Chem.* **250**:4007–4021.
54. Greengard, P., 1982, *Harvey Lect.* **75**:277–331.
55. Shashoua, V. E., 1977, *Proc. Natl. Acad. Sci. U.S.A.* **74**:1734–1747.
56. Bock, E., 1978, *J. Neurochem.* **30**:7–14.
57. Morrell, J. I., and Pfaff, D. W., 1982, *Science* **217**:1273–1276.
58. Harlan, R. E., Shivers, B. D., Kow, L. M., and Pfaff, D. W., 1982, *Brain Res.* **238**:153–167.
59. McEwen, B. S., and Grafstein, B., 1968, *J. Cell Biol.* **38**:494–508.
60. Davis, P. G., Chaptal, C. V., and McEwen, B. S., 1978, *Horm. Behav.* **12**:12–19.
61. Goy, R. W., and McEwen, B. S. (eds.), 1980, *Sexual Differentiation of the Brain*, MIT Press, Cambridge.
62. McEwen, B. S., Lieberburg, I., Chaptal, C. V., and Krey, L. C., 1977, *Horm. Behav.* **9**:249–263.
63. McEwen, B. S., 1982, *Molecular Approaches to Neurobiology* (I. Brown, ed.), Academic Press, New York, pp. 195–219.
64. Toran-Allerand, C., 1976, *Brain Res.* **106**:407–412.
65. Truman, J. W., and Schwartz, L. M., 1982, *Neurosci. Comment.* **1**:66–72.
66. Luine, V. N., and McEwen, B. S., 1983, *Neuroendocrinology* **36**:475–482.
67. Fischette, C. F., Biegon, A., and McEwen, B. S., 1983, *Science* **222**:333–335.
68. Vito, C. C., and Fox, T. O., 1979, *Science* **204**:517–519.
69. MacLusky, N. J., Lieberburg, I., and McEwen, B. S., 1979, *Brain Res.* **178**:129–142.
70. Lieberburg, I., MacLusky, N. J., and McEwen, B. S., 1980, *Brain Res.* **196**:125–138.
71. McGinnis, M. Y., Meany, M. J., Davis, P. J., and McEwen, B. S., 1982, *Soc. Neurosci. Abstr.* **8**:424.
72. Kobayashi, F., and Gorski, R. A., 1970, *Endocrinology* **86**:285–289.
73. Roffi, J., Corbler, P., and Rhodes, J., 1982, *The Ontogenesis of the Endocrine System* (J. Bertramo and J. Saeg, eds.), INSERM, Paris.
74. Gorski, R. A., Harlan, R. E., Jacobson, C. D., Shryne, J. E., and Southam, A. M., 1980, *J. Comp. Neurol.* **193**:529–539.
75. Greenough, W. T., Carter, C. S., Steerman, C., and DeVoogd, T. J., 1977, *Brain Res.* **126**:63–72.
76. Raisman, G., and Field, P. M., 1973, *Brain Res.* **54**:1–29.
77. Matsumoto, A., and Arai, Y., 1981, *Neuroendocrinology* **33**:166–169.
78. Wilkinson, M., Herdon, H., Pearce, M., and Wilson, C., 1979, *Brain Res.* **168**:652–655.
79. Luine, V. N., and McEwen, B. S., 1977, *J. Neurochem.* **28**:1221–1227.

3

Adrenocortical Hormone Action

E. Ronald de Kloet and H. Dick Veldhuis

1. INTRODUCTION

The major secretory products of the adrenal cortex (see Fig. 1) are classified as glucocorticoids [cortisol (F) and corticosterone (B)], mineralocorticoids [aldosterone (ALDO) and deoxycorticosterone (DOC)] and weak androgens [dehydroepiandrosterone (DHEA)]. Small amounts of progesterone (PROG) and estrogenic steroids are secreted as well. The principal glucocorticoid is F in some species (hamster, pig, primates), but is B in mouse and rat.

The corticoids have a potent influence on the brain. Corticoid action is ultimately expressed in behavior and neuroendocrine regulation. Glucocorticoids affect mood,[1] and sleep,[2] sensitivity to sensory stimuli,[3] and growth differentiation[4] and play an important role in adaptation of the animal to the environment.[5-7] Mineralocorticoids regulate NaCl appetite[8] and under certain conditions adaptive behavior.[9] Glucocorticoids feed back on the brain to regulate the release of CRH and on the pituitary to regulate the release of hormones of the opiocortin family of peptides, in particular, corticotrophin (ACTH) (see for reviews refs. 10–12).

Corticoids exert their action on behavior via modulation of electrolyte balance,[13] energy and neurotransmitter metabolism,[14,15] and electrophysiological responses.[16,17]

If a steroid hormone affects cell metabolism, the cell usually contains intracellular receptor sites. Such receptor sites have been identified in nerve cells for gluco- and mineralocorticoids.[18] Their properties and localization have given the impetus for a molecular approach in understanding of the modulatory action of corticoids on neuronal processes. On the other hand, corticoids may evoke direct cellular responses not mediated by the intracellular receptor system. These actions are less well understood, and the cell membrane seems the predominant site of action.

This chapter surveys the progress on the cellular mechanism of corticoid action in the brain, the sites of corticoid action, and the circuitry involved.

E. Ronald de Kloet and H. Dick Veldhuis • Rudolf Magnus Institute for Pharmacology, Medical Faculty, University of Utrecht, 3521 GD Utrecht, The Netherlands.

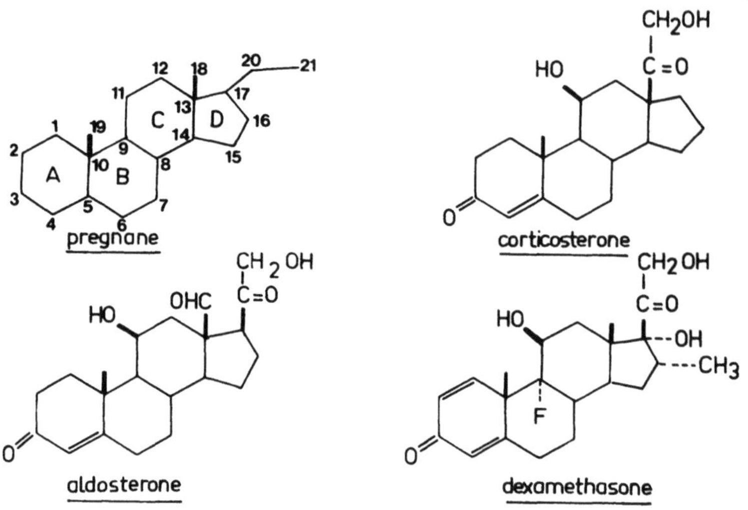

Fig. 1. Structure of the pregnane skeleton showing the numbering of the steroid molecule and the structure of the major steroids discussed in this chapter.

The last section summarizes effects of adrenocortical hormones in neuroendocrine regulation and behavior.

2. FUNCTION OF ADRENOCORTICAL HORMONES

The secretion of glucocorticoids is stimulated by stress. On the other hand, these steroids do protect against stressful influences. In the brain the glucocorticoids control neuronal circuits that become activated by stress, and in the pituitary the stress-induced release of ACTH is suppressed. Apart from the central effects that are directed to control the psychic and neuroendocrine components of adaptive processes (see Section 7), the glucocorticoids modify the flow of substrates through intermediary metabolism. They act to redistribute metabolic energy from lipids, proteins, and nucleic acids to carbohydrates. Carbohydrates constitute a readily available energy resource, which is rapidly utilized in cases of stress.[19,20]

In many tissues such as fat, skin, muscle, and lymphoid tissue, glucocorticoids inhibit glucose uptake, depress macromolecular synthesis, and enhance protein degradation. In fat cells lipolysis is stimulated, and in muscle glycogen stores are depleted. This catabolic action of glucocorticoids results in release of amino acids and free fatty acids into the circulation, which serve as substrates of gluconeogenesis in the liver, where excess glucose is stored as glycogen. As a consequence of gluconeogenesis, the blood glucose level rises, but this effect is counteracted by insulin. Since enzymes are induced and macromolecular synthesis is enhanced in liver, glucocorticoids have an anabolic action in this tissue.

In addition to the liver, the brain, heart, and red blood cells escape the catabolic action of glucocorticoids. These tissues are provided with more glucose after stress-induced release of glucocorticoids. A coordinated mechanism thus ensures maximal availability of substrates for energy metabolism in tissues that have an essential function in response to or coping with stress. However, the brain itself is a target tissue for glucocorticoids as well. Glucocorticoid receptors are found in those regions, such as the limbic structures, that have a function in the interpretation of sensory stimuli and the expression of the behavioral response serving adaptation or emotion.

Although some of the actions of the glucocorticoids are direct, in certain cases the glucocorticoid-induced metabolic responses permit or amplify the action of other hormones.[21] The term permissive action has been introduced by Ingle.[22] Examples are found in cyclic-cAMP- or Ca^{2+}-dependent hormones such as epinephrine, which is released prior to the adrenocortical hormones after stress.

Mineralocorticoids have an important function in the transport of ions via epithelial cells that is manifest in Na^+ retention and loss of K^+. The steroids catalyze the Na^+,K^+-ATPase that serves as an electrolyte pump. The principal site of action is in the epithelia of the renal distal tubules, but it is also present in the cells of the salivary and sweat glands and in the intestinal tract.[23] Mineralocorticoids and glucocorticoids have several influences on the cardiovascular system and regulate the blood pressure under stress.[24]

3. STEROID–CELL INTERACTION

Steroid hormones circulate in blood and because of their lipophilic nature, have easy access to all tissues including the brain. They are thought to freely enter most cells via a process that is best described as passive diffusion. On entering the cells, the current consensus is that the hormones bind to intracellular receptor sites and that the hormone–receptor complex interacts with cell nuclear chromatin. The cell nucleus is the principal site of action, where steroid hormones stimulate the synthesis of macromolecular synthesis (see Fig. 2). There is also evidence that steroid hormones interact with the plasma membrane of the cell. This section describes the various molecular events that occur in the cellular uptake and binding process of a steroid hormone.

3.1. Cellular Uptake

The naturally occurring glucocorticoid hormones F and B circulate about 75% bound to corticosteroid-binding globulin (transcortin or CBG) (K_D 10^{-8} M). Fifteen percent is bound to serum albumin, and the rest circulates as free steroid.[25] Corticosteroids bound to CBG are biologically inactive.[26] The CBG-bound corticosteroids serve as a pool of hormone that is protected against metabolism and excretion but makes physiologically active steroid available when steroid production is increased. The capillary transit time is the rate-limiting step in the cellular uptake process and depends on the polarity of the steroid hormone.[27,28]

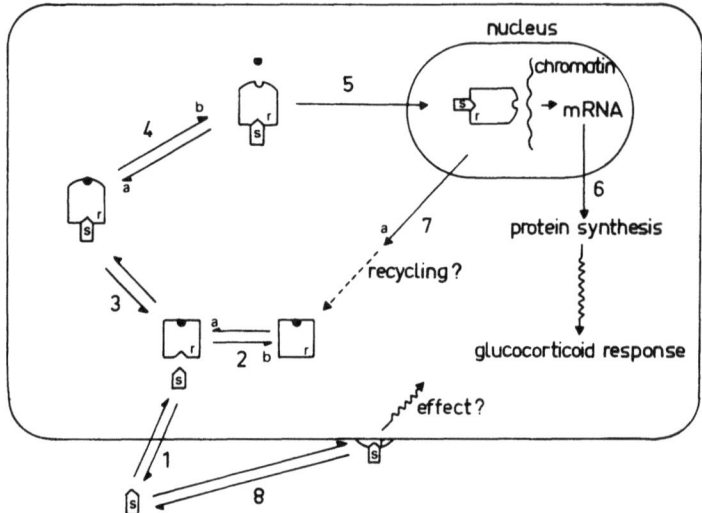

Fig. 2. Schematic representation of the possible steps in glucocorticoid hormone action: (1) penetration of the target cell by the steroid (s), which does not appear to be rate limiting; (2) "activation" of free receptor (r); (3) binding of the steroid to the "activated" receptor, which may cause a conformational change; (4) "transformation" of the steroid–receptor complex so that it shows enhanced affinity towards cell nuclear chromatin and DNA—during this "transformation," a low-molecular-weight substance is thought to dissociate from the receptor; (5) nuclear association of the steroid receptor complex in some way modulating gene expression; (6) changes in amino acid incorporation, specific protein formation, and induction of characteristic enzymes finally bringing about the target cell response to glucocorticoid hormone action; (7) a phosphorylation step, which might be involved in the possible "recycling" of the dephosphorylated receptor protein; (8) membrane interaction of the steroid hormone, which, at the moment, is only poorly understood; a, phosphorylation step; b, dephosphorylation step.

In some tissues, corticosteroids bound to CBG enter the cells. This is the case for liver and pituitary,[29] and CBG or CBG-like proteins actually have been found associated with membranes and elsewhere in the cell and have been shown to participate actively in the cellular uptake and distribution of corticoids.[30-34] The CBG-bound steroid is excluded from most brain regions with the possible exception of the areas outside the blood–brain barrier (see Section 4.8).

The involvement of the plasma membrane in the interaction of steroid hormones with target cells is at present unclear. There is some evidence that an "entry" step exists that is specific for a particular steroid hormone and involves binding.[35-37] It is possible that at this level the steroid hormone triggers a response related to membrane functioning prior to or independently of the hormone–receptor effects mediated via gene expression. That membrane interaction of a steroid hormone may occur is elegantly illustrated in the reinitiation of meiotic division by PROG in *Xenopus laevis* oocyte. A progesterone analogue (androsta-4,1-3,1 carboxylic acid), when coupled to a soluble polymer (aminopropylpolyethylene oxide) cannot enter the cell but is able to promote oocyte meiosis[38,39] Electrophysiological and biochemical effects with an extremely short latency may also find their molecular basis in membrane effects. Steroid binding that actually occurs to nerve cell membranes has only rarely been reported.[40,41] Part of these findings are explained by binding to CBG-like protein (Section 4.8).

The search for membrane binding sites is also complicated, since the initial localization of the intracellular receptor protein at present is unknown. It has been proposed that the receptor may be attached to or be a component of the plasma membrane,[42] from which it is easily detached during homogenization and hence found in the cytosol. On the other hand, free receptor could be present in the cell nuclear compartment as well, being dependent on the free water content of this cellular compartment.[43]

3.2. Intracellular Biotransformation of Steroids

Some steroid hormones that enter cells are prohormones for the biologically active steroid moiety. This is the case for testosterone, which can be converted to dihydrotestosterone or estradiol in certain pituitary and brain cells,[44–47] and for estradiol, which is converted to catecholestrogens.[48] Acetylation and sulfation of corticoids have been reported,[49–51] but there is no evidence that these conversions provide biologically active hormones. In fact, more than 90% of unchanged [^{3}H]steroid is recovered from the hippocampal cell nuclear fraction 1 h after administration of [^{3}H]B. In contrast, extensive metabolism of DOC has been reported in the brain.[52]

3.3. Activation of the Receptor and Transformation to its DNA-Binding State

Steroid receptors display target cell specificity and have a high degree of stereoselectivity in binding the appropriate steroid hormone ligands or synthetic analogues. The receptor proteins are thermolabile proteins and contain sulfhydryl groups. Oxidation or alkylation destroys the steroid binding activity. The receptors are large molecules with sedimentation coefficients in the range of 6–9 S and a molecular weight of several hundred thousands. This section describes the sequence of molecular events that takes place once the steroid interacts with the receptor (see for review refs. 53,54). These events are given in a somewhat speculative scheme in Fig. 2.

Before the receptor protein binds the steroid, the inactive form (aporeceptor) may have to undergo an energy-dependent alteration.[53] This "activation" of free receptor exposes a high-affinity binding site to the steroid and requires phosphorylation as well as reduction. Such a mechanism has been inferred from *in vitro* studies using ATP, molybdate, and dithiothreitol; Ca^{2+} and Mg^{2+} seem to be involved as well.[54–59]

Binding of the steroid to the "activated" receptor may cause a conformational change.[60] The steroid receptor complex may be "transformed" to a complex that shows an enhanced affinity towards cell nuclear chromatin and DNA. In cell-free systems "transformation" of the complex is induced by temperature increase, changes in ionic strength, or removal of low-molecular-weight substances after gel filtration.[61-65] "Transformation" involves changes in size, shape, and other physicochemical characteristics. A general feature is exposure of cationic sites at the cell surface of the receptor that confer on the molecule an enhanced affinity towards DNA.[63,66] Separation of receptor complexes showed that under physiological conditions these changes in receptor properties are sequential events.[67-69]

"Transformation" of the steroid–receptor complex appears to involve dephosphorylation. Molybdate blocks not only the temperature-dependent inactivation of unoccupied receptors but also the temperature-dependent transformation of occupied receptors to the DNA-binding state.[58] Such an action could be achieved by a phosphoprotein phosphatase intimately associated with the receptor or by an enzyme not associated with the receptor but present in the cytosol.[58,68] Another line of evidence suggests that a low-molecular-weight substance (structure unknown) associated with the receptor prevents receptor transformation.[64,65,70,71] Moreover, this low-molecular-weight substance becomes phosphorylated and dephosphorylated, and *in vitro* temperature and ionic strength promote dissociation, yielding the DNA-binding form of the receptor.[58,68]

These findings suggest that an internal phosphorylation–dephosphorylation mechanism is responsible for switching "on" or "off" the steroid-binding site. This would provide the cell with a mechanism to regulate the number of sites capable of interacting with steroid and with DNA. In the target cell, therefore, a substantial pool of latent or "cryptic" binding sites may be present.[54] In fact, such "cryptic" receptor sites for B have been detected in the rat hippocampus.[72]

A number of models have been postulated to explain agonistic and antagonistic steroid actions on the basis of the receptor. The free receptor is thought to be represented by "active" and "inactive" forms in the cell. Agonists preferably bind to the "active" conformation and help to stabilize the complex in this state. Antagonists associate in particular with the "inactive" form and would prevent agonist binding. The models can be depicted as an allosteric equilibrium or by induced fit of subunits of the receptor with multiple interacting binding sites. The issue is discussed at length elsewhere.[60,73-76]

3.4. Translocation of the Steroid–Receptor Complex and Initiation of Genomic Effect

The "transformed" steroid–receptor complex displays high affinity to cell nuclear chromatin. The nature of these nuclear "acceptors" is still poorly defined. The nuclear chromatin contains high-affinity sites that show tissue and receptor specificity. Following the binding of the steroid–receptor complex to cell nuclear chromatin, the chromatin structure becomes altered in a subtle way that leads to changes in the rate of transcription of spceific mRNA sequences.[77] It is speculated that the steroid–receptor complex acts on a certain site of the genome by causing α-helix destabilization followed by increased RNA polymerase activity. The specific mRNA molecules are then translated to produce specific proteins that serve the hormone-induced physiological re-

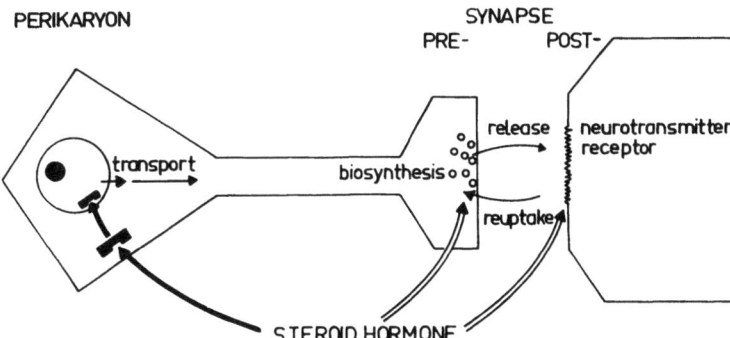

Fig. 3. Schematic representation of genomic and nongenomic effects of steroid hormones on pre- and postsynaptic events.

sponses. These may be proteins that are involved in a wide variety of cellular functions, e.g., functional components of membrane processes, enzymes participating in energy or neurotransmitter metabolism, protein kinases, protein hormones. For a detailed review on genomic action of adrenocortical hormones, the reader is referred to ref. 20.

The signals that terminate steroid hormone action are poorly understood. Decline in steroid hormone concentration and decay of the "activation" and "transformation" process may be involved. Also, pyridoxal phosphate has been postulated as a factor to block cell nuclear uptake of the steroid.[68] The receptor may return to the cytoplasm and be added to the pool of newly synthetized receptors, or it may be degraded.[78,79] Proteolytic enzymes have been shown to generate steroid receptor sites and DNA binding sites as independent moieties in the receptor molecule.[78,80] These processes may be coupled to cellular metabolism and require ATP for recycling of the receptor. In other words, after recycling to the cytoplasm, the dephosphorylated receptor protein (the activated unoccupied receptor) is ready for steroid binding and phosphorylation (the transformed steroid–receptor complex) followed by translocation and interaction with the genome.

3.5. Steroid–Nerve Cell Interaction

There are no experimental arguments to exclude that steroid receptor interaction and involvement of the genome in neural tissue are distinct from that in non neuronal target tissue. The uptake and intracellular transport of steroid hormones followed by the modulation of genomic expression are time-consuming processes. Steroid action involving the genome may induce long-lasting changes in cell metabolism that persist well after disappearance of the hormone from the tissue. In addition, in the brain certain gene products have to be transported by axonal transport to distant sites (see Fig. 3).

There is also evidence that steroid hormones may interact with the nerve cell membrane. Such actions of steroid hormones that do not involve the genome are rapid in onset and short in duration. They occur in parallel with rise and decline of hormone levels in the tissue.[81]

4. ADRENAL STEROID RECEPTORS IN BRAIN

4.1. Cellular Localization of Corticosterone

When [^{3}H]B is administered in a tracer dose to adrenalectomized (ADX) animals, the radioactive labeled hormone is retained preferentially by cell nuclei of the limbic system, e.g., hippocampus and parts of septum and amygdala.[18,82,83] Subsequent studies have extended this finding to the rhesus monkey, suggesting that such limbic regions are common target cells for corticosteroids in mammals irrespective of whether the principal adrenal glucocorticoid is B or F.[84] Endogenous B abolishes in intact rats the retention of the tracer dose of radioactive labeled steroid, which suggests that the B receptor system has a low capacity. In intact rats, B extracted from cell nuclei localizes with essentially the same preference in the hippocampus. This is an important finding, since it demonstrates that the B receptor system operates in the presence of the whole spectrum of adrenocortical secretions.[85]

The principal retention of [^{3}H]B in rat brain is in cell nuclei of neurons, as shown by autoradiography[86-88] (Figs. 4,5). A similar localization was seen after administration of [^{3}H]F,[89] although the amount of [^{3}H]F that accumulated was only 5% of B. Most heavily labeled were the neurons of the hippocampal CA_I and CA_{II} neurons of the Ammon's horn, and the granular neurons of the dentate gyrus, CA_{III}, CA_{IV}, the subiculum, the induseum griseum (supracallosal hippocampus), and the anterior (preseptal) hippocampus. Intense labeling was found in the dorsolateral part of the septum close to the lateral ventricle and under the corpus callosum. The medial septum was practically devoid of label. Somewhat less but still considerable labeling was found in the amygdala and parts of the cortex. In the amygdala, most label was concentrated in neurons of the cortical and basal amygdala. In the cortex, the entorhinal, suprarhinal, pyriform, and cingulate cortex retained some [^{3}H]B. Other regions that showed labeled neurons were more diffusely distributed in the olfactory nucleus, habenular nucleus, and red nucleus. Motor neurons of the cranial nerve nuclei and the spinal cord were heavily labeled. Some glial cells also retained label. In the hypothalamus, only a few scattered neurons contained [^{3}H]B in the cell nuclei, and there was no pronounced labeling observed in the paraventricular nucleus, recently shown to be the site of synthesis of corticotrophin-releasing factor (CRF).[90,91]

[^{3}H]Corticosterone can be extracted from isolated brain cell nuclei with 0.4 M KCl. A considerable amount of [^{3}H]B remains bound to macromolecules after passage through a sephadex G100 column.[92] Maximal cell nuclear labeling *in vivo* in the hippocampus occurs 1 h after administration of [^{3}H]B. Labeling of cytoplasmic sites precedes the cell nuclear labeling.[72,87,93]

4.2. Cellular Localization of Mineralocorticoids and Dexamethasone

[^{3}H]Aldosterone administered to adrenalectomized animals displays a neuroanatomic distribution pattern that closely resembles that of [^{3}H]B.[94-96] Accordingly, the highest concentration of label was observed in the neurons of

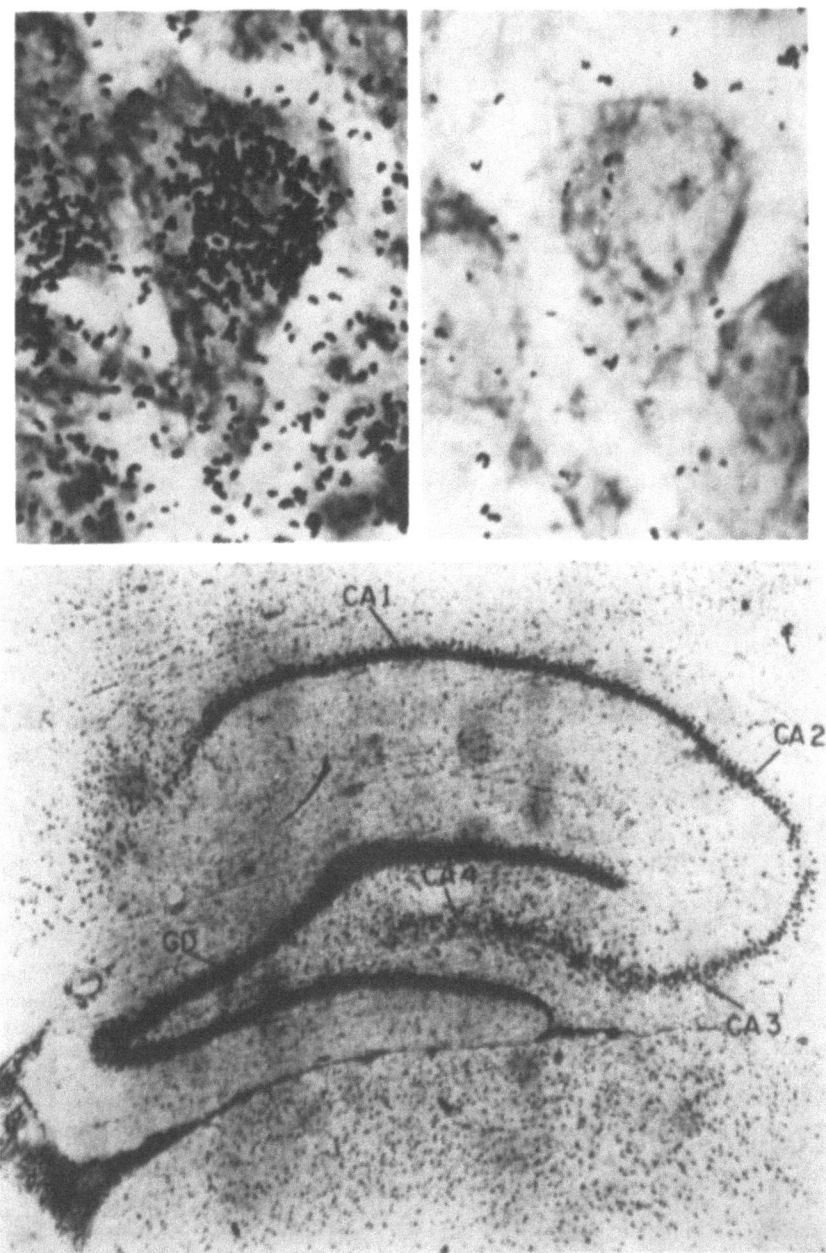

Fig. 4. Autoradiographs of [^{3}H]B uptake by hippocampal neurons of ADX male rats. Upper left, control uptake of [^{3}H]B; upper right, uptake of [^{3}H]B in the presence of 3 mg unlabeled B. (From McEwen et al.[328] reprinted with permission.) Below, unstained autoradiograph of the dorsal hippocampus of an ADX male rat showing uptake of systemically injected [^{3}H]B as black silver grains. The figure shows the longitudinal fields CA1 to CA4 of the pyramidal neuron layer in Ammon's horn and the granular neuron layer in the dentate gyrus. (From McEwen et al.,[328] reprinted with permission.)

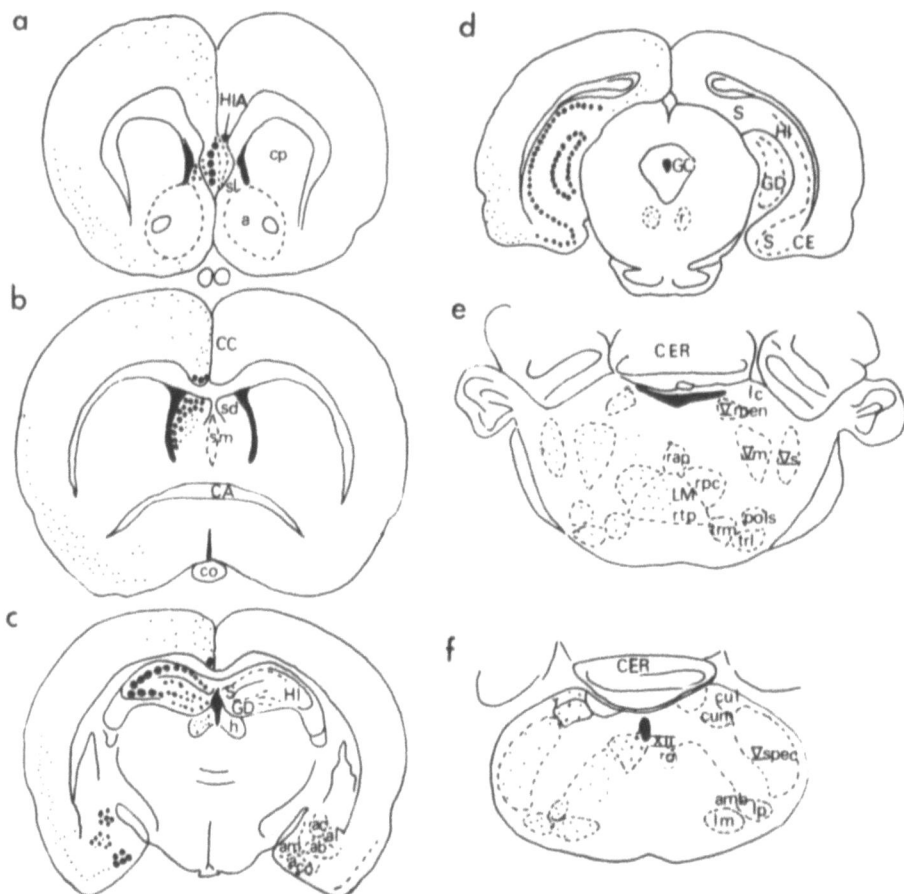

Fig. 5. Transverse sections of rat brain depicting cell nuclear localization of [^{3}H]B administered 60 min previously. The size and number of dots (left half) indicate the intensity of cell nuclear uptake of [^{3}H]B and the frequency of occurrence of B-concentrating cells. The schematic drawings of e and f are 1.5 times the proportional size of a–d. Abbreviations: a, n. accumbens; ab, n. amygdaloideus basalis; ac, n. amygdaloideus centralis; aco, n. amygdaloideus corticalis; al, n. amygdaloideus lateralis; am, n. amygdaloideus medialis; amb, n. ambiguus; C, cingulum; CA, commissura anterior; CE, cortex entorhinalis; CER, cerebellum; CO, chiasma opticum; Cul, n. cuneatus lateralis; Cum, n. cuneatus medialis; GC, griseum centrale; GD, gyrus dentatus; h, n. habenularis; HI, hippocampus; HIA, hippocampus anterior; lc, locus coeruleus; lm, n. reticularis lateralis magnocellularis; LM, lemniscus medialis; Lp, n. reticularis lateralis parvocellularis; pols, n. paraolivaris superior; r, n. ruber; ro, n. roller; rpc, n. reticularis pontis caudalis; rtp, n. reticularis tegmenti; S, subiculum; sd, n. septalis dorsalis; sl, n. septalis lateralis; sm, n. septalis medialis; trl, n. trapezoidus lateralis; trm, n. trapezoidus medialis; Vm, n. motorius nervi trigemini; Vmes, n. tractus mesencephali nervi trigemini; Vs, n. sensible nervi trigemini; Vspec, n. caudalis tractus spinalis nervi trigemini; XII, n. nervi hypoglossi.

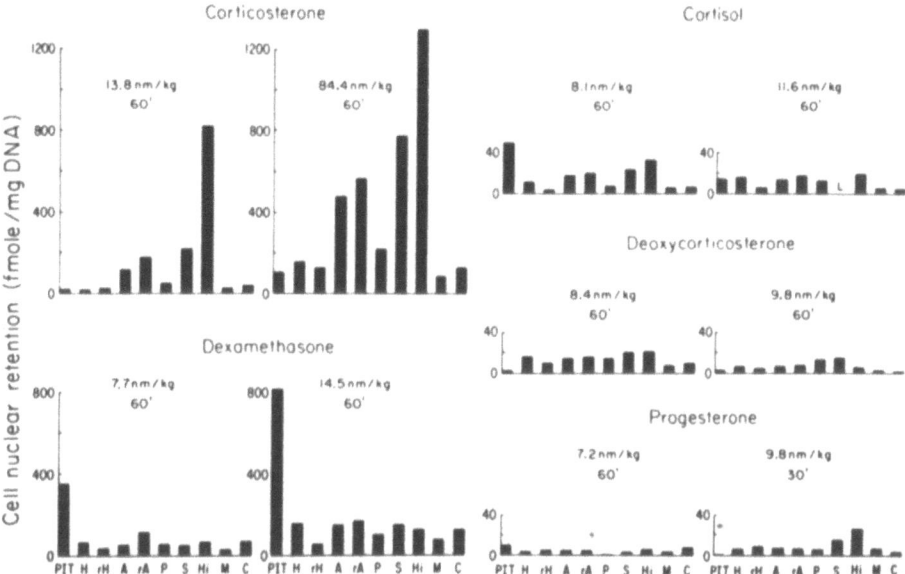

Fig. 6. Cell nuclear retention of [³H]steroids by various brain regions after their infusion via tail vein in ADX–OVX rats. Time between injection and sacrifice and the dose of [³H]steroid used (nmol/kg body weight) are indicated at the top of each experiment. Tissue from three to four identically treated rats was pooled for cell nuclear isolation. PIT, pituitary; H, basomedial hypothalamus; rH, rest of hypothalamus; A, corticomedial amygdala; rA, rest of amygdala; P, medial preoptic nucleus; S, septum; Hi, hippocampus; M, midbrain; C, cerebral cortex. (From McEwen et al.,[97] reprinted with permission.)

the limbic system and the motor neurons of cranial nerves as well as dispersed in the reticular formation. One difference from [³H]B was the relative intensity of the label. [³H]Aldosterone was more intense in the induseum griseum, whereas the reverse was true for [³H]B. Another difference was the time course and extent of cell nuclear uptake. [³H]Aldosterone uptake reaches maximal levels within 15 min after administration, but the amount is still half that of [³H]B 1 h later.[95,96]

[³H]Deoxycorticosterone uptake in cell nuclei is very low. About 1% of the amount that accumulates after [³H]B in equal doses of 5 nmol.[97] [³H]Progesterone cell nuclear uptake is negligible. However, the use of a synthetic progestagen, [³H]RU 5020, showed that progestagen-sensitive sites are localized predominantly in hypothalamic regions and in the preoptic areas.[98]

Synthetic glucocorticoids are not retained in a similar way by brain cell nuclei as B and F (Fig. 6). [³H]Dexamethasone is only very little retained by hippocampal neurons and weakly concentrated in an even way over cell brain regions.[87,97,99–101] Cell nuclear concentration was about 10% of that of B. The heaviest labeling occurs in endothelial cells around blood vessels and in epithelial cells lining choroid plexus and ventricles. Cells of the circumventricular organs and the medial basal hypothalamus also concentrate [³H]DEX, in particular in cells of the ventral caudal part of the arcuate nucleus. In one study it was noted that [³H]DEX labeling was heavy near the ventricle, as if the

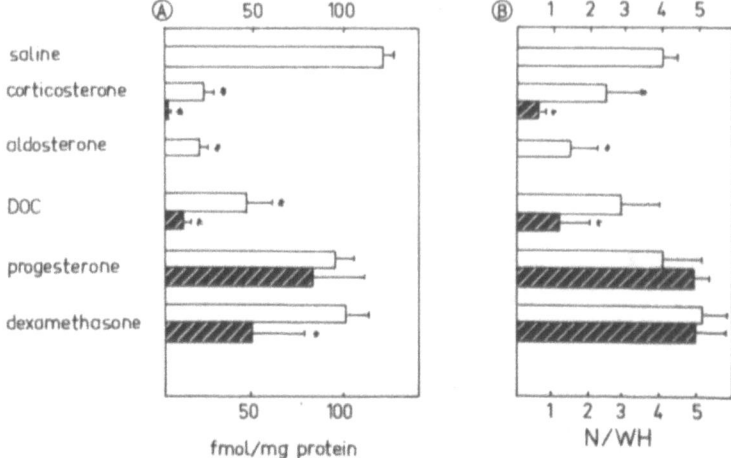

Fig. 7. Blockade of hippocampal cell nuclear uptake of [³H]B *in vivo* by prior administration of unlabeled steroids in a dose of 30 μg (open bars) or 300 μg (hatched bars) per 100 g rat. The unlabeled steroids were administered 30 min prior to administration of a tracer dose of [³H]B (50 μCi) to rats adrenalectomized 3 days previously. The animals were sacrificed by decapitation 1 h after administration of the tracer dose of [³H]B. (A) Cell nuclear retention of [³H]B expressed as fmol/mg nuclear protein. Saline: ADX rats receiving saline 30 min prior to the tracer. (B) Ratio of cell nuclear uptake to tissue uptake of [³H]B. *$P < 0.05$ *vs.* saline (Newman–Keuls multiple-range test). (From Veldhuis et al.,[96] reprinted with permission.)

synthetic steroid seemed to enter the brain via the liquor cerebrospinalis.[100] The intensity and regional distribution pattern are the same in ADX rats as in intact rats with circulating endogenous adrenal steroids.

4.3. Competition for Cell Nuclear Localization of Corticosterone

Endogenous or previously administered B (30 μg/100 g body weight) completely suppresses the cell nuclear retention of [³H]B.[87,96,102] Pretreatment with the same dose of DOC or ALDO similarly suppresses uptake of the tracer. Cortisol is a weaker suppressor, and in this dose range DEX and PROG lack any competitive effect. Only a tenfold higher dose of DEX leads to a significant suppression.[96] Some further insight into the way the steroid hormones interact with the cell nuclear retention mechanism is derived from a comparison of the amount of steroid localized in the whole tissue and in the cell nuclear compartment. This ratio (N/WH) is greatly depressed after pretreatment with B and ALDO, indicating that these steroids interact with the same receptor system for cell nuclear translocation. Dexamethasone and PROG suppress uptake in tissue and cell nuclei equally well (N/WH unchanged), which does not exclude that the two steroids have affinity for the B binding sites (Fig. 7).

4.4. Cell Nuclear Uptake in Vitro in Tissue Slices

In tissue slices of the hippocampus, it was shown that the cell nuclear uptake process of [³H]B is temperature dependent[93] as was shown in cytosol

Table I
Binding Constants of [³H]Corticosterone in
Cytosol of Brain Regions

Tissue	b_{max}	K_D	R
Hippocampus	858 ± 140	3.80 ± 0.8	0.984
Hypothalamus	344 ± 12	2.50 ± 0.2	0.998
Raphe area	Not detectable		

nuclear transfer experiments in other systems. However, the large differences observed in cell nuclear retention *in vivo* between B and steroids such as DEX, DOC, and F disappear.[98,100] Although [³H]B uptake in the hippocampus at saturating concentration (2×10^{-8} M) is still highest and comparable to that observed *in vivo*, the uptake of [³H]DEX and [³H]DOC has increased tenfold. Also, the clear regional differences in cell nuclear uptake disappear. The [³H]B uptake in cell nuclei of hypothalamus and anterior pituitary have increased considerably.[99] In addition, [³H]B uptake *in vitro* is efficiently blocked by corticoids and PROG.[93] Similar observations were reported for the brains of pigs and cats.[103,104]

These comparisons of *in vivo* and *in vitro* data stress that under *in vitro* conditions a specificity-conferring mechanism is lost. Such a mechanism may reflect different cellular uptake kinetics and be related to the rate of permeation of brain cells, metabolism, and binding to blood CBG. Difference in permeability may be of even greater relevance when a bolus injection is used. For instance, the binding of B to CBG causes more prolonged exposure of brain cells to this steroid, in contrast to ALDO, which does not bind. Metabolism may have reduced the uptake of [³H]DOC, since this steroid undergoes extensive metabolism.[52] Dexamethasone has a biological half-life of 4 h[105] and shows no affinity for blood CBG. Thus, if DEX had a high affinity towards the cell nuclear uptake and retention mechanism in the hippocampal neurons, it would show substantial accumulation.

4.5. Binding of Adrenal Steroids to Soluble Receptors

The binding properties of the soluble receptor sites have been characterized on the basis of affinity constants and specificity of the receptor sites for various steroid ligands. The animals were previously adrenalectomized for depletion of endogenous adrenal steroids and extensively perfused with saline through the heart for removal of blood CBG.

[³H]Corticosterone-labeled receptors were clearly distinct from plasma CBG on the basis of physicochemical properties and steroid specificity.[93,102] The distribution of these B-labeled sites parallels the regional differences observed in cell nuclear uptake *in vivo* and *in vitro* in tissue slices. Thus, the largest numbers of sites were found in the hippocampus. However, in the hypothalamus and anterior pituitary (see Section 4.8.), a considerable number of high-affinity binding sites were also determined *in vitro*[99,102,106] (Table I).

Other regions such as parts of the midbrain are practically devoid of soluble corticoid receptor sites, whereas the spinal cord contains such binding activity.[107] Similar corticoid receptors were found in human brain cytosol[108,109] and brain cytosol of other species.[103,104]

The binding activity is very labile. In the absence of ligands, the sites decay rapidly the first hour after tissue disruption. The disappearance of binding activity does not proceed at the same rate for each ligand. Whereas binding of [^{3}H]B 1½ h after homogenization is 20% less, that of [^{3}H]DEX is reduced 50%.[99]

Scatchard analysis of [^{3}H]B and [^{3}H]DEX provided linear plots.[96,110,111] The relative binding affinity (RBA) of various steroid ligands for [^{3}H]B-labeled sites was B = DOC > PROG = DEX > ALDO. The same order of potency was measured with [^{3}H]DEX as the ligand. Inclusion of "pure" glucocorticoid, RU 26988, displayed a different competition pattern. Excess of this steroid displaced both [^{3}H] steroids to a maximum of 45%. Apparently, some fractions of the B- and DEX-labeled receptor proteins lack glucocorticoid-binding properties.[96]

Scatchard analysis of [^{3}H]ALDO binding in hippocampus or hypothalamus gave a curvilinear plot (see Fig. 8). Such a curvilinear pattern is indicative of two populations of binding sites, which can be distinguished on the basis of the affinity constants. Inclusion of a very small amount (0.6 × amount) of unlabeled B or a 100-fold excess of RU 26988 linearizes the ALDO Scatchard plot and leaves only the high-affinity sites available.[96,111] Another study, however, showed a linear Scatchard plot for [^{3}H]ALDO.[112] In their study, the low-affinity component was absent, presumably because of the different experimental conditions. These authors found capacities for ALDO-labeled sites ranking hippocampus > pituitary, septum > amydala > cortex, hypothalamus > preoptic area = 0.

Increasing the excess B tenfold also depresses the high-affinity ALDO sites. The relative binding affinity for [^{3}H]ALDO-labeled sites is in the order B = DOC > PROG > DEX > ALDO. Inclusion of a 100-fold excess of RU 26988, which leaves only mineralocorticoid sites, gave the order B = DOC > PROG > ALDO > DEX.

4.6. Heterogeneity in Adrenal Steroid Receptor Systems

The *in vivo/in vitro* uptake and binding studies have provided sufficient evidence to enable one to state that the rat brain contains heterogeneous sets of receptor sites for corticoids. Research in recent years has focused on two major issues. First, are there separate mineralo- and glucocorticoid receptor systems? Second, is the glucocorticoid receptor system compartmentalized in neurons and glial cells? A serious drawback in these studies was that the steroids employed were defined in biological activity on the basis of their classical effects on metabolism in peripheral target tissues. The extensive "cross talk" of ligands between the various receptor systems complicated the issue even more. It appeared, therefore, necessary ultimately to define the type of receptor on the basis of the steroid effects on brain and behavior (see Section 7).

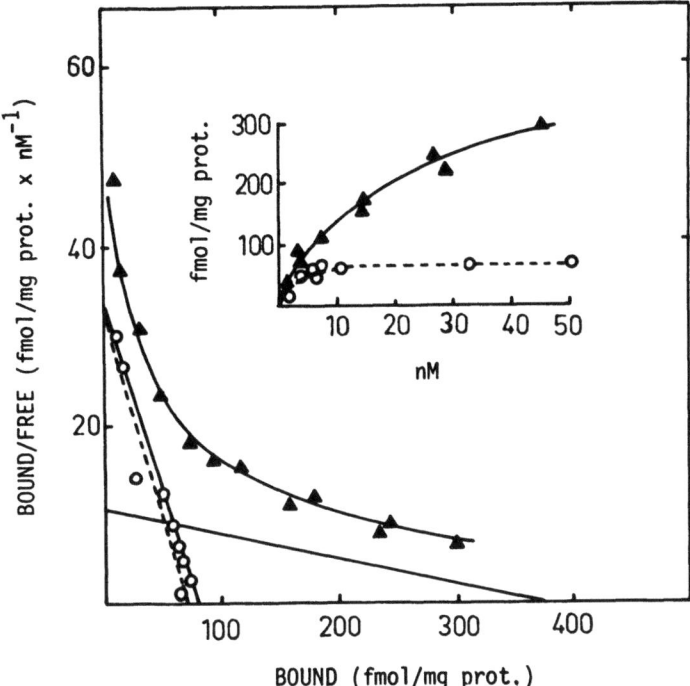

Fig. 8. [³H]Aldosterone binding *in vitro* to soluble macromolecules in hippocampal cytosol and competition with unlabeled B. Hippocampal cytosol was incubated with increasing concentrations of [³H]ALDO in the absence (▲——▲) or presence (○- - -○) of B in a concentration 0.6 times that of [³H]ALDO. Nonspecific binding was assessed by inclusion of a 500-fold excess of unlabeled ALDO. Inset, saturation analysis in the absence (▲——▲) or presence (○- - -○) of unlabeled B. Scatchard plot of [³H]ALDO binding in the absence of unlabeled B. Site I: K_D 2.2 nM; B_{max} 70 fmol/mg protein. Site II: K_D 30.3 nM; B_{max} 367 fmol/mg protein. Inclusion of unlabeled B: K_D 2.2 nM; B_{max} 69 fmol/mg protein. (From Veldhuis et al.,[96] reprinted with permission.)

4.6.1. Mineralo- and Glucocorticoid Receptor Sites

The strongest arguments are derived from curvilinear Scatchard analysis of [³H]ALDO and the inability of potent and "pure" glucocorticoids to displace [³H]ALDO binding *in vitro* in cytosol.[96,110–113] However, B appears to display the highest affinity, and ALDO and B apparently may use the same receptor system to be retained in neuronal cell nuclei of the hippocampus. Yet B has agonistic effects with stringent specificity on brain and behavior (see Section 7); ALDO is ineffective on these behaviors, and prior administration of this steroid or of PROG or DEX reveals an antagonistic action[114] (see Section 7.2).

In hippocampus neurons, mineralo- and glucocorticoid receptors may coexist in the same neurons. The sites may be part of the same molecule[115] and influence via an allosteric mechanism the affinity for each other's site.[75] Such an interrelationship has been established between PROG and glucocorticoid receptors. It appeared that PROG enhanced the dissociation of [³H]DEX from its binding site.[116]. On the other hand, DEX enhanced the dissociation rate of

[³H]promegestone.[116] These findings are reminiscent of those in the kidney, where a similar heterogeneity in adrenal steroid receptor sites has been noted.[73,117-122]

4.6.2. Glial/Neuronal Glucocorticoid Receptor Sites

Biochemical studies have clearly shown that glial cells are target cells for glucocorticoids. Glycerol-3-phosphate dehydrogenase is an enzyme that is under control of glucocorticoids,[123,124] and immunocytochemical studies have shown that this enzyme is localized in oligodendroglial cells.[125] Another enzyme, ornithine decarboxylase (ODC), is also under glucocorticoid control[126] (see Section 6.4). Moreover, glucocorticoid-induced hypertrophy of glial cells (astrocytes) has also been reported.[127] Glucocorticoid receptors have been found in glial cells, as shown in studies with enucleated optic nerves[128] and glial tumor cell line C_6, which derives from rat brain.[129] Interestingly, the *in vivo* retention of [³H]DEX in surviving tissue after optic nerve enucleation is larger than that of [³H]B.[130,131] It therefore seems likely that the uniform uptake of [³H]DEX in rat brain cell nuclei represents glial cell labeling by [³H]DEX in contrast to predominant labeling of neuronal cells by [³H]B. The difference is not absolute. There is some cell nuclear localization by [³H]DEX in neurons, in particular in the medial basal hypothalamus, and [³H]B will also bind to glial cell receptors.

4.7. Fractionation of Adrenal Steroid Receptor Sites

Proof for different types of receptors for adrenal steroids could come from isolation by analytical chemical techniques. So far only little progress has been made. One study reported a 200-fold purification of the receptor.[132] Labeled receptor complexes could be fractionated into two components by DEAE cellulose chromatography,[133] as was also observed with other steroid cell systems.[77] The ratio of the two labeled receptor components appeared dependent on the steroid ligand. With [³H]DEX, the majority of the complex eluted at low ionic strength, suggesting an overall more negative charge of the molecule. With [³H]DOC, the majority of the complex eluted at high ionic strength, whereas equal amounts were isolated with B.[134].

Two components could also be fractionated with isoelectric focusing (p*I* = 5.9 and 6.8).[135] Warming of the cytosol for 15 min at 25°C results in an enhanced affinity for DNA but does not change the overall DEAE elution of isoelectrofocusing profile.[134,135] However, Wrange[136] found that when conditions were chosen such that proteolysis was prevented, only one peak eluted, whereas two components again evolved after tryptic digestion. Although this finding may point to artificial formation of the two moieties *in vitro*, it certainly does not exclude that formation of one of the components may be of functional significance. So far, however, there is little evidence that the glucocorticoid receptors from optic nerve, pituitary, or hippocampus are different in physiochemical properties,[130] and neither has a difference been observed between those of hippocampus and liver.[136]

4.8. Corticoid Receptors and CBG-like Proteins

Cells of the anterior pituitary and not those of the intermediate or posterior lobe retain [^{3}H]B.[100] Whereas in cell nuclei of hippocampal neurons [^{3}H]B was retained in much higher quantities than [^{3}H]DEX, anterior pituitary cell nuclei preferentially accumulated the synthetic glucocorticoid.[97,99–101] [^{3}H]Dexamethasone concentrates in a tenfold higher amount than [^{3}H]B, and the localization is primarily in the corticotrophs of the anterior lobe.[137] [^{3}H]Dexamethasone was associated with pituicytes in the posterior lobe.[137]

Glucocorticoid and mineralocorticoid receptor sites have been identified in the soluble cell fraction of the pituitary.[111,112,138–141] There is also firm evidence for a population of high-affinity binding sites that selectively bind B. These binding sites are found in pituitary cytosol and resemble plasma CBG in molecular weight, isoelectric focusing profile, ammonium sulfate precipitability, and immunologic properties.[30,142] The proteins lack the ability to transport B to cell nuclei[30] or to DNA.[142] The CBG-like binding system is not removed by extensive perfusion and is also found in isolated pituitary cells.[31,33] Intracellular sites resembling CBG have also been noted in other cells such as human lymphocytes.[34,144–146]

There is evidence that the CBG molecules actively participate in the cellular uptake and distribution process of B; CBG is present at the plasma cell membrane and probably can be internalized in a tissue such as the pituitary, which has a protein-permeable vascular bed.[147,148] That membrane components may be involved in glucocorticoid uptake also became evident in work with pituitary tumor cell line AtT-20/0-1.[36] The membrane components of the latter study differ, however, from CBG in that they also bind synthetic glucocorticoid with high affinity. The proportion that enters the cell as free steroid or as protein-bound hormone is not known.

Although the entry of B is facilitated, possibly by mediation of CBG,[32] the ultimate extent of cell nuclear localization is much lower *in vivo* than *in vitro* in tissue slices[32,99,149,150] and pituitary tumor cells.[140] The difference not only results from its intrinsic ability to promote cell nuclear translocation but also seems to be a consequence of the participation of the CBG system. The CBG system is thought to compete with the glucocorticoid receptor for binding of B, and as a consequence less of the steroid is made available to the receptor for cell nuclear translocation.[32,33,112] This role is accounted for by the extracellular as well as the intracellular pool of CBG, since both regulate the amount of free steroid. Dexamethasone does not bind to CBG and bypasses this mechanism, which contributes to the extensive cell nuclear localization in the pituitary. These aspects of glucocorticoid cell interaction may help to explain the potent suppressive effect of the synthetic glucocorticoid on stress-induced pituitary ACTH release (see Section 7.1). The brain virtually lacks CBG-like protein[33,135] with the possible exception of regions outside the blood–brain barrier. The brain, therefore, may be more susceptible to small changes in B level than the pituitary.

5. REGULATION OF THE ADRENAL STEROID RECEPTOR SYSTEM

One of the factors determining the magnitude of the biological response to corticoids is the amount of corticoid–receptor complex that is bound to cell nuclear chromatin. The amount of receptor complexes in the cell nucleus depends on the concentration of the steroid hormones and the number of utilized receptor molecules. The steroid concentration that reaches the cell nucleus is effective, as agonists may have to compete with circulating partial agonists or antagonists for binding to the receptor. The receptor occupancy varies with the circadian variation in plasma and brain corticoid level,[151–153] ranging from 50% occupancy at the morning trough to 80% at the afternoon peak.[151,154,155] A circadian variation was also noted in receptor levels of ADX rats.[156,157]

The number of receptor sites is subject to two types of regulation, which we refer to as short-term and long-term regulation. Short-term regulation refers to the intracellular processes that take place during steroid–receptor interaction. These have been defined as receptor activation, receptor transformation to the cell nuclear binding state, and the actual binding process of the receptor complex to cell nuclear constituents (see Section 3.3). Influences on this chain of events have been reported *in vitro* to be dependent on pH, ionic strength, the low-molecular-weight inhibitor, and the internal phosphorylation/dephosphorylation cycle. The role of these *in vitro* established facts for *in vivo* modulation of receptor-mediated responses is still poorly understood. Long-term changes in receptor number have been induced by endocrine and neural manipulations. Such manipulations have helped to identify factors that are involved in control of receptor number. These factors are neurotropic substances or are intrinsic to the receptor-containing cells.

5.1. Neurotropic Substances

Three categories of neurotropic substances can be distinguished that are involved in control of corticoid receptor capacity.

The first category is the autoregulation by B. The number of receptors increases after bilateral removal of the adrenals, and this is reflected *in vitro* in cytosol as well as in cell nuclear uptake in tissue slices.[154,157–159] The first 2 h after ADX, the binding capacity rises rapidly, which is probably a result of removal of endogenous corticoids. The gradual increase during the subsequent hours may represent newly synthesized receptors. Excess corticoids or chronic stress reduces the number of receptor sites.[156,160–162]

In pituitary tumor cells, prolonged incubation with DEX leads to a progressive diminution in the number of glucocorticoid receptors.[163] The authors found evidence that chronic exposure to DEX results in enhanced degradation

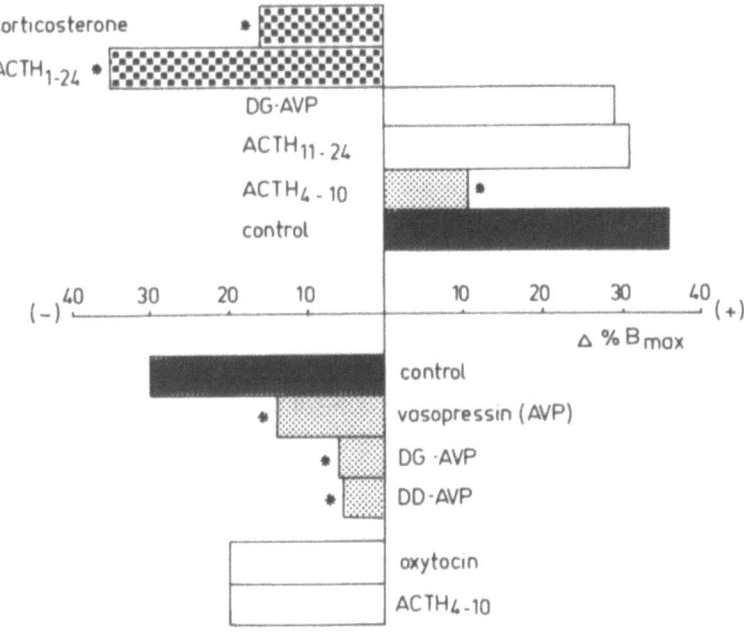

Fig. 9. Effect of chronically administered neuropeptides and B on the apparent maximal binding capacity (B_{max}) for [^{3}H]B in hippocampal cytosol of hypophysectomized and homozygous diabetes insipidus (HO-DI) rats. B_{max} of sham-operated and of non-diabetes-insipidus rats of the Brattleboro strain was taken as the reference value and indicated as 0%; all data are expressed as the percent difference from these values. The upper part of the figure represents the effect on B_{max} of hypophysectomy (control) and of the various chronic treatments. The lower part depicts the effect on B_{max} in the HO-DI rats without (control) and with peptide treatment. *$P < 0.05$ vs. hypophysectomized and vs. HO-DI animals. (According to Veldhuis and De Kloet,[161] reprinted with permission.)

of receptor sites. Since the repletion is sensitive to cycloheximide, this process represents synthesis of new receptor molecules.

The second category are neuropeptides related to ACTH or vasopressin. That such peptides are involved became apparent after a classical endocrine approach. A hormone imbalance was induced after removal of the pituitary, or rats were used that have a genetically determined defect in the synthesis of vasopressin (diabetes insipidus rats of the Brattleboro strain[164]). The disturbed hormone balance is reflected in an altered receptor capacity for B in such target tissues as hippocampus and pituitary. Receptor capacity was restored after replacement with the appropriate hormone or neuropeptide (see Fig. 9).

Removal of the pituitary results in a gradual increase in receptor capacity for B in all brain regions, e.g., 60%, 36%, and 72% for hippocampus, hypothalamus, and septum, respectively. Plasma CBG levels showed a fall in capacity. The increase in receptor capacity exceeded that following bilateral

ADX.[161,165] This indicates that there are more factors involved in receptor control than solely B. One of these factors is the neuropeptide $ACTH_{4-10}$, a sequence with potent behavioral activity.[166] Replacement with $ACTH_{4-10}$ every other day reduced the receptor number to that observed in the sham-operated animals. This effect was specific for the hippocampus. $ACTH_{11-24}$ and DGAVP were ineffective. Since $ACTH_{4-10}$ lacks corticotropic activity, the peptide action is not mediated via the adrenocortical secretion. Cell nuclear uptake *in vivo* parallels the changes seen in receptor capacity.[165]

The diabetes insipidus animals had about 35% less receptor capacity for B in hippocampus and anterior pituitary, whereas the number of receptors was twice as great in the neurointermediate lobe.[157,168] Treatment of diabetes insipidus animals with arginine vasopressin (AVP), des-glycinamide arginine vasopressin (DGAVP), or 1-desamino-8-D-arginine vasopressin (DDAVP) daily for 1 week resulted in elevation of receptor capacity in hippocampus and anterior pituitary to near the nondiabetic control levels. It should be noted that DGAVP is a peptide that lacks peripheral action on diuresis but has retained its behavioral activity. No effects were observed on the other brain regions or on the neurointermediate lobe. The effect was also not observed after administration of oxytocin or $ACTH_{4-10}$ fragment.[168]

These two examples of regulation of receptor number for B reveal a distinct specificity in tissue and neuropeptide. The principal sites of action are the typical corticosteroid target tissues hippocampus and anterior pituitary. The peptides not only restored the altered receptor number but also normalized the behavior in those animals. The receptor population affected are the glucocorticoid receptor sites in the anterior pituitary and the B receptor in the hippocampus.

The third category of receptor regulators are the neurotransmitter systems that innervate the regions with steroid-sensitive and receptor-containing cells. An example is the neurotransmitter serotonin (5-HT), which projects from the cell bodies in the raphe nuclei via the cingulum, the fornix bundle, and the entorhinal cortex into the hippocampus. Administration of the neurotoxic drug 5,6-dihydroxytryptamine in the median raphe nuclei or in the fornix–fimbria bundle caused a marked depletion of 5-HT in the hippocampus. At the same time, the number of B receptors in the hippocampus increased twofold.[169,170]

The number of B receptor sites was not altered after removal of the cholinergic input by lesioning the medial septal nucleus[171] or by destruction of the noradrenergic innervation after administration of the neurotoxic drug 6-OH-dopamine in the locus coeruleus.[170] Changes in noradrenergic neurotransmission alter the concentration of progesterone receptors in the hypothalamus[172] and of sex steroid receptors in the pineal gland.[173] These data suggest that modulation of target cell responsiveness to steroids is an important mechanism by which neurotransmitters influence steroid-dependent processes.[170,172,174]

5.2. Compensatory Changes in Number of Steroid Receptor Sites

In the course of our studies we have noted that removal of a part of the receptor-containing cells in the septal–hippocampal complex included a compensatory increase in number of B receptors in the remaining cells. It first became apparent in a study in which discrete septal lesions were applied.[171] Only lesions in the dorsolateral septum, a region rich in B receptors, increased the number of B receptor sites in the hippocampus. In subsequent studies, unilateral removal of the hippocampus evoked a considerable increase in B receptor number in the remaining contralateral lobe, which further substantiates the notion of a compensatory mechanism in regulation of the number of receptor sites.[175]

An increase in receptor capacity is also observed after partial destruction of hippocampal neurons following local kainic acid administration. These hippocampus-lesioned animals became hyperactive in exploratory activity and at the same time displayed a supersensitivity to B. The findings emphasize the significance of an increase in receptor number for enhanced steroid responsiveness.[176]

5.3. Ontogeny and Aging

Receptor capacity for corticoids in the hippocampus shows large changes during the life-span. On postnatal day 1 of the rat, the lowest number of receptor sites is measured in the hippocampus. The cytosol binding capacity increases about fourfold and gradually reaches adult levels around 4 weeks of age.[159,177] With the "pure" glucocorticoid RU 26988[112] used to measure mineralo- and glucocorticoid receptors separately, it appeared that both types of receptor sites are present early in the rat's life.[111] Nuclear retention *in vivo* following administration of a B tracer followed an identical pattern during development.[155] Autoradiography showed that receptor-containing cells appeared in parallel with the genesis of dentate gyrus and pyramidal neurons.[178] Hypothalamic cytosol binding and cell nuclear uptake of B are changed to a much smaller extent.[159]

Strikingly different is the ontogenetic pattern of gluco- and mineralocorticoid receptors and CBG-like molecules in the pituitary. Glucocorticoid receptors do not change greatly in capacity from day 1 up to adult levels[111,159]; the mineralocorticoid receptors appear only at day 6 and reach adult levels after 4 weeks.[111,179] After birth, a rapid decline occurs in CBG-like molecules and this binding system is undetectable with labeled B between day 6 and day 10 of postnatal age.[111,180] Thereafter, the levels show a rapid rise and reach adult levels at 4 weeks of age.[111,159] In parallel with these changes are the variations in plasma CBG levels.[25,111,180]

It is of interest to note that the pituitary–adrenal system displays changes in activity that can be related to the animals' age. A stress response can be evoked at any time in the newborn animals, but the responsiveness is not fully developed before the second postnatal week.[181–184] Circadian rhythmicity gradually appears and reaches the magnitude of adult animals at 3 weeks of age.[182,185] It has been found that the development of pituitary–adrenal function is accelerated by exposing the neonatal rats to other stressful events. These procedures also have consequences for behavior and pituitary–adrenal function in adulthood.[185,186]

Feedback action on pituitary ACTH release is also less well developed the first 10 days after birth.[186] This steroid resistance can be explained in part by a deficiency in the steroid receptor molecule. The steroid–receptor complex was hampered in its ability to become associated with the cell nuclear chromatin.[111,180] This inability to translocate to the cell nucleus was somehow related to an impairment in transformation of the receptor molecule to the DNA-binding state.

However, although there is a deficiency in cell nuclear uptake *in vitro*, the *in vivo* administration of a tracer amount of [^{3}H]B resulted in a much more extensive uptake in pituitary cell nuclei than in the case of the adult animals. This apparent *in vivo/in vitro* discrepancy can be explained by the absence of CBG and CBG-like molecules at that time. Because of the absence of these molecules, there is more steroid made available for cell nuclear localization. In spite of the receptor deficiency, this leads to more extensive cell nuclear uptake. These deficiencies in binding, however, coincide with the period that the animals display a reduced stress responsiveness and an increased steroid resistance.

Senescent rats, alternatively, have elevated plasma B levels and heavier adrenals.[187] This may be related to progressive astrogliosis, which was observed in aging rats and which presumably leads to disinhibitory influences in hippocampus function.[188] Interestingly, certain indices for aging are reversed after removal of the adrenals or treatment with neuropeptides such as ACTH $_{4-9}$ analogue.[189] The number of receptors for B declines in the senescent rats and mice.[190-192] The decline exceeds the percentage of neuronal loss during aging.[192] Factors that control the number of receptor sites during aging are poorly understood. One of them may be B itself, which circulates in higher levels, exerting long-term down-regulatory influences on receptor number.

In naive mature rats of the same age maintained under apparently homogeneous conditions, there was a large variability in B receptor capacity. Individual binding values ranged between 220 and 550 fmol/mg protein in hippocampus cytosol. Classification of these animals in groups displaying good or poor avoidance behavior as well as acquisition of a conditioned avoidance response shows that such animals have significant differences in number of receptors. Animals that have been qualified as being good avoiders have the highest number of hippocampal B receptors.[193] It may well be that these individual differences not only are genetically determined but also are the consequence of environmental influences during life. Such environmental influences on the number of B receptors may be expressed via some of the hormones, neurotransmitters, and neuropeptide discussed in Section 5.1.

6. EFFECTS OF ADRENAL STEROIDS ON BRAIN CHEMISTRY

Numerous steroid hormone effects have been reported on morphology, chemistry, and electrical activity of the brain (see reviews[13,194,195]). In this chapter we have avoided exhaustive review of all available data but rather have emphasized those studies that have been designed to understand the consequences of steroid–receptor interaction. This implies that we have focused in particular on studies with ADX subjects that have been replaced with low doses of particular steroid hormones (see Table II). In order to assign receptor interaction as the first step in a given response to the steroid, this response should obey the criteria as prescribed by the properties of the receptor. These include the localization (septal–hippocampal complex) and steroid specificity (B action in neuron, synthetic glucocorticoid in glial) of the effect. Studies should be performed with doses that are within the range of the receptor capacity.

6.1. Neurotransmitters

Among the neurotransmitter systems that have been identified in the hippocampus are the septal cholinergic input,[217] the ascending serotonin (5-HT) from the raphe nuclei,[14] and the norepinephrine (NE) from locus coeruleus.[218] In addition, there are the excitatory amino acids glutamate via the entorhinal cortex and aspartate of hippocampal commissural fibers and the inhibitory transmitter γ-aminobutyric acid (GABA).[219]

Serotonin (5-HT) metabolism is under the influence of the adrenocortical hormones. Adrenalectomy decreased the turnover* of 5-HT in hippocampus, midbrain, and hypothalamus by reducing the activity of the rate-limiting biosynthetic enzyme tryptophan hydroxylase (TR-OH).[203,208,220] Adrenalectomy reduces tryptophan uptake[207,221] as well as 5-HT uptake and release in brain synaptosomes *in vitro*.[206] It also blocks the developmental rise in TR-OH activity.[207] The 5-HT response is rapid and is observed in parallel with the disappearance of B in the first hours after ADX[209,211] and the resultant increase of pituitary ACTH release.[210]

Replacement treatment of ADX animals with B restores 5-HT turnover and all other indices of 5-HT metabolism in the aforementioned brain regions[203,204,206,208,209,211,220,222] including monoamine oxidase (MAO) activity.[223] Corticosterone is necessary for the elevation of TR-OH activity by exposure of the animals to footshock, cold, or ether stress[220] or reserpine treatment.[207]

The 5-HT response to B occurs simultaneously in hippocampus (terminal area) and in the raphe nuclei (cell body region). The B-receptor-containing cells are, however, soley located postsynaptically in the hippocampal neurons, since the raphe area is devoid of such receptors.[220,224] The raphe–hippocampal system, therefore, is probably activated by B via a transsynaptic action. Surgical isolation of the hippocampus prevents the midbrain TR-OH response,[226] whereas electrical stimulation stimulates 5-HT turnover.[227,228] The nature of this hippocampal–raphe connection is not known but may well be GABAergic.

Normalization of 5-HT response in the raphe–hippocampal 5-HT system is specific for B. Replacement with the same dose of ALDO was ineffective, and DEX even decreased 5-HT turnover.[224] Pretreatment of B-substituted ADX

---

* Turnover rate has been estimated from the rate of accumulation of 5-HT and decline of 5-hydroxyindoleacetic acid (5-HIAA) on administration of pargyline, which is a MAO blocker.[225]

Table II
Effect of Adrenalectomy and Hormone Replacement on Brain Chemistry

Biochemical parameter	Effect of ADX	Hormone replacement Effect	Steroid	Dose/duration	Brain region	Reference
Catecholamines						
NE turnover	Increased	Decreased	B	Endogenous[b]	Whole brain	196
	Increased	Decreased	F	25 mg/kg, 2 d.[b]	Whole brain	197
TH activity	Decreased	Increased	DEX	0.3 mg/kg, 7 d.[b]	Median eminence	198
DBH activity	Decreased	Increased	B	100 mg/kg, 4 h[a]	Hypothalamus	199
MAO activity	Increased	Decreased	DEX	2 × 30 μg/kg, 10 d.[b]	Hypothalamus	200
Sensitivity of NE-receptor-coupled adenylate cyclase	Increased	Decreased	B	10 mg/kg, 5 d.[b]	Frontal cortex	201
B_{max} of β-adrenergic receptor	Increased	Decreased	B	1 mg/kg, 12 h[a], 7 d.[b]	Hippocampus/dorsal NE bundle lesion	202
Serotonin						
TR-OH activity	Decreased	Increased	B	0.4–3 mg/kg, 4 h[a], 5 d.[b]	Midbrain	203
5-HT content	Decreased	Increased	B	1 mg/kg, 30–60 min[a]	Hypothalamus	204
	Decreased	Increased	B	1 mg/kg, 15 min[a]	Mesencephalon, amygdala, hypothalamus, hippocampus, septum	205
5-HT uptake and release	Decreased	Increased	B	1 mg/kg, 30 min[a]	Hypothalamus	206
Developmental rise TR-OH activity	Blocked	Increased	B	5 mg/kg, 3 d.[b]	Whole brain, neonate	207
Reserpine-induced rise TR-OH activity	Blocked	Increased	B	20 mg/kg, 6 d.[b]	Whole brain, adult	207

Try content	Decreased	Increased	B	10 mg/kg, 3–7 d.[b]	Brainstem	208
Tr-OH activity, 5-HT content	Decreased	Increased	B	10 mg/kg, 3–7 d.[b]	Brainstem	208
5-HIAA content	Increased	Decreased	B	10 mg/kg, 3–7 d.[b]	Brainstem	208
5-HT turnover	Decreased	Increased	B	2 × 0.3–0.4 mg/kg, 2h[a], 3 d.[b]	Hypothalamus, brainstem	209, 210
	Decreased	Increased	DEX	0.25 mg/kg, 2 h[a]	Hypothalamus, brainstem	209, 210
	Decreased	Increased	B	0.3 mg/kg, 1 h[a]	Hippocampus, raphe area	211
5-HT turnover	Decreased	Ineffective	DEX	0.3 mg/kg, 1 h[a]	Hippocampus, raphe area	211
GABA						
GABA uptake	Increased	Decreased	B	Solid implant, 4–10 d.[b]	Hippocampal synaptosomes	212
GABA receptor binding	Increased	Decreased	B	5 mg/kg, 24 h[a]	Midbrain, striatum	213
Miscellaneous						
GPDH-activity	Decreased	Increased	F	2–3 mg/kg, 13–14 d.[b]	Brainstem, cerebellum	123
VIP content	Decreased	Increased	B	160 μg/ml in drinking water, 14 d.[b]	Dorsal hippocampus	214
	Decreased	Increased	DEX	173 μg/ml in drinking water, 10 d[b]	Dorsal hippocampus	214
Angiotensin content	Decreased	Increased	B	1 mg/kg, 4 d.[b]	Preoptic area, anterior hypothalamus, periaqueductal gray, area postrema	215
Neurophysin and vasopressin content in terminals	Increased	Decreased	B	Solid implant, 14 d.[b]	Median eminence	216

[a] Acute injection.
[b] Chronic treatment; the dose of steroid given was injected daily.

animals with either DEX or ALDO abolished normalization of the 5-HT response by the naturally occurring glucocorticoid. Aldosterone blockade may well be exerted by competition for the B receptor (see Section 4.3). Since DEX does not bind very well to this receptor, the antagonizing action of this steroid reveals another aspect of steroid influence on 5-HT metabolism. Dexamethasone apparently blocks the availability or uptake of the tryptophan precursor.[207] Low and high doses of B have also been shown to act in opposite ways on 5-HT metabolism.[229]

Yet, the B effect on 5-HT metabolism displays a remarkable correlation with the receptor specificity. It will be shown in Section 7.2 that this correlation can be extended to the influence of B on certain behaviors.

Some aspects of GABA chemistry are also under control of B. Although the enzymes GAD and GABA-T[230-232] are unaffected by ADX, this procedure in the hippocampus leads within 24 h to an increase in GABA uptake by hippocampal synaptosomes *in vitro*.[212] The effect is specific for the hippocampus, and GABA uptake is restored again after chronic replacement with B.[212] The number of GABA receptors increases in the midbrain of ADX rats, but this effect seems to be related to ACTH or ACTH-related neuropeptides.[213]

Changes in norepinephrine (NE), epinephrine (E), or dopamine (DA) metabolism after ADX were mostly restricted to hypothalamic regions. The effects are, however, contradictory and may well depend on the time interval and the particular hypothalamic nucleus investigated. Short-term ADX increased NE* turnover in median eminence (EM), arcuate nucleus (na), paraventricular nucleus (npv), and dorsomedial nucleus (ndm), E turnover in npv only, and DA turnover in na and EM.[233] It may be related to the corticosterone effect on prolactin release.[234] All other changes in catecholamine turnover were not different shortly after ADX or S-ADX. Long-term ADX reduced NE turnover in hypothalamus, and B replacement in general had the opposite effect. Synthesis *in vitro* from [^{3}H]tyrosine precursor was also not affected by ADX,[233] and neither was catecholamine uptake in synaptosomes.[206]

Although there were no effects on NE metabolism in the hippocampus, corticosterone can influence the number of β-adrenergic receptor sites[202] and regulates the sensitivity of NE-receptor-coupled adenylate cyclase.[201,235] The effect on β-adrenergic receptor sites becomes apparent after lesioning of the dorsal bundle: ADX potentiates the consequent rise in NE receptor sites, and a reduction occurs after B replacement. Involvement of pituitary ACTH release cannot be excluded, however.

Enzymes involved in catecholamine metabolism are also affected by B in some cells. The presence of glucocorticoids is necessary for maximal adrenal PNMT activity.[236] In the superior cervical ganglion, glucocorticoids greatly amplify induction of TH by stress, by nerve growth factor, or acetylcholine.[237] Tyrosine hydroxylase, however, was also sensitive to E, which is removed as well on ADX.[238]

---

* Norepinephrine, E, DA turnover were calculated from the rate of disappearance of the catecholamines after administration of the synthesis inhibitor α-methyl-*p*-tyrosine (α-MPT).

6.2. Neuropeptides

Corticosterone inhibits hypothalamic CRF release *in vitro*.[12,239] The CRFs include the recently discovered 41-amino-acid peptide claimed to be "the CRF"[240] as well as vasopressin. "The CRF" is synthetized in the same nucleus as vasopressin, i.e., the paraventricular nucleus.[91] More aspects of interaction of B with CRF will be revealed soon.

Vasopressin is also under control of B. The first 24 h after ADX, the peptide is depleted from hypothalamus and hippocampus (E. R. de Kloet unpublished data), but at longer time intervals the concentration is increased in median eminence, as shown by immunocytochemistry.[216] Corticosterone replacement reduces this increase in vasopressin concentration, but the mineralocorticoid DOC is ineffective.[216]

ACTH is one of the biosynthetic end products of proopiomelanocortin. Other known end products are β-LPH, β-endorphin, α-MSH, and related peptides.[241] The precursor protein is synthesized not only in the anterior pituitary but also in cells of the intermediate lobe and in the periarcuate region of the brain. These sources differ in pattern of peptides released, ACTH and β-LPH being predominant in the anterior pituitary, but α-MSH and β-endorphin (or the acetylated peptides) predominating in intermediate lobe and brain.[242] Pituitary peptides are secreted in the vascular system, whereas distribution of the brain peptides is anatomically funneled to distant brain regions via axonal transport.[243]

Adrenalectomy results in changes in opiomelanocortin level in pituitary and brain. ACTH immunoreactivity in the anterior pituitary is minimal 12 h after ADX and in the hypothalamus and hippocampus after 24 h and 72 h, respectively. The depletion cannot be restored by B replacement. Rather, the changes in level seem primarily related to the stress. However, glucocorticoids block release of pituitary ACTH.[12] Glucocorticoids may affect intra- or extracellular processing of opiocortins[244] and the number of opiate receptors.[245]

Cell bodies containing vasoactive intestinal peptide (VIP) have been found in the hippocampus.[246] The peptide is suggested to have a function in energy metabolism. Adrenalectomy decreases VIP concentration, and the level is restored with chronic B as well as DEX treatment. This lack of specificity makes the finding difficult to interpret in terms of neuronal B receptors.[214]

6.3. Proteins

In hippocampal tissue slices B stimulates [^{3}H]uridine incorporation into cell RNA,[247] and [^{3}H]leucine incorporation into a species of 54k soluble proteins.[248] In this latter study, PROG was ineffective, but preincubation with PROG inhibits the B effect on protein synthesis. Steroid specificity and dose (10^{-9} M) suggest that B receptors are involved.

Corticosterone increases the amount of radioimmunoassayable protein I in the hippocampus.[249] Protein I is a phosphoprotein that is present only in neurons and is concentrated in most, and possibly all, presynaptic terminals.[250,251] The association of protein I with neurotransmitter vesicles suggests that it has an important role in functioning of those vesicles. The effect of B is specific and occurs only in the hippocampus. The changes in protein I may be interpreted as an indication for increased production in the vesicles of neurotransmitters such as 5-HT.

6.4. Enzymes

In Table II a number of enzymes have been listed that were tested for responsiveness to ADX and steroid replacement. Of these enzymes, only four respond to glucocorticoids. These are, in addition to the previously mentioned neurotransmitter enzymes TR-OH and TH, glycerolphosphate dehydrogenase (GPDH) and ornithine decarboxylase (ODC). Glycerolphosphate dehydrogenase is involved in energy metabolism and has an exclusive glial cell localization; DEX is a more potent inducer than B.[123,125,231]

Ornithine decarboxylase also responds better to DEX than to B. Its induction time is 4 to 6 h, and since there is no regional specificity, the enzyme presumably represents a glial response.[126] Ornithine decarboxylase is a rate-limiting enzyme in the biosynthesis of polyamines[252] and is known to play a role in protein synthesis.[253] Lesioning of the hippocampus is also a potent stimulus for ODC induction.[254] This lesion-induced response is, however, more pronounced in ADX animals and lasts for 5 days. Replacement with B normalizes the ODC response to the lesion; ODC thus appears to be a highly sensitive index for glucocorticoid action localized primarily in glial cells.

6.5. Morphology

Corticosteroids have inhibitory effects on growth and development and facilitate the appearance of morphological signs of aging. Neonatal administration of glucocorticoids causes long-lasting deficits in myelination and dendritic branching.[255,256] These effects are long lasting and have consequences for pituitary–adrenal function and behavior in later life.

Recovery of brain tissue after damage is also affected by glucocorticoids. Lesions in the hippocampus are followed by glial cell proliferation serving the removal of degenerating nerve terminals.[257] Glial cell proliferation lasts for about 5 days, which agrees well with the duration of elevated ODC (see Section 6.4) before axonal growth takes place. Glucocorticoids act on both processes: they enhance glial cell proliferation and decrease the amount of axonal sprouting that occurs in response to a lesion in the entorhinal cortex.[258]

Hypertrophy of astrocytes, neuronal loss, and retarded axonal sprouting are age-related phenomena[258,259] and correlate well with increased adrenocortical activity during aging.[188] The morphological characteristics of aging were reversed after ADX, although performance in behavioral tests did not improve. Interestingly, treatment of aged rats for 6 months with an ACTH4–9 analogue also reduced the aging process in morphology as well as in behavior.[189] On the basis of these findings, it was concluded that hormones may gradually damage neural target sites and, on the way, contribute to some aspects of aging. The beneficial effect of the ACTH analogue may well be exerted on the level of the B receptor (see Section 5.1).

6.6. Other Actions

Adrenalectomy reduced brain excitability and the ratio of intra- to extracellular Na$^+$ but increased this ratio for K$^+$. This effect is reversed by mineralocorticoids.[13] Cerebral blood flow and oxygen consumption are reduced by ADX and restored after F but not DOC replacement.[13] The concentration of free amino acids is reduced after ADX; F increased the concentration of most amino acids except for glutamine and GABA.[13]

Clinically, synthetic glucocorticoids are widely used to reduce brain edema.[260] As described in Section 4.2, the synthetic glucocorticoid concentrates in particular in cerebral capillary endothelial cells,[100] and this localization suggests that the steroids act on these cells to reduce the rate of formation of CSF. Direct effects on transport of electrolytes through brain cell membranes might also contribute to the clinical efficacy of the steroids.

7. CONTROL OF BRAIN FUNCTION BY ADRENAL STEROIDS

Glucocorticoids are secreted by adrenocortical cells in response to ACTH from the anterior pituitary, which in turn is released in response to CRFs of hypothalamic or even extrahypothalamic origin. Glucocorticoids feed back on brain and pituitary to block the release and synthesis of ACTH (and other opiomelanocortins) and CRFs and have effects on their metabolism. Removal of the adrenals results in a high rate of synthesis and release of these hormones.

The activity of the system at rest is on the one hand a function of the net results of all inhibitory and stimulatory neural influences converging on the hypothalamic CRF cells and, on the other hand, the magnitude of the corticoid feedback signal (Fig. 10). The feedback by glucocorticoids can easily be overridden by excitatory impulses, which are nonspecifically termed stress: environmental changes, pain, blood loss, loud noise, anxiety, or frustration. In fact, every internal or external disturbance in homeostasis evokes a response of the pituitary–adrenal system.[261]

It has been suggested that the essential property of all stressors is the ability to elicit arousal that results in a behavioral response aimed to fight, flight, or cope with the threat.[262] Glucocorticoids are released in response to stress and serve to restore depleted energy resources and help to facilitate extinction of aversive experiences and retard extinction based on an appetitive behavioral paradigm. Glucocorticoids also protect the organism against stressful experiences and in this way serve the response to the threat. The neuroendocrine and adrenal responses to stress are rather slow. This contrasts with the autonomic responses to stress that produce stimulus specific rapid adaptive changes in many organs including the brain. All of these responses serve homeostasis and have been termed by Selye[263] the "general adaptation

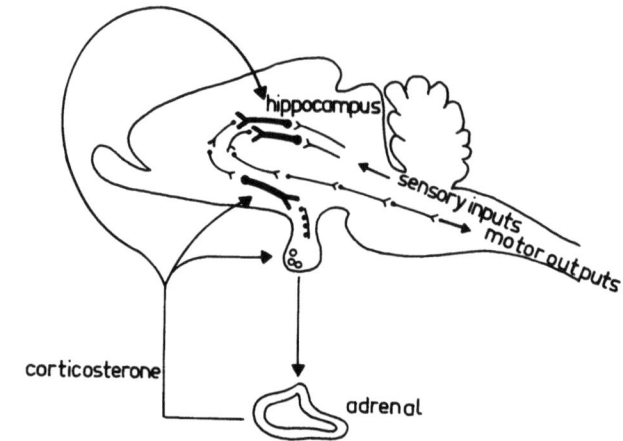

Fig. 10. Feedback action of B on brain and pituitary. Corticosterone controls hypothalamic CRH and pituitary ACTH release. In higher brain regions, B exerts an influence on phase and timing of stress-induced neuroendocrine response and modulates aspects of adaptive behavior. Arrows in brain depict the predominant putative sites of corticosterone action and presumptive neural circuits involved in processing of sensory information with appropriate neuroendocrine and behavioral motor responses.

syndrome." In other words, life is a continuous adaptation to changes in internal and external conditions.

The next two sections are concerned with modulatory influences of adrenocortical hormones on neuroendocrine and behavioral responses to environmental challenges.

7.1. Neuroendocrinology

Elevated plasma B level in response to stress or corticosterone infusion brings about two subsequent phases in suppression of pituitary adrenal activity. The first phase occurs during the first minutes, when plasma B level is rising at its fastest rate.[264–266] This rate-sensitive or fast feedback control is observed on ACTH and CRF release[239,267] and disappears between 5 and 20 min after the stress response. Then a gradual inhibition of ACTH release develops. The level-sensitive or delayed feedback may last several hours. Onset and duration depend on the magnitude of the glucocorticoid feedback signal and the severity of the stress.[268] With a very potent synthetic glucocorticoid, it may still take 30–90 min before suppression of stress-induced ACTH release is obtained.[105] Fast and delayed feedback also occur in man.[240] The only agonists in fast feedback are the naturally occurring glucocorticoids B and F.[266] The other steroids secreted by the adrenal gland are not active in this way, as is the case for synthetic glucocorticoids.[239,269] Some of these steroids (DOC) even antagonize the fast feedback action of B. In contrast, many steroids suppress pituitary ACTH release several hours after administration. Extremely potent are synthetic glucocorticoids, but the mineralocorticoids ALDO and DOC are effective.[111,269,270] Temporal aspects and difference in steroid specificity suggest different mechanisms and receptor systems for fast and delayed feedback.

There is considerable evidence that synthetic glucocorticoids such as DEX have their principal site of action in the anterior pituitary.[105,271,272] The potency to suppress ACTH release *in vitro* from pituitary cells has been correlated with the extent of cell nuclear localization of the glucocorticoid–receptor complex.[139,140,143] The use of protein synthesis inhibitors revealed that genomic events were involved in the suppression.[273] Recently, putative DNA sequences were found that seem to mediate glucocorticoid control of opiocortin gene expression.[274] *In vitro*, ACTH release from pituitary glands is also inhibited immediately by DEX and other glucocorticoids under conditions in which ACTH synthesis is not altered. Accordingly, such steroids *in vitro* suppress the release directly and the new synthesis subsequently. Corticosterone is less potent in suppression of pituitary ACTH release. This diminished efficacy seems partly to result from the presence of an intracellular CBG-like binding system in the anterior pituitary (see Section 4.8). That ALDO also suppresses ACTH release has been ascribed to the affinity of the glucocorticoid receptor for this mineralocorticoid.[111]

Besides acting on the pituitary to suppress ACTH release, glucocorticoids also inhibit the release of CRF from the hypothalamus or hypothalamic synaptosomes *in vitro*.[239,275,276] As is the case for pituitary ACTH release *in vitro*, the glucocorticoids have an immediate effect on CRF release by adding them to the medium and a delayed effect in suppression of CRF synthesis and release after pretreatment of the animals 24 h earlier.[12] Calcium ions, acetylcholine, and 5-HT are stimuli for CRF release that can be blocked by glucocorticoids.[275]

It is tempting to speculate that the immediate membrane-mediated effects in pituitary and hypothalamus *in vitro* reflect the fast feedback effect. However, fast feedback is present in animals with the hypothalamus isolated.[12,277] This favors a hypothalamic site of fast feedback action. Receptor sites that are the most likely candidates for mediating glucocorticoid action on hypothalamic CRF release are the soluble receptor molecules that can be detected in appreciable amounts in this region (see Section 4.5).[99]

In contrast to the direct blocking effect of glucocorticoids on pituitary–adrenal activity in hypothalamus and pituitary, the action of B on its target cells in extrahypothalamic limbic structures (hippocampus, septum, and amygdala) seems to be implicated in the modulation of the influence these structures have on CRF synthesis and release. It has been suggested that the excitatory amygdaloid pathway and the inhibitory hippocampal input act in sequence,[262] the amygdala being involved in initiation of the stress response and the hippocampus in termination. Some authors have pointed out that the inhibitory influence might be "state dependent." Such state dependence implies that the extent of inhibitory action on CRF release provoked by the hippocampus is proportional to the severity of the stress the animal is exposed to.[262,278] Corticosterone may modulate the timing and magnitude of the limbic influence on pituitary–adrenal activity.

An interesting example is the stress responsiveness during the circadian variation in B level. Adrenocortical activation after a light emotional stress such as handling has a rapid onset and short duration at the diurnal trough; the same stimulus causes much slower adrenocortical activation and is more persistent when applied during the diurnal peak of B secretion.[279] It is as if the initiating and terminating mechanisms in stress-induced ACTH release taking place via B-sensitive amygdaloid and hippocampal neurons work less efficiently at the diurnal crest.[102] This line of reasoning implicates B as a hormonal factor of peripheral origin in state dependence. Another example is the steroid influence on circadian variation. Although the circadian variation in pituitary–adrenal activity is of neural origin, B implants in the dorsal hippocampus abolish the rhythm.[280]

If hippocampal influence on hypothalamic CRF release is indeed state dependent, this may explain the many contradictory results of hippocampal function in neuroendocrine regulation that are based on electrical stimulation or lesioning of the structure. Most of these studies point to an inhibitory influence of the hippocampus,[281,282] but this may depend on the time of day and other variables.[102,283] Moreover, electrical stimulation or lesions may alter the sensitivity of other neural structures or the pituitary to corticosteroid feedback.[284,285]

With the immunocytochemical localization of CRF cells in the paraventricular nucleus, the neurotransmitter systems that terminate on these neurons can be identified. Previous work has already implicated acetylcholine, 5-HT, and amino acid transmitters. The 5-HT system may play an important role in modulation of the stress response.[14] The system has extensive projections to forebrain regions including the hypothalamus and the hippocampus. Moreover, it needs B for expression of maximal activity[209,211,220] and is implicated in control of pituitary–adrenal function.

7.2. Behavior

Corticoids have little if any effect on learning of active and passive avoidance or approach behaviors. The steroids improve performance and have a profound effect on retention of such behaviors.[5,6] Corticoids usually facilitate extinction of active avoidance behaviors.[301,314] High doses of corticoid suppress passive avoidance retention,[5,314] but low doses (100 µg corticosterone/100 g) facilitate.

The early behavioral studies with corticoid administration to intact rats are, however, in conflict with the properties of B receptor system in the hippocampus and other limbic structures. First, steroid effects in intact animals lack the structural specificity that is required for interaction with the receptor. Second, implantation studies have indicated behavioral responsive sites in extensive limbic–midbrain areas, which are also outside the areas containing the receptor. Third, the doses used to elicit behavioral response are far beyond the capacity of the receptor system. Thus, such behavioral effects employ a mechanism that seems distinct from the one that responds to intracellular receptor interaction.

In order to resolve the physiological function of B receptor interaction in behavioral adaptation, studies have been designed based on the properties of these receptors. This includes the finding that deficiencies occurring after bilateral removal of the adrenals are restored specifically with B in a dose range not exceeding the physiological level. Two further aspects require consideration after adrenalectomy. First is the implication of medullary catecholamines in disturbed behaviors. Second is the high level of circulating ACTH, since ACTH (and other opiomelanocortins) often has effects on behavior opposite to those of corticosteroids.[315] Moreover, behaviors have been studied that are known to be associated with hippocampal function.[316,317]

In Table III behavioral observations after ADX and hormone replacement have been listed. Few of these studies meet the criteria prescribed by the hippocampal B receptor. Accordingly, ADX facilitated extinction of a food-reinforced straight runway response,[304,305] impaired the forced extinction of a learned passive avoidance response,[6,114,309] and reduced exploratory behavior.[298] Replacement with physiological doses of B reversed the ADX effects and restored the behavioral paradigm as observed in the sham-operated control animals; DEX administration to ADX rats was ineffective.[114,298,304,305]

The studies on extinction of a conditioned appetitive response were performed with chronic B replacement via subcutaneous implantation of steroid-containing pellets. The same authors found that after chronic B substitution the amount of paradoxical sleep was suppressed in the 3-h interval preceding light.[287] The direction of these B effects was the same as observed after ablation of the hippocampus.[318] Thus, it seems that B acts to suppress hippocampal function when tested in this behavioral paradigm.[15] Moreover, lesioning of the dorsal NE bundle also retarded extinction of appetitive behavior.[304,319-321] The dorsal NE bundle input normally facilitates selective attention.[319-321] These correlates among B action, hippocampectomy, and NE lesion have led McEwen[15] to speculate that B actually suppresses mechanisms underlying selective attention to a particular sensory stimulus.

The other two behavioral tests were performed with acute B treatment. The forced extinction paradigm was used in a one-trial passive avoidance condition.[322] Confinement of the rats for 5 min, 3 h after the learning trial, to the compartment where they had experienced punishment, resulted in a complete extinction of the passive avoidance response (forced extinction).[323,329] Forced extinction was impaired when the adrenals were removed 1 h previously, but was maintained in sham-operated animals when rats were given B at the time of bilateral adrenal removal. Dexamethasone or mineralocorticoids not only were ineffective, but pretreatment with these steroids antagonized the effect of B replacement.[6,114,309] For study of exploratory behavior, the same steroid specificity was observed. In these experiments, rats were adrenalectomized 10 days previously. Rearing and ambulation were increased to the level of sham-operated controls 1 h after a single B injection. Again, DEX and ALDO were ineffective and inhibited the B effect.[298]

Serotonin turnover in the raphe area and hippocampus is subject to essentially the same stringent criteria of steroid specificity (see Section 6.1). That B activates 5-HT whereas DEX and ALDO antagonize this B action reveals a striking correlation between behavioral expression and 5-HT metabolism in the raphe hippocampal projection.[211] It seems that this 5-HT system reflects the

Table III
Effect of Adrenalectomy and Hormone Replacement on Behavior

Behavioral paradigm	Species	Effect ADX or adrenocortical insufficiency	Steroid specificity[a]	References
Sensory perception of olfactory, auditory stimuli	Man, rat	Detection increased, recognition decreased	Gluco	3
Mood and affect	Man	Disturbed	Gluco	1
Sleep	Man	REM sleep decreased, attenuation of circadian variation decreased 3 h before light	Gluco	2
			Gluco	286
	Rat		B	287
Feeding	Rat	Food intake reduced	Gluco	288
Drinking	Rat	Salt appetite increased	Mineralo	289
			Gluco	290
Motor activity	Rat	Running wheel activity reduced	B	291
			Gluco	292
	Rat	General activity increased (gross body movements)	Gluco	293
Stress-induced analgesia	Rat	Latency of tail-flick response reduced	B	294
Aggression	Rat	Reduced	Gluco	295
Social interaction	Rat	Reduced	B	296
Exploration of novel environment	Rat	Reduced	B	297,298
Acquisition fear-motivated (avoidance) behavior	Rat	No effect	—	299–303
Intertrial responses of active avoidance	Rat	Increased	Gluco	300
Free operant avoidance, effect of prestimulation	Rat	Absent	Mineralo	9
Extinction of active avoidance behavior	Rat	Delayed	Gluco	300,301
Extinction of food-reinforced response	Rat	Facilitated	B	304,305
Retention of passive avoidance	Rat	Facilitation	Gluco	303
Acquisition and retention of passive avoidance	Rat	With 6-OHDA lesions: impaired	Not tested	306–308
Forced extinction of passive avoidance	Rat	Impaired	B	309
Consolidation of memory	Rat	Inhibition of drug- or surgery-induced amnesia	Gluco	310–313

[a] Steroid specificity in normalization of disturbed behavior after ADX: B, specific for corticosterone; Gluco, responsive to synthetic and naturally occurring glucocorticoids; Mineralo, specific for mineralocorticoids.

actual state of hippocampal functioning than being involved in mediation of the behavioral response.[14]

The antagonism displayed by ALDO in the abovementioned behavioral responses to B could be explained on the basis of competitive binding to the receptor. Alternatively, there are behaviors specifically influenced by ALDO. An example is the enhancement of an avoidance response following prestimulation by a weak stressful event (handling or an air blast). The effect of prestimulation is absent in ADX rats but is restored by ALDO replacement and not by B.

The facilitated extinction of an avoidance response has been interpreted as if B suppresses behavior that is of no further relevance. This does not necessarily contradict the notion of B suppression of selective attention[15] and could well be dependent on conditions that determine whether the cue is a rewarding or an aversive stimulus. Hypo- and hypersecretion of adrenocortical hormones disrupt sensory processes in man[3] and in rats.[325] Patients with Addison's disease have a heightened sensitivity to detection of olfactory and auditory stimuli that could be restored with glucocorticoid administration and not with mineralocorticoids. The relationship of this finding to B effects on animal behavior and the aforementioned aspects of hippocampal chemistry remain to be established.

Mental disturbances are frequently observed in patients with Cushing's or Addison's disease.[1] On the other hand, depressed patients often have abnormalities in adrenal steroid secretion. Repetitious blood sampling (every 20 min) showed that there are more frequent episodes of F secretion in depressed patients.[326] More than 50% of the depressed patients show an early escape from the DEX suppression of pituitary ACTH release.[327] The DEX suppression test is one of the most widely used tests in psychiatric diagnosis. The primary site of action of DEX is in the pituitary. The question thus remains whether such mental disturbances are intrinsic to a disturbed pituitary–adrenal system and/or manifestations of disturbed neuronal process, perhaps in the limbic structure, that underlie emotion and adaptation.

8. CONCLUDING REMARKS

Receptor systems for adrenal steroids in rat brain are heterogeneous. Limbic neurons, particularly in hippocampus, dorsal septum, and the amygdala region, contain receptor sites specific for the naturally occurring glucocorticoid B as well as for the mineralocorticoid ALDO. Synthetic glucocorticoids such as DEX have lower affinity and less access to these sites. In contrast, DEX has a potent influence on glial cell function, and the glucocorticoid receptor system in glial cells resembles that in peripheral target tissues, including the anterior pituitary. These characteristics of binding specificity and regional or cellular localization represent the criteria for study of receptor-mediated steroid responses. In such studies a classical endocrine design of the experiments was used: replacement of ADX subjects with a particular steroid hormone in amounts sufficient to reach a physiological level of receptor occupation should normalize disturbed neurochemical, neuroendocrinological, and behavioral response.

Such an approach clearly demonstrated that aspects of sleep, food and water intake, aggression, locomotion, and extinction of certain limbic-associated learned behaviors were selectively modulated by B in the rat. Corticosterone-specific changes occurred in 5-HT turnover, NE receptors, GABA uptake, and protein synthesis in hippocampus or regions innervated by hippocampal neurons.

That ALDO also binds with high affinity to B receptors and acts as an antagonist in some parameters presents a puzzling problem. There are specific mineralocorticoid actions on brain and behavior. The conclusive evidence for independent ALDO receptors that coexist with the B receptor in the same neurons must await purification of the two receptor populations.

The knowledge gained of the properties of glucocorticoid receptors in rodents may have one further implication. Provided that DEX also has poor affinity and access to the limbic B receptor in humans, it is tempting to speculate that in regard to limbic processes therapy with synthetic glucocorticoids would result in the opposite goal. Although such a therapy effectively blocks pituitary ACTH release and affects glial processes, it would reduce rather than replace the action of the naturally occurring glucocorticoids in the limbic brain. Ongoing research is aimed to test this hypothesis in man.

Corticosterone or F action in limbic brain regions seems primarily involved in interpretation of sensory stimuli and the expression of the behavioral and neuroendocrine response (Fig. 10). In man, adrenocortical insufficiency has been associated with a decrease in threshold for sensory stimuli and a decreased ability of sensory recognition. Excess of adrenocortical hormones also leads to disturbances in detection, interpretation, and response to environmental stimuli. It promotes psychopathology and aspects of brain aging. The number of glucocorticoid receptors in limbic brain is dramatically reduced at senescence and is subject to control by specific endocrine and neural factors. It is conceivable that the receptor system not only participates in homeostasis by transmitting the steroid signal to the genome but also adjusts its capacity or activity to environmental needs. Such a regulatory mechanism exerted via modulation of receptor activation, transformation, recycling, or new synthesis promises to be an intriguing avenue of research.

ACKNOWLEDGMENT. The work from our laboratory described in this chapter was supported in part by the Foundation for Medical Research (FUNGO) and the European Training Program in Brain and Behavior Research (ETPBBR). We express our appreciation to Dr. Bela Bohus for stimulating discussions in the course of our research.

REFERENCES

1. Von Zerssen, D., 1976, *Psychotropic Actions of Hormones* (I. M. Itil, G. Laudahn, and W. M. Hermann, eds.), Spectrum, New York, pp. 195–222.

2. Gillin, J. C., Jacobs, L. S., Fram, D. H., and Snyder, F., 1972, *Nature* **237**:398–399.
3. Henkin, R. I., 1979, *Prog. Brain Res.* **32**:270–293.
4. Doupe, A. J., and Patterson, P. H., 1982, *Current Topics in Neuroendocrinology*, Volume 2 (D. Ganten and D. Pfaff, eds.), Springer-Verlag, Berlin, Heidelberg, New York, pp. 23–44.
5. Bohus, B., 1975, *The Hippocampus*, Volume I (R. L. Isaacson and K. H. Pribram, eds.), Plenum Press, New York, pp. 323–353.
6. Bohus, B., De Kloet, E. R., and Veldhuis, H. D., 1982, *Current Topics in Neuroendocrinology*, Volume 2 (D. Ganten and D. Pfaff, eds.), Springer-Verlag, Berlin, Heidelberg, New York, pp. 107–148.
7. McEwen, B. S., and Micco, D. J., Jr., 1980, *The Brain as an Endocrine Target Organ in Health and Disease* (P. A. Van Keep and D. De Wied, eds.), MTP Press, Lancaster, pp. 11–28.
8. Richter, C. P., 1936, *Am. J. Physiol.* **115**:481–487.
9. Gray, P., 1976, *J. Comp. Physiol. Psychol.* **90**:1–17.
10. Ganong, W. F., 1970, *The Hypothalamus* (L. Martini, M. Motta, and F. Fraschini, eds.), Academic Press, New York, pp. 313–333.
11. De Kloet, E. R., and McEwen, B. S., 1976, *Molecular and Functional Neurobiology* (W. H. Gispen, ed.), Elsevier, Amsterdam, pp. 257–307.
12. Jones, M. T., Gillham, B., Greenstein, B. D., Beckford, U., and Holmes, M. C., 1982, *Current Topics in Neuroendocrinology*, Volume 2 (D. Ganten and D. Pfaff, eds.), Springer-Verlag, Berlin, Heidelberg, New York, pp. 45–68.
13. Woodbury, D. M., 1972, *Handbook of Neurochemistry*, Volume 7 (A. Lajtha, ed.), New York, Plenum Press, pp. 225–287.
14. Azmitia, E. C., 1978, *Handbook of Psychopharmacology*, Volume 9 (L. L. Iversen, S. D. Iversen, and S. H. Snyder, eds.), Plenum Press, New York, pp. 233–314.
15. McEwen, B. S., 1982, *Current Topics in Neuroendocrinology*, Volume 2 (D. Ganten and D. Pfaff, eds.), Springer-Verlag, Berlin, Heidelberg, New York, pp. 1–22.
16. McEwen, B. S., 1979, *Monographs in Endocrinology*, Volume 12 (J. D. Baxter and G. G. Rousseau, eds.), Springer-Verlag, Berlin, Heidelberg, New York, pp. 467–492.
17. Riker, W. F., Baker, T., and Sastre, A., 1982, *Current Topics in Neuroendocrinology*, Volume 2 (D. Ganten and D. Pfaff, eds.), Springer-Verlag, Berlin, Heidelberg, New York, pp. 69–105.
18. McEwen, B. S., Weiss, J. M., and Schwartz, L. S., 1968, *Nature* **220**:911–912.
19. Baxter, J. D., and Rousseau, G. G. (eds.), 1979, *Glucocorticoid Hormone Action*, Springer-Verlag, Berlin.
20. Baxter, J. D., and Rousseau, G. G., 1979, *Monographs in Endocrinology*, Volume 12 (J. D. Baxter and G. G. Rousseau, eds.), Springer-Verlag, Berlin, Heidelberg, New York, pp. 1–24.
21. Granner, D. K., 1979, *Monographs in Endocrinology*, Volume 12 (J. D. Baxter and G. G. Rousseau, eds.), Springer-Verlag, Berlin, Heidelberg, New York, pp. 593–612.
22. Ingle, D. J., 1954, *J. Clin. Endocrinol.* **14**:1272–1274.
23. Forman, B. H., and Mulrow, P. J., 1975, *Handbook of Physiology*, Section 7, Volume VI (R. O. Greep and E. B. Astwood, eds.), American Physiological Society, Washington, pp. 179–190.
24. Christy, A. (ed.), 1971, *The Human Adrenal Cortex*, 1st ed., Harper & Row, New York.
25. Westphal, U. (ed.), 1971, *Steroid Protein Interaction*, Springer-Verlag, Berlin.
26. Ballard, P. L., 1979, *Monographs in Endocrinology*, Volume 12 (J. D. Baxter and G. G. Rousseau, eds.), Springer-Verlag, Berlin, Heidelberg, New York, pp. 25–48.
27. Pardridge, W. M., and Mietus, L. J., 1979, *J. Clin. Invest.* **64**:145–154.
28. Giorgi, E. P., and Stein, W. D., 1981, *Endocrinology* **108**:688–697.
29. Pardridge, W. M., 1981, *Diabetologica* **20**:246–254.
30. Koch, B., Lutz, B., Briaud, B., and Mialhe, C., 1976, *Biochim. Biophys. Acta* **444**:497–507.
31. Koch, B., Lutz-Bucher, B., Briaud, B., and Mialhe, C., 1978, *J. Endocrinol.* **79**:215–222.
32. Koch, B., Sakly, M., and Lutz-Bucher, B., 1981, *Mol. Cell. Endocrinol.* **22**:169–178.
33. De Kloet, E. R., Burbach, J. P. H., and Mulder, G. H., 1977, *Mol. Cell. Endocrinol.* **7**:261–273.

34. Al-Khoury, H., and Greenstein, B. D., 1980, *Nature* **287**:58–60.
35. Milgrom, E., Atger, M., and Baulieu, E. E., 1973, *Biochim. Biophys. Acta* **320**:267–274.
36. Harrison, R. W., Fairfield, S., and Orth, D. N., 1977, *Biochim. Biophys. Acta* **466**:357–364.
37. Rao, G. S., 1981, *Mol. Cell. Endocrinol.* **21**:97–108.
38. Baulieu, E. E., 1978, *Mol. Cell. Endocrinol.* **12**:247–254.
39. Sadler, S. E., and Maller, J. L., 1982, *J. Biol. Chem.* **257**:355–361.
40. Finidorf-Lépicard, J., Schorderet-Slatkine, S., Hanoune, J., and Baulieu, E. E., 1981, *Nature* **292**:255–257.
41. Towle, A. C., and Sze, P. Y., 1978, *Neurosci. Abstr.* **4**:356.
42. Jackson, V., and Chalkley, R., 1974, *J. Biol. Chem.* **249**:1615–1626.
43. Martin, P. M., and Sheridan, P. J., 1982, *J. Steroid Biochem.* **16**:215–229.
44. Denef, C., Magnus, C., and McEwen, B. S., 1973, *J. Endocrinol.* **59**:605–621.
45. Naftolin, F., Ryan, K. J., Davies, I. J., Reddy, V. V., Flores, F., Petro, Z., and Kuhn, M., 1975, *Recent Prog. Horm. Res.* **31**:295–315.
46. Lieberburg, I., and McEwen, B. S., 1977, *Endocrinology* **100**:588–597.
47. Karavolas, H. J., and Nuti, K. M., 1976, *Subcellular Mechanisms in Reproductive Neuroendocrinology* (F. Naftolin, K. J. Ryan, and J. Davies, eds.), Elsevier, Amsterdam, pp. 305–326.
48. Pane, S. M., and Axelrod, J., 1977, *Science* **197**:657–659.
49. Purdy, R. H., and Axelrod, L. R., 1968, *Steroids* **11**:851–862.
50. Hampel, M. R., Peng, L.-H., Pearlman, M. R. J., and Pearlman, W. H., 1978, *J. Biol. Chem.* **253**:8545–8553.
51. Ueda, M., 1978, *J. Steroid. Biochem.* **9**:1261–1262.
52. Kraulis, I., Foldes, G., Traikov, H., Dubrovsky, B., and Birmingham, M. K., 1975, *Brain Res.* **88**:1–14.
53. Cidlowski, J. A., and Munck, A., 1980, *Pharmacological Modulation of of Steroid Action* (E. Genazzani, ed.), Raven Press, New York, pp. 205–216.
54. Milgrom, E., 1981, *Biochemical Actions of Hormones*, Volume 8 (G. Litwack, ed.), Academic Press, New York, pp. 465–493.
55. Sando, J. J., Hammond, N. D., Stratford, C. A., and Pratt, W. B., 1979, *J. Biol. Chem.* **254**:4779–4789.
56. Moudgil, V. K., and John, J. K., 1980, *Biochem. J.* **190**:799–808.
57. Nielsen, C. J., Sando, J. J., Vogel, W. M., and Pratt, W. B., 1977, *J. Biol. Chem.* **252**:7568–7578.
58. Leach, K. L., Dahmer, M. K., Hammond, M. D., Sando, J. J., and Pratt, W. B., 1979, *J. Biol. Chem.* **254**:11884–11890.
59. Noma, K., Nakao, K., Sato, B., Nishizawa, Y., Matsumoto, K., and Yamamura, Y., 1980, *Endocrinology* **107**:1205–1212.
60. Munck, A., and Leung, K., 1977, *Receptors and Mechanism of Action of Steroid Hormones*, Part II (J. R. Pasqualini, ed.), Marcel Dekker, New York, pp. 311–397.
61. Jensen, E. V., and De Sombre, E. R., 1973, *Science* **182**:126–134.
62. Higgins, S. J., Rousseau, G. G., Baxter, J. D., and Tomkins, G. M., 1973, *J. Biol. Chem.* **248**:5866–5872.
63. Mainwaring, W. I. P., and Irving, R., 1973, *Biochem. J.* **134**:113–127.
64. Cake, M. H., Goidl, J. O., Parchmann, G. L., and Litwack, G., 1976, *Biochem. Biophys. Res. Commun.* **71**:45–52.
65. Bailly, A., Sallas, N., and Milgrom, E., 1977, *J. Biol. Chem.* **252**:858–863.
66. Disorbo, D. M., Phelps, D. S., and Litwack, G., 1980, *Endocrinology* **106**:922–929.
67. Munck, A., and Foley, R., 1979, *Nature* **278**:752–754.
68. Litwack, G., Schmidt, T. J., Marković, R. D., Eisen, H. J., Barnett, C. A., Disorbo, D. M., and Phelps, D. S., 1980, *Perspectives in Steroid Receptor Research* (F. Bresciani, ed.), Raven Press, New York, pp. 113–131.
69. Atger, M., and Milgrom, E., 1976, *Biochemistry* **15**:4298–4304.
70. Goidl, J. A., Cake, M. H., Dolan, K. P., Parchman, L. G., and Litwack, G., 1977, *Biochemistry* **16**:2125–2130.
71. Sato, M., Noma, K., Nishizawa, Y., Nakao, N., Matsumoto, K., and Yamamura, Y., 1980, *Endocrinology* **106**:1142–1148.

72. Turner, B. B., and McEwen, B. S., 1980, *Brain Res.* **189**:169–182.
73. Rousseau, G. G., and Baxter, J. D., 1979, *Monographs in Endocrinology*, Volume 12 (J. D. Baxter and G. G. Rousseau, eds.), Springer-Verlag, Berlin, Heidelberg, New York, pp. 49–78.
74. Sherman, M. R., 1979, *Monographs in Endocrinology*, Volume 12 (J. D. Baxter and G. G. Rousseau, eds.), Springer-Verlag, Berlin, Heidelberg, New York, pp. 123–134.
75. Suthers, M. B., Pressley, L. A., and Funder, J. W., 1976, *Endocrinology* **99**:260–269.
76. Samuels, H. H., and Tomkins, G. M., 1970, *J. Mol. Biol.* **52**:57–74.
77. O'Malley, B. W., and Schrader, W. T., 1976, *Sci. Am.* **234**:32–43.
78. Sherman, M. R., Pickering, L. A., Rollwagen, F. M., and Miller, L. K., 1978, *Fed. Proc.* **37**:167–173.
79. Horwitz, K. B., and McGuire, W. L., 1978, *J. Biol. Chem.* **255**:9699–9705.
80. Vedekis, W. V., Schrader, W. T., and O'Malley, B. W., 1980, *Biochemistry* **191**:343–349.
81. McEwen, B. S., Krey, L. C., and Luine, V. N., 1978, *The Hypothalamus* (S. Reichlin, R. J. Baldessarini, and J. B. Martin, eds.), Raven Press, New York, pp. 255–266.
82. McEwen, B. S., Weiss, J. M., and Schwartz, L. S., 1970, *Brain Res.* **17**:471–482.
83. Knizley, M., Jr., 1972, *J. Neurochem.* **19**:2737–2745.
84. Gerlach, J. L., McEwen, B. S., Pfaff, D. W., Moskovitz, S., Ferin, M. Carmel, P. W., and Zimmermann, E. A., 1976, *Brain Res.* **103**:603–612.
85. McEwen, B. S., Stephenson, B. S., and Krey, L. C., 1980, *J. Neurol. Sci. Methods* **3**:57–65.
86. Gerlach, J. L., and McEwen, B. S., 1972, *Science* **175**:1133–1136.
87. Rhees, R. W., Grosser, B. I., and Stevens, W., 1975, *Brain Res.* **83**:293–300.
88. Warembourg, M. Y., 1975, *Brain Res.* **89**:61–70.
89. Stumpf, W. E., and Sar, M., 1979, *J. Steroid Biochem.* **11**:801–807.
90. Stumpf, W. E., and Sar, M. (eds.), 1975, *Anatomical Neuroendocrinology*, Karger, Basel.
91. Bugnon, C., Fellmann, D., Gouget, A., and Cardot, J., 1982, *Neurosci. Lett.* **30**:25–30.
92. McEwen, B. S., and Plapinger, L., 1970, *Nature* **226**:263–264.
93. McEwen, B. S., and Wallach, G., 1973, *Brain Res.* **57**:373–386.
94. Ermisch, A., and Rühl, H.-J., 1978, *Brain Res.* **147**:154–158.
95. De Nicola, A. F., Tornello, S., Weisenberg, L., Fridman, O., and Birmingham, M. K., 1981, *Horm. Metab. Res.* **13**:103–106.
96. Veldhuis, H. D., Van Koppen, C., Van Ittersum, M., and de Kloet, E. R., 1982, *Endocrinology* **110**:2044–2051.
97. McEwen, B. S., De Kloet, E. R., and Wallach, G., 1976, *Brain Res.* **105**:129–136.
98. Moguilevski, M., and Raynaud, J. P., 1977, *Steroids* **30**:99–109.
99. De Kloet, E. R., Wallach, G., and McEwen, B. S., 1975, *Endocrinology* **96**:598–609.
100. Rees, H. D., Stumpf, W. E., and Sar, M., 1975, *Anatomical Neuroendocrinology* (W. F. Stumpf and L. D. Grant, eds.), Karger, Basel, pp. 262–269.
101. Coutard, M., Osborne-Pellegrin, P., and Funder, J. W., 1978, *Endocrinology* **103**:1144–1152.
102. McEwen, B. S., Zigmond, R. E., and Gerlach, J. L., 1972, *Structure and Function of the Nervous Tissue*, Volume 5 (G. H. Bourne, ed.), Academic Press, New York, pp. 205–291.
103. Stith, R. D., and Weingarten, D., 1978, *Neuroendocrinology* **26**:129–140.
104. Stith, R. D., and Weingarten, D., 1979, *Neuroendocrinology* **29**:363–373.
105. De Kloet, E. R., Van der Vies, J., and De Wied, D., 1974, *Endocrinology* **94**:61–73.
106. Grosser, B. I., Stevens, W., and Reed, D. J., 1973, *Brain Res.* **57**:387–396.
107. Clark, C. R., MacLusky, N. J., and Naftolin, F., 1981, *Brain Res.* **217**:412–415.
108. Tsuboi, S., Kawashima, R., Tomioka, O., Nakata, M., Sakamoto, N., and Fujita, T., 1977, *Brain Res.* **179**:181–185.
109. Yu, Z. Y., Wrange, Ö., Boethius, J., Gustafsson, J. Å., and Granholm, L., 1981, *Brain Res.* **223**:325–333.
110. Krozowsky, Z., Hamilton, C. A., and Funder, J. W., 1982, *64th Annual Meeting of the Endocrine Society, USA*, Endocrine Society, Bethesda, p. 238.
111. Sakly, M., 1982, *Interactions Multiples au Cours de l'Ontogénèse entre Récepteurs et Corticoides dans le Contrôle de l'Activité Corticotrope de l'Hypofyse*, Thesis, Université Louis Pasteur, Strassbourg.

112. Moguilevsky, M., and Raynaud, J. P., 1980, *J. Steroid Biochem.* **12**:309–314.
113. Anderson, N. S., and Fanestil, D. D., 1976, *Endocrinology* **98**:676–684.
114. Bohus, B., and De Kloet, E. R., 1981, *Life Sci.* **28**:433–440.
115. Loose, D. S., Schurman, D. J., and Feldman, D., *Nature* **293**:477–479.
116. Moguilevsky, M., and De Raedt, R., 1981, *J. Steroid Biochem.* **15**:329–335.
117. Funder, J. W., Feldman, D., and Edelman, I. S., 1973, *Endocrinology* **92**:994–1004.
118. Funder, J. W., Feldman, D., and Edelman, I. S., 1973, *Endocrinology* **92**:1005–1013.
119. Feldman, D., Funder, J., and Loose, D., 1978, *J. Steroid Biochem.* **9**:141–145.
120. Strum, J. M., Feldman, D., Taggart, B., Marver, D., and Edelman, I. S., 1975, *Endocrinology* **97**:505–516.
121. Marver, D., 1980, *Endocrinology* **106**:611–618.
122. Lan, N. C., Graham, B., Bartter, F. C., and Baxter, J. D., 1982, *J. Clin. Endocrinol. Metab.* **54**:332–342.
123. De Vellis, J., and Inglish, D., 1968, *J. Neurochem.* **15**:1961–1970.
124. McGinnis, J. F., and De Vellis, J., 1978, *J. Biol. Chem.* **253**:8483–8492.
125. Leveille, P. J., McGinnis, J. F., Maxwell, D. S., and De Vellis, J., 1980, *Brain Res.* **196**:287–305.
126. Cousin, M. A., Lando, D., and Moguilevski, M., 1982, *J. Neurochem.* **38**:1296–1304.
127. Scheff, S. W., Benardo, L. S., and Cotman, C. W., 1980, *Exp. Neurol.* **68**:195–201.
128. Meyer, J. S., Leveille, P. J., McEwen, B. S., and De Vellis, J., 1978, *60th Annual Meeting Endocrine Society USA*, Endocrine Society, Bethesda, p. 276.
129. De Vellis, J., McEwen, B. S., Cole, R., and Inglish, D., 1974, *J. Steroid Biochem.* **5**:392–393.
130. Meyer, J. S., Luine, V. N., Khylchevskaya, R. I., and McEwen, B. S., 1982, *J. Neurochem.* **39**:435–442.
131. Meyer, J. S., Leveille, De Vellis, J., Gerlach, J. L., and McEwen, B. S., 1982, *J. Neurochem.* **39**:423–434.
132. De Kloet, E. R., and Burbach, J. P. H., 1978, *J. Neurochem.* **30**:1505–1507.
133. De Kloet, E. R., and McEwen, B. S., 1976, *Biochim. Biophys. Acta* **421**:124–132.
134. De Kloet, E. R., Dam, C. W., and Bohus, B., 1977, *Multiple Molecular Forms of Steroid Hormone Receptors*, Elsevier, Amsterdam, pp. 65–79.
135. MacLusky, N. J., Turner, B. B., and McEwen, B. S., 1977, *Brain Res.* **130**:564–571.
136. Wrange, Ö., 1979, *Biochim. Biophys. Acta* **582**:346–357.
137. Rees, H. D., Stumpf, W. E., Sar, M., and Petrusz, P., 1977, *Cell Tissue Res.* **182**:347–356.
138. Krozowsky, Z., and Funder, J. W., 1981, *Endocrinology* **109**:1221–1224.
139. Watanabe, H., Orth, D. N., and Toft, D. O., 1973, *J. Biol. Chem.* **248**:7625–7630.
140. Svec, F., and Harrison, R. W., 1979, *Endocrinology* **104**:1563–1568.
141. Lan, N. C., Matulick, T., Morris, J. A., and Baxter, J. D., 1981, *Endocrinology* **109**:1963–1970.
142. De Kloet, E. R., and McEwen, B. S., 1976, *Biochim. Biophys. Acta* **421**:115–123.
143. Koch, B., Lutz-Bucher, B., Briaud, B., and Mialhe, C., 1979, *Neuroendocrinology* **28**:169–177.
144. Werthamer, S., Samuels, A. J., and Amaral, L., 1973, *J. Biol. Chem.* **248**:6398–6407.
145. Milgrom, E., and Baulieu, E. E., 1970, *Endocrinology* **87**:276–287.
146. Rosenthal, H. E., Paul, M. A., and Sandberg, A. A., 1974, *J. Steroid Biochem.* **5**:219–225.
147. Raymoure, W. J., Siiteri, P. K., Green, A., and Kuhn, R., 1982, *64th Annual Meeting Endocrine Society USA*, Endocrine Society, Bethesda, p. 80.
148. De Kloet, E. R., Voorhuis, T. H., Leunissen, J., and Koch, B., 1984, *J. Steroid Biochem.* **20**:367–372.
149. Rotsztejn, W. H., Normand, M., Lalonde, J., and Fortier, C., 1975, *Endocrinology* **97**:223–230.
150. Koch, B., Lutz-Bucher, B., Briaud, B., and Mialhe, C., 1978, *Horm. Metab. Res.* **10**:174.
151. Stevens, W., Reed, D. J., Erickson, J., and Grosser, B. I., 1973, *Endocrinology* **93**:1152–1156.
152. Carroll, B. J., Heath, B., and Jarrett, D. B., 1975, *Endocrinology* **97**:290–300.
153. Butte, J. C., Kakihana, R., and Noble, E. P., 1976, *J. Endocrinol.* **68**:235–239.

154. McEwen, B. S., Wallach, G., and Magnus, C., 1974, *Brain Res.* **16**:227–241.
155. Turner, B. B., 1980, *Am. Zool.* **18**:461–475.
156. Angelucci, L., Valeri, P., Palméry, M., Pattachioli, F. R., and Catalani, A., 1980, *Receptors for Neurotransmitters and Peptide Hormones*, Raven Press, New York, pp. 391–406.
157. De Kloet, E. R., and Veldhuis, H. D., 1980, *Neurosci. Lett.* **16**:187–192.
158. Stevens, W., Reed, D. J., and Grosser, B. I., 1975, *J. Steroid Biochem.* **6**:521–527.
159. Olpe, H. R., and McEwen, B. S., 1976, *Brain Res.* **105**:121–128.
160. Valeri, P., Angelucci, L., and Palmery, M., 1978, *Neurosci. Lett.* **9**:249–254.
161. Veldhuis, H. D., and De Kloet, E. R., 1981, *Proceedings of the XXVIII International Congress of Physiological Sciences*, Volume 13, Akadémiai Kiado, Budapest, pp. 61–65.
162. Tornello, S., Fridman, O., Weisenberg, L., Coirini, H., and De Nicola, A. F., 1981, *J. Steroid Biochem.* **14**:77–81.
163. Svec, F., and Rudis, M., 1981, *J. Biol. Chem.* **256**:5984–5987.
164. Valtin, H., and Schroeder, H. A., 1964, *Am. J. Physiol.* **206**:425–430.
165. Veldhuis, H. D., and De Kloet, E. R., 1982, *Endocrinology* **110**:153–157.
166. De Wied, D., 1969, *Frontiers of Neuroendocrinology*, Volume I (L. Martini and W. F. Ganong, eds.), Oxford University Press, London, pp. 97–140.
167. De Kloet, E. R., Veldhuis, H. D., and Bohus, B., 1980, *Receptors for Neurotransmitters and Peptide Hormones* (G. Pepeu, M. J. Kuhar, and S. J. Enna, eds.), Raven Press, New York, pp. 373–382.
168. Veldhuis, H. D., and De Kloet, E. R., 1982, *Neuroendocrinology* **34**:374–380.
169. De Kloet, E. R., Berkers, J., Veldhuis, H. D., and Bohus, B., 1982, *J. Endocrinol* **11** suppl. 438.
170. Angelucci, L., Patacchioli, F. R., Bohus, B., and de Kloet, E. R., 1982, *Typical and Atypical Antidepressants* (E. Costa and G. Racagni, eds.), Raven Press, New York, pp. 365–370.
171. Nyakas, C., De Kloet, E. R., and Bohus, B., 1979, *Neuroendocrinology* **29**:301–312.
172. Nock, B., Blaustein, J. D., and Feder, H., 1981, *Brain Res.* **207**:371–396.
173. Cardinali, D. P., Nagle, C. A., and Rosner, J. M., 1975, *Life Sci.* **16**:93–106.
174. De Kloet, E. R., Angelucci, L., Kovács, G. L., Versteeg, D. H. G., and Bohus, B., 1983, *Integrative Neurohumoral Mechanisms* (E. Endröczi, D. DeWied, L. Angelucci, and U. Scapagnini, eds.), Elsevier Biomedical Press, Amsterdam, pp. 147–156.
175. Nyakas, C., De Kloet, E. R., Veldhuis, H. D., and Bohus, B., 1981, *Neurosci. Lett.* **21**:339–343.
176. Nyakas, C., De Kloet, E. R., Veldhuis, H. D., and Bohus, B., 1983, *Brain Res.* **288**:219–228.
177. Clayton, C. J., Grosser, B. I., and Stevens, W., 1977, *Brain Res.* **134**:445–453.
178. Altman, G., and Bayer, S., 1975, *The Hippocampus*, Volume 1 (R. L. Isaacson and K. H. Pribram, eds.), Plenum Press, New York, pp. 95–122.
179. Sakly, M., and Koch, B., 1981, *Neuroendocrinol. Lett.* **3**:375–378.
180. Sakly, M., and Koch, B., 1981, *Endocrinology* **108**:591–596.
181. Shapiro, S., 1968, *Gen. Comp. Endocrinol.* **10**:214–218.
182. Hiroshige, T., and Sato, T., 1970, *Endocrinology* **86**:1184–1186.
183. Levine, S., 1970, *Prog. Brain Res.* **32**:79–85.
184. Gray, P., 1971, *Endocrinology* **89**:1126–1128.
185. Ader, R., 1968, *J. Comp. Physiol. Psychol* **66**:264–268.
186. Levine, S., Haltmeyer, G. C., and Kara, G. C., 1967, *Physiol. Behav.* **2**:55–59.
187. Tang, F., and Phillips, J. G., 1978, *J. Endocrinol.* **75**:325–326.
188. Landfield, P. W., Waymire, J. C., and Lynch, G., 1978, *Science* **202**:1098–1101.
189. Landfield, P. W., Baskin, R. K., and Pitter, T. A., 1981, *Science* **214**:581–583.
190. Finch, C., and Latham, K., 1974, *56th Annual Meeting Endocrine Society*, Endocrine Society, Bethesda, p. 236.
191. Roth, G. S., 1976, *Brain Res.* **107**:345–354.
192. Roth, G. S., 1980, *Adv. Exp. Med. Biol.* **129**:157–169.
193. Angelucci, L., Valeri, P., Grossi, E., Veldhuis, H. D., Bohus, B., and De Kloet, E. R., 1981, *Progress in Psychoneuroendocrinology* (F. Brambilla, G. Racagni, and D. De Wied, eds.), Elsevier, Amsterdam, pp. 177–185.

194. McEwen, B. S., Davis, P. G., Parsons, B., and Pfaff, D. W., 1979, *Annu. Rev. Neurosci.* **2**:65–112.
195. Rees, H. D., and Gray, H. F., 1984, *Behavioral Endocrinology* (C. B. Nemeroff and A. J. Dunn, eds.), Spectrum, New York (in press).
196. Javoy, F., Glowinski, J., and Kordon, C., 1968, *Eur. J. Pharmacol.* **4**:103–104.
197. Fuxe, K., Corrodi, H., Hökfelt, T., and Jonsson, G., 1970, *Prog. Brain Res.* **32**:42–56.
198. Kizer, J. S., Palkovits, M., Zivin, J., Brownstein, M., Saavedra, J. M., and Kopin, I. J., 1974, *Endocrinology* **95**:799–812.
199. Shen, J. T., and Ganong, W. F., 1976, *Neuroendocrinology* **20**:311–318.
200. Clarke, P. E., and Sampath, S. S., 1975, *Experientia* **31**:1098–1100.
201. Mobley, P. L., and Sulser, F., 1980, *Nature* **286**:608–609.
202. Roberts, D. C. S., and Bloom, F. E., 1981, *Eur. J. Pharmacol.* **74**:37–41.
203. Azmitia, E. C., and McEwen, B. S., 1969, *Science* **166**:1274–1276.
204. Vermes, I., Telegdy, G., and Lissak, K., 1973, *Acta Physiol. Acad. Sci. Hung.* **43**:33–42.
205. Telegdy, G., and Vermes, I., 1975, *Neuroendocrinology* **18**:16–26.
206. Vermes, I., Smelik, P. G., and Mulder, A. H., 1976, *Life Sci.* **19**:1719–1726.
207. Sze, P. Y., Neckers, C., and Towle, A. C., 1976, *J. Neurochem.* **26**:169–173.
208. Rastogi, R. B., and Singhal, R. L., 1978, *J. Neural Transm.* **42**:63–71.
209. Van Loon, G. R., Shum, A., and Sole, M. J., 1981, *Endocrinology* **108**:1392–1402.
210. Van Loon, G. R., Shum, A., and De Souza, E. B., 1981, *Endocrinology* **108**:2269–2276.
211. De Kloet, E. R., Kovács, G. L., Szabó, G., Telegdy, G., Bohus, B., and Versteeg, D. H. G., 1982, *Brain Res.* **239**:659–663.
212. Miller, A. L., Chaptal, C., McEwen, B. S., and Peck, E. J., 1978, *Psychoneuroendocrinology* **3**:155–164.
213. Kendall, D. A., McEwen, B. S., and Enna, S. J., 1982, *Brain Res.* **236**:365–374.
214. Rotsztejn, W. H., Besson, J., Briaud, B., Gagnant, L., Rosselin, G., and Kordon, C., 1980, *Neuroendocrinology* **31**:287–291.
215. Wallis, C. J., and Printz, M. P., 1980, *Endocrinology* **106**:337–342.
216. Silverman, A. J., Hoffman, D., Gadde, C. A., Krey, L. C., and Zimmerman, E. A., 1981, *Neuroendocrinology* **32**:129–133.
217. Kuhar, M. J., 1975, *The Hippocampus* (R. L. Isaacson and K. H. Pribram, eds.), Plenum Press, New York, pp. 269–285.
218. Moore, R. Y., and Bloom, F. E., 1979, *Annu. Rev. Neurosci.* **2**:113–168.
219. Nadler, J. V., Vaca, K. W., White, W. F., Lynch, G. S., and Cotman, C. W., 1976, *Nature* **260**:538–540.
220. Azmitia, E. C., and McEwen, B. S., 1974, *Brain Res.* **78**:291–302.
221. Hillier, J., Hillier, J. G., and Redfern, P. H., 1975, *Nature* **253**:566–567.
222. Millard, S. A., Costa, E., and Gall, E. M., 1972, *Brain Res.* **40**:545–551.
223. Parvez, H., and Parver, S., 1973, *J. Neurochem.* **20**:1011–1020.
224. De Kloet, E. R., Kovács, G. L., and Versteeg, D. H. G., 1983, *Brain Res.* **264**:323–327.
225. Tozer, T. N., Neff, N. H., and Brodie, B. B., 1966, *J. Pharmacol. Exp. Ther.* **153**:177–182.
226. Azmitia, E. C., and Conrad, L. C. A., 1976, *Neuroendocrinology* **21**:338–349.
227. Kostowski, W., and Giacalone, E., 1969, *Eur. J. Pharmacol.* **7**:176–179.
228. Gumulka, W., Samanin, R., and Valzelli, L., 1970, *Eur. J. Pharmacol.* **12**:276–279.
229. Kovács, G. L., Kishonti, J., Lissák, K., and Telegdy, G., 1977, *Neurochem. Res.* **2**:311–322.
230. Dunn, A. J., Gildersleeve, N. B., and Gray, H. E., 1978, *J. Neurochem.* **31**:977–982.
231. Meyer, J. S., Luine, V. N., Khylchevskaya, R. I., and McEwen, B. S., 1979, *Brain Res.* **166**:172–175.
232. Acz, Z., Palkovits, M., and Stark, E., 1980, *Neurosci. Lett.* **19**:97–101.
233. Versteeg, D. H. G., Van Zoest, I., and De Kloet, E. R., 1979, *Neurosci. Lett. (Suppl.)* **3**:S182.
234. Gala, R. R., Kothari, L. S., and Haisenleder, D. J., 1981, *Life Sci.* **29**:2113–2117.
235. Mobley, P. L., and Sulser, F., 1980, *Eur. J. Pharmacol.* **65**:321–322.
236. Wurtman, R. J., and Axelrod, J., 1966, *J. Biol. Chem.* **241**:2301–2305.
237. Otten, U., and Thoenen, H., 1977, *J. Neurochem.* **29**:69–75.
238. Markey, K. A., and Sze, P. Y., 1980, *Brain Res.* **202**:347–356.

239. Jones, M. T., Hillhouse, E. W., and Burden, J. L., 1977, *J. Endocrinol.* **74**:415–424.
240. Vale, W., Spiess, J., Rivier, C., and Rivier, J., 1981, *Science* **213**:1394–1397.
241. Gramsch, C., Kleber, G., Höllt, V., Pari, A., Mehraein, P., and Herz, A., 1980, *Brain Res.* **192**:109–119.
242. O'Donohue, T. L., and Chappell, M. C., 1982, *Peptides* **3**:69–75.
243. De Kloet, E. R., Mezey, E., and Palkovits, M., 1981, *Pharmacol. Ther.* **12**:321–351.
244. Van Dijk, A. M. A., Van Wimersma Greidanus, T. B., Burbach, J. P. H., De Kloet, E. R., and De Wied, D., 1981, *J. Endocrinol.* **88**:243–253.
245. Roosevelt, S., Wolfsen, A. R., and Odell, W. D., 1979, *Endocrinology* **104**(Suppl.):A125.
246. Larsson, L. I., Fahrenkrug, J., Schaffalitsky de Muckadell, O. B., Sundler, F., Hakanson, R., and Rehfeld, J. F., 1976, *Proc. Natl. Acad. Sci. U.S.A.* **73**:3197–3200.
247. Dokas, L. A., 1979, *Soc. Neurosci. Abstr.* **5**:443.
248. Etgen, A. M., Martin, M., Gilbert, R., and Lynch, G., 1980, *J. Neurochem.* **35**:598–602.
249. Nestler, E. J., Rainbow, T. C., McEwen, B. S., and Greengard, P., 1981, *Science* **212**:1162–1164.
250. Ueda, T., and Greengard, P., 1977, *J. Biol. Chem.* **252**:5155–5159.
251. Bloom, F. E., Ueda, T., Battenberg, E., and Greengard, P., 1979, *Proc. Natl. Acad. Sci. U.S.A.* **76**:5982–5987.
252. Maudsley, D. V., 1979, *Biochem. Pharmacol.* **28**:153–161.
253. Goertz, B., 1979, *Brain Res.* **173**:125–135.
254. De Kloet, E. R., Cousin, M. A., Veldhuis, H. D., Voorhuis, T. D., and Lando, D., 1983, *Brain Res.* **275**:91–98.
255. Balász, R., and Cotterell, M., 1972, *Nature* **236**:348–360.
256. Bohn, M. C., 1980, *Neuroscience* **5**:2003–2012.
257. McWilliams, R., and Lynch, G., 1980, *J. Comp. Neurol.* **187**:191–198.
258. Scheff, S. W., Benardo, L. S., and Cotman, C. W., 1978, *Science* **202**:775–778.
259. Cotman, C. W. (ed.), 1978, *Neuronal Plasticity*, Raven Press, New York.
260. Reulen, H. J., 1976, *Br. J. Anaesthesiol.* **48**:741–752.
261. Yuwiler, A., 1971, *Handbook of Neurochemistry*, Volume VI (A. Lajtha, ed.), Plenum Press, New York, pp. 103–171.
262. Mason, J. W., 1958, *Reticular Formation of the Brain* (H. H. Jasper, L. O. Procter, R. S. Knighton, C. W. Noshay, and R. T. Costello, eds.), Little, Brown, Boston, pp. 645–652.
263. Selye, H., 1950, *Stress: The Physiology and Pathology of Exposure to Stress*, Acta Medica, Montreal.
264. Dallman, M. F., and Yates, F. E., 1969, *Ann. N.Y. Acad. Sci.* **156**:696–721.
265. Zimmerman, E., and Critchlow, V., 1972, *Neuroendocrinology* **9**:235–243.
266. Jones, M. T., Tiptaft, E. M., Brush, F. R., Ferguson, D. A. N., and Neame, R. L. B., 1974, *J. Endocrinol.* **60**:223–233.
267. Sato, T., Sato, M., Shinsako, S., and Dallman, M. F., 1975, *Endocrinology* **97**:265–274.
268. Smelik, P. G., and Papaikonomou, E., 1973, *Prog. Brain Res.* **39**:99–110.
269. Jones, M. T., and Tiptaft, E. M., 1977, *Br. J. Pharmacol.* **59**:35–41.
270. Birmingham, M. K., Kraulis, I., Traikov, A., Bartova, M. P., Lichan, T. H., and Possanza, G., 1974, *J. Steroid Biochem.* **5**:789–794.
271. De Wied, D., 1964, *J. Endocrinol. (Kbh.)* **29**:29–37.
272. Kendall, J. W., 1971, *Frontiers in Neuroendocrinology*, Volume 2 (L. Martini and W. F. Ganong, eds.), Oxford University Press, London, pp. 209–235.
273. Portonova, R., and Sayers, G., 1974, *Biochem. Biophys. Res. Commun.* **56**:928–933.
274. Cochet, M., Chang, A. C. Y., and Cohen, S. N., 1982, *Nature* **297**:335–339.
275. Jones, M. T., Hillhouse, E. W., and Burden, J. L., 1977, *Frontiers in Neuroendocrinology*, Volume 4 (L. Martini and W. F. Ganong, eds.), Raven Press, New York, pp. 195–226.
276. Edwardson, J. A., and Bennett, G. W., 1974, *Nature* **251**:425–427.
277. Abe, K., and Critchlow, V., 1977, *Endocrinology* **101**:498–505.
278. Kawakami, M., Sato, L., and Yoshida, K., 1968, *Neuroendocrinology* **3**:349–353.
279. Ader, R., and Friedman, S. B., 1968, *Neuroendocrinology* **3**:378–386.
280. Slusher, M. A., 1966, *Exp. Brain Res.* **1**:184–194.
281. Moberg, G. A., Scapagnini, U., De Groot, J., and Ganong, W. F., 1971, *Neuroendocrinology* **7**:11–15.

282. Fischette, C. T., Komisaruk, B. R., Edinger, H. M., Feder, H. H., and Siegel, A., 1980, *Brain Res.* **195**:373–387.
283. Van Hartesveldt, C., 1975, *The Hippocampus* (R. L. Isaacson and K. H. Pribram, eds.), Plenum Press, New York, pp. 375–391.
284. Feldman, S., and Conforti, N., 1980, *Neuroendocrinology* **30**:52–55.
285. Wilson, M. M., Greer, S. E., Greer, M. A, and Roberts, L., 1980, *Brain Res.* **197**:433–441.
286. Terkel, J., Johnson, J. M., Whitmoyer, D. I., and Sawyer, C. M., 1974, *Neuroendocrinology* **14**:103–113.
287. Micco, D. J., Jr., Meyer, J. S., and McEwen, B. S., 1980, *Brain Res.* **200**:206–212.
288. Yukimara, Y., Bray, G. A., and Wolfsen, A. R., 1978, *Endocrinology* **103**:1924–1928.
289. Fregly, M. J., and Waters, I. W., 1966, *Physiol. Behav.* **1**:65–74.
290. Weisinger, R. S., Denton, D. A., McKinley, M. J., and Nelson, J. F., 1978, *Pharmacol. Biochem. Behav.* **8**:339–342.
291. Leshner, A. I., 1971, *Physiol. Behav.* **6**:551–558.
292. Kendall, J. W., 1970, *Horm. Behav.* **1**:327–336.
293. Katz, R. J., and Carroll, B. J., 1978, *Physiol. Behav.* **20**:25–30.
294. MacLennan, A. J., Drugan, R. C., Hyson, R. L., Maier, S. F., Madden, J., and Barchas, J. D., 1982, *Science* **215**:1530–1532.
295. Brain, P. F., 1972, *Behav. Biol.* **7**:453–477.
296. File, S. E., Velucci, S. V., and Wendlandt, S., 1979, *J. Pharm. Pharmacol.* **31**:300–305.
297. McIntyre, D. C., 1976, *Horm. Behav.* **17**:789–795.
298. Veldhuis, H. D., De Kloet, E. R., Van Zoest, I., and Bohus, B., 1982, *Horm. Behav.* **16**:191–199.
299. Fuller, J. L., Chambers, R. M., and Fuller, R. P., 1956, *Psychosom. Med.* **18**:234–242.
300. Bohus, B., and Lissak, K., 1968, *Int. J. Neuropharmacol.* **7**:301–306.
301. De Wied, D., 1967, *Proceedings of the Second International Congress Hormonal Steroids* (L. Martini, F. Fraschini, and M. Motta, eds.), Excerpta Medica Foundation, Amsterdam, pp. 945–951.
302. Van Delft, A., 1970, *Conditioned Avoidance Behavior and the Pituitary–Adrenal System in the Rat*, Thesis, University of Utrecht, Utrecht.
303. Weiss, J. M., McEwen, B. S., Silva, M. T., and Kalkut, M., 1970, *Am. J. Physiol.* **218**:864–868.
304. Micco, D. J., McEwen, B. S., and Shein, W., 1979, *J. Comp. Physiol. Psychol.* **93**:323–329.
305. Micco, D. J., and McEwen, B. S., 1980, *J. Comp. Physiol. Psychol.* **94**:624–633.
306. Ögren, S.-O., and Fuxe, K., 1974, *Med. Biol.* **52**:399–405.
307. Roberts, D. C. S., and Fibiger, H. C., 1977, *Pharmacol. Biochem. Behav.* **7**:191–194.
308. Wendlandt, S., and File, S. E., 1979, *Behav. Neurol. Biol.* **26**:189–201.
309. Bohus, B., and De Kloet, E. R., 1977, *J. Endocrinol.* **72**:64P–65P.
310. Flexner, J. B., and Flexner, L. B., 1970, *Proc. Natl. Acad. Sci. U.S.A.* **66**:48–52.
311. Nakajima, S., 1975, *J. Comp. Physiol. Psychol.* **88**:378–385.
312. McIntyre, D. C., and Wann, P. D., 1978, *Physiol. Behav.* **20**:469–474.
313. Bookin, H. B., and Pfeiffer, W. D., 1978, *Behav. Biol.* **24**:527–532.
314. Kovács, G. L., and Telegdy, G., 1978, *Results in Neuroendocrinology, Neurochemistry and Sleep Research*, Volume 7 (K. Lissák, ed.), Akadémiai Kiadó, Budapest, pp. 31–97.
315. Bohus, B., and De Wied, D., 1980, *General, Comparative and Clinical Endocrinology of the Adrenal Cortex*, Volume 3 (I. Chester-Jones and I. W. Henderson, eds.), Academic Press, London, pp. 256–347.
316. Douglas, R. J., 1967, *Psychol. Bull.* **67**:416–422.
317. Kimble, D. P., 1968, *Psychol. Bull.* **70**:285–295.
318. Iuvone, P. M., and Van Hartesveldt, C., 1977, *Behav. Biol.* **19**:228–237.
319. Roberts, D. C. S., Price, M. T. C., and Fibiger, H. C., 1976, *J. Comp. Physiol. Psychol.* **90**:363–372.
320. Mason, S. T., and Iversen, S. D., 1975, *Nature* **258**:422–424.
321. Mason, S. T., and Iversen, S. D., 1979, *Brain Res.* **150**:135–148.
322. Ader, R., Weijnen, J. A. W. M., and Moleman, P., 1972, *Psychoneurol. Sci.* **26**:125–128.
323. Robustelli, F., Geller, A., and Jarvik, M. E., 1972, *J. Comp. Physiol. Psychol.* **81**:472–482.

324. Bohus, B., 1974, *Brain Res.* **66**:366–367.
325. Sakellaris, P. C., 1972, *Physiol. Behav.* **9**:495–521.
326. Sachar, E. J., Hellman, L., Roffwarg, H., Halpern, F. S., Fukushima, D. K., and Gallagher, T. F., 1973, *Arch. Gen. Psychiatry* **28**:19–24.
327. Carroll, B. J., Curtis, G. C., and Mendels, J., 1976, *Arch. Gen. Psychiatry* **33**:1051–1058.
328. McEwen, B. S., Gerlach, J. L., and Micco, D. J., 1975, *The Hippocampus* (R. L. Isaacson and K. Pribram, eds.), Plenum Press, New York, pp. 285–322.

4

Neurotensin

François B. Jolicoeur, Francis Rioux, and Serge St.-Pierre

1. INTRODUCTION

It has now been a decade since neurotensin was first isolated from bovine hypothalami by Carraway and Leeman.[1] The peptide was so named because of its presence in neural tissue and its marked hypotensive effect in the rat. Shortly after this, neurotensin was found to be a tridecapeptide having the following amino acid sequence: pGlu-Leu-Tyr-Glu-Asn-Lys-Pro-Arg-Arg-Pro-Tyr-Ile-Leu-OH.[2] Later, neurotensin was also detected in the gastrointestinal tract of animals[3] and, therefore, is now considered part of a growing list of brain–gut peptides. Since its discovery, neurotensin has attracted considerable experimental attention, and the vast amount of information that has already accumulated on this peptide has been the subject of several reviews[4-10] and, recently, the focus of an international symposium.[11] In the present chapter, we attempt to summarize the findings on the distribution, metabolism, and actions of neurotensin in the CNS as well as discuss the possible neurophysiological roles of this neuropeptide.

2. DISTRIBUTION OF NEUROTENSIN IN THE CENTRAL NERVOUS SYSTEM

By use of both radioimmunoassay and immunohistochemical techniques, neurotensinlike immunoreactivity (NTLI) has been detected in the nervous system of organisms spanning the phylogenetic scale.[12,13] The localization of NTLI has been most extensively studied in the CNS of a variety of mammalian species including man,[3,14-17] and, in general, similar patterns of distribution have been found. The central distribution of NTLI in rat CNS has been ex-

---

François B. Jolicoeur • Department of Psychiatry, Faculty of Medicine, University of Sherbrooke, Sherbrooke, Quebec J1H 5N4, Canada. *Francis Rioux and Serge St.-Pierre* • Department of Physiology and Pharmacology, Faculty of Medicine, University of Sherbrooke, Sherbrooke, Quebec J1H 5N4, Canada.

Table I
Regional Distribution of Neurotensinlike Immunoreactivity in Rat Brain[a]

	NTLI (pmol neurotensin/g wet weight)
Cerebral cortex and hippocampus	
Frontal cortex	4.1 ± 1.2
Cingulate cortex	5.3 ± 1.5
Parietal cortex	5.2 ± 2.8
Pyriform cortex	12.8 ± 4.1
Hippocampus	7.1 ± 2.8
Subiculum	9.2 ± 2.8
Basal forebrain	
Olfactory bulb	3.5 ± 1.8
Olfactory tubercle	25.7 ± 8.8
Nucleus accumbens	29.9 ± 3.8
Lateral septum	73.0 ± 25.9
Nucleus caudate–putamen	10.2 ± 3.8
Globus pallidus	12.6 ± 5.8
Bed nucleus of the stria terminalis	171.8 ± 18.6
Central amygdaloid nucleus	106.0 ± 31.2
Hypothalamus	
Lateral preoptic area	124.0 ± 12.0
Medial preoptic area	143.5 ± 13.2
Lateral hypothalamic nucleus	76.0 ± 12.9
Anterior hypothalamic nucleus	113.4 ± 5.1
Arcuate nucleus/median eminence	128.1 ± 11.6
Dorsomedial hypothalamic nucleus	123.9 ± 9.5
Ventromedial hypothalamic nucleus	87.7 ± 22.3
Mammillary body	128.1 ± 11.6
Brainstem	
Ventral tegmental area/ interpeduncular nucleus	35.5 ± 16.5
Substantia nigra	13.3 ± 4.9
Dorsal raphe	48.1 ± 9.5
Medial raphe	6.7 ± 4.6
Locus coeruleus	54.0 ± 11.4
Periaqueductal gray	42.2 ± 7.2
Cerebellum	0.9 ± 0.2
Nucleus of the solitary tract	49.8 ± 6.4
Trigeminal nucleus	35.16 ± 2.2
Spinal cord	
Cervical dorsal horn	23.7 ± 5.6
Cervical ventral horn	2.0 ± 0.8
Thoracic dorsal horn	17.1 ± 2.7
Thoracic ventral horn	2.6 ± 1.3
Lumbar dorsal horn	22.7 ± 2.5
Ventral horn	3.3 ± 0.2

[a] Determinations were carried out using NH_2-terminal specific antiserum. Values are means ± SE.

amined recently by members of our group and collaborators,[18,19] and the results of these studies are summarized in Table 1.

It can be seen that although NTLI can be detected in many cortical, subcortical, and spinal areas of the CNS, regional concentrations vary enormously.

Very high concentrations are localized in the stria terminalis, central amygdaloid nucleus, and in several hypothalamic nuclei. Significant amounts of NTLI are found in regions of the limbic system and of the basal ganglia as well as in areas giving rise to important monoaminergic pathways such as the ventral tegmental area, substantia nigra, dorsal raphe, and locus coeruleus. Relatively smaller concentrations are present in cortical regions and in cerebellum. In the spinal cord, NTLI is found mostly in the dorsal horn section, where, as other studies have shown, neurotensin is contained in cell bodies intrinsic to this area.[12-15]

At the subcellular level, NTLI has been localized in neuronal cell bodies, fibers, and terminals.[24-28] There is evidence indicating that some neurotensin-containing fibers are organized into distinct pathways in the CNS.[23] Already, two such pathways have been described: one originating from the central amygdaloid nucleus and projecting via the stria terminalis into the bed nucleus[29] and another arising from cell bodies in hippocampus and terminating in the anterior cingulate cortex.[30] The presence of NTLI in nerve terminals suggests that the peptide may be released as a neurotransmitter substance. Consistent with this hypothesis is the finding that neurotensin can be excreted by depolarized brain tissue in a calcium-dependent process.[31,32]

Also supporting a possible neurotransmitter role for neurotensin is the existence of binding sites for the peptide in the CNS. The binding of neurotensin to neuronal membranes displays all of the characteristics of affinity, specificity, saturability, and reversibility usually associated with classical neurotranmitter receptors.[24,33-35] The localization of neurotensin's binding sites in the CNS has been examined with autoradiographic procedures,[36-38] and, in general, the regional distribution of these putative receptors parallels that of NTLI. For instance, high densities of receptors are present in areas containing relatively large concentrations of NTLI, such as lateral septum, central amygdaloid nucleus, habenula, ventral tegmentum, substantia nigra, and various hypothalamic nuclei. Interestingly, high densities of binding sites are also found in cingulate cortex and in bed nucleus of stria terminalis, areas thought to be respective end points of the two aforementioned neurotensin pathways.[29,30]

3. METABOLISM OF NEUROTENSIN IN THE CENTRAL NERVOUS SYSTEM

Although considerable attention has been given recently to the biosynthesis of various neuropeptides, particularly the opioid peptides, to our knowledge, the synthetic processes yielding neurotensin have not been examined. If the synthesis of neurotensin is similar to that of most other peptides studied so far,[39] it is expected that neurotensin will be shown to be derived from post-translational modifications of larger peptide precursors synthesized in the cell body of neurons.[40]

The catabolism of neurotensin in neuronal tissue has been examined in several reports.[18,19,41,42] The results of these studies demonstrate that neurotensin can be degraded by proteases located in both soluble and particulate fractions of cortex, thalamus, pituitary, and hypothalamus, with the most rapid biotransformation occurring in soluble fractions of the latter structure. The nature of the metabolites obtained after degradation was studied with high performance liquid chromatography, amino acid analysis, and radioimmunoassay using antisera directed towards the amino- or carboxy-terminal of neurotensin.[18,19] It appears that the major metabolic event consists of the hydrolysis of the Arg^8–Arg^9 bond, yielding the fragments neurotensin$_{1-8}$ (pGlu-Leu-Try-Glu-Asn-Lys-Pro-Arg) and neurotensin$_{9-13}$ (Arg-Pro-Tyr-Ile-Leu). The fragment neurotensin$_{1-8}$ seems to be relatively impervious to further degradation, whereas neurotensin$_{9-13}$ is rapidly hydrolyzed, leading ultimately to free tyrosine. The fragment neurotensin$_{1-10}$ is also found after degradation, which suggests that an enzyme acting at the postproline bond also participates in the catabolism of the peptide. It should be mentioned that the same preparation that breaks down neurotensin also degrades a variety of other neuropeptides.[43] This obviously raises doubts concerning the existence of specific "neurotensinases" in the CNS.

4. ELECTROPHYSIOLOGICAL STUDIES

The effects of direct application of neurotensin on electrical activity of single neurons have been examined in numerous studies, but the results obtained so far are often discordant and difficult to reconcile. Both excitation and depression of neuronal activity have been observed following application of neurotensin in dorsal horn of the spinal cord of anesthetized cats.[44,45] On the other hand, *in vitro* studies report only excitatory effects of neurotensin in dorsal horn of isolated amphibian[46] and mammalian[47] spinal cords. Variable effects were also obtained in locus coereleus,[48,49] stria terminalis,[50-52] septum,[50] preoptic area,[50] and cerebellum.[50,53,54] In this latter brain region, the effects appear to depend on procedures, since pressure-ejected neurotensin markedly depressed neuronal activity, whereas iontophoretically released neurotensin did not.[54] Contrary to the discrepant results obtained in the aforementioned CNS sites, consistent stimulatory effects of neurotensin in cerebral cortex have been documented.[50,55] Also, initial studies in nucleus accumbens[53] and substantia nigra[56] indicate inhibitory and stimulatory actions of neurotensin, respectively.

Taken together, the results of these studies reveal that neurotensin has no consistent electrophysiological effects in the CNS and that, often, the peptide produces opposite effects on neurons localized in the same brain region. Whether these discrepancies are mainly attributable to methodological differences or reflect a multifaceted influence of neurotensin on neuronal activity remains to be ascertained.

5. NEUROBEHAVIORAL EFFECTS OF CENTRAL ADMINISTRATION OF NEUROTENSIN

The administration of neurotensin into the CNS results in a variety of neurophysiological and behavioral effects, which are reviewed in the present section. Central injection of the peptide also produces a diversity of autonomic[57,58] and endocrine reactions,[59,60] which have been summarized and discussed elsewhere.[61-64]

5.1. Body Temperature

Of all central effects of neurotensin, the hypothermic action of the peptide has been the most extensively studied. Neurotensin decreases body temperature in a majority of mammalian species.[65] Although the hypothermia is seen when animals are tested in both normal (23°C) and cold environments (4°C), the effect is more pronounced in the latter situation.[66] Under normal temperature conditions, we have observed that intraventricular administration of neurotensin in doses ranging from 1.8 to 120 µg/rat significantly decreases rectal temperature in a dose-related fashion.[67] The hypothermic effect is maximal at 60 min after injections and then gradually dissipates over a 2-h period. When animals are placed in warmer environments (30–34°C), the hypothermic effect has been shown to disappear completely in mice[68] and actually to be replaced by hyperthermia in rats.[69] This latter finding suggests that neurotensin may be a poikilothermic agent in some animal species.

The exact mechanisms underlying neurotensin's influence on body temperature are still unknown, but there is sufficient evidence that this influence is centrally mediated. First, peripheral administration of neurotensin, even in high doses, does not alter body temperature of animals.[70] This negative finding probably reflects the inability of the peptide to cross the blood–brain barrier. Secondly, the results of a recent study indicate that neurotensin's hypothermia is accompanied by lower metabolic heat production but not by peripheral heat loss as a result of cutaneous vasodilation.[69] Finally, direct injections of neurotensin into specific brain sites have been shown to replicate the hypothermic effect of cerebroventricular administration of the peptide.[71,72] Brain sites in which neurotensin is active are numerous and include the medial preoptic area, the anterior hypothalamus, and the floor of the fourth ventricle, three regions that have been implicated in central thermoregulation and in which high concentrations of NTLI are found. However, the presence of NTLI within its boundaries does not necessarily render a region sensitive to the hypothermic action of neurotensin, since direct administration of the peptide into nucleus accumbens and bed nucleus of the stria terminalis does not affect body temperature of animals.[72]

Whether neurotensin has a direct effect in the aforementioned sensitive areas or its influence results from the modulation of neurotransmitter systems in these regions remains to be determined. In this respect, it has been shown that neurotensin's hypothermic effect was not altered by different pharmacological manipulations of the cholinergic, noradrenergic, and serotoninergic systems but was significantly enhanced by selective depletion of dopamine with 6-hydroxydopamine or by administration of the dopamine antagonist haloperidol.[73] Because of these results, it has been suggested that an interaction of neurotensin with dopaminergic processes may underlie the hypothermic effect of the peptide. However, the nature of this interaction remains to be clearly established. Both 6-hydroxydopamine and haloperidol produce significant falls in body temperature,[73,74] so that the greater hypothermia observed after the combination of neurotensin with any of these treatments may only reflect an additive effect of two hypothermic agents. Moreover, we have shown recently that neurotensin's hypothermia is also enhanced by administration of apomorphine and n-propylnorapomorphine.[75] Since neurotensin's hypothermia appears to be enhanced by both inhibition and stimulation of dopaminergic activity, it is difficult to envisage that a functional interaction of neurotensin and dopamine mediates the influence of the peptide on body temperature.

The administration of another neuropeptide, thyrotropin-releasing hormone (TRH), has been shown to completely inhibit the hypothermic effect of neurotensin.[73] This inhibition is obtained with relatively high doses of TRH in that a minimum of 20 μg is needed to obtain the effect.[73,76] The significance of this finding is unclear. Neurotensin and TRH have opposite effects or can antagonize each other's action in a multiplicity of physiological and behavioral situations.[77–81] It remains to be established that these opposing effects are the result of functional pharmacological interactions and not the consequence of physiological antagonism of two substances acting on different brain mechanisms.

In summary, it is clear that neurotensin has a prominent influence on body temperature of animals. Although the effects of neurotensin on thermoregulation are most likely centrally mediated, the exact mechanisms underlying these effects remain to be determined.

5.2. Muscular Tone

Studies on the possible effects of neurotensin on muscle tone have relied on indirect behavioral measures and have yielded inconclusive results. In mice, intracisternal injection of 30 μg neurotensin caused muscular relaxation in the Julou–Courvoisier test, a procedure that has been used to screen compounds with neurolepticlike activity.[82] In our laboratory we have utilized the traction and grasping tests to assess muscle tone.[67] We have found that with these procedures, which were described in detail elsewhere,[84] the effects of various drugs and neurotoxicants known to affect muscular tone could be detected reliably.[84,85] Starting at an intraventricular dose of 1.8 μg/rat, neurotensin significantly decreased muscle tone in the traction test; however, the grasping response of animals was not altered even with relatively high doses (120 μg).[83] Because of the limitations of such indirect measures, it is not possible to determine if and how neurotensin functionally affects mechanisms controlling muscular tone. This is especially evident in view of a recent study in which centrally administered neurotensin had no effect on the EMG activity recorded from the biceps femoris muscle of anesthetized rats.[79]

5.3. Motor Activity

The effects of neurotensin on locomotor activity had received little attention until recently. In an earlier study, the intraventricular administration of a single dose of 30 μg was found to decrease locomotor activity in rats. Recently, we observed that doses as small as 1.8 μg injected by the same route decrease locomotor activity of rats.[67,87] The hypokinetic effect is maximal at 7.5 μg, dissipates gradually with higher doses, and disappears entirely at 60 μg. The biphasic nature of the dose–effect curve is interesting since it indicates that the decrements in activity seen at small doses do not reflect some nonspecific toxic effect of the peptide. Also, this suggests that the hypokinetic effect of neurotensin cannot be attributed to its effects on body temperature, since this measure decreases linearly as a function of dose. The decreases in activity were statistically significant at 15 and 30 min but not at 60, 90, and 120 min after injection.

In these studies, the methods utilized to measure activity could not discriminate which aspects of motor activity were affected by neurotensin. However, in a recent study in which specific components of activity, locomotion, rearing, and grooming, were assessed following intraventricular administration of neurotensin (10 μg), all three measures were found to be decreased by the peptide.[88] Thus, it appears that neurotensin has a general depressant influence on motor activity in animals. The results of all of the above studies were obtained after intraventricular administration in rats. It is noteworthy that intracisternal injection of doses as small as 25 pg of neurotensin in mice has been shown to markedly lower motor activity.[89] Whether this finding reflects a greater sensitivity of this species to the hypokinetic effect of neurotensin and/or to the route of administration utilized remains to be investigated.

How neurotensin affects motor activity is still unknown. Relatively high concentrations of NTLI are found in ventral tegmental area, substantia nigra, nucleus accumbens, striatum, and globus pallidus. These regions have all been implicated in the complex regulation of activity by the CNS.[90-92] Investigations on the motor effects of direct administration into specific brain sites have been sparse. Motor activity was not found to be affected by injection of 2 to 5 μg into the nucleus accumbens but was reported significantly decreased following administration of a higher dose (10 μg) in the same region.[8,81] On the other hand, a marked stimulation of motility was observed following application of 2–5 μg into the ventral tegmental area (VTA).[93] Surprisingly, this stimulatory effect of intra-VTA administration of neurotensin could be blocked by intraaccumbens injection of 4.0 μg of the peptide.[94] This somewhat intriguing but interesting finding suggests that neurotensin's influence on motor activity is multifaceted and may involve differential actions at various brain sites. However, the results of the above studies should be interpreted prudently in view of the size of doses utilized to obtain these effects. Caution seems particularly warranted in view of the fact that administration of 1.8 μg into the whole brain via the cerebroventricles decreases motor activity and that the dose–effect relationship obtained with this route is a biphasic one.

5.4. Reactivity to Noxious Stimuli

Although the influence of neurotensin on reactivity of animals to aversive stimuli has been examined in numerous studies, the antinociceptive properties of the peptides are still difficult to ascertain. Possible antinoceptive effects have been studied in both rats and mice, and the results appear to be more consistent in the latter species than in the former. Neurotensin was first shown to decrease responsiveness to noxious stimuli in mice subjected to the hot plate and acetic acid writhing procedures.[89,93] These effects, which were obtained following intracisternal administration of doses as small as 250 pg in the writhing and 25 ng in the hot plate tests, were not reversed by the opiate antagonist naloxone. Similar effects were seen in a later study in which intracisternal administration of at least 300 ng neurotensin lowered reactivity of mice submitted to the tail immersion test.[96] This effect was also found to be naloxone resistant but was completely blocked by administration of TRH.[97]

In rats, the picture is more complicated. Intracisternal administration of 25 ng neurotensin in this animal species was reported to exert analgesic activity in the hot plate test.[89] In a subsequent study in which identical procedures were used, this effect was not observed until dosage was increased to 1.7 µg.[98] On the other hand, in the tail flick situation, intracisternal administration of up to 2.5 µg neurotensin was devoid of any analgesic effect.[89] Intraventricular injections of neurotensin in doses of 20 to 60 µg were reported to decrease reactivity of rats in the hot plate test.[98] However, using identical procedures and doses, we have failed to replicate this finding (unpublished observations). The results of a more recent study also point to the ineffectiveness of similar doses of neurotensin administered intraventricularly to alter response of rats on a hot plate.[99] In this study, some analgesic effects appeared with larger doses, and the ED_{50} for neurotensin's antinociception in this situation was estimated to be 93.5 µg/rat. Finally, intraventricular injection of 10 to 100 µg neurotensin was found to be incapable of modifying the responsiveness of rats subjected to the tail compression and the saline writhing tests.[78]

When neurotensin is administered in the spinal subarachnoid space, the reported effects are also conflicting. In one study, intrathecal administration of a wide range of neurotensin doses (1–80 µg) was not found to affect the reactivity of rats in the hot plate test.[98] However, in another experiment using the same route of administration and similar doses, significant analgesic effects were obtained in the hot plate and in the acetic acid writhing tests.[100] Surprisingly, these effects could be reversed by naloxone. In the same study, intrathecal administration of neurotensin did not alter the heat-induced tail flick response of rats.

Adding to the confusion are the findings of a recent study on the effects of intraventricular administration of antisera to neurotensin.[101] Results of that study indicate that the putative immunologic blockade of endogenous neurotensin produces both analgesic and hyperalgesic effects in tail flick responses depending on the intensity of the thermal stimulus. However, no effect was obtained in the flinch jump responses to electric footshocks. The effects of direct administration of neurotensin into specific brain sites were also examined, and the data gathered so far are somewhat more consistent. Significant increases in hot plate response latencies were found in rats following application of neurotensin into the central amygdaloid nucleus, medial preoptic area, ventral thalamus, and medial pontine reticular formation.[72] Direct injections of neurotensin into the mesencephalic gray area, a region implicated in the central processing of noxious stimuli,[102,103] produced analgesic effects in 38% of animals after unilateral administration of 2.5 µg in one study[104] an in 70% of animals after bilateral infusions of the same dose in another report.[72]

In summary, the influence of neurotensin on nociception is ambiguous. The analgesic properties of the peptide are elusive in that they appear to be very sensitive to experimental parameters utilized and probably to other, still undefined independent variables. One aspect of this research that deserves more attention is the possible relationship between neurotensin's effects on nociception and other central actions of the peptide. Although the reported analgesic effects of neurotensin have been clearly dissociated from its hypothermic action,[72,104] the possible influence of the decreasing effects of the peptide on motor activity and, in some instances, on muscular tone has not been explored. This is particularly crucial considering that the tests utilized to detect analgesia in animals require normal motor activity and intact muscle tone. In this respect the results of a recent report are noteworthy since it was observed that, although the motor responsiveness of rats given electric footshocks was decreased, the vocalization induced by the painful stimulation was not.[88]

5.5. Other Central Effects

In addition to the aforementioned actions of neurotensin, other neurobehavioral effects have been reported. One of the first described central effects of neurotensin was a significant potentiation of pentobarbital-induced sleeping time.[86] This effect was shown to be partly caused by a decreased elimination rate of the barbiturate in the brain, liver, and blood. Later, neurotensin was also found to enhance narcosis produced by ethanol, although this effect was not accompanied by changes in the metabolism of the drug.[105] In another domain, neurotensin was reported to induce catalepsy in mice,[106] although we have not seen this effect in rats.[67] Again, this discrepancy may reflect the apparent greater sensitivity of mice to the central actions of neurotensin as was noted earlier for the hypokinetic and antinociceptive effects of the peptide.

Significant reductions of food intake have been observed following intraventricular administration of neurotensin in food-deprived animals.[107] Recently, direct administration of the peptide into paraventricular hypothalamus was shown to replicate this effect in food-deprived animals and was also shown to decrease food consumption induced by infusion of norepinephrine into this brain region.[108] Interestingly, intrahypothalamic administration did not affect water ingestion in liquid-deprived animals,[108] pointing to a specific influence of the peptide on feeding mechanisms.

The possible effects of neurotensin on conditioned behavior have been only sparsely documented. Neurotensin was first reported to decrease conditioned active avoidance performance in rats.[109] However, in contrast to this finding, more recent results indicate that the peptide significantly retards the extinction of active avoidance and markedly enhances avoidance behavior in a conditioned passive avoidance paradigm.[88] Since the strategies of behavioral pharmacology have proven to be useful for the delineation of the central actions of other neuroactive substance,[110] it can be expected that the effects of neurotensin on conditioned behavior will be more extensively studied in the future.

6. STRUCTURE–ACTIVITY STUDIES

Aside from its actions following central administration, neurotensin injected peripherally exerts a wide variety of pharmacological effects on the cardiovascular, gastrointestinal, and endocrine systems. The peripheral pharmacology of neurotensin has been reviewed by us[61,64] and others[4,5,11] elsewhere. Early structure–activity studies on peripheral effects of neurotensin have demonstrated convincingly that the carboxy-terminal portion (Arg^9-Pro^{10}-Tyr^{11}-Ile^{12}-Leu^{13}) of the peptide is primarily responsible for its biological activity.[111,112] In more recent studies we have examined the stimulatory effects of neurotensin, several fragments, and analogues in rat stomach strips and portal vein as well as in isolated, spontaneously beating atria of guinea pigs.[113,114] Results of these studies confirmed the determinant role of residues 9–13 and indicated that whereas Arg^9, Pro^{10}, Ile^{12}, and Leu^{13} mainly contribute to the affinity of neurotensin to its receptors, Tyr^{11} appears to be implicated specifically in the activation of these receptors. In all preparations, substitution of Tyr^{11} with its D-isomer or an aliphatic amino acid yielded compounds with very low potency unable to attain neurotensin's maximum stimulatory effects.

Structure–activity investigations of neurotensin's neurobehavioral effects have been relatively few and have concentrated on the hypothermic effect of the peptide.[112,115] A consistent finding of these studies is that certain analogues such as [D-Tyr^{11}]NT and [D-Phe^{11}]NT were much more potent than neurotensin in producing hypothermia. Until recently, structure–activity studies on other central effects of the peptide had not been performed. In a series of recent experiments, we examined and compared the effects of several fragments of neurotensin and various structural analogues in which tyrosine in position 11 was replaced by other amino acids. These experiments have been described in detail elsewhere.[67,87,116] A partial list of fragments and analogues that were studied is given in Table II.1 Substances were injected intraventricularly in doses incremented from 0.1 μg/rat to a dose that produced significant effects on body temperature, motor activity, and muscular tone, up to a maximum of 240 μg/rat. The salient findings of these studies are presented in Table II, where relative potencies of fragments and analogues in producing statistically significant changes in motor activity and body temperature are presented. Results for muscular tone are not included in the table, but relative potencies of fragments and analogues in decreasing muscular tone were essentially identical to those found for hypothermia. For comparison purposes, relative potencies of the same compounds in brain binding assays and in two *in vitro* smooth muscle assays are also presented in the table.

An examination of the results obtained with the various fragments reveals that the group of amino acids responsible for neurotensin's central effects resides in the C-terminal. This is particularly exemplified by the fact that NT_{1-10} was devoid of any activity. On the other hand, the two fragments NT_{8-13} and

Table II
Relative Potencies of Neurotensin, Its Fragments, and Analogues in
Neurobehavioral Tests, Binding Assays, and Smooth Muscle Preparations[a]

Peptide	Central effects[b]		Binding[c]		Smooth muscles[d]	
	Motor activity	Hypothermia	Synaptic membranes	Brain tissue slices	Rat portal vein	Rat stomach strip
Neurotensin (NT)	100 ↓	100	100	100	100	100
NT_{8-13}	24 ↓	48	15	39	94	39
NT_{9-13}	12 ↓	12	1.0	11	2.4	11
NT_{10-13}	<0.7	<0.7	0.05	0.65	0	0
NT_{1-10}	<0.7	<0.7	NI[e]	0.01	0	0
NT_{1-12}	6 ↓	6	<0.05	NI	0.02	0.65
[Trp¹¹]NT	450 ↓	900	NI	91	115	130
[D-Trp¹¹]NT	200 ↑	900	NI	1	0.2	0.7
[D-Tyr¹¹]NT	48 ↑	200	1	0.8	<0.01	0.05
[D-Phe¹¹]NT	24 ↑	200	0.1	0.08	<0.01	0.02
[Phe¹¹]NT	3 ↓	3	19	20	94	16
[Ala¹¹]NT	<0.7	<0.7	NI	0.05	NI	0.01

[a] Potency of neurotensin in each test is arbitrarily fixed at 100.
[b] Relative potencies in producing significant decreases in body temperature and significant changes, increase (↑) or decrease (↓), in locomotor activity.[67,87,116]
[c] Relative binding potencies to brain synaptic membranes[118] and to brain tissue slices.[38]
[d] Relative potencies for contracting rat portal vein[114] and stomach strips.[113]
[e] NI, not investigated.

NT_{9-13} were able to significantly decrease activity and body temperature, although, when compared to neurotensin, larger doses were required to produce these effects. The results also reveal that the fragment NT_{9-13} constitutes the minimal effective sequence of the peptide since NT_{10-13} was without effect. Therefore, the C-terminal residues appear to be primordial for the central actions of the peptide. However, this interpretation should be limited to the central effects examined in these studies, since a recent report indicates that an amino terminal fragment, NT_{1-8}, injected centrally can inhibit the so-called "wet dog shakes" induced by TRH in rats.[117]

Results obtained with the various analogues confirm the previously reported greater potency of [D-Tyr¹¹]NT and [D-Phe¹¹]NT in producing hypothermia and extend this attribute to [Trp¹¹]NT and [D-Trp¹¹]NT. A surprising finding of these studies was the opposite effects of neurotensin and some analogues on motor activity of animals. In contrast to neurotensin, various analogues such as [D-Tyr¹¹]NT, [D-Phe¹¹]NT, and [D-Trp¹¹]NT markedly enhanced motor activity (Table II). Subsequent studies revealed that another analogue, [DOPA¹¹]NT, also increased activity, whereas [D-Leu¹¹]NT did not (data not shown). Therefore, the opposite effects of certain analogues are not caused solely by D-isomer replacement of the tyrosine in position 11. All analogues that increased activity were also very potent in producing hypothermia (Table II). However, the analogue [Trp¹¹]NT, which was very effective in lowering body temperature, had, similarly to neurotensin, a depressing effect on activity.

Thus, these data demonstrate that there is no functional relationship between these two central effects of neurotensin and its analogues.

The differential effects of neurotensin and certain structural analogues on motor activity are difficult to interpret. One of many possible explanations is that these analogues act as specific antagonists to endogenous neurotensin on those putative receptors mediating the modulation of motor activity by the peptide. In agreement with this hypothesis, we have observed that [D-Tyr^{11}]NT can completely block neurotensin-induced hypoactivity.[87] This blockade can be obtained with very small doses of the analogue, doses that in themselves do not affect motor activity.

However, results obtained in binding assays with these analogues militate against the hypothesized antagonistic properties of these analogues.[38,118] As can be seen in Table II, the analogues that have opposite effects on motor activity display weak potencies in displacing radiolabeled neurotensin from neuronal membranes. On the other hand, it would be interesting to see in another set of binding experiments if neurotensin could displace these analogues from binding sites. It should also be mentioned that the binding characteristics of the various analogues do not correspond to their efficacy in lowering body temperature (Table II). Many analogues that were found to be more potent than neurotensin in inducing hypothermia are very weak in dislodging bound neurotensin. The reasons for these discrepancies are unclear. It is possible that the greater *in vivo* potencies of some analogues are attributable to the pharmacokinetic characteristics of these substances in the CNS. These very potent analogues may be absorbed more efficiently from the ventricular space, pervade the brain more easily, and/or be more resistant to catabolic processes. Supporting this latter possibility is our observation of a positive correlation between the potency of analogues and their duration of action.

When the central effects of fragments and analogues are compared to their peripheral effects (Table II), some similarities as well as prominent differences are revealed. Results indicate that in both the CNS and periphery the carboxy-terminal portion is crucial for neurotensin's pharmacological activity. Also, aromaticity in position 11 appears to be essential for biological activity in the brain as well as in peripheral tissue. Substitution of the phenolic group with nonaromatic residues such as alanine (Table II) or leucine, serine, lysine, and others (data are shown) yields mostly inactive analogues. On the other hand, the distribution of potencies of analogues in the brain and in periphery are markedly divergent. For example, as can be seen in Table II, the analogues [D-Tyr^{11}]NT, [D-Trp^{11}]NT, and [D-Phe^{11}]NT are very potent in their central effects but relatively weak in their actions on smooth muscles. Although more work is needed to delineate the exact reasons for the differences, these results might indicate that the peripheral receptors for neurotensin are pharmacologically distinct from those subserving the central actions of the peptide.

7. INTERACTIONS OF NEUROTENSIN WITH DOPAMINE AND OTHER NEUROTRANSMITTERS

For the most part, studies on possible interactions of neurotensin with known neurotransmitters have concentrated on the dopaminergic system.

7.1. Dopamine

Data accumulated so far certainly point to neurotensin–dopamine interrelationships in the CNS. As seen earlier, neurotensin's immunoreactivity as well as putative receptors for the peptide are present in many areas containing dopamine cell bodies or terminals.[18,38] These areas include the VTA, substantia nigra, nucleus accumbens, and striatum, thus indicating that neurotensin may interact with both major dopaminergic fiber tracts: the nigrostriatal and the so-called mesolimbic pathways. Recent studies have shown that chemical lesioning of both pathways by means of 6-hydroxydopamine results in a loss of neurotensin binding sites in VTA, substantia nigra, and striatum, suggesting that in these areas neurotensin makes direct synaptic contact with dopamine neurons.[119,120] However, the same treatment does not affect receptors densities in nucleus accumbens, indicating that neurotensin does not act directly on dopamine cells in this region.[120]

As mentioned earlier, initial studies indicate that iontophoretically applied neurotensin affects neuronal activity in nucleus accumbens and VTA.[53,56] Neurotensin has also been shown to affect the synthesis and utilization of dopamine in various brain regions.[121-125] The two principal metabolites of dopamine, 3,4-dihydroxyphenylacetic acid (DOPAC) and homovanillic acid (HVA), are markedly elevated in accumbens, striatum, and olfactory tubercles following central administration of neurotensin.[123,124] This apparent increase in turnover rate is accompanied by a stimulatory action of neurotensin on dopamine synthesis, as evidenced by the enhanced accumulation of DOPA induced by DOPA decarboxylase inhibitors.[123,124] In vitro release of neurotensin from hypothalamic tissue is increased by dopamine and, conversely, decreased by the antagonist haloperidol.[32] However, the concentration of immunoreactive neurotensin in nucleus accumbens was found to be increased following acute and chronic administration of various dopamine antagonists.[126] Finally, neurotensin has recently been reported to facilitate in vitro release of dopamine from striatal tissue.[127]

Together, the above studies support the existence of interactions between neurotensin and dopamine in the CNS. However, the results from these pioneer studies are too disparate to allow a coherent definition of the nature of this interaction. One experimental strategy that has been fruitful in the past for delineating peptide–transmitter interactions involves the study of peptide effects on behaviors elicited by drugs known to affect relatively specific facets of a neurotransmitter.[83] In a series of recent experiments, we examined the effects of neurotensin on the hyperactivity and stereotypy induced by drugs that stimulate dopaminergic activity by either pre- or postsynaptic mechanisms. Drugs selected were amphetamine, nomifensine, apomorphine, and n-propylnorapomorphine. Amphetamine and nomifensine are thought to act presynaptically mainly by releasing dopamine and blocking its reuptake, respectively.[90,128,129] Apomorphine and n-propylnorapomorphine are two direct agonists of dopamine.[90,130,131] The inclusion of a second agonist was prompted by reports indicating that the hyperactivity and stereotypy induced by n-propylnorapomorphine can be more easily dissociated than those produced by apomorphine.[130] The methods and results of these studies have been detailed elsewhere.[75,87] Briefly, results obtained with each drug treatment were identical: hyperactivity was significantly decreased by intraventricular administration of neurotensin, whereas stereotypy was not affected.

The hyperactivity induced by the drugs in these studies is generally considered to be mediated mostly by mesolimbic dopaminergic fibers terminating in the nucleus accumbens, whereas stereotypy is thought to involve more the nigrostriatal dopaminergic pathway.[90,130,133,139] The fact that neurotensin affected only hyperactivity induced by these drugs obviously points to a specific action of the peptide on mesolimbic processes. In agreement with this view is the demonstration that direct intraaccumbens administration of neurotensin significantly reduces amphetamine-induced hyperactivity, whereas intrastriatal injections of the peptide do not alter stereotypy produced by this drug.[134] Moreover, we have observed that intraventricular administration of neurotensin can significantly decrease hyperactivity elicited by direct administration of dopamine into the accumbens.[75] Others have reported that combined administration of neurotensin and dopamine into the accumbens results in lower levels of activity than those produced by dopamine alone.[135] Although neurotensin appears to have a specific inhibitory influence in the nucleus accumbens, the precise mechanisms by which the peptide interacts with dopamine synapses in this region are unknown.

The results of the studies just described indicate that neurotensin can affect hyperactivity caused by direct stimulation of dopamine receptors, suggesting that the peptide may operate at the level of the receptors for this neurotransmitter. However, it has been shown that neurotensin does not alter the binding of [^{3}H]spiroperidol to various membranes, including those in nucleus accumbens.[124,136] Furthermore, neurotensin does not seem to affect dopamine-stimulated adenylate cyclase activity.[136] If neurotensin does not directly affect dopamine receptors, other mechanisms of action must be envisaged that could explain the influence of this peptide on postsynaptic events of dopamine stimulation. In recent years, evidence has accumulated that the accumbens is modulated in a complex fashion by inputs from several brain regions involving different neurotransmitters.[137-139] Of particular interest are the findings that the hyperactivity produced by intraaccumbens administration of dopamine is inhibited by pharmacological activation of serotoninergic as well as GABAergic systems.[139] It is thus conceivable that the modulatory influence of neurotensin on dopamine in the accumbens may be mediated by stimulation of one or both of these inputs.

7.2. Neuroleptic Properties of Neurotensin

The profile of neurotensin's neuropharmacological effects shares many similarities with those of neuroleptics, and because of this the suggestion has been made that neurotensin might act as an endogenous neuroleptic.[8] Similarly to neuroleptics, neurotensin in animals induces hypothermia, hypoactivity, and decreased muscular tone as assessed by indirect behavioral measures.[67,73,82] Also like the neuroleptics, neurotensin has been reported to decrease rates of self-stimulation in rats.[136] However, analgesic effects of neurotensin have been reported, whereas neuroleptics are not known to produce antinociception. But, as discussed earlier, the analgesic action of the peptide is not always seen and seems to depend on a variety of experimental variables. Neuroleptics typically decrease conditioned avoidance behavior in animals.[140] As mentioned earlier, the reported effects of neurotensin on such behavior are still too fragmentary to be conclusive.[88,109]

Contrary to typical neuroleptics,[140] neurotensin does not affect amphetamine- or apomorphine-induced stereotypy.[75] Furthermore, the induction of catelepsy in rats is a typical effect of neuroleptics,[140] but we have never seen such an action of neurotensin in this species.[67] On the other hand, these discrepancies may be irrelevant, since it has been shown that certain clinically effective neuroleptics such as clozapine, sulpiride, and thioridazine have little effect on drug-induced stereotypy and are rather impotent in producing catalepsy.[140-142] For these reasons these substances are referred to as "atypical" neuroleptics.[143,144] It has been proposed that the reduction of amphetamine-induced hyperactivity constitutes a better predictive index of neuroleptic action since this effect was initially found with both typical and atypical neuroleptics.[141] This effect is produced by neurotensin as well. However, in a recent study, the clinically effective neuroleptic thioridazine did not significantly attenuate amphetamine-induced hyperactivity.[143] Thus, it appears that this action cannot always predict the therapeutic usefulness of neuroleptics.

One effect that does seem to be shared by all typical and atypical neuroleptics is their ability to increase synthesis and turnover of dopamine, especially in striatal tissue, probably as a result of facilitatory feedback mechanisms.[145,146] As we have seen, neurotensin produces similar effects on dopamine metabolism. Even more pertinent to the present discussion is the finding that, similarly to what has been described for neurotensin, certain atypical neuroleptics such as clozapine and sulpiride will increase striatal DOPAC levels without affecting the dopamine-sensitive adenylate cyclase system or altering spiroperidol binding to neuronal membranes.[146-150.]

Another effect that is displayed by both atypical and typical neuroleptics is their attenuation of hyperactivity induced by direct stimulation of the accumbens by dopamine,[144,145] again an effect observed with neurotensin. In addition, it appears that, like neurotensin, the atypical neuroleptics will produce this effect at doses that do not reduce drug-induced stereotypy or induce catalepsy in animals.[140,144]

In summary, it appears that neurotensin does have a neuropharmacological profile akin to the so-called "atypical" neuropleptics. From the literature on neuroleptic drugs it appears that all therapeutically effective compounds share two neuropharmacological properties: an increase in dopamine turnover and an inhibition of hyperactivity produced by intraaccumbens administration of dopamine. Neurotensin also produces these two effects. However, before neuroleptogenic properties can be ascribed to the peptide, some important questions must be addressed. At the present time the data derived from the above studies can only help to establish a correlative relationship between clinical efficacy and neuropharmacological effects in animals. It remains to be determined if these effects are functionally responsible for the therapeutic action of drugs and, as importantly, to what extent they are implicated in the production of the serious side effects associated with neuroleptic treatment.

Evidently, the proposition that neurotensin might act as a neuroleptic has spurred considerable interest and has already prompted the study of neurotension levels in psychiatric patients labeled as schizophrenics. In two recent studies, the concentration of immunoreactive neurotensin in various brain regions of schizophrenic patients was not found to be significantly different from control values.[152,153] However, since schizophrenia is probably not a uniform entity and can be produced by a diversity of neurobiochemical defects, it is unlikely that significant differences in the levels of neurotensin, or probably any other neurochemical, can be found using between-group comparisons. This is especially pertinent in view of another study that showed that in a subgroup of schizophrenic patients the levels of neurotensin in cerebral spinal fluid were markedly low.[154] Of further interest was the finding that CSF concentrations of neurotensin returned to normal values following treatment and dissipation of symptoms. Although the status of neurotensin in schizophrenia cannot be established from such limited data, it can be anticipated that it will be the subject of considerable experimental attention in the near future.

7.3. Interactions with Other Neurotransmitters

The possible interactions of neurotensin with neurotransmitters other than dopamine have not been examined thoroughly. High concentrations of immunoreactive neurotensin are found in the dorsal raphe nucleus,[18] which is constituted mainly of serotoninergic cell bodies, suggesting that the peptide might influence this neurotransmitter. The effects of neurotensin on serotonin metabolism have been examined in two studies, and significant elevations of serotonin turnover have been reported.[122,123] However, this effect was only obtained at relatively large doses of the peptide, since no changes were seen following dose smaller than 100 μg.[123]

The influence of neurotensin on noradrenergic systems also warrants further experimental attention. Neurotensin is found in the pons and medulla oblongata, where cell bodies of the ventral noradrenergic bundle originate.[18] The locus coeruleus, the point of departure for the dorsal noradrenergic bundle, also contains significant concentrations of immunoreactive neurotensin.[18] Intraventricular administration of neurotensin has been shown to increase the rate of utilization of norepinephrine in one report.[122] Also, it has been found that the inhibitory effect of the peptide on the ongoing firing rates of cerebellar cells was probably caused by stimulation of presynaptic noradrenergic inhibitory influences on these cells.[54] In addition, the results of a recent study indicate a marked stimulatory effect of the peptide on both the spontaneous and potassium-induced release of norepinephrine.[155]

Finally, the possible effect of neurotensin on acetylcholine has been the subject of only one study. Significant augmentation in acetylcholine concentrations were found in striatum and parietal cortex following central administration of 10 μg neurotensin. The same treatment was found to increase the turnover rate of acetylcholine in the diencephalic region.[156]

8. SUMMARY

In the 10 years since its discovery, a vast amount of data has emerged on the presence, metabolism, and actions of neurotensin in the CNS. These data were obtained with a multiplicity of experimental techniques involving the disciplines of neuroanatomy, biochemistry, neuropharmacology, neurophysiology, and behavioral science. Although the multiplicity of approaches has succeeded in unveiling many important facets of this neuropeptide, it is difficult to organize into a coherent whole the diversity of the information gleaned so far.

The presence of neurotensin in the CNS in a variety of animal species is now well established, and it is interesting that similarities in the patterns of regional distribution of the peptide have been observed. The proximity of binding sites to nerve terminals containing neurotensin as well as the demonstrated release of the peptide under physiological conditions suggest strongly that the actions of neurotensin in the brain are mediated via synaptic chemical transmission. When administered into the CNS, neurotensin produces a wide variety of effects. As we have seen, many of the electrophysiological and neurobehavioral effects are variable and appear to be very sensitive to the methodologies employed to investigate them. The availability of a genuine neurotensin antagonist would certainly help to verify and characterize the actions of this peptide in the CNS.

Another important issue pertains to the physiological significance of the observed effects following central administration of neurotensin. One perennial concern with these studies is that the injected peptide may be interacting in a pharmacological way with various central processes and not functionally simulate the neurophysiological activity of the endogenous peptide at specific receptors. The observed neurobehavioral effects occur following intracranial injection of doses that are much larger than the endogenous immunoreactive neurotensin. However, this does not necessarily mean that doses administered are "pharmacological," since it is not known how much of the injected peptide is degraded by peptidases and what concentration is available at the receptor level. On the other hand, even if some of these effects are eventually shown to be pharmacological in nature, these efforts will not be wasted, since characterization of these effects could lead to the development of specific, therapeutically useful substances derived from neurotensin.

At the neurochemical level, research emphasis has been placed on the interaction of neurotensin with dopamine, and the data accumulated do suggest an interrelationship of the peptide with this neurotransmitter system. However, it is unlikely that the neurochemical influence of the peptide is restricted to dopamine, as early studies point to interactions with other neurotransmitters.

Finally, considerable attention has focused on the neurolepticlike actions of neurotensin. As we have seen, the peptide shares some but not all of the properties of classical neuroleptic drugs. However, direct comparisons are complicated by the fact that these drugs as a class have very few common neuropharmacological features. At the present time, it appears that neurotensin most closely resembles the so-called "atypical" neuroleptics.

Although it is too early to define the neurophysiological role or functions of neurotensin in the CNS, the interest generated by the accumulated data will undoubtedly continue to spur research efforts so that in the not too distant future the ways that neurotensin interacts with other neurochemicals in the genesis of normal and perhaps abnormal behaviors will be elucidated.

REFERENCES

1. Carraway, R. E., and Leeman, S. E., 1973, *J. Biol. Chem.* **248**:6854–6861.
2. Carraway, R. E., and Leeman, S. E., 1975, *J. Biol. Chem.* **250**:1907–1911.
3. Carraway, R. E., and Leeman, S. E., 1976, *J. Biol. Chem.* **251**:7045–7052.
4. Bissette, G., Manberg, P. J., Nemeroff, C. B., and Prange, A. J., Jr., 1978, *Life Sci.* **23**:2173–2182.
5. Brown, D. R., and Miller, R. J., 1982, *Br. Med. Bull.* **38**:239–245.
6. Carraway, R. E., 1978, *Methods of Hormone Radioimmunoassay*, 2nd ed. (B. M. Jaffe and H. R. Behrman, eds.), Academic Press, New York, pp. 139–169.
7. Leeman, S. E., 1982, *Recent Prog. Horm. Res.* **38**:93–132.
8. Nemeroff, C. B., 1980, *Biol. Psychiatry* **15**:283–302.
9. Nemeroff, C. B., Luttinger, D., and Prange, A. J., Jr., 1980, *Trends Neurosci.* **3**:212–215.
10. Uhl, G. R., and Snyder, S. H., 1981, *Neurosecretion and Brain Peptides* (J. B. Martin, S. Reichlin, and K. L. Bick, eds.), Raven Press, New York, pp. 87–106.
11. Nemeroff, C. B., and Prange, A. J., Jr. (eds.), 1982, *Neurotensin, a Brain and Gastrointestinal Peptide*, The New York Academy of Sciences, New York.
12. Falkner, S., Carraway, R. E., El-Salhy, M., Emdin, S. O., Grimelius, L., Rechfeld, J. F., Reinecke, M., and Schwartz, T. F. W., 1981, *UCLA Forum Med. Sci.* **23**:21–42.
13. Grimmelikhuijzen, C. J. P., Carraway, R. E., Rokaeus, A., and Sundler, F., 1981, *Histochemistry* **72**:199–209.
14. Kobayashi, R., Brown, M., and Vale, W., 1977, *Brain Res.* **126**:584–588.
15. Uhl, G. R., and Snyder, S. H., 1976, *Life Sci.* **19**:1827–1832.
16. Kataoko, K., Mizuno, N., and Frohman, L., 1980, *Brain Res. Bull.* **4**:57–60.
17. Cooper, P., Fernstrom, M., Rorstad, O., Leeman, S. E., and Martin, J., 1981, *Brain Res.* **218**:219–232.
18. Emson, P. C., Goedert, M., Horsfield, P., Rioux, F., and St.-Pierre, S., 1982, *J. Neurochem.* **38**:992–999.
19. Emson, P. C., Goedert, M., Benton, H., St.-Pierre, S., and Rioux, F., 1982, *Adv. Biochem. Psychopharmacol.* **35**:477–485.
20. Seybold, U., and Elde, R., 1980, *J. Histochem. Cytochem.* **28**:367–370.
21. Ainsworth, A., Hall, P., Wall, P. D., Allt, G., Mackensie, L., Gibson, S., and Polak, J. M., 1981, *Pain* **2**:379–388.
22. Hunt, S. P., Kelly, J. S., Emson, P. C., Kimmel, J. R., Miller, R. J., and Wu, J. Y., 1981, *Neuroscience* **6**:1883–1898.
23. Polak, J. M., and Bloom, S. R., 1982, *Neurotensin, a Brain and Gastrointestinal Peptide* (C. B. Nemeroff and A. J. Prange, Jr., eds.), The New York Academy of Sciences, New York, pp. 75–93.
24. Uhl, G. R., and Snyder, S. H., 1977, *Eur. J. Pharmacol.* **41**:89–91.
25. Fernstrom, M. H., Garraway, R. E., and Leeman, S. E., 1980, *Frontiers in Neuroendocrinology*, Volume 6 (L. Martini, ed.), Raven Press, New York, pp. 103–127.
26. Khan, D., Abrams, G. M., Zimmerman, E. A., Carraway, R. E., and Leeman, S. E., 1980, *Endocrinology* **107**:47–53.
27. Uhl, G. R., Kuhar, M., and Snyder, S. E., 1977, *Proc. Natl. Acad. Sci. U.S.A.* **74**:4059–4063.
28. Khan, D., Hou-Yu, A., and Zimmerman, E. A., 1982, *Neurotensin, a Brain and Gastrointestinal Peptide* (C. B. Nemeroff and A. J. Prange, Jr., eds.), New York Academy of Sciences, New York, pp. 117–131.

29. Uhl, G. R., and Snyder, S. H., 1979, *Brain Res.* **161**:522–526.
30. Roberts, G. W., Crow, T. J., and Polak, J. M., 1981, *Peptides* **2**:37–43.
31. Iversen, L. L., Iversen, S. D., Bloom, F., Douglas, C., Brown, M., and Vale, W., 1978, *Nature* **273**:161–163.
32. Maeda, K., and Frohman, L. A., 1981, *Brain Res.* **210**:261–269.
33. Kitabji, P., Carraway, R. E., Van Rietschoten, F., Morgat, J. L., Menez, A., Leeman, S. E., and Freychet, P., 1977, *Proc. Natl. Acad. Sci. U.S.A.* **74**:1846–1850.
34. Lazarus, L. H., Brown, M. R., and Perrin, M. H., 1977, *Neuropharmacology* **16**:625–629.
35. Uhl, G. R., Bennett, J. P., and Snyder, S. H., 1977, *Brain Res.* **130**:299–313.
36. Young, W. S. III, and Kuhar, M. J., 1981, *Brain Res.* **206**:273–285.
37. Young, W. S. III, and Kuhar, M. J., 1979, *Eur. J. Pharmacol.* **59**:161–163.
38. Quirion, R., Gaudreau, P., St.-Pierre, S., Rioux, F., and Pert, C. B., 1982, *Peptides* **3**:757–763.
39. Vale, W., Spiess, J., Rivier, C., and Rivier, J., 1981, *Science* **213**:1394–1397.
40. Habener, J. F., 1981, *Neurosecretion and Brain Peptides* (J. B. Martin, S. Reichlin, and K. L. Bick, eds.), Raven Press, New York.
41. Dupont, A., and Merand, Y., 1978, *Life Sci.* **22**:1623–1630.
42. McDermott, J. R., Smith, A. I., Biggins, J. A., and Edwardson, J. A., 1982, *Regul. Peptides* **3**:397–404.
43. Edwardson, J. A., and McDermott, J. R., 1982, *Br. Med. Bull.* **38**:259–264.
44. Henry, J. L., 1982, *Neurotensin, a Brain and Gastrointestinal Peptide* (C. B. Nemeroff and A. J. Prange, Jr., eds.), New York Academy of Sciences, New York, pp. 216–227.
45. Miletic, U., and Randig, M., 1979, *Brain Res.* **169**:600–604.
46. Phillis, J. W., and Kirkpatrick, J. R., 1979, *Can. J. Physiol. Pharmacol.* **57**:887–889.
47. Suzue, T., Yanaihara, N., and Otsuka, M., 1981, *Neurosci. Lett.* **26**:137–142.
48. Guyenet, P. G., and Aghajanian, G. K., 1977, *Brain Res.* **136**:178–184.
49. Young, W. S., Uhl, G. R., and Kuhar, M. J., 1978, *Brain Res.* **150**:431–435.
50. Dao, W. P. C., Yajima, H., Kitagawa, K., and Walker, R. J., 1981, *Advances in Physiological Sciences*, Volume 14 (E. Stark, G. B. Makara, B. Halasz, and G. Y. Rappay, eds.), Raven Press, New York, pp. 249–254.
51. Sawada, S., Takada, S., and Yamamoto, C., 1980, *Brain Res.* **188**:578–581.
52. Sawada, S., and Yamamoto, C., 1980, *Neurosi. Lett.* **4**:58.
53. McCarthy, P. S., Walker, R. J., Yajima, H., Kitagawa, K., and Woodruff, G. N., 1979. *Gen. Pharmacol.* **10**:331–333.
54. Marwaha, J., Hoffer, B., and Freedman, R., 1980, *Regul. Peptides* **1**:115–125.
55. Phillis, J. W., and Kirkpatrick, J. R., 1980, *Can. J. Physiol. Pharmacol.* **58**:612–623.
56. Andrade, R., and Aghajanian, G. K., 1981, *Soc. Neurosci. Abstr.* **7**:573.
57. Rioux, F., Quirion, R., St-Pierre, S., Regoli, D., Jolicoeur, F. B., Bélanger, F., and Barbeau, A., 1981, *Eur. J. Pharmacol.* **69**:241–247.
58. Osumi, Y., Nagasaka, Y., Want, L. H. F., and Fuziwara, M., 1978, *Life Sci.* **23**:2275–2280.
59. Maeda, K., and Frohman, L. A., 1978, *Endocrinology* **103**:1903–1909.
60. Vijayan, E., and McCann, S. M., 1979, *Endocrinology* **105**:64–68.
61. Rioux, F., Kérouac, R., Quirion, R., and St-Pierre, S., 1982, *Neurotensin, a Brain and Gastrointestinal Peptide* (C. B. Nemeroff and A. J. Prange, Jr., eds.), The New York Academy of Sciences, New York, pp. 56–74.
62. McCann, S. M., Vijayan, E., Koenig, J., and Krulich, L., 1982, *Neurotensin, a Brain and Gastrointestinal Peptides* (C. B. Nemeroff and A. J. Prange, Jr., eds.), The New York Academy of Sciences, New York, pp. 160–171.
63. Frohman, L. A., Maeda, K., Berelowitz, M., Szabo, M., and Thominet, J., 1982, *Neurotensin, a Brain and Gastrointestinal Peptide* (C. B. Nemeroff, and A. J. Prange, Jr., eds.), New York Academy of Sciences, New York, pp. 172–191.
64. St-Pierre, S., Kérouac, R., Quirion, R., Jolicoeur, F. B., and Rioux, F., 1984, *Peptide and Protein Rev.* **2**:83–171.
65. Prange, A. J., Jr., Nemeroff, C. B., Bissette, G., Manberg, P. J., Oshbar, A. J., Burnett, G. B., Loosen, P. T., and Kraemer, G. W., 1979, *Pharmacol. Biochem. Behav.* **11**:473–477.
66. Bissette, G. B., Manberg, P. J., Nemeroff, C. B., Loosen, P. T., Prange, A. J., Jr., and Lipton, M. A., 1976, *Nature* **262**:607–609.

67. Jolicoeur, F. B., Barbeau, A., Rioux, F., Quirion, R., and St-Pierre, S., 1980, *Peptides* **2**:171–175.
68. Mason, G. A., Nemeroff, C. B., Luttinger, D., Hatley, O. L., and Prange, A. J., Jr., 1980, *Regul. Peptides* **1**:53–60.
69. Chandra, A., Chou, H. C., and Lin, M. T., 1981, *Neuropharmacology* **20**:715–718.
70. Dorsa, D. M., De Kloet, E. R., Mezey, E., and De Wied, D., 1979, *Endocrinology* **104**:1663–1666.
71. Martin, G. E., Bacino, C. B., and Pap, N. L., 1981, *Peptides* **1**:333–339.
72. Kalivas, P. W., Jennes, L., Nemeroff, C. B., and Prange, A. J., Jr., 1982, *J. Comp. Neurol.* **210**:225–238.
73. Nemeroff, C. B., Bissette, G., Manberg, P. J., Oshbar, A. J., Breese, G. R., and Prange, A. J., Jr., 1980, *Brain Res.* **195**:69–84.
74. Clark, W. G., 1979, *Neurosci. Biobehav. Rev.* **3**:179–231.
75. Jolicoeur, F. B., De Michele, G., Barbeau, A., and St-Pierre, S., 1983, *Neurosci. Biobehav. Rev.* **7**:385–390.
76. Morley, J. E., Levine, A. S., Oken, M. M., Grace, M., and Kneip, J., 1982, *Peptides* **3**:1–6.
77. Nemeroff, C. B., Loosen, P. T., Bissette, G., Manberg, P. J., Wilson, I. C., Lipton, M. A., and Prange, A. J., Jr., 1979, *Psychoneuroendocrinology* **3**:279–310.
78. Cowan, A., and Gmerek, D. E., 1982, *Neurotensin, a Brain and Gastrointestinal Peptide* (C. B. Nemeroff and A. J. Prange, Jr., eds.), The New York Academy of Sciences, New York, 438–439.
79. Yarbrough, G. G., and McGuffin-Clineschmidt, J. J., 1979, *Eur. J. Pharmac.* **60**:41–46.
80. Dunn, A. J., Snijders, R., Hurd, R. W., and Kramarcy, N. R., 1982, *Neurotensin, a Brain and Gastrointestinal Peptide* (C. B. Nemeroff and A. J. Prange, Jr., eds.), The New York Academy of Sciences, New York, pp. 345–353.
81. Griffiths, E. C., Slater, P., and Webster, A. D., 1981, *J. Physiol. (Lond.)* **320**:90.
82. Oshbar, A. J., Nemeroff, C. B., Manberg, P. J., and Prange, A. J., Jr., 1979, *Eur. J. Pharmacol.* **54**:299–302.
83. Jolicoeur, F. B., Rondeau, D., St.-Pierre, S., Rioux, F., and Barbeau, A., 1981, *Clinical Pharmacology of Apomorphine and Other Dopaminomimetics* (J. L. Gessa and G. U. Corsini, eds.), Raven Press, New York, pp. 19–25.
84. Jolicoeur, F. B., Rondeau, D., Hamel, E., Butterworth, R. F., and Barbeau, A., 1979, *Can. J. Neurol. Sci.* **6**:209–214.
85. Jolicoeur, F. B., Rondeau, D., Belanger, F., Fouriezos, G., and Barbeau, A., 1980, *Peptides* **1**:103–106.
86. Nemeroff, C. B., Bissette, G., Prange, A. J., Jr., Loosen, P. T., Barlow, T. S., and Lipton, M. A., 1977, *Brain Res.* **128**:485–496.
87. Jolicoeur, F. B., Rioux, F., St.-Pierre, S., and Barbeau, A., 1981, *Brain Neurotransmitters and Hormones* (R. Collu, R. J. Ducharme, A. Barbeau, and G. Tollis, eds.), Raven Press, New York, pp. 171–178.
88. Van Wimersma Greidanius, T. J. B., Van Praag, M. C. G., Kalmann, R., Rinkel, G. J. E., Croiset, G., Hoeke, E. C., Van Egmond, M. A. H., and Fekete, M., 1982, *Neurotensin, a Brain and Gastrointestinal Peptide* (C. B. Nemeroff and A. J. Prange, eds.), The New York Academy of Sciences, New York, pp. 319–329.
89. Clineschmidt, B. U., McGuffin, J. C., and Bunting, P. B., 1979, *Eur. J. Pharmacol.* **54**:129–139.
90. Costall, B., Naylor, R. J., Marsden, C. B., and Pycock, C. J., *Brain Res.* **123**:89–97.
91. Jones, D. L., and Mogenson, G. J., 1980, *Brain Res.* **188**:93–96.
92. Mogenson, G. J., Wu, M., and Manchanda, S. K., 1979, *Brain Res.* **161**:311–319.
93. Kalivas, P. W., Nemeroff, C. B., and Prange, A. J., Jr., 1981, *Brain Res.* **229**:525–529.
94. Kalivas, P. W., Nemeroff, C. B., and Prage, A. J., Jr., 1982, *Eur. J. Pharmacol.* **78**:471–474.
95. Clineschmidt, B. U., and MacGuffin, J. C., 1977, *Eur. J. Pharmacol.* **46**:395–396.
96. Nemeroff, C. B., Oshbar, A. J., Manberg, P. J., Ervin, G. N., and Prange, A. J., Jr., 1979, *Proc. Natl. Acad. Sci. U.S.A.* **76**:5368–5371.

98. Martin, G. E., Naruse, T., and Pap, N. L., 1981, *Neuropeptides* **1**:447–454.
99. Kalivas, P. W., Gau, B. A., Nemeroff, C. B., and Prange, A. J., Jr., 1982, *Brain Res.* **243**:279–286.
100. Yaksh, T. L., Schmauss, C., Miceviych, P. E., Abay, E. O., and Go, V. L. W., 1982, *Neurotensin, a Brain and Gastrointestinal Peptide* (C. B. Nemeroff, and A. J. Prange, Jr., eds.), The New York Academy of Sciences, New York, pp. 228–243.
101. Bodnar, R. J., Wallace, M. M., Nilaver, G., and Zimmerman, E. A., 1982, *Neurotensin, a Brain and Gastrointestinal Peptide* (C. B. Nemeroff and A. J. Prange, Jr., eds.), The New York Academy of Sciences, New York, pp. 244–258.
102. Bennett, G. J., and Mayer, P. J., 1979, *Brain Res.* **172**:243–257.
103. Yaksh, T. L., and Rudy, T. A., 1978, *Pain* **4**:299–359.
104. Clineschmidt, B. U., Martin, G. E., and Veber, D. F., 1982, *Neurotensin, a Brain and Gastrointestinal Peptide* (C. B. Nemeroff and A. J. Prange, Jr., eds.), The New York Academy of Sciences, New York, pp. 283–306.
105. Luttinger, D., Nemeroff, C. B., Mason, G. D., Frye, G. A., Breese, G. R., and Prange, A. J., Jr., 1981, *Neuropharmacology* **20**:305–309.
106. Snijders, R., Kramarcy, N. R., Hurd, R. W., Nemeroff, C. B., and Dunn, A. J., 1982, *Neuropharmacology* **21**:465–468.
107. Luttinger, D., King, R. A., Sheppard, D., Strupp, J., Nemeroff, C. B., and Prange, A. J., Jr., 1982, *Eur. J. Pharmacol.* **81**:499–503.
108. Stanley, B. G., Eppel, N., and Hoebel, B. G., 1982, *Neurotensin, a Brain and Gastrointestinal Peptide* (C. B. Nemeroff and A. J. Prange, Jr., eds.), The New York Academy of Sciences, New York, pp. 425–427.
109. Luttinger, D., Nemeroff, C. B., and Prange, A. J., Jr., 1982, *Brain Res.* **237**:83–82.
110. Seiden, L. S., and Dykstra, L. A., 1977, *Psychopharmacology*, Van Nostrand Reinhold, New York.
111. Carraway, R. E., and Leeman, S. E., 1975, *Peptides: Chemistry, Structure and Biology* (R. Watter and J. Meienhofer, eds.), Ann Arbor Science, Ann Arbor, pp. 679–685.
112. Rivier, J. E., Lazarus, L. H., Perrin, M. H., and Brown, M. R., 1977, *J. Med. Chem.* **20**:1409–1412.
113. Quirion, R., Regoli, D., Rioux, F., and St.-Pierre, S., 1980, *Br. J. Pharmacol.* **68**:83–91.
114. Rioux, F., Quirion, R., Regoli, D., Leblanc, M. A., and St.-Pierre, S., 1980, *Eur. J. Pharmacol.* **66**:273–279.
115. Loosen, P. T., Nemeroff, C. B., Bissette, G., Burnett, G. B., Prange, A. J., and Lipton, M. A., 1978, *Neuropharmacology* **17**:109–113.
116. Jolicoeur, F. B., Barbeau, A., Quirion, R., Rioux, F., and St.-Pierre, S., 1982, *Neurotensin, a Brain and Gastrointestinal Peptide* (C. B. Nemeroff and A. J. Prange, Jr., eds.), The New York Academy of Sciences, New York, pp. 440–441.
117. Griffiths, E. C., Widdowson, P. S., and Slater, P., 1982, *Neurosci. Lett.* **31**:171–174.
118. Kitabgi, P., Poustis, C., Granier, C., Van Rietschoten, J., Rivier, J., Morgat, J. L., and Freychet, P., 1980, *Mol. Pharmacol.* **18**:11–19.
119. Palacios, J. M., and Kuhar, M. J., 1981, *Nature* **294**:587–589.
120. Quirion, R., Everist, H. D., and Pert, A., 1982, *Soc. Neurosci. Abstr.* **8**:582.
121. Haubrich, D. R., Martin, G. E., Pflueger, A. B., and Williams, M., 1982, *Brain Res.* **231**:216–221.
122. Garcia-Sevilla, J. A., Magnusson, T., Carlsson, A., Leban, J., and Folkers, K., 1978, *Naunyn Schmiedebergs Arch. Pharmacol.* **305**:213–218.
123. Widerlöv, E., and Breese, G. R., 1982, *Neurotensin, a Brain and Gastrointestinal Peptide* (C. B. Nemeroff and A. J. Prange, Jr., eds.), The New York Academy of Sciences, New York, pp. 428–430.
124. Reches, A., Burke, R. E., Jiang, D. H., Wagner, H. R., and Fahn, S., 1982, *Neurotensin, a Brain and Gastrointestinal Peptide* (C. B. Nemeroff and A. J. Prange, Jr., eds.), The New York Academy of Sciences, New York, pp. 420–421.
125. Magnusson, T., Garcia-Sevilla, J. A., and Carlsson, A., 1979, *Catecholamines, Basic and Clinical Frontiers* (E. Usdin, I. J. Kopin, and J. Barchas, eds.), Pergamon Press, New York, pp. 1209–1211.

126. Govoni, S., Hong, J. S., Yang, H. Y. T., and Costa, E., 1980, *J. Pharmacol. Exp. Ther.* **215**:413–417.
127. Starr, M. S., 1982, *Neurochem. Int.* **4**:233–240.
128. Creese, I., and Iversen, S. D., 1975, *Brain Res.* **83**:419–436.
129. Algeri, S., Ponzio, F., Achilli, G., and Perego, G., 1982, *Typical and Atypical Antidepressants* (E. Costa and G. Racagni, eds.), Raven Press, New York, pp. 219–228.
130. Costall, B., Hui, S. C. G., and Naylor, R. J., 1980, *Neuropharmacology* **19**:1039–1048.
131. Anden, N. E., Rubenson, A., Fuxe, K., and Hokfelt, T., 1967, *J. Pharm. Pharmacol.* **19**:627–629.
132. Iversen, L. L., Kelley, P. H., Miller, R. J., and Seviour, P., 1975, *Br. J. Pharmacol.* **54**:244–249.
133. Wilcox, R. E., Riffee, W. H., and Smith, R. U., 1979, *Pharmacol. Biochem. Behav.* **11**:653–659.
134. Ervin, G. N., Birkemo, L. S., Nemeroff, C. B., and Prange, A. J., Jr., 1981, *Nature* **291**:73–76.
135. Kalivas, P. W., Nemeroff, C. B., and Prange, A. J., Jr., 1982, *Neurotensin, a Brain and Gastrointestinal Peptide* (C. B. Nemeroff and A. J. Prange, Jr., eds.), The New York Academy of Sciences, New York, pp. 307–318.
136. Nemeroff, C. B., Hernandez, D. E., Luttinger, D., Kalivas, P. W., and Prange, A. J., Jr., 1982, *Neurotensin, a Brain and Gastrointestinal Peptide* (C. B. Nemeroff and A. J. Prange, Jr., eds.), The New York Academy of Sciences, New York, pp. 330–344.
137. Carter, C. J., and Pycock, C. J., 1973, *Life Sci.* **23**:953–958.
138. Costall, B., Naylor, R. J., Marsden, C. B., and Pycock, C. J., 1976, *J. Pharm. Pharmacol.* **28**:523–530.
139. Jones, D. L., Mogenson, G. J., and Wu, M., 1981, *Neuropharmacogy* **20**:29–37.
140. Niemergeers, C. J. E., and Janssen, P. A. J., 1979, *Life Sci.* **24**:2201–2216.
141. Honda, F., Satoh, Y., Shimomura, K., Satoh, H., Noguchi, H., Uchida, S., and Katol, R., 1977, *Jpn. J. Pharmacol.* **27**:397–411.
142. Tagliamonte, A., De Montis, G., Olianas, M., Vargin, L., Corsini, G. U., and Gessa, G. L., 1975, *J. Neurochem.* **24**:707–710.
143. Bentall, A. C. C., and Herberg, L. J., 1980, *Neuropharmacogy* **19**:699–703.
144. Costall, B., and Naylor, R. J., 1976, *Eur. J. Pharmacol.* **40**:9–19.
145. Westerink, B. H. C., 1979, *Neurobiology of Dopamine* (A. S. Horn, J. Korf, and B. H. C. Westerink, eds.), Academic Press, London, pp. 255–273.
146. Stanley, M., and Wilk, S., 1979, *Life Sci.* **24**:1907–1922.
147. Peringer, E., Jenner, P., Donaldson, I. M., Marsden, C. D., and Miller, R., 1975, *Neuropharmacogy* **15**:463–469.
148. Roufogalis, B. D., Thornton, M., and Wade, D. N., 1976, *Life Sci.* **19**:927–934.
149. Venner, P., Chow, A., Reavill, C., Theodorou, A., and Marsden, C. D., 1978, *Life Sci.* **23**:545–550.
150. Howard, J. L., Large, B. T., Wedley, S., and Pullar, I. A., 1978, *Life Sci.* **23**:599–604.
151. Costall, B., and Naylor, R. J., 1976, *Eur. J. Pharmacol.* **35**:161–170.
152. Kleinmann, J. E., Karoum, F., Rosenblatt, J., Gillin, J. C., Hong, J., Bridge, T. P., Zalchman, S., Storck, F., Carmen, R., and Wyatt, R. J., 1981, *Biological Psychiatry* (C. Perris, G. Struwe, and B. Jansson, eds.), Elsevier North-Holland Biomedical Press, Amsterdam, pp. 711–714.
153. Roberts, G. W., Ferrier, I. N., Lee, Y. C., Adrian, T. E., O'Shaughnessy, D. J., Crow, T., Polak, J. M., and Bloom, S. R., 1982, *Regul. Peptides* **3**:81–86.
154. Widerlöv, E., Lindström, L. H., Besev, G., Manberg, P. J., Nemeroff, C. B., Breese, G. R., Kizer, J. S., and Prange, A. J., Jr., 1982, *Am. J. Psychiatry* **139**:1122–1126.
155. Okuma, Y., and Osumi, Y., 1982, *Life Sci.* **30**:77–84.
156. Malthe-Sorenssen, D., Wood, P. L., Cheney, D. L., and Costa, E., 1978, *J. Neurochem.* **31**:685–691.

5

Cholecystokinin

John E. Morley and Allen S. Levine

1. INTRODUCTION

The universal distribution of most of the small peptides is now well recognized. Many of the peptides originally isolated from the brain are now also known to be present in the gastrointestinal tract, and many peripheral peptides are now also found in the brain. Recent evidence suggests that cholecystokinin, a classical gastrointestinal polypeptide hormone, plays an equally important role as a neurotransmitter.

Cholecystokinin (CCK) was first isolated as a 33-amino-acid hormone from the porcine gastrointestinal tract by Mutt and Jorpes.[1] Historically, the existence of CCK was first suspected when Okada, working in Starling's laboratory, found that acidification of the dog intestine led to expulsion of gallbladder bile as well as increased secretion of hepatic bile flow.[2] In 1928, Ivy and Oldberg[3] described a substance that was released from the upper intestine to produce gallbladder contraction and suggested that this hormone be named cholecystokinin. Between 1941 and 1943, Harper, Vass, and Raper[4,5] described a series of experiments showing that porcine intestinal tissue contained a hormone distinct from secretin that stimulated pancreatic enzyme secretion and suggested that this hormone be named pancreozymin. Originally, CCK and pancreozymin were thought to be separate hormones, but after they had been purified and characterized, they proved to be identical.[1] Cholecystokinin has numerous effects on the gastrointestinal tract and biliary systems, as outlined in Table I. Details of these effects have been reviewed recently[6,7] and are not further discussed here. The effects on the gallbladder, pancreas, and gastrointestinal tract have been clearly shown to be physiological, whereas many of the other effects are thought only to be pharmacological. Following the original isolation and characterization of CCK, it has been demonstrated that besides being present in the intestine, CCK can also be extracted from brain tissue.

John E. Morley and Allen S. Levine • Neuroendocrine Research Laboratory, Minneapolis Veterans Administration Medical Center, Minneapolis, Minnesota 55417; and Departments of Medicine and Food Science and Nutrition, University of Minnesota, Minneapolis–St. Paul, Minnesota 55455.

Table I
Gastrointestinal Effects of Cholecystokinin

Motility
   Stimulation of gallbladder contraction and relaxation of sphincter of Oddi[8]
   Inhibition of lower esophageal sphincter pressure[9]
   Inhibition of gastric emptying[10]
   Stimulation of intestinal motility[11]
Exocrine secretion
   Stimulation of pancreatic enzyme secretion and enhancement of secretin-induced bicarbonate secretion[12,13]
   Low doses stimulate gastric acid secretion[14] and competitively inhibit gastrin-stimulated gastric acid secretion[15]
   Increase hepatic bile flow[16]
Pancreatic endocrine secretion
   Insulinotropic[17,18]
   Stimulate glucagon release[19]
Miscellaneous
   Trophic effect on pancreas[20]
   Increase mucosal blood flow[21]
   Inhibition of water and electrolyte absorption[22]

The presence of gastrinlike immunoreactivity in the central nervous system of humans and several other species was first described by Vanderhaegen *et al.*[23] in 1975. The following year, the gastrinlike immunoreactivity in extracts of cerebral tissue was shown to reside in peptides more closely resembling CCK than gastrin. This chapter deals predominantly with the rapidly expanding literature on the neuropharmacological effects of cholecystokinin.

2. CENTRAL NERVOUS SYSTEM DISTRIBUTION

Since the original observations in 1975 of Vanderhaegen *et al.*[23] of gastrinlike immunoreactivity in the brain, a number of workers in Liverpool,[24] New York,[25] and Aaarhus, Denmark[26,27] have extended these observations and shown that this new peptide most resembles the carboxyterminal octapeptide of cholecystokinin (CCK-8), which cross reacts with some gastrin antisera because of the common carboxy terminus of gastrin and CCK. There have been numerous reports describing the heterogeneity of cholecystokinin in the brain and the gut. It has been demonstrated that CCK exists in the brain in at least five forms: a component larger than CCK-39, a component similar in size to CCK-39 or CCK-33, a component similar to CCK-12, and one similar to CCK-8, as well as some smaller molecular forms.[28–33] Although controversy exists concerning the major forms of CCK in the brain (mainly because of differences in the extraction techniques used), there seems generally to be agreement that the predominant form is CCK-8 (60–70%), with approximately 15% being CCK-33. The structures of the various forms of CCK are depicted in Fig. 1. Besides having immunoreactivity similar to CCK, brain extracts containing CCK have

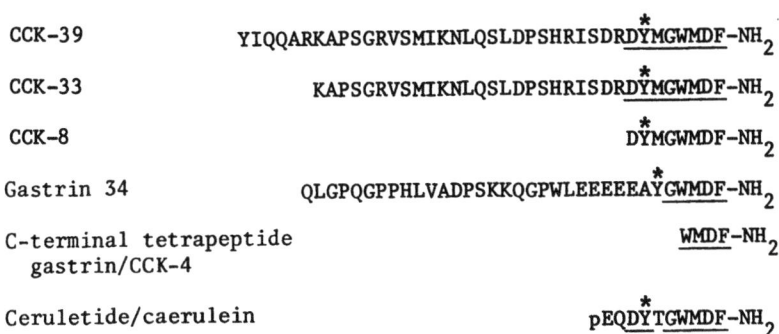

Fig. 1. The structure of cholecystokinin and related peptides. Italic sequences represent sequence homology with CCK-8. Amino acids: A, alanine; R, arginine; N, asparagine; D, aspartic acid; Q, glutamine; E, glutamic acid; G, glycine; H, histidine; I, isoleucine; L, leucine; K, lysine; M, methionine; F, phenylalanine; P, proline; S, serine; T, threonine; W, tryptophan; Y, tyrosine; V, valine; p, pyro-; *, SO₃H. Note that gastrins can exist in either sulfated or unsulfated forms.

been shown to have bioactivity similar to that of gut CCK *in vitro* in the guinea pig gallbladder bioassay[30] and *in vivo* on pancreatic exocrine secretion in the anesthetized rat.[34] Recently, Dockray et al.[35] have isolated and reported the exact amino acid sequence from sheep brain, showing that it is identical to CCK-8.

Besides the widespread existence of CCK in mammalian brains, a CCK-like substance has been demonstrated by radioimmunoassay and chromatography[36] to be present in the brain and gut of the lamprey, suggesting that the dual distribution of CCK in brain and gut is widespread throughout the vertebrates. Cholecystokinin has also been localized by immunocytochemistry in the brain of the blowfly,[37] and radioimmunoassayable CCK-like activity is present in the nervous system of insects,[38] showing that phylogenetically the gastrin/CCK peptides were established early in evolution within the neuronal elements of invertebrates. Cholecystokinin has been localized by immunohistochemistry in the nerve cells of the coelenterate *Hydra*, which has the most primitive nervous system in the animal kingdom.[39] These data show that the gastrin/CCK family emerged very early in evolution, with CCK most probably being phylogenetically older than gastrin.[40]

The quantitative distribution of CCK has been shown to be the same in the rat,[41] swine,[27] and guinea pig[31] brain. The distribution of CCK in the brain is different from those of all other known peptides except for vasoactive intestinal peptide (VIP) in that both CCK and VIP[42] are predominately cerebral cortical peptides, though both have wide noncortical distributions. Within the cerebral cortex of the rat, the highest CCK concentrations are found in the cingulate, pyriform, and entorhinal areas.[41] Substantial CCK concentrations are found in all other brain areas with the exception of the pons, medulla, and cerebellum. Outside of the cerebral cortex the highest concentration is in the caudate nucleus.[41] By histochemical localization, dense collections of CCK-8-containing cells have been found in the periaqueductal gray and in the dorsomedial hypothalamus, with more limited groups of cells in the pyramidal layer of the hippocampus and the dorsal raphe.[43,44] Hokfeldt et al.[45] have shown that in the ventral tegmental area (AlO) and in the pars lateralis of the substantia nigra (zona compacta area), the CCK immunoreactivity coexists in a subpopulation of dopamine-containing neurons projecting to the limbic areas.

However, the need for caution in interpreting immunohistochemical data needs to be emphasized, particularly in view of the finding by Schultzberg et al.[46] that capsaicin can lead to the depletion of CCK-like immunoreactivity in the rat dorsal spinal cord when it is detected by immunochemistry but not when it is measured by radioimmunoassay. One substance that may be responsible for aberrant immunohistochemical results is the molluscan cardioexcitatory peptide (FMRFamide). This neuropeptide has recently been demonstrated to be present in the brain and gut of several vertebrate species including the rat, cow, dog, frog, and chicken.[47,48] Its amino acid sequence (Phe-Met-Arg-Phe-NH_2) is closely related to the C-terminal tetrapeptide of CCK and gastrin (Trp-Met-Asp-Phe-NH_2).

Yamada et al.[49] have established the presence of CCK-like immunoreactivity in frog retina. They also demonstrated in vitro synthesis of CCK in cultured frog retina. The unstimulated synthesis rate was 2.5 fmol/h, which, when compared to the total CCK content in the retina (2.2 pmol/g or approximately 23.6 pmol/retina), suggests a very low basal turnover of this peptide in the retina. The major form of CCK in frog retina appears to be the octapeptide. Immunohistochemistry demonstrated that the CCK was present in amacrine cells that give rise to single primary processes that branch to form a dense plexus of fine processes within the distal two-thirds of the inner plexiform layer. The distribution of CCK in the retina is thus similar to that of somatostatin.[50] The distribution of these multiple peptides in the retina and the relatively easy accessibility of the retina compared to other neural tissue suggests that this will prove to be an exciting area in which to study the pathophysiology of the neuropeptides.

Substantial CCK-like immunoreactivity is also found in the posterior lobe of the pituitary.[51,52] The majority of CCK in the posterior lobe originates in the paraventricular nucleus of the hypothalamus.[51] Preliminary studies by our group in collaboration with Gary Robertson in Chicago have suggested that centrally administered CCK may inhibit the release of arginine vasopressin (antidiuretic hormone) from the posterior pituitary. In contrast to the abundant CCK-like immunoreactivity in the posterior lobe, there is minimal CCK-like activity in the anterior and intermediate lobes of the pituitary. However, immunoreactive gastrins are present in pituitary extracts[51] and have been localized by immunohistochemistry in scattered cells of the anterior lobe and all cells of the intermediate lobes of cat, pig, dog, and human pituitaries.[52,53] The gastrins in the pituitary include a large component as well as G-34 and G-17. The gastrin-containing cells also contain ACTH- and α-MSH-related peptides.[52,53]

3. CHOLECYSTOKININ-CONVERTING ENZYMES

The presence of multiple but well-defined forms of CCK in the brain and gastrointestinal tract suggested the possibility that there may exist specific enzyme(s) capable of converting the large-molecular-weight forms of CCK to the smaller carboxy-terminal fragments that appear to be the predominant forms in the brain. Straus and Yalow[33,54–56] have extracted two cholecystokinin-converting enzymes from porcine and bovine brain. These enzymes convert CCK-33 to CCK-12 and CCK-8, respectively (Fig. 1). They are not simply hormone-specific enzymes but are capable of splitting the synthetic dipeptides Arg-Ile (or other bonds between arginine and neutral amino acids) and Arg-Asp, respectively. The Arg-Ile hydrolase activity is associated with a protein larger in molecular weight than that associated with the Arg-Asp hydrolase activity, and both proteins are of a molecular weight intermediate between albumin and γ-globulin in molecular weight. The activation of cholecystokinin-converting enzyme in the brain is dependent on the formation of a complex with $Mg^{20,52}$ Eng et al.[58] have suggested that the CCK enzyme converting system is much more efficient in the brain than in the gut.

4. CHOLECYSTOKININ RECEPTORS IN THE BRAIN

Specific high-affinity cholecystokinin-binding sites have been reported in rat and guinea pig brain.[59–61] The reported dissociation constant (K_D) is in the range of 0.65 to 2 nM. The highest areas of specific binding for CCK are in the cerebral cortex, olfactory bulb, and caudate nucleus, with appreciable binding also being present in the hippocampus and hypothalamus.[59] These specific binding sites for CCK correlate well with the distribution of immunoreactive CCK in the rat brain. The CCK receptors demonstrated in the rat brain show differences in binding affinities of CCK, gastrin, and desulfated analogues compared to the CCK receptors in the pancreas.[61,62] In addition to the presence of CCK receptors in the brain and pancreas, CCK receptors have been identified in the rat vagus nerve.[63] Of interest is the fact that some of these receptors appear to be being transported distally along the nerve by fast axonal flow. These receptors most probably originate from cell bodies in the brainstem nuclei. At present there have been no *in vitro* correlations of CCK binding activity with CCK biological effects on intact nerve tissue.

Because of the role of CCK as a putative satiety factor (*vide infra*), the effect of fasting on CCK receptors has been investigated. Saito et al.[64] found that 42-h fasting in mice significantly increased CCK binding through an increase in the number of CCK receptors in the olfactory bulb and hypothalamus but not in other brain regions. However, Hays et al.[65] found no difference in CCK binding to hypothalamic and cerebral cortical membranes in 96-h-fasted rats. They did, however, report an increase in specific CCK binding in the hypothalami of both Zucker Fatty Rats and genetically obese mice (C57B1/6J-ob).

Hays et al.[65] also studied the ontogeny of CCK receptors and found that the levels were extremely low in neonatal rats, with an increase to maximal density at 12 days of age followed by a decline to adult levels by day 26. These developmental changes reflect alterations in receptor number with the apparent affinity (K_D) remaining constant. These findings differ from the ontogeny of

CCK peptide in the brain, which increases linearly from low levels in the neonate to adult levels at 25–30 days post-partum.[66] Chronic reserpine treatment leads to an increase in CCK receptor number,[65] which could be secondary to the associated weight loss.[64]

Kainic acid injection into the caudate nucleus reduces CCK receptor density by approximately 75%,[56,57] whereas wide knife cuts severing all caudal afferents reduced caudate CCK receptors by only about 25%.[67] This suggests that some of the receptors in the caudate lie on axons and/or terminals of neurons originating outside the caudate but that the majority of CCK receptors are located on neuronal cell bodies intrinsic to the caudate nucleus. Specific knife cuts suggest that the extrinsic CCK receptors come predominantly from processes arising in the thalamus and possibly from caudal connections arising laterally in the amygdala and ventrooccipital cortical areas. No effect on caudate CCK receptors could be found following lesions of the nigrostriatal dopaminergic tracts, suggesting that the dopamine terminals of this pathway do not contain CCK receptors. As kainic acid lesions of the striatum have been proposed as a model of Huntington's disease,[68] the authors also studied CCK binding in caudate–putamen postmortem specimens from patients with Huntington's disease and other postmortem controls.[56] There was a marked reduction in CCK binding in caudate–putamen samples of Huntington's patients, suggesting a role for CCK in the regulation of extrapyramidal function. This finding is consistent with the report of the coexistence of dopamine and CCK in mesolimbic neurons[45] and the reduction of CCK levels in the basal ganglia in Huntington's disease[69] and suggests that CCK may be involved in the regulation of extrapyramidal motor function.

5. DIRECT EFFECT OF CHOLECYSTOKININ ON NEURONS

Ishibashi *et al.*[70] using the multibarreled microelectrode technique, showed that CCK had a facilitatory effect on 18% of cerebral cortical neurons and an inhibitory effect on 3%. They could show no significant effect on hypothalamic neurons. Stimulation of thalamic neurons occurred in 8% of cases. Phillis and Kirkpatrick[71] found an excitatory effect on 28% of tested cerebral cortical neurons. Oomura *et al.*[72] found enhancement of firing of 22% of cortical neurons tested with CCK and little or no effect on neurons in the venteromedial and lateral hypothalamus. Dodd and Kelly[73] found that CCK-8 and CCK-4 are potent excitants of CA1 pyramidal neurons and evoke a substantial depolarization accompanied by a decrease in membrane resistance. Cholecystokinin also has been shown to have a depolarizing action on motoneurons and dorsal root terminals of the isolated toad spinal cord.[74]

6. NEUROPHARMACOLOGY OF CHOLECYSTOKININ

Cholecystokinin has been reported to produce a variety of effects on the central nervous system[75] (Table II).

Table II
Neuropharmacological Effects of CCK

Hyperglycemia
Hypothermia
Analgesia
Rotational syndrome
Ptosis
Catalepsy
CNS depression
Reduction of exploratory behavior
Satiety factor
Anterior pituitary hormone effects
↑ Growth hormone
↓ Thyrotropin
↑ ACTH and corticosteroids

6.1. Hyperglycemia

Morley and Levine[76] have demonstrated that intracerebroventricular administration of CCK-8 produces hyperglycemia in rats. Hyperglycemia was produced over a dose range of 2.5 to 250 ng. Only the highest concentrations of CCK administered altered glucagon, and none of the concentrations altered insulin levels. Intraventricular CCK-8 did not alter glucose clearance. Adrenalectomy abolished the CCK-8-induced hyperglycemia. Thus, the hyperglycemia induced by CCK appears to be similar to that induced by other brain peptides such as bombesin[77] and β-endorphin.[78]

6.2. Hypothermia

Intraventricular administration of CCK-8 (20–250 ng) induces a dose-related hypothermia in rats at room temperature.[79–81] This hypothermia is antagonized by thyrotropin-releasing hormone and prostaglandin E_2.[80,82] CCK-8 potentiates pentobarbital-induced hypothermia but does not alter ethanol-induced hypothermia.[82] Nonsulfated CCK-8 was ineffective at lowering body temperature, indicating that the sulfated tyrosine in the CCK molecule is indispensable for its hypothermic action. Caerulein produced minimal hypothermia.[82] The degree of hypothermia produced by CCK is less than that produced by bombesin[83] and approximately equivalent to that produced by neurotensin[84] at equimolar doses.

6.3. Analgesia

Parenteral administration of CCK-8 (50–1000 μg/kg) and the frog skin decapeptide caerulein (which displays remarkable homology with CCK-8) produced analgesia in mice in the hot-plate and the writhing test but not in the tail-flick test.[85] This analgesia was antagonized by low doses of naloxone but was not affected by tolerance to morphine. Both the sulfated and unsulfated forms are analgesic after central administration, with the sulfated forms being slightly more potent.[105] In view of the high concentration of CCK in the periaqueductal gray,[43] an area known to be important for the generation of analgesia, the possibility that CCK plays a physiological role in the control of analgesia needs to be considered.

The mechanism by which CCK induces analgesia is uncertain. It is possible that CCK releases endogenous opioid peptides as appears to be the case for substance P[86] and somatostatin.[87] Preliminary results in humans show that caerulein releases β-endorphin.[106] Alternatively, the analgesic activity could be related to the *in vitro* opiate activity of unsulfated CCK-7,[88] which can be explained on the basis of structural similarities between CCK-7 and methionine-enkephalin.[89] In the guinea pig ileum assay, CCK and the opioid peptides have been shown to be physiological antagonists of one another.[90] Of interest in this regard is that morphine causes a dose-dependent decrease in hypothalamic and thalamic CCK-33[30] and that morphine addiction leads to chronically low levels of CCK-33 in the cortex.[91] Caerulein does not bind directly to the opiate receptor.[105] Further studies are indicated to elucidate the facinating mechanism of interaction between CCK and the endogenous opioid peptides. An additional mechanism by which CCK and related analogues may produce analgesia is through the release of calcitonin,[107,108] which is known to be a potent antinociceptive agent.[109]

6.4. Rotational Syndrome

Mann *et al.*[92] have described a syndrome of barrel rotations accompanied by a distorted head and body position, lack of spontaneous motor activity other than rotations, characteristic limb flexion, and extension and loss of some reflexes after central injection of the sulfated C-terminal heptapeptide of cholecystokinin (4×10^{-8} mol in 10 μl).

6.5. Central Nervous System Depression and Exploratory Behaviors

Pharmacological doses (150–1000 μg/kg) of parenteral CCK-8 lead to sedation, catalepsy, ptosis, and inhibition of spontaneous rearing activity in mice.[85,93,94] CCK-8 and caerulein also delay the onset of convulsions following strychnine, pentylenetetrazole, picrotoxin, and bicuculline[95,96] and prolong hexobarbital-induced sleeping time.[93] The methylphenidate-induced stereotyped behavior of biting and gnawing is reduced by CCK-8, suggesting that CCK may antagonize dopaminergic functions.[96] Further evidence for the antagonism of dopaminergic systems by CCK comes from the observation of Itoh and Katsuura[94,97] that marked suppression of the L-DOPA behavioral effects occurs after administration of intraventricular CCK-8 (400–1600 ng) to rats. They also demonstrated that CCK antagonizes the potentiation of L-DOPA by the ergotropic peptide TRH.[110,111] In addition, TRH antagonized the prolongation of pentobarbital- and ethanol-induced sleeping time produced by CCK-8,[97] and CCK-8 suppresses the serotonin-potentiating effect of TRH.[98] Microgram doses of CCK-8, the nonsulfated forms, and caerulein all diminished the duration of β-endorphin-induced catalepsy in rats.[99] CCK-8 and the nonsulfated form impair the acquisition of active avoidance and facilitate extinction of active avoidance after both central and peripheral administration.[100] CCK-8 also increases the latency of passive avoidance behavior.[101]

Crawley and her colleagues[102,103] have carefully investigated the effect of CCK on exploratory behaviors. They have shown that CCK-8 administered intraperitoneally decreased investigation of environmental objects and interactions with a female mouse while increasing the amount of time spent in the corners of the test arena and the duration of nonexploratory pauses.[102] Similar results were obtained in rats.[103] The effects on exploratory behavior are only seen with the sulfated form of CCK. Subsequently, they have shown that vagotomy eliminates the effect of CCK-8 on exploratory behavior, suggesting that CCK-8 exerts its effects on exploratory behavior by stimulating peripheral CCK receptors.

7. EFFECTS OF CHOLECYSTOKININ ON PITUITARY HORMONE RELEASE

Both CCK-8 and CCK variant (CCK-39) have been demonstrated to release growth hormone and inhibit thyrotropin release from an acute anterior pituitary explant system.[112] CCK-8 also induced growth hormone release from monolayer cultures of a GH_3 tumor cell line and reversed the inhibitory effect of somatostatin on growth hormone release in GH_3 cells.[112] Based on these results, it appears possible that CCK is one of a family of physiologically important growth hormone-releasing hormones, although a recent study by Samson et al.[113] failed to duplicate these results using different sources of CCK. Further evidence favoring this concept came from the experiments of Vijayan et al.,[114] who found that injection of CCK (4–500 ng) into the third ventricle of conscious ovariectomized rats led to a significant increase of growth hormone within 15 min of injection. Central CCK also produced a suppression of thyrotropin and luteinizing hormone release. Preliminary studies in humans have shown that CCK-8 can decrease the thyrotropin response to TRH.[115] Central administration of gastrin also leads to an increase in growth hormone and a suppression of basal thyrotropin levels in rats.[116] Intravenous administration of gastrin tetrapeptide to man has been shown to increase growth hormone secretion,[117] as has intravenous administration of CCK-8 to dogs.[118] Thus, a large body of evidence is now available suggesting that the gastrin/CCK family may play a role in the modulation of growth hormone and thyrotropin release from the anterior pituitary. The paucity of CCK in the anterior pituitary and the presence of gastrin[51–53] suggest that small-molecular-weight gastrinlike peptides may play a more important role in the regulation of these two anterior pituitary hormones.

Itoh et al.[119] have found that intracerebroventricular and intraperitoneal injection of a crude CCK extract (Boots Co.) leads to pronounced elevations in plasma corticosterone levels and that this effect is not present in hypophysectomized animals. A similar increase in corticosterone levels has also been reported after central administration of CCK-8[122] and intraperitoneal administration of CCK-33.[123] The possibility exists that the corticosterone release after peripheral administration represents a stress response. The finding that high doses of the CCK extract produced an increase in ACTH release from pituitary explant cultures[119] suggests that this may be a specific effect of CCK. In addition, Porter and Sander[120,121] showed that CCK-8 significantly stimulates basal ACTH output from hemipituitaries. Both CCK-8 and CCK-33 inhibited ACTH stimulation evoked by high-level stimulation of CRF-containing hypothalamic stalk median eminence and enhanced low levels of stimulation. It is well accepted that corticotrophin-releasing factor activity in crude stalk median eminence resides in a multifactor system,[124,125] and available evidence suggests that a CCK-like peptide may be one of these factors. This is particularly attractive in view of the high concentrations of CCK in the intermediate lobe of the pituitary.[41,44] In addition, the simultaneous release of ACTH and GH hormone is well recognized to occur under a number of conditions.[126]

Intravenous administration of CCK to the rat produced a dose-related increase in prolactin levels within 5 min.[114] Malarkey et al.[127] found prolactin release from rat monolayer cultures exposed to CCK-33 and CCK-8 but not from rat (GH_3) cultures or from normal human pituitary and a human prolactinoma. Morley et al.[112] found no release of prolactin from rat pituitary explants. The reason for the conflicting results is not known. A possible explanation for these varying results is that pharmacological concentrations of CCK can on occasion antagonize the inhibitory effect of dopamine on prolactin release. Whether CCK plays a major physiological role in anterior hormone release remains to be determined.

8. CHOLECYSTOKININ AND SATIETY

More than a decade ago, Schally and his associates[128] demonstrated that intravenous and subcutaneous injection of "enterogastrone," a preparation now known to be rich in CCK, caused reduced food intake in mice. In 1971, Glick et al.[129] reported that this satiety effect was not caused by CCK or secretin. However, perusal of their data shows a clear-cut, albeit nonsignificant, tendency for CCK to reduce meal size after intraperitoneal injection. Subsequently, Smith and Gibbs at Cornell University have shown that exogenous CCK produces satiety in the rat and the monkey.[130] The doses administered were in most cases so large that they resulted in circulating levels likely to be in excess of those occurring postprandially. However, they did show that *l*-phenylalanine, a potent releaser of CCK,[131] produces large reductions in food intake, whereas *d*-phenylalanine, which does not release CCK, fails to alter appetite regulation.[132] In addition, administration of the endogenous trypsin inhibitor trasylol decreases food intake in lean and obese rats.[208] Trypsin inhibitors are postulated to increase CCK by inhibiting the negative feedback signal for its release.

These data support the concept that endogenous CCK may act as a satiety factor. The satiety effect of CCK has now been observed in chickens, rabbits, pigs, sheep, rhesus monkeys, lean mice, genetically obese mice and rats, and neurologically obese rats.[133] Recently, Kissileff et al.[134] have shown that CCK-8 infusions in lean men decrease food intake, although previous studies had produced conflicting data.[135,136] Stacher et al.[137] have shown that ceruletide (caerulein) decreases food intake in humans by about 17% and that this effect is coupled with less hunger and activation and prolonged reaction times. A number of recent reviews have evaluated the supporting and conflicting evidence for the role of CCK as a short-term satiety factor.[138-140] A recent study in our laboratory (C. J. Billington, A. S. Levine, and J. E. Morley, unpublished observations) has shown that increasing the length of starvation decreases the satiety potency of CCK, strongly suggesting that nausea is not a factor in the mechanism by which CCK produces satiety. Based on the available evidence, it appears that CCK may be one of a number of factors that plays a role in short-term appetite regulation following the ingestion of a meal.

There is controversy concerning whether or not CCK produces its anorectic effect through a peripheral or central locus of action. It appears that in the rat the major effect of CCK is seen after peripheral administration, whereas in sheep and pigs central CCK appears to play an important role in satiety. At present, it is generally accepted that central administration of CCK has little effect on "normal" feeding in the rat,[141-143] although earlier studies have suggested the possibility of a small effect predominately on intermeal interval.[144-146] Central administration of CCK appears to have a marked effect on stress-induced eating as tested using the mild tail-pinch paradigm.[147-149] This effect is most probably secondary to the hyperglycemia produced following central administration of CCK, as the suppressive effect of CCK on stress-induced feeding is abolished by adrenalectomy, a procedure that reduces the hyperglycemia.[149] It is well recognized that hyperglycemia alters the affinity of the opiate receptor both in vivo and in vitro.[150,151] As tail-pinch-induced feeding represents a nociceptive phenomenon,[152-155] it is perhaps not surprising that it should be modulated by hyperglycemia. In fact, we have shown that tail-pinch-induced feeding is reduced in diabetic mice,[156] further confirming the integral interrelationship of glucose with the opiate receptors involved in the initiation of this phenomenon. McCaleb and Myers[157] have found that microinjections of CCK-8 into the hypothalamus significantly attenuated norepinephrine-induced feeding in the rat. It thus appears clear that CCK can, in a few instances, suppress feeding in the rat after cental injection, though this ability appears to be high state dependent.

In sheep and pigs there now appear to be unequivocal data showing that the major effect of CCK is through a direct effect on the brain. Continuous injections of picomole quantities of CCK-8 into the cerebral ventricles of sheep decrease feeding.[141,158] This suppression is not related to changes in water intake or temperature.[158] Central CCK-8 did, however, reduce the amplitude of rumen contractions to a degree similar to that found during feeding.[158] Recently, cholecystokinin antibody injected into the cerebral ventricles has been found to stimulate feeding in sheep but not in rats.[159] Parrott and Baldwin[160] have shown that intracerebroventricular injection of CCK-8 reduces feeding in food-deprived pigs but not drinking in water-deprived animals. The concentrations of CCK-8 given centrally were less than those given intravenously (but not intraportally, *vide infra*) to reduce feeding.[160,161]

Studies by Smith and his colleagues[162,163] have suggested that peripherally administered CCK-8 acts in the abdomen through vagal fibers and not directly on the brain to produce satiety. They showed that total abdominal vagotomy and selective gastric (but not celiac or hepatic) vagotomy reduced the satiety effect of CCK. These findings were confirmed by Lorenz and Goldman[164] but not by Anika *et al.*[165] Further, Smith *et al.*[163] found no effect of atropine on CCK-induced satiety, suggesting that the effect is mediated by afferent rather than efferent vagal fibers.

The presence of CCK receptors in the vagus nerve[63] provides further circumstantial evidence supporting the possibility that CCK may, in part, work by activating vagal fibers. Peripheral CCK reduces gastric emptying in monkeys, suggesting that this may be the primary effect leading to secondary activation of vagal afferents.[166,167] Further evidence for the attenuation of gastric emptying playing a role in the satiety effect of CCK comes from the attenuation of glucose and insulin responses to meals it produces in free feeding baboons.[168] Based on their monkey studies, Moran and McHugh[167] have suggested that the satiety effect of CCK is an indirect one that is dependent on inhibition of gastric emptying, which results in gastric distension with further food ingestion. In their view, CCK is a link in a chain of physiological elements[169] producing the short-term satiety that cumulates in the appropriate interruption of a meal or a bout of feeding behavior.

The recent discovery that calcitonin is a potent anorectic agent in rats, mice, monkeys, and humans[170-173] coupled with the fact that calcitonin is released following a meal[173] and by CCK[107,108] produces the possibility that CCK may partially produce its effects by release of calcitonin. Calcitonin appears to be a far more active centrally acting satiety factor than CCK.[173,175] Further studies are necessary to evaluate this possibility. In addition, in pigs, Anika *et al.*[161] have demonstrated that intraportal infusion of CCK produces a greater depression of feeding than intraduodenal, intrajugular, or intraventricular infusion. This suggests the possibility that in this species, CCK may activate some hepatic satiety mechanism.[176]

Another potential mechanism of CCK is its interaction with the hedonic qualities (orosensory feedback) of food. Antin, Gibbs, and Smith[177] have shown that the suppressive effect of CCK on sham feeding increases with increasing amounts of pregastric stimulation prior to injection. Gosnell and Hsiao,[178] using the food-seeking operant model, have extended these findings to show that with a total lack of the food-related orosensory stimulation the CCK satiety effect is absent. Waldbillig and O'Callaghan[179] have shown that CCK reduces sucrose-flavored water ingestion while not altering normal water ingestion. This finding suggests that CCK may produce part of its effect on food intake by modulating the behavior-controlling characteristics of taste. The role of flavor in modulating the antidipsogenic effect of another brain peptide, substance P, has been documented in greater detail, showing the potential importance of orosensory mechanisms in appetite regulation.[180] Further evidence for the importance of the orosensory reflex in the pathogenesis of CCK-induced satiety comes from findings in rats with bilateral lesions of the ventromedial hypothalamus (VMH). Rats with VMH lesions fed a normal rat chow display appetite suppression after CCK[163,181] but when fed a highly palatable high-fat diet failed to respond to CCK.[182]

Studies by McCaleb and Myers[157,183] and our laboratory[175,184–186] have focused on the satiety effect of CCK on a variety of pharmacological inducers of feeding. Parenterally administered CCK-8 suppressed norepinephrine-induced feeding[183–185] but not feeding induced by the GABA agonist muscimol.[175] The suppressive effect of CCK-8 in the tail-pinch model is reversed by a number of well-known appetite stimulants, e.g., [D-Ala]methionine enkephalin, diazepam, muscimol, and propranolol.[186] Based on these results, it appears that the end result of CCK administration is most probably to inhibit the stimulatory effect of α agonists on GABA-induced feeding. The vagal afferent input to the medial hypothalamus[187] represents a possible pathway by which the action of CCK may be mediated.

Before we conclude the section on CCK and appetite, it is necessary to briefly comment on the confusion introduced by the measurement of CCK in the brain. Early studies reported that genetically obese mice (ob/ob) had decreased brain immunoreactive CCK compared to their lean littermates[188] and that starvation decreased CCK levels.[189] A number of subsequent studies have failed to confirm these findings.[30,190–192] Similar discrepancies exist as to potential changes in CCK receptors during starvation (*vide supra*). In view of the multiple problems with CCK assays and the lack of availability of simple techniques to measure synthesis and release of brain peptides, further elucidation of these discrepancies must await the introduction of suitable technical advances.

To conclude this section, it appears that the major effect of CCK on satiety is mediated through activation of gastric vagal afferents, which leads to the inhibition of the stimulatory effect of the α agonists on GABA-induced feeding (Fig. 2). This effect of CCK is modulated by orosensory inputs. Brain CCK may produce satiety indirectly as an epiphenomenon secondary to its effects on glucose metabolism and gastric motility. The possibility that the satiety effect of CCK may also be mediated, in part, by calcitonin release and/or activation of hepatic satiety mechanisms needs to be further investigated (Table III). Some of these mechanisms may be species specific. A final twist to be added to the saga of CCK is the recent finding by Kraly *et al.*[193] that at certain times of the day CCK may actually enhance appetite. We (C. J. Billington, J. E. Morley, and A. S. Levine, unpublished observations) have also noted a mild enhancement of feeding following low doses of CCK under certain circumstances. In addition, tolerance to the satiety effect of CCK develops extremely rapidly during constant infusions.[194]

9. MONOAMINES AND CHOLECYSTOKININ

Because a subpopulation of mesolimbic dopamine neurons contain CCK-like immunoreactivity,[195] a number of groups have investigated the role of CCK

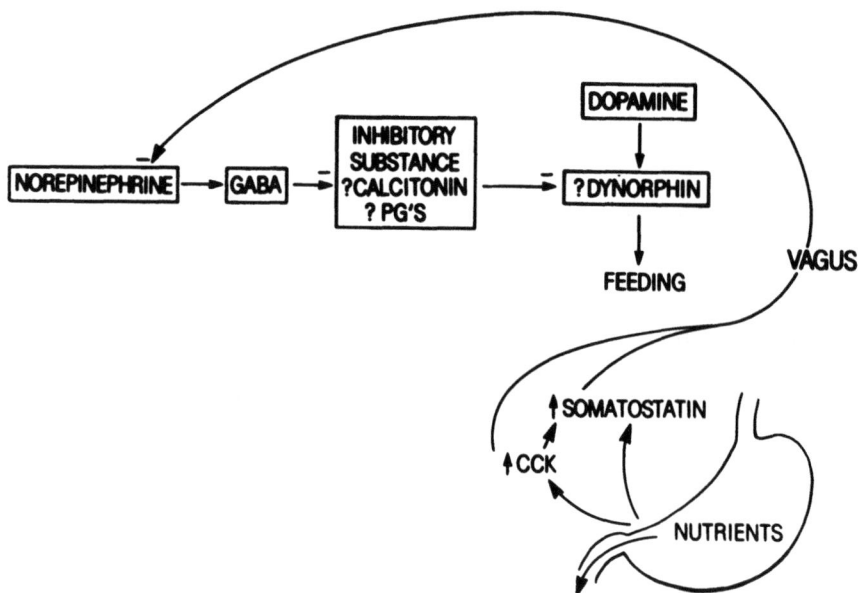

Fig. 2. Putative model of the mechanism by which peripheral CCK inhibits feeding in the rat. Nutrients cause the release of CCK, which then either directly or via somatostatin release stimulates the vagus. Vagal projections to the paraventricular nucleus then inhibit the firing of norepinephrine-responsive neurons, resulting in a decrease of the GABA inhibitory input. This leads to disinhibition of the major inhibitory substance. Increased activity of the inhibitory substance decreases the dopamine–dynorphin (opiate) feeding drive in the area of the lateral hypothalamus, leading to satiety.

on dopamine metabolism in the central nervous system. Fuxe et al.[196] found that CCK reduced dopamine turnover in the nucleus caudatus and in the anterior nucleus accumbens of the rat. Kovacs et al.[197] found that CCK-8 results in an increased release and disappearance of striatal dopamine in rats. Telegdy[198] has found that CCK-8 injected into the lateral ventricle increased dopamine levels in the hypothalamus, mesencephalon, and septum while decreasing it in the amygdala and striatum. Norepinephrine levels were increased in the mesencephalin, amygdala, septum, and striatum. In general, they found that CCK antiserum produced the opposite effects on dopamine and norepinephrine levels.[199] Cholecystokinin tended to increase serotonin levels.[198] Katsuura et al.[200] found a decrease in dopamine content in α-methyl-p-tyrosine-pretreated animals after intraperitoneal injection of a crude commercial preparation of CCK. They also noticed an increased norepinephrine turnover rate after CCK. Myers and McCaleb[183] have found increased norepinephrine levels in the brain after parenteral administration of CCK. These data would be compatible with the hypothesis that CCK is a competitive antagonist of norepinephrine (see Section 8).

The effects on dopamine metabolism are compatible with the concept of CCK playing a role as a physiological dopamine antagonist in some areas.

Table III
Mechanisms Incriminated in Satiety
Caused by CCK

Peripheral effects
  Activation of vagal afferents
  Inhibition of stomach emptying
  Interaction with orosensory feedback
  ? Calcitonin release
  ? Activation of hepatic satiety system
CNS effects
  Hyperglycemia
  Altered gastric motility
  Direct inhibition of norepinephrine-induced
    feeding

Few data are available on the effect of monoamines on brain CCK levels. The serotonin precursor L-tryptophan has been demonstrated to decrease brain CCK levels acutely.[30] Serotonin is a well-recognized satiety substance,[139] and it is possible that its anorectic action may involve the release of CCK. Serotonin also results in ACTH[201] and growth hormone[126] release, and there are other effects that could be mediated by CCK release (see Section 7).

10. CONCLUSION

Cholecystokinin is now firmly established as a central and peripheral neurotransmitter or neuromodulator. Evidence establishing CCK as a neurotransmitter includes (1) its regional distribution with localization within neuronal cell bodies and axons (*vide supra*), (2) the demonstration that it can be synthesized in neuronal tissue,[202,203] (3) the fact that it can be released from nerve tissue by depolarizing stimuli *in vitro*,[204,205] (4) the presence of specific high-affinity receptors for CCK in the brain (*vide supra*), and (5) the finding that it can activate isolated neurons (*vide supra*). Evidence for its role as a peripheral neurotransmitter derives from the studies described above linking it to the vagus nerve and to the studies by Rehfeld *et al.*[206] showing its presence in pancreatic nerve fibers. In view of the recent finding by Passaro *et al.*[207] that intraventricularly administered CCK rapidly appears in the peripheral circulation, some caution needs to be observed when claiming that effects seen after central administration of a high dose of CCK are caused by central actions and are not secondary to peripheral leakage.

With the discovery of the high concentrations of CCK in the central nervous system, CCK began a rapid ascent from an orphan gut hormone to a star in the rapidly growing galaxy of neuropeptides. The recent findings that a number of the central effects of CCK are in fact secondary to peripheral stimulation of the vagus suggest that before the CCK saga is finally put in perspective we may well have to descend to the gut again. Finally, the high concentrations of CCK in the cerebral cortex suggest that future studies will produce further surprises concerning the physiological role of the gallbladder-contracting hormone that came of age with the discovery of its wide distribution in the central nervous system.

REFERENCES

1. Mutt, V., and Jorpes, J. E., 1971, *Biochem. J.* **125**:57–58.
2. Okada, S., 1914/15, *J. Physiol. (Lond.)* **49**:457–482.
3. Ivy, A. C., and Oldberg, E., 1928, *Am. J. Physiol.* **86**:599–613.
4. Harper, A. A., and Vass, C. C. N., 1941, *J. Physiol. (Lond.)* **99**:415–435.
5. Harper, A. A., and Raper, H. S., 1943, *J. Physiol. (Lond.)* **102**:115–125.
6. Mutt, V., 1980, *Gastrointestinal Hormones* (G. B. Jerzy, ed.), Raven Press, New York, pp. 169–221.
7. Walsh, J. H., 1978, *Gastrointestinal Disease*, Volume 1 (M. H. Sleisenger and J. S. Fordtran, eds.), W. B. Saunders, Philadelphia, pp. 107–155.
8. Lin, T.-M., 1975, *Gastroenterology* **69**:1006–1022.
9. Resin, H., Stern, D. H., Sturdevant, R. A. L., and Isenberg, J. I., 1973, *Gastroenterology* **64**:946–950.
10. Debas, H. T., Farooq, O., and Grossman, M. I., 1975, *Gastroenterology* **68**:1211–1217.
11. Gutierrez, J. G., Chey, W. Y., and Dinoso, V. P., 1974, *Gastroenterology* **67**:35–41.
12. Debas, H. T., and Grossman, M. I., 1978, *Digestion* **9**:469–481.
13. Meyer, J. H., Spingola, L. J., and Grossman, M. I., 1971, *Am. J. Physiol.* **221**:742–747.
14. Preshaw, R. M., and Grossman, M. I., 1965, *Gastroenterology* **48**:36–44.
15. Gillespie, I. E., and Grossman, M. I., 1964, *Gut* **5**:342–345.
16. Jones, R. S., and Grossman, M. I., 1970, *Am. J. Physiol.* **219**:1014–1018.
17. Pfeiffer, E. F., 1969, *Excerpta Med. Int. Conf. Ser.* **172**:419–424.
18. Glick, Z., Baile, C. A., and Mayer, J., 1970, *Endocrinology* **86**:927–931.
19. Unger, R. H., Ketterer, H., Dupre, J., and Eisentraut, A. M., 1967, *J. Clin. Invest.* **46**:630–645.
20. Mainz, D. L., Black, O., and Webster, P. D., 1973, *J. Clin. Invest.* **52**:2300–2304.
21. Thulin, L., and Samnegard, H., 1978, *Acta Chir. Scand. [Suppl]* **482**:73–74.
22. Moritz, M., Finkelstein, G., Mechkinpour, H., Fingerhut, J., and Lorber, S., 1973, *Gastroenterology* **64**:76–81.
23. Vanderhaegen, J. J., Signeau, J. C., and Gepts, W., 1975, *Nature* **257**:604–605.
24. Dockray, G. J., 1977, *Nature* **270**:359–361.
25. Mueller, J. E., Straus, E., and Yalow, R. S., 1977, *Proc. Natl. Acad. Sci. U.S.A.* **74**:3035–3041.
26. Rehfeld, J. F., 1977, *Acta Pharmacol. Toxicol. (Kbh.)* **41**:24–35.
27. Rehfeld, J. F., 1978, *J. Biol. Chem.* **253**:4022–4030.
28. Rehfeld, J. F., Goltermann, N., Larsson, L.-I., Emson, P. M., and Lee, C. M., 1979, *Fed. Proc.* **38**:2325–2329.
29. Straus, E., and Yalow, R. S., 1978, *Proc. Natl. Acad. Sci. U.S.A.* **75**:486–489.
30. Lamers, C. B., Morley, J. E., Poitras, P., Sharp, B., Carlson, H. E., Hershman, J. M., and Walsh, J. H., 1980, *Am. J. Physiol.* **239**:E232–E235.
31. Larsson, L.-I., and Rehfeld, J. F., 1979, *Brain Res.* **165**:201–218.
32. Ryder, S. W., Eng, J., Straus, E., and Yalow, R. S., 1981, *Proc. Natl. Acad. Sci. U.S.A.* **78**:3892–3896.
33. Straus, E., Ryder, S. W., Eng, J., and Yalow, R. S., 1981, *Recent Prog. Horm. Res.* **37**:447–475.
34. Varro, A., Berger, Z., Hajnal, F., Lonovies, J., and Pap, A., 1981, *Scand. J. Gastroenterol.* **16**:611–614.
35. Dockray, G. J., Gregory, R. A., Hutchinson, J. B., Harris, J. I., and Runswick, M. J., 1978, *Nature* **274**:711–713.

36. Holmquist, A. L., Dockray, G. J., Rosenquist, G. L., and Walsh, J. H., 1979, *Gen. Comp. Endocrinol.* **37**:474–481.
37. Duve, H., and Thorpe, A., 1981, *Gen. Comp. Endocrinol.* **43**:381–391.
38. Kramer, K. J., Spiers, R. D., and Childs, C. N., 1977, *Gen. Comp. Endocrinol.* **32**:423–426.
39. Grimmelikhuijzen, C. J. P., Sundler, F., and Rehfeld, J. F., 1980, *Histochemistry* **69**:61–68.
40. Larsson, L.-I., and Rehfeld, J. F., 1977, *Nature* **269**:335–338.
41. Beinfeld, M. C., Meyer, D. K., Eskay, R. L., Jensen, R. T., and Brownstein, M. J., 1981, *Brain Res.* **212**:51–57.
42. Besson, J., Rotsztejn, W., Labruthe, M., Epelbaum, J., Beaudet, A., Kordon, C., and Rosselin, G., 1979, *Brain Res.* **165**:79–85.
43. Innis, R. B., Correa, F. M. A., Uhl, G. R., Schneider, B., and Snyder, S. H., 1979, *Proc. Natl. Acad. Sci. U.S.A.* **76**:521–525.
44. Beinfeld, M. C., Meyer, D. K., and Brownstein, M. J., 1980, *Nature* **288**:376–378.
45. Hokfeldt, T., Skirboll, L., Rehfeld, J. F., Goldstein, M., Markey, K., and Dann, O., 1980, *Neuroscience* **12**:2093–2124.
46. Schultzberg, M., Dockray, G. J., and Williams, R. G., 1982, *Brain Res.* **235**:198–204.
47. Dockray, G. J., Vaillant, C., and Williams, R. G., 1981, *Nature* **293**:656–657.
48. Weber, E., Evans, C. J., Samuelsson, S. J., and Barchas, J. D., 1981, *Science* **214**:1248–1251.
49. Yamada, T., Brecha, N., Rosenquist, G., and Bassinger, S., 1982, *Peptides* **2**:93–97.
50. Yamada, T., Marshak, D., Bassinger, S., Walsh, J., Morley, J. E., and Stell, W., 1980, *Proc. Natl. Acad. Sci. U.S.A.* **77**:1691–1695.
51. Rehfeld, J. F., 1978, *Nature* **271**:771–773.
52. Larsson, L.-I., and Rehfeld, J. F., 1981, *Science* **213**:768–770.
53. Vanderhaegen, J. J., 1981, *Horm. Cell Regul.* **5**:149–157.
54. Straus, E., Malesci, A., and Yalow, R. S., 1978, *Proc. Natl. Acad. Sci. U.S.A.* **75**:5711–5714.
55. Malesci, A., Straus, E., and Yalow, R. S., 1980, *Proc. Natl. Acad. Sci. U.S.A.* **77**:597–599.
56. Ryder, S. W., Straus, E., and Yalow, R. S., 1980, *Proc. Natl. Acad. Sci. U.S.A.* **77**:3669–3671.
57. Lee, T. H., and Lee, M. S., 1980, *Fed. Proc.* **39**:1864;1353A.
58. Eng, J., Shilna, Y., Straus, E., and Yalow, R. S., 1982, *Endocrinology* **110**:612A.
59. Saito, A., Sankaran, H., Goldfine, I. D., and Williams, J. A., 1980, *Science* **208**:1155–1156.
60. Hays, S. E., Beinfeld, M. C., Jensen, R. T., Goodwin, F. K., and Paul, S. M., 1980, *Neuropeptides* **1**:53–62.
61. Innis, R. B., and Snyder, S. H., 1980, *Eur. J. Pharmacol.* **65**:123–124.
62. Jensen, R. T., and Gardner, J. D., 1980, *Polypeptide Hormones* (R. F. Beers, Jr. and E. G. Bassett, eds.), Raven Press, New York, pp. 295–412.
63. Zarbin, M. A., Wamsley, J. K., Innis, R. B., and Kuhar, M. J., 1981, *Life Sci.* **29**:697–705.
64. Saito, A., Williams, J. A., and Goldfine, I. D., 1981, *Nature* **289**:599–600.
65. Hays, S. E., Goodwin, F. K., and Paul, S. M., 1981, *Peptides* **2**:21–26.
66. Noyer, M., Bai, N. D., Deschodt-Lanckman, M., Robberecht, P., Noussen, M.-C., and Christophe, J., 1980, *Life Sci.* **27**:2197–2203.
67. Hays, S. E., Meyer, D. K., and Paul, S. M., 1981, *Brain Res.* **219**:208–213.
68. Mason, S. T., and Fibiger, H. C., 1978, *Brain Res.* **155**:313–329.
69. Emson, P. C., Rehfeld, J. F., Langevin, H., and Rosor, M., 1980, *Brain Res.* **198**:497–500.
70. Ishibashi, S., Oomura, Y., Okajima, T., and Shibata, S., 1979, *Physiol. Behav.* **23**:401–403.
71. Phillis, J. W., and Kirkpatrick, J. R., 1980, *Can. J. Physiol. Pharmacol.* **58**:612–623.
72. Oomura, Y., Ohta, M., Kita, H., Ishibashi, S., and Okajima, T., 1978, *Iontophoresis and Transmitter Mechanisms in the Mammalian Central Nervous System*, Elsevier, New York, pp. 120–123.
73. Dodd, J., and Kelly, J. S., 1981, *Brain Res.* **205**:337–350.
74. Phillis, J. W., and Kirkpatrick, J. R., 1979, *Can. J. Physiol. Pharmacol.* **57**:887–889.
75. Morley, J. E., 1982, *Life Sci.* **30**:479–493.
76. Morley, J. E., and Levine, A. S., 1981, *Life Sci.* **28**:2187–2190.
77. Brown, M., Tache, Y., and Fisher, D., 1979, *Endocrinology* **105**:660–665.

78. Brown, M., and Vale, W., 1978, *Endocrinology* **102**:779A.
79. Morley, J. E., and Levine, A. S., 1980, *Clin. Res.* **28**:721A.
80. Katsuura, G., Hirota, R., and Itoh, S., 1981, *Experientia* **37**:60.
81. Morley, J. E., Levine, A. S., and Lindblad, S., 1981, *Eur. J. Pharmacol.* **74**:249–251.
82. Itoh, S., and Katsuura, G., 1982, *Jpn. J. Physiol.* **32**:145–148.
83. Brown, M., Rivier, J., and Vale, W., 1977, *Science* **196**:998–999.
84. Yehuda, S., and Kastin, A. J., 1980, *Neurosci. Biobehav. Rev.* **4**:459–471.
85. Zetler, G., 1980, *Neuropharmacology* **19**:415–422.
86. Krivoy, W. A., Kroeger, D. C., and Zimmerman, E., 1977, *Psychoneuroendocrinology* **2**:43–52.
87. Rezek, M., Havlicek, V., Leybin, L., Labella, F. S., and Friesen, H., 1978, *Can. J. Physiol. Pharmacol.* **56**:227–231.
88. Schiller, P. W., Lipton, A., Horrobin, D. F., and Bodanszky, M., 1978, *Biochem. Biophys. Res. Commun.* **85**:1332–1338.
89. Schiller, P. W., Natarajan, S., and Bodanszky, M., 1978, *Int. J. Peptide Protein Res.* **12**:139–142.
90. Zetler, G., 1979, *Eur. J. Pharmacol.* **60**:67–77.
91. Morley, J. E., Yamada, T., Walsh, J. H., Lamers, C. B., Wong, H., Shulkes, A., Damassa, D. A., Carlson, H. E., and Hershman, J. M., 1980, *Life Sci.* **26**:2239–2244.
92. Mann, J. F. E., Boucher, R., and Schiller, P. W., 1980, *Pharmacol. Biochem. Behav.* **13**:125–127.
93. Zetler, G., 1980, *Eur. J. Pharmacol.* **65**:133–139.
94. Itoh, S., and Katsuura, G., 1982, *Jpn. J. Pharmacol.* **32**:83–91.
95. Zetler, G., 1980, *Eur. J. Pharmacol.* **65**:297–300.
96. Zetler, G., 1981, *Neuropharmacology* **20**:277–283.
97. Katsuura, G., and Itoh, S., 1982, *Jpn. J. Physiol.* **32**:83–91.
98. Itoh, S., and Katsuura, G., 1982, *Jpn. J. Physiol.* **32**:145–148.
99. Itoh, S., and Katsuura, G., 1981, *Eur. J. Pharmacol.* **74**:381–384.
100. Fekete, M., Szabo, A., Balázs, M., Penke, B., and Telegdy, G., 1981, *Acta Physiol. Acad. Sci. Hung.* **58**:39–45.
101. Fekete, M., Penke, B., and Telegdy, G., 1981, *Neuropeptides* **1**:301–307.
102. Crawley, J. N., Hays, S. E., Paul, S. M., and Goodwin, F. K., 1981, *Physiol. Behav.* **27**:407–411.
103. Crawley, J. N., Hays, S. E., O'Donohue, T. L., Paul, S. M., and Goodwin, F. K., 1981, *Peptides* **2**:123–129.
104. Crawley, J. N., Hays, S. E., and Paul, S. M., 1981, *Eur. J. Pharmacol.* **73**:379–380.
105. de Castiglione, R., 1981, *Peptides* **2**:61–63.
106. Basso, N., Materia, A., D'Intinosante, V., Bianchi, E., and Speranza, V., 1980, *Regul. Peptides* **1**:58.
107. Case, A. D., Bruce, J. B., Boelkins, J., Kenny, A. D., Conaway, H., and Arast, C. S., 1971, *Endocrinology* **89**:262–271.
108. Passeri, M., Carapezzi, C., Seccato, S., Monica, C., Strozzi, D., and Palummeri, E., 1975, *Experientia* **31**:1234–1235.
109. Braga, P., Ferri, S., Santagostino, A., Olgiati, V. R., and Pecile, A., 1978, *Life Sci.* **22**:971–978.
110. Metcalf, G., and Dettmar, P. W., 1981, *Lancet* **1**:586–589.
111. Morley, J. E., 1979, *Life Sci.* **25**:1539–1550.
112. Morley, J. E., Melmed, S., Briggs, J., Carlson, H. E., Hershman, J. M., Solomon, T. E., Lamers, C., and Damassa, D. A., 1979, *Life Sci.* **25**:1201–1206.
113. Samson, W. K., Koenig, J. I., and McCann, S. M., 1982, *Fed. Proc.* **41**:1488A.
114. Vijayan, E., Samson, W. K., and McCann, S. M., 1979, *Brain Res.* **172**:295–302.
115. Morley, J. E., 1981, *Endocrine Rev.* **2**:396–436.
116. Vijayan, E., Samson, W. K., and McCann, S. M., 1978, *Life Sci.* **23**:2225–2232.
117. Domschke, W., Lux, G., and Domschke, S., 1980, *N. Eng. J. Med.* **303**:458.
118. Baranetsky, N. G., Modlin, I., Morley, J. E., Lamers, C., and Carlson, H. E., 1980, *Clin. Res.* **28**:719A.

119. Itoh, S., Hirota, R., Katsuura, G., and Odaguchi, K., 1979, *Life Sci.* **25**:1725–1730.
120. Porter, J. R., and Sander, L. D., 1981, *Fed. Proc.* **40**:1285A.
121. Porter, J. R., and Sander, L. D., 1981, *Regul. Peptides* **2**:245–252.
122. Fekete, M., Boker, M., Penke, B., Kovais, K., and Telegdy, G., 1981, *Neurochem. Int.* **3**:165–169.
123. Sander, L. D., and Porter, J. R., 1982, *Fed. Proc.* **41**:1111.
124. Gillies, G., and Lowry, P., 1979, *Nature* **278**:463–464.
125. Yasuda, N., Greer, M. A., and Aizawa, T., 1982, *Endocrine Rev.* **3**:123–140.
126. Martin, J. B., Brazeau, P., Tannenbaum, G. S., Willoughby, J. O., Pelbaum, J. E., Terry, L. C., and Durand, D., 1978, *The Hypothalamus*, Raven Press, New York.
127. Malarkey, W. B., O'Dorisio, T. M., Kennedy, M., and Cataland, S., 1981, *Life Sci.* **28**:2489–2495.
128. Schally, A. V., Redding, T. W., Lucien, H. W., and Meyer, J., 1967, *Nature* **157**:210–212.
129. Glick, Z., Thomas, D. W., and Mayer, J., 1971, *Physiol. Behav.* **6**:5–8.
130. Gibbs, J., Young, R. C., and Smith, G. P., 1973, *Nature* **245**:323–325.
131. Meyer, J. H., and Grossman, M. I., 1972, *Am. J. Physiol.* **222**:1058–1065.
132. Gibbs, J., and Smith, G. P., 1977, *Am. J. Clin. Nutr.* **30**:758–761.
133. Smith, G. P., Gibbs, J., Jerome, C., Pi-Sunyer, F. X., Kissileff, H. R., and Thornton, J., 1981, *Peptides* **2**:57–59.
134. Kissileff, H. R., Pi-Sunyer, F. X., Thornton, J., and Smith, G. P., 1981, *Am. J. Clin. Nutr.* **34**:154–160.
135. Sturdevant, R., and Goetz, H., 1976, *Nature* **261**:713–715.
136. Greenway, F. L., and Bray, G. A., 1977, *Life Sci.* **21**:769–771.
137. Stacher, G., Steinringer, H., Schmierer, G., Schneider, C., and Winklehner, S., 1982, *Peptides* **3**:607–612.
138. Mueller, K., and Hsiao, S., 1978, *Neurosci. Biobehav. Rev.* **2**:79–87.
139. Morley, J. E., 1980, *Life Sci.* **27**:355–368.
140. Smith, G. P., and Gibbs, J., 1981, *Neurosecretion and Brain Peptides* (J. B. Martin, ed.), Raven Press, New York, pp. 389–395.
141. Della-Fera, M. A., and Baile, C. A., 1979, *Science* **206**:471–473.
142. Grinker, J. A., Schneider, B. S., Ball, G., Cohen, A., Strohmayer, A., and Hirsh, J., 1980, *Fed. Proc.* **39**:501, 1234A.
143. Smith, G. P., and Gibbs, J., 1979, *Progress in Psychobiology and Physiological Psychology*, Volume 8 (J. M. Sprague and A. N. Epstein, eds.), Academic Press, New York, pp. 224–230.
144. Maddison, S., 1977, *Physiol. Behav.* **19**:819–824.
145. Stern, J. J., Cudillo, C. A., and Kruper, J., 1976, *J. Comp. Physiol. Psychol.* **90**:484–490.
146. Stern, J. J., and Page, P., 1977, *Psychol. Rep.* **40**:3–8.
147. Nemeroff, C. B., Osbahr, A. J., Bissette, G., Jahnke, G., Lipton, M. A., and Prange, A. J., 1978, *Science* **200**:793–794.
148. Morley, J. E., and Levine, A. S., 1980, *Clin. Res.* **28**:721A.
149. Levine, A. S., and Morley, J. E., 1981, *Regul. Peptides* **2**:353–357.
150. Morley, J. E., Levine, A. S., Brown, D. M., and Handwerger, B. S., 1981, *Soc. Neurosci. Abstr.* **7**:854.
151. Levine, A. S., Morley, J. E., Brown, D. M., and Handwerger, B. S., 1982, *Physiol. Behav.* **28**:987–989.
152. Rowland, N. E., and Marques, D. M., 1980, *Appetite* **1**:225–228.
153. Levine, A. S., and Morley, J. E., 1982, *Appetite* **3**:135–138.
154. Morley, J. E., and Levine, A. S., 1981, *Science* **214**:1150–1151.
155. Levine, A. S., Wilcox, G. L., Grace, M., and Morley, J. E., 1982, *Physiol. Behav.* **28**:959–962.
156. Levine, A. S., Morley, J. E., Wilcox, G., Brown, D. M., and Handwerger, B. S., 1982, *Physiol. Behav.* **28**:39–43.
157. McCaleb, M. L., and Myers, R. D., 1980, *Peptides* **1**:47–49.
158. Della-Fera, M. A., and Baile, C. A., 1980, *Physiol. Behav.* **24**:943–950.
159. Della-Fera, M. A., Baile, C. A., Schneider, B. S., and Grinker, J. A., 1981, *Science* **212**:687–689.

160. Parrott, R. F., and Baldwin, B. A., 1981, *Physiol. Behav.* **26**:419–422.
161. Anika, S. M., Houpt, T. R., and Houpt, K. A., 1981, *Am. J. Physiol.* **240**:R310–318.
162. Smith, G. P., and Cushin, B. J., 1978, *Soc. Neurosci. Abstr.* **4**:180.
163. Smith, G. P., Jerome, C., Cushin, B. J., Eterno, R., and Simansky, K. J., 1981, *Science* **213**:1036–1037.
164. Lorenz, D. N., and Goldman, S. A., 1978, *Soc. Neurosci. Abstr.* **4**:178.
165. Anika, S. M., Houpt, T. R., and Houpt, K. A., 1977, *Physiol. Behav.* **19**:761–766.
166. Moran, T. H., and McHugh, P. R., 1979, *Fed. Proc.* **38**:1131.
167. Moran, T. H., and McHugh, P. R., 1982, *Am. J. Physiol.* **242**:R491–497.
168. Woods, S. C., and Stein, L. J., 1982, *Diabetes* **31**:121A.
169. McHugh, P. R., 1979, *Johns Hopkins Med. J.* **144**:147–155.
170. Freed, W. J., Perlow, M. J., and Wyatt, R. D., 1979, *Science* **206**:471–473.
171. Perlow, M. J., Freed, W. J., Carman, J. S., and Wyatt, R. J., 1980, *Pharmacol. Biochem. Behav.* **12**:609–612.
172. Morley, J. E., and Levine, A. S., 1981, *Clin Res.* **29**:297A.
173. Levine, A. S., and Morley, J. E., 1981, *Brain Res.* **222**:187–191.
174. Talmage, R. V., Doppelt, S. H., and Cooper, C. W., 1975, *Proc. Soc. Exp. Biol. Med.* **149**:855–859.
175. Morley, J. E., Levine, A. S., and Kneip, J., 1981, *Life Sci.* **29**:1213–1218.
176. Sawchenko, P. E., and Freidman, M. I., 1979, *Am. J. Physiol.* **236**:R5–20.
177. Antin, J., Gibbs, J., and Smith, G. P., 1978, *Physiol. Behav.* **20**:67–70.
178. Gosnell, B., and Hsiao, S., 1981, *Physiol. Behav.* **27**:153–156.
179. Waldbillig, R. J., and O'Callaghan, 1980, *Physiol. Behav.* **25**:25–30.
180. Morley, J. E., Levine, A. S., and Murray, S. S., 1981, *Brain Res.* **226**:334–338.
181. Kulkosky, P. J., Beckenridge, C., Krinsky, R., and Woods, S. C., 1976, *Behav. Biol.* **18**:227–234.
182. Krinsky, R., Lotter, E. C., and Woods, S. C., 1979, *Physiol. Psychol.* **7**:67–69.
183. Myers, R. D., and McCaleb, M. L., 1981, *Neuroscience* **6**:645–655.
184. Morley, J. E., Levine, A. S., Murray, S. S., and Kneip, J., 1981, *Pharmacol. Biochem. Behav.* **16**:225–228.
185. Morley, J. E., Levine, A. S., Murray, S. S., and Kneip, J., 1981, *Endocrinology* **108**:884A.
186. Levine, A. S., and Morley, J. E., 1981, *Peptides* **2**:261–264.
187. Ricardo, J. A., and Koh, E. T., 1978, *Brain Res.* **153**:1–26.
188. Straus, E., and Yalow, R. S., 1979, *Science* **203**:69–70.
189. Straus, E., and Yalow, R. S., 1980, *Life Sci.* **26**:969–970.
190. Schneider, B. S., Monahan, J. W., and Hirsch, J., 1979, *J. Clin. Invest.* **64**:1348–1356.
191. Oku, J., Glick, Z., Shimonura, Y., Inoue, S., Bray, G. A., and Walsh, J., 1980, *Clin. Res.* **28**:25A.
192. Ho, P., and Hansky, J., 1979, *Gastroenterology* **76**:1155.
193. Kraly, F. S., 1981, *Appetite* **2**:177–191.
194. Crawley, J. N., and Beinfield, M. C., 1983, *Nature* **302**:703–706.
195. Hokfeldt, T., Rehfeld, J. F., Skirboll, L., Ivemark, B., Goldstein, M., and Markey, K., 1980, *Nature* **285**:476–478.
196. Fuxe, K., Andersson, K., Locatelli, V., Agnati, L. F., Hokfeldt, T., Skirboll, L., and Mutt, V., 1980, *Eur. J. Pharmacol.* **67**:329–331.
197. Kovacs, G. L., Szabo, G., Penke, B., and Telegdy, G., 1981, *Eur. J. Pharmacol.* **69**:313–319.
198. Telegdy, G., 1980, *Acta Physiol. Acad. Sci. Hung.* **55**:273–281.
199. Fekete, M., Kadar, T., and Telegdy, G., 1981, *Acta Physiol. Acad. Sci. Hung.* **57**:177–183.
200. Katsuura, G., Hirota, R., and Itoh, S., 1980, *Jpn. J. Pharmacol.* **30**:811–814.
201. Ganong, W. F., 1971, *Excerpta Medica Int. Cong. Ser.* **210**:61.
202. Goltermann, N. R., Rehfeld, J. F., and Roigaard-Petersen, H., 1980, *J. Biol. Chem.* **255**:6181–6185.
203. Goltermann, N. R., Rehfeld, J. F., and Roigaard-Petersen, H., 1980, *J. Neurochem.* **35**:479–483.
204. Dodd, P. R., Edwardson, J. A., and Dockray, G. J., 1980, *Regul. Peptides* **1**:17–29.

205. Pinget, M., Straus, E., and Yalow, R. S., 1979, *Life Sci.* **25**:339–342.
206. Rehfeld, J. F., Larsson, L.-I., Goltermann, N. R., Schwartz, T. W., Holst, J. J., Jensen, S. L., and Morley, J. S., 1980, *Nature* **284**:33–38.
207. Passaro, E., Dabas, H., Oldendorf, W., and Yamada, T., 1982, *Brain Res.* **241**:338–340.
208. Perkin, S. R., and McLaughlin, C. L., 1982, *Fed. Proc.* **41**:388,692.

6

β-Lipotropin

László Gráf

1. INTRODUCTION

In 1964, Birk and Li[1] reported the discovery of a new polypeptide of the pituitary gland. Because of the considerable lipolytic activity of this polypeptide on rabbit adipose tissue *in vitro*, Li[2] named it lipotropic hormone (LPH) or lipotropin. Fourteen years later, when Dr. Li was asked about his reason for the designation, he explained[3]: "The designation of a new natural substance is often a matter of convenience. Lipotropic hormone had been named by me on the basis of its biological effect first observed. If the melanocyte-stimulating activity of lipotropin had been noticed first, it would have been called γ-melanotropin."

Shortly after the isolation of β-LPH, the largest molecular form of ovine lipotropins, the amino acid sequence of the polypeptide was elucidated,[4] revealing that it contains the complete structure of β-melanotropin (β-MSH) in residues 41–58 (Fig. 1). This structural feature rather than the designation itself directed interest, between 1964 and 1975, to some biological properties of the hormone also shown by melanotropins. Studies with natural and synthetic β-LPH fragments have indeed shown that the active center of the molecule for both lipolytic and melanocyte-stimulating effects resides in region 41–58 of the sequence.[5–7]

This biased view about the possible biological significance of lipotropins is well illustrated by the historical fact that although a lipolytically active 58-residue natural β-LPH fragment (designated as γ-LPH; Fig. 1) was isolated and sequenced from both ovine[8] and porcine pituitaries,[9] no attention was paid to the complementary C-terminal fragment of β-LPH. This 31-residue fragment, actually isolated as a "side product" in both Li's and the author's laboratories but not published till 1976, is known today as the most potent natural analgesic peptide isolated so far.[10–12]

The discovery of enkephalins by Hughes *et al.*[13] in 1975 and the local sequence identity between β-LPH and Met-enkephalin (Fig. 1) suggested a new area in which to look for further biological functions of β-LPH. As a result of

---

László Gráf • The Institute for Drug Research, 1325 Budapest, Hungary.

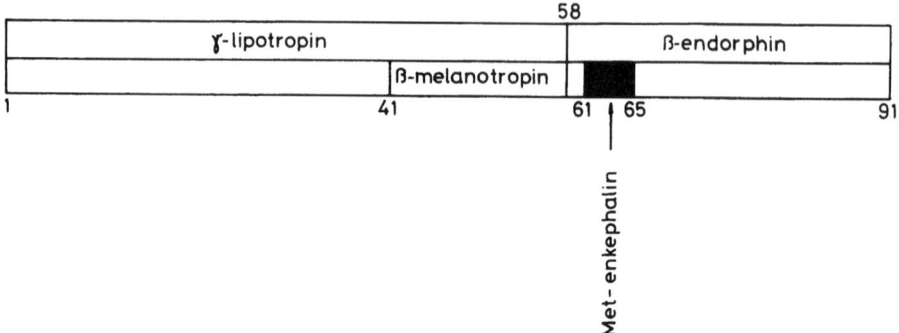

Fig. 1. Schematic representation of the ovine and porcine β-LPH structure; structural relationships with γ-LPH, β-MSH, Met-enkephalin, and β-endorphin.

this approach, it was first reported from the author's laboratory that although porcine β-LPH is practically inactive as an opioid agonist, its digestion with trypsin generates a nonapeptide, β-LPH,$_{61-69}$ with opiate agonist activity comparable to that of Met-enkephalin.[14] Independently, a series of opioid peptides, all β-LPH fragments with Tyr$_{61}$ at their N-termini, were isolated from hypothalamic and pituitary extracts.[10-12,15,16] However, only one of these, β-endorphin[10] (Fig. 1), has significant morphinelike properties in vivo.[10-12] Furthermore, adequate biosynthetic evidence is now available showing that γ-LPH and β-endorphin are formed in situ from β-LPH in the anterior pituitary,[17-19] and it is also likely that these two latter peptides are further processed in the pars intermedia to β-MSH,[20] β-endorphin,$_{1-27}$, and its α-N-acetyl derivative.[21]

Although β-LPH is the biological precursor of at least three biologically active peptides, γ-LPH, β-MSH, and β-endorphin (Fig. 1), it is in turn a product of a large biosynthetic precursor, proopiomelanocortin.[17,18,22,23] This precursor molecule consists of and is enzymatically processed into three main fragments, a large N-terminal glycoprotein, corticotropin (ACTH), and β-LPH. The fact that β-LPH and its fragments are generated and released concomitantly with different biologically active peptides strongly suggests that its physiological significance cannot be evaluated without considering the contribution of the other biologically active components of the system. One is tempted to speculate that these peptides of common biosynthetic origin act all in a cooperative fashion in the organism. Thus, the elucidation of β-LPH biosynthesis, more than any other investigation, has considerably changed our view of the biological importance of β-LPH.

2. STRUCTURE

2.1. Amino Acid Sequence

Ovine β-LPH was the first lipotropin isolated and sequenced by Li and co-workers.[1,4] Reexamination of the primary structure of ovine β-LPH revealed that it contains 91 amino acids[24] instead of 90 as proposed originally.[4] The revised amino acid sequence of ovine β-LPH is shown in Fig. 2. Porcine β-LPH was first isolated by Gráf and Cseh,[25] and its amino acid sequence was elucidated in the same laboratory.[26] Further studies[27–29] confirmed this structure. Based on comparative peptide mapping of ovine and bovine β-LPHs, the two species homologues were originally thought to have identical structures,[30] but recent reinvestigation of the sequence[31] shows that the bovine hormone is composed of 93 instead of 91 residues (Fig. 2). Recently, the complete amino acid sequence of ostrich β-LPH[32] and that of the γ-LPH segment of mouse β-LPH[33] were reported (Fig. 2). Partial sequence data for whale β-LPH are also available[34] (not included in Fig. 2).

Although our knowledge of the chemistry of lipotropins, including several exotic species homologues, has steadily grown, the primary structure of human β-LPH has remained controversial until very recently. Human β-LPH was one of the first species homologues isolated.[35] The C-terminal sequence region, residues 41–91 of the hormone, was determined by Cseh et al.[36] Li and Chung[37] confirmed these results and proposed a complete sequence for human β-LPH. However, doubt arose about the validity of this sequence in both Li's and the author's laboratory, and reinvestigation of the most disputed sequence region, residues 9–24, was carried out by these two groups independently. The studies led to similar sequence proposals (see versions I and II in Table I) which were, however, different from the human β-LPH structure deduced from nucleotide sequence data.[39] Although the DNA sequence of human proopiomelanocortin was revised,[40] the human β-LPH structure derived from this sequence still differed in five positions from the sequence proposed by Li et al.[31,38] The first protein sequence data confirming the revised DNA sequence[40] were reported by Hsi et al.[41] (Table I, sequence III).

To resolve the remaining contradictions among the protein sequence data published,[31,37,41] a joint study was carried out by the Hungarian and Canadian workers. Human β-LPH and several papaic and thermolysin fragments were prepared in Hungary,[35,36,42] and the sequence analysis was done in the Canadian laboratory. The results[43] confirmed the sequence previously reported by Hsi et al.,[41] thus suggesting that the differences between the sequences published probably arose from some technical difficulties in peptide sequencing rather than from a possible polymorphism of the human proopiomelanocortin gene.[38,39] To illustrate the uncertainty of locating enzymic fragments on the basis of amino acid composition and partial sequence data, the compositions of a few fragments derived from sequence region 9–24 are shown in Table II. Sequence proposals I and II[31,38] were partly based on the assumption that fragments P1 and 2S9 contained 7 and 8 residues, respectively (column a in Table II). The composition of the same fragments of double size, however, is consistent with sequence III (Table I) proposed by Hsi et al.[41] and confirmed by Seidah et al.[43] and Spiess et al.[44]

The primary structures of six species homologues of β-LPH are aligned in Fig. 2. The comparison of the sequences to the ovine β-LPH structure, the one first elucidated,[4,24] is arbitrary and does not take into account the actual evolutionary events that have occurred in the gene coding for β-LPH. From

```
                    1                                              10                                  20
Ovine     Glu-Leu-Thr-Gly-Glu-Arg-Leu-Glu-Gln-Ala-Arg-Gly-Pro-Glu-Ala-Gln-Ala-Glu-Ser-Ala-Ala-Ala-Arg-Ala-Glu(———)-
Porcine                     Ala        Ala-Pro-Pro            Pro        Asp                                 Gly       (———)-
Bovine
Ostrich   Ala             Pro-Pro-Ala-Ala-Met-Leu-Pro         Ala-Ala-Glu                Glu-Glu-Glu-Gly-Glu(———)
Mouse                                        Glu    Pro(———————————————————)
Human                                               Gln       Arg-Glu-Gly-Asp            Asp-Asp-Gly        Gly-Ala-Gln-Ala-Asp-

                              30                              40                                      50
Ovine     Leu-Glu-Tyr-Gly-Leu-Val-Ala-Glu(———)Ala-Glu-Ala-Ala-Glu-Lys-Asp-Ser-Gly-Pro-Tyr-Lys-Met-Glu-His-Phe-
Porcine
Bovine                                  (———)                                                 Glu
Ostrich                             Ala-Glu
Mouse                    Gly-Leu        Gln-Val-Leu                    Glu                           Gly    Ser    Arg    Arg
Human                    His-Ser        Leu-Val(———————————)Ser-Asp-Ala-Glu                          Asp           Arg-Val
                                                                                                     Glu           Arg

                              60                                      70
Ovine     Arg-Trp-Gly-Ser-Pro-Pro-Lys-Asp-Lys-Arg-Tyr-Gly-Gly-Phe-Met-Thr-Ser-Glu-Lys-Ser-Gln-Thr-Pro-Leu-Val-Thr-
Porcine
Bovine
Ostrich                   Gln-Ala     Leu                              Ser                           Arg-Gly-Arg-Ala
Mouse                     Ser-Asn              Total residues (γ-LPH) = 38
Human

                              80                              90                              Total residues
Ovine     Leu-Phe-Lys-Asn-Ala-Ile-Ile-Lys-Asn-Ala-Ala-His-Lys-Lys-Gly-Gln                           91
Porcine                       Val                                                                   91
Bovine                                                                                              93
Ostrich                   Val     Ser                      Tyr                                      79
Human                                                      Tyr                 Glu                  89
```

Fig. 2. Comparison of the amino acid sequences of porcine β-LPH,[26] ostrich β-LPH,[32] mouse γ-LPH,[33] and human β-LPH[41,43] to that of ovine β-LPH.[4,24] Lines within parentheses indicate deletions in the corresponding sequences. Invariable sequence portions are underlined.

β-Lipotropin

Table I
An Analysis of Different Sequence Proposals for Residues 9–24 in Human β-LPH

		10 15 20
I.	Li et al.[38]	Gln-Gly-Asp-Gly-Pro-Asn-Ala-Gly-Ala-Asp-Asp-Gly-Pro-Gly-Ala-Gln
II.	Gráf et al.[38]	Gln-Gly-Asp-Gly-Pro-Asp-Ala-Gly-(Ala,Asp,Gly,Pro,Asp,Gly)Ala-Gln
III.	Hsi et al.[41]	Glu-Gly-Asp-Gly-Pro-Asp-Gly-Pro-Ala-Asp-Asp-Gly-Ala-Gly-Ala-Gln

the comparison it is evident, however, that the N-terminal sequence portions (corresponding to residues 1–48 in the ovine β-LPH structure) are extremely variable, and homologous sections are only present at the C-terminal part of the molecule. The size variability of the N-terminal sequence is worth mentioning. Whereas the γ-LPH portions (residues 1–58 of the ovine and porcine β-LPH) in mouse, ostrich, and human lipotropins are, respectively, 20, 12, and two residues shorter than that of ovine β-LPH, bovine β-LPH has a two-residue insertion within the same region (Fig. 2). It is of particular interest that the paired basic residues, which are thought to direct the proteolytic cleavage at the N-terminal end of β-MSH in ovine, porcine, and bovine β-LPHs,[8,9,23] are absent from the mouse and ostrich lipotropin molecules (Fig. 2).

The conservation of some sequence regions in the β-LPH molecule indicates that they may carry important biological information. Indeed, the His-Phe-Arg-Trp tetrapeptide (residues 49–52 in ovine β-LPH) present in all species homologues sequenced so far (Fig. 2) has long been known to be the active core of many related pituitary hormones for steroidogenic, melanocyte-stimulating, and lipolytic activities. Another homologous sequence region, Lys-Asp-Lys-Arg-Tyr-Gly-Gly-Phe-Met (residues 57–65 in ovine β-LPH), contains a proteolytic cleavage site (Arg-Tyr) for the generation of β-endorphin[18,23] and also the biologically quintessential N-terminus of β-endorphin.[10–16] Further conserved sections of the β-endorphin portion in the lipotropin molecule (see Fig. 2) have important structural roles in a proposed model of the β-endorphin conformation at the functional receptor site.[45]

Table II
Amino Acid Compositions of Some Enzymic Fragments of Human β-LPH from Region 9–24

	Th1[31]	2S8[43]	PI[31] a	PI[31] b	2S9[43] a	2S9[43] b
Asp	4.4 (4)	4.0 (4)	1.8 (2)	3.6 (4)	2.2 (2)	4.4 (4)
Glu	1.4 (1)	1.9 (2)	0.7 (1)	1.4 (1)	0.6 (1)	1.2 (1)
Pro	1.9 (2)	1.6 (2)	0.9 (1)	1.8 (2)	0.8 (1)	1.6 (2)
Gly	5.0 (5)	5.4 (5)	2.0 (2)	4.0 (5?)	2.6 (3)	5.2 (5)
Ala	2.8 (3)	2.7 (3)	1.3 (1)	2.6 (3)	1.2 (1)	2.4 (2)
Residues	9–23(I,II,III)	9–24(I,II,III)	9–15(I,II)	9–23(I,II,III)	9–16(I,II)	9–22(I,II,III)

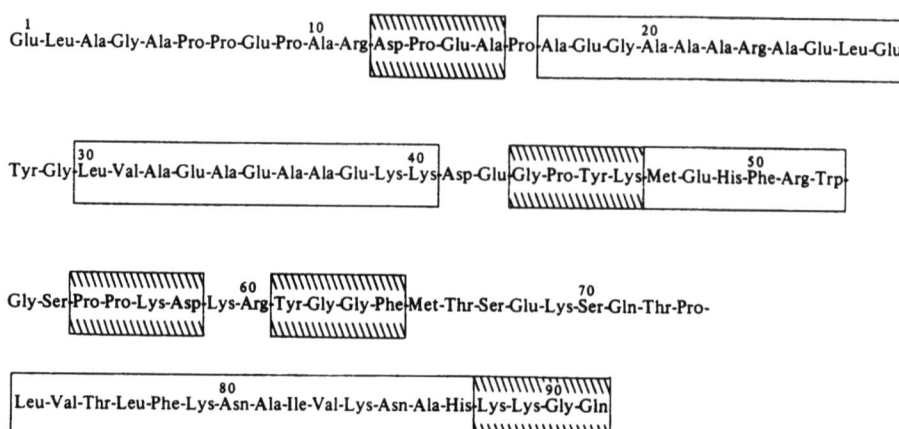

Fig. 3. α-Helix (empty rectangles) and β-turn (hatched rectangles) regions predicted for the porcine β-LPH structure. (From Gráf and Hollósi.[48])

2.2. Conformation

Circular dichroism studies have shown that ovine[46,47] and porcine β-LPH[48] have a low amount of α-helix in aqueous solutions. However, secondary-structure-promoting organic solvents such as dioxan, methanol, and trifluoroethanol considerably increase the proportion of helical structure in lipotropins.[46-48] Comparative studies on the secondary structure of β-LPH, γ-LPH,[48] and β-endorphin[49] showed that the α-helix content of the three peptides increased in a similar fashion in response to an increase in the trifluoroethanol concentration in the solutions. Interestingly, the highest percentage of α-helix measured in pure trifluoroethanol was about the same, 50%, for all three peptides.[48,49] This is, however, in agreement with the results of the predictive analysis of the secondary structure[48,49] (Fig. 3) by the method of Chou and Fasman.[50,51] The α-helical stretches predicted between residues 17–27, 30–40, 47–52, and 74–87 in the porcine β-LPH structure (Fig. 3) may account for the approximately 50% α-helix of either β-LPH[48] or β-endorphin[49] in trifluoroethanol.

Based on extensive secondary structure–activity studies on β-endorphin, Gráf and his co-workers[45,49,52] have proposed that the conformation of the peptide in trifluoroethanol is related to its biologically active conformation, i.e., the conformation of β-endorphin at its functional receptor site. As for the predicted secondary structure of β-LPH, it has been speculated that such a conformation adopted in the secretory granules might direct the specificity of the conversion of β-LPH into β-endorphin by even a common trypsinlike proteinase.[48]

3. BIOLOGICAL PROPERTIES

Li and co-workers[1,5,8] reported first that ovine lipotropins had lipolytic, melanocyte-stimulating, and adrenal-stimulating activities (Table III). Further

Table III
In Vitro Biological Activities of β-MSH, ACTH, and Lipotropins

Hormones	MSH activity (U/g)	Lipolytic activity (MED^a)		Adrenal-stimulating activity (U/mg)
		Rabbit adipose tissue	Rat adipose tissue	
Ovine β-MSH[8]	1.2×10^9	0.0006	Inactive	Not tested
Ovine ACTH[5]	4×10^7	0.008	0.001	179
Ovine β-LPH[5]	2×10^7	0.05	0.8	1.1
Ovine γ-LPH[8]	1.6×10^7	0.01–0.1	Inactive	0.4
Porcine β-LPH[7,25,53]	Not tested	0.005	1–20	Not tested
Human β-LPH[35]	Not tested	0.005	10	Not tested
Porcine γ-LPH[7]	Not tested	0.06	Inactive	Not tested

^{a} MED, minimal effective dose (μg).

in vitro biological studies on other species homologues[7,25,35,53] supported these results. As Table III shows, lipotropins as lipolytic agents are much more active in rabbit than in rat adipose tissue. The same species specificity was demonstrated *in vivo* in comparisons of the effects of porcine β-LPH on serum free fatty acid in different species.[53] Human β-LPH was also shown to be active *in vitro* in human adipose tissue[35] (data not included in Table III).

In view of the low values of adrenal-stimulating activity of lipotropins (Table III and ref. 25), it is possible that this activity resides in ACTH present as a contaminant in the lipotropin preparations. This would also explain the low and variable lipolytic activity of β-LPH in rat adipose tissue (Table III). Unfortunately, synthetic ovine β-LPH[54] has not been tested for the above biological affects, and the question of ACTH contamination in natural lipotropin preparations is still open. Natural ovine and human β-LPH preparations were found to stimulate the production of aldosterone in collagenase-dispersed rat adrenal capsular cells.[55] The effect of β-LPH was comparable to that of ACTH; thus, it could not be attributable to contaminant ACTH. In addition, in a single experiment with synthetic ovine β-LPH, aldosterone stimulation was similar to that caused by natural products.[55]

As already noted in Section 1, the lipolytic and melanocyte-stimulating activities of lipotropins reside in the β-MSH region, residues 41–58, of their structures. Originally, this was suggested by the sequence homology itself[4,8,26] (Fig. 1). Adequate evidence to prove that the β-MSH region is indeed the active core of lipotropins was provided by structure–activity studies with a large series of natural and synthetic lipotropin fragments.[5–7,56] The C-terminal cyanogen bromide fragment of porcine γ-LPH,[7] residues 48–58, has a significant lipolytic activity as tested on renal adipose tissue of rabbit *in vitro*. An extension of this molecule by four N- and two C-terminal residues to form β-LPH$_{44-60}$ (a fragment produced by chemical synthesis[6]) resulted in a dramatic increase in lipolytic activity.[7,57] Further addition of the N-terminal 43 residues of porcine β-LPH to form β-LPH$_{1-60}$ did not affect the activity.

To determine the effect of the C-terminal sequence portion, residues 59–91, of porcine β-LPH on lipolytic activity, a series of N-terminal plasmin frag-

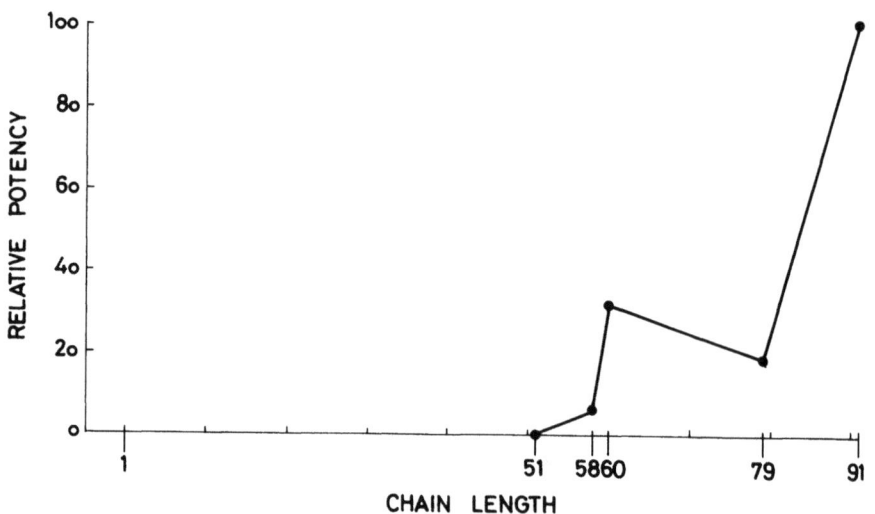

Fig. 4. Relative lipolytic activities of N-terminal fragments of porcine β-LPH consisting of residues 1–51, 1–58, 1–60, 1–79, and 1–91 (β-LPH). (From Gráf et al.[57])

ments were tested on rabbit adipose tissue.[7,57] The results are demonstrated in Fig. 4. As the figure shows, the introduction of Lys^{59}-Arg^{60} into the C-terminus of the γ-LPH molecule considerably enhances the lipolytic activity. The largest increment of activity was observed in response to the addition of residues 80–91 to the structure of β-LPH$_{1-79}$. The question was raised by Gráf et al.[57] whether the positive contribution of the C-terminal 12 residues to the lipolytic activity of β-LPH (Fig. 4) had something to do with the slight lipolytic potency of natural porcine β-endorphin.[57] Two groups[58,59] have recently confirmed our earlier finding on the intrinsic lipolytic activity of β-endorphin,[57] and Schwandt et al.[58] now suggest that in addition to the active core within residues 47–53, β-LPH contains a second message sequence within region 78–91 for lipolytic activity.

Porcine β-LPH was not found to cause any significant change in the blood sugar, serum triglyceride, cholesterol, or phospholipid levels of rabbit.[53] From in vitro studies in rat epididymal fat pad, however, ovine β-LPH appeared to stimulate glucose oxidation and incorporation into fatty acids.[60] Other weak in vivo biological activities of ovine β-LPH such as hypocalcemic[61] and blood coagulation-stimulating effects in rabbit[62] have also been reported.

The discovery of the structural relatedness of enkephalins to β-LPH[13] initiated interest in the possible opiatelike properties of β-LPH itself. Both ovine and human β-LPH were reported to show a slight opiate activity in vitro,[63] and highly purified porcine β-LPH was found to have a weak analgesic effect[14,64] on intracerebroventricular administration. The latter finding has recently been confirmed.[65]

These properties of β-LPH, however, apparently disagree with our recent knowledge of the structural requirements of a peptide for opiate activity. Namely, the peptide has to have an N-terminal tyrosine residue with a free amino group to elicit opiate activity, a basic criterion clearly not fulfilled by lipotropins. There may be two explanations to resolve the apparent controversy: all natural β-LPH preparations tested so far[63-65] were contaminated by about 1% (by weight) β-endorphin—an amount that cannot easily be detected by chemical methods—or some β-endorphin (or other opioid peptide) may have been formed from β-LPH in response to an appropriate enzymic cleavage in the brain. The latter possibility is an interesting one, which, however, should be investigated by using synthetic β-LPH[54] rather than natural preparations.

4. BIOSYNTHESIS IN THE PITUITARY

In 1973 two independent immunohistochemical studies[66,67] indicated that a group of pituitary hormones (ACTH, lipotropins β-MSH) were contained in the same cells of the anterior and intermediate lobes of bovine, ovine, and porcine pituitaries. The cooccurrence of two or more different secretory peptides in the same endocrine cells would clearly suggest, today, a common biosynthetic pathway of these peptides. Ten years ago, however, the concept of the common precursor to different hormones was not established yet, and the above immunohistochemical findings,[66,67] later confirmed by more detailed studies,[68,69] were not interpreted in terms of biosynthesis until 1977.

As has already been noted in Section 1, before the discovery of enkephalins[13] and endorphins,[10-12,15,16] interest in lipotropins centered around their lipolytic activities. Although *in vitro* labeling experiments by Chrétien *et al.*[70] demonstrated the biosynthesis of both β-LPH and γ-LPH in bovine pituitary slices, the formation of the C-terminal fragment of β-LPH (missing from γ-LPH) was not looked for. In a separate series of experiments, Mains and Eipper[71,72] provided evidence that a large glycoprotein with a molecular weight of about 31,000 (31 K) was the biosynthetic precursor for three smaller forms of ACTH. In 1977, Mains *et al.*[17] extended their pulse-labeling studies to investigate the possibility that the large precursor of ACTH could also serve as a biosynthetic precursor to lipotropins and endorphins, and, in fact, this has been found to be the case.

4.1. Pulse–Chase Studies: The Pathway of the Biosynthesis

It is now generally accepted that all secretory proteins and peptides are synthesized in the form of large precursors and that these precursors undergo intracellular proteolytic conversions into the biologically active products (for reviews see refs. 18,22,73). The pulse-labeling and -chase technique applied to *in vitro* slice and cell culture systems has been a useful tool in studying the pathway of peptide biosynthesis. In the case of an endocrine gland producing a series of hormones, as in the case of the pituitary, specific chemical or immunologic techniques are necessary to detect the formation of the radiolabeled precursor protein in question (pulse), and to follow its conversion into the smaller peptides (chase). It has been a piece of luck that the common precursor for ACTH and β-endorphin, and all the biosynthetic intermediates formed from

1. 28.5K ACTH/β-endorphin

2. 31K and 29K ACTH/β-endorphin

3. 23K and 20K ACTH
   β-LPH

4. 16K fragment
   13K and 4.5K ACTH

5. γ-LPH
   β-endorphin

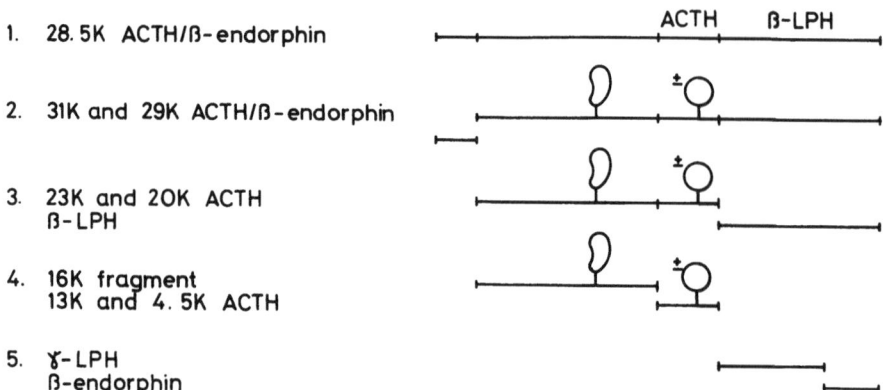

Fig. 5. Model for the structure and processing of mouse pituitary tumor prepro-ACTH/β-endorphin according to Mains and Eipper[18,76] and Roberts and Herbert.[74]

it, could be immunoprecipitated by either ACTH or β-endorphin antibodies.[17,18] In fact, double-antibody immunoprecipitation (with the two antibodies) combined with sodium dodecylsulfate gel electrophoresis provided the first evidence that ACTH and β-endorphin originate from a common biosynthetic precursor in a mouse pituitary tumor cell line.[17] In addition, translation of mRNA isolated from the same anterior pituitary tumors in a cell-free system produced a similar precursor molecule with ACTH and β-LPH immunoreactivity.[74]

Though the immunologic approach has been criticized by Chrétien's group,[22] in the case of the ACTH/β-endorphin synthesizing system at least, it has been successfully used not only to explore the pathway of proteolytic processing[17,18,75] but also to locate the main products of the processing in the structure of the precursor.[18,76,77] Based on the above studies, a schematic model for the structure of mouse pituitary tumor prepro-ACTH/β-endorphin and for the main steps of proteolytic processing is shown in Fig. 5. Accordingly, the initial translation product (28.5 K ACTH/β-endorphin) detected in a cell-free system[74] is processed by the cleavage of the N-terminal signal peptide and glycosylation of the protein to yield two forms (31 K and 29 K) of pro-ACTH/β-endorphin,[18,76] also designated proopiomelanocortin.[22] A further proteolytic cleavage generates β-LPH and the glycosylated and nonglycosylated ACTH intermediates (23 K and 20 K ACTH). Subsequent cleavages give rise to 13 K and 4.5 K ACTH, fragment 16 K, γ-LPH, and β-endorphin[18] (Fig. 5). Interestingly, the conversion of β-LPH into γ-LPH and β-endorphin was found to be partial in the anterior pituitary tumor cell line.[18] There is now general agreement in the literature that β-LPH and not β-endorphin (and γ-LPH) is the predominant precursor product in the anterior pituitary.[19,78,79]

Although Mains and Eipper[17,18,71,72,76] and Roberts and Herbert[74,75,77] used a mouse pituitary tumor cell line as a model system for studying ACTH and β-endorphin biosynthesis, Chrétien and his group applied the pulse-labeling and -chase technique to isolated cells from pars intermedia of beef[80,81] and rat pituitaries.[82,83] The above authors identified the biosynthetic products chemically (rather than immunologically) by using a combination of microsequencing and peptide-mapping analyses.

It is clear from these studies that although the main features of the precursor structure and the initial steps of the processing in the pars intermedia cells are similar to those in the anterior pituitary (see Fig. 5), the peptides are processed to a different extent in the two lobes.[18,22] Whereas ACTH, β-LPH, and β-endorphin are the final products in the anterior pituitary,[18,19,78,79] in the pars intermedia they serve a short-lived biosynthetic intermediates in the production of α-MSH and corticotropinlike intermediate lobe peptide (CLIP; from ACTH) and γ-LPH (or β-MSH) and β-endorphin (from β-LPH).[20,81,83] Zakarian and Smyth[21,84] have shown that the main endorphin-related products in the pars intermedia of porcine pituitary are β-endorphin$_{1-27}$ and α-N-acetyl-β-endorphin$_{1-27}$, suggesting that even β-endorphin is further processed in the intermediate pituitary. Kinetic studies on the processing of β-endorphin in rat pars intermedia cells have confirmed this view.[85,86] According to this work, β-endorphin is first acetylated to form α-N-acetyl-β-endorphin, which is subsequently cleaved to produce α-N-acetyl-β-endorphin$_{1-27}$.[85,86]

4.2. Nucleotide Sequencing of Cloned cDNA for the Precursor and Protein Sequencing of the Biosynthetic Intermediates

Theoretically, there are three methodological approaches to determine the precise primary structure of the precursor proteins: (1) amino acid sequencing of the biosynthetically labeled precursor isolated from tissue slices or cell cultures, (2) amino acid sequencing of the cell-free translation product of the corresponding mRNA, and (3) nucleotide sequencing of cloned cDNA for the precursor. Because of the very small amounts of the biosynthetically labeled precursor proteins isolated and of some technical difficulties in sequencing of proteins of such size, the use of methods 1 and 2 provided only partial sequence data on mouse[33,74,87] and rat (pre)proopiomelanocortin.[88] No doubt, DNA cloning together with nucleotide sequence analysis has been the most powerful technique to elucidate the whole preproopiomelanocortin structure. By this approach, the complete DNA sequence of bovine[23] and human proopiomelanocortin[39,40,89] has been determined. Nucleotide sequence analyses have also been reported for large portions of rat[90] and mouse genomic DNA[91] coding for proopiomelanocortin.

The structure deduced from the revised DNA sequence of human proopiomelanocortin[40,89] is shown in Fig. 6. This agrees perfectly with the amino acid sequences of four peptides isolated from human pituitaries, including a 76-residue glycopeptide,[92] the so-called joining peptide,[93] ACTH,[94] and β-LPH[41,43,44] (Fig. 6). The relative positions of ACTH and β-LPH within this precursor molecule and also in the known structures of other species homologues of proopiomelanocortin[23,90,91] are the same as in the schematic precursor model suggested by Mains and Eipper's pulse-chase experiments and partial product characterization[18,76] (see Fig. 5). Residues 1–109 of the human proopiomelanocortin structure (Fig. 6) correspond to the 16 K fragment formed from mouse pro-ACTH/β-endorphin[17,18] (Fig. 5). In the human pituitary gland,

1
Trp-Cys-Leu-Glu-Ser-Ser-Gln-Cys-Gln-Asp-Leu-Thr-Thr-Glu-Ser-Asn-Leu-Leu-Glu-Cys-Ile-Arg-Ala-Cys-Lys-
          10   20
|——N-terminal glycopeptide——

Pro-Asp-Leu-Ser-Ala-Glu-Thr-Pro-Met-Phe-Pro-Gly-Asn-Gly-Asp-Glu-Gln-Pro-Leu-Thr-Glu-Asn-Pro-Arg-Lys-
      30   40   50
                                                                    Thr
                                                                    CHO

Tyr-Val-Met-Gly-His-Phe-Arg-Trp-Asp-Arg-Phe-Gly-Arg-Arg-Asn-Ser-Ser-Ser-Ser-Gly-Ser-Ser-Gly-Ala-Gly-
          60   70

Gln-Lys-Arg-Glu-Asp-Val-Ser-Ala-Gly-Glu-Asp-Cys-Gly-Pro-Leu-Pro-Glu-Gly-Gly-Pro-Glu-Pro-Arg-Ser-Asp-
      80   90   100
                                    Asn
                                    CHO
        |——"Joining peptide"——|

Gly-Ala-Lys-Pro-Gly-Pro-Arg-Glu-Gly-Lys-Arg-Ser-Tyr-Ser-Met-Glu-His-Phe-Arg-Trp-Gly-Lys-Pro-Val-Gly-
          110   120
        |——Glu-NH₂ |——Adrenocorticotropin——

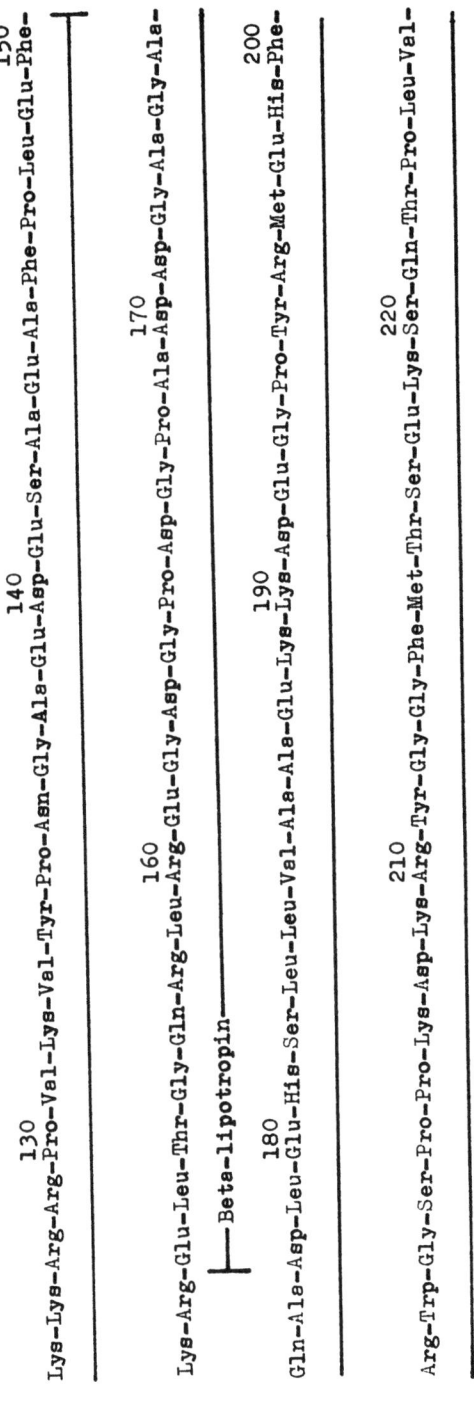

Fig. 6. The amino acid sequence of human proopiomelanocortin derived from the nucleotide sequence of cloned cDNA[40,89] with the isolated and completely sequenced component peptides, the N-terminal glycopeptide,[92] the so-called joining peptide,[93] ACTH,[94] and β-lipotropin[41,43,44] shown.

however, this large N-terminal segment appears to be further processed into two fragments at least, as shown in Fig. 6. In view of the presence of an MSH-like peptide sequence within this region of the precursor molecule (residues 51–62 in the human proopiomelanocortin structure, Fig. 6), there has been considerable interest in the possible generation of such a peptide, designated as γ-MSH by Nakanashi et al.[23] Although several authors have reported the isolation and biological properties of γ-MSH-like peptides from different sources,[95,97] biosynthetic evidence for their generation from proopiomelanocortin is not, as yet, available.

Thus, it is easy to see that the different methodological approaches are complementary to each other, and only their combined use can reveal a complete picture on proopiomelanocortin processing.

4.3. Proteinases Involved in the Processing of the Precursor and in the Release of the Final Products

Sequence comparison of proopiomelanocortin and its products identified to date indicates that all the component peptides in the precursor structure are flanked by pairs of basic residues on both the N- and C-terminals (Fig. 6). Thus, the specificity of processing is apparently directed by the positions of paired basic amino acid residues in the prohormone structure. Theoretically, sequential actions of trypsin and carboxypeptidase B might generate all of the peptides shown in Fig. 6 and also those isolated from the pars intermedia of other species, provided that cleavages of further trypsin-sensitive bonds within their sequences are prevented by a specific conformation of the precursor.[48,98,99]

Although the involvement of trypsinlike proteinases in intracellular proteolytic processing has been suggested for a long time (for reviews see refs. 73,100), earlier attempts at their detection and characterization by conventional biochemical methods were not successful. It was first reported from the author's laboratory[101,102] that crude homogenates from porcine and rat anterior and intermediate pituitaries (but not from neurohypophysis and brain) contain proteinase(s) that are able to split a few lysyl and arginyl peptide bonds of exogenous β-LPH. The pH optimum of the activity is about 8,[101,102] and it can be inhibited by pancreatic trypsin inhibitor[101] and some tripeptide aldehydes,[57] indicating that the enzyme activity is attributable to serine proteinase(s). The specific regional location of the activity together with its subcellular localization to the small secretory granule fraction from porcine anterior pituitaries[100,103] has suggested that the enzyme activity may have something to do with the biosynthesis or release of some pituitary hormones. Supporting this view, a positive correlation has been found between the trypsinlike and biosynthetic–secretory activities of rat anterior pituitaries[104]: dexamethasone- and ACTH-induced suppression of ACTH/β-endorphin biosynthesis and secretion decreased, whereas adrenalectomy stimulating the biosynthesis and secretion increased the trypsinlike activity in the pituitary.[104]

There is now a renewed interest in proteinases involved in proopiomelanocortin[105] and proenkephalin processing.[106–109] These studies, except one,[107] report secretory granule-associated endo- and exopeptidases with pH optima of 5–6. This low pH optimum of the enzymes is consistent with the internal pH of about 5.5 found in secretory granules (ref. 105 and refs. therein). Contamination of the secretory granule fractions with lysosomes, however, is a potential source of error in these studies. For example, a proenkephalin-processing carboxypeptidase characterized by Hook et al.[106] appears to be a lysosomal rather than a secretory granule enzyme.[108] The proopiomelanocortin-converting activity associated with a secretory granule fraction from rat intermediate pituitary[105] also has some overlapping properties with lysosomal cathepsin B. This, however, does not necessarily indicate that the enzyme activity detected actually resides in lysosomal enzyme(s) contaminating the secretory granule preparation. Similarities between the lysosomal and secretory granule enzymes are conceivable in view of the cytochemical observation that the membranes of both lysosomes and secretory granules originate from the same membrane of the inner Golgi cisternae in the secretory cells (for review see ref. 110).

Even such a brief discussion of the recent data on secretory granule-associated proteinases may indicate the technical difficulties of this approach. The apparent contradiction between reports on serine proteinases with a pH optimum of about 8 (refs. 100,103,107) and proteinases of low pH optimum[105,108,109] in the secretory granules may be resolved by assuming that these enzymes are involved in different steps of the complex mechanism of precursor processing and product release. Recently, we have found that a serine proteinase inhibitor, BOC-D-Phe-Pro-Arg-H, inhibits the stimulated release of ACTH- and β-endorphinlike peptides from cultured rat anterior pituitary cells.[111]

5. OCCURRENCE IN THE PITUITARY, PERIPHERY, AND BRAIN

It is now well established that although both the anterior pituitary and pars intermedia contain "opiomelanocorticotroph" cells, only the former lobe is a source of circulating β-LPH.[18,19,21,78,79,112–114] Data on the amount of β-LPH relative to that of β-endorphin in the anterior pituitary, however, vary to a large extent. Although several groups[78,112,114] have reported that the molar concentration of β-endorphin is less than 5% relative to that of β-LPH, others have found that this concentration is about 20–40%.[18,19,21,79,113] Liotta et al.[78] have proposed that the higher β-endorphin-to-β-LPH ratios obtained by the latter group of authors might reflect artificial conversion of β-LPH to β-endorphin during the extraction procedure. It is an intriguing possibility that proteinase(s) involved in the secretion of the hormones (see Section 4.3 and ref. 111) may also mediate the conversion of β-LPH into β-endorphin. Although this idea is a speculative one, it is interesting to note that the molar ratios of β-endorphin to β-LPH measured from culture media rather than tissue extracts of the anterior pituitary are consistently 1:3–4 (refs 18,19,79,111).

Release studies with primary anterior pituitary cell cultures,[79,115,116] mouse pituitary tumor cells,[18,117] and intact anterior pituitary[19] have shown that the known ACTH secretagogues, such as norepinephrine, vasopressin, and corticotropin-releasing factor, stimulate the release of ACTH, β-endorphin, and β-LPH in parallel. Physiological or pharmacological stimulatory effects such as acute stress,[118] long-term adrenalectomy,[119] and acute treatment with serotonin-acting drugs[120] also resulted in a more or less parallel increase in the plasma levels of ACTH, β-endorphin, and β-LPH in rats. In studies with rats, however, one has to consider the contribution of both anterior pituitary and pars intermedia to the secretion of β-endorphin-like peptides, i.e., peptides with β-endorphinlike immunoreactivity and the size of β-endorphin (see ref. 21).

The human pituitary, unlike rat pituitary, does not have an intermediate lobe. Therefore, it has been of interest to study the plasma levels of β-LPH and β-endorphin in human subjects. After some controversy in the literature concerning the absence[121,122] or presence of β-endorphin in human plasma,[123,124] there is now a general agreement that both β-LPH (5–15 fmol/ml) and β-endorphin (3–7 fmol/ml) are present in plasma from normal human subjects.[124,125] Similarly, elevated plasma levels of β-LPH and β-endorphin have been found in patients with pituitary disease (Cushing's disease, Nelson's syndrome) or excessive ACTH production (Addison's disease).[121,122,124] According to Yamaguchi et al.,[126] however, following acute stimulation with vasopressin or insulin in normal human subjects, there is a two- to threefold disproportionate rise in the plasma levels of β-LPH as compared to those of β-endorphin. The greater metabolic clearance rate and fractional rate of disappearance and the shorter plasma half-life of β-endorphin as compared to those of β-LPH may account, at least in part, for the above phenomenon.[125]

In addition to the pituitary, other peripheral organs such as human placenta,[127–129] human[130] and rat pancreas,[114] rat gut, pineal, kidney, and adrenal gland[114] are known to contain a peptide(s) with β-endorphinlike immunoreactivity. Fractionation of the tissue extracts by gel filtration and high-performance liquid chromatography revealed that 60–100% of the immunoreactive material resided in a peptide behaving like synthetic β-endorphin.[114,129] Thus, β-LPH, if present at all, may be only a minor component in these tissues.

In view of the various effects on the central nervous system of centrally administered β-endorphin, it has been of particular interest whether β-endorphin and its related peptides are present in the brain. By use of gel filtration of the tissue extracts and radioimmunoassay, significant amounts of β-endorphinlike substances have been detected in the brains of several species.[131–135] It has been reported that neither adrenalectomy[133] nor hypophysectomy[133,136] significantly alters the concentrations of these peptides in the rat brain.

These data suggest that β-endorphin is synthesized in the brain and not simply derived by transport from the pituitary gland. However, a study by Ogawa et al.[137] on the brain concentration of β-endorphin-immunoreactive peptides in intact and hypophysectomized rats is not entirely consistent with this view. These authors[137] have found that the brain concentrations of immunoreactive β-endorphin in rats killed by microwave irradiation are about ten times higher than those in rats killed by decapitation[133] and that hypophysectomy causes a dramatic decrease of β-endorphin immunoreactivity in the brain. The reduced β-endorphin concentration of the brain, which remains stable as long as 3 months after hypophysectomy,[137] is about the same as that reported by Rossier et al.[133] Thus, it may well be that this "conserved" fraction of β-endorphin immunoreactivity in the brain is indeed of brain origin, whereas the excess amount of brain endorphin found in microwave-irradiated rats[137] was transported from the pituitary by retrograde blood flow.[138]

The strongest evidence for the central nervous system origin of β-endorphinlike peptides was provided by their intraneuronal localization immunohistochemically[139–141] and by demonstration of the *in vitro* biosynthesis of a precursor to ACTH- and β-endorphinlike materials by hypothalamic cells.[142,143] This is further strengthened by the observation that monosodium-glutamate-induced lesions of arcuate nucleus (the only brain region in which β-endorphin- or β-LPH-containing cell bodies were detected[139–141]) resulted in a marked decrease in the ACTH and β-endorphin contents of the brain.[144]

As to the chemical identity of the β-endorphin-immunoreactive peptides in the brain, gel filtration of rat and human brain extracts revealed that the great majority of the immunoreactive material eluted as β-endorphin from the columns.[135] Swann and Li[145] have isolated four β-endorphin-immunoreactive peptides from bovine brain with similar physicochemical properties to those of β-endorphin. The amino acid composition of one of these peptides is practically identical to that of bovine β-endorphin from the pituitary gland. More recently, Zakarian and Smyth[146] have succeeded in identifying, in rat brain, six β-endorphin-related peptides that are also present in the pars intermedia (also see ref. 84). These are the following: β-endorphin$_{1-31}$, β-endorphin$_{1-27}$, and β-endorphin$_{1-26}$ in both α-N-acetyl and free forms.[146] Rat brain extracts have been reported to contain even smaller β-endorphin fragments such as β-endorphin$_{1-17}$, β-endorphin$_{1-16}$, and their des-Tyr derivatives.[147]

6. PHYSIOLOGICAL SIGNIFICANCE

According to the classical view on endocrine hormone function, the hormone is released from the endocrine cells and transported by the circulation to the target organ, where it exerts its effect by specifically interacting with the receptors. ACTH provides a good example of this mode of action. Until 1976, this view strongly influenced investigations on the physiological role of β-LPH (for review see ref. 148). The lipolytic activity of β-LPH, on the basis of which it was designated as lipotropic hormone,[1–3] has not been a specific property of β-LPH. On the contrary, almost all adenohypophyseal hormones (including ACTH and β-MSH; see Table III) have lipolytic side effects in addition to their main biological functions. Similarly, neither the stimulatory effect on aldosterone secretion[55] nor the effect on the secretion of the pancreas[149] are unique biological properties of β-LPH. Similar biological actions are carried out by other component peptides of the same biosynthetic precursor such as the 76-residue N-terminal fragment of human proopiomelanocortin[150] (aldosterone-stimulating activity) and β-endorphin[151] (insulin- and glucagon-releasing action).

Structurally unrelated fragments of proopiomelanocortin (α-MSH, β-endorphin) have been reported to elicit homologous actions in the central nervous system too.[152] Melanotropins, ACTH, and β-endorphin induce grooming behavior in rats that can be blocked by opiate antagonists.[153,154] On the other hand, some studies suggest that ACTH and β-endorphin might interact antagonistically in the central nervous system.[155,156] Székely et al.[157] have reported that α-MSH moderates the development of morphine tolerance and dependence in rats.

The above pharmacological observations are only a few characteristic ones picked out from a large quantity of data supporting our recent view that the multiple secretory peptides from the same biosynthetic precursor may act in a cooperative fashion in either the periphery or the brain. Cooperativity can become effective through either mutual modulatory actions of the coreleased peptides at the same receptor site or their coordinated effects at different receptor sites. In any case, the physiological significance of β-LPH and its intra- or extracellularly formed fragments cannot be understood without viewing the peripheral and central nervous system opiomelanotropinergic systems as an integral whole. These systems appear to be complex enough to be targets of extensive research for a few more decades, at least.

REFERENCES

1. Birk, Y., and Li, C. H., 1964, *J. Biol. Chem.* **239**:1048–1052.
2. Li, C. H., 1964, *Nature* **201**:924–925.
3. Li, C. H., 1978, *Endorphins '78* (L. Gráf, M. Palkovits, and A. Z. Rónai, eds.), Akadémiai Kiadó, Budapest, p. 269.
4. Li, C. H., Barnafi, L., Chrétien, M., and Chung, D., 1965, *Nature* **208**:1093–1094.
5. Lohmar, P., and Li, C. H., 1968, *Endocrinology* **82**:898–904.
6. Medzihradszky, K., Bajusz, S., S.-Vargha, H., and Láng, Z., 1969, *Acta Chim. Acad. Sci. Hung.* **59**:165–168.
7. Cseh, G., Gráf, L., and Sutka, K., 1973, *Acta Biochim. Biophys. Acad. Sci. Hung.* **8**:245–251.
8. Chrétien, M., and Li, C. H., 1967, *Can. J. Biochem.* **45**:1163–1174.
9. Gráf, L., Cseh, G., and Medzihradszky-Schweiger, H., 1969, *Biochim. Biophys. Acta* **175**:444–447.
10. Li, C. H., and Chung, D., 1976, *Proc. Natl. Acad. Sci. U.S.A.* **73**:1145–1148.
11. Bradbury, A. F., Smyth, D. G., Snell, C. R., Birdsall, N. J. M., and Hulme, E. C., 1976, *Nature* **260**:793–795.
12. Gráf, L., Székely, J. I., Rónai, A. Z., Dunai-Kovács, Z., and Bajusz, S., 1976, *Nature* **263**:240–242.
13. Hughes, J., Smith, T. W., Kosterlitz, H. W., Fothergill, L. A., Morgan, B. A., and Morris, H. R., 1975, *Nature* **258**:577–579.
14. Gráf, L., Rónai, A. Z., Bajusz, S., Cseh, G., and Székely, J. I., 1976, *FEBS Lett.* **64**:181–184.
15. Guillemin, R., Ling, N., and Burgus, R., 1976, *C.R. Acad. Sci. [D] (Paris)* **283**:783–785.
16. Ling, N., Burgus, R., and Guillemin, R., 1976, *Proc. Natl. Acad. Sci. U.S.A.* **73**:3942–3946.
17. Mains, R. E., Eipper, B. A., and Ling, N., 1977, *Proc. Natl. Acad. Sci. U.S.A.* **74**:3014–3018.
18. Mains, R. E., Eipper, B. A., 1978, *Endorphins '78* (L. Gráf, M. Palkovits, and A. Z. Rónai, eds.), Akadémiai Kiadó, Budapest, pp. 79–126.
19. Przewlocki, R., Höllt, V., Voigt, K. H., and Herz, A., 1979, *Life Sci.* **24**:1601–1608.

20. Scott, A. P., Ratcliffe, J. G., Rees, L. H., Landon, J., Bennett, H. J. P., Lowry, P. J., and McMartin, C., 1973, *Nature [New Biol.]* **244**:65-69.
21. Smyth, D. G., and Zakarian, S., 1980, *Nature* **288**:613-615.
22. Chrétien, M., and Seidah, N. G., 1981, *Mol. Cell Biochem.* **34**:101-127.
23. Nakanashi, S., Inoue, A., Kita, T., Nakamura, M., Chang, A. C. Y., Cohen, S. N., and Numa, S., 1979, *Nature* **278**:423-427.
24. Gráf, L., and Li, C. H., 1973, *Biochem. Biophys. Res. Commun.* **53**:1304-1309.
25. Gráf, L., and Cseh, G., 1968, *Acta Biochim. Biophys. Acad. Sci. Hung.* **3**:175-177.
26. Gráf, L., Barát, E., Cseh, G., and Sajgó, M., 1971, *Biochim. Biophys. Acta* **229**:276-278.
27. Gilardeau, C., and Chrétien, M., 1972, *Chemistry and Biology of Peptides* (J. Meinhofer, ed.), Ann Arbor Science, Ann Arbor, MI, pp. 609-611.
28. Pankov, Yu. A., and Yudaev, N. A., 1972, *Biokhimia* **37**:991-1004.
29. Gráf, L., Kenessey, ., Patthy, A., Grynbaum, A., Marks, N., and Lajtha, A., 1979, *Arch. Biochem. Biophys.* **193**:101-109.
30. Li, C. H., Tan, L., and Chung, D., 1977, *Biochem. Biophys. Res. Commun.* **77**:1088-1093.
31. Li, C. H., and Chung, D., 1981, *Int. J. Peptide Protein Res.* **17**:131-142.
32. Naudé, R. J., Chung, D., Li, C. H., and Oelofsen, W., 1981, *Int. J. Peptide Protein Res.* **18**:138-147.
33. Keutmann, H. T., Lampman, G. W., Mains, R. E., and Eipper, B. A., 1981, *Biochemistry* **20**:4148-4155.
34. Kawauchi, H., Chung, D., and Li, C. H., 1980, *Int. J. Peptide Protein Res.* **15**:171-176.
35. Cseh, G., Gráf, L., and Góth, E., 1968, *FEBS Lett.* **2**:42-44.
36. Cseh, G., Barát, E., Patthy, A., and Gráf, L., 1972, *FEBS Lett.* **21**:344-346.
37. Li, C. H., and Chung, D., 1976, *Nature* **260**:622-624.
38. Li, C. H., Chung, D., and Yamashiro, D., 1980, *Proc. Natl. Acad. Sci. U.S.A.* **77**:7214-7217.
39. Chang, A. C. Y., Cochet, M., and Cohen, S. N., 1980, *Proc. Natl. Acad. Sci. U.S.A.* **77**:4890-4894.
40. Takahashi, H., Teranishi, Y., Nakanashi, S., and Numa, S., 1981, *FEBS Lett.* **135**:97-102.
41. Hsi, K. L., Seidah, N. G., Lu, C. L., and Chrétien, M., 1981, *Biochem. Biophys. Res. Commun.* **103**:1329-1335.
42. Barát, E., Patthy, A., and Gráf, L., 1979, *Proc. Natl. Acad. Sci. U.S.A.* **76**:61220-6123.
43. Seidah, N. G., Hsi, K. L., Chrétien, M., Barát, E., Patthy, A., and Gráf, L., 1982, *FEBS Lett.* **147**:267-272.
44. Spiess, J., Mount, C. D., Nicholson, W. E., and Orth, D. N., 1982, *Proc. Natl. Acad. Sci. U.S.A.* **79**:5071-5075.
45. Gráf, L., Hollósi, M., Barna, I., Hermann, I., Borvendég, J., and Ling, N., 1980, *Biochem. Biophys. Res. Commun.* **95**:1623-1627.
46. St.-Pierre, S., Gilardeau, C., and Chrétien, M., 1976, *Can. J. Biochem.* **54**:992-998.
47. Makarov, A. A., Esipova, N. G., Pankov, Y. A., Grishkovsky, B. A., Lobachev, V. M., and Sukhomudrenko, A. G., 1976, *Mol. Biol. Mosk.* **10**:704-712.
48. Gráf, L., and Hollósi, M., 1980, *Biochem. Biophys. Res. Commun.* **93**:1089-1093.
49. Hollósi, M., Kajtár, M., and Gráf, L., 1977, *FEBS Lett.* **74**:185-189.
50. Chou, P. Y., and Fasman, G. D., 1974, *Biochemistry* **13**:210-221.
51. Chou, P. Y., and Fasman, G. D., 1974, *Biochemistry* **13**:222-245.
52. Gráf, L., and Hollósi, M., 1982, *Hormonally Active Brain Peptides* (K. W. McKerns, and V. Pantic, eds.), Plenum Press, New York and London, pp. 25-43.
53. Tamási, G., Cseh, G., and Gráf, L., 1969, *Experientia* **25**:360-361.
54. Yamashiro, D., and Li, C. H., 1978, *J. Am. Chem. Soc.* **100**:5174-5179.
55. Matsuoka, H., Mulrow, P. J., and Li, C. H., 1980, *Science* **209**:307-308.
56. Yamashiro, D., and Li, C. H., 1974, *Proc. Natl. Acad. Sci. U.S.A.* **71**:4945-4949.
57. Gráf, L., Cseh, G., Barát, E., Rónai, A. Z., Székely, J. I., Kennessey, Á., and Bajusz, S., 1977, *Ann. N.Y. Acad. Sci.* **297**:63-83.
58. Schwandt, P., Richter, W., and Morley, J. S., 1981, *Neuropeptides* **1**:211-216.
59. Jean-Baptiste, E., and Rizack, M. A., 1980, *Life Sci.* **27**:135-141.
60. Bielmann, P., Chrétien, M., and Gattereau, A., 1972, *Horm. Metab. Res.* **4**:22-25.
61. Lis, M., Gilardeau, C., and Chrétien, M., 1972, *Acta Endocrinol.* **69**:507-516.

62. Chrétien, M., Dufault, C., Gratton, J., and Gilardeau, C., 1971, *Horm. Metab. Res.* **3:**355–356.
63. Chrétien, M., Benjannnet, S., Dragon, N., Seidah, N. G., and Lis, M., 1976, *Biochem. Biophys. Res. Commun.* **72:**472–478.
64. Rónai, A. Z., Székely, J. I., Gráf, L., Dunai-Kovács, Z., and Bajusz, S., 1976, *Life Sci.* **19:**733–738.
65. Garzón, J., Sanchez-Franco, F., Kapcala, L., and Del Rio, J., 1981, *Neuropharmacology* **20:**627–632.
66. Dubois, M. P., and Gráf, L., 1973, *Horm. Metab. Res.* **5:**229.
67. Moon, H. D., Li, C. H., and Jennings, B. M., 1973, *Anat. Rec.* **175:**529–537.
68. Pelletier, G., Leclerc, R., Labrie, F., Cote, J., Chrétien, M., and Lis, M., 1977, *Endocrinology* **100:**770–776.
69. Weber, E., Martin, R., and Voigt, K. H., 1979, *Life Sci.* **25:**1111–1118.
70. Chrétien, M., Benjannet, S., Bertagna, X., Lis, M., and Gilardeau, C., 1974, *Clin. Res.* **22:**730–735.
71. Eipper, B. A., and Mains, R. E., 1975, *Biochemistry* **14:**3836–3844.
72. Mains, R. E., and Eipper, B. A., 1976, *J. Biol. Chem.* **254:**4115–4120.
73. Steiner, D. F., Kemmler, W., Tager, H. S., and Peterson, J. D., 1974, *Fed. Proc.* **33:**2105–2115.
74. Roberts, J. L., and Herbert, E., 1977, *Proc. Natl. Acad. Sci. U.S.A.* **74:**4826–4830.
75. Roberts, J. L., Phillips, M., Rosa, P. A., and Herbert, E., 1978, *Biochemistry* **17:**3609–3618.
76. Eipper, B. A., and Mains, R. E., 1978, *J. Biol. Chem.* **253:**5732–5744.
77. Roberts, J. L., and Herbert, E., 1977, *Proc. Natl. Acad. Sci. U.S.A.* **74:**5300–5304.
78. Liotta, A. S., Suda, T., and Krieger, D. T., 1978, *Proc. Natl. Acad. Sci. U.S.A.* **75:**2950–2954.
79. Vale, W., Rivier, C., Yang, L., Minick, S., and Guillemin, R., 1978, *Endocrinology* **103:**1910–1915.
80. Crine, P., Benjannet, S., Seidah, N. G., Lis, M., and Chrétien, M., 1977, *Proc. Natl. Acad. Sci. U.S.A.* **74:**1403–1406.
81. Crine, P., Benjannet, S., Seidah, N. G., Lis, M., and Chrétien, M., 1977, *Proc. Natl. Acad. Sci. U.S.A.* **74:**4276–4280.
82. Seidah, N. G., Gianoulakis, C., Crine, P., Lis, M., Benjannet, S., Routhier, R., and Chrétien, M., 1978, *Proc. Natl. Acad. Sci. U.S.A.* **75:**3153–3157.
83. Crine, P., Gossard, F., Seidah, N. G., Blanchette, L., Lis, M., and Chrétien, M., 1979, *Proc. Natl. Acad. Sci. U.S.A.* **76:**5085–5089.
84. Smyth, D. G., Massey, D. E., Zakarian, S., and Finnie, M. D. A., 1979, *Nature* **279:**252–254.
85. Mains, R. E., and Eipper, B. A., 1981, *J. Biol. Chem.* **256:**5683–5688.
86. Eipper, B. A., and Mains, R. E., 1981, *J. Biol. Chem.* **256:**5689–5695.
87. Herbert, E., Budarf, M., Phillips, M., Rosa, P., Policastro, P., Oates, E., Roberts, J. L., Seidah, N. G., and Chrétien, M., 1980, *Ann. N.Y., Acad. Sci.* **343:**79–93.
88. Gossard, F., Seidah, N. G., Crine, P., Routhier, R., and Chrétien, M., 1980, *Biochem. Biophys. Res. Commun.* **92:**1042–1051.
89. Cochet, M., Chang, A. C. Y., and Cohen, S. N., 1982, *Nature* **297:**335–339.
90. Drouin, J., and Goodman, H. M., 1980, *Nature* **288:**610–613.
91. Roberts, J. L., Seeburg, P. H., Shine, J., Herbert, E., Baxter, J. D., and Goodman, H. M., 1979, *Proc. Natl. Acad. Sci. U.S.A.* **76:**2153–2157.
92. Seidah, N. G., and Chrétien, M., 1981, *Proc. Natl. Acad. Sci. U.S.A.* **78:**4236–4240.
93. Seidah, N. G., Rochemont, J., Hamelin, J., Benjannet, S., and Chrétien, M., 1981, *Biochem. Biophys. Res. Commun.* **102:**710–716.
94. Gráf, L., Bajusz, S., Patthy, A., Barát, E., and Cseh, G., 1971, *Acta Biochim. Biophys. Acad. Sci. Hung.* **6:**415–417.
95. Böhlen, P., Esch, F., Shibasaki, T., Baird, A., Ling, N., and Guillemin, R., 1981, *FEBS Lett.* **128:**67–70.
96. Browne, C. A., Bennett, H. P. J., and Solomon, S., 1981, *Biochem. Biophys. Res. Commun.* **100:**336–343.

97. Shibasaki, T., Ling, N., and Guillemin, R., 1980, *Nature* **285**:416–417.
98. Geisow, M. J., 1978, *FEBS Lett.* **87**:111–114.
99. Loh, P. Y., and Gainer, H., 1978, **96**:269–272.
100. Gráf, L., and Kenessey, Á., 1981, *Hormonal Proteins and Peptides,* Volume X (C. H. Li, ed.), Academic Press, New York, pp. 35–63.
101. Gráf, L., Kenessey, Á., Berzétei, I., and Rónai, A. Z., 1977, *Biochem. Biophys. Res. Commun.* **78**:1114–1123.
102. Kenessey, Á., Gráf, L., and Palkovits, M., 1977, *Brain Res. Bull.* **2**:247–250.
103. Kenessey, Á., Gráf, L., Páldi-Haris, P., Hermann, I., and Borvendég, J., 1978, *Endorphins '78* (L. Gráf, M. Palkovits, and A. Z. Rónai, eds.), Akadémiai Kiadó, Budapest, pp. 139–154.
104. Kenessey, Á., Páldi-Haris, P., Makara, G. B., and Gráf, L., 1979, *Life Sci.* **25**:437–444.
105. Loh, Y. P., and Gainer, H., 1982, *Proc. Natl. Acad. Sci. U.S.A.* **79**:108–112.
106. Hook, V. Y. H., Eiden, L. E., and Brownstein, M. J., 1982, *Nature* **295**:341–342.
107. Lindberg, I., Yang, H.-Y. T., and Costa, E., 1982, *Biochem. Biophys. Res. Commun.* **106**:186–193.
108. Fricker, L. D., and Snyder, S. H., 1982, *Proc. Natl. Acad. Sci. U.S.A.* **79**:3886–3890.
109. Mizuno, K., Miyata, A., Kangawa, K., and Matsuo, H., 1982, *Biochem. Biophys. Res. Commun.* **108**:1235–1242.
110. Smith, R. E., and Van Frank, R. M., 1975, *Lysosomes in Biology and Pathology,* Volume 4 (J. T. Dingle and R. T. Dean, eds.), North Holland, Amsterdam, pp. 193–249.
111. Barna, I., Gráf, L., Makara, G. B., and Rappay, G., 1982, *Neuropeptides* **3**:65–70.
112. Rubinstein, M., Stein, S., Gerber, L. D., and Udenfriend, S., 1977, *Proc. Natl. Acad. Sci. U.S.A.* **74**:3052–3055.
113. Lissitsky, J.-C., Morin, O., Dupont, A., Labrie, F., Seidah, N. G., Chrétien, M., Lis, M., and Coy, D. H., 1978, *Life Sci.* **22**:1715–1722.
114. Vuolteenaho, O., Vakkuri, O., and Leppaluoto, J., 1980, *Life Sci.* **27**:57–65.
115. Raymond, V., Lepine, J., Lissitsky, J. C., Cole, J., and Labrie, F., 1979, *Mol. Cell. Endocrinol.* **16**:113–122.
116. Lis, M., Lariviere, N., Maurice, G., Julesz, J., Seidah, N. G., and Chrétien, M., 1982, *Life Sci.* **30**:1159–1164.
117. Allen, R. G., Herbert, E., Hinman, M., Shibuya, H., and Pert, C. B., 1978, *Proc. Natl. Acad. Sci. U.S.A.* **75**:4972–4976.
118. Rossier, J., French, E. D., Rivier, C., Ling, N., Guillemin, R., and Bloom, F. E., 1977, *Nature* **270**:618–620.
119. Guillemin, R., Vargo, T., Rossier, J., Minick, S., Ling, N., Rivier, C., Vale, W., and Bloom, F. E., 1977, *Science* **197**:1367–1369.
120. Petraglia, F., Penalva, A., Genazzani, A. R., and Müller, E. E., 1982, *Life Sci.* **31**:2809–2817.
121. Suda, T., Liotta, A. S., and Krieger, D. T., 1978, *Science* **202**:221–223.
122. Krieger, D. T., 1978, *Endorphins '78* (L. Gráf, M. Palkovits, and A. Z. Rónai, eds.), Akadémiai Kiadó, Budapest, pp. 275–294.
123. Nakai, Y., Nakao, K., Oki, S., Imura, H., and Li, C. H., 1978, *Life Sci.* **23**:2292–2298.
124. Höllt, V., Müller, O. A., and Fahlbusch, R., 1979, *Life Sci.* **25**:37–44.
125. Aronin, N., Wiesen, M., Liotta, A. S., Schussler, G. C., and Krieger, D. T., 1981, *Life Sci.* **29**:1265–1269.
126. Yamaguchi, H., Liotta, A. S., and Krieger, D. T., 1980, *J. Clin. Endocrinol. Metab.* **51**:1002–1008.
127. Nakai, Y., Nakao, K., Oki, S., and Imura, H., 1978, *Life Sci.* **23**:2013–2018.
128. Odagiri, E., Sherrell, B. J., Mount, C. D., Nicholson, W. E., and Orth, D. N., 1979, *Proc. Natl. Acad. Sci. U.S.A.* **76**:2027–2031.
129. Liotta, A. S., Houghten, R., and Krieger, D. T., 1982, *Nature* **295**:593–595.
130. Bruni, J. F., Watkins, W. B., and Yen, S. S. C., 1979, *J. Clin. Endocrinol. Metab.* **49**:649–651.
131. LaBella, F., Queen, G., Senyshyn, J., Lis, M., and Chrétien, M., 1977, *Biochem. Biophys. Res. Commun.* **75**:350–357.
132. Krieger, D. T., Liotta, A., Suda, T., Palkovits, M., and Brownstein, M. J., 1977, *Biochem. Biophys. Res. Commun.* **76**:930–936.

133. Rossier, J., Vargo, T. M., Minick, S., Ling, N., Bloom, F. E., and Guillemin, R., 1977, *Proc. Natl. Acad. Sci. U.S.A.* **74**:5162–5165.
134. Borvendég, J., Gráf, L., Hermann, I., Palkovits, M., and Merétey, K., 1978, *Endorphins '78* (L. Gráf, M. Palkovits, and A. Z. Rónai, eds.), Akadémiai Kiadó, Budapest, pp. 177–186.
135. Gramsch, C., Kleber, G., Höllt, V., Pasi, A., Mehraein, P., and Herz, A., 1980, *Brain Res.* **192**:109–119.
136. Krieger, D. T., Liotta, A., and Brownstein, M. J., 1977, *Proc. Natl. Acad. Sci. U.S.A.* **74**:648–652.
137. Ogawa, N., Panerai, A. E., Lee, S., Forsbach, G., Havlicek, V., and Friesen, H. G., 1979, *Life Sci.* **25**:317–326.
138. Oliver, C., Mical, R. S., and Porter, J. C., 1977, *Endocrinology* **101**:598–604.
139. Watson, S. J., Barchas, J. D., and Li, C. H., 1977, *Proc. Natl. Acad. Sci. U.S.A.* **74**:5155–5158.
140. Bloom, F., Battenberg, E., Rossier, J., Ling, N., and Guillemin, R., 1978, *Proc. Natl. Acad. Sci. U.S.A.* **75**:1591–1595.
141. Pelletier, G., Désy, L., Lissitsky, J.-C., Labrie, F., and Li, C. H., 1978, *Life Sci.* **22**:1799–1804.
142. Liotta, A. S., Gildersleeve, D., Brownstein, M. J., and Krieger, D. T., 1979, *Proc. Natl. Acad. Sci. U.S.A.* **76**:1448–1452.
143. Liotta, A. S., Loudes, C., McKelvy, J. F., and Krieger, D. T., 1980, *Proc. Natl. Acad. Sci. U.S.A.* **77**:1880–1884.
144. Krieger, D. T., Liotta, A. S., Nicholsen, G., and Kizer, J. S., 1979, *Nature* **278**:562–563.
145. Swann, R. W., and Li, C. H., 1980, *Proc. Natl. Acad. Sci. U.S.A.* **77**:230–233.
146. Zakarian, S., and Smyth, D. G., 1982, *Nature* **296**:250–252.
147. Verhoef, J., Loeber, J. G., Burbach, J. P. H., Gispen, W. H., Witter, A., and de Wied, D., 1980, *Life Sci.* **26**:851–859.
148. Gráf, L., 1976, *Pharmacol. Ther.* **2**:753–769.
149. Schwandt, P., Richter, W. O., Kerscher, P., and Bottermann, P., 1981, *Life Sci.* **29**:345–349.
150. Seidah, N. G., Rochemont, J., Hamelin, J., Lis, M., and Chrétien, M., 1981, *J. Biol. Chem.* **256**:7977–7984.
151. Ipp, E., Dobbs, R., and Unger, R. H., 1978, *Nature* **276**:190–191.
152. Walker, J. M., Akil, H., and Watson, S. J., 1980, *Science* **210**:1247–1249.
153. Gispen, W. H., and Wiegant, V. M., 1976, *Neurosci. Lett.* **2**:159–164.
154. Gispen, W. H., Wiegant, V. M., Bradbury, A. F., Hulme, E. C., Smyth, D. G., Snell, C. R., and de Wied, D., 1976, *Nature* **264**:794–795.
155. Jacquet, Y. F., 1978, *Science* **201**:1032–1034.
156. Fratta, W., Rosetti, Z. L., Poggioli, R., and Gessa, G. L., 1981, *Neurosci. Lett.* **24**:71–74.
157. Székely, J. I., Miglécz, E., Dunai-Kovács, Z., Tarnawa, I., Rónai, A. Z., Gráf, L., and Bajusz, S., 1979, *Life Sci.* **24**:1931–1938.

7

Prolactin

Priscilla S. Dannies

1. INTRODUCTION

Prolactin is a hormone secreted from the anterior pituitary gland that has been studied extensively. Early studies showed prolactin to have a variety of effects on animals, including actions on reproduction, development, and regulation of osmolarity.[1] Interest in prolactin continued to grow after human prolactin was recognized as a hormone distinct from human growth hormone.[2] Prolactin has served as a useful model for cell biologists and molecular biologists to study synthesis and secretion of proteins. The large number of factors that affect prolactin production has made it an important tool for investigating the mechanisms of neurotransmitter and hormone action and the regulation of hormone production. This chapter summarizes the control of prolactin production using data primarily from rat and human studies.

2. SYNTHESIS AND PROCESSING OF PROLACTIN

The anterior pituitary gland contains several different types of cells that secrete different hormones. These cells were first distinguished by chemical stains; cells synthesizing and storing prolactin stained with acidic dyes. The morphology of the cells was later characterized using the electron microscope.[3] Prolactin-secreting cells differ from other hormone-producing cells in that they may contain irregularly shaped granules and round granules that are larger than those found in other cell types; the large granules can range from 500 to 900 nm in diameter. Subsequently, immunocytochemical stains for prolactin were developed that confirmed that the cells with large granules are prolactin-producing cells. In addition, use of these stains has revealed that some prolactin-producing cells do not contain the large secretory granules,[4] indicating that all the prolactin cells in the pituitary gland do not have the same morphology.

Prolactin is synthesized on membranes of the rough endoplasmic reticulum,[5] and cells that actively synthesize prolactin contain large amounts of rough

---

Priscilla S. Dannies • Department of Pharmacology, Yale University School of Medicine, New Haven, Connecticut 06510.

endoplasmic reticulum, often in a distinct concentric pattern.[3] Prolactin resembles most other secreted proteins in that the hormone is made in a precursor form larger than the final product. The precursor is larger because it has an extra sequence on the N-terminal end that is hydrophobic.[5] After this N-terminal end is synthesized, factors in the cell recognizing this sequence cause it to bind to and insert into membranes of the rough endoplasmic reticulum. Translocation across the membranes occurs only as the protein is synthesized; the precursor of prolactin is not translocated if membranes are added after the protein is synthesized. The precursor of prolactin is rapidly cleaved into prolactin in the intact cell, so that preprolactin is only detectable using short labeling periods and is not normally found stored in granules.

The course of prolactin through the cell after synthesis on membranes of the rough endoplasmic reticulum has been followed by Farquhar and co-workers[6] using autoradiographic studies of tritiated leucine incorporation. Cells from estrogen-treated rats incorporate most of their amino acids into prolactin after a brief labeling period, so that the location in the cell of incorporated radioactive leucine primarily indicates the location of prolactin. Dispersed cells were incubated with tritiated leucine for 5 min, at which time the incorporated leucine is mainly associated with the rough endoplasmic reticulum. Fifteen minutes after this 5-min labeling period, autoradiographic grains are located over elements of the Golgi zone, and there are a few grains in secretory granules. One hour after the 5-min labeling period, the amount of radioactivity in the granules increases and continues to increase for the next 2 h. Small granules and irregularly shaped granules have more radioactivity at early times; at later times (2 to 3 h after the pulse), the large rounded granules are labeled, indicating that these are the last to be formed.

Biochemical studies using continuous labeling give results that are consistent with the results of the autoradiographic studies. Labeled prolactin from primary cells in culture begins to appear in the medium in large quantities 2 h after tritiated leucine is added to cultures,[7] which is enough time for prolactin to get packaged into secretory granules. All of the newly synthesized prolactin, however, does not get out of the cells with the same kinetics. Swearingen[8] showed that pituitary glands that had incorporated [^{3}H]leucine subsequently secreted prolactin with a higher specific activity than the prolactin that remained in the glands, and similar studies have been performed in cultures of monolayer cells.[9] These studies show that the cells have a slowly turning over pool and that release of this pool can be preferentially stimulated by agents that release prolactin. The presence of more than one processing route may complicate studies on mechanisms of prolactin release.

Prolactin is tightly packaged in the mature forms of the secretory granules; Farquhar et al.[6] estimate that prolactin in the mature granule is 50–150 times more concentrated than prolactin in the rough endoplasmic reticulum, based on relative grain density after autoradiography. Once packaged, the contents of granules are relatively stable; the inner cores remain visible by electron microscopy after the surrounding membranes are removed by detergent treatment, and intact and membraneless granules are stable at low pH.[10] Purified prolactin granules have molecules that can be labeled with [^{35}S]sulfate and appear to be proteins.[11] The granules also have antigenic sites in common with secretory vesicles in neurons.[12] The anterior of the granules has an acidic pH,[13] so presumably there is an ion pump to maintain the pH gradient. However, the mechanisms by which prolactin is concentrated are not understood, and not much is known about the function of other components of the granules.

Prolactin granules are released from the cell membrane by exocytosis, that is, fusion of the granule membrane to the outer membrane of the cells to release the contents. This process has been observed in several laboratories and has been reviewed by Farquhar.[3] If the granules are not released, they may be digested by lysosomes, a process termed crinophagy.[3] This process was demonstrated in suckling rats, which synthesize large amounts of prolactin. Such rats show a drop in the release of prolactin when deprived of their young, and large granules accumulate in cells in the pituitary gland for the next 12 h. At later times, lysosomes that contain the secretory granule inner cores are seen, indicating the excess granule contents are eliminated by digestion in the lysosomes.

3. REGULATION OF PROLACTIN PRODUCTION

3.1. Agents That Regulate Prolactin Production by Acting Directly at the Pituitary Gland

A variety of hormones and neuropeptides regulate prolactin production and do so at several steps in the processing and release of prolactin. Prolactin synthesis, storage, degradation, and release, as well as the number of prolactin-producing cells can all be affected. Four different systems have been used to investigate which agents have effects at the level of the pituitary gland and what the mechanisms of action of these agents are. The systems are: (1) perfusion of the anterior pituitary gland while still in the animal; (2) incubation of anterior pituitary halves or pieces in culture; (3) dispersed cell cultures of anterior pituitary glands, either in perfused columns or in monolayer cultures; (4) cultures of cell strains derived from pituitary tumors.

The choice of system depends on the questions to be asked. To determine if an agent can directly affect normal prolactin cells, the perfusion of the gland and the incubation of pituitary halves are the best systems, since cells remain in their original environment undisturbed by enzymatic or mechanical processes. The perfusion system may most closely resemble the gland in the intact animal but is the most technically difficult. The use of pituitary halves is easier; a disadvantage is that substances may slowly diffuse into the interior of the gland, since access to medium by all cells is not uniform. To study the synthesis and processing of prolactin as well as the mechanisms by which these processes are regulated, monolayer cultures of cells, normal or tumor, are easiest to use to get the large number of replicate samples required, and in both cases all cells are equally exposed to the medium. Dispersed cultures of normal glands are more likely to resemble normal cells in the intact gland than tumor cells are, but the enzymatic process used to disperse the cells as well as adaptations to culture conditions may change the properties of the cells. Cultures of normal cells also have a heterogeneous population of cells, a disadvantage for biochemical studies such as measurements of cyclic AMP levels, since one cannot be sure in which cell type the changes occur. Clonal strains of pituitary tumor cells provide a relatively homogeneous population and can be obtained in large quantities but may have lost some of the normal processes and responses that occur in normal cells and gained others that normal cells do not have.

Detection of a response to a given agent in the first three systems need not mean that the compound tested is acting directly on the prolactin-producing cell itself. The pituitary gland has recently been shown to contain endorphins; structures similar to gastrin, secretin, calcitonin, and vasoactive intestinal peptide (VIP) have been detected by immunocytochemical techniques. These substances may exist in addition to the six hormones traditionally regarded as anterior pituitary hormones (prolactin, growth hormone, thyrotropin, follicle-stimulating hormone, luteinizing hormone, and adrenocorticotropic hormone). Some of the other peptides, such as VIP, can stimulate prolactin secretion, so agents that affect prolactin production may do so indirectly by affecting secretion of VIP or other factors that affect prolactin release. The demonstration of a direct action on pituitary tumor cells may not be what occurs physiologically, since the cells may have developed inappropriate responses.

The compounds discussed below affect prolactin production by acting at the level of the anterior pituitary gland in normal or tumor cells. The number of factors known to affect prolactin production has been growing, so the list reported here may not be complete. In some cases, there is disagreement among laboratories as to whether a compound has an effect, and the discrepancy often cannot be simply explained by the use of different systems. One possible reason for these discrepancies is that basal release can vary from one experiment to another from low to high amounts of release. If basal release is low, it may be easier to see stimulation of release and more difficult to see inhibition. A failure, therefore, to see a direct action at the level of the pituitary gland in one set of experiments need not mean that such an action cannot exist. For this reason, we have listed compounds as causing effects when the data are conflicting. The references listed below are not intended to be complete, so that not all studies that show effects are listed.

3.1.1. Growth of Prolactin-Producing Cells

Estrogen has been known since 1947 to increase pituitary gland size[14]; it does this by causing hypertrophy and hyperplasia of the prolactin-producing cells. Prolonged treatment with estrogen induces formation of prolactin-secreting pituitary tumors in rats.[15] Estrogen can increase cell number by a direct effect at the level of the pituitary gland. Estradiol stimulates growth of mammotrophs in primary cultures of pituitary glands[16] and stimulates growth of GH cells, which are strains of rat pituitary tumor cells.[17] There are problems with seeing reproducible estrogen effects in culture; the medium used to maintain the cells usually contains serum with significant amounts of estrogens. The stimulation of growth of pituitary cells is quite sensitive to estradiol, occurring

Table I
Agents That Affect the Synthesis of Prolactin

Agent	Reference
Stimulate	
Estrogen	5
Thyrotropin-releasing hormone	20
Epidermal growth factor	21
Thyroid hormone	22
1,25-Dihydroxyvitamin D_3	23
Inhibit	
Hydrocortisone	24
Bromocriptine	5
Thyroid hormone	25
1,25-Dihydroxyvitamin D_3	26

at concentrations as low as 5×10^{-12} M, so that endogenous estrogens in the serum can fully stimulate cell growth.[17] Serum treated with charcoal to remove steroids, serum-free medium, or serum from castrated animals has been used to overcome this problem.

Dopaminergic agonists such as bromocriptine can inhibit pituitary tumor development,[18] and estrogen-induced mitosis is also inhibited.[19] As yet, there are no studies to indicate whether this growth-inhibiting ability is a direct effect at the level of the pituitary gland. Both estrogen and dopaminergic agents could affect pituitary growth by changing the output of hypothalamic factors that affect the gland as well as by direct effects.

3.1.2. Synthesis

The agents that affect prolactin synthesis are listed in Table I. Estradiol, thyrotropin-releasing hormone (TRH), and the inhibitory effects of thyroid hormone and bromocriptine have been shown to be mediated by changes in messenger RNA levels. Thyroid hormone and vitamin D_3 have been reported to both increase and decrease prolactin synthesis; the factors that control the direction of the responses are not determined. Dopamine presumably decreases synthesis, since dopaminergic agonists such as bromocriptine do. Other factors that have been studied in less detail may also influence prolactin synthesis. Insulin, for example, causes a long-term increase in prolactin production that probably reflects increased synthesis.[27] Progesterone inhibits estrogen stimulation of prolactin production, and this also is probably an effect on synthesis.[28]

3.1.3. Storage

Estrogen, insulin, and epidermal growth factor can synergistically increase the capacity of GH cells to store prolactin, an effect that is independent of the stimulation of hormone synthesis and release.[27] In animals and man, estrogen

Table II
Agents That Affect Prolactin Release

Agent	Reference
Stimulate	
Thyrotropin-releasing hormone (TRH)	29
Vasoactive intestinal peptide (VIP)	30
Epidermal growth factor	21
Bombesin	31
Opiate peptides	32
Calcitonin	33
Inhibit	
Dopamine	34
Somatostatin	35
Acetylcholine	7
γ-Aminobutyric acid (GABA)	36
Sauvagine	37
Histidyl-proline-diketopiperazine	38

increases the amount of prolactin that can be released by stimulating agents[18]; the increased release may result in part from larger stores of prolactin induced by estrogen.

3.1.4. Release

Factors that both stimulate and inhibit prolactin release are listed in Table II. These include hypothalamic factors such as TRH as well as peptides from frog skin such as bombesin and sauvagine. Peptides similar to frog skin peptides can be found in mammalian brain and may have a role in prolactin regulation. Many of these agents may affect synthesis as well as release of prolactin from glands, but the effects on synthesis have not been examined. Bombesin, however, has been shown only to affect prolactin release and not synthesis from GH cells.

3.1.5. Degradation

Bromocriptine, a dopaminergic agonist, causes degradation of newly synthesized prolactin in pituitary cells in culture in addition to inhibiting release and synthesis of prolactin.[7] Maurer[39] demonstrated that cycloheximide prevents the degradation of prolactin if it is added with bromocriptine but not if the cells are pretreated with bromocriptine. These data suggest that protein synthesis may be necessary for the induction of prolactin degradation. Dopamine also has been reported to elevate activity of lysosomal enzymes.[40] These experiments therefore indicate that dopamine regulates prolactin degradation and that it is an active process, not just secondary to the inhibition of release.

3.1.6. Interactions Among Agents That Regulate Prolactin Production

In addition to directly affecting prolactin production, hormones can regulate the ability of other hormones to release prolactin. Estrogen is the best studied example; it increases the amount of prolactin released by TRH and makes the cells more sensitive to the peptide.[18] It is difficult to see a TRH-induced release of prolactin in male rats unless they have been primed with estrogen. Estrogen also reduces the inhibitory effects of dopamine, both by making the cells less sensitive to dopaminergic agents and by reducing the maximum amount of inhibition of release caused by such agents.[41,42] At least part of this regulation appears to be caused by regulation of receptor number, since there is evidence that the amount of dopamine receptors in the pituitary gland changes during the estrous cycle,[43] and estrogen can increase the number of TRH receptors in pituitary tumor cells.[44] Estrogen probably also affects release through other means; for example, estrogen can increase the amount of prolactin released by TRH by affecting stores of the hormone in the gland. Thyroid hormones can also modulate the response of prolactin release to TRH by changing the number of TRH receptors.[45] Other factors, such as somatostatin, can also decrease TRH receptors, indicating that peptide hormones as well as estrogen and thyroid hormone may regulate the ability of agents to influence prolactin production.

3.2. Regulation at Levels Other Than the Pituitary Gland

The agents listed above may exert their effects not only at the level of the pituitary gland but also at the hypothalamus and other parts of the brain. Factors that affect production of prolactin indirectly may not cause the same effects that they do when added directly to the gland. For example, γ-aminobutyric acid (GABA) and acetylcholine agonists increase serum prolactin levels in some experiments, although the direct effects on the pituitary gland are inhibitory. The stimulatory effects are presumably mediated through the central nervous system. There are other compounds that also affect prolactin secretion by acting through the central nervous system (for reviews, see refs. 18,29,46). Two neurotransmitters that have marked effects are serotonin and histamine. Serotonin appears important in suckling-induced prolactin release. Activation of the serotonergic system by inhibiting serotonin uptake or stimulating serotonergic receptors results in release of prolactin. Histaminergic agents also affect prolactin release. Intraventricular injection of histamine in animals increases prolactin release, and the increase can be blocked by H_1 antagonists. On the other hand, H_2 antagonists such as cimetidine enhance prolactin release, suggesting that histamine acts at an H_1 site to increase prolactin secretion and an H_2 site to decrease prolactin secretion.

4. PHYSIOLOGICAL PATTERNS OF PROLACTIN SECRETION

4.1. Prolactin Secretion in Man

The patterns of prolactin secretion have been closely studied in humans.[18,29] Men and women have similar basal levels of prolactin with considerable variability in the range of normal values (up to 25 ng/ml). In both men and women prolactin is secreted in a pusatile fashion—that is, small bursts occur over periods of 2–3 h. This pattern of episodic secretion is not unique to prolactin but is true of all anterior pituitary hormones studied, although the timing of the pulses vary. In addition to these frequent bursts, prolactin secretion rises during sleep; the rise begins after sleep starts and continues to increase through the duration of sleep. Prolactin levels in the serum may take several hours after waking to return to basal daytime levels. The rise in prolactin is sleep related; it does not occur at night if the subject does not sleep.

Prolactin levels also increase in men and women with physical stress, such as surgical procedures, and with noninvasive procedures such as proctoscopy. Prolactin levels are elevated during exercise and insulin-induced hypoglycemia. Levels also rise with emotional stress.

Females that are pregnant or lactating have high prolactin concentrations in their serum. Prolactin values rise during pregnancy and drop after birth. Subsequently, there is a large burst of prolactin release that occurs after suckling. Unlike oxytocin release, prolactin release does not occur through auditory or visual cures; it is the actual act of suckling that triggers prolactin release. Nipple stimulation alone in some nonlactating females and males can stimulate prolactin release.

4.2. Prolactin Secretion in Rats

Prolactin release in the rat has recently been reviewed in detail.[47] Female rats have a distinctive pattern of prolactin release that occurs during the estrous cycle. Prolactin levels are low throughout diestrus and the morning of proestrus; in the afternoon of proestrus, there is a large increase in prolactin release. At night, the levels slowly decline until they have reached basal values by the morning of estrus. This increase appears to be caused by the increase in estrogen in the serum during proestrus. The effect of estrogen on prolactin secretion differs in males and females. In males, there is a stimulatory effect of estrogen on basal prolactin levels, but the surges of prolactin that occur when estrogen is implanted in a female rat do not occur. At least part of the action of estrogen in this case appears to be on the hypothalamus; a lesion behind the optic chiasm prevents the estrogen-induced surges of prolactin.

Prolactin release occurs in two other stages in the female rat: in pregnancy and pseudopregnancy. For the first 8 days of pregnancy, rats have two bursts of prolactin release in each 24 h, a diurnal and a nocturnal surge. The diurnal surge is not observed after day 8; the nocturnal surge lasts one more day than the diurnal surge, so on day 9, only a nocturnal surge is seen. Serum prolactin levels do not rise again until a day or two before birth. Pseudopregnancy is caused by cervical stimulation or sterile mating. It results in two bursts of release a day, as is found in normal pregnancy. This pattern of release continues for about 10 days, until finally the rat returns to a normal estrous cycle. There is a large increase in serum prolactin after suckling, as there is in man, and the initiation of suckling is necessary to release prolactin. Stress can also increase prolactin levels in the serum in rats as well as man, especially in female rats at estrus.

4.3. Factors That Regulate Physiological Changes in Prolactin Secretion

Releasing factors and neurotransmitters released from the hypothalamus travel through the portal blood system that connects the hypothalamus to the pituitary gland. In view of the large number of factors that can regulate prolactin secretion and the different circumstances under which prolactin is released, it is probably not valid to discuss "the" releasing hormone or inhibitory factor—almost certainly a combination of factors will be involved in all circumstances. Studies to determine the ways in which prolactin release are physiologically regulated will have to deal with two complications.

One requirement for showing that an agent is involved in physiological control of release is to show that the levels of the compound in the portal blood change appropriately as prolactin levels change. Since so many agents affect prolactin release, changes in release may occur because changes in several factors in the blood have changed, and measuring one compound will not be sufficient. For example, TRH may stimulate prolactin release because levels of dopamine drop.

A second requirement is that infusion of the agents postulated to cause the response should mimic prolactin release patterns that occur physiologically. Experiments to demonstrate this may be complicated because the response of the gland or animal depends on its previous exposure to other factors as indicated by two experiments described below.

Grosvenor and Mena[48] have shown a change in the response of the pituitary gland before and after a brief period of suckling. They used as a model system lactating rats that had been deprived of their young for hours and had, therefore, built up large stores of prolactin. Compounds were tested using rats either in this state or after they had been allowed to suckle their young for just 10 min. Prolactin release rose during this 10-min period but dropped to basal release when suckling was discontinued. Using this system, Grosvenor and Mena have found that TRH stimulated prolactin release only slightly in rats that had not suckled, but there was a larger elevation of release from rats that had first suckled and then received TRH after prolactin release had returned to basal levels. Fagin and Neill have done a study on the effect of dopamine on TRH-induced prolactin release using perfused pituitary glands.[49] They found that when cells were exposed to dopamine, prolactin release was decreased to about 35% of untreated glands. When dopamine infusion was stopped, prolactin release rose to control levels, and when it was added back, prolactin release was inhibited to the same degree as before. When TRH infusion was begun 18 min after dopamine infusion was reinitiated, TRH stimulated prolactin release more than when TRH was administered to glands that had been continuously subjected to dopaminergic inhibition.

These two sets of experiments show that the order and the time of administration may affect the response to different agents. At present, we can not say which factors that can release prolactin are involved in the physiological patterns of prolactin release.

5. INTRACELLULAR MECHANISMS BY WHICH PROLACTIN PRODUCTION IS CONTROLLED

5.1. Release

Investigations into the mechanisms by which the agents mentioned in Table II control prolactin release have centered primarily on two components—one is calcium, and the other, cyclic AMP. these experiments have been reviewed in detail[7,18]; the following is a brief summary of the evidence for the involvement of these two components.

Douglas[50] proposed a general model for the role of calcium in secretion from neuronal, exocrine, and endocrine cells. According to this model, depolarization of the cell membrane leads to a calcium influx; the increased intracellular concentrations of calcium trigger hormone release. Several kinds of evidence support a role of intracellular calcium in prolactin release. (1) Calcium in the medium often stimulates hormone release. (2) Depolarization of the cell membrane by high potassium concentrations causes release of prolactin that is dependent on the presence of calcium in the medium. (3) Divalent cation ionophores that carry calcium through the membrane into the cell stimulate release of prolactin. (4) Pituitary cells have action potentials that contain a calcium component, and TRH, which stimulates release of prolactin, stimulates the frequency of these action potentials, whereas dopamine, which inhibits prolactin release, reduces the frequency.[51] (5) Calcium mobilization has been associated with phosphatidylinositol turnover, although the exact relationship is still a matter of debate. TRH rapidly increases phosphatidylinositol turnover,[52] and dopamine inhibits turnover of this compound.[53]

Cyclic AMP has also been proposed to mediate prolactin release for the following reasons. (1) Agents that increase cyclic AMP, either by activating adenylate cyclase, such as cholera toxin, or by inhibiting phosphodiesterase activity, such as isobutylmethylxanthine, often increase prolactin release. (2) Agents that mimic the effects of endogenous cyclic AMP, such as 8-bromo cyclic AMP, often stimulate prolactin release. (3) Vasoactive intestinal peptide stimulates adenylate cyclase directly and increases prolactin release, whereas dopamine inhibits activation of adenylate cyclase and inhibits prolactin release.

After the increase of intracellular mediators such as cyclic AMP and calcium, a next step is almost certainly an activation of one or more kinases. There are at least three kinases that can be affected. Cyclic AMP activates a cyclic-AMP-dependent protein kinase that will phosphorylate some proteins. Calcium will bind to the calcium-binding protein calmodulin, which has the capacity to activate calcium-dependent protein kinases as well as other enzymes. Although studies using inhibitors of calmodulin, such as trifluoroperazine, have been used to ask whether calmodulin is involved, such studies are not valid, since these agents have so many side effects. One side effect is inhibition of the third kinase, phospholipid-activated kinase. When phosphatidylinositol turnover is activated, there is increased accumulation of the product diacylglycerol, which can also activate a protein kinase.[54] This phospholipid-activated kinase requires calcium for activity but does not require calmodulin, although the kinase is inhibited by agents that also inhibit calcium ion binding to calmodulin.

The steps involved after calcium mobilization and cyclic AMP accumulation will be proven by isolating and characterizing the components involved. The mechanisms are apt to be quite complicated for several reasons. (1) At least some of the factors that control prolactin release appear to act by more than one mechanism. Dopamine, for example, appears to inhibit phosphatidylinositol turnover and prevent calcium entry as well as to decrease the activation of adenylate cyclase. These may be two separate effects. (2) Changes in calcium and cyclic AMP are not isolated, static effects but trigger compensatory changes in the cell, and one factor can induce changes that regulate the others. For example, calcium can activate a phosphodiesterase that breaks down cyclic AMP. Numerous examples of such interactions in other systems have been summarized by Rasmussen.[55] (3) As discussed above, there is more than one route of transport of prolactin through the cells, since part of the newly synthesized prolactin goes into at least one pool that turns over more slowly than the newly synthesized prolactin, which is rapidly released from these cells. In addition, there is morphological evidence for heterogeneity in the pituitary gland. Whether prolactin in these different pools is controlled by the same agents and released by the same mechanisms is not known. (4) Mechanisms other than effects on calcium, phosphatidylinositol, and cyclic AMP may be involved. GABA, for example, changes chloride conductance in the central nervous system, and it is not known how it inhibits prolactin production from the pituitary gland.

5.2. Synthesis

Prolactin synthesis can also be regulated by calcium and cyclic AMP. Increases of both agents can increase synthesis of prolactin through an increase in prolactin messenger RNA.[56,57] Agents other than polypeptide hormones, however, presumably regulate synthesis by direct interactions of the receptor with the nucleus of the cell.

6. EFFECTS OF PROLACTIN

Prolactin is present not only in mammals that lactate but in other vertebrates down through fish. Prolactin has many different effects depending on the species, and these varieties of effects have been classified and summarized by Nicoll[7] into several main groups:

1. Nurturing the young. The involvement in this state can be expressed in several ways. In mammals, it causes development of the breast for

Table III
Classification of Hormones and Receptors Based on Observed Affinities[a]

Hormones		Receptors		
		Type 1 for lactogenic hormones	Type 2 for nonprimate GH	Type 3 for primate GH
Type I	Primate GH	100	100	100
	Bovine PL	100	100	100
Type II	Nonprimate PL	0	100	0.03
Type III	Human PL	100	0	0.03
	Ovine PRL	100	0	0.03
	Human PRL	100	0	0

[a] From Lesniak et al.[59]

lactation; in doves, it causes secretion of crop milk; in some fish, it causes secretion of mucus on which the young feed.
2. Gonadotrophic or antigonadotrophic effects.
3. Osmoregulatory actions.
4. Effects on metamorphosis.
5. Effects on parental behavior.

A brief summary of the known effects in man and rat is listed below; these effects are covered more extensively in other reviews.[7,18,58]

6.1. Prolactin Receptors

The first step in prolactin action is binding to receptors. Receptors on the membranes of cells have been characterized using iodinated hormones.[29,58,60] In primates, growth hormone can bind as well as prolactin can to the lactogenic receptor, but there are specific receptors for growth hormone that do not bind prolactin. Lesniak et al.[59] have analyzed work done with receptors and hormones from various species and suggested three types of receptors, as shown in Table III. The following discussion is limited to lactogenic hormones, but one can observe from the table that these lactogenic receptors interact with growth hormone; this is true of primates as well as nonprimates such as rats. It is also apparent that suppressing prolactin secretion in primates with drugs such as dopaminergic agonists will not necessarily prevent all lactogenic activity, because growth hormone release will not be affected.

Prolactin receptors have been found in several different organs, including mammary gland, ovaries, testes, seminal vesicles, prostate, adrenal cortex, kidney, liver, and lung.[58,60] The number of these receptors can be regulated in the different tissues by hormones such as estrogens and androgens, insulin, and prolactin itself. This receptor is internalized into the cell, as other surface membrane receptors are, and transported into the Golgi. At this point, at least some of the receptor may be transferred to the lysosomes. As yet there is no evidence that the internalization plays a role in mediating the actions of prolactin; internalization may be a means of recycling or emptying the receptor.

6.2. Effects of Prolactin

As described above, prolactin has a variety of actions in different species.[7] The most obvious are perhaps the effects on the mammary gland. The growth and development of breast cells are dependent on the hormonal environment, and the general pattern has been elucidated, although it varies in detail from species to species and even from one strain of rat to another.[58] A simplified overall view is that estrogen promotes duct growth, which occurs during adolescence and during pregnancy. Lobuloalveolar development occurs at the ends of the ducts, and this development is most pronounced during pregnancy, when it is stimulated by estrogen, progesterone, and prolactin. The levels of these three hormones are elevated in pregnancy. In addition, thyroid hormone, insulin, hydrocortisone, and growth hormone are required. After birth, levels of sex steroids drop, and lactation begins. Milk formation is stimulated by prolactin; until birth, the high progesterone or estrogen concentrations or both prevent prolactin from causing milk formation. Other hormones such as insulin, growth hormone, and thyroid hormone must usually be present to see these changes. When breast feeding ceases, the gland involutes, and the alveolar epithelium is reduced, so that some breast cells are programmed to regress and die in response to changes in the environment. In many species, including man and rat, the presence of prolactin is necessary for the maintenance of lactation after birth.

Prolactin also has effects on the gonads in both sexes.[29] In female rats, prolactin maintains the corpus luteum, although under some conditions it can also cause regression of the corpus luteum. Less work has been done with primate ovaries, but prolactin has been reported to have positive and negative effects in primates as well. The dose–response curve of prolactin is biphasic under some conditions when assayed in breast cell cultures.[61] If this concentration dependence is also true in the ovary, it will make the effects of prolactin more difficult to completely elucidate.

In male rodents, prolactin is present in the seminal fluid and is required for spermatogenesis. Prolactin appears to exert part of its action directly on the Leydig cells. In men, the role is less clear, perhaps in part because human growth hormone also has lactogenic effects. Since hyperprolactinemia is associated with impotence and low testosterone levels, there may be a biphasic effect in men as well.

The effects of prolactin on other organs is less simple and involves complex interactions with other hormones and changes in physiological states. Prolactin is reported to influence calcium and vitamin D metabolism,[62] at least in part by regulating the activity of 1-hydroxylase in the mitochondria of the kidney, thereby controlling the formation of the active metabolite of vitamin D, 1,25-dihydroxyvitamin D. There is also evidence that prolactin is involved in osmoregulation,[63] although some of the studies on prolactin have used preparations contaminated by vasopressin, which has confused the interpretation of these studies.

Finally, recent studies have begun to examine the effects of prolactin in the brain. Prolactin had been shown to influence dopamine turnover in the hypothalamus,[64] a process that may be part of a feedback mechanism to regulate prolactin release. Other studies have shown that prolactin can increase dopamine release from additional parts of the brain, such as the striatum,[65] and can affect other neurotransmitters, such as acetylcholine,[66] indicating that the brain may also be a target organ for prolactin.

7. PATHOLOGY

At present, the only pathological effects known to occur with changing levels of prolactin are infertility in women and impotence in men, both associated with high serum levels of prolactin.[29] The cause or causes of these disturbances are not known, and they may be a result of actions at the level of the pituitary gland, the ovary, or at the hypothalamus or higher levels in the central nervous system. The high prolactin levels may result from hypothyroidism, treatment with drugs such as the antipsychotic drugs, pituitary tumors, or unknown causes. In the last two cases, prolactin levels can often be successfully reduced with long-lasting dopaminergic agonists such as bromocriptine.

Prolactin has also been suggested to have a role in the development of breast cancers. There is no question that prolactin is necessary in the development of many breast tumors in model animal systems, but these findings may not apply to human cancer.[58] Reduction of prolactin levels with bromocriptine was not useful in treating breast cancer, but interpretation of these studies may be difficult because of the lactogenic action of human growth hormone. Interest in the role of prolactin has been revived with the finding that women in families with breast cancer have higher than normal serum levels of prolactin when blood levels are monitored for 24 h.[67] At this time, however, the role of prolactin in human breast cancer is not resolved.

8. CONCLUSION

Prolactin is a hormone that can be directly regulated by many factors and indirectly by more, so it is probably more unusual to find a compound that does not affect prolactin secretion than one that does. The release process is heterogeneous with more than one component, so that elucidating the mechanisms that control prolactin production is complicated. Finally, prolactin apparently has actions at many sites in the body, including gonads, breast, kidney, and brain, but the biphasic effects of prolactin and the variety of interactions with other hormones will also make these actions difficult to elucidate.

REFERENCES

1. Nicoll, C. S., 1974, *Handbook of Physiology, Section 7: Endocrinology,* Volume IV (E. Knobil and W. H. Sawyer, eds.), American Physiological Society, Washington, pp. 253–292.

2. Frantz, A. G., 1973, *Frontiers in Neuroendocrinology* (W. F. Ganong and L. Martini, eds.), Oxford University Press, New York, pp. 337-374.
3. Farquhar, M. G., 1977, *Advances in Experimental Medicine and Biology: Comparative Endocrinology of Prolactin*, Volume 80 (H. D. Dellman, J. A. Johnson, and D. M. Klachko, eds.), Plenum Press, New York, pp. 37-94.
4. Nogami, H., and Yoshimura, F., 1982, *Anat. Rec.* **202**:261-274.
5. Maurer, R. A., 1982, *Cellular Regulation of Secretion and Release* (P. M. Conn, ed.), Academic Press, New York, pp. 267-300.
6. Farquhar, M. G., Reid, J. J., and Daniell, L. W., 1978, *Endocrinology* **102**:296-311.
7. Dannies, P. S., 1982, *Cellular Regulation of Secretion and Release* (P. M. Conn, ed.), Academic Press, New York, pp. 529-568.
8. Swearingen, K. C., 1971, *Endocrinology* **89**:1380-1388.
9. Walker, A. M., and Farquhar, M. G., 1980, *Endocrinology* **107**:1095-1104.
10. Giannattasio, G., Zanini, A., and Meldolesi, J., 1975, *J. Cell Biol.* **64**:246-251.
11. Matthew, W. D., Reichardt, L. F., and Tsavaler, L., 1981, *Cold Spring Harbor Reports in Neurosciences*, Volume 2: *Monoclonal Antibodies to Neural Antigens* (R. McKay, M. C. Raff, and L. F. Reichardt, eds.), Cold Spring Harbor Laboratory, Cold Spring Harbor, New York, pp. 163-180.
12. Matthew, W. D., Reichardt, L. F., and Travaler, L., 1981, *Cold Spring Harbor Reports in Neurosciences*, Volume 2: *Monoclonal Antibodies to Neural Antigens* (R. McKay, M. C. Raff, and L. F. Reichardt, eds.), Cold Spring Harbor Laboratory, Cold Spring Harbor, New York, pp. 163-180.
13. Carty, S. E., Johnson, R. G., and Scarpa, A., 1982, *J. Biol. Chem.* **257**:7269-7273.
14. Baker, B. L., and Everett, N. B., 1947, *Endocrinology* **41**:144-157.
15. Furth, J., Clifton, K. N., Gadsen, E. L., and Buffet, R. F., 1956, *Cancer Res.* **16**:608-616.
16. Lieberman, M. E., Maurer, R. A., Claude, P., and Gorski, J., 1982, *Mol. Cell. Endocrinol.* **25**:277-294.
17. Amara, J. F., and Dannies, P. S., 1983, *Endocrinology* **112**:1141-1143.
18. Flückiger, E., del Pozo, E., and von Werder, K., 1982, *Prolactin: Physiology, Pharmacology and Clinical Findings*, Springer-Verlag, New York.
19. Lloyd, H. M., Meares, J. D., and Jacobi, J., 1975, *Nature* **255**:497-498.
20. Potter, E., Nicolaisen, A. K., Ong, E. S., Evans, R. M., and Rosenfeld, M. G., 1981, *Proc. Natl. Acad. Sci. U.S.A.* **78**:6662-6666.
21. Schonbrunn, A., Krasnoff, M., Westendorf, J. M., and Tashjian, A. H., Jr., 1980, *J. Cell Biol.* **85**:786-797.
22. Perrone, M. H., Greer, T. L., and Hinkle, P. M., 1980, *Endocrinology* **106**:600-605.
23. Wark, J. D., and Tashjian, A. H., Jr., 1982, *Endocrinology* **111**:1755-1757.
24. Dannies, P. S., and Tashjian, A. H., Jr., 1973, *J. Biol. Chem.* **248**:6174-6179.
25. Maurer, R. A., 1982, *Endocrinology* **110**:1507-1513.
26. Murdoch, G. H., and Rosenfeld, M. G., 1981, *J. Biol. Chem.* **256**:4050-4055.
27. Kiino, D. R., and Dannies, P. S., 1982, *Yale J. Biol. Med.* **55**:409-420.
28. Haug, E., 1979, *Endocrinology* **104**:429-437.
29. Martin, J. B., Reichlin, S., and Brown, G. M., 1977, *Clinical Neuroendocrinology*, F. A. Davis, Philadelphia.
30. Gourdji, D., Bataille, D., Vauclin, N., Grouselle, D., Rousselin, G., and Tixier-Vidal, A., 1979, *FEBS Lett.* **104**:165-168.
31. Westendorf, J. M., and Schonbrunn, A., 1982, *Endocrinology* **110**:352-358.
32. Enjalbert, A., Ruberg, M., Arancibra, S., Fiore, L., Priam, M., and Kordon, C., 1979, *Endocrinology* **105**:823-826.
33. Iwasaki, Y., Chihara, K., Iwasaki, J., Abe, H., Fujita, T., 1979, *Life Sci.* **25**:1243-1248.
34. MacLeod, R. M., 1976, *Frontiers in Neuroendocrinology*, Volume IV (L. Martini and W. F. Ganong, eds.), Raven Press, New York, pp. 169-194.
35. Vale, W., Rivier, C., Brazeau, P., and Guillemin, R., 1974, *Endocrinology* **95**:968-977.
36. Grandison, L., and Guidotti, L., 1979, *Endocrinology* **105**:754-759.
37. Falaschi, P., D'Urso, R., Negri, L., Rocco, A., Montecucchi, P. C., Henschen, A., Melchiorri, P., and Erspamer, V., 1982, *Endocrinology* **111**:693-695.

38. Peterkofsky, A., Battaini, F., Koch, Y., Takahara, Y., and Dannies, P., 1981, *Mol. Cell. Biochem.* **42**:45–63.
39. Maurer, R. A., 1980, *Biochemistry* **19**:3573–3579.
40. Nansel, D. D., Gudelsky, G. A., Reymond, M. J., Neaves, W. B., and Porter, J. C., 1981, *Endocrinology* **108**:896–902.
41. Raymond, V., Beaulieu, M., Labrie, F., and Boissier, J., 1978, *Science* **200**:1173–1175.
42. West, B., and Dannies, P. S., 1980, *Endocrinology* **106**:1108–1113.
43. Heiman, M. L., and Ben-Jonathan, N., 1982, *Endocrinology* **111**:37–41.
44. Gershengorn, M. C., Marcus-Samuels, B. E., and Geras, E., 1979, *Endocrinology* **105**:171–176.
45. Hinkle, P. M., Perrone, M. H., and Schonbrunn, A., 1981, *Endocrinology* **108**:199–205.
46. Weiner, R. I., and Ganong, W. F., 1978, *Physiol. Rev.* **58**:905–976.
47. Neill, J. D., 1980, *Frontiers in Neuroendocrinology*, Volume 6 (L. Martini and W. F. Ganong, eds.), Raven Press, New York, pp. 129–156.
48. Grosvenor, C. E., and Mena, F., 1980, *Endocrinology* **107**:863–868.
49. Fagin, K. D., and Neill, J. D., 1981, *Endocrinology* **109**:1835–1840.
50. Douglas, W. W., 1968, *Br. J. Pharmacol.* **34**:451–474.
51. Vincent, J. D., and Dufy, B., 1982, *Cellular Regulation of Secretion and Release* (P. M. Conn, ed.), Academic Press, New York, pp. 107–146.
52. Rebecchi, M. J., Kolesnick, R. N., and Gershengorn, M. C., 1983, *J. Biol. Chem.* **258**:227–234.
53. Canonico, P. L., Valdenegro, C. A., and MacLeod, R. M., 1982, *Endocrinology* **111**:347–349.
54. Kishimoto, A., Takai, Y., Mori, T., Kikkawa, U., and Nishizuka, Y., 1980, *J. Biol. Chem.* **255**:2273–2276.
55. Rasmussen, H., and Goodman, D. B. P., 1977, *Physiol. Rev.* **57**:421–509.
56. White, B., Bauerle, L. R., and Bancraft, F. C., 1982, *J. Biol. Chem.* **256**:5942–5945.
57. Murdoch, G. H., Rosenfeld, M. G., and Evans, R. M., 1982, *Science* **218**:1315–1317.
58. Cowie, A. T., Forsyth, I. A., and Hart, I. C., 1980, *Hormonal Control of Lactation*, Springer-Verlag, Berlin, Heidleberg, New York.
59. Lesniak, M. A., Gorden, P., and Roth, J., 1977, *J. Clin. Endocrinol. Metab.* **44**:838–849.
60. Nagasawa, H., Sakai, S., and Banerjee, M. R., 1979, *Life Sci.* **24**:193–208.
61. Djiane, J., Houdebine, L. M., and Kelly, P. A., 1982, *Endocrinology* **110**:791–795.
62. Fraser, D. R., 1980, *Physiol. Rev.* **60**:588–607.
63. Loretz, C. A., and Bern, H. A., 1982, *Neuroendocrinology* **35**:292–304.
64. Perkins, N. A., and Westfall, T. C., 1978, *Neuroscience* **3**:59–63.
65. Chen, Y. F., and Ramirez, V. D., 1982, *Endocrinology* **111**:1740–1742.
66. Wood, P. L., Cheney, D. L., and Costa, E., 1980, *J. Neurochem.* **34**:1053–1057.
67. Levin, P. A., and Malarkey, W. B., 1982, *J. Clin. Endocrinol. Metab.* **53**:179–183.

8

Thyrotropin-Releasing Hormone

Chandan Prasad

1. INTRODUCTION

Although the hypothalmic control of anterior pituitary function through endogenous chemical substances was postulated more than a generation ago, not until 1955 did the search begin for the hypothalamic release-modulating substances.[1] Thyrotropin-releasing hormone (TRH) (Fig. 1) was the first hypothalamic hypophysiotropic hormone to be chemically characterized in the laboratories of Drs. Andrew Schally and Roger Guillemin.[2,3] Soon after this discovery, synthetic TRH became available, and TRH radioimmunoassays were developed.[4–6] This prompted a great flurry of research activity in various aspects of TRH endocrinology, biochemistry, physiology, and pharmacology. It has become obvious now that thyrotropin (TSH) release from the pituitary is only one of numerous actions of TRH, and this tripeptide has been associated with a large variety of other endocrine and nonendocrine central nervous system-related functions (see Section 4, Tables IV and V).

The purpose of this chapter is to review data on the distribution of this peptide within and outside the central nervous system, its regional and subcellular distribution, its metabolism and multiple biological actions, and then, finally, its possible function as a central neurotransmitter/neuromodulator. Detailed consideration is not given to the issue of clinical effects of TRH, which have been reviewed elsewhere.[7,8]

2. DISTRIBUTION AND MODULATION OF ENDOGENOUS TRH CONCENTRATION

2.1. In Central Nervous System

TRH-like immunoreactivity or biological activity is widely distributed throughout the nervous system of vertebrate species, including man, rat, guinea pig, mouse, frog, salamander, chicken, snake, and salmon (Table I).[9,10] TRH-

---

Chandan Prasad • Departments of Medicine (Section of Endocrinology) and Biochemistry, Louisiana State University Medical Center, New Orleans, Louisiana 70112.

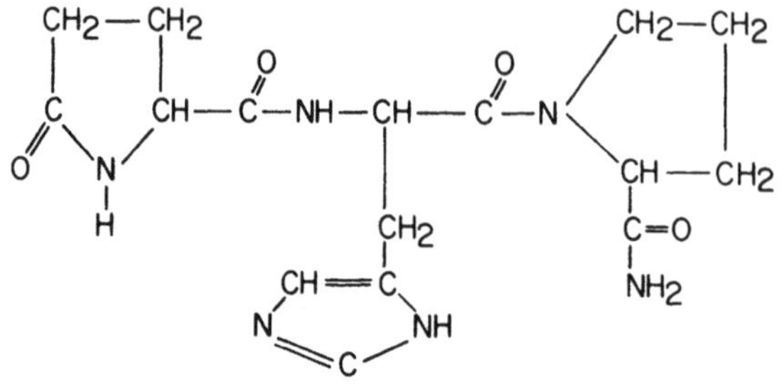

Fig. 1. Structure of thyrotropin-releasing hormone (pGlu-His-Pro-NH$_2$). The three constituent amino acids, from left to right, are pyroglutamic acid, histidine, and prolineamide.

like immunoreactivity has also been shown to be present in the whole brain of the larval lamprey, in the head end of the amphioxus, and in the circumesophageal ganglia of the invertebrate snail.[11] Over 70% of the total TRH in the central nervous system is found outside the hypothalamus, although the concentrations are lower than those within the hypothalamus or posterior pituitary.[12]

TRH-like immunoreactivity has also been demonstrated in the rat spinal cord[13] and the retina of the rat[14] and the human.[15] Rat retinal TRH-like activity

Table I
Distribution of TRH in Tissues and Body Fluids of Various Animal Species

Species	Tissue/body fluid	Concentration (fmol/mg protein)	Reference
Human	Cerebellum	15 ± 7	214
	Hypothalamus	1,093 ± 312	214
	Placenta	54.6	215
	Pineal gland	71.7	32
	Peripheral blood	78 pg/ml	48
Rat	Cerebellum	31.5 ± 4.0	73
	Hypothalamus	2,208.5 ± 155.3	73
	Caecum	91 ± 27	27
	Rectum	27.6 ± 5.5	27
	Pancreas	93.8 ± 24.8	27
	Retina	828	16
	Peripheral blood	77 ± 21 pg/ml	48
	Hypophyseal portal plasma	801 ± 110 pg/ml	48
Frog	Brain	6.89 ± 1.56	172
	Skin	2.56 ± 0.69	172
	Blood	227 ± 57 ng/ml	216
Monkey	Cerebellum, anterior	2.3 ± 0.6	73
	Hypothalamus, anterior	150.8 ± 19.2	73
Chicken	Hypothalamus	938–1,352	9
Snake	Hypothalamus	10,844–20,171	9
Salmon	Hypothalamus	5,188–7,284	9

has been found to be low during the night and high during the day.[16] Similarly, constant exposure to light has been shown to increase TRH content of frog pineal in spring and midwinter and to decrease it in autumn months.[17] In contrast, hypothalamic content of TRH exhibited seasonal variation (autumn > spring), but light or darkness had no effect.[17]

The hypothalamic content of TRH in rat is very low at birth, but it increases progressively during the postnatal period, approaching the adult level by the 16th day.[18] This may in part explain the reduced basal levels of serum TSH in neonatal compared with adult rats. Interestingly, neonatal rats have increased numbers of pituitary TRH receptors and exaggerated TSH response to exogenous TRH.[19] In contrast, the concentration of TRH in the human fetal cerebellum is much higher than that of the adult cerebellum.[20]

To explore the potential physiological and pharmacological functions of endogenous TRH, several studies have been conducted to alter its endogenous concentration.[17,18,21,26] Changes in hypothalamic TRH content have been observed with photoillumination and seasonal changes in the frog[17] and nyctohemerally[21] and following insulin-induced hypoglycemia[22] in the rat. However, hypophysectomy, thyroidectomy, and treatment with several hormones and neurotransmitter-modulating drugs have failed to bring about significant changes in rat brain or hypothalmic TRH content.[23] Caloric intake restriction in neonatal rats has been shown to produce a decrease in hypothalamic TRH content leading to tertiary hypothyroidism.[18] Recently, Winokur et al.[24] have reported substantial increases in TRH content in the forebrain and posterior cortex, but not in other regions, of rat brain after intracisternal administration of 6-hydroxydopamine, a catecholamine-depleting agent. The levels of TRH in the caudate nucleus and putamen but not hippocampus were elevated threefold in Huntington's chorea patients compared to control subjects,[25] suggesting a possible involvement of this peptide in the pathophysiology of Huntington's disease. Consistent with an overall excitatory activity of TRH on the central nervous system is a recent observation by Kubek et al.[26] that electroconvulsive shock increased TRH content in the brainstem but not the hypothalamus or cortex of rats.

2.2. In Gastrointestinal Tract and Pancreas

Morley et al.[27] were the first to present immunologic, biological, and chromatographic evidence for the existence of TRH in the rat gastrointestinal tract, which was confirmed a year later by Leppäluoto et al.[28] The distribution of TRH is uneven throughout the gastrointestinal tract, with the highest concentrations in the pancreas and the cecum and the lowest concentrations in the rectum, jejunum, and duodenum.[29] Pancreatic islet TRH content is increased by starvation and decreased by streptozotocin, whereas hypothyroidism has no effect.[29] These authors[29] also demonstrated very high concentrations of TRH in the pancreas of the neonatal rats. A detailed comparative study of the ontogenetic development of pancreatic and hypothalamic TRH contents in the rat followed this observation,[30] and the data show that the fetal and newborn rats of up to 4–8 days contain higher TRH concentrations in pancreas than in hypothalamus, after which the hypothalamic TRH steadily rises and pancreatic TRH declines to adult levels. Recently, Koivusalo[31] has identified and characterized TRH-like immunoreactivity in the human fetal pancreas.

2.3. In Body Fluids and Other Tissues

TRH-like immunoreactivity has been characterized in many other tissues including ovine pineal gland,[32] rat and human placenta,[33,34] the reproductive system of the male rat,[35] and frog skin.[36] Uncharacterized immunoreactivity has been reported to occur in kidney, liver, and lung[27] and in the adrenals.[37] Recently, the presence of TRH-like immunoreactivity has been confirmed by affinity column chromatography and high-pressure liquid chromatography in gallbladder and colonic carcinomas, leiomyosarcoma, and, to a much lesser extent, normal lung.[38]

Among body fluids, TRH-like immunoreactivity has been demonstrated in amniotic fluid,[39] cerebrospinal fluid,[40,41] breast milk,[42] urine,[12] and blood.[44–48] Although the nature of the urinary TRH-like activity is not certain, large quantities of material resembling acid TRH (deamidated TRH) have been characterized from human urine.[43] Amniotic fluid TRH appears to be immunologically similar to synthetic TRH, and its activity increases with gestational age.[39] TRH-like material has been shown to be present both in cisternal fluid and in lumbar cerebrospinal fluid,[40] and its levels are increased in depressed patients.[44]

Considerable effort has been devoted to measurement of the changes in TRH levels in blood in relation to the functional changes in the hypothalamic–pituitary–thyroid axis. Early attempts to measure TRH levels in human plasma extracts were made by Oliver et al.[44] and Lombardi et al.[45] Studies of the possible relationship between thyroid status and plasma levels of TRH have suggested an inhibitory role of thyroid hormones on the levels of plasma TRH in man.[46,47] However, the experimental protocol utilized in these studies failed to address one or more of the problems that are generally associated with the radioimmunoassay of peptide hormones in biological fluids. These include metabolism, the necessity for chemical and chromatographic characterization of the immunoreactivity, and, finally, the recovery and the change in the specific activity during purification. More recently, Mallik et al.[48] reinvestigated the presence of TRH in human peripheral blood and rat peripheral and hypothalamohypophyseal portal blood using immunologic, chromatographic, and enzymatic criteria. They found no significant change in the plasma concentration of TRH (78 pg/ml) during altered thyroid status. Moreover, their TRH measurements in selected venous compartments in the rat suggest that most of the peripheral blood TRH is of nonhypothalamic origin.[48] Therefore, caution should be exercised in interpreting peripheral TRH concentration as a reliable reflection of the changes in the hypothalamo–pituitary–thyroid axis.

3. METABOLISM OF TRH

3.1. Biochemical Pathways

Incubation of [^{3}H-Pro]TRH with rat, porcine, bovine, human, and hamster serum[49] as well as extracts of various tissues such as brain, eye, heart, lung, kidney, liver, diaphragm, skeletal muscle, and stomach[50,51] abolishes its biological and immunologic activity, leading to the formation of radioactive proline. However, because of its unique chemical structure, containing blocked N- and C-termini, TRH is resistant to degradation by a number of well-characterized proteases and peptidases such as pepsin, substilisin, carboxypeptidase B, aminopeptidase, prolidase, and pancreatins.[52-54] Understanding the mechanisms of TRH inactivation by brain tissues, pituitary, and blood is important because (1) enzymatic inactivation of TRH by blood may regulate its activity by determining the number of molecules available to the pituitary cells, (2) the metabolism of TRH may be essential for the termination of its action as a putative neurotransmitter, and (3) a clear knowledge of the TRH catabolic pathway in the brain may allow investigators to inhibit TRH metabolism during biosynthetic studies.

Both brain extracts and plasma or serum metabolize TRH by similar but independent pathways—pyroglutamyl aminopeptidase and TRH amidase (Fig. 2). In the pyroglutamyl aminopeptidase pathway, N-terminal pyroglutamic acid is cleaved from TRH to yield histidyl-prolineamide, which then cyclizes (probably nonenzymatically) to histidyl-proline diketopiperazine or cyclo(His-Pro). Deamidation of TRH by TRH-amidase leads to the formation of acid TRH, which is probably further metabolized to individual amino acids by the combined actions of carboxypeptidase and pyroglutamate aminopeptidase. Brain extracts also metabolize histidyl-prolineamide to histidine and prolineamide by action of the enzyme imidopeptidase.[55] The presence of imidopeptidase activity has not yet been reported in serum or plasma.

Both the anterior and posterior regions of the pituitary gland actively metabolize TRH to yield acid TRH, cyclo(His-Pro), and the constituent amino acids, though the anterior pituitary contains the highest activity.[56,57] The formation of acid TRH from TRH has also been demonstrated in lung and liver homogenates.[51,58]

The metabolism of TRH by rat and human blood leading to the loss of its biological and immunologic activity is well documented.[49,51] Marked differences, however, in the rate of TRH metabolism by peripheral and pituitary portal plasma have been reported.[59] The products of the metabolism of radioactive TRH by blood have been reported to be acid TRH, cyclo(His-Pro), and the constituent amino acids.[49,51,58,60-63] Acid TRH is further metabolized by rat and human serum,[64] whereas cyclo(His-Pro) appears to be resistant to further metabolism.[65] Recently, Emerson et al.[66] studied the metabolism of TRH and acid TRH by human serum using RIA. Using a 2-h incubation period and a substrate concentration of 50 ng/450 µl reaction volume, they reported metabolism of 46.3 ± 1.3 ng TRH and 14.2 ± 5.1 ng acid TRH.[66] This observation, coupled with the lack of detectable (<3.1 ng) formation of acid TRH from TRH,

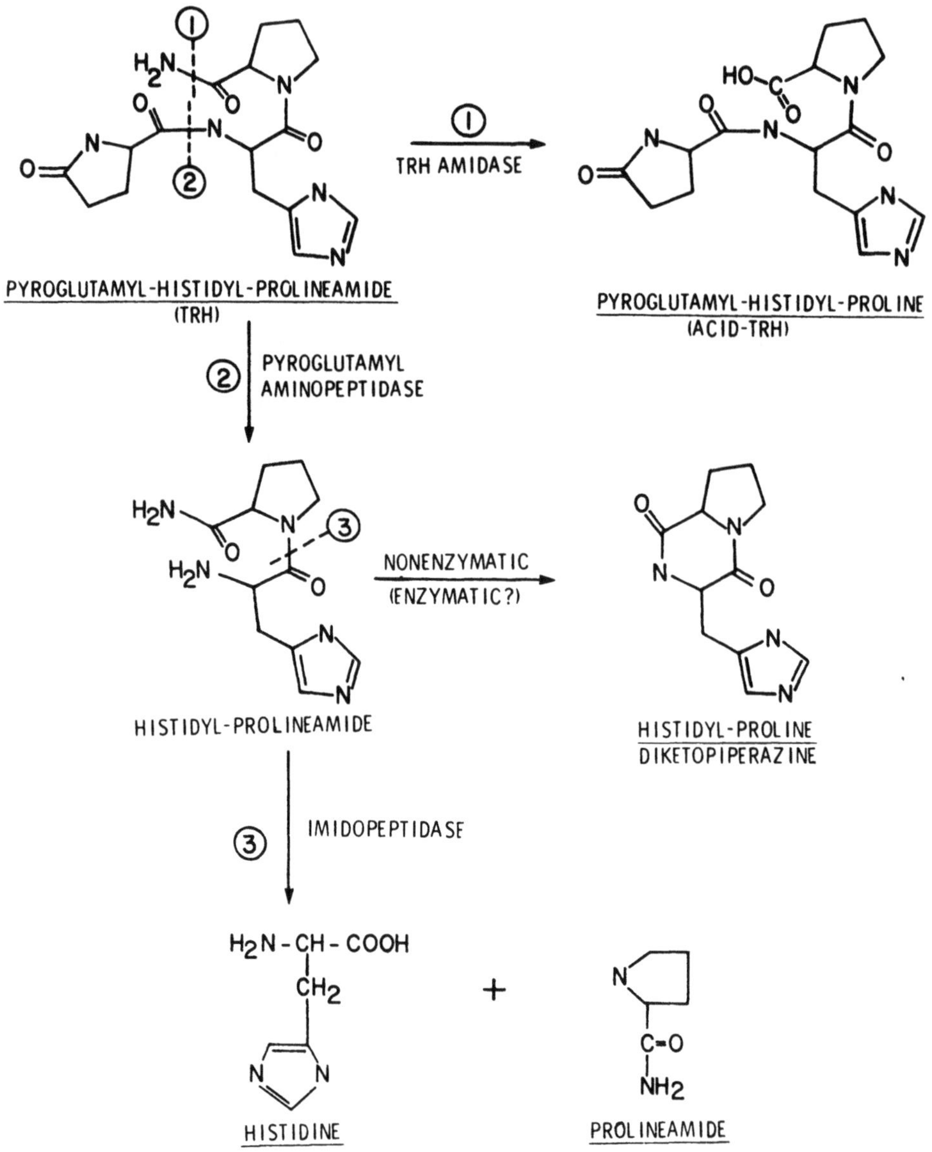

Fig. 2. Pathways of thyrotropin-releasing hormone metabolism. The sites of cleavage by three enzymes are shown by dashed lines. (1) TRH amidase, (2) pyroglutamate aminopeptidase, and (3) imidopeptidase.

led these workers to conclude that TRH deamidation does not occur in human serum. However, this conclusion should be considered tentative, since the rate of the metabolism of acid TRH may be saturated at the concentration of acid TRH achieved by deamidation of TRH, and, thus, acid TRH might not accumulate in the incubation medium.

3.2. Properties of the Enzymes

3.2.1. TRH-Amidase

The existence of TRH-amidase activity, first suggested in 1971 by Nair et al.,[67] was subsequently confirmed by in vitro formation of acid TRH from TRH by serum and hypothalamic extracts.[51,61] In 1976, for the first time, we separated and characterized TRH-amidase activity from crude hamster hypothalamic extract.[51] This enzyme has since been partially purified and characterized from the brains and the pituitaries of a number of species. The enzymes from various sources are similar with regard to molecular weight, pH optima, and substrate specificity.[71,72] For example, the pituitary enzyme (molecular weight 76,000), with a K_m of 410 μM for TRH, exhibits optimal activity between pH 7.4 and 7.6 and has an absolute -SH and -OH group requirement for the maximal activity. The enzymatic deamidation of radioactive TRH is competitively inhibited by synthetic and biologically active natural peptides containing a Pro-X bond, suggesting TRH-amidase to be a postproline-cleaving endopeptidase.[69] TRH-amidase activity is almost evenly distributed throughout the rat and mouse brain with slightly higher activity in the hypothalamus.[70,71] The activity of this enzyme in rat pituitary and hamster hypothalamus resides exclusively in the soluble fraction.[51,57]

3.2.2. Pyroglutamate Aminopeptidase

Recently, the pyroglutamate aminopeptidase activities from hamster hypothalamus,[51] rat serum,[63] and bovine pituitary have been characterized.[72] The serum and the tissue enzymes appear to be entirely different. For example, the serum enzyme of 260,000 molecular weight is inhibited by EDTA and does not accept acid TRH as a substrate, whereas the tissue enzyme has a molecular weight of less than 100,000, accepts acid TRH as substrate, and is not affected by EDTA.[51,63,72]

Although rat pituitary pyroglutamate aminopeptidase activity is mostly associated with the particulate fraction,[57] about 50% of the total brain enzyme activity is in 100,000 × g supernatant fraction, and the other half is almost equally distributed between the crude mitochondrial and microsomal fractions (Table II). The enzyme activity is distributed ubiquitously throughout the rat brain.[73] Although the enzyme activity in the cerebellum (650 ± 31 fmol/min per mg protein) was the highest, it was statistically similar ($P > 0.05$) to that from the hypothalamus, pons–medulla, or midbrain.[73] The ratio of the lowest activity (cortex, 431 ± 48) to the highest activity (cerebellum) was only 0.72 ± 0.12. In contrast, pyroglutamate aminopeptidase activity in rat gastrointestinal tract is unevenly distributed (M. Mori, C. Prasad, and J. F. Wilber, unpublished observations). Duodenum (12.8 ± 5.6 fmol/min per mg protein) contained the lowest enzyme activity, whereas the colon (149.6 ± 18.2) contained the highest. However, the colon enzyme activity was statistically similar to that of the stomach (76.9 ± 15.9), ileum (62.8 ± 6.9), or jejunum (59.0 ± 5.9). The low activity in duodenal extract did not result from the presence of en-

Table II
Subcellular Distribution of Rat Brain Pyroglutamate Aminopeptidase Activity

Subcellular fraction	Pyroglutamate aminopeptidase activity[a] (fmol cyclo(His-Pro) formed/min per mg protein)		
	Whole brain	Cortex	Hypothalamus
Nuclear	86.7 ± 18.5	30.1 ± 7.9	40.7 ± 10.8
	(2.1 ± 0.7)	(1.8 ± 1.1)	(1.5 ± 0.3)
Crude mitochondrial	127.5 ± 34.2	228.4 ± 18.6	81.8 ± 16.3
	(19.1 ± 2.1)	(42.7 ± 4.1)	(19.8 ± 1.4)
Microsomal	213.2 ± 23.3	347.1 ± 30.2	191.2 ± 46.3
	(19.5 ± 0.8)	(32.5 ± 1.3)	(17.2 ± 3.0)
Soluble	440.3 ± 18.4	144.7 ± 22.3	357.0 ± 61.4
	(51.3 ± 4.5)	(18.2 ± 1.1)	(54.3 ± 5.9)

[a] Mean ± S.E.M. ($n = 5$). The data presented in the parentheses represents percent of the total enzymatic activity.

dogenous enzyme inhibitor(s), since a mixture of the colon and the duodenal extracts gave activity that was equal to the sum of activities in the two extracts assayed individually.

3.3. Regulation of TRH Metabolism

Several investigators have measured the total activities of the enzyme metabolizing TRH in human and rat serum or plasma and reported it to decrease with hypothyroidism and to increase during hyperthyroid status.[74,76,77] However, these observations have not been confirmed by others.[48,66,75] Although the reason for this apparent discrepancy is not clear, it is possible that the differences observed in TRH metabolism during altered thyroid status may simply be a reflection of the altered hormonal (e.g., TSH and ACTH) composition of the serum or plasma, since both pituitary and adrenal hormones are known to inhibit one or more of the enzymes associated with TRH metabolism.[51,55] Age-dependent changes in total TRH-degrading activity in both rat and human serum have been reported.[78,79] TRH-inactivating activity is absent in the serum of 2- to 15-day-old rats, but it reaches up to 75% of the adult level by the age of 40 days.[78] Neary et al.[80] have shown lower levels of TRH degradation in human cord and in maternal sera than in the serum of euthyroid, nonpregnant adults. Measurements of the activities of individual TRH-degrading enzymes are limited only to mouse brain, where the specific activities of both pyroglutamate aminopeptidase and TRH-amidase were found to be the highest on the 13th fetal day and then to decline to the adult level by the 20th to 22nd postnatal day.[71] Total TRH-degrading activity in the hypothalamus of dams but not fetuses has also been shown to be reduced by protein deprivation or food restriction during gestation.[81]

3.4. Possible Physiological Significance of TRH Metabolism

Known mechanisms for the termination of the biological actions of neurotransmitters include enzymatic hydrolysis, increased molecular weight, or a change in the intrinsic structure of the transmitter. This is best exemplified by the hydrolysis of acetylcholine by acetylcholinesterase at the muscle endplate and the methylation of catecholamine by catechol-O-methyltransferase at the synaptic junction. The inactivating enzymes could also modulate the action of the transmitter by regulating the levels of the circulating transmitter. Such a control would require modulation of the synthesis and/or the activity of the inactivating enzymes.[82,83]

Data concerning the localization of TRH-degrading enzymes at the synaptic junction or preferably in close proximity to TRH receptor are yet unavailable. However, Joseph-Bravo et al.[84] have shown that rat brain synaptosome-associated total TRH-degrading activity (about 10% of the total brain activity) resides mainly in the membrane fraction. Studies from my laboratory have recently shown about 20% of the total brain pyroglutamate aminopeptidase to be associated with P_2 fraction containing mitochondria, synaptosomes, and myelin (Table II). Indirect evidence for the role of TRH-inactivating enzymes in the biological actions of TRH comes from a series of studies in which increased biological activities of various TRH analogues have been positively correlated with decreased metabolism.[60,85,86]

Another possible role for the degradative enzymes may be the generation of new biologically active compounds through limited proteolysis. The typical examples are the conversion of chymotrypsinogen to chymotrypsin,[87] generation of malanocyte-stimulating hormone release-inhibiting factor (Pro-Leu-Gly-NH_2) from oxytocin,[88] and the formation of angiotensin from angiotensinogen.[89] In the case of TRH, the support for this mechanism comes from the demonstration of the formation of cyclo(His-Pro) by the limited proteolysis of TRH by the brain enzyme pyroglutamate aminopeptidase.[49] Studies into the biological effects of this metabolite have shown it to possess a number of interesting properties, some of which are similar to those of TRH (Table III).

Attempts to associate biological activity with acid TRH, a product of limited proteolysis of TRH by enzyme TRH-amidase, have been uniformly negative.[90] However, recently, Boschi et al.[91] reported acid TRH to be as potent as TRH in eliciting apomorphine-reversible "wet-dog" shakes in rats. The action of TRH-amidase, a postproline cleavage enzyme acting on the Pro-X bond,[69] on oxytocin may generate Leu-Gly-NH_2, which then could cyclize to cyclo(Leu-Gly), a peptide known to modulate the development of morphine tolerance.[92]

The significance of the enzymatic degradation of TRH or the role of TRH metabolites may not be restricted to physiological circumstances alone and could be applicable to pathological conditions as well. In a study on the effect of thyroid status on the metabolic clearance rate of TRH, Jackson et al.[77] found that the half-life of the first component was similar (2.1–2.2 min) in both control and hypothyroid rats, but the second component was prolonged in the hypothyroid rats (4.8 vs. 3.6; $P < 0.01$). The changes in the half-life were attributed to the changes in the plasma TRH degradation rates rather than distribution space or renal clearance.[77] Recently, we have demonstrated significant increases in the hypothalamic contents of cyclo(His-Pro) but not TRH during primary hypothyroidism,[73] chronic alcohol consumption,[93] and acute starvation.[42]

Table III
A Comparison of Various Biological Activities Ascribed to TRH and Cyclo(His-Pro)

Biological activities	Reference
TRH-related activities of cyclo(His-Pro)	
TRH-like activities	
Antagonism of ethanol narcosis	90
Elevation of brain cyclic GMP levels	217
Inhibition of food intake	218
Inhibition of cholesterol synthesis	151
Inhibition of abstinence syndrome in opiate-dependent mice	219
Attenuation of ketamine-induced anesthesia	220
TRH-opposite activities	
Hypothermia in rats	99
Inhibition of *in vitro* prolactin secretion by rat pituitary gland	221
Inhibition of *in vivo* prolactin secretion in monkeys	222
TRH-unrelated activity of cyclo(His-Pro)	
Inhibition of dopamine uptake	223
Cyclo(His-Pro)-unrelated activities of TRH	
Stimulation of thyrotropin secretion	221
Interaction with TRH-receptor	221
Behavioral effects, including piloerection, body tremor, and tail lifting	224
Inhibition of pentobarbital-induced sleep	90

4. MULTIPLE BIOLOGICAL EFFECTS OF TRH

4.1. Behavioral Effects

Plotnikoff and his co-workers[95] were the first to demonstrate the action of TRH in the central nervous system. These authors showed TRH potentiation of the stimulant properties of L-dihydroxyphenylalanine in pargyline-treated mice. Furthermore, this action of TRH was reported to persist in both hypophysectomized and thyroidectomized animals, suggesting a noninvolvement of the pituitary–thyroid axis in its behavioral effects.[96] Since this major discovery, TRH has been shown to elicit numerous centrally mediated behavioral effects in man and animals (see refs. 7,8,97,98 for reviews). However, it is largely unknown whether endogenous TRH participates in these behavioral manifestations. Studies from my laboratory have shown that exogenous administration of TRH produces hyperthermia in rats, and, conversely, immunologic blockade of endogenous TRH produces hypothermia.[99,100]

Several lines of investigation[38] have been pursued in the past to attempt to elucidate the mechanism(s) of behavioral effects of TRH. These include (1) comparison of the behavioral effects of TRH with drugs such as morphine and amphetamine, (2) the effect of TRH on the behavioral effects of other CNS-active drugs, and (3) the effect of other CNS-active agents (such as agonists

Table IV
Multiple Behavioral Effects of TRH[a]

Behavioral effect	Species	TRH response
Sedative effects of ethanol, barbiturates, chloral hydrate, ether, diazepam, chlorpromazine, and reserpine	Rat	Attenuation
Hypothermic effects of neurotensin, ethanol, barbiturates, oxotremorine, and bombesin	Rat	Attenuation
Thermoregulation	Rat	Hyperthermia
	Cat	Hypothermia
Increased locomotor activity following pargyline plus L-DOPA or 5-HTP administration	Rat	Potentiation
Pentobarbital anticonvulsive potency	Rat	Increased
Strychnine and pentylenetetrazole lethality	Rat	Increased
Food consumption	Rat	Decreased
Muscle tremor	Rat	Increased
Modulation of analeptic effects of TRH by antimuscarinic agents	Mouse	Attenuation
	Rabbit	Attenuation
	Rat	No effect

[a] Data from refs. 7,8,97,98,103–106.

and antagonists of putative neurotransmitters) on the behavioral action of TRH. A critical examination of the results of these studies, which are summarized in Table IV, suggests that the major effects of TRH, i.e., its arousing, analeptic, hyperthermic, and respiratory stimulant effects, are mediated through different neuronal mechanisms and that a unifying hypothesis concerning the mode of action of TRH in eliciting its stimulant actions is not available. Formulation of a mechanism of action of TRH based on the effect of various pharmacological agents to modulate the behavioral effects of TRH is precluded by numerous problems, including, in particular, the species differences in the behavioral effects of TRH.[97,101–105] For example, antimuscarinic agents block the analeptic effects of TRH in mice and rabbits[103,104] but not in rats.[106]

4.2. Endocrine Effects

Although TRH was initially isolated and characterized as the hypothalamic peptide responsible for the synthesis and release of thyrotropin,[107–110] subsequent research on the endocrine effects of this tripeptide exhibited a striking exception to the concept of neurohormonal specificity (Table V). Jacobs et al.[111] were the first to present multiple endocrine effects of TRH in man when they showed TRH to be at least equipotent in releasing prolactin or TSH and prolactin release to precede the release of TSH. Simultaneously, Tashjian et al.[112] demonstrated TRH stimulation of prolactin release from cultured rat pituitary tumor cells. This finding was also confirmed with hypothyroid rat

Table V
Multiple Endocrine Effects of TRH

Target	Species	Experimental or clinical condition	Effect	Reference
Pituitary (anterior)	Rat	In vitro	Stimulation of TSH release	221
		In vitro	Stimulation of TSH release	2,3
	Human	Normal	Stimulation of TSH release	111
	Rat	In vivo	Stimulation of prolactin release	113,114
		In vitro	Stimulation of prolactin release	112,221
	Sheep	In vitro	Stimulation of prolactin release	119
	Bovine	In vitro	Stimulation of prolactin release	117
	Frog	In vitro	Stimulation of prolactin release	118
	Human	Normal	Stimulation of prolactin release	111
	Sheep	In vitro	Stimulation of growth hormone release	116
	Bovine	In vitro	Stimulation of growth hormone release	120
	Human	Acromegaly	Stimulation of growth hormone release	121–124
		Uremia	Stimulation of growth hormone release	121–124
		Mental depression	Stimulation of growth hormone release	125
		Liver cirrhosis	Stimulation of growth hormone release	127
		Normal	Stimulation of FSH release	225
		Corticotropin adenoma cells	Stimulation of ACTH release	226
Pituitary (posterior)	Rat	In vivo	Stimulation of oxytocin and arginine-vasopressin release	129
Pancreas		In vivo	Stimulation of Arg-induced glucagon release	130
	Human	Normal	Attenuation of pancreatic popypeptide release during insulin-induced hypoglycemia	133
Gastrointestinal tract	Dog	In vivo	Inhibition of gastrin-stimulated acid secretion	131
	Human	In vivo	Inhibition of gastrin-stimulated acid secretion	132
Pineal	Cat	In vivo	Stimulation of arginine vasotocin release into cerebrospinal fluid	134

pituitary[113] and in proestrous rats *in vivo*.[114] The prolactin-releasing activity of TRH has also been shown in other species, including humans,[115] sheep,[116] cattle,[117] and frog.[118]

An increased growth hormone release following TRH has been reported in sheep[116] or bovine[120] pituitaries *in vitro* and in human subjects during acromegaly and uremia,[121–124] mental depression,[125] anorexia nervosa,[126] and liver cirrhosis.[127] In the intact rat, however, TRH has been shown to produce only a slight or no increase in growth hormone release.[128]

The effects of TRH on other endocrine tissues include the stimulation of arginine vasopressin and oxytocin release from posterior pituitary,[129] enhancement of arginine-induced glucagon release from the rat pancreas,[130] inhibition of gastrin-stimulated acid secretion in dog[131] and human,[132] and attenuation of pancreatic polypeptide release during insulin-induced hypoglycemia.[133] The release of arginine vasotocin from the cat pineal into the CSF has also been shown to be stimulated by TRH.[134]

4.3. Cardiovascular Effects

Central administration of TRH in rabbits produces a pressor response[135,136] that is not blocked by chlorpromazine, scopolamine, pentobarbital, phenoxybenzamine, propranolol, hexamethonium, guanethidine, reserpine, or atropine. This effect of TRH is centrally mediated since the response was not altered in animals with spinal transection below T_1, but transections in the cervical region of the spinal cord abolished the response. Administration of TRH produces a transient increase in blood pressure in man[137,138] as well as cats[139]; TRH also induces tachycardia in dogs[140] and in rats.[141] Recently, TRH has been shown to be effective in improving cardiovascular function and survival in experimentally induced endotoxic and hypovolemic shock[142] and spinal injury[143] in cats.

4.4. Gastrointestinal Effects

Thyrotropin-releasing hormone produces a number of pronounced nonendocrine effects on the gastrointestinal tract that appear to be centrally mediated and expressed through the enhancement of parasympathetic outflow. Intraventricular administration of TRH has been shown to activate the extrinsic neural input to the gastrointestinal tract, leading to increased colonic and duodenal activity.[144,145] TRH stimulates the guinea pig ileum[146] as well as rat antrum, pyloric sphincter, and colon *in vitro*.[147] The TRH effect on antral motility is inhibited by histamine antagonists.[148] Additionally, TRH may act directly to modulate cholinergic receptor responsiveness in the gut, since in the isolated guinea pig ileum, high concentrations of TRH enhanced the contraction produced by the addition of acetylcholine.[149] In the dog, Morley *et al.*[131] have shown that TRH (8 μg/kg) markedly increased myoelectric contraction in the antrum of unanesthetized dogs. In man, TRH infusion has been shown to inhibit glucose and xylose absorption.[150] Recently, low concentrations of TRH and its metabolite cyclo(His-Pro) have been shown to have an inhibitory effect on the activation of 3-hydroxy-3-methylglutaryl-CoA reduction, which occurs during organ culture of canine intestinal mucosa.[151]

4.5. Respiratory and Other Effects

Administration of TRH to pregnant rabbits results in increased pulmonary surfactant production by the fetus.[152] Central administration of TRH produces tachypnea.[153] In normal men[154] and paraplegic patients,[155] exogenous TRH has been reported to produce acute urinary urgency. It has also been shown to produce an increase in skeletal muscle tonus in several species, probably by a direct action on spinal motor neurons.[156–158] Recently, Koss[139] found TRH to cause a dose-dependent stimulation of tonic ciliary nerve activity. This increase in nerve activity appeared to be independent of the excitatory light reflex. Sobue *et al.* have shown that the ataxia and eye movement abnormalities of spinocerebellar degeneration improve markedly following TRH treatment.[159,160]

5. THYROTROPIN-RELEASING HORMONE AS A POSSIBLE NEUROTRANSMITTER OR NEUROMODULATOR

Information processing in the brain largely involves chemical communication among neurons through substances called neuroregulators. These substances may be subdivided into those that convey information between adjacent nerve cells (neurotransmitters) and those that amplify or dampen neuronal activity (neuromodulators). During the last decade, TRH and a dozen or more other brain peptides have been added to the growing list of neuronal substances that may subserve a neuroregulator function.[161] The criteria by which a compound is classified as a neurotransmitter or neuromodulator have been extensively reviewed and debated in recent years.[82,83] Catecholamines have been regarded as the model neurotransmitters of the central nervous system based on the regional and subcellular distribution, mechanism of formation, release, reuptake, metabolism, and effects on pre- and postsynaptic receptors. However, even catecholamines fail to satisfy all the criteria set for a neurotransmitter. The concept of a neuromodulator compound, however, is relatively new, and it comes from the discovery of compounds that affect general communication between nerve cells by acting in a hormonelike manner rather than in a transsynaptic fashion.[162] Unlike neurotransmitters, a neuromodulator need not have specific receptors; instead, it might affect neurotransmitter synthesis, release, receptor interaction, reuptake, or metabolism.

5.1. Uneven Distribution in Mammalian Brain and Spinal Cord

Thyrotropin-releasing hormone is ubiquitously but unevenly distributed throughout the rat, monkey, and human brain.[10,73,163] Although TRH concentration is highest in the hypothalamus, about 70–80% of the total immunoreactive TRH in brain resides in extrahypothalamic areas such as brainstem, midbrain, septum, basal ganglia, cerebral cortex, and spinal cord.[10,73,163] Of all the extrahypothalamic areas, cerebellar TRH content appears to be the lowest in all the species (human, monkey, and rat) examined.[10,64,65,73] However, Pacheco et al.[164] have recently observed a significant increase in rat cerebellar TRH concentration when the crude extract was assayed following rigorous purification. Furthermore, these authors reported an uneven distribution of TRH within cerebellum, with high concentrations in paraflocculi, flocculi plus nodulus, and deep cerebellar nuclei. The concentration of TRH in cerebellar paraflocculi (786 pg/mg) was much higher than that in the hypothalamus (465 pg/mg). Thus, 19% of the cerebellar mass contained 94% of the total cerebellar TRH, whereas the concentration of TRH in the cerebellar hemispheres and vermis (which represent most of the total cerebellar mass) was very low. Consistent with the radioimmunoassay analysis of extrahypothalamic TRH are the immunohistochemical studies that revealed TRH-containing fibers in many areas of the brain, including motor nuclei of the trigeminal, facial, and hypoglossal nerves, the nucleus accumbeus, the lateral septal nuclei, and the bed nucleus of the stria terminalis.[165-168]

However, in two recent reports, Youngblood and his co-workers[169,170] have claimed a nonidentity between synthetic TRH and TRH-like immunoreactivity from urine, serum, extrahypothalamic brain, pancreas, eye, pineal gland, and placenta but not hypothalamus and frog skin. To explain the failure of these investigators to show the similarities of thin-layer chromatographic behavior between immunoreactive and synthetic TRH, Kreider et al.[171] studied the effect of different extraction procedures on the recovery of TRH and found that the extraction procedure (acid–ethanol followed by acetone–water) used by Youngblood et al.[170] was highly inefficient in extracting TRH from tissue as well as silica gel plates. However, in the case of frog skin, Youngblood et al. used methanol extraction, which has been shown by us and others[172,173] to be very effective for TRH.

Thyrotropin-releasing hormone has been detected by means of radioimmunoassay throughout the hypothalamus.[174] It is present in especially high concentrations in the median eminence,[174] where it is found in nerve endings.[165-168] The location(s) of the cell bodies that give rise to these endings have been a matter of conjecture, since TRH-containing neuronal perikarya have been visualized immunocytochemically in several hypothalamic nuclei, including the suprachiasmatic, dorsomedial, paraventricular, and periventricular nuclei.[165-168] Recently, Brownstein et al.[175] have shown that lesions that destroy the hypothalamic paraventricular nuclei cause a significant decrease in TRH levels in the median eminence, arcuate nucleus, and posterior pituitary but not in the medial forebrain bundle. Thus, TRH in the median eminence and arcuate nucleus seems to be provided by neurons in or immediately adjacent to the paraventricular nuclei.

Fibers staining for TRH in the anterior horn of the rat spinal cord have been demonstrated by the use of an indirect immunofluescence technique,[176] and Wilber et al.[13] observed immunoreactive TRH in transverse sections of the rat spinal cord. Recently, spinal TRH has been shown to be unevenly distributed, with highest concentrations in the anterior horn and central canal areas.

5.2. Synaptosomal Localization

Both hypothalamic and extrahypothalamic[177] synaptosomes are richly endowed with immunoreactive TRH. Immunocytochemically, TRH has been shown to be present in nerve endings.[166–168] In the hypothalamus, immunoreactive TRH is found in two subcellular populations, consisting of large and small particles, with sedimentation characteristics similar to those of synaptosomes containing norepinephrine and dopamine.[178,179] Following hypoosmotic shock or exposure to solubilizing agents, most of the TRH appears near the top of the gradient. The remaining fraction of TRH was associated with small particles with a sedimentation coefficient similar to that of synaptic vesicles containing acetylcholine. These properties suggest that both TRH-containing particles are of synaptosomal origin. Barnea et al.[180] have characterized the distribution of TRH-containing particles during ontogenesis. In newborn rats, TRH is associated exclusively with the subpopulation of small particles. An age-dependent increase in the TRH associated with the subpopulation of large particles was observed. By day 7 after birth, TRH was equally distributed between the two subpopulations of particles.

5.3. Uptake and Release

The termination of neurotransmitter action at the synapse may be regulated by removal of the neurotransmitter from the synaptic cleft. Reuptake of neurotransmitters is one mechanism by which neurotransmitters are removed. Such an uptake process may take place either at the pre- or postsynaptic neuronal elements or by glial cells that form the synaptic capsule. Therefore, the demonstration of an active uptake process that may be involved in the rapid and specific removal of TRH from the extracellular space may suggest a neurotransmitter role for this tripeptide.

Early studies on the uptake of TRH suggested that a metabolite of TRH, proline, but not TRH itself is taken up by crude synaptosomal fractions from the hypothalamus and cerebral cortex.[181] In these studies no effort was made to inhibit TRH metabolism during uptake. However, a recent investigation[182] of the *in vitro* uptake of exogenous TRH by rat cerebellar slices in the presence of bacitracin, a protease/peptidase inhibitor, suggest that it is taken up via a process sharing many of the properties of a high-affinity transport system: (1) saturation kinetics; (2) high-affinity kinetic constants ($K_{m_1} = 1.06 \times 15^5$ M, $K_{m_2} = 5.6 \times 10^{-6}$ M); (3) temperature dependence ($Q_{10} = 1.48$); sodium dependence; and a high tissue/medium ratio. After 1 h of incubation at 27°C, the tissue/medium ratio was 5:1, and 70% of the total radioactivity was recovered as TRH.

During the last 7 years, numerous contradictory data have been published with regard to the role of neurotransmitters or neuropeptides such as dopamine, norepinephrine, serotonin, and somatostatin in the modulation of TRH release *in vitro* from rat brain slices, fragments, or synaptosomes.[183–190] For example, serotonin has been shown to stimulate,[190] inhibit,[183] or have no effect[188] on TRH release. A summary of these data is presented in Table VI. Although the

Table VI
In Vitro Release of TRH from Neural Tissues: Some Contradictory Data

Releasing agent	Tissue	Type of preparation	Effect	Reference
55–60 mM KCl	Sheep hypothalamus	Synaptosomes	No effect	183
	Rat hypothalamus	Synaptosomes	Stimulation	184,185
	Rat hypothalamus	Synaptosomes	No effect	188
	Rat hypothalamus	Fragments	Stimulation	189
Histamine	Rat hypothalamus	Slices or fragments	Stimulation	186,188
Norepinephrine	Rat hypothalamus	Fragments in culture	Stimulation	187
	Rat hypothalamus	Fragments	No effect	188
	Sheep hypothalamus	Synaptosomes	No effect	183
Somatostatin	Rat hypothalamus	Fragments in culture	Inhibition	187
Dopamine	Sheep hypothalamus	Synaptosomes	No effect	187
	Rat septal area	Synaptosomes	No effect	185
	Rat hypothalamus	Fragments	Inhibition	188
	Rat hypothalamus	Fragments	Stimulation	189
Serotonin	Sheep hypothalamus	Synaptosomes	Inhibition	183
	Rat hypothalamus	Synaptosomes	No effect	188
	Rat hypothalamus	Fragments	Stimulation	190
Acetylcholine, GABA, or epinephrine	Sheep hypothalamus	Synaptosomes	No effect	183
	Rat hypothalamus	Fragments	No effect	188
Electrical stimulation	Sheep hypothalamus	Synaptosomes	Stimulation	183

reasons for these discrepancies are not apparent, the differences in the experimental procedures may be the most likely cause. For instance, the release of TRH from hypothalamic synaptosomes following chemical depolarization was shown by some[184,185,189] but not all[183,188] investigators to be stimulated. A review of the experimental protocol used by these investigators[183–190] clearly indicates that in order to observe a depolarization-induced TRH release, a synaptosomal preparation must be thoroughly washed or a protease inhibitor be used to inhibit the metabolism of TRH. Schaeffer et al.[185] have clearly shown a requirement for bacitracin, an inhibitor of TRH metabolism, in the incubation medium. Additional factors may include the variation in the alteration of synaptosomal plasma membrane microchemistry as a result of activation of endogenous phospholipses after decapitation-induced ischemia[191] and the time elapsed between the preparation of synaptosomes and the release experiment, since synaptosomes apparently become metabolically labile with increased time of incubation.[192]

5.4. The TRH Receptor

Although TRH is widely distributed in various tissues (Table I), the presence of high-affinity stereospecific binding of TRH to membranes is limited to relatively few tissues, including brain,[193] pituitary,[19,194] retina,[195] and spinal cord.[227] Burt et al.[193] were the first to demonstrate the stereospecific binding of [³H]TRH to rat brain membranes and to establish the presence of both high-

Table VII
Subcellular Distribution of TRH Receptor in Rat Cerebellum[198]

Fraction	TRH (fmol bound/mg protein)[a]
Nuclear fraction (P_1)	3.57 ± 1.37
Crude mitochondrial (P_2)	24.24 ± 5.51
Synaptic plasma membrane	44.83 ± 3.32
Light synaptic plasma membrane	29.96 ± 6.26
Myelin	13.75 ± 4.95
Mitochondria	4.05 ± 0.32
Microsomal membrane (P_3)	20.90 ± 4.37

[a] Mean ± S.E.M. ($n = 3$).

and low-affinity binding sites. Subsequent studies revealed a close similarity between the TRH receptor from the central nervous system (cortex, nucleus accumbens, and retina) and that in the anterior pituitary gland.[19,193–196]

Recently, however, a dissimilarity in the response of brain and pituitary binding sites to guanine nucleotides has been observed.[197] In the presence of micromolar concentrations of GTP and related nucleotides, the affinity of [^{3}H]TRH binding exhibited a twofold reduction in the sheep pituitary gland but not brain.[197] The regional distribution of TRH receptor in sheep and rat central nervous system has been reported to be similar except for the absence of the receptor in rat cerebellum.[193] However, a recent study from this laboratory has shown the presence of TRH receptor in rat cerebellum.[198] Moreover, we have also demonstrated a relative enrichment of TRH binding sites in the crude mitochondrial fraction (P_2). Subfractionation of P_2 yielded an increase in the specific binding as the purity of the synaptic component in each subfraction increased (Table VII). Although the precise number of TRH receptors at the synaptic junction could not be determined (the detergent used for synaptic junction preparation inactivates the TRH receptor), as many as 30% of the synaptic plasma-membrane-associated receptors may reside at the synaptic junction. However, the presence of a significant number of TRH receptors in light synaptic plasma and microsomal membranes suggests the existence of extrajunctional receptors as well.

5.5. Action of TRH on Neurons

A considerable amount of research effort has been devoted to studying the possible effects of TRH on neuronal excitability. The available data are consistent with a neurotransmitter or a neuromodulator role of TRH in several areas of the CNS.

By microionotophoretic techniques, TRH has been shown to inhibit a significant portion of hypothalamic, cerebral cortical, and cerebellar neurons.[199,200] However, a number of investigators fail to discern any direct actions of TRH on certain populations of cerebral cortical neurons.[201,202] Moss[203] observed both excitation and inhibition of hypothalamic neurons. Yarbrough[202] observed TRH enhancement of the excitatory actions of acetylcholine and carbachol (but not glutamate) on the firing of spontaneously active cerebral cortical neurons in anesthetized rats. However, some recent studies[200,204] not only have failed to substantiate Yarbrough's observation but have shown TRH to cause a selective reduction in neuronal excitation evoked by L-glutamate but not by acetylcholine in rat cerebral cortex. In the cat spinal cord, TRH increases the frequency of spontaneous ventral root action potentials and increases the amplitude of the monosynaptic reflex and the amplitude and onset of polysynaptic potentials, leading to a general activation of muscle tonus.[156]

5.6. Effect on Putative Neurotransmitter Metabolism

Several laboratories have examined the effects of TRH on central catecholamine and acetylcholine metabolism to elucidate its possible role as a neuromodulator. Although acute TRH administration does not alter the levels of endogenous serotonin, dopamine, or norepinephine, it consistently increases the rate of depletion of norepinephrine following α-methyl-*p*-tyrosine treatment.[205,206] It has also been shown to increase the concentration of a norepinephrine metabolite (3-methoxy-4-hydroxyphenylethyleneglycol) in brain and to accelerate the rate of conversion of tyrosine to norepinephrine.[207] However, in pentobarbital-anesthetized mice, TRH failed to enhance the incorporation of tritiated tyrosine into norepinephrine.[103] The TRH enhancement of norepinephrine turnover has also been shown by fluorescence histochemistry.[208] Consistent with TRH potentiation of norepinephrine turnover, Reigle *et al.*[209] have shown that TRH stimulates the rate of efflux of intracisternally administered tritiated norepinephrine from the brain. However, TRH attenuated the conversion of intraventricularly administered tritiated dopamine to norepinephrine and also inhibited dopamine-β-hydroxylase activity *in vitro*.[210] Recently, centrally administered TRH was reported to decrease acetylcholine levels and to increase the turnover rate of acetylcholine the parietal cortex,[211] whereas TRH was found to antagonize pentobarbital-induced regional increases in acetylcholine levels.[212]

6. DISCUSSION

Thyrotropin-releasing hormone is a biologically active endogenous tripeptide molecule that is ubiquitously distributed throughout the central nervous system and other tissues of both vertebrates and invertebrates (Table I). Exogenous administration of this tripeptide in animals, including man, elicits a number discrete changes related to behavioral, endocrine, cardiovascular, gastrointestinal, and respiratory functions (see Section 4; Tables IV and V). Although the mechanism(s) by which TRH elicits these multiple biological effects is presently unknown or at the best poorly understood, as in the case of anterior pituitary hormone secretion, three general mechanisms could be advanced. (1) Action of TRH on its receptor at different neural loci may modulate the function of only local circuits leading to different biological outcomes. In line with this hypothesis is the fact that the majority of CNS neurons do not project great distances but instead are members of local redundant circuits. (2) The TRH may elicit some of its various biological activities by amplifying or dampening the activity of putative neurotransmitters such as acetylcholine and catecholamines. (3) Finally, a role of limited proteolysis in the biological actions of TRH has been shown by our studies on the formation and the actions of cyclo(His-Pro) (Table III). The range of the biological activities of TRH could be further amplified if cyclo(His-Pro) modulates local neuronal circuitry.

The presence of TRH in different tissues from evolutionarily distant species from plants to primates raises important questions about its evolutionary and embryologic origin. It is possible that most cells have a constant low level of expression of many peptides including TRH, which will argue against the necessity to invoke a common embryologic origin for tissues that express TRH. Pearse[213] originally postulated that all tissues containing similar peptides were of neural crest origin. However, it is now known that the pineal gland, anterior pituitary, and hypothalamus (all containing TRH) arise not from the neural crest but from the neuroectoderm or specialized ectodermal placodes. The origin of TRH-producing cells of the gastrointestinal tract remains uncertain.

There are a number of areas in which future research into TRH neurobiology may be directed. For example neither the cellular or anatomic origin nor the TRH biosynthetic mechanism is clear. A knowledge of the neuroanatomic loci that mediate different biological effects of TRH would greatly facilitate the understanding of its mechanism of action. The question of the pharmacology of TRH secretion from the hypothalamus should be reexamined by measuring its *in vivo* secretion into hypophyseal portal plasma. Finally, to evaluate its possible role as a neurotransmitter or neuromodulator, data are needed concerning the subsynaptosomal localization of TRH, its receptor and metabolic enzymes, and its effect on the regional turnover of calecholamines, serotonin, acetylcholine, histamine, and various neuropeptides.

ACKNOWLEDGMENTS. The author thanks Mr. Charles F. Chapman and Ms. Margo Johnson of the Editorial Office for excellent editorial and secretarial assistance, Dr. Robert E. Wehmann for many helpful discussions, and Mrs. Ruth M. Edwards for her help with the preparation of bibliography. Research from the author's laboratory was supported in part by LSU Medical Center, Environmental Protection Agency (CR-8071401-01-2), Office of Naval Research (N 00014-80-C-0416), Louisiana Heart Foundation, American Diabetes Association–Louisiana affiliate, and NIH (BRSG 507-RR-5376).

REFERENCES

1. Harris, G. W., 1955, *Neural Control of the Pituitary Gland*, Edward Arnold, London.
2. Boler, J., Enzmann, F., Folkers, K., Bowers, C. Y., and Schally, A. V., 1969, *Biochem. Biophys. Res. Commun.* **37**:705–710.
3. Burgus, R., Dunn, T. F., Desiderio, D., and Guillemin, R., 1969, *C. R. Acad. Sci. [D] (Paris)* **269**:1870–1873.
4. Bassiri, R. M., and Utiger, R. D., 1972, *Endocrinology* **90**:722–727.

5. Bassiri, R. M., and Utiger, R. D., 1973, *J. Clin. Invest.* **52**:1616–1619.
6. Montoya, E., Seibel, M. J., and Wilber, J. F., 1975, *Endocrinology* **96**:1413–1418.
7. Nemeroff, C. B., Loosen, P. T., Bissette, G., Manberg, P. J., Wilson, I. C., Lipton, M. A., and Prange, A. J., Jr., 1979, *Psychoneuroendocrinology* **3**:279–310.
8. Morley, J. E., 1979, *Life Sci.* **25**:1539–1550.
9. Jackson, I. M. D., and Reichlin, S., 1974, *Endocrinology* **95**:854–862.
10. Kubek, M. J., Lorincz, M. A., and Wilber, J. F., 1977, *Brain Res.* **126**:196–200.
11. Jackson, I. M. D., 1979, *Hormones and Evolution* (E. J. W. Barrington, ed.), Academic Press, New York, pp. 723–789.
12. Jackson, I. M. D., 1979, *Central Nervous System Effects of Hypothalamic Hormones and other Peptides* (R. Collu, A. Barbeau, J. R. Ducharme, and J. G. Rochefort, eds.), Raven Press, New York, pp. 3–54.
13. Wilber, J. F., Montoya, E., Plotnikoff, N. A., White, W. F., Gendrich, R., Renaud, L., and Martin, J. B., 1976, *Recent Prog. Horm. Res.* **32**:117–159.
14. Martino, E., Seo, H., Lernmark, A., and Refetoff, S., 1980, *Proc. Natl. Acad. Sci. U.S.A.* **77**:4345–4348.
15. Martino, E., Nardi, M., and Vaudagna, G., 1980, *J. Endocrinol. Invest.* **3**:267–271.
16. Schaffer, J. M., Browstein, M. J., and Axelrod, J., 1977, *Proc. Natl. Acad. Sci. U.S.A.* **74**:3579–3581.
17. Jackson, I. M. D., Saperstein, R., and Reichlin, S., 1977, *Endocrinology* **100**:97–100.
18. Shambaugh, G. E. III, and Wilber, J. F., 1974, *Endocrinology* **94**:1145–1149.
19. Banerji, A., and Prasad, C., 1982, *Endocrinology* **110**:663–664.
20. Winters, A. J., Eskay, R. L., and Porter, J. C., 1974, *J. Clin. Endocrinol. Metab.* **39**:960–963.
21. Collu, R., Du Ruisseau, P., Tache, Y., and Ducharme, J. R., 1977, *Endocrinology* **100**:1391–1393.
22. Leung, Y., Guansing, A. R., Ajlouni, K., Hagen, T. C., Rosenfeld, P. S., and Barboriak, J. J., 1975, *Endocrinology* **97**:380–384.
23. Kardon, F., Marcus, R. J., Winokur, A., and Utiger, R. D., 1977, *Endocrinology* **100**:1604–1609.
24. Winokur, A., Kreider, M. S., Dugan, J., and Utiger, R. D., 1978, *Brain Res.* **152**:203–208.
25. Spindel, E. R., Wurtman, R. J., and Bird, E. D., 1980, *N. Engl. J. Med.* **303**:1235–1236.
26. Kubek, M. J., Etchison, D., and Sattin, A., 1981, *Abstr. Soc. Neurosci. Mtg.* **7**:379.
27. Morley, J. E., Garvin, T. J., Pekary, A. E., and Hershman, J. M., 1977, *Biochem. Biophys. Res. Commun.* **79**:314–318.
28. Leppäluoto, J., Koivulsalo, F., and Kraama, R., 1978, *Acta Physiol. Scand.* **104**:175–179.
29. Martino, E., Lernmark, A., Seo, H., Steiner, D. F., and Refetoff, S., 1978, *Proc. Natl. Acad. Sci. U.S.A.* **75**:4265–4267.
30. Engler, D., Scanlon, M. F., and Jackson, I. M. D., 1981, *J. Clin. Invest.* **67**:800–808.
31. Koivusalo, F., 1981, *J. Clin. Endocrinol. Metab.* **53**:734–736.
32. White, W. F., Hedlund, M. T., Weber, G. F., Rippel, R. H., Johnson, E. S., and Wilber, J. F., 1974, *Endocrinology* **94**:1422–1426.
33. Gibbons, J. M., Mitnick, M., and Chifco, V., 1975, *Am. J. Obstet. Gynecol.* **121**:127–131.
34. Shambaugh, G. E., Kubek, M., and Wilber, J. F., 1978, *Clin. Res.* **26**:495A.
35. Pekary, A. E., Meyer, N. V., Vaillant, C., and Hershman, J. M., 1980, *Biochem. Biophys. Res. Commun.* **95**:993–1000.
36. Jackson, I. M. D., and Reichlin, S., 1977, *Science* **198**:414–415.
37. Leppaluoto, J., Koivusalo, F., and Kraama, R., 1977, *Int. U. Physiol. Sci.* **13**:1300A.
38. Wilber, J. F., and Spinella, P., 1981, *Clin. Res.* **29**:509A.
39. Morley, J. E., Bayshore, R. A., Reed, A., Carlson, H. E., and Hershman, J. M., 1979, *Am. J. Obstet. Gynecol.* **134**:581–584.
40. Oliver, C., Charvet, J. P., Codaccioni, J. L., Vague, J., and Porter, J. C., 1974, *Lancet* **1**:873.
41. Kirkegaard, C., Faber, J., Hummer, L., and Rogowski, P., 1979, *Psychoneuroendocrinology* **4**:227–235.
42. Baram, T., Koch, Y., Hazum, E., and Fridkin, M., 1977, *Science* **198**:300–302.
43. Bhandaru, L., and Emerson, C. H., 1980, *J. Clin. Endocrinol. Metab.* **51**:410–412.

44. Oliver, C., Charvet, J., Codaccioni, J., and Vague, J., 1974, *J. Clin. Endocrinol. Metab.* **38**:406–409.
45. Lombardi, G., Lupoli, G., Scopacasa, F., Ponza, R., and Minozzi, M., 1978, *J. Endocrinol. Invest.* **1**:69–75.
46. Mitsuma, T., Hirooka, Y., and Nihei, N., 1976, *Acta Endocrinol.* **83**:225–235.
47. Gignier, F., 1981, *Eur. J. Nuclear Med.* **6**:73–77.
48. Mallik, T. K., Wilber, J. F., and Pegues, J., 1982, *J. Clin. Endocrinol. Metab.* **54**:1194–1198.
49. Redding, T. W., and Schally, A. V., 1969, *Proc. Soc. Exp. Biol. Med.* **131**:415–419.
50. Bauer, K., Sy, J., and Lipmann, F., 1973, *Fed. Proc.* **32**:489.
51. Prasad, C., and Peterkofsky, A., 1976, *J. Biol. Chem.* **251**:3229–3234.
52. Schally, A. V., Arimura, A., Bowers, C. Y., Kastin, A. J., Sawano, S., and Redding, T. W., 1968, *Recent Prog. Horm. Res.* **24**:497–503.
53. Burgus, R., Ward, D. W., Sakiz, E., and Guillemin, R., 1966, *C.R. Acad. Sci. [D.] (Paris)* **262**:2643–2645.
54. Masson, M. A., Moreau, O., Debuire, W., Han, K. K., Morier, E., and Rips, R., 1979, *Biochimie* **61**:847–854.
55. Matsui, T., Prasad, C., and Peterkofsky, A., 1979, *J. Biol. Chem.* **254**:2439–2445.
56. Marks, N., 1976, *Subcellular Mechanisms in Reproductive Endocrinology* (F. Naftolin, K. J. Ryan, and J. Davies, eds.), Elsevier, Amsterdam, pp. 129–134.
57. Bauer, K., Gräf, K. J., Faivre-Baumann, A., Beier, S., Tixier-Vidal, A., and Keinkauf, H., 1978, *Nature* **274**:174–176.
58. Visser, T. J., Klootwijk, W., Docter, R., and Henneman, G., 1976, *Neuroendocrinology* **21**:204–213.
59. Knigge, K. M., Schock, D., and Ching, M., 1976, *Acta Endocrinol.* **83**:449–453.
60. Benuck, M., and Marks, N., 1976, *Life Sci.* **19**:1271–1276.
61. Knigge, K. M., and Schock, D., 1975, *Neuroendocrinology* **19**:277–287.
62. Loudes, C., Joseph-Bravo, P., LeBlanc, P., and Kordon, C., 1978, *Biochem. Biophys. Res. Commun.* **83**:921–926.
63. Taylor, W. L., and Dixon, J. E., 1976, *J. Biol. Chem.* **253**:6934–3940.
64. Visser, T. J., Klootwijk, W., Docter, R., and Hennemann, G., 1977, *Acta Endocrinol.* **86**:449–456.
65. Prasad, C., Jacobs, J. J., Edwards, R. M., and Banerji, A., 1980, *Clin. Res.* **28**:884A.
66. Emerson, C. H., Mishal, A., Mahabeer, H. L., and Currie, B. L., 1979, *J. Clin. Endocrinol. Metab.* **49**:138–140.
67. Nair, R. M. G., Redding, T. W., and Schally, A. V., 1971, *Biochemistry* **10**:3621–3624.
68. Tate, S., 1978, *Fed. Proc.* **37**:1780.
69. Knisatschek, H., and Bauer, K., 1979, *J. Biol. Chem.* **254**:10936–10943.
70. Taylor, W. L., and Dixon, J. E., 1976, *Biochim. Biophys. Acta* **444**:428–434.
71. Faivre-Baumann, A., Knisatschek, H., Tixier-Vidal, A., and Bauer, K., 1981, *J. Neurosci. Res.* **6**:63–74.
72. Mudge, R. W., and Fellows, R. E., 1973, *Endocrinology* **93**:1428–1434.
73. Prasad, C., Mori, M., Wilber, J. F., Pierson, W., Pegues, J., and Jayaraman, A., 1982, *Peptides* **3**:591–598.
74. Redding, T. W., and Schally, A., 1969, *Proc. Soc. Exp. Biol. Med.* **131**:420–425.
75. Vale, W., Burgus, R., Dunn, T., and Guillemin, R., 1971, *Hormones* **2**:193–203.
76. Bauer, K., 1976, *Nature* **259**:591–593.
77. Jackson, I. M. D., Papapetrou, P. D., and Reichlin, S., 1979, *Endocrinology* **104**:1292–1298.
78. Neary, J. T., Kieffer, J. D., Nakamura, C., Mover, H., Soodak, M., and Maloof, F., 1978, *Endocrinology* **103**:1849–1854.
79. Griffiths, E. C., Jeffcoate, S. L., Thorne, J., and White, N., 1978, *J. Physiol. (Lond.)* **275**:28P.
80. Neary, J. T., Nakamura, C., Davies, I. J., Soodak, M., and Maloof, F., 1978, *J. Clin. Invest.* **62**:1–5.
81. Hastings-Roberts, M. M., and Zeman, F. J., 1979, *Neuroendocrinology* **29**:9–13.
82. Werman, R., 1966, *Comp. Biochem. Physiol.* **18**:745–766.
83. Barchas, J. D., Akil, H., Elliott, G. R., Holman, R. B., and Watson, S. J., 1978, *Science* **200**:964–973.

84. Joseph-Bravo, P., Loudes, C., Charli, J. L., and Kordon, C., 1979, *Brain Res.* **166**:321–329.
85. Dvorak, J. C., and Utiger, R. D., 1977, *J. Clin. Endocrinol. Metab.* **44**:582–585.
86. Brewster, D., and Rance, M. J., 1980, *Biochem. Pharmacol.* **29**:2619–2623.
87. Sigler, P. B., Blow, D. M., Matthews, B. W., and Henderson, R., 1968, *J. Mol. Biol.* **35**:143–164.
88. Celis, M. E., Taleisnik, S., and Walter, R., 1971, *Prac. Natl. Acad. Sci. U.S.A.* **68**:1428–1433.
89. Yang, H. Y. T., and Neff, N. H., 1972, *J. Neurochem.* **19**:2443–2450.
90. Prasad, C., Matsui, T., and Peterkofsky, A., 1977, *Nature* **268**:142–144.
91. Boschi, G., Launay, N., and Rips, R., 1980, *Neurosci. Lett.* **16**:209–212.
92. Walter, R., Ritzmann, R. F., Bhargava, H. N., and Flexner, L. B., 1979, *Proc. Natl. Acad. Sci. U.S.A.* **76**:518–520.
93. Mori, M., Prasad, C., and Wilber, J. F., 1982, *J. Neurochem.* **38**:1785–1786.
94. Mori, M., Pegues, J., Prasad, C., and Wilber, J. F., 1982, *Abstr. Endocrinol. Soc. Mtg.* **64**:211.
95. Plotnikoff, N. P., Prange, A. J., Breese, G. R., Anderson, M. S., and Wilson, I. C., 1972, *Science* **178**:417–418.
96. Plotnikoff, N. P., Prange, A. J., Breese, G. R., and Wilson, I. C., 1974, *Life Sci.* **14**:1271–1278.
97. Yarbrough, G. G., 1979, *Prog. Neurobiol.* **12**:291–312.
98. Manberg, P. J., Nemeroff, C. B., and Prange, A. J., Jr., 1979, *Prog. Neuropsychopharmacol.* **3**:303–314.
99. Prasad, C., Matsui, T., Williams, J., and Peterkofsky, A., 1978, *Biochem. Biophys. Res. Commun.* **85**:1582–1587.
100. Prasad, C., Jacobs, J. J., and Wilber, J. F., 1980, *Brain Res.* **193**:580–583.
101. Metcalf, G., 1974, *Nature* **252**:310–311.
102. Prange, A. J., Jr., Breese, G. R., Cott, J. M., Vartin, B. R., Cooper, B. R., Wilson, I. C., and Plotnikoff, N. P., 1974, *Life Sci.* **14**:447–455.
103. Breese, B. R., Cott, J. M., Cooper, B. R., Prange, A. J., Jr., and Lipton, M. A., 1975, *J. Pharmacol. Exp. Ther.* **193**:11–22.
104. Horita, A., Carino, M. A., and Smith, J. R., 1976, *Pharmacol. Biochem. Behav.* **5**:111–116.
105. Kraemer, G. W., Mueller, R., Breese, G. R., Prange, A. J., Jr., Lewis, J. K., Morrison, H., and McKinney, W. T., Jr., 1976, *Pharmacol. Biochem. Behav.* **4**:709–712.
106. Santori, E. M., Schmidt, D. E., Kalivas, P. W., and Horita, A., 1981, *Psychopharmacology* **74**:13–16.
107. Schally, A. V., Bowers, C. Y., Redding, T. W., and Barrett, J. F., 1969, *J. Biol. Chem.* **244**:4077–4088.
108. Burgus, R., Dunn, T. F., Desiderio, D., Ward, D. N., Vale, W., and Guillemin, R., 1970, *Nature* **226**:321–325.
109. Folkers, K., Enzmann, F., Böler, J., Bowers, C. Y., and Schally, A. V., 1969, *Biochem. Biophys. Res. Commun.* **37**:123–126.
110. Gillesen, D., Felix, A. M., Lergier, W., and Stüder, R. O., 1970, *Helv. Chim. Acta* **53**:63–72.
111. Jacobs, L. S., Snyder, P. H., Wilber, J. F., Utiger, R. D., and Daughaday, W. H., 1971, *J. Clin. Endocrinol. Metab.* **33**:996–998.
112. Tashjian, A. H., Jr., Barowsky, N. J., and Jensen, D. K., 1971, *Biochem. Biophys. Res. Commun.* **43**:516–523.
113. Vale, W., Blackwell, R., Grant, G., and Guillemin, R., 1973, *Endocrinology* **93**:26–33.
114. Mueller, G. P., Chen, H. J., and Meites, J., 1977, *Proc. Soc. Exp. Biol. Med.* **144**:613–615.
115. Bowers, C. Y., Friesen, H. G., and Folkers, K., 1973, *Biochem. Biophys. Res. Commun.* **51**:512–521.
116. Takahara, J., Arimura, A., and Schally, A. V., 1974, *Endocrinology* **95**:1490–1494.
117. Convey, E. M., Tucker, H. A., Smith, V. G., and Zolman, J., 1973, *Endocrinology* **92**:471–476.
118. Clemons, G. K., Russell, S. M., and Nicoll, C. S., 1979, *Gen. Comp. Endocrinol.* **38**:62–67.
119. Fell, L. R., Findlay, J. K., Cumming, I. A., and Goding, J. R., 1973, *Endocrinology* **93**:487–491.

120. Machlin, L. J., Jacobs, L. S., Cirulis, N., Kimes, R., and Miller, R., 1974, *Endocrinology* **95**:1350–1358.
121. Irie, M., and Tsushima, T. J., 1972, *J. Clin. Endocrinol. Metab.* **35**:97–100.
122. Schalch, D. S., Gonzales-Barcena, D., Kastin, A. J., Schally, A. V., and Lee, L. A., 1972, *J. Clin. Endocrinol. Metab.* **35**:609–615.
123. Faglia, G., Beck-Peccoz, P., Travaglini, P., Paracchi, A., Spada, A., and Lewin, A., 1973, *J. Clin. Endocrinol. Metal.* **37**:338–340.
124. Gonzales-Barcena, D., Kastin, A. J., Schalch, D. S., Torres-Zamora, M., Perez-Pasten, E., Kato, A., and Schally, A. V., 1973, *J. Clin. Endocrinol. Metab.* **36**:117–120.
125. Maeda, K., Kato, Y., Ohgo, S., Chihara, K., Yoshimoto, Y., Yamaguchi, N., Kuromaru, S., and Imura, H., 1975, *J. Clin. Endocrinol. Metab.* **40**:501–505.
126. Maeda, K., Kato, Y., Yamaguchi, N., Chihara, K., Ohgo, S., Iwasaki, Y., Yoshimoto, Y., Moridera, K., Kuromaru, S., and Imura, H., 1976, *Acta Endocrinol.* **81**:1–8.
127. Panerai, A. E., Salerno, F., Manneschi, M., Cocchi, D., and Müller, E. E., 1977, *J. Clin. Endocrinol. Metab.* **45**:134–140.
128. Udeschini, G., Cocchi, D., Panerai, A. E., Gil-Ad, I., Rossi, G. L., Chiodini, P. G., Liuzzi, A., and Müller, E. E., 1976, *Endocrinology* **98**:807–814.
129. Weitzman, R. E., Firemark, H. M., Glatz, T. H., and Fisher, D. A., 1979, *Endocrinology* **104**:904–907.
130. Morley, J. E., Levin, S. R., Phelevanian, M., Adachi, R., Pedary, A. E., and Hershman, J. M., 1979, *Endocrinology* **104**:137–139.
131. Morley, J. E., Steinbach, J. H., Feldman, E. J., and Solomon, T. E., 1979, *Life Sci.* **24**:1059–1066.
132. Dolva, L. Ö., Hanssen, K. F., and Berstad, A., 1979, *Scand. J. Gastroenterol.* **14**:33–34.
133. Dolva, L. O., Flaten, O., Hanssen, K. F., Schrumpf, E., and Lundquist, G., 1979, *Acta Endocrinol. (Supplement)* **225**:322.
134. Goldstein, R., and Pavel, S., 1977, *J. Endocrinol.* **75**:175–176.
135. Beale, J. S., White, R. P., and Huang, S., 1977, *Neuropharmacology* **16**:499–506.
136. Horita, A., and Carino, M. A., 1977, *Proc. West. Pharmacol. Soc.* **20**:303–304.
137. Tuck, M., Morley, J. E., Mayes, D., Rosenblatt, S., and Hershman, J. M., 1979, *Clin. Res.* **27**:261A.
138. Hall, R., Amos, J., Garry, R., and Buxton, R. L., 1970, *Br. Med. J.* **2**:274–277.
139. Koss, M. C., 1980, *Eur. J. Pharmacol.* **65**:105–108.
140. Hine, B., Sanghvi, I., and Gershon, S., 1973, *Life Sci.* **13**:1789–1792.
141. Tonoue, T., 1977, *Endocrinol. Jpn.* **24**:271–274.
142. Holaday, J. W., D'Amato, R. J., and Faden, A. I., 1981, *Science* **213**:216–218.
143. Faden, A. I., Jacobs, T. P., and Holaday, J. W., 1981, *N. Engl. J. Med.* **305**:1063–1067.
144. Smith, J. R., LaHann, T. R., Chesnut, R. M., Carino, M. A., and Horita, A., 1977, *Science* **196**:660–662.
145. Tonoue, T., and Nomoto, T., 1979, *Eur. J. Pharmacol.* **58**:369–377.
146. Almqvist, S., 1972, *Front. Horm. Res.* **1**:38–44.
147. Bruce, L. A., Behsudi, F. M., and Fawcett, C. P., 1977, *IRCS Med. Sci.* **5**:469.
148. Bruce, L. A., Behsudi, F. M., and Fawcett, C. P., 1979, *Gastroenterology* **76**:908–912.
149. Ormston, B. J., 1972, *Front. Horm. Res.* **1**:45–75.
150. Dolva, O., Hanssen, K. F., and Frey, H. M. M., 1978, *Scand. J. Gastroenterol.* **13**:599–604.
151. Gebhard, R. L., Morley, J. E., Prigge, W. F., Goodman, M. W., and Prasad, C., 1981, *Peptides* **2**:137–140.
152. Rooney, S. A., Marino, P. A., Gobran, L. I., Gross, I., and Warshaw, J. B., 1979, *Pediatr. Res.* **13**:623–625.
153. Kruse, H., 1974, *Naunyn Schmiedebergs Arch. Pharmacol.* **282**:R51.
154. Fleisher, N., 1972, *J. Clin. Endocrinol. Metab.* **34**:617–622.
155. Martin, J. B., Reichlin, S., and Brown, G., 1977, *Clinical Neuroendocrinology*, F. A. Davis, Philadelphia, p. 218.
156. Cooper, B. R., and Boyer, C. E., 1978, *Neuropharmacology* **17**:153–156.
157. Nicoll, R. A., 1978, *J. Pharmacol. Exp. Ther.* **207**:817–824.
158. Yarbrough, G. G., and McGuffin-Clineschmidt, J. C., 1979, *Eur. J. Pharmacol.* **60**:41–46.

159. Sobue, I., 1977, *Clin. Neurol.* **17**:791-799.
160. Sobue, I., Yamamoto, H., Konagaya, M., Lida, M., and Takayanagi, T., 1980, *Lancet* **1**:418-419.
161. Snyder, S. H., 1980, *Science* **209**:976-983.
162. Levine, S. (ed.), 1972, *Hormones and Behavior*, Academic Press, New York.
163. Winokur, A., and Utiger, R. D., 1974, *Science* **185**:265-267.
164. Pacheco, M. F., McKelvy, J. F., Woodward, D. J., Loudes, C., Joseph-Bravo, P., Krulich, L., and Griffin, W. S. T., 1981, *Peptides* **2**:277-282.
165. Hökfelt, T., Fuxe, K., Johansson, O., Jeffcoate, S., and White, N., 1975, *Eur. J. Pharmacol.* **34**:389-392.
166. Johansson, O., and Hökfelt, T., 1980, *J. Histochem. Cytochem.* **28**:364-366.
167. Johansson, O., Hökfelt, L. T., Jeffcoate, S. L., White, N., and Sternberger, L. A., 1980, *Exp. Brain Res.* **38**:1-10.
168. Choy, V. J., and Watkins, W. B., 1977, *Cell Tissue Res.* **177**:371-374.
169. Youngblood, W. W., Lipton, M. A., and Kizer, J. S., 1978, *Brain Res.* **146**:95-107.
170. Youngblood, W. W., Humm, J., and Kizer, J. S., 1979, *Brain Res.* **163**:101-110.
171. Kreider, M. S., Winokur, A., and Utiger, R. D., 1979, *Brain Res.* **171**:161-165.
172. Prasad, C., Wilber, J. F., and Amborski, R. L., 1982, *Dev. Neurosci.* **5**:293-297.
173. Kellokumpu, S., Vuolteenaho, O., and Leppäluoto, J., 1980, *Life Sci.* **26**:475-480.
174. Brownstein, M. J., Palkovits, M., Saavedra, J. M., Bassiri, R. M., and Utiger, R. D., 1974, *Science* **185**:267-269.
175. Brownstein, M. J., Eskay, R. L., and Palkovits, M., 1982, *Neuropeptides* **2**:197-201.
176. Hökfelt, T., Fuxe, K., Johansson, O., Jeffcoate, S., and White, N., 1975, *Neurosci. Lett.* **1**:133-139.
177. Winokur, A., Davis, R., and Utiger, R. D., 1977, *Brain Res.* **120**:423-434.
178. Barnea, A., Ben-Jonathon, N., Colston, C., Johnston, J. M., and Porter, J. C., 1975, *Proc. Natl. Acad. Sci. U.S.A.* **72**:153-157.
179. Barnea, A., Ben-Jonathan, N., and Porter, J. C., 1976, *J. Neurochem.* **27**:477-484.
180. Barnea, A., Neaves, W. B., and Porter, J. C., 1977, *Endocrinology* **100**:1068-1079.
181. Parker, C. R., Neaves, W. B., Barnea, A., and Porter, J. C., 1977, *Endocrinology* **101**:66-75.
182. Pacheco, M. F., Woodward, D. J., McKelvy, J. F., and Griffin, W. S. T., 1981, *Peptides* **2**:283-288.
183. Bennett, G. W., Edwardson, S. L., Holland, D., Jeffcoate, S. L., and White, N., 1975, *Nature* **257**:323-325.
184. Warberg, J., Eskay, R. L., Barnea, A., Reynolds, R. C., and Porter, J. C., 1977, *Endocrinology* **100**:814-825.
185. Schaeffer, J. M., Axelrod, J., and Brownstein, M. J., 1977, *Brain Res.* **138**:571-574.
186. Charli, J. L., Joseph-Bravo, P., Palacios, J. M., and Kordon, C., 1978, *Eur. J. Pharmacol.* **52**:401-403.
187. Hirooka, Y., Hollander, C. S., Suzuki, S., Ferdinand, P., and Juan, S. I., 1978, *Proc. Natl. Acad. Sci. U.S.A.* **75**:4509-4513.
188. Joseph-Bravo, P., Charli, J. L., Palacios, J. M., and Kordon, C., 1979, *Endocrinology* **104**:801-806.
189. Maida, K., and Frohman, L. A., 1980, *Endocrinology* **106**:1837-1842.
190. Chen, Y. F., and Ramirez, V. D., 1981, *Endocrinology* **108**:2359-2366.
191. Bazan, N. G., Jr., 1971, *J. Neurochem.* **18**:1379-1385.
192. Wheeler, D. D., 1978, *J. Neurochem.* **30**:109-120.
193. Burt, D. R., and Snyder, S. H., 1975, *Brain Res.* **93**:309-328.
194. Banerji, A., and Prasad, C., 1982, *Life Sci.* **30**:2293-2299.
195. Burt, D. R., 1979, *Exp. Eye Res.* **29**:353-365.
196. Burt, D. R., and Taylor, R. L., 1980, *Endocrinology* **106**:1416-1423.
197. Taylor, R. L., and Burt, D. R., 1981, *Mol. Cell. Endocrinol.* **21**:85-91.
198. Banerji, A., and Prasad, C., 1982, *Fed. Proc.* **41**:1507.
199. Renaud, L. P., and Martin, J. B., 1975, *Brain Res.* **86**:150-154.
200. Winokur, A., and Beckman, A. L., 1978, *Brain Res.* **150**:205-209.

201. Dyer, R. G., and Dyball, R. E., 1974, *Nature* **252**:486–488.
202. Yarbrough, G. G., 1976, *Nature* **263**:523–524.
203. Moss, R. L., 1977, *Fed. Proc.* **36**:1978–1983.
204. Renaud, L. P., Blume, H. W., Pittman, Q. J., Lamour, Y., and Tan, A. T., 1979, *Science* **205**:1275–1277.
205. Horst, W. D., and Spirt, N., 1974, *Life Sci.* **15**:1073–1082.
206. Marek, K., and Haubrich, D. R., 1977, *Biochem. Pharmacol.* **26**:1817–1818.
207. Keller, H., Bartholini, G., and Pletscher, A., 1974, *Nature* **248**:528–529.
208. Constantinidis, J., Geissbuhler, F., Gaillard, J. M., Hovaguimian, T., and Tissot, R., 1974, *Experientia* **30**:1182.
209. Reigle, T. G., Auni, J., Platz, P. A., Schildkraut, J., and Plotnikoff, N. P., 1974, *Psychopharmacologia* **37**:1–6.
210. Stolk, J. M., and Nisula, B. C., 1975, *Prog. Brain Res.* **42**:47–56.
211. Malthe-Sorenssen, D., Wood, P. L., Cheney, D. L., and Costa, E., 1978, *J. Neurochem.* **31**:685–691.
212. Schmidt, D. E., 1977, *Psychopharmacol. Commun.* **1**:469–473.
213. Pears, A. G. E., 1969, *J. Histochem. Cytochem.* **17**:303–313.
214. Emson, P. C., Bennett, G. W., and Rossov, M. N., 1981, *Neuropeptides* **2**:115–122.
215. Shambaugh, G. III, Kubek, M., and Wilber, J. F., 1979, *J. Clin. Endocrinol. Metab.* **48**:483–486.
216. Jackson, I. M. D., and Reichlin, S., 1979, *Endocrinology* **104**:1814–1818.
217. Yanagishawa, T., Prasad, C., Williams, J., and Peterkofsky, I. A., 1979, *Biochem. Biophys. Res. Commun.* **86**:1146–1152.
218. Morley, J. E., Levine, A. S., and Prasad, C., 1981, *Brain Res.* **210**:475–478.
219. Bhargava, H. N., 1980, *Life Sci.* **28**:1261–1267.
220. Bhargava, H. N., 1981, *Neuropharmacology* **20**:699–702.
221. Prasad, C., Wilber, J. F., Akerstrom, V., and Banerji, A., 1980, *Life Sci.* **27**:1979–1983.
222. Brabant, G., Wicking, E. J., and Neischlag, E., 1981, *Acta Endocrinol.* **98**:189–194.
223. Battiaini, F., and Peterkofsky, A., 1980, *Biochem. Biophys. Res. Commun.* **94**:240–247.
224. Schenkel-Hulliger, L., Koella, W. P., Hartman, A., and Maitre, L., 1974, *Experientia* **100**:751–754.
225. Hall, R., Besser, G. M., Schally, A. V., Coy, D. H., Evered, D., Goldie, D. J., Kastin, A. J., McNeilly, A. S., Mortimer, C. H., Turnbridge, W. M. G., Phenedos, C., and Weightman, D., 1973, *Lancet* **2**:581–584.
226. Ishibashi, M., and Yamaji, T., 1981, *J. Clin. Invest.* **68**:1018–1027.
227. Prasad, C., and Edwards, R. M., 1983, *Clin. Res.* **31**:870A(Abstract).

9

Role of Calmodulin in the Regulation of Neuronal Function

M. Billingsley, I. Hanbauer, and D. Kuhn

1. INTRODUCTION

Calcium and cyclic nucleotides have long been recognized to act as second messengers operative in the amplification of extracellular signals controlling cellular responses in eucaryotic cells.[1] The mode by which the free intracellular Ca^{2+} concentration is coupled to cellular regulatory mechanisms has been elucidated by the discovery of structurally related proteins that possess high-affinity binding sites for Ca^{2+}.[2] Among this group of proteins, only calmodulin (CaM) has a broad distribution within the cell and throughout different tissues and species.[3] Moreover, in the nervous system CaM has been found in both pre- and postsynaptic neuronal elements. The ability of CaM to bind Ca^{2+} and subsequently to interact with a large number of enzymes and proteins has established a pivotal role of this regulatory protein in neuronal function.

The scope of this chapter is to review the role of CaM in the regulation of pre- and postsynaptic biochemical processes (see Table 1). The presynaptic biochemical events that are mediated by CaM include the regulation of neurotransmitter-synthesizing enzymes and neurotransmitter release from storage vesicles and motility of these vesicles. In postsynaptic sites, CaM is involved in the regulation of catecholaminergic receptors through modulation of the activities of adenylate cyclase, protein kinases, phosphodiesterase, and ATPases.

2. NEUROTRANSMITTER SYNTHETIC ENZYMES

From the work of several laboratories it has been well established that calcium entry into nerve terminals may affect the biosynthesis of neurotransmitters.[4,5] Recent research has indicated that the activity of tryptophan hy-

Table I
Neuronal Processes in Which Ca^{2+}–CaM Has Been Implicated

Process	System
Presynaptic events	Protein kinase system
Neurotransmitter release	Protein I kinase
	Tubulin kinase
Vesicle translocation	Tubulin kinase
	Protein I kinase
Vesicle fusion and recycling	Protein I kinase
Neurotransmitter biosynthesis	
Tryptophan hydroxylase/5-HT	Ca^{2+} · CaM-dependent kinase
Tyrosine hydroxylase/NE, DA	Unknown Ca^{2+}-dependent kinase
Postsynaptic events	Ca^{2+} · CaM-activated enzyme or protein
Catabolism of cyclic AMP/cyclic GMP	Ca^{2+} · CaM-dependent phosphodiesterase
Formation of cyclic AMP	Ca^{2+} · CaM-stimulated adenylate cyclase
Membrane binding of CaM	Fodrin (230,000 daltons)
Postsynaptic protein phosphorylation	Membrane-bound Ca^{2+} · CaM-dependent protein kinase

droxylase, the enzyme that catalyzes the initial and rate-limiting step in the biosynthesis of serotonin (5-HT), can be altered by CaM-dependent phosphorylation reactions.[10,11]

Although a large body of evidence indicates that tyrosine hydroxylase is phosphorylated by a cyclic-AMP-dependent mechanism,[6,7] there is evidence available showing that this enzyme can also be activated by a cyclic-nucleotide-independent, possibly Ca^{2+}-dependent phosphorylation.[8,9]

2.1. Tryptophan Hydroxylase

It was first demonstrated by Hamon and colleagues[10] that the *in vitro* activity of tryptophan hydroxylase could be increased by conditions that promote protein phosphorylation. Similar observations were published from our laboratory almost simultaneously.[11] Briefly, the activity of tryptophan hydroxylase was increased two- to threefold by ATP and Mg^{2+}, as demonstrated in Fig. 1, and this effect was attributed to a phosphorylation effect and not to an allosteric influence on the enzyme for the following reasons: (1) the nonhydrolyzable analogue of ATP, AMPPNP, could not substitute for ATP in the enzyme activation, and (2) the activated state of the enzyme was retained after chromatography on Sephadex G-25 to remove ATP and Mg^{2+} from the enzyme solution. Another interesting feature was that the ATP-dependent activation of tryptophan hydroxylase required a Mg^{2+}-dependent protein kinase for catalyzing the phosphorylation reaction but was not dependent on cyclic nucleotides.

Numerous attempts were made to insure that cyclic AMP or cyclic GMP was not mediating the activation of tryptophan hydroxylase, including the use of cyclic-AMP-dependent protein kinase inhibitors and removal of endogenous cyclic AMP by gel filtration, and even under these conditions activation still

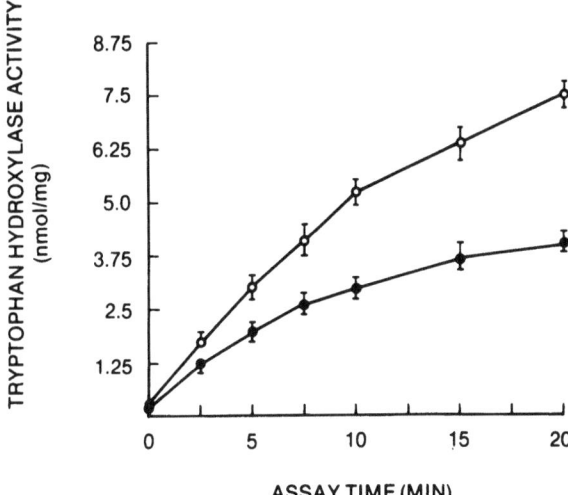

Fig. 1. Effects of phosphorylating conditions on tryptophan hydroxylase activity. Enzyme was assayed for the length of time indicated on the abscissa in the presence (○——○) or absence (●——●) of 0.5 mM ATP, 5.0 mM Mg^{2+}, and 100 μM Ca^{2+}. The reaction is not linear over time since the concentration of the pterin cofactor used (6-methyltetrahydropterin) was subsaturating.

occurred.[10,11] However, it was consistently observed that the activation of tryptophan hydroxylase by phosphorylating conditions was dependent on Ca^{2+}. The activation was blocked by very low concentrations of EGTA, and the addition of low concentrations of Ca^{2+} into the assay led to slight increases in the ATP–Mg^{2+} effect.[10,11] Although this Ca^{2+} stimulation has been somewhat difficult to verify,[12] the careful control of the free Ca^{2+} concentration by the use of Ca^{2+}·EGTA buffers makes the Ca^{2+} influence easily observable.[13]

The dependence of the activation of tryptophan hydroxylase by ATP and Mg^{2+} on Ca^{2+} suggested a possible role for CaM in this process. In fact, soon after the appearance of reports in the literature describing the activation of tryptophan hydroxylase, other papers were forthcoming that described CaM as an essential component.[8,14] Since CaM is distributed throughout the nervous system, it was important to remove CaM from the extracts containing hydroxylase activity to carefully assess its role. A solution to this problem was provided by two related observations. First, Levin and Weiss[15] observed that various drugs commonly referred to as antipsychotics (e.g., trifluoperazine, pimozide, fluphenazine) bound to CaM with high affinity in a Ca^{2+}-dependent fashion. Second, Charbonneau and Cormier[16] devised an affinity chromatographic method for the isolation of CaM by covalently binding fluphenazine to Sepharose 6B. We observed that chromatography of tryptophan hydroxylase-containing extracts over fluphenazine–Sepharose very effectively removed CaM, and such chromatographed preparations of tryptophan hydroxylase no longer responded to ATP–Mg^{2+}.[14] Only on readdition of CaM did tryptophan hydroxylase respond to ATP–Mg^{2+}. Furthermore, the CaM reversal was dependent on Ca^{2+} and not on other divalent cations.

These experiments with fluphenazine–Sepharose chromatography also revealed that the basal activity of tryptophan hydroxylase was not altered by CaM, since removal of CaM did not change the enzyme activity. Furthermore, in separate experiments, we have observed that rather high concentrations of

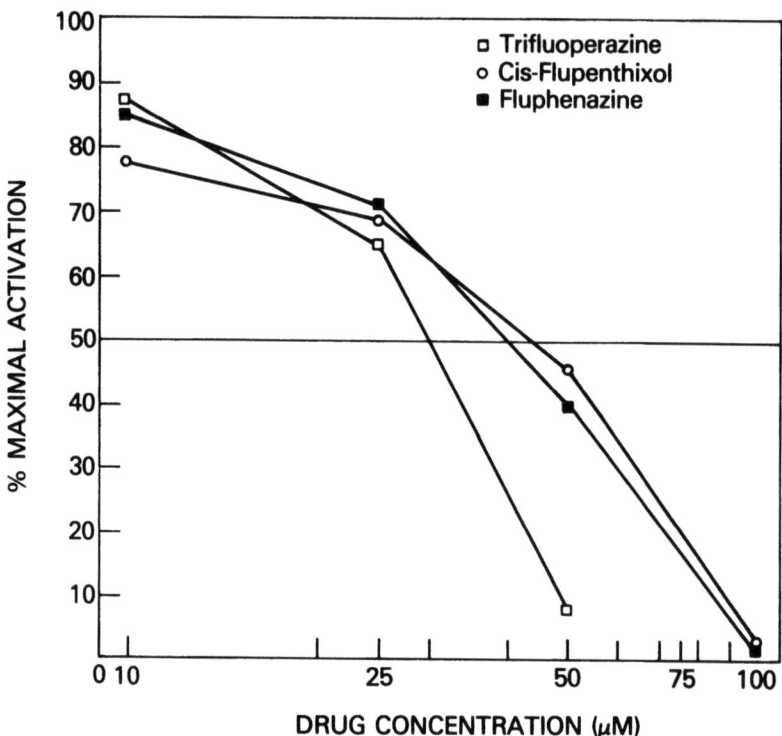

Fig. 2. Drug-induced inhibition of the effects of phosphorylating conditions on tryptophan hydroxylase. Drugs were added in the final concentrations listed on the abscissa. (From Kuhn et al.[14])

CaM did not change tryptophan hydroxylase activity when added into the enzyme assay (D. Kuhn and W. Lovenberg, unpublished observations). However, tryptophan hydroxylase appeared to interact with CaM in a yet undetermined fashion, since the enzyme binds to CaM–Sepharose in the presence of Ca^{2+} (D. Kuhn and W. Lovenberg, unpublished observations).

A role for CaM in the activation of tryptophan hydroxylase could also be seen by adding the CaM-binding drugs referred to above into the assay mixture. By adding a range of concentrations of such drugs as trifluoperazine, fluphenazine, haloperidol, pimozide, fluphenazine, and flupenthixol into the incubation medium, it was possible to observe that these agents produced concentration-related antagonism of the effect of ATP–Mg^{2+} on tryptophan hydroxylase. These results are shown in Fig. 2. The ability of these drugs to block the activation of tryptophan hydroxylase was closely related to their binding capacities to CaM and suggested that these effects were one and the same. The demonstration that the effects of these drugs could be specifically overcome by additional CaM indicated that antipsychotic drugs may block the ATP·Mg^{2+}-induced activation of tryptophan hydroxylase by binding to CaM.[14]

The results discussed above led to the conclusion that tryptophan hydroxylase was activated by a cyclic-nucleotide-independent, Ca^{2+}·CaM-dependent protein kinase. While this *in vitro* work was progressing, similar research was being carried out to determine if the *in vivo* activity of tryptophan hydroxylase was altered by a CaM-dependent phosphorylation reaction. It had already been demonstrated that the activation of tryptophan hydroxylase was mediated by an endogenous Ca^{2+}-stimulated protein kinase and that this activation occurred in many different brain areas.[17] Hamon and co-workers[17] demonstrated the physiological relevance of this activation of tryptophan hydroxylase in brain slices. Depolarization with potassium ions (K^+) of slices from various brain areas was positively correlated with the activation of tryptophan hydroxylase by ATP–Mg^{2+} (*in vitro*). Furthermore, once activated in slices by K^+ depolarization, tryplophan hydroxylase could not be further aclinated *in vitro* by ATP–Mg^{2+}.[17] Finally, Hamon and co-workers indicated that the Ca^{2+}-dependent, K^+-induced activation of tryptophan hydroxylase was reduced in slice preparations incubated in the presence of fluphenazine,[18] a drug that was shown to block the CaM-mediated activation of tryptophan hydroxylase.[14]

Additional studies have supported the role of a Ca^{2+}-dependent mechanism in the activation of tryptophan hydroxylase in slice preparations. For example, Boadle-Biber[19,20] and Elks *et al.*[21] demonstrated that Ca^{2+} ionophores or agents that release Ca^{2+} from intraneuronal sites (e.g., metabolic poisons such as rotenone, dicoumarol) but do not depolarize the slices also activate tryptophan hydroxylase. Boadle-Biber[22] has further demonstrated that the Ca^{2+}-dependent activation of tryptophan hydroxylase in slice preparations is blocked by haloperidol and fluphenazine. These experiments showed that antipsychotic drugs were acting intracellularly to block the depolarization-induced enzyme activation, and it was concluded that CaM played an important role in this process. The similarity between the activation of tryptophan hydroxylase in slices and *in vitro* (by ATP–Mg^{2+}) led to speculations that increased neuronal depolarization facilitated hydroxylase activation through a $Ca^{2+}\cdot$CaM-dependent phosphorylation reaction. These results together with the finding that a Ca^{2+}-dependent protein kinase was present in tryptophan hydroxylase-containing cells[17] were also consistent with this idea. It is also interesting, in this regard, that Haycock *et al.*[28] recently demonstrated that the increased production of catecholamines in adrenal medullary cells in response to K^+ depolarization was accompanied by the phosphorylation of tyrosine hydroxylase. Unfortunately, the direct phosphorylation of tryptophan hydroxylase under similar conditions has not been demonstrated.

2.2. Tyrosine Hydroxylase

Early findings by Lovenberg *et al.*[6] and Morgenroth *et al.*[7] demonstrated that a slight activation of brain tyrosine hydroxylase could be elicited by ATP and Mg^{2+} in the absence of cyclic AMP.

Andrews and Weiner[24] observed that the activation of tyrosine hydroxylase was only partially blocked when the inhibitor of cyclic-AMP-dependent protein kinase was present in the incubation medium. This "residual" activation was similar in magnitude to the activation of tyrosine hydroxylase produced by ATP and Mg^{2+} in the absence of cyclic AMP. In agreement with these findings, Raese and co-workers[25] demonstrated that highly purified tyrosine hydroxylase from bovine corpus striatum was phosphorylated by a cyclic-AMP-independent protein kinase. These studies clearly showed that the activation of tyrosine hydroxylase is associated with a phosphorylation mechanism that involves a cyclic-AMP-independent protein kinase. It has to be noted that although a lack of dependence on cyclic nucleotides is evident, there is no unequivocal support for a phosphorylation of tyrosine hydroxylase mediated by CaM.

Yamauchi and Fujisawa[9] provided data showing that tyrosine hydroxylase is activated by a CaM-dependent mechanism. These investigators demonstrated in the supernatant fraction of brainstem homogenates that tyrosine hydroxylase stimulation by ATP and Mg^{2+} was prevented when CaM was removed from the enzyme solution by differential ammonium sulfate fractionation. Readdition of CaM to the incubation medium restored the activation properties. In contrast, in rat adrenal gland, tyrosine hydroxylase does not appear to be activated by a $Ca^{2+}\cdot CaM$-dependent mechanism. The reason for this lack of responsiveness was shown to be the absence of CaM-dependent protein kinase. The same authors showed that adrenal tyrosine hydroxylase can also be a substrate for $Ca^{2+}\cdot CaM$-dependent protein kinase purified from brain.[26] Although there is some evidence for the role of CaM in the activation of tyrosine hydroxylase, the CaM-mediated phosphorylation of this enzyme may be tissue specific and may not regulate tyrosine hydroxylase in every tissue. Although, Ca^{2+} is essential for the activation of tyrosine hydroxylase in adrenergic nerve terminals during depolarization,[27] a Ca^{2+}-dependent, phospholipid-sensitive protein kinase may also be operative in the regulation of tyrosine hydroxylase, as suggested by Haycock et al.[28]

In conclusion, there is strong evidence that a $Ca^{2+}\cdot CaM$-dependent phosphorylation mechanism plays an important role in the stimulation of tryptophan hydroxylase. A direct role for CaM in the activation of tyrosine hydroxylase has not yet been demonstrated with certainty. However, CaM might well interact with other systems apart from a soluble protein kinase (receptors, membrane phosphorylation) where Ca^{2+} is also involved to produce an alteration in tyrosine hydroxylase activity.

3. CALCIUM- AND CALMODULIN-STIMULATED PHOSPHORYLATION OF MEMBRANE PROTEINS IN NERVOUS TISSUE

In neuronal tissue, the mechanisms underlying synaptic function and neurotransmitter release have been closely associated with the modulation of Ca^{2+} channels. The mechanisms shown to be operative in the modulation of Ca^{2+} channels include phosphorylation and methylation processes, which confer a certain degree of fluidity to the lipid bilayer of neuronal membranes associated with either closing or opening of ion channels. Among the wide variety of protein kinases that have been described to be sensitive to various second messengers, the Ca^{2+}·CaM-dependent protein kinases represent a new class of enzymes.

Reports in the literature suggested that in synaptosomes, Ca^{2+} may modulate the phosphorylation of membrane proteins.[29] Thus, much of the work in presynaptic events has been directed towards understanding the Ca^{2+}-CaM involvement in regulating protein kinase activity. This particular emphasis is undoubtedly related to the realization that a voltage-sensitive inward flux of Ca^{2+} is intimately involved in neurotransmitter secretion.[4,5] Other investigators have examined the role of CaM as an activator of Ca^{2+}-sensitive protein kinase(s) in postsynaptic densities (PSDs). Pharmacological studies exploiting the interaction of phenothiazines with CaM have attempted to alter the state of membrane phosphorylation and correlate such alterations with known biochemical and physiological effects of these drugs.

3.1. Presynaptic Protein Phosphorylation

3.1.1. Protein I (Synapsin I)

Several distinct classes of protein kinases appear to coexist in the neuron; these exhibit cyclic AMP, cyclic GMP, Ca^{2+}-CaM, and Ca^{2+}-phospholipid dependencies. It appears that these classes of enzymes can be further differentiated by their selectivity of endogenous protein substrates.

Greengard and co-workers[29-33] characterized synaptosomal proteins that were phosphorylated in a Ca^{2+}-dependent manner. They found that a doublet protein, termed protein I, was prominently phosphorylated in synaptosomes depolarized with K^+ or veratridine. Protein I consists of two subunits (M_r 86,000 and 80,000), which have pI values of 10.2 and 10.3, respectively. Both subunits were phosphorylated at distinct sites by both cyclic-AMP- and Ca^{2+}·CaM-dependent protein kinases.[31] Studies on the cellular and subcellar distribution of protein I revealed that this protein occurs exclusively in neurons, where it can be detected on the outside surface of synaptic vesicles.[34] Protein I is associated with neurotransmitter vesicles in peripheral nerves but, interestingly, is not found on adrenal chromaffin granules.[35] The state of phosphorylation of synaptosomal protein I was increased by depolarizing agents. Incubation with 60 mM K^+ enhanced protein I phosphorylation, whereas in the presence of 5 mM K^+ (nondepolarizing), protein I was dephosphorylated.[32] This depolarization-induced phosphorylation was dependent on Ca^{2+} influx into presynaptic nerve terminals. Further studies indicated that the Ca^{2+} dependence of the depolarization-induced phosphorylation was dependent on a heat-stable protein that was identified as CaM.[33,36]

The phosphorylation state of protein I was facilitated by electrical stimulation, incubation with dopamine in the neurohypophysis[37] and superior cervical ganglia,[38] or incubation with 5-HT of sections of the facial motor nucleus.[39] Protein I content in the pineal shows a diurnal rhythm that is mediated by β-adrenergic agents.[40] Thus, there is considerable evidence suggesting that protein I may be intimately involved in events controlling neurotransmitter release. Several of the events associated with the function of synaptic vesicles that appear to be modulated by protein I are summarized in Table I. However, the precise role of protein I in the dynamics of neurotransmitter release, such as promotion of translocation or membrane fusion after Ca^{2+} influx, remains to be determined.

3.1.2. Tubulin

The work of DeLorenzo and co-workers[41] suggested that Ca^{2+}, via activation of a Ca^{2+}–CaM protein kinase system, may stimulate the release of neurotransmitters and induce the phosphorylation of several proteins. These phosphoproteins were identified as tubulin.[42] Further studies indicated that tubulin and a Ca^{2+}·CaM-dependent protein kinase were present in isolated presynaptic vesicles.[43] This system, also termed Ca^{2+}–CaM tubulin kinase, was quite labile to proteolytic degradation, since rapid homogenization in buffers containing protease inhibitors was necessary to preserve activity.[44] Thus, Ca^{2+}–CaM tubulin kinase may be distinguished from protein I kinase by its substrate specificity for tubulin and its poor resistance to proteases. During membrane depolarization, a temporal relationship between tubulin phosphorylation and neurotransmitter release appears to exist, which suggests a role for this system in neurosecretion.[43]

Recently, Fukunaga and co-workers[45] purified a Ca^{2+}·CaM-dependent protein kinase from brain that could phosphorylate several substrates including myosin light chain, microtubule-associated protein (MAP), tubulin, myelin basic protein, and casein. This protein kinase has an apparent molecular weight of 640,000 and consists of multiple subunits of 49,000 daltons. This kinase was strictly dependent on Ca^{2+} and CaM and was not affected by phospholipids. Since, in general, Ca^{2+}–CaM protein kinases have been proven difficult to isolate, it was not yet possible to decide whether this particular kinase is the same as or related to those phosphorylating protein I or tubulin.

3.2. Calmodulin-Stimulated Phosphorylation in Postsynaptic Densities

Grab and co-workers reported that a Ca^{2+}–CaM protein kinase[46,47] and several proteins that bound [^{125}I]CaM[48] are present in postsynaptic densities (PSDs). Two of these proteins, with molecular weights of 51,000 and 62,000, were preferentially phosphorylated over protein I when the endogenous kinase system of the PSD was used.[47] Cyclic AMP did not stimulate the phosphorylation of either the 51,000- or the 62,000-dalton protein band. This is in contrast to protein I, which can be partially phosphorylated by a cyclic-AMP-dependent protein kinase. The function of the Ca^{2+}·CaM-dependent protein kinase present in PSDs depended on exogenous addition of CaM,[47] implying that activation of the protein kinase in the PSD may depend on the availability of cytoplasmic CaM. Binding studies carried out with [^{125}I] CaM showed that CaM binds with high affinity to protein bands with the apparent molecular weights of 51,000, 60,000, 140,000, and 230,000.[48] The latter protein corresponds to the ubiquitous CaM-binding protein (fodrin) described by Carlin and co-workers.[49] Table II lists some of the neuronal processes that may, in part, be mediated by Ca^{2+}-CaM.

3.3. Phosphorylation of Myelin

Recently, several laboratories have described a Ca^{2+}-CaM and a Ca^{2+}-phospholipid-sensitive protein kinase in brain myelin fractions which primarily phosphorylated myelin basic proteins (MBP). Endo and Hidaka[50] showed that Ca^{2+} and exogenous CaM enhanced the phosphorylation of MBP. Turner and co-workers[51] indicated that the phosphorylation of MBP was independent of CaM, but dependent on Ca^{2+} and phospholipids. Recently, Sulakhe and coworkers[52] reported that cAMP-dependent kinase activity appears to be associated with myelinlike brain membranes, whereas the Ca^{2+}-stimulated kinase is associated with myelin itself. The physiological function of Ca^{2+}-induced phosphorylation of MBP is presently not known, but it may play a role in the interaction between MBP and other proteins.

3.4. Drug Effects on Ca^{2+}·CaM-Dependent Phosphorylation of Membrane Proteins

Phenothiazines, particularly trifluoperazine, inhibit various CaM-dependent protein kinase systems.[53] In addition to the phenothiazines, several napthalene-sulfonamide derivatives have been shown to inhibit the actions of CaM in various phosphorylating systems.[54] Thus, phosphorylation of membrane proteins is blocked by treating homogenates or cells with inhibitors of CaM. Moreover, in hippocampal slices trifluperazine could block the phosphorylation of a 40,000-dalton protein elicited by electrical stimulation.[55] This membrane protein band was identified as a subunit of pyruvate dehydrogenase, which is located in mitochondria.

In a similar hippocampal slice preparation, electrical stimulation or Met-enkephalin enhanced the phosphorylation of a 52,000-dalton protein.[56] Furthermore, several pharmacological studies indicated that when rats were chronically treated with haloperidol,[57] the Ca^{2+}·CaM-dependent protein kinase activity was increased, whereas chronic morphine treatment stimulated the phosphorylation of specific proteins.[58] Similarly, rats treated with triethyltin show altertions in protein phosphorylation in synaptic plasma membranes.[59] For additional reading concerning the interaction of drugs with CaM, the reader is directed to other sources.[60]

In conclusion, Ca^{2+}·CaM-dependent phosphorylations of membrane-bound proteins have been demonstrated in both pre- and postsynaptic membranes and in myelin. The CaM-dependent protein kinases are differentiated primarily on the basis of their substrate specificity, requirements for phospholipids, and location in tissue. So far, experiments have suggested that Ca^{2+}·CaM-induced phosphorylations are intimately involved in processes related to neurotransmitter release and may be related to postsynaptic events as well. More evidence is needed to conclusively establish distinct physiological roles for Ca^{2+}·CaM-sensitive protein kinases.

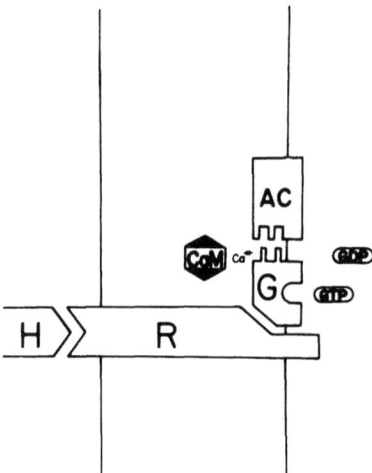

Fig. 3. Scheme for molecular organization of the striatal dopamine receptor. H, dopamine receptor agonist; R, recognition site; G, G protein; CaM, calmodulin; AC, adenylate cyclase.

4. POSTSYNAPTIC RECEPTOR REGULATION

The discovery that the enzymes involved in the synthesis[61] and breakdown of cyclic nucleotides[62] require Ca^{2+} and CaM for maximal functional expression directed considerable attention on the function of CaM in the regulation of neurotransmitter receptors. The concept of specific receptors for hormones and drugs has been popular among endocrinologists and pharmacologists throughout this century. Only since the discovery of adenylate cyclase activation by epinephrine or glucagon[63] has a substantial elucidation of the function and structure of neurotransmitter receptors occurred. Receptors are now viewed as supramolecular entities consisting of various components that are embedded in the membrane lipid bilayer and interact in a specified manner. Figure 3 depicts in a simplified scheme the various components that constitute the receptor complex on the postsynaptic membrane. The activation of the postsynaptic receptor can be visualized as an interaction of various proteins that is facilitated by changes in lipid fluidity and allosteric modifications of these proteins.

4.1. Regulatory Proteins

The response of catecholamine receptors to their physiological agonists is initiated by the binding of the agonist to a specific recognition site. This occupancy of recognition sites is coupled to adenylate cyclase. In the coupling process, the guanine nucleotide regulatory (G) protein and CaM play an important functional role. The G protein was first discovered by Rodbell and co-workers[64] and is now considered to be the regulatory subunit of the adenylate cyclase complex (for review see ref. 65). Independently of the G-protein-mediated process, adenylate cyclase activity is also regulated by CaM. Two different laboratories[61,66] simultaneously reported that CaM increases brain adenylate cyclase activity in the presence of Ca^{2+}. The molecular interaction of CaM with adenylate cyclase is not completely resolved. The available experimental evidence on the regulatory function of CaM leads to the conclusion that it may modulate the Ca^{2+} content in the postsynaptic membrane. Such an interaction is supported by findings from Gnegy and co-workers[67] showing that addition of Ca^{2+} decreases adenylate cyclase stimulation by dopamine. Furthermore, it was suggested that the association between GPP(NH)P or GTP and adenylate cyclase may be impaired in the presence of Ca^{2+}, whereas CaM, by binding free Ca^{2+}, facilitates the activated state of the enzyme.

The regulation of cyclic nucleotide phosphodiesterase, the only enzyme that catalyzes the hydrolytic degradation of cyclic nucleotides, was shown to depend on Ca^{2+} and CaM.[62] In brain tissue, there exist various molecular forms of this enzyme, which differ in regard to affinity for the substrate and in the ability to be regulated by CaM in the presence of Ca^{2+}.[68] Formation of an enzyme–CaM–Ca^{2+} complex increases the affinity of the enzyme for cyclic nucleotides. Various laboratories have therefore suggested that this mechanism may be instrumental in the down-regulation of the cytosolic cyclic nucleotide content.[69,70]

An important aspect to be considered in conjunction with the activation of adenylate cyclase or phosphodiesterase is whether Ca^{2+} uptake or CaM availability is the rate-limiting event. Calmodulin is contained in both membrane and cytosol compartments. Although it may seem that this protein is abundant in both compartments, there exists evidence that CaM translocates from its membrane binding sites to the cytosol. This inference was based on findings showing that phosphorylation of synaptosomal membrane preparations *in vitro* released membrane-bound CaM into the incubation medium and diminished the response of adenylate cyclase to subsequent stimulation by dopamine.[71] Further support for a physiological role of CaM translocation was obtained by experiments showing that prolonged incubation of striatal slices with dopamine receptor agonists increased the cytosolic CaM content and that the membrane-bound CaM content was diminished.[72]

The interaction of CaM with adenylate cyclase and cyclic nucleotide phosphodiesterase in brain tissue has been of great interest in regard to understanding the mechanism of action of neuroleptics and central stimulants. Particular attention was focused on the role of CaM in dopamine receptor function in rat corpus striatum during desensitization or development of supersensitivity.

4.2. Desensitized and Supersensitive Dopamine Receptors

Prolonged stimulation of a neurotransmitter receptor with an agonist results in desensitization of this receptor to subsequent stimulation. The degree of desensitization is usually documented by the decrease in adenylate cyclase stimulation elicited by the agonist.

Desensitization of striatal dopamine receptors is characterized by a number of changes that occur at the level of the synaptosomal membrane and the cytosol compartment (Table II).[72] Exposure of striatal slices to dopamine for at least 30 min uncouples the dopamine recognition sites from dopamine-sen-

Table II
Modifications of Biochemical Processes during Desensitization or Supersensitivity of Striatal Dopamine Receptors

Biochemical events	Desensitized	Supersensitive
Membrane		
Recognition sites (B_{max})	↓	↑
Adenylate cyclase stimulation by dopamine	↓	↑
CaM content	↓	↑
Protein phosphorylation	↑	↔
Cytosol		
Cyclic AMP phosphodiesterase activity	↑	↔
CaM content	↑	↔
CaM translocation	Yes	No

sitive adenylate cyclase.[72] This phenomenon is associated with a decreased affinity of N-propylnorapomorphine for its membrane binding sites and with a reduced efficiency of the G protein. In addition to the reduced function of the regulatory components, a change in the compartmentation of CaM was observed. Prolonged exposure of striatal slices to dopamine receptor agonists elicited an increase in the cytosolic calmodulin and a decrease in the membrane-bound CaM pool.[74] The increase of cytosolic calmodulin content was shown to be associated with an activation of cyclic AMP phosphodiesterase and may constitute an important step in the down-regulation of cytosolic cyclic AMP content. During desensitization of β-adrenergic receptors, the uncoupling of the recognition sites was shown to be associated with a decreased response of the regulatory and catalytic component of the adenylate cyclase complex.[73] It is important to note that in desensitized β-adrenergic receptors, recognition sites translocate from the membrane to the cytosol compartment,[74] whereas neither CaM nor the regulatory or catalytic subunits of adenylate cyclase were observed to change compartments. The translocation of striatal CaM may depend on increased phosphorylation of membrane proteins. Recently, experimental evidence obtained in striatal slice preparations indicated that prolonged exposure to dopamine increased the phosphorylation of several membrane protein bands.[75] This phosphorylation appeared to be catalyzed by a cyclic-AMP-dependent protein kinase.

Supersensitivity of neurotransmitter receptors has been achieved by two methods: (1) by decreasing the amount of neurotransmitter released into the synaptic cleft through inhibition of synthesis or storage or destruction of synaptic terminals and (2) by limiting the occupancy of available postsynaptic recognition sites by neurotransmitters through the use of competitive antagonists. The changes of molecular mechanisms during dopaminergic supersensitivity elicited by either transection of the nigrostriatal fiber bundles[76] or by long-term treatment of rats with neuroleptics[77] appear to be restricted to the membrane-located processes (Table II). Long-term treatment of rats with haloperidol increases the density of dopamine recognition sites, the adenylate cyclase stimulation by dopamine, and the amount of CaM contained in striatal membranes.[77] Studies of the turnover rate of CaM in chronic haloperidol-treated rats ruled out an increased rate of synthesis as the reason for the increased CaM content in the striatal membrane fraction.[78] Therefore, it can be inferred that the affinity of CaM for its membrane binding sites may be increased. This inference was supported by data showing that in chronically haloperidol-treated rats the amount of CaM that can be released from striatal membranes by phosphorylation was reduced by 70%.[79] Similarly, transection of the nigrostriatal fibers increases the responsiveness of adenylate cyclase to dopamine and the CaM content in striatal membranes. These changes were prevented when rats were repeatedly injected with apomorphine immediately after the transection.[76] From these results, it may be inferred that the occupancy of dopamine recognition sites by agonists may be required in order to promote the translocation of CaM.

In conclusion, CaM was shown to play a role in the regulation of dopamine receptors located at postsynaptic sites in corpus striatum. The adaptive changes of this receptor in either sub- or supersensitive states appears to be determined by the degree of CaM translocation from the membrane to the cytosol fraction.

REFERENCES

1. Rasmussen, H., 1970, *Science* **170**:404–412.
2. Vamaman, T. C., Sharief, F. S., and Watterson, D. M., 1977, *Calcium Binding Protein and Function* (R. H. Wasserman, R. A. Corradino, E. Carafoli, R. H. Kretsinger, D. H. MacLennan, and F. L. Seigel, eds.), American Elsevier, New York, pp. 107–116.
3. Cheung, W. Y., 1980, *Calcium and Cell Function*, Volume 1, Academic Press, New York.
4. Katz, B., and Miledi, R., 1967, *J. Physiol. (Lond.)* **192**:407–436.
5. Douglas, W. W., 1968, *Br. J. Pharmacol.* **34**:451–474.
6. Lovenberg, W., Bruckwick, E. A., and Hanbauer, I., 1975, *Proc. Natl. Acad. Sci. U.S.A.* **72**:2955–2958.
7. Morgenroth, V. H., Hegstrand, L. R., Roth, R. H., and Greengard, P., 1975, *J. Biol. Chem.* **250**:1946–1948.
8. Yamauchi, T., and Fujisawa, H., 1979, *Biochem. Biophys. Res. Commun.* **90**:28–35.
9. Yamauchi, T., and Fujisawa, H., 1980, *Biochem. Int.* **1**:98–104.
10. Hamon, M., Bourgoin, S., Hery, F., and Glowinski, J., 1978, *Mol. Pharmacol.* **14**:99–110.
11. Kuhn, D. M., Vogel, R. L., and Lovenberg, W., 1978, *Biochem. Biophys. Res. Commun.* **82**:759–766.
12. Vitto, A., and Mandell, A. J., 1981, *J. Neurochem.* **37**:601–607.
13. Kuhn, D. M., 1982, *Psychopharmacol. Bull.* **18**:161–165.
14. Kuhn, D. M., O'Callaghan, J. P., Juskevich, J., and Lovenberg, W., 1980, *Proc. Natl. Acad. Sci. U.S.A.* **77**:4688–4691.
15. Levin, R. M., and Weiss, B., 1979, *J. Pharmacol. Exp. Ther.* **208**:454–459.
16. Charbonneau, H., and Cormier, M. J., 1979, *Biochem. Biophys. Res. Commun.* **90**:1039–1047.
17. Hamon, M., Bourgoin, J. S., Artaud, F., and Glowinski, J., 1979, *J. Neurochem.* **33**:1031–1042.
18. Mestikevay, S. L., Bourgoin, S., Artaud, F., and Hamon, M., 1982, *Function and Regulation of Monoamine Enzymes: Basic and Clinical Aspects* (E. Usdin, N. Weiner, and M. Youdim, eds.), Macmillan, London, New York, pp. 175–186.
19. Boadle-Biber, M. C., 1978, *Biochem. Pharmacol.* **27**:1069–1079.
20. Boadle-Biber, M. C., 1979, *Biochem. Pharmacol.* **28**:2129–2138.

21. Elks, M. L., Youngblood, W. W., and Kizer, J. S., 1979, *Brain Res.* **172**:471–486.
22. Boadle-Biber, M. C., 1982, *Biochem. Pharmacol.* **31**:2495–2503.
23. Haycock, J. W., Meligeni, J. A., Bennett, W. F., and Waymire, J. C., 1982, *J. Biol. Chem.* **257**:12641–12648.
24. Andrews, D. N., and Weiner, N., 1979, *Proc. West. Pharmacol. Soc.* **22**:163–167.
25. Raese, J. D., Edelman, A. M., Makk, G., Bruckwick, E. A., Lovenberg, W., and Barchas, J. D., 1979, *Commun. Psychopharmacol.* **3**:295–301.
26. Yamauchi, T., Nakata, H., and Fujisawa, H., 1981, *J. Biol. Chem.* **256**:5404–5409.
27. Weiner, N., 1970, *Annu. Rev. Pharmacol.* **10**:273–290.
28. Haycock, J. W., Bennett, W. F., George, R. J., and Waymire, J. C., 1982, *J. Biol. Chem.* **257**:13699–13703.
29. Greengard, P., 1981, *Harvey Lect.* **75**:277–331.
30. Ueda, T., and Greengard, P., 1977, *J. Biol. Chem.* **252**:5155–5163.
31. Kennedy, M. B., and Greengard, P., 1981, *Proc. Natl. Acad. Sci. U.S.A.* **78**:1293–1297.
32. Forn, J., and Greengard, P., 1978, *Proc. Natl. Acad. Sci.* **75**:5195–5199.
33. Schulman, H., and Greengard, P., 1978, *Nature* **271**:478–479.
34. Bloom, F. E., Ueda, T., Battenberg, E., and Greengard, P., 1979, *Proc. Natl. Acad. Sci. U.S.A.* **76**:5982–5986.
35. Fried, G., Nestler, E. J., De Camilli, P., Stjarne, L., Olson, L., Lundberg, J. M., Hokfelt, T., Ouimet, C. C., and Greengard, P., 1982, *Proc. Natl. Acad. Sci. U.S.A.* **79**:2717–2721.
36. Schulman, H., and Greengard, P., 1978, *Proc. Natl. Acad. Sci. U.S.A.* **75**:5432–5436.
37. Tsou, K., and Greengard, P., 1982, *Proc. Natl. Acad. Sci. U.S.A.* **79**:6075–6079.
38. Nestler, E. J., and Greengard, P., 1982, *Nature* **296**:452–454.
39. Dolphin, H. C., and Greengard, P., 1981, *Nature* **289**:76–79.
40. Nestler, E. J., Zatz, M., and Greengard, P., 1982, *Science* **217**:357–359.
41. DeLorenzo, R. J., Freedman, S. D., Yohe, W. B., and Maurer, S. C., 1979, *Proc. Natl. Acad. Sci. U.S.A.* **76**:1838–1842.
42. Burke, B. E., and DeLorenzo, R. J., 1981, *Proc. Natl. Acad. Sci. U.S.A.* **78**:991–995.
43. Burke, B. E., and DeLorenzo, R. J., 1982, *J. Neurochem.* **38**:1205–1218.
44. Burke, B. E., and DeLorenzo, R. J., 1982, *Brain Res.* **236**:393–415.
45. Fukunaga, K., Yamamoto, H., Matsui, K., Higashi, K., and Miyamoto, E., 1982, *J. Neurochem.* **39**:1607–1617.
46. Grab, D. J., Carlin, R. K., and Siekevitz, P., 1981, *J. Cell Biol.* **89**:433–439.
47. Grab, D. J., Carlin, R. K., and Siekevitz, P., 1981, *J. Cell. Biol.* **89**:440–448.
48. Carlin, R. K., Grab, D. J., and Siekevitz, P., 1981, *J. Cell. Biol.* **89**:449–455.
49. Carlin, R. K., Bartelt, D. C., and Siekevitz, P., 1983, *J. Cell. Biol.* **96**:443–448.
50. Endo, T., and Hidaka, H., 1980, *Biochem. Biophys. Res. Comm.* **97**:553–558.
51. Turner, R. S., Chou, C. H. J., Kibler, R. F., and Kuo, J. F., 1982, *J. Neurochem.* **39**:1397–1404.
52. Sulakhe, P. V., Petrali, E. H., Davis, E. R., and Thiessen, B. J., 1980, *Biochemistry* **19**:5363–5371.
53. Prozialeck, W. C., and Weiss, B., 1982, *J. Pharmacol. Exp. Ther.* **222**:509–516.
54. Tanaka, T., Ohmura, T., And Hidaka, H., 1982, *Mol. Pharmacol.* **22**:403–407.
55. Finn, R. C., Browning, M., and Lynch, G., 1980, *Neurosci. Lett.* **19**:103–108.
56. Bar, P. R., Tielen, A. M., Lopes, D., Silva, F. H., Zwiers, H., and Gispen, W. H., 1982, *Brain Res.* **245**:69–79.
57. Lau, Y. S., and Gnegy, M. E., 1982, *Life Sci.* **30**:21–28.
58. O'Callaghan, J. P., Juskevich, J. C., and Lovenberg, W., 1982, *J. Pharmacol. Exp. Ther.* **220**:696–702.
59. O'Callaghan, J. P., Miller, D. B., and Reiter, L. W., 1983, *J. Pharmacol. Exp. Ther.* **224**:466–472.
60. Weiss, B., Prozialeck, W. C., and Wallace, T. L., 1982, *Biochem. Pharmacol.* **31**:2217–2226.
61. Brostrom, C. O., Huang, Y. C., Breckenridge, B.McL., and Wolff, D. J., 1975, *Proc. Natl. Acad. Sci. U.S.A.* **72**:64–68.
62. Chueng, W. Y., 1971, *J. Biol. Chem.* **246**:2859–2869.
63. Robison, G. A., Butcher, R. W., and Sutherland, E. W., 1967, *Ann. N.Y. Acad. Sci.* **139**:107.

64. Rodbell, M., Birnbaumer, L., Pohl, S. L., and Krans, M. J., 1971, *J. Biol. Chem.* **246**:1877–1882.
65. Stadel, J. M., Lean, A. D., and Lefkowitz, R. J., 1982, *Adv. Enzymol* **53**:1–43.
66. Chueng, W. Y., Bradham, L. S., Lynch, T. J., Lin, Y. M., and Tallant, E. A., 1975, *Biochem. Biophys. Res. Commun.* **66**:1055–1062.
67. Gnegy, M. E., and Treisman, G., 1981, *Apomorphine and Other Dopaminometics*, Volume 1, *Basic Pharmacology* (G. L. Gessa and G. V. Corsini, eds.), Raven Press, New York, pp. 161–170.
68. Uzunov, P., and Weiss, B., 1972, *Biochem. Biophys. Acta* **284**:220–226.
69. Uzunov, P., Revuelta, A., and Costa, E., 1975, *Mol. Pharmacol.* **11**:506–510.
70. Hanbauer, I., Gimble, J., and Lovenberg, W., 1979, *Neuropharmacology* **18**:851–857.
71. Gnegy, M. E., Nathanson, J. A., and Uzunov, P., 1977, *Biochim. Biophys. Acta* **497**:75–85.
72. Memo, M., Lovenberg, W., and Hanbauer, I., 1982, *Proc. Natl. Acad. Sci. U.S.A.* **79**:4456–4460.
73. Su, Y.-F., Harden, T. K., and Perkins, J. P., 1979, *J. Biol. Chem.* **254**:38–41.
74. Chuang, D. M., and Costa, E., 1979, *Proc. Natl. Acad. Sci. U.S.A.* **76**:3024.
75. Memo, M., and Hanbauer, I., 1982, *Fed. Proc.* **41**:6163.
76. Lucchelli, A., Guidotti, A., and Costa, E., 1978, *Brain Res.* **155**:130–135.
77. Gnegy, M. E., Uzunov, P., and Costa, E., 1977, *J. Pharmacol. Exp. Ther.* **202**:558–564.
78. Hanbauer, I., Pradhan, S., and Yang, H.-Y. T., 1980, *Ann. N.Y. Acad. Sci.* **356**:292–303.
79. Gnegy, M. E., 1979, *Catecholamines: Basic and Clinical Frontiers*, Volume 1 (E. Usdin, J. Barchas, and I. J. Kopin, eds.), Pergamon Press, New York, pp. 577–579.

10

Effects of Gastrointestinal Peptides on the Nervous System

Gyula Telegdy

1. INTRODUCTION

The discovery that some gastrointestinal peptides are present not only in the gastrointestinal tract but also in the central and the peripheral nervous system has made this topic a very exciting one in current neurobiology.

The classical gastrointestinal hormones gastrin, secretin, cholecystokinin, etc. were considered to be hormones because they are produced in cells of the gastrointestinal tract and exert their action via the circulation. The system has been characterized as the "diffuse endocrine system." However, a growing amount of evidence indicates that some of the cells producing these hormones (peptides) may release their products within their own vicinity, thereby influencing the cell function in this area. This development has given rise to the term "paracrine system."

As a result of developments in peptide chemistry as well as immunochemical methods, gastrointestinal peptide staining has been detected in a number of neural cells and in fibers in the gastrointestinal tract and in the central nervous system. Some peptides, such as vasoactive intestinal peptide and cholecystokinin, were first demonstrated in the gastrointestinal tract and later discovered to be present in the central nervous system. Other peptides were first described in the central nervous system and later found in the gastrointestinal tract as well (enkephalin and somatostatin).

The exact role of these peptides in the neural system is not known. Certain evidence indicates that some of the peptides may behave as transmitters and others as neuromodulators in the sense that their primary action is to change the activity of other "classical" transmitters or metabolic events. The modulatory role of some of these peptides is also supported by morphological evidence. Peptides may coexist with different transmitters or even with peptides in the same neurons. It is possible that certain peptides may act at the same time as a hormone in an endocrine or paracrine manner and as a transmitter

---

Gyula Telegdy • Institute of Pathophysiology, University Medical School, Szeged, Hungary.

Table I
*Peptides Present in the
Gastroenteropancreatic System*

Secretin
Vasoactive Intestinal Polypeptide /VIP/
Gastrin
Cholecystokinin
Motilin
Gastric Inhibitory Polypeptide /GIP/
Angiotensin-II
Bombesin /Gastrin-releasing Peptide?/
/Entero/glucagon
Substance P
Neurotensin
Somatostatin
Enkephalin /methionin-/leucin/
β-endorphine
ACTH
Pancreatic Polypeptide /PP/
TRH
Insulin
Bradykinin

in some function and a neuromodulator in some other function. About 20 peptides have so far been described in the gastrointestinal tract. A number of them have been identified only by immunohistochemical techniques. However, until they have been identified chemically, their chemical nature must be accepted with some reservation. A list of the peptides known to be present in the gastroenteropancreatic system is given in Table I.

This chapter does not deal with those peptides that are covered in detail in other chapters of this volume, such as neurotensin, cholecystokinin, somatostatin, etc., or with those whose actions on the neural system have not yet been investigated.

2. SECRETIN

Bayliss and Starling[1] observed that acid instillation into the duodenum or an intravenous injection of a small bowel extract causes stimulation of pancreas secretion from the denervated pancreas. Their conclusion was that the duodenum must contain a chemical messenger that stimulates pancreas secretion via the circulation; this was named secretin. The peptide was purified from hog small intestine by Jorpes and Mutt[2] in 1961, and the amino acid sequence was identified in 1966.[3] The peptide includes sequences of many amino acid groups identical with those in other gastrointestinal hormones [vasoactive intestinal polypeptide (VIP), glucagon, gastric inhibitory polypeptide (GIP)]. It is possible that these hormones evolved from a common origin (Fig. 1).

Gastrointestinal Peptides

	1	2	3	4	5	6	7	8	9	10	11	12	13	14	15	16	17	18	19	20	21	22
Secretin	His	Ser	Asp	Gly	Thr	Phe	Thr	Ser	Glu	Leu	Ser	Arg	Leu	Arg	Asp	Ser	Ala	Arg	Leu	Gln	Arg	Leu
Porcine VIP (bovine)	His	Ser	Asp	Ala	Val	Phe	Thr	Asp	Asn	Tyr	Thr	Arg	Leu	Arg	Lys	Gln	Met	Ala	Val	Lys	Lys	Tyr
Chicken VIP	His	Ser	Asp	Ala	Val	Phe	Thr	Asp	Asn	Tyr	Ser	Arg	Phe	Arg	Lys	Gln	Met	Ala	Val	Lys	Lys	Tyr
Glucagon	His	Ser	Gln	Gly	Thr	Phe	Thr	Ser	Asp	Tyr	Ser	Lys	Tyr	Leu	Asp	Ser	Arg	Arg	Ala	Gln	Asp	Phe
GIP	Tyr	Ala	Glu	Gly	Thr	Phe	Ile	Ser	Asp	Tyr	Ser	Ile	Ala	Met	Asp	Lys	Ile	Arg	Gln	Gln	Asp	Phe

	23	24	25	26	27	28	29	30	31	32	33	34	35	36	37	38	39	40	41	42	43
Secretin	Leu	Gln	Gly	Leu	Val	NH_2															
Porcine VIP (bovine)	Leu	Asn	Ser	Ile	Leu	Asn	NH_2														
Chicken VIP	Leu	Asn	Ser	Val	Leu	Thr	NH_2														
Glucagon	Val	Gln	Trp	Leu	Met	Asn	Thr														
GIP	Val	Asn	Trp	Leu	Leu	Ala	Gln	Gln	Lys	Gly	Lys	Lys	Ser	Asp	Trp	Lys	His	Asn	Ile	Thr	Gln

Fig. 1. Amino acid sequences of secretin group peptides.

2.1. Distribution

Secretin has been localized by immunohistochemical techniques in the endocrine S cells in the small bowel,[4] and with a similar technique secretin immunoreactivity has recently been detected in the central nervous system in the hypothalamus, cerebral cortex, septum, olfactory tubercle, medulla, pons, thalamus, caudate nucleus, midbrain, pineal gland, adenohypophysis, and neurohypophysis.[5,6]

2.2. Effect on Brain Function

Injection of 17 nmol secretin i.c.v. increases dopamine (DA) stores and turnover in the medial and lateral palisade zones. Although it has no effect on the norepinephrine stores or turnover in the same brain regions, it causes a weak acceleration of the norepinephrine turnover in the parvocellular parts of the paraventricular hypothalamic nucleus. In the forebrain, the dopamine turnover is reduced in the nucleus accumbens and in the tuberculum olfactorium.[7]

Secretin has very little effect on the electrical activity of the brain,[8] but it evokes a tetrodotoxin-insensitive depolarization of the dorsal and ventral roots.[9]

2.3. Effect on Endocrine System

Secretin has been reported to increase the release of LH from the anterior pituitary *in vitro*,[10] although other authors did not detect this action.[11,12] A similar stimulation has been seen for prolactin *in vitro*.[11,12] Intracerebroventricular administration causes an inhibition of prolactin release *in vivo*.[7,11]

2.4. Conclusion

The presence of secretin-immunoreactive material in the central nervous system has been demonstrated only recently; its chemical identification remains to be performed. The data so far available suggest that the neural action of secretin is mainly brought about by its modulating the action of other transmitters, particularly those of the dopaminergic system. This action might also be responsible for the endocrine effects.

3. VASOACTIVE INTESTINAL POLYPEPTIDE

Vasoactive intestinal polypeptide was isolated in 1970 from porcine intestine[13] and was named after its biological vasodilatator activity. In 1976, Larson and his co-workers[14] demonstrated VIP immunoreactivity in neurons in the central nervous system and peripheral nerves.

Vasoactive intestinal polypeptide is a peptide containing 28 amino acids. Chemically, bovine and porcine VIP are identical, whereas VIP isolated from chicken differs in four amino acids (Fig. 1).

3.1. Distribution

Nerve fibers containing VIP are present in all layers of the gastrointestinal wall; their cell bodies are localized in the submucosal and myenteric plexuses. In the sphincter regions, the smooth muscle layer is especially rich in VIP fibers.[15]

In other organs, in the upper respiratory tract, pancreas, and salivary and thyroid glands, VIP terminals can be found in the smooth muscle and glandular epithelium and around blood vessels.[16-18]

In the CNS, VIP-positive neurons are found in all cerebral cortical areas, amygdaloid nuclei, hippocampus, anterior olfactory area, in the hypothalamus in the suprachiasmatic and supraoptic nuclei, in the mesencephalon, in the periaqueductal gray matter, in the superior colliculus, and in the caudal thalamus.[19]

In addition, VIP-positive fibers are found in the cortical area originating from local cortical neurons in layers II–IV, their processes extending vertically through layers I–VI. Abundant nerve fibers can be found in the striatum, the origins of the hippocampus, the bed nucleus of the stria terminalis and the n. amygdala centralis, originating from the mesencephalon via the medial forebrain bundle.[20] Another projection of the VIP system has been proposed from the amygdala to the bed nucleus and anterior hypothalamic area.[21,22] In the spinal cord, primary afferent fibers may originate in the dorsal root ganglia to the dorsal horn.[23]

Vasoactive intestinal polypeptide is localized in vesicles of presynaptic terminals[24]; VIP is synthesized in the neuronal cell bodies and exported along the axons or dendrites to refill the terminal vesicles.[25,26] In some neuronal elements it seems that VIP coexists with other transmitters, e.g., with acetylcholine. In cat, VIP-immunoreactive ganglion cells are present in the paravertebral sympathetic ganglia,[26] almost overlapping with acetylcholinesterase. More typical is the VIP coexistence with acetylcholine in the sphenopalatine ganglia, which are considered to be cholinergic.[26] Exocrine glands are also supplied with VIP and cholinergic fibers.[27]

3.2. Peripheral Action

In the autonomic nervous system, VIP may play a role as a nonadrenergic, noncholinergic neurotransmitter.

In the gastrointestinal tract, VIP causes smooth muscle relaxation, for example, that of the lower esophageal sphincter; this can be antagonized by VIP antiserum.[28] In the stomach, VIP infusion causes relaxation[29] in the corpus fundus part and pyloric region.[30] During digestion, VIP is probably involved in the local vasodilatation.[29,31]

In the pancreas, VIP present in the nerves surrounding pancreatic exocrine and endocrine cells[32] stimulates the juice and bicarbonate secretion[33] and also pancreatic insulin and glucagon secretion.

In the salivary gland, VIP is probably involved in the vasodilatation associated with salivary secretion.[27] Neurally induced vasodilatation can be reduced by intraarterial administration of VIP antiserum.[27] The VIP antiserum neutralizes the released VIP and hence prevents its action on vascular receptors.

In the urogenital tract, VIP inhibits the spontaneous muscle activity in myometrial strips.[34] It inhibits the prostaglandin $F_{2\alpha}$-induced response in rabbit myometrial autografts.[35]

3.3. Central Effects

Binding sites for VIP have been demonstrated in synaptic membrane from central and peripheral nervous tissues[36]; VIP receptor binding is higher in the synaptosomal fraction than in the nuclear, crude mitochondrial, myelin sheet, and separated mitochondrial fractions.[37] It seems that cyclic AMP is involved as a second messenger in the action of VIP in brain slices[38]; in synaptosomes[39] VIP stimulates cyclic AMP formation.

Vasoactive intestinal polypeptide stimulates adenylate cyclase activity in homogenates of rat cerebral cortex, cerebellar cortex, hypothalamus, and hippocampus. In the brainstem and caudate nucleus no stimulation occurs. The enzyme stimulation is inhibited by Ca^{2+} but unaffected by guanine nucleotides.[40]

Potassium releases VIP from synaptosomal preparations obtained from rat cerebral cortex, hypothalamus, and striatum[41] and from rat hypothalamic slices.[42] This release is calcium dependent. Stimulation of preganglionic parasympathetic nerves by electric current causes a frequency-dependent increase in VIP release. This release cannot be blocked by atropine or by adrenoreceptor-blocking agents, but it can be blocked completely by the nicotinic receptor blocking drug hexamethonium.[31,43]

Given intraventricularly in a 17-nmol dose, VIP significantly depletes catecholamine stores in the peri- and paraventricular hypothalamic nuclei and increases the turnover in these brain regions as well as in the subependymal layer. The DA level and turnover are not affected in these brain regions. However, in the forebrain, the tuberculum olfactorium, and the marginal zone of the n. caudatus, the DA turnover increases.[7]

When applied into single cells in the cerebral cortex or hippocampus or spinal cord neurons, VIP causes depolarization and excitation.[44-46] The VIP effect is not blocked by tetrodotoxin or low Ca^{2+} or high Mg^{2+} ion concentration. Morphologically, the close association of VIP with pial and cerebral blood vessels (VIP dilates cerebral blood vessels) suggests that VIP may play a physiological role in the local regulation of cortical blood flow associated with specific functions.

Intracerebroventricular administration of VIP in cat causes hyperthermia.[47]

3.4. Effect on Endocrine System

The high concentration of VIP in certain hypothalamic nuclei suggests a specific role of VIP in hypothalamic neuroendocrine function. The median eminence and the anterior preoptic, supraoptic, and suprachiasmatic areas are densely innervated with VIP terminals.[48]

Deafferentation of the mediobasal hypothalamus (MBH) causes a 40% decrease in the VIP concentrations of the caudal MBH, the stalk, part of the ventromedial nucleus, and premammillary region. This finding suggests that hypothalamic nerve endings of a peptidergic nature originate from cell bodies located outside the MBH.[49]

The fact that VIP can be released into the hypophyseal portal blood[50] suggests that VIP may influence pituitary function; VIP significantly increases prolactin secretion, whereas FSH, LH, GH, and TSH secretion are not affected *in vitro*.[12,51,52] When administered *in vivo* into the third ventricle, VIP stimulates the release of PRL, LH, and GH.[53] The *in vivo* effect of VIP on prolactin release might be caused by its action on the median eminence, inhibiting DA release from nerve terminals in this region, whereas the effect on LH may be a direct action on LH-RH neurons. Indeed, VIP significantly stimulates the LH-RH release from a hypothalamic synaptosomal preparation.[54] The GH-stimulating effect of VIP may be a result of inhibition of somatostatin release from hypothalamic neurons. *In vitro* evidence supports this concept.[55]

3.5. Conclusion

It seems that VIP fulfills a number of the prerequisites of a neurotransmitter. In the autonomic system it acts as a noncholinergic, nonadrenergic neurotransmitter, causing vasodilatation, acting mainly as an inhibitory transmitter; on the exocrine glands it behaves as a stimulator. In the central nervous system, it may play a role in the local vasodilatation associated with cortical function or as a neuromodulator, regulating a specific endocrine function.

4. GASTRIN

The first observation concerning the possible presence in the antral mucosa of a humoral factor that stimulates acid secretion was published by Edkins in 1905.[55] However, the true gastrin epoch started with the isolation of pure gastrin from hog antral mucosa.[56]

Gastrin exists in three different forms, all of which have biological activity, e.g., stimulating gastric acid secretion. These are "big gastrin" (G-34), "little gastrin" (G-17), and "minigastrin" (G-14).[57] All three gastrins also exist in a form sulfated on the tyrosine residue.

G-17 is present mainly in the antral mucosa, comprising about 90% of the extractable gastrin. In the duodenal and jejunal extracts, more G-34 is present than G-17.[58]

In 1975, Vanderhaeghen *et al.*[59] reported that the brains of different mammals contain a gastrinlike immunoreactive peptide. This peptide was later shown to be cholecystokinin octapeptide; gastrin and cholecystokinin have the same C-terminal pentapeptide sequence, and the immune sera could not differentiate between gastrin and cholecystokinin (Fig. 2).

Porcine gastrin-17

Glp-Gly-Pro-Trp-Met-Glu-Glu-Glu-Ala-Tyr-Gly-Trp-Met-Asp-Phe-NH$_2$
 1   2   3   4   5   6   7   8   9 10 11 12 13 14 15 16 17

with SO$_3$H on Tyr (position 12)

Porcine CCK-33

Lys-Ala-Pro-Ser-Gly-Arg-Val-Ser-Met-Ile-Lys-Asn-Leu-Gln-Ser-Leu-Asp-
 1   2   3   4   5   6   7   8   9 10 11 12 13 14 15 16 17

-Pro-Ser-His-Arg-Ile-Ser-Asp-Arg-Asp-Tyr-Met-Gly-Trp-Met-Asp-Phe-NH$_2$
 18 19 20 21 22 23 24 25 26 27 28 29 30 31 32 33

with SO$_3$H on Tyr (position 27)

Fig. 2. Amino acid sequences of gastrin–cholecystokin peptides.

Phe-Val-Pro-Ile-Phe-Thr-Tyr-Gly-Glu-Leu-Gln-Arg-Met-Gln-Glu-Lys-Glu-Arg-Asn-Lys-Gly-Gln

Fig. 3. Amino acid sequence of motilin.

With more specific antisera and with chemical separation, gastrin has so far been detected only in pig anterior and posterior pituitary[60] and in the peripheral nerve in the vagus.[61] G-17 and G-34 have been found in human cerebrospinal fluid.[62]

4.1. Peripheral Effects

Stimulation of the sciatic nerves causes gastrin release.[63] Gastrin stimulates gastric secretion and contracts the stomach and the jejunum.[56]

Gastrin and pentagastrin are able to release acetylcholine from the Auerbach plexus of a guinea pig longitudinal muscle strip.[64] The sulfate group on the tyrosine molecule is important for this action. Desulfated gastrin has a considerably lower biological activity. Shortening of the gastrin molecule reduces its activity, but even the C-terminal dipeptide still retains some biological activity.[65]

4.2. Central Effects

The dopamine, norepinephrine, and serotonin levels in the hypothalamus, mesencephalon, septum, amygdala, and striatum have been studied following administration of human [Leu^{15}]G-17 analogue of gastrin and its fragments into the lateral brain ventricle. The dopamine levels are increased in all brain areas except the striatum, in which an opposite action is observed. The C-terminal and N-terminal fragments as well as the total molecule exhibit a similar action. The [Leu^{15}]G-17 analogue has a similar effect on norepinephrine levels. However, the N-terminal fragments generally increase the NE levels in the brain areas studied, whereas the C-terminal molecule has a similar action to the whole [Leu^{15}]G-17 peptide. The effect exerted on serotonin levels is rather weak. The [Leu^{15}]G-17 analogue shows only a slight biphasic effect; the N- and C-terminal sequences tend to increase the serotonin levels in different brain areas. The central portion of the molecule decreases the serotonin in the same brain areas.[66-68]

Pentagastrin i.p. decreases the acetylcholinesterase activity in the cerebrum but is ineffective in the cerebellum.[69]

The C-terminal tetrapeptide of G-17 (which is identical with CCK-4, the C-terminal tetrapeptide of cholecystokinin) applied by pressure ejection into the vicinity of the CA1 region of the rat hippocampus causes a short-latency depolarization accompanied by increased excitability. During application, desensitization occurs, the excitation disappears, and a relatively long latency is needed for the next response. The leucine analogues of G-14 and G-13 without a sulfate group on the tyrosine also cause excitation, depolarization, and a decrease in membrane resistance. However, when its action is compared with that of the C-terminal tetrapeptide of G-17, this excitation lasts only for seconds after the application of the peptide.[70]

A high dose of the C-terminal fragment of G-17 (pentagastrin) administered into the lateral ventricle in sheep is able to inhibit food intake[71] and decrease gastric motility.[72]

4.3. Endocrine Effects

Gastrin in a high dose inhibits TSH release from the paired hemipituitary.[53] Intraventricular administration of gastrin inhibits LH, TSH, and prolactin and stimulates growth hormone release *in vivo*.[73] The C-terminal tetrapeptide sequence of G-17, which is identical with CCK C-terminal tetrapeptide, also stimulates growth hormone release *in vitro*.[10]

The C-terminal hexa-, penta-, and tetrapeptide fragments of [Leu^{15}]G-17 analogues administered into the lateral brain ventricle in rats increase plasma corticosterone,[74] whereas the original pentapeptide of G-17 causes a decrease in the activity of the pituitary adrenal axis as measured by the plasma corticosterone levels.[75]

4.4. Conclusion

As a gastrointestinal hormone, gastrin is undoubtedly important in the regulation of the gastrointestinal function via a hormonal mechanism. Its significance as a neurotransmitter or neuromodulator is questionable. The data presented concerning its action on the neural functions suggest mainly its pharmacological rather than its physiological importance. However, it might also be important in conditions involving high gastrin levels in the peripheral blood, such as, for example, the Zollinger–Ellison syndrome.

The presence of gastrin in the pituitary and its physiological significance await further elucidation.

5. MOTILIN

Brown *et al.*[76] first reported that the duodenum contains motilin, which stimulates contraction of the canine gastric fundus. It is a polypeptide consisting of 22 amino acid residues.[77] In structure, motilin does not display a molecular similarity with either the gastrin or the secretin family (Fig. 3).

5.1. Distribution in the Nervous System

By means of immunocytochemical techniques, motilin-containing neurons have been found in the submucosa and muscular layer of the esophagus, stomach, small intestine, colon, and gallbladder.[78]

Yanaihara *et al.*[79] demonstrated motilin immunoreactivity in the mammalian hypothalamus, pituitary, and pineal gland, as well as in the cortex and cerebellum.[80–83] The highest concentration of motilin has been observed in the cerebellum[84] and in the cerebellar Purkinje cells.[85] A high concentration of immunoreactivity is found in the dog cerebral cortex,[82] but no motilin-containing neurons were found in the rat cerebral cortex.[86] A high concentration of motilin immunoreactivity has been detected in the hypothalamic nuclei, organum vasculosum laminae terminalis, amygdala, mammillary body, and central gray matter.[84,86]

Glp-Gln-Arg-Leu-Gly-Asn-Gln-Trp-Ala-Val-Gly-His-Leu-Met-NH$_2$

Fig. 4. Amino acid sequence of bombesin.

The distribution of motilin in the central nervous system suggests that it may play a role in regulating both neuroendocrine and neurological processes.

5.2. Peripheral Effects

Motilin causes a stimulation of an intestinal muscle strip that is about 50 times stronger than that of acetylcholine on a molar basis. This action is not mediated through nervous pathways.[87] Motilin increases the gastric emptying of food,[88] stimulates pepsin secretion,[76,89] stimulates pancreatic protein and bicarbonate secretion, and inhibits stimulated pancreatic secretion.[90]

5.3. Central Action

Phillis and Kirkpatrick[9] have shown that motilin causes excitation of the isolated toad spinal cord, depolarizing the dorsal root terminals and motor neurons. Tetrodotoxin reduces this effect, suggesting a primary action on spinal cord interneurons. In the cerebral cortex, iontophoretic application of motilin excites most of the identified corticospinal neurons and about 50% of the unidentified neurons.[8]

5.4. Endocrine Action

Motilin increases the release of growth hormone from anterior pituitary tissue *in vitro*.[6]

5.5. Conclusion

Motilin is one of the most recently found members of the family of gastrointestinal peptides. Even in the gastrointestinal tract, its physiological role is not well understood. Its presence in both the gut mucosa and the neurons in the gastrointestinal tract suggests that it might have a dual role. Locally, it may act as an endocrine or paracrine hormone, and in the neurons as a neurotransmitter or neuromodulator.

Still less is known about the central role of motilin. Although its localization hints at a special function, elucidation of this function awaits further experimentation.

6. BOMBESIN

Bombesin is a 14-amino-acid peptide and was originally extracted from the skin of the diglossid frog *Bombina bombina*[91] (Fig. 4).

6.1. Distribution

With immunocytochemical techniques, bombesin has been shown to be present in brain and in nerves in the lung and gut in man and other mammals.[92] Bombesinlike immunoreactivity (BLI) reveals the highest concentration in the gastric mucosa, but significant amounts are present in other parts of the gut.

In the brain, the highest concentration is observed in the hypothalamus, followed by the cortex and midbrain. None has been found in the cerebellum and pons–medulla in rats. In developing rats aged 1–2 days, a moderate amount is found in the gastric fundus, but it is undetectable in other parts of the gut and in the brain. By 8–16 days of life, the concentration in the gastrointestinal tract is near the adult level. The brain BLI increases more gradually, reaching the adult levels by 26 days.[93] No BLI has been detected in the pineal gland, anterior pituitary gland, adrenal, liver, or kidney.[94]

6.2. Peripheral Effects

Bombesin infusion in man stimulates gastric and pancreatic secretion, releasing different gut hormones such as gastrin, neurotensin, pancreatic polypeptide, motilin, and enteroglucagon.[95]

Bombesin stimulates gastrointestinal motility, gallbladder contraction, myoelectric activity, and smooth muscle contraction.[94]

6.3. Central Effects

In the dog, bombesin increases the blood pressure and produces antidiuresis.[94] Intracisternal administration causes hyperglycemia and lowers the body temperature.[96,97] The hyperglycemic effect seems to be mediated by a central mechanism leading to adrenal medullary epinephrine secretion and increased glucagon and decreased insulin levels. Bombesin causes satiety without apparent toxicity.[98] Bombesin given intracerebroventricularly suppresses stress-induced eating. Adrenalectomy, which abolishes the hyperglycemic effect of bombesin, does not influence stress-induced eating.[99]

Bombesin caused excitation in four of 66 CA1 pyramidal neurons; this was accompanied by depolarization, although there was no detectable change in membrane resistance. After the initial excitation, a long (10-min) latency period followed.[100]

6.4. Endocrine Effects

Bombesin releases prolactin[101] and causes hypothermia[96]; both can be reversed by naloxone, suggesting that the action of bombesin might be mediated at least in part by endorphins.[102]

Bombesin causes the release not only of prolactin but also of growth hormone. These effects are absent in cultured pituitary cells, suggesting that bombesin acts centrally.[103]

Acting centrally, bombesin prevents the rise in plasma TSH normally observed following cold exposure. Also, TRH reverses the hypothermic effects of bombesin.[104] The effects of bombesin, bombesin analogues, and fragments on hypothermia have been studied in detail by Brown and Vale.[105] The shortest peptide that retains the biological activity of bombesin on hypothermia is the bombesin octapeptide. The most potent analogues are those in which positions 1 and 5 are altered. Glutamine at position 7 and glycine at position 11 can be replaced by D-Gln and D-Ala without change in potency. When methionine at position 14 is replaced by the D-isomer, the biological activity is only one-tenth that of the original peptide. Any other alteration in the C terminal greatly reduces the activity of these peptides.

6.5. Conclusion

Although bombesin is present in the central and peripheral nerves, its function is not well understood. It has been proposed that bombesin may play a role within the brain via a mechanism that is able to increase the blood sugar and decrease the temperature. In this way, the organism is able to conserve energy and to maintain central nervous system function even in a good-deprived state.[105] Others have suggested that bombesin may serve as a satiety signal for the organism.[99] The common C-terminal amino acid sequence of gastrin-releasing peptide (GRP) and bombesin[106] has led to the supposition that GRP might be the mammalian bombesin.[107]

7. ANGIOTENSIN

The renin–angiotensin system is essentially linked to kidney function and is involved in the homeostasis of the peripheral vascular resistance and in water and salt balance.[108,109]

Immunohistochemical methods have recently been used to demonstrate angiotensin-II-like immunoreactivity in the plexus myentericus of the Auerbach neurons in the gastrointestinal tract and in the central nervous system.[110,111]

The classical renin–angiotensin system is able to form biologically active angiotensin and to degrade it. Renin, which is formed in the kidney, is a proteolytic endopeptidase. In the bloodstream it hydrolyzes the Leu–Leu bond in the N-terminal part of angiotensinogen, and α_2-globulin. This cleavage results in a decapeptide, angiotensin I (Ang-I). Converting enzyme splits off a dipeptide (His-Leu) from the C-terminal end of Ang-I and produces the bioactive octapeptide angiotensin II (Ang-II). A further aminopeptidase hydrolyzes an asparagine from the N-terminal end of Ang-II and forms another biologically active metabolite, Ang-III. The blood, kidney, peripheral organs, etc. contain angiotensinases that further degrade Ang-III to biologically inactive metabolites (Fig. 5).

Renin can be found in a number of peripheral organs (when it is called isorenin), which allows the formation of angiotensin locally. The origin, chemical identity, and function of Ang-II immunoreactivity in the gastrointestinal tract are not clear. The isorenin–angiotensin system (iRAS) in the brain has been extensively studied in recent years (see below).

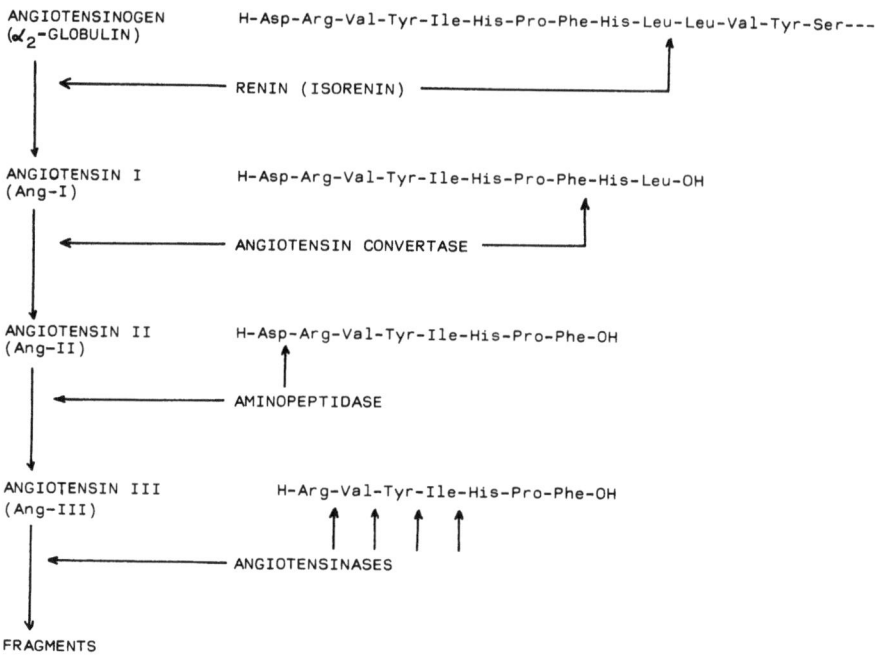

Fig. 5. Renin–angiotensin system.

7.1. Distribution

In the gastrointestinal tract, Ang-II-immunopositive material has been found in the plexus myentericus of Auerbach.[110,111]

In the CNS, all components of the RAS are present and are independent of the peripheral RAS.[108,112]

Angiotensin II can be produced in the brain by renin present in the brain. Free renin has been separated from human brain as well as from dog brain proteases.[112,113] Endogenous renin has also been isolated from the pituitary gland.[114]

Brain renin is different from peripheral renin.[115] Both active and inactive renin are present in the brain, the amounts varying in the different brain regions.[115] Renin has been found in most of the brain tissues studied, e.g., hypothalamus, cerebellum, amygdala, brainstem, caudate, and hippocampus, but the highest renin concentrations are observed in the pineal gland, anterior pituitary, and choriod plexus.[115]

Renin-positive cells have been described in the Purkinje cells in the cerebellum, in the thalamic and hypothalamic nuclei, including the paraventricular, supraoptic, suprachiasmatic, and septal nuclei, and in the hippocampal cells.[116,117] In the rat pituitary, renin coexists with luteinizing hormone in ovoid cells staining to anti-LH.[118] Similar cells have been reported to contain angiotensin II.[119] Renin-positive cells have been found in the hypothalamus, in the paraventricular, periventricular, and supraoptic nuclei, Purkinje cells, and medulla oblongata. Renin may coexist with oxytocin in the same cells.[120] In the human brain, glia cells also show a positive reaction to renin.[121] Renin and angiotensin-I- and II-converting enzymes are present in neuroblastoma cell lines, indicating an intracellular mechanism for angiotensin II formation in neuronal tissues.[122] Whether this mechanism operates in normal tissue too remains to be seen. The presence of angiotensinogen in the brain was first described by Ganten et al.[123] The brain and CSF angiotensinogen is independent of the plasma angiotensinogen.[124]

The angiotensinogen-converting enzyme shows a regional distribution in the brain. It has been found to be relatively concentrated in the striatum, cerebellum, and pituitary.[125] In the posterior lobe, the activity is higher than in the anterior lobe.[126]

Angiotensin can be extracted from the brains of nephrectomized animals,[123,127] which reveals that angiotensin is formed in the brain.

Angiotensin II has been observed in the substantia gelatinosa of the spinal cord, the nucleus tractus spinalis nervi trigemini, the eminentia mediana, and the nucleus amygdaloideus centralis dense plexi. A moderate density occurs in the locus coeruleus. Immunofluorescence is absent from the cerebellar and cerebral cortex.[110,111,124]

A high-affinity binding site for Ang-II has been identified in bovine and rat brain membranes.[128,129] Different analogues or fragments all compete for binding sites. [des-Asp]Angiotensin II (Ang-III) is more potent than Ang-II, whereas Ang-I, the 3–8 hexapeptide, and the 4–8 pentapeptide are even less active. In rat receptors, Ang-II displays the highest activity in the thalamus, hypothalamus, and midbrain.[128,129] In calf brain, the highest receptor binding is found in the cerebellar cortex and deep nuclei of the cerebellum; the binding is lower in the chorioid plexus of the fourth ventricle and superior colliculus.[128]

By means of electrophysiological methods Ang-II-sensitive neurons have been localized in the subfornical organ in cat[130,131] and also in the supraoptic nucleus *in vivo*[132] and *in vitro*.[133]

7.2. Peripheral Action

Circulating Ang-II causes vasoconstriction; by potentiating sympathetic nerve stimulation in the mesenteric vascular bed,[134] it stimulates the sympathetic nerve cell bodies both directly[135] and indirectly by increasing acetylcholine release from preganglionic terminals.[136] Whether Ang-II produced locally in the gastrointestinal tract has a similar action to the circulating Ang-II remains to be discovered.

7.3. Central Action

Angiotensin II given intracerebroventricularly causes a dose-dependent drinking response in rats.[137] The receptive sites for this Ang-II action are the subfornical organ,[138,139] the anterior hypothalamus, the median and supraoptic portions of the preoptic region, and the organum vasculosum laminae terminalis.[140] In the cat, the most sensitive brain areas are the septal region, the anterior hypothalamus, and the preoptic area,[141] whereas in the rat, it is the ventricle.[142]

It seems that the dopaminergic system may play a role in mediating the dipsogenic activity of Ang-II.[143,144] The dopamine receptor blocker haloperidol, given prior to Ang-II, blocks thirst, whereas α- or β-adrenergic blockers have no action.

Angiotensin II injected into the lateral and third ventricles produces a dose-related response in increasing blood pressure.[145] It is suggested that this effect is mediated by Ang-II receptors in the mesencephalon and in the anterior ventral portion of the third ventricle.[146,147] The blood pressure-increasing effect of Ang-II might be mediated by increased sympathetic tone,[146] but other humoral factors such as vasopressin might also be considered.[138,147]

Intracerebroventricularly given Ang-II increases the sodium appetite[148] and releases vasopressin[149,150] and oxytocin.[151] It has been shown that Ang-II causes ADH release from isolated posterior lobes[152] *in vitro*. On the other hand, an angiotensin blocker reduces the blood pressure and decreases thirst.[124]

7.4. Conclusion

The data indicate that two renin–angiotensin systems exist. The first one, described many years ago, is associated with the kidney function and regulates blood pressure and salt and water metabolisms. The second one, the isorenin–angiotensin system, operates in the central and in some peripheral nerves, modulating drinking behavior and the central blood pressure and perhaps altering the visceral vascular tone. In the latter functions, Ang-II might be regarded as a neural modulator, but its role as a neural transmitter remains to be clarified.

8. PANCREATIC POLYPEPTIDE

Pancreatic polypeptide (PP) was first discovered in chicken pancreas[153] followed by the isolation of bovine, porcine, ovine, and human forms.[154,155] Since PP shows species differences, it is designated in accordance with its origin: avian PP (aPP), bovine PP (bPP), porcine PP (pPP), and human PP (hPP).

Pancreatic polypeptide contains 36 amino acids. There are up to four differences in amino acid composition among bovine, porcine, ovine, and human PP, whereas aPP differs in 20 positions from the others (Fig. 6).

There are PP-immunoreactive cells present in the pancreas—endocrine-type cells (PP) that differ from the glucagon (A) and insulin (B) cells.[156]

8.1. Distribution

Avian PP immunoreactivity has been observed in the central and peripheral nervous system.[157,158]

	1	2	3	4	5	6	7	8	9	10	11	12	13	14	15	16	17	18	19	20
hPP	Ala	Pro	Leu	Glu	Pro	Val	Tyr	Pro	Gly	Asp	Asp	Ala	Thr	Pro	Glu	Gln	Met	Ala	Gln	Tyr
bPP	Ala	Pro	Leu	Glu	Pro	Gln	Tyr	Pro	Gly	Asp	Asp	Ala	Thr	Pro	Glu	Gln	Met	Ala	Gln	Tyr
aPP	Gly	Pro	Ser	Gln	Pro	Thr	Tyr	Pro	Gly	Asp	Asp	Ala	Pro	Val	Glu	Asp	Leu	Ile	Arg	Phe

	21	22	23	24	25	26	27	28	29	30	31	32	33	34	35	36
hPP	Ala	Ala	Asp	Leu	Arg	Arg	Tyr	Ile	Asn	Met	Leu	Thr	Arg	Pro	Arg	Tyr
bPP	Ala	Ala	Glu	Leu	Arg	Arg	Tyr	Ile	Asn	Met	Leu	Thr	Arg	Pro	Arg	Tyr
aPP	Tyr	Asp	Asn	Leu	Glu	Gln	Tyr	Leu	Asn	Val	Val	Thr	Arg	His	Arg	Tyr

Fig. 6. Amino acid sequences of pancreatic polypeptides.

In the gastrointestinal tract, immunoreactive perikaria have been detected in the submucosal and myenteric plexuses, and immunoreactive fibers in smooth muscle and around blood vessels and in fibers in the core of the villi.[157]

By immunohistochemical methods, PP has been demonstrated in the brain and peripheral nervous system in chickens, mice, and cats. Only antiserum raised against aPP reveals immunopositive cell bodies and nerves, whereas antiserum raised against bPP or hPP is ineffective. Nerve cell bodies have been found in the cortex, n. accumbens, neostriatum, and septum and in nerve fibers in the nucleus accumbens, interstitial nucleus of the stria terminalis, para- and periventricular hypothalamic nuclei, medial preoptic area, and cortex.[157] AvianPP immunoreactivity is also present in some adrenergic nerve cells.

In the rat peripheral nervous system, immunoreactivity has been observed in the superior cervical, stellate, and celiac ganglia. Immunoreactive plexuses have been seen in the vas deferens and the heart and some fibers in the iris and in the submaxillary gland. The cell bodies and fibers also contain tyrosin hydroxylase (TH). In the central nervous system, aPP immunoreactivity has been observed together with TH and phenylethanolamine-N-methyltransferase (PNMT) in cell bodies in the medulla oblongata, nucleus tractus solitarii, and aPP and TH in the locus coeruleus, nerve terminals in the sympathetic lateral column of the spinal cord, nucleus tractus solitarii, locus coeruleus, nucleus periventricularis thalami, and some hypothalamic areas.[159]

By immunohistochemical methods, aPP has been shown to coexist with catecholamine in neurons of the locus coeruleus and with Met-enkephalin in the sacral parasympathetic system. Avian-PP-like immunoreactivity has been found in axonal and nerve terminals in the dorsal horn of the spinal cord, originating in part from intrinsic cell bodies, in terminals around sacral and caudal lumbar motor neurons and the lateral sympathetic column neurons, and in fibers around neurons of the periaqueductal gray, caudate raphe, locus coeruleus, and arcuate nucleus.[160]

8.2. Peripheral Effect

Avian PP given intravenously to chickens causes increased proventricular secretion of fluid, H^+, and pepsin and hepatic glycogenolysis without accompanying hyperglycemia.[161] Bovine PP in dog stimulates gastric acid secretion but inhibits pentagastrin-induced secretion, inhibits pancreas secretion, causes gallbladder relaxation, increases the choledochal tone, and decreases intestinal motility.[162]

Bovine PP in human inhibits the basal secretion of trypsin in the pancreatic juice[163] and the bilirubin output into the duodenum.[164,165]

8.3. Central Effects

Avian PP lowers the respiration rate in rats and enhances the clonidine-induced reduction of the respiration frequency. It is suggested that aPP-like peptides in the adrenergic neurons may play a modulatory role in the epinephrine synapses in the medulla oblongata involved in the regulation of the respiratory function.[166]

Avian PP reversibly inhibits the atropine-resistant vasodilatation induced by parasympathetic nerve stimulation in the submandibular and salivary glands.[167]

8.4. Conclusion

The physiological significance of PP in the nervous system is almost unknown. Its function in the gastrointestinal tract seems to be an inhibitory one. In the central nervous system, the few data available so far suggest that it might modulate epinephrine neurons in certain functions and interact with atropine-resistant vasodilatation in the salivary gland (VIP interaction?).

9. INSULIN

Insulin is produced by pancreatic β cells. More recently, however, insulin has been found in a number of extrapancreatic tissues, such as the heart, kidney, skeletal muscle, lung, brain, small intestine, liver, etc., in humans and rats.[168]

9.1. Distribution

Insulin has been identified in the brains of humans,[168] rats,[169,170] rabbits, and dogs.[171] The highest concentrations are found in the hypothalamus, olfactory bulb, and cerebellum, and the lowest in the brainstem and cerebral cortex.

By immunofluorescence, insulin has been shown to be present in neuronal perikarya of the olfactory bulb and central cortex.[172] However, it is estimated that insulin present in the pancreas is 10,000 times more concentrated than that in the extrapancreatic tissues.[168]

Insulin receptors are widely distributed on cell membranes from different regions of the rat brain. The highest level is observed in the olfactory bulb, followed by the cerebral cortex. The hippocampus, anterior hypothalamus including the preoptic area, amygdala, septum, and posterior hypothalamus also exhibit high binding. A lower binding is observed in the striatum, mesencephalon, thalamus, pons, and cerebellum. The lowest binding is to be seen in the cervical spinal cord, medulla, retina, and pituitary.[168]

With autoradiographic studies utilizing *in vivo* administration of labeled insulin, localization of insulin binding is observed in the medial basal hypothalamus, circumventricular organ,[173] blood vessels,[174] and nerve terminals in the median eminence and arcuate nucleus of the rat.[175] In mouse, immunofluorescence techniques reveal insulin binding in cell bodies, tanycytes lining the third ventricle, and the chorioid plexus.[176] The findings are essentially similar following administration of insulin into the brain ventricle or intracardially.[177]

9.2. Conclusion

It seems that extrapancreatic insulin does not differ chemically from pancreatic insulin, but its physiological role is different. In genetic obese hyperglycemic mice with hyperinsulinemia, intracellular insulin is chemically the same as in nonobese littermates. Under extreme conditions, such as fasting or streptozotocin-induced diabetes with low insulin, the brain insulin levels do not change.[168] It appears that intracellular insulin is not regulated by the same factors as extrapancreatic insulin. Intracellular insulin probably plays little if any role in carbohydrate metabolism. The exact role of intracellular insulin, especially in the brain, is not known.

REFERENCES

1. Bayliss, W. M., and Starling, E. H., 1902, *J. Physiol. (Lond.)* **28**:325–335.
2. Jorpes, J. E., and Mutt, V., 1961, *Acta Chem. Scand.* **15**:1790–1791.
3. Jorpes, J. E., and Mutt, V., 1966, *Acta Physiol. Scand.* **66**:316–325.
4. Polak, J. M., Coulling, I., Bloom, S. R., and Pearse, A. G. E., 1973, *Scand. J. Gastroenterol.* **6**:739–744.
5. O'Donohue, T. L., Charlton, C. G., Miller, R. L., Boden, G., and Jacobowitz, D. M., 1981, *Proc. Natl. Acad. Sci. U.S.A.* **78**:5221–5224.
6. Samson, W. K., Lumpkin, M. D., Vijayan, E., and McCann, S. M., 1982, *Endocrinol. Exp.* **16**:177–189.
7. Fuxe, K., Anderson, K., Hökfelt, T., Mutt, V., Ferland, L., Agnati, L. F., Ganten, D., Said, S., Eneroth, P., and Gustafsson, J. A., 1979, *Fed. Proc.* **38**:2333–2340.
8. Phillis, J. W., and Kirkpatrick, J. R., 1980, *Can. J. Physiol. Pharmacol.* **58**:612–623.
9. Phillis, J. W., and Kirkpatrick, J. R., 1979, *Can. J. Physiol. Pharmacol.* **57**:887–899.
10. Morley, J. E., Melmed, S., Briggs, J., Carlson, H. E., Hershman, J. M., Solomon, T. E., Lamers, C., and Damassa, D. A., 1979, *Life Sci.* **25**:1201–1206.
11. Samson, W. K., and Lumpkin, M. D., 1981, *Abstract, 63rd Annual Meeting of the Endocrine Society,* Endocrine Society, Bethesda, p. 320.
12. Enjalbert, A., Arancibia, S., Ruberg, M., Priam, M., Bluet-Pajot, M. T., Rotsztejn, W. H., and Kordon, C., 1980, *Neuroendocrinology* **31**:200–204.
13. Said, S. I., and Mutt, V., 1970, *Science* **169**:1217–1218.
14. Larsson, L.-I., Fahrenkrug, J., Schaffalitzky de Muckadell, O. B., Sundler, F., Hakanson, R., and Rehfeld, J. F., 1976, *Proc. Natl. Acad. Sci. U.S.A.* **73**:3197–3200.
15. Alumets, J., Fahrenkrug, J., Hakanson, R., Schaffalitzky de Muckadell, O. B., Sundler, F., and Uddman, R., 1979, *Nature* **280**:155–156.
16. Fahrenkrug, J., Schaffalitzky de Muckadell, O. B., Holst, J. J., and Lindkeer Jensen, S., 1979, *Am. J. Physiol.* **237**:E535–E540.
17. Ahrén, B., Alumets, J., Ericsson, M., Fahrenkrug, J., Fahrenkrug, L., Hakanson, R., Hedner, P., Lorén, I., Melander, A., Rerup, C., and Sundler, F., 1980, *Nature* **287**:343–345.
18. Lundberg, J. M., Hökfelt, T., Anggard, A., Uvnas-Wallensten, K., Brimijoin, S., Brodin, E., and Fahrenkrug, J., 1980, *Neural Peptides and Neuronal Communication* (E. Costa and M. Trabucchi, eds.), Raven Press, New York, pp. 25-36.
19. Loren, I., Emson, P. C., Fahrenkrug, J., Björklund, A., Alumets, J., Hakanson, R., and Sundler, F., 1979, *Neuroscience* **4**:1953–1976.
20. Marley, P. D., Emson, P. C., Hunt, S. P., and Fahrenkrug, J., 1981, *Neurosci. Lett.* **27**:261–266.
21. Roberts, G. W., Woodhams, P. L., Crow, T. J., and Polak, J. M., 1980, *Brain Res.* **195**:471–475.
22. Palkovits, M., Besson, J., and Rotsztejn, W., 1981, *Brain Res.* **213**:455–459.

23. Lundberg, J. M., Hökfelt, T., Anggard, A., Uvnas-Wallensten, K., Brimijoin, S., Brodin, E., and Fahrenkrug, J., 1980, *Adv. Biochem. Psychopharmacol.* **22**:25–36.
24. Larsson, L.-I., 1977, *Histochemistry* **54**:173–176.
25. Gilbert, R. F. T., Emson, P. C., Fahrenkrug, J., Dee, G. M., Penman, E., and Wass, J., 1980, *J. Neurochem.* **34**:108–113.
26. Lundberg, J. M., 1981, *Acta Physiol. Scand. (Suppl.)* **496**:1–57.
27. Lundberg, J. M., Anggard, A., Fahrenkrug, J., Hökfelt, T., and Mutt, V., 1980, *Proc. Natl. Acad. Sci. U.S.A.* **77**:1651–1655.
28. Goyal, R. K., Rattan, S., and Said, S. I., 1980, *Nature* **288**:378–380.
29. Eklund, S., Jodal, M., Lundgren, O., and Sjöqvist, A., 1979, *Acta Physiol. Scand.* **105**:461–468.
30. Edin, R., Lundberg, J. M., Ahlman, H., Dahlström, A., Fahrenkrug, J., Hökfelt, T., and Kewenter, J., 1979, *Acta Physiol. Scand.* **107**:185–187.
31. Fahrenkrug, J., Hanglund, U., Jodal, M., Lundgren, O., Olbe, L., and Schaffalitzky de Muckadell, O. B., 1978, *J. Physiol. (Lond.)* **284**:291–305.
32. Larsen, L.-I., Fahrenkrug, J., and Holst, J. J., 1978, *Life Sci.* **22**:773–780.
33. Lindkaer, Jensen, S., Fahrenkrug, J., Holst, J. J., Vagn Nielsen, O., and Schaffalitzky de Muckadell, O. B., 1978, *Am. J. Physiol.* **235**:387–391.
34. Ottesen, B., Larsen, J. J., Fahrenkrug, J., Stjernquist, M., and Sundler, F., 1981, *Am. J. Physiol.* **240**:E32–E36.
35. Ottesen, B., Wagner, G., and Fahrenkrug, J., 1980, *Prostaglandins* **19**:427–435.
36. Taylor, D. P., and Pert, C. B., 1979, *Proc. Natl. Acad. Sci. U.S.A.* **76**:660–664.
37. Taylor, D. P., Pert, C. B., and Herkenham, M., 1979, *Peptides: Structure and Biological Functions* (E. Gross and J. Meidhalfer, eds.), Pierce Chemical, Rockford, IL, pp. 917–920.
38. Quick, M., Iversen, L. L., and Bloom, S. R., 1978, *Biochem. Pharmacol.* **27**:2209–2213.
39. Kerwin, R. W., Pay, S., Bhoola, K. D., and Pycock, C. J., 1980, *J. Pharm. Pharmacol.* **32**:561–566.
40. Borghi, C., Nicosia, S., Giachetti, A., and Said, S. I., 1979, *Life Sci.* **24**:65–70.
41. Giachetti, A., Said, S. I., Reynolds, R. C., and Koniges, F. C., 1977, *Proc. Natl. Acad. Sci. U.S.A.* **74**:3424–3428.
42. Emson, P. C., Fahrenkrug, J., Schaffalitzky de Muckadell, O. B., Jessell, T. M., and Iversen, L. L., 1978, *Brain Res.* **143**:174–178.
43. Bloom, S. R., and Edwards, A. V., 1980, *J. Physiol. (Lond.)* **299**:437–452.
44. Phillis, J. W., Kirkpatrick, J. R., and Said, S. I., 1978, *Can. J. Physiol. Pharmacol.* **56**:337–340.
45. Dodd, J., Kelly, J. S., and Said, S. I., 1979, *Br. J. Pharmacol.* **66**:125P–126P.
46. Dingledine, R., Dodd, J., and Kelly, J. S., 1980, *J. Neurosci. Methods* **2**:323–362.
47. Clark, W. G., Lipton, J. M., and Said, S. I., 1978, *Neuropharmacology* **17**:883–885.
48. Samson, W. K., Said, S. I., and McCann, S. M., 1979, *Neurosci. Lett.* **12**:265–269.
49. Besson, J., Rotsztejn, W., Laburthe, M., Epelbaum, J., Beaudet, A., Kordon, C., and Rosselin, G., 1979, *Brain Res.* **165**:79–85.
50. Said, S. I., and Porter, J. C., 1979, *Life Sci.* **24**:227–230.
51. Samson, W. K., Said, S. I., Snyder, G., and McCann, S. M., 1980, *Peptides* **1**:325–352.
52. Shaar, C. J., Clemens, J. A., and Dininger, N. B., 1979, *Life Sci.* **25**:2071–2074.
53. Vijayan, E., Samson, W. K., Said, S. I., and McCann, S. M., 1979, *Endocrinology* **104**:53–57.
54. Samson, W. K., Burton, K. P., Reeves, J. P., and McCann, S. M., 1981, *Regul. Peptides* **2**:253–264.
55. Epelbaum, J., Tapia-Arancibia, L., Besson, J., Rotsztejn, W. H., and Kordon, C., 1979, *Eur. J. Pharmacol.* **58**:493–495.
56. Gregory, R. A., and Tracy, H. J., 1964, *Gut* **5**:103–117.
57. Walsh, J. H., and Grossman, M. I., 1975, *N. Engl. J. Med.* **292**:1324–1332.
58. Berson, S. A., and Yalow, R. S., 1971, *Gastroenterology* **60**:215–222.
59. Vanderhaeghen, J. J., Signeau, J. C., and Gepts, W., 1975, *Nature* **257**:604–605.
60. Rehfeld, J. F., 1978, *Nature* **271**:771–773.
61. Uvnas-Wallensten, K., Rehfeld, J. F., Larsson, L. I., and Uvnas, B., 1977, *Proc. Natl. Acad. Sci. U.S.A.* **74**:5707–5710.

62. Rehfeld, J. F., and Kruse-Larsen, C., 1978, *Brain Res.* **155**:19–26.
63. Uvnas-Wallensten, K., and Uvnas, B., 1978, *Acta Physiol. Scand.* **103**:349–351.
64. Vizi, E. S., Bertaccini, G., Impicciatore, M., and Knoll, J., 1972, *Eur. J. Pharmacol.* **17**:175–178.
65. Vizi, E. S., Bertaccini, G., Impicciatore, M., Mantovani, P., Zséli, J., and Knoll, J., 1974, *Naunyn Schmiedebergs Arch. Pharmacol.* **284**:233–243.
66. Telegdy, G., 1980, *Acta Physiol. Acad. Sci. Hung.* **55**:273–281.
67. Várszegi, M., Fekete, M., Penke, B., Kovács, K., and Telegdy, G., 1981, *Endocrinology, Neuropeptides-I* (E. Stark, G. B. Makara, Z. Ács, E. Endröczi, eds.), Pergamon Press, London, pp. 215–219.
68. Várszegi, M., Fekete, M., Telegdy, G., Penke, B., and Kovács, K., 1980, *Modulation of Neurochemical Transmission* (E. S. Vizi, ed.), Pergamon Press, London, Akadémiai Kiadó, Budapest, pp. 439–449.
69. Nandi Majumder, A. P., and Nakhla, A. M., 1978, *Experientia* **34**:974–975.
70. Kelly, J. S., and Dodd, J., 1981, *Neurosecretion and Brain Peptides,* (J. B. Martin, S. Reichlin, and K. J. Blick, eds.), Raven Press, New York, pp. 133–144.
71. Della-Fera, M. A., and Baile, C. A., 1979, *Science* **206**:471–473.
72. Goltermann, N. R., Stengaard-Pedersen, K., Rehfeld, J. F., and Christensen, N. J., 1981, *Neurochemistry* **36**:959–965.
73. Vijayan, E., Samson, W. K., and McCann, S. M., 1978, *Life Sci.* **23**:2225–2232.
74. Telegdy, G., Fekete, M., Várszegi, M., and Kádár, T., 1980, *Advances in Pharmacological Research and Practice* (J. Knoll, E. S. Vizi, and M. Wolleman, eds.), Pergamon Press, pp. 169–185.
75. Itoh, S., Hirota, R., Katsuura, G., and Odaguchi, K., 1979, *Endocrinol. Jpn.* **26**:741–744.
76. Brown, J. C., Mutt, V., and Dryburgh, J. R., 1971, *Can. J. Physiol. Pharmacol.* **49**:399–405.
77. Brown, J. C., Cook, M. A., and Dryburgh, J. R., 1973, *Can. J. Biochem.* **51**:533–537.
78. Chey, W. Y., and Lee, K. Y., 1980, *Clin. Gastroenterol.* **9**:645–656.
79. Yanaihara, C., Sato, H., Yanaihara, N., Naruse, S., Forssman, W. G., Helmstaedter, V., Fujita, T., Yamaguchi, K., and Abe, K., 1978, *Adv. Exp. Biol. Med.* **106**:269–283.
80. O'Donohue, T. L., Beinfeld, M., Chey, W. Y., and Jacobowitz, D. M., 1981, *Soc. Neurosci. Abstr.* **7**:508.
81. O'Donohue, T. L., Charlton, C. G., Miller, R. L., Boden, G., and Jocobowitz, D. M., 1981, *Proc. Natl. Acad. Sci. U.S.A.* **78**:5221–5224.
82. Chey, W. Y., Escoffery, R., Roth, F., Chang, T. M., and Yajima, H., 1980, *Peptides (Suppl.)* **1**:S19.
83. Nagai, K., Yanaihara, C., Yanaihara, N., Shimuzu, F., Kobayashi, S., and Fujita, T., 1980, *Peptides (Suppl)* **1**:S78.
84. O'Donohue, T. L., Beinfeld, M. C., Chey, W. Y., Chang, T. M., Nilaver, G., Zimmerman, E. A., Yajima, H., Adachi, H., Poth, M., McDevitt, R. P., and Jacobowitz, D. M., 1981, *Peptides* **2**:467–477.
85. Nilaver, G., Defendini, R., Zimmerman, E. A., Beinfeld, M. C., and O'Donohue, T. L., 1981, *Soc. Neurosci. Abstr.* **7**:99.
86. Jacobowitz, D. M., O'Donohue, T. L., Chey, W. Y., and Chang, T. M., 1981, *Peptides* **2**:479–487.
87. Strunz, U., Domschke, W., Mitznegg, P., Domschke, S., Schubert, E., Wunsch, E., Jaeger, E., and Demling, L., 1975, *Gastroenterology* **68**:1485–1491.
88. Christofides, N. D., Modlin, I. M., Fitzpatric, M. D., and Bloom, S. R., 1978, *Gut* **19**:A436–437.
89. Koch, J., Domschke, S., Belohlavek, D., Domschke, W., Wunsch, E., Jaeger, E., and Demling, L., 1976, *Scand. J. Gastroenterol.* **11**:93–96.
90. Konturek, S. J., Krol, R., Dembinski, A., and Wunsch, E., 1976, *Pflugers Arch.* **364**:297–300.
91. Erspamer, V., and Melchiorri, P., 1973, *Pure Appl. Chem.* **35**:463–494.
92. Polak, J. M., Ghatel, M. A., Wharton, J., Bishop, A. E., Bloom, S. R., Solcia, E., Brown, M. R., and Pearse, A. G. E., 1978, *Scand. J. Gastroenterol.* **13**:S49–148.
93. Walsh, J. H., Wong, H. C., and Dockray, G. J., 1979, *Fed. Proc.* **38**:2315–2319.

94. Brown, M., and Vale, W., 1979, *Trends Neurosci.* **2**:95-97.
95. Lezoche, F., Ghatei, M. A., Carlei, F., Blackburn, A. M., Basso, N., Adrian, T. E., Speranza, V., and Bloom, S. R., 1979, *Gastroenterology* **76**:1185.
96. Brown, M. R., Rivier, J., and Vale, W., 1977, *Life Sci.* **21**:1729.
97. Rivier, J. E., and Brown, M. R., 1978, *Biochemistry* **17**:1766.
98. Gibbs, J., Fauser, D. J., Rowe, E. A., Rolls, B. J., Rolls, E. T., and Maddison, S. P., 1979, *Nature* **282**:208-210.
99. Morley, J. E., and Levine, A. S., 1980, *Pharmacol. Biochem. Behav.* **14**:149-151.
100. Dodd, J., and Kelly, J. S., 1981, *Brain Res.* **205**:337-350.
101. Pederson, R. A., Dryburgh, J. R., and Brown, J. C., 1975, *Can. J. Physiol. Pharmacol.* **53**:1200.
102. Brown, M., Rivier, J. E., Kobayashi, R., and Vale, W. W., 1978, *Gut Hormones*, Churchill Livingston, Edinburgh, pp. 550-558.
103. Vale, W., Rivier, C., Rivier, J., and Brown, M., 1978, *Psychopharmacology: A Generation Progress* (M. A. Lipton, ed.), Raven Press, New York, pp. 403-421.
104. Brown, M., Rivier, J., Wolfe, A., and Vale, W., 1977, *Endocrinology* **100**:279.
105. Brown, M. R., and Fisher, D. A., 1980, *Peptides: Integrators of Cell and Tissue Function* (F. E. Bloom, ed.), Raven Press, New York, pp. 81-97.
106. McDonald, T. J., Jörnvall, H., Nilsson, G., Vagne, M., Ghatei, M., Bloom, S. R., and Mutt, V., 1979, *Biochem. Biophys. Res. Commun.* **90**:227-233.
107. Sundler, F., Hakanson, R., and Leander, S., 1980, *Clin. Gastroenterol.* **9**:517-543.
108. Peach, M. J., 1977, *Physiol. Rev.* **57**:313-370.
109. Regoli, D., Park, W. K., and Rioux, F., 1974, *Pharmacol. Rev.* **26**:69-129.
110. Fuxe, K., Ganten, D., Hökfelt, T., and Bolme, P., 1976, *Neurosci. Lett.* **2**:229-234.
111. Hökfelt, T., Schultzberg, M., Johansson, O., Ljungdahl, A., Elfvin, L., Elde, R., Terenius, L., Nilsson, G., Said, S., and Goldstein, M., 1978, *Gut Hormones* (S. R. Bloom, ed.), Churchill Livingstone, Edinburgh, pp. 423-433.
112. Ganten, D., and Speck, G., 1978, *Biochem. Pharmacol.* **27**:2379-2389.
113. Osman, M. Y., Smeby, R. R., and Sen, S., 1979, *Hypertension* **1**:53-60.
114. Hirose, S., Ohsawa, T., Inagami, T., and Murakami, K., 1982, *J. Biol. Chem.* **257**:6316-6321.
115. Hirose, S., Yokosawa, H., Inagami, T., and Workman, R. J., 1980, *Brain Res.* **91**:489-499.
116. Celio, M. R., Clemens, D. L., and Inagami, T., 1980, *Biomed. Res.* **1**:427-431.
117. Inagami, T., Celio, M. R., Clemens, D. L., Lau, D., Takii, T., Kasselberg, A. G., and Hirose, S., 1980, *Clin. Sci.* **59**:49-51.
118. Naruse, K., Takii, Y., and Inagami, T., 1981, *Proc. Natl. Acad. Sci. U.S.A.* **78**:7579-7583.
119. Ganten, D., Fuxe, K., Ganten, U., Hökfelt, T., and Bolme, P., 1977, *Hypertension and Brain Mechanisms* (W. de Jong, A. P. Provoost, and A. P. Shapiro, eds.), Elsevier/North Holland Biomedical Press, Amsterdam, pp. 155-160.
120. Fuxe, K., Ganten, D., Andersson, K., Calza, L., Agnati, L. F., Lang, R. E., Poulsen, K., Hökfelt, T., and Bernardis, P., 1982, *The Renin-Angiotensin System in the Brain* (D. Ganten, M. P. Printz, M. I. Philips, and B. A. Scholkens, eds.), Springer-Verlag, Berlin, pp. 208-232.
121. Slater, E. E., Defendini, R., and Zimmerman, E., 1980, *Proc. Natl Acad. Sci. U.S.A.* **77**:5458-5460.
122. Inagami, T., 1982, *Neuroendocrinology* **35**:475-482.
123. Ganten, D., Marques-Julio, A., Granger, P., Hayduk, K., Karsunky, K. P., Boucher, R., and Genest, J., 1971, *Am. J. Physiol.* **221**:1733-1737.
124. Ganten, D., Fuxe, K., Philips, M. I., Mann, J. F. E., and Ganten, U., 1978, *Frontiers in Neuroendocrinology*, Volume 5 (W. F. Ganong, and L. Martini, eds.), Raven Press, New York, pp. 61-99.
125. Yang, H.-Y., and Neff, N. H., 1972, *J. Neurochem.* **19**:2443-2450.
126. Yang, H.-Y., and Neff, N. H., 1973, *J. Neurochem.* **21**:1035-1036.
127. Fischer-Ferraro, C., Nahmod, V. E., Goldstein, D. J., and Finkielman, S., 1971, *J. Exp. Med.* **133**:353-361.
128. Bennet, J. B., and Snyder, S. H., 1976, *J. Biol. Chem.* **251**:7423-7430.
129. McLean, A. S., Sirett, N. E., Bray, J. J., and Hubbard, J. I., 1975, *Proc. Univ. Otago Med. School* **53**:19-20.

130. Philips, M. I., and Felix, D., 1976, *Brain Res.* **109**:531-540.
131. Philips, M. I., Felix, D., Hoffman, W. E., and Ganten, D., 1977, *Neuroscience Symposium* (W. M. Cowan, and J. A. Ferrendelli, eds.), Society for Neuroscience, Bethesda, pp. 308-339.
132. Nicoll, R. A., and Barker, J. L., 1971, *Nature (New Biol.)* **233**:172-173.
133. Sakai, K. K., Marks, B. H., George, J., and Koestner, A., 1974, *Life Sci.* **14**:1337-1344.
134. Zimmerman, B. G., 1967, *J. Pharmacol. Exp. Ther.* **158**:1-10.
135. Reit, E., 1972, *Fed. Proc.* **31**:1338-1343.
136. Panisset, J. C., 1967, *Can. J. Physiol. Pharmacol.* **45**:313-317.
137. Fitzsimons, J. T., 1971, *Proc. R. Soc. Med.* **64**:1074.
138. Severs, W. B., Summy-Long, J., Taylor, J. S., and Connor, J. P., 1970, *J. Pharmacol. Exp. Ther.* **174**:27-34.
139. Simpson, J. B., 1981, *Neuroendocrinology* **32**:248-256.
140. Buggy, J., and Johnson, A. K., 1977, *Am. J. Physiol.* **233**:R44-R52.
141. Swanson, L. W., and Sharpe, L. G., 1973, *Am. J. Physiol.* **225**:566-572.
142. Johnson, A. K., and Epstein, A. N., 1975, *Brain Res.* **86**:399-418.
143. Fitzsimons, J. T., and Setler, P. E., 1975, *J. Physiol. (Lond.)* **250**:613-631.
144. Setler, P. E., 1973, *The Neuropsychology of Thirst* (A. N. Epstein, H. R. Kissileff, and E. Stellar, eds.), H. V. Winston, Washington, pp. 279-291.
145. Buckley, J. P., and Jandhyala, B. S., 1977, *Life Sci.* **20**:1485-1494.
146. Buckley, J. P., 1981, *Trends Pharmacol. Sci.* **22**:161-162.
147. Hutchinson, J. S., Schelling, P., Möhring, J., and Ganten, D., 1976, *Endocrinology* **99**:819-823.
148. Fitzsimons, J. T., 1978, *Fed. Proc.* **37**:2669-2675.
149. Keil, L. C., Summy-Long, J., and Severs, W. B., 1975, *Endocrinology* **96**:1063-1065.
150. Simonnet, G., Rodriguez, F., Fumoux, F., Czernichow, P., and Vincent, J. D., 1979, *Am. J. Physiol.* **237**:R20-R25.
151. Lang, R. E., Rascher, W., Heil, J., Unger, T., Wiedemann, G., and Ganten, D., 1981, *Life Sci.* **29**:1425-1428.
152. Gagnon, D. J., Sirois, P., and Boucher, P. J., 1975, *Clin. Exp. Pharmacol. Physiol.* **2**:305-313.
153. Kimmel, J. R., Pollock, H. G., and Hazelwood, R. L., 1968, *Endocrinology* **83**:1323-1330.
154. Chance, R. E., 1972, *Diabetes* **21**(Suppl. 2):536.
155. Chance, R. E., and Jones, W. E., 1974, *United States Patent Office* **842**:/3/, 063.
156. Orci, L., Malaisse-Lagae, F., Baetens, D., and Perrelet, A., 1978, *Lancet* **2**:1200.
157. Loren, I., Alumets, J., Häkansson, R., and Sundler, F., 1979, *Cell Tissue Res.* **200**:179-186.
158. Hökfelt, T., Lundberg, J. M., Terenius, L., Jansco, G., and Kimmel, J., 1981, *Peptides* **2**:81-87.
159. Lundberg, J., Hökfelt, T., Anggard, A., Kimmel, J., Goldstein, M., and Markey, K., 1980, *Acta Physiol. Scand.* **110**:107-109.
160. Hunt, S. P., Emson, P. C., Gilbert, R., Goldstein, M., and Kimmel, J. R., 1981, *Neurosci. Lett.* **21**:125-130.
161. Hazelwood, R. L., and Turner, S. D., 1973, *Gen. Comp. Endocrinol.* **21**:485-497.
162. Lin, T.-M., and Chance, R. E., 1978, *Gut Hormones*, Churchill Livingstone, Edinburgh, pp. 242-246.
163. Greenberg, G. R., McCloy, R. F., Chadwick, V. S., Adrian, T. E., Baron, J. H., and Bloom, S. R., 1979, *Am. J. Dig. Dis.* **24**:11-14.
164. Greenberg, G. R., McCloy, R. F., Adrian, T. E., Baron, J. H., and Bloom, S. R., 1978, *Acta Hepato-Gastroenterol.* **25**:384-387.
165. Greenberg, G. R., McCloy, R. F., Adrian, T. E., Chadwick, V. S., Baron, J. H., and Bloom, S. R., 1978, *Lancet* **2**:1280-1282.
166. Fuxe, K., Agnati, L. F., Härfstrand, A., Lundberg, J., Hökfelt, T., Calza, L., Kimmel, J., and Bernardi, P., 1982, *Acta Physiol. Scand.* **115**:381-384.
167. Lundberg, J., Anggard, A., Hökfelt, T., and Kimmel, J., 1980, *Acta Physiol. Scand.* **110**:199-201.

168. Rosenzweig, J. L., Havrankova, J., Brownstein, M., and Roth, J., 1980, *Endocrinology, Neuroendocrinology, Neuropeptides-I* (E. Stark, G. B. Makara, Z. Ács, and E. Endröczi, eds.), Pergamon Press, London, pp. 175–186.
169. Havrankova, J., Schmechel, D., Roth, J., and Brownstein, M., 1978, *Proc. Natl. Acad. Sci. U.S.A.* **75**:5737–5741.
170. Havrankova, J., Brownstein, M. J., and Roth, J., 1979, *J. Clin. Invest.* **64**:636–642.
171. Eng, J., and Yalow, R. S., 1980, *Diabetes* **29**:105–109.
172. Havrankova, J., Roth, J., and Brownstein, M., 1978, *Nature* **272**:827–829.
173. van Houten, M., Posner, B. I., Kopriwa, B. M., and Brawer, J. R., 1979, *Endocrinology* **105**:666–673.
174. van Houten, M., and Posner, B. I., 1979, *Nature* **282**:623–625.
175. van Houten, M., Posner, B. I., Kopriwa, B. M., and Brawer, J. R., 1980, *Science* **207**:1081–1083.
176. Pansky, B., and Hatfield, J. S., 1978, *Am. J. Anat.* **153**:459–467.
177. Cruz, L., and Antonetty, C., 1980, *Endocrinology, Neuroendocrinology, Neuropeptides-I* (E. Stark, G. B. Makara, Z. Ács, and E. Endröczi, eds.), Pergamon Press, London, pp. 187–192.

11

Peptidergic Systems

Peter Schotman, Loes H. Schrama, and Philippa M. Edwards

1. INTRODUCTION

A peptidergic system is a functionally relevant complex consisting of a cell that synthesizes and releases the peptide, a cell that responds to that peptide by some change in function, and a means whereby the peptide is transferred from its site of synthesis to its site of action. We have included in our considerations cases in which the source of the peptide is a peptidergic neuron and the target cell is nonneuronal (neuroendocrine function), in which a nonneuronally synthesized peptide acts on a peptide-sensitive neuron, and in which both the source of and target for the peptide are neuronal.

The study of peptidergic systems is still in its infancy. A large number of peptides have been demonstrated to be present in the mammalian brain, and in most cases, neuronal localization has been indicated. However, complete definition of a peptidergic system requires knowledge of the site and mode of synthesis of a peptide, of release mechanisms and their regulation, of the mode of transport of the peptide to its target cell, of the interaction with target cell receptors, of the response of the target cell, of the mechanisms for terminating the peptide action, and of the physiological role of the system. In a few instances [e.g., some hypophysiotrophic hormones and substance P (*SP*) in the peripheral nervous system], some of these criteria have been satisfied, but in many (e.g., *insulin* and *bombesin*), little more is known than that the peptides are present. From the information currently available, a common pattern is observed: synthesis of a large precursor by the protein synthetic machinery in the cell body, packaging in granules, transport to and release from the nerve terminals, and processing and destruction by specific and nonspecific enzymes. Small peptides (such as *carnosine*) that are synthesized by specific amino-acid-linking enzymes have not been considered here.

Peter Schotman, Loes H. Schrama, and Philippa M. Edwards • Division of Molecular Neurobiology, Institute of Molecular Biology, Laboratory of Physiological Chemistry and Rudolf Magnus Institute for Pharmacology, State University of Utrecht, 3508 TB Utrecht, The Netherlands.

A large majority of neuropeptides are also peripheral hormones. Brain peptidergic systems can thus be seen as a combination of the two major modes of intracellular communication: the endocrine and the nervous systems. The concept unifying these two systems was first formulated by Pearse in the so-called APUD (amine content and/or amine precursor uptake and decarboxylation) concept to explain the possession by neurons and endocrine cells producing polypeptide hormones of a set of common characteristics.[1] Pearse proposed that the cells of the APUD series all originate from the embryonic neuroectoderm and migrate to the nervous system, endocrine glands, gastrointestinal tract, and skin during development. Thus, the same or similar peptides are found in association with catecholamines in these diverse organs, in which different functions (neurotransmitter or hormone) are served by catecholamine release.

Neuroactive peptides may have either effects on the electrical activity of neurons or long-term trophic influences that may, for instance, be involved in the stability of synaptic contacts or cell survival. The peptides considered in this chapter all (*insulin* is an exception) have effects on the electrical activity of neuronal circuits. Apart from nerve growth factor (*NGF*) in the peripheral system (see Chapter 20, Vol. 5), very little is known about peptides that have purely trophic effects, although studies on cultured cells indicate that such factors may exist. Some peptides that have electrophysiological effects also induce changes that indicate an additional, trophic influence on nerve cells.

Neuroregulatory substances have been classified according to their proposed actions as neurotransmitters, neuromodulators, or neurohormones[2,3] (see Section 4). Transmitters and hormones have many features in common,[4,5] and one substance may act both as a neurotransmitter and a hormone, as has been shown for the catecholamines.[4]

The frequent combined occurrence of peptides and classical neurotransmitters argues against a role for the peptide as a neurotransmitter in these instances, since this is at odds with Dale's concept of one cell–one transmitter. However, only 40% of the known synapses present in the CNS can be accounted for by the classic neurotransmitters,[6] so the scope for neuropeptides as putative neurotransmitters is large even without questioning Dale's concept. Most neuropeptides must be considered as putative neurotransmitters, neuromodulators, and neurohormones,[3,4] and assignment to one category can only be made with respect to a particular action at a particular site.

Separate chapters have been devoted to several of the most extensively investigated peptides [opiate peptides, *SP*, neurotensin (*NT*), hypothalamic releasing factors, and prolactin (*PRL*)] and to the effects of peptides on behavior. We have therefore used this chapter to give an overview of the localization of peptidergic systems (site of synthesis, site of action, and route from former to latter), the relationships between peptides and classical neurotransmitters, interactions with target cell membranes, and modulation of target cell function. More detailed descriptions are given for peptides that are not covered in other chapters.

2. ANATOMIC CONSIDERATIONS RELEVANT TO THE ROLE OF PEPTIDES AS NEUROHORMONES

Until recently, the blood–brain barrier was considered to be impermeable to peptides. However, in spite of their short half-life (of the order of minutes),[7] many peptides have been reported to be present in the brain following peripheral injection.[7-9] Further evidence that blood-borne peptides can reach brain nerve cells comes from behavioral studies. Extirpation of the pituitary gland (which is peripheral to the blood–brain barrier) or one of its lobes leads to behavioral deficiencies that were restored to normal by peripheral administration of corticotropin (*ACTH*), melanotropin (*MSH*), or vasopressin (*VP*), or with synthetic analogues lacking the classical endocrine activities.[10] Although specific transport systems across the blood–brain, blood–CSF, and/or CSF–brain barriers cannot be excluded, it is clear that brain regions lacking these barriers in their normal form are very important in neurohormonal communication. Such areas are the choroid plexi, the median eminence (ME) of the hypothalamus and circumventricular organs (CVOs) including the organum vasculosum of the lamina terminalis (OVLT), the area postrema, the subfornical organ (SFO), and the pineal. In all of these regions, the capillary endothelium is fenestrated (tight junctions exist in the rest of the brain), and the permeability barrier is shifted to a thin layer of ependymal cells, which may be permeable to or specifically transport many peptides.

All of the areas are highly vascularized. The ME is of particular interest since it is the area that has the highest concentration of neuropeptides in the brain and because it is linked by a portal system to the pituitary (Fig. 1). In addition, the presence of two capillary beds (in the primary and subependymal plexi) with fenestrated epithelia provides the anatomic basis for all permutations of transfer of peptides between neurons, blood, and CSF. The primary plexus represents the site of entry of hypothalamic peptides into the hypothalamic–pituitary portal system. The subependymal plexus is in close association with the third ventricle (CSF). Specialized ependymal cells (tanycytes), which contact both the pericapillary space and the CSF and may also contact nerve terminals, have been proposed to play a special role in transport functions.

It has been suggested[11] that the vascular connections between the hypothalamus and pituitary may also serve to transport pituitary hormones to the hypothalamus. The concentrations of pituitary peptides are one or two orders of magnitude higher in portal than in peripheral blood. Neuropeptides injected into the pituitary appear almost immediately in the CNS, and this transfer is markedly reduced by stalk section.[7] The portal system shows extremely fine dynamics, and local changes in pressure can influence the direction of blood flow. A similar reverse flow may occur between the pituitary and the hypothalamic subependymal plexus. This plexus is characterized by a high degree of anastomization, and both arteries and veins pass from this plexus to the arcuate nucleus (ARC). Thus, a reversed flow between the pituitary and the

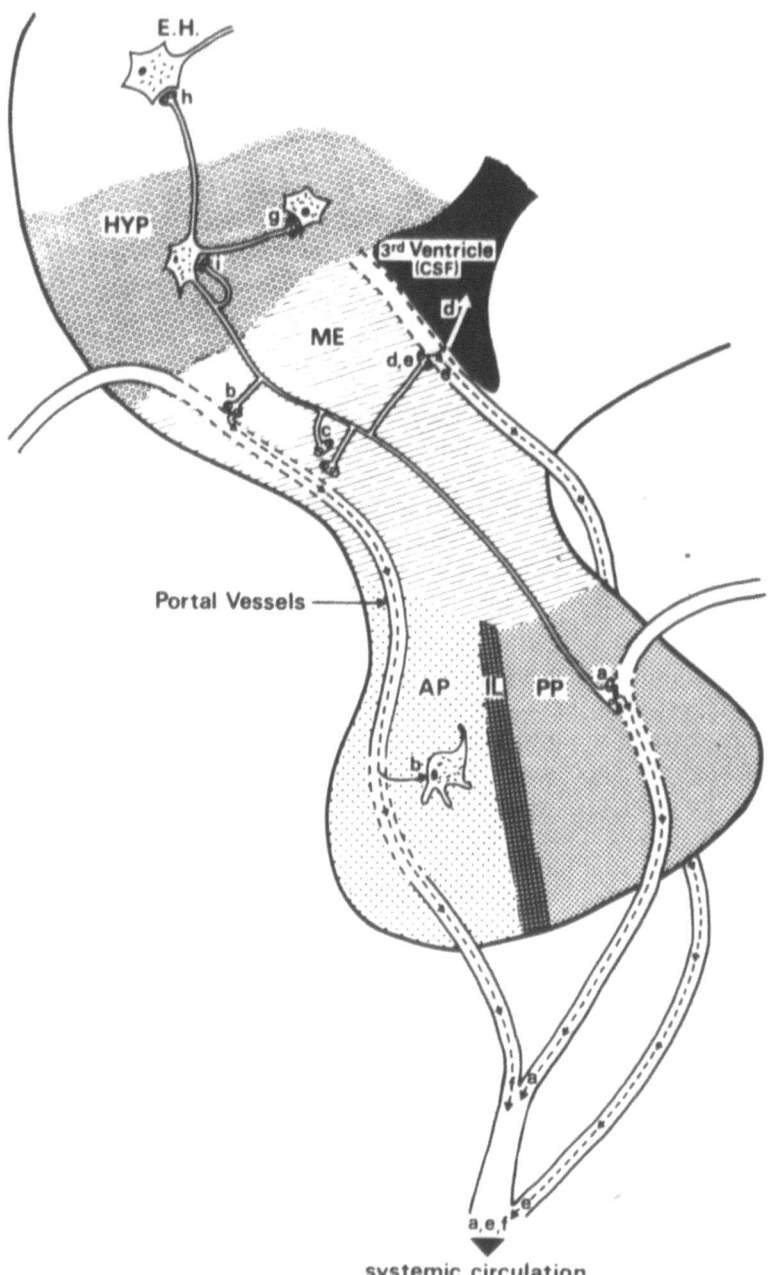

Fig. 1. Schematic diagram of avenues of communication between hypothalamic neurons and target cells. The various regions indicated are: HYP, hypothalamus; ME, median eminence of hypothalamus; AP, anterior pituitary; IL, intermediate lobe of pituitary; PP, posterior pituitary; EH, extrahypothalamic brain. The various pathways shown are: (a) access to systemic circulation via release in PP, (b) entry to portal vessels via primary plexus in ME and access to target pituicytes, (c) regulation of release of other substances from terminals within ME, (d) access to CSF via subependymal and primary plexi and other circumventricular organs, (e) access to systemic circulation via same areas as d, (f) access to systemic circulation via portal system, (g) intrahypothalamic neural connection, (h) extrahypothalamic neural connection, and (i) ultrashort feedback. Fenestrated capillaries are indicated by a broken line in the blood vessel walls.

subependymal plexus could also serve to deliver pituitary peptides to the ARC.[11] However, the physiological occurrence of backflow from the pituitary to the hypothalamus is still open to question. It must be borne in mind that the procedures used to demonstrate this flow (blood sampling, injection of peptides) cause local pressure changes that may not occur physiologically, thus generating an abnormal transport system.

3. LOCALIZATION OF PEPTIDERGIC SYSTEMS

3.1. General Considerations

Rarely can the presence of peptides be traced from the nerve cell body along the complete fiber tract to the nerve terminals. The source of peptides is often deduced from its disappearance from terminals following lesioning of the areas suspected to contain the synthesizing cell bodies. Further supporting evidence is obtained by interrupting the axoplasmic transport by lesion, ligation, or microtubule-disrupting drugs. Direct demonstration of local synthesis has only been given in a few instances. In some cases, evidence from brain/blood concentration ratios is the major indication of peptide synthesis within the brain. It is still more difficult to be certain of pathways in cases in which CSF or blood is believed to play a role, as exemplified by the discussion above on the role of reversed blood flow between the pituitary and the hypothalamus.

Information on the localization of peptides within the CNS mainly arises from immunologic techniques and is only as reliable as the specificity of the antibodies used. The specificity is particularly critical, since almost all neuropeptides are formed by proteolytic cleavage of larger precursor molecules, which include most of the antigenic determinants of the final product. In some cases [*SP*, somatostatin (*SRIF*), *insulin*], only one active derivative is known to be formed from the precursor, so the distribution of precursor plus product will accurately trace the pathway. However, the situation with peptides derived from proopiomelanocortin (*POMC*) requires much greater caution, as different active fragments can be formed (Section 3.2). Recently, recombinant-DNA techniques have made it possible to trace the distribution of *POMC*-like mRNA in the brain.[12]

Evidence for the localization of target cells derives from their juxtaposition to neuropeptide-containing terminals, from binding studies, and from observed effects of the peptide on cell function.

3.2. Peptides of Pituitary Origin

Three groups of peptides are synthesized in the endocrine cells of the anterior and intermediate lobes of the pituitary: (1) *POMC*-derived peptides, (2) growth hormone (*GH*) and prolactin (*PRL*), and (3) thyroid-stimulating hormone (*TSH*), luteinizing hormone (*LH*), and follicle-stimulating hormone (*FSH*). The release of hormones from the anterior pituitary is regulated by hypothalamic releasing factors via the portal circulation. There is evidence that each of the pituitary hormones exerts feedback regulation (referred to as internal or short feedback) in the hypothalamus and also has effects on brain function unrelated to neuroendocrine regulation.

Proopiomelanocortin itself has hardly any biological activity but contains a linear arrangement of active principles, which can elicit a variety of physiological responses. The expression of those principles is dependent on cleavage of the precursor and other modifications such as acetylation. All *POMC*-producing cells studied to date produce and secrete more than one of the active principles, but there is selective processing, since a different mixture is formed in different cell types.[13] The corticotropes in the anterior pituitary secrete *ACTH*, β-lipotropin (β-*LPH*), *pro*-γ-*MSH*, β-endorphin (β-*END*), and corticotropinlike intermediate peptide (*CLIP*). *ACTH*-producing cells are also present in the intermediate lobe, but the processing of *POMC* in the majority of cells in this lobe is far more extensive, and there is considerable interspecies variation. β-*Endorphin*, α-*MSH*, and *CLIP* are common to various species, whereas β-*LPH*, β-*MSH*, and *pro*-γ-*MSH* have been found in only some.[14]

Further processing of $ACTH_{1-39}$ into $ACTH_{1-38}$ and into smaller N-terminal fragments and several forms of *CLIP* has been described more recently.[15] Also, β-*LPH* and $β\text{-}END_{1-31}$ seem to be further processed into $β\text{-}END_{1-27}$ and $β\text{-}END_{1-26}$ and *glycylglutamine*.[16] There are indications that not only α-*MSH*, but also β-*END* is α-N-acetylated before release.[17]

The stimulatory influence of the hypothalamus on peptide release from corticotropes is mediated via the portal system and is thought to involve at least two factors, vasopressin (*VP*)[18] and a recently purified peptide with potent corticotropin-releasing factor (*CRF*) activity.[19] Corticotropes represent a stage in the hypothalamo–pituitary–adrenal system, which is instrumental in the physiological adaptation of the organism in response to a variety of stressful stimuli.[20] Peripherally, *ACTH* stimulates the adrenal gland to synthesize and release glucocorticoids; γ-*MSH* potentiates this effect.[21] Glucocorticoids exert a negative feedback at the level of the corticotropes and via regulation of hypothalamic *CRF* release.[11]

A separate release mechanism, triggered by neurogenic (as opposed to systemic) stress and circulating catecholamines, has been proposed to regulate the *ACTH*- and β-*END*-producing cells of the intermediate lobe.[22] This lobe is poorly vascularized, and blood that reaches the melanotropes is independent of the hypothalamo–hypophyseal portal system, which may explain why release is not under *CRF* control.[11] The intermediate lobe receives a rich innervation by aminergic fibers; a few peptidergic fibers are also observed, but contacts are seldom established.[13] Two factors originating from the paraventricular nucleus (PVN) of the hypothalamus are thought to stimulate (melanotropin-releasing factor, *MRF*) and to inhibit (melanotropin-release inhibiting factor, *MIF*) the release of α-*MSH*.[14] Peptides from the intermediate lobe are presumed to reach the systemic circulation by diffusion from the site of release via the extracellular channels between the parenchymal cells to the capillary networks in the anterior and posterior lobes.

Pituitary hormones may also take part in behavioral adaptational processes, reaching target neurons either via the systemic or the portal circulation (see Section 2). The peripheral responses may be a poor model for these actions, since, for example, γ-*MSH* and *ACTH* peptides have synergistic corticotropic activity[21] but opposite effects on avoidance behavior.[23] Although behavioral effects of *MSH*-like peptides have been documented,[10] no clear evidence for a central role for pituitary–as opposed to centrally—synthesized peptides has been found. The intermediate lobe is absent from adult man but may be important in the fetus for brain development.[24]

Growth hormone and *PRL* are structurally related hormones with molecular weights of the order of 20,000. There is considerable overlap of activity and species variation in both hormone structure and peripheral receptor specificity. The two hormones are synthesized in separate cells in the anterior pituitary, and their release is regulated by independent hypothalmic factors. Prolactin has been implicated in certain types of behavior (maternal, migration, feeding patterns), but the anatomic correlates of these effects are unknown. *Prolactin* is present in the CSF and can reach the CSF following intravenous (i.v.) injection[25]; *PRL* in the systemic circulation can influence neuronal function, since it exerts negative feedback via modulation of dopaminergic neurons originating in hypothalamic nuclei (see Section 5.1.1). *Growth hormone* has an analogous dual role in the CNS: poorly understood effects on learning,[26] sleep, and CNS metabolism[25] and feedback regulation of the hypothalamic neurons that regulate *GH* release from the pituitary. Little is known about effects of the other group of pituitary hormones on the brain except that they also exert feedback inhibition. Presumably, the peripheral circulation plays the major role in delivering the peptides to their target cells.

3.3. Peptides of Hypothalamic Origin

3.3.1. General

The classic role for peptidergic neurons in the hypothalamus is as neuroendocrine transducers that integrate neuronal signals from the brain and translate them into hormone messages in the periphery. The peptide may either act itself as a peripheral hormone (*VP* and oxytocin, *OT*) or act by regulating the release of pituitary hormones. It is becoming increasingly evident that many, if not all, hypothalamic peptides have actions additional to those that were originally discovered.

Peptides synthesized in hypothalamic neurons show a wide diversity of target cells and of pathways for reaching them, as shown schematically in Fig. 1. Most peptides seem to participate in several routes even from a single neuron via axon collaterals, and a multiplication to 2000 endings from one neuron have been observed. Connections between neurons within the hypothalamus indicate complex feedback and coregulation mechanisms. Two major pathways to the pituitary exist: a direct neural pathway passing through the inner zone of the ME to terminals predominantly in the posterior pituitary and transport via the portal system. The external zone of the ME consists of a mosaic of capillaries, axons, and terminals containing different peptides and amines, which provides the anatomic basis for hypophysiotropic and other neurohormonal actions and for interactions between terminals (see also Section 2). Hypothalamic nuclei also receive innervation from and project to extrahypothalamic brain regions. Thus, not only do hypothalamic peptidergic neurons integrate incoming signals, they also orchestrate a complex integrated response.

3.3.2. Oxytocin and Vasopressin

The neurohypophyseal hormones *VP* and *OT* are synthesized mainly in the magnocellular neurons in the hypothalamic PVN and supraoptic nucleus (SON). Recent immunochemical studies using purified antibodies have indicated that similar percentages of both peptides are present in both nucleic.[27] The perikarya are, however, distinct, those synthesizing *OT* being located mainly in the anterior part of the nuclei, whereas *VP*-ergic cells occur mainly in the posterior regions. *Vasopressin* is also present in the small (parvicellular) neurons of the suprachiasmatic nucleus (SCN), where it may be involved in biological rhythms (see Section 5.3).

For each peptide, a high-molecular-weight precursor containing the hormone and its carrier protein (neurophysin) is synthesized and packed in granules in the neuronal perikarya. The enzymes responsible for cleavage of the precursor are probably also contained in the same granules. The granules are transported by fast axoplasmic flow to the nerve terminals. Axons from the PVN and SON pass through the ME to the posterior pituitary. Terminals are observed in close association with the capillary networks both within the ME and in the posterior pituitary.

The terminals in the posterior pituitary are the release site for the entry of *VP* and *OT* into the systemic circulation (site a, Fig. 1), whereby they exert their classical hormonal effects (antidiuresis; parturition and lactation, respectively). The physiological significance of the terminals within the ME may be related to regulation of anterior pituitary hormone release (site b) via the portal system. Indeed, *VP* has been proposed as one of the factors that regulate *ACTH* release by stress.[18,28] *Vasopressin* and *OT* could also modulate the release of other hormones within the ME (site c) or have more widespread effects via entry into the blood or CSF (sites d and e; see also Section 2). In addition to the terminals in the ME and posterior pituitary, projections to the OVLT on the wall of the third ventricle and to the choroid plexus are potential sites for release of *VP* and *OT* into the blood and CSF (sites d and e). Fibers containing *VP* and *OT* also project to the pineal gland.

The *VP* and *OT* systems thus described are largely neurohumoral. Morphological evidence for several neurotransmitter and/or neuromodulator systems has also been described. Fibers arising from hypothalamic neurons make synapselike connections with neural structures both within and outside the hypothalamus (sites g and h).[29] Probably most, if not all, of the extrahypothalamic *VP* and *OT* is of hypothalamic origin. Several pathways have been more or less well described: *VP* and *OT* fibers from the magnocellular neurons in the PVN project to the ventral hippocampus and entorhinal cortex, to nuclei in the amygdala, to the medulla oblongata, and to the substantia gelatinosa of the spinal cord, and *VP* fibers from parvicellular neurons in the SCN project to the lateral septum and lateral habenular nucleus.[27] In diencephalic and mesencephalic subcortical brain areas, more *VP* than *OT* fibers are present, but the reverse situation exists in the medulla oblongata and spinal cord. There seems also to exist direct projection from *VP* and *OT* neurons in the PVN to preganglionic cell groups of both the parasympathetic and sympathetic divisions of the autonomic nervous system and to the peripheral sensory nucleus of the vagus and glossopharyngeal nerves, the nucleus solitarii.[30] Along this pathway, *VP* might interact directly with the autonomic nervous system and influence such parameters as blood pressure and the level of arousal.[31]

Central terminals containing *VP* and *OT* may function in memory consolidation and retrieval processes.[32-34] Various lines of evidence have connected the existence of *VP* terminals and their depletion in the dorsolateral septum with the action of this peptide on consolidation[33] and in the central amygdala with the action on retrieval.[34] The dorsal raphe nucleus (see also Section 5.2) and dentate gyrus of the hippocampus are also involved in the facilitation by *VP* of memory consolidation (see also Section 5.3). These actions most probably are exerted by metabolites of *VP*.[35] The hexapeptide [pGlu^4,Cyt^6]AVP_{4-9}, a major metabolite in the brain, and its desglycinamide derivative are considerably more potent than *VP* with respect to their central action, although they lack vasopressor activity.[35] Interestingly, *VP* binding sites could be localized on synaptosomal plasma membranes prepared from septum and hippocampus but not other brain regions.[36]

Microinjection techniques have implicated the central amygdala and dentate gyrus of the hippocampus as the site action of *VP* on retrieval. As has already been discussed for the pituitary peptides, *VP* and *OT* may also exert behavioral effects by reentering the brain following release into pituitary or hypothalamic blood or CSF. Antibodies against *VP* and *OT* have severe effects on memory processes when injected intracerebroventricularly (i.c.v.). However, this effect could occur through an interaction with CNS receptors rather than by sequestration of *VP* and *OT* from the CSF (where they are normally present).[32]

3.3.3. Hypophysiotropic Hormones

The sources of hypothalamic releasing hormones [gonadotropin-releasing hormone (*GnRH*), *SRIH*, thyrotropin-releasing hormone (*TRH*), *MIF, MRF*, and *CRF*] present in the ME of the hypothalamus are predominantly, but not exclusively, hypothalamic. Release in the region of the primary plexus and transport via the portal system form the major mode of access to target pituicytes. Terminals present in the posterior pituitary may also deliver peptides to the same cells or serve different, unknown functions. Projections to the OVLT (described for *SRIH* and *GnRH*) provide a further site for entry into blood and into the CSF. Since a separate chapter is devoted exclusively to these hormones, only a brief consideration is given here.

Somatostatin has been shown to be localized in neurons in several hypothalamic nuclei (ARC, SCN, periventricular, ventral premammillary, and ventromedial). *De novo* synthesis of *SRIH* and larger-molecular-weight precursors has been demonstrated in hypothalamic tissue *in vitro*. Four molecular weight forms that are biologically (inhibition of *GH* release) and immunologically active have been found, and several are present in hypothalamic portal blood. The only *SRIH* pathway that has been anatomically defined is from neurons in the anterior periventricular area and SON via the ME to the posterior pituitary.[37]

Considerable technical problems have been encountered in tracing *GnRH*-containing pathways. The *GnRH* perikarya are not restricted to a single cell mass but rather are found in bands or patches across the hypothalamus. Relatively large numbers of perikarya were found in the preoptic area and the ARC. The ME and OVLT are the two major sites of terminals.[38] Fibers containing *GnRH* have also been traced to the interpeduncular nucleus and the midbrain central gray region. This latter pathway may be important in the behavioral effects of *GnRH*, since local injection of the peptide in the preoptic area, ARC, or midbrain central gray area potentiates mating behavior.[39] Releasable *GnRH* and *GnRH*-containing fibers are found in other areas both within and outside the hypothalamus, and *GnRH*-sensitive cells have been shown throughout the hypothalamus, preoptic area, septum, midbrain central gray, and in the cerebral and cerebellar cortices.[25,39]

The localization of neurons supplying other releasing factors (*TRH*, *MRF*, *MIF*, and *CRF*) is still under investigation. Fibers and terminals containing *TRH* have been described in a number of regions[25]; *MRF* and *MIF* are believed to originate in the PVN and regulate the secretion of α-*MSH*.[23] All of these peptides have effects on behavior, but the relative importance of hypothalamic as against central sources of peptide cannot be assessed. *Thyrotropin-releasing hormone* also exhibits ultrashort feedback of its own release, but whether this is a direct loop as indicated in Fig. 1 (site i) is unknown.

Recently, a growth hormone-releasing factor (*GRF*) has been identified in fibers and cell bodies of the hypothalamus. This *GRF* consists of 44 amino acid residues and structurally belongs to the group of pancreatic–intestinal peptides (*glucagon, secretin, VIP*).[40]

3.3.4. Proopiomelanocortin-Derived Peptides

Presence of a 31K precursor of proopiomelanocortin (*POMC*) similar to that found in the pituitary has been indicated in a single major cell group (sometimes referred to as the periarcuate nucleus) that overlaps the anatomic borders of the ARC, ventromedial, and premammillary nuclei.[41] Processing of the 31K precursor in this system seems similar to that of the melanotropes (Section 3.2), i.e., to pro-γ-*MSH*, β-*LPH*, β-*END*, and α-*MSH*. Further processing, for example, during axonal transport of β-END_{1-31} to give β-END_{1-27}, β-END_{1-26}, glycylglutamine,[16] α-*END*, γ-*END*, destyrosyl DT α-*END*, and destyrosyl DT γ-*END* has been suggested.[42]

The potential of synaptic membranes to cleave $ACTH_{1-39}$ into $ACTH_{1-38}$, $ACTH_{1-16}$, and several forms of *CLIP*[43], as has been found in the pituitary, opens the possibility of processing of the *ACTH* part of the *POMC* in the vicinity of putative target sites. The information on the brain fiber systems containing these peptides has been mainly derived from studies aimed at the localization of individual peptides. The extrahypothalamic projections, however, are so similar that it is likely that there is a common system. Within the hypothalamus, α-*MSH*/β-*END*-containing fibers from the ARC have been shown in the ME. Branches from these fibers form terminals close to the capillaries in the primary plexus (site b, Fig. 1). An extensive intranuclear network innervating non-opiomelanocortin-containing ARC cells is apparently formed by axon collaterals (site g).[11] A dense plexus of immunopositive axons can be found along the whole length of the hypothalamus from the rostral part of the PVN to the mammillary body, with a large number of fibers in the SON, PVN, SCN, dorsomedial and periventricular nuclei, and in the lateral hypothalamus. α-*Melanotropin* is also found in the ependyma of the third ventricle, where it is probably located in supraependymal nerve terminals (sites d and e).

Fibers from the hypothalamus pass via a variety of tracts to anatomically divergent areas outside the hypothalamus (site h). Only two pathways have been fully traced[11]: (1) caudally, via the medial preoptic and lateral septal nuclei, to the bed nucleus of the stria terminalis and (2) dorsally, to the periventricular thalamic nucleus. Fibers traced to the ansa lenticularis, ventral amygdofugal pathway, and median forebrain bundle may be the source of some of the α-*MSH* found in various structures associated with the limbic and extrapyramidal systems.

Apart from the periarcuate group of cells, perikarya containing α-*MSH* have been described in the dorsolateral hypothalamus. Fibers from these cells project to the caudate nucleus, cortex, and hippocampus.[41] α-*Melanotropin* terminals of unknown origin are also found in dorsal root ganglia and in the pineal gland.[24]

Although behavioral effects of *ACTH*/β-*END*/*MSH* group of peptides are well known,[10,26] it is not yet possible to ascribe particular functions to the pathways that have been localized, although the predominant presence of β-, α-, and γ-*ENDs* in hypothalamic and septal areas corroborates the view that these peptides have a regulatory function in brain processes.[42]

3.3.5. Enkephalins

Evidence is accumulating that brain enkephalins (*ENKs*) are not derived from the 31K *POMC* precursor but from a distinct pro-*ENK* molecule.[44] This molecule contains the sequences for *Met-ENK* and *Leu-ENK* in similar relative proportions (4:1)[45] to those found in brain (3:1). Two additional peptides, a heptapeptide and an octapeptide, that have potent opiatelike activity and are also present in appreciable quantities in the brain are found within the pro-*ENK* sequence. Immunotechniques used to delineate *ENK*ergic pathways do not always distinguish between *Leu-* and *Met-ENK*,[46] but there is some evidence that they exist, at least in part, in separate terminals. This suggests selective processing as has been described for *POMc*.

Unlike β-*END*, *ENKs* are not restricted to any single cell group but have been shown in widespread individual perikarya in almost all hypothalamic nuclei. Fibers also project to a large number of nuclei, suggesting that *ENK*ergic cells function as interneurons (site g, Fig. 1). *Enkephalin* is also present in fibers terminating in the posterior pituitary, where they seem to play a role in regulating the secretion of *VP*.[47]

Another group of opioidlike peptides comprising *dynorphins* (*DYN*) (a 17-amino-acid peptide and its N-terminal octapeptide), α-*neo-END*, and β-*neo-END* has recently been discovered.[41] *Dynorphin* and α-*neo-END* fiber systems traced by antibody techniques show a high degree of parallelism, suggesting that a common precursor may exist.[48] Presence of these peptides in the magnocellular neurons of the SON and PVN has been demonstrated.[48] *Dynorphin* and *AVP* occur in the same cells of the SON and fibers of the pars nervosa of the pituitary.[49]

3.3.6. Other Peptides

Cholecystokinin (pancreozymin, *CCK*) is a gut peptide synthesized as a 33-amino-acid precursor, which is then cleaved to give a number of biologically active N-terminal fragments; *CCK*-8 is the predominant form in the brain. *Gastrin* has also been shown to be present in the brain, but localization methods using immunologic techniques do not distinguish it from *CCK* (the five N-terminal amino acids are identical).[50] Fibers-containing *CCK–gastrin* derive from magnocellular neurons in the PVN and SON and terminate in a dense plexus in the region of the ARC, SON, SCN, periventricular, ventromedial, and dorsomedial nuclei.[46] This is identical to an *ENK* pathway. Immunoreactive cell bodies are also present in the dorsomedial and periventricular hypothalamic nuclei.[50] The hypothalamus appears to play a role in the satiety signal actions of these peptides, since the effect of peripheral peptides is abolished by lesions in the ventromedial hypothalamus. Direct application of *CCK* to this area results in decreased feeding behavior. However, the role of intrinsic pathways in the hypothalamus is questionable, since the doses required for activity are higher via direct application to the brain than by peripheral injection.

Vasointestinal polypeptide (*VIP*) is present in relatively high concentrations in the SCN, anterior hypothalamic area, and ME. Terminals present in the ME may be involved in regulating the release of pituitary hormones including *GH* and *PRL*.[51] This could be a direct effect on the pituicytes (type b, Fig. 1) or indirect (type c) by modulating the release of hypophysiotropic factors, since it has been shown that i.c.v. *VIP* decreases *SRIH* secretion. Other peptides present in the hypothalamus may serve similar functions. For example, angiotensin II (*ATII*), which is present in the dorsal medial hypothalamus, increases the firing rate of SON neurons and alters the release of *VP*.[52] Interestingly, *GRF*, recently isolated from hypothalamus, shows 43% homology with *VIP*.[40] *Vasointestinal peptide* itself, however, does not exhibit detectable *GH*-releasing activity.

Bradykinin has been found in nerve cell bodies exclusively in the hypothalamus (mostly in the medial, but also in the anterior, posterior, and lateral regions) and thalamus.[50] Defined pathways have not been described, but fibers are found throughout the hypothalamus, thalamus, mesencephalon, lateral septal area, caudate/pallidum, and cortex. Local injections of *bradykinin* have implicated a role for these fibers in the lateral septal area in blood pressure regulation, in the periaqueductal gray in pain perception, and in the hypothalamus in body temperature regulation.

Neurotensin-containing neurons are also found in the hypothalamus in relatively high concentrations in a variety of nuclei and in nerve fibers in the external zone of the ME[53] but, unlike *bradykinin*, are also found in many other brain regions.[54,55] Fibers and terminals are present, and calcium-dependent release and the ability of *NT* to influence cell firing have been shown. A C-terminal peptide of *NT*, [Lys^8, Asn^9]NT_{8-13}, has also been found in rat brain and intestine[53]; the C-terminal part of *NT* is required for both receptor binding and biological activity. Little is known of the biological function of *NT*; it appears to be the most potent anticonceptive substance known so far.[53] *Neurotensin* alters the release of pituitary hormones (*PRL*, *GH*, *FSH*, and probably *ACTH*). This action could be direct, of paracrine nature at the level of the anterior pituitary[53] (site b, Fig. 1), or via hypophysiotrophic hormone release (site c). A direct effect on hypothalamic cell bodies is unlikely, since although *NT* binding in the hypothalamus is high, no electrophysiological effects could be found.[56] The pathways involved may be complex, since i.v. and i.c.v. routes of administration give opposite effects on pituitary hormone release.[54]

3.4. Peptides Originating in Extrahypothalamic Brain and of Uncertain Origin

The presence of projections from the hypothalamus to various brain regions makes it difficult to be certain about the proportions of many peptides that originate outside the hypothalamus. In some cases, the relatively high amounts of peptide may be indicative of extrahypothalamic synthesis. For example, the bulk of brain *TRH* and *SRIH* are extrahypothalamic and seem to be independent of hypothalamic sites of synthesis.[25,37] *Somatostatin* is present in nerve cell bodies in the zona incerta, bed nucleus of the stria terminalis, cortical amygdaloid nucleus, hippocampus, and in the pyriform, entorhinal, and neocortices. Immunoreactive *SRIH*-like material has also been shown in synaptic-terminal-like structures around the pyramidal neurons of the hippocampus and neurons throughout the neocortex. A more diffuse localization is seen around neurons in the thalamus. In some of these areas (cerebral cortex, hippocampus, caudate–putamen), cells responding to *SRIH* by electrophysiological or biochemical criteria have been described. *Somatostatin* elicits a spectrum of behavioral effects when injected into the CNS, but attempts using stereotactic injection to localize the areas involved have not been very successful. In addition to the nerve–nerve connections suggested by its presence in perikarya and synaptic structures, humoral routes may also be involved in *SRIH* modulation of CNS function, as *SRIH* is present in CVOs and CSF.[37]

Most brain *GnRH* originates in the hypothalamus, but neurons containing this peptide have also been described in extrahypothalamic areas.[39] The connections of these neurons with blood vessels suggest that they function in regulation of brain blood supply. The pathways involved in the CNS actions of

GnRH (on memory and mating behavior) have not been defined but seem to be independent of the effects of the peptide on the pituitary.[39]

Eighty percent of immunoreactive *TRH* present in the brain is extrahypothalamic, but detailed localization of pathways has not yet been achieved. The discovery of both high concentrations of *TRH*-immunoreactive varicosities and high *TRH* binding in the nucleus accumbens indicates that this is a target site for *TRH* action.[57] It has been suggested that certain behavioral and biochemical effects of *TRH* are mediated by presynaptic modulation of DA release in this nucleus. Coincidence of high concentrations of *TRH* and *TRH*-binding sites in the septum suggests that this may also be a target area. Local injection of *TRH* in this region results in antagonism of barbiturate-induced narcosis.[58]

ENK-containing neurons, fibers, and terminals are found scattered throughout the nervous system,[46] but details of pathways are limited, perhaps reflecting a divergent interneuronlike role for *ENK*ergic neurons. Only a fraction of the total *ENK* present in neuronal structures has been assigned to pathways: from perikarya in the amygdala to terminals in the bed nucleus of the stria terminalis, intrinsic neurons in the substantia nigra (SN), and from cell bodies in the striatum to the terminals in the globus pallidus.[44,46] A transmitter role for *ENK* present in structures involved in extrapyramidal function has been proposed. It has been proposed that *ENK*-containing terminals in relay stations of the sensory system such as the dorsal horn, periaqueductal gray, raphe, and trigeminal nucleus may influence pain perception.[51]

Substance P is present within numerous intrinsic pathways throughout the CNS. Its synaptosomal localization and the presence of specific inactivating enzymes suggest a role of *SP* as a neuromodulator.[25] *Substance P* may be an important modulator of sensory input, for high densities of *SP* receptors have been traced in the olfactory bulb and tubercule (olfaction), in the superior and inferior colliculi (vision, audition, and pain), and in the periaqueductal gray (pain).[59]

The gut peptides *CCK/gastrin* and *VIP* are both found in the highest concentrations in the cerebral cortex, where they are present in both perikarya and terminals.[60,61] Both *CCK* and *VIP* have excitatory actions on neurons in the cortex and hippocampus (where both types of peptidergic terminals have also been observed); *CCK* also exerts an activating effect in midbrain, where it appears to coexist with DA in a subpopulation of dopaminergic neurons (A9 area).[62] Depolarization-stimulated release of both *CCK* and *VIP* from neural structures has been described.[60] The distribution of *CCK* is rather similar to that of *ENK*, but the functional significance of this association is unknown. *Vasointestinal polypeptide* may also act as a local hormone, since fibers have been described in the walls of cerebral blood vessels, and the peptide has been shown to be capable of altering cerebral blood flow.

Neurotensin has been shown to be present in perikarya, fibers, and terminals in many brain regions, and in the amygdala, the localization parallels that of *ENK*.[60]

Some of the peptides present in brain have posed serious problems for localizing the systems involved; *insulin* and *ATII* are examples. *Insulin* has been shown to be present in neuronal perikarya, fibers, and terminals, but *de novo* synthesis has not been established. The concentration of *insulin* in the brain is ten times that in plasma, and variations in the latter are not reflected in the former.[63] *In vitro* methods have shown high *insulin* binding in brain tissue distributed over various brain regions and not restricted to neurons. *In vivo* methods label predominantly areas in which fenestrated epithelia are present. It has been proposed that hypothalamic glucoreceptor neurons are a target site for *insulin*,[64] but otherwise the significance of brain *insulin* content and binding sites is completely unknown.

De novo synthesis of *ATII* in neurons has also not been unequivocally established, but all of the enzymes necessary for the processing of angiotensinogen to *ATII* and its subsequent destruction are present. Although *ATII* has been shown to be present in neurons in the dorsal medial hypothalamus, locus coeruleus, and caudate nucleus, humoral sources appear to play a part in the best-described central effects of *ATII* (induction of drinking behavior). The removal of the SFO (a CVO) abolishes the behavioral response to i.c.v. *ATII*.[65] Local processing of angiotensinogen also seems to be important, since injection of renin induces drinking behavior, and the dipsogenic actions of exogenous angiotensinogen can be blocked by inhibitors of enzymic cleavage.[66]

The role of CSF in peptidergic systems is not fully understood. In some cases, i.c.v. administration of the peptide has effects markedly different from those resulting from peripheral dosing (for instance, excessive grooming behavior after i.c.v. but not after systemic application of *ACTH*- and *END*-like peptides).[64] Antibodies against peptides can block several of their actions when applied i.c.v.,[32] and different circadian rhythms in plasma, CSF, and brain have been shown, for instance, for α-*MSH*[68] and *VP*[69] (see also Section 5.3).

3.5. Peptides in Spinal Cord and Peripheral Nerves

A large number of peptides have been found in the peripheral nervous system, if which the best described is *SP*.[25] Synthesis, transport, release, and target cells for this peptide have been localized in a functionally significant pathway. *Substance P* released from the central branch of a sensory neuron acts as a neurotransmitter and neuromodulator; *SP* is also transported to and released from the peripheral branch and here exerts paracrine effects (e.g., vasodilation). It is also present in interneurons in the spinal cord and in fibers and terminals descending from higher centers (as are also *TRH* and *ENK*).

Somatostatin has also been shown to be localized in primary afferent nerves and synthesized in cells in spinal ganglia, although the cells are distinct from those containing *SP*. A similar role for both peptides in pain perception has been proposed.[25,37]

Vasointestinal polypeptide and *gastrin/CCK* and an *ATII*-like peptide have also been shown in axons and terminals from the peripheral sensory nerves. *Enkephalin*ergic and *NT*ergic neurons seem to be localized predominantly in interneurons in the spinal cord (as was the case in the CNS). *Enkephalins* inhibit *SP* release (see Section 6.3) and may play a role in the processing of sensory information. It has been suggested that *CCK*, *NT*, *SP*, *SRIH*, and *ENK* all play a role in pain perception.

The autonomic ganglia also contain networks of peptide-immunoreactive fibers originating from various sources.[70] In the inferior mesenteric ganglion, dense plexi of *ENK* fibers derive from descending preganglionic fibers whose cell bodies are in the spinal cord. *Substance P* fibers probably arise from collaterals of ascending sensory nerves, the cell bodies of which are located in the spinal ganglion. The peripheral and central terminals of these neurons are in the gut and spinal cord, respectively. Probably, *VIP* and *gastrin/CCK* fibers arise from neurons in the enteric plexi in the gut. *Somatostatin* is the only peptide that has been clearly demonstrated in neurons of the autonomic ganglia. Projections from these cells represent the postganglionic peptidergic innervation. Gut peptides (*SP, NT, CCK, VIP, ENK, TRH, SRIH, ACTH, insulin, secretin,* and *pancreatic polypeptide*) may be synthesized either in intrinsic neurons of the enteric plexi or in the mucosal, endocrinelike cells. The peptides have local action on gut target cells and may also contribute to the circulating hormone concentration and have target cells in the CNS reached by neuronal connections.[70]

4. PEPTIDES AS NEUROREGULATORS

Of the three modes of neuroregulation, neurotransmission, neuromodulation, neurohormonal, only the first is electrophysiologically clearly defined. Neurotransmitters are released by calcium-sensitive, depolarization-induced mechanisms within specialized synaptic structures and interact with specific receptors on the postsynaptic membranes to cause a transient change in specific, voltage-independent conductance channels. Other characteristics of neurotransmission are desensitization of the postsynaptic cell, the presence of high-affinity uptake or degradation systems, and antagonism by specific agents. Neuromodulation is said to occur between juxtaposed cells when the initial response is other than a change in voltage-independent conductance channels. Neurohormonal influences occur between cells in the absence of juxtaposition. In reality, the firing rate of an individual neuron represents the integration of a number of incoming signals. Neurotransmitters that have subthreshold excitatory effects or inhibitory, hyperpolarizing actions (e.g., probably *SRIH* in several areas[37]) alter the background excitability of the target neuron and thus "modulate" the effect of other incoming signals. Conversely, modulators can precipitate nerve cell firing by altering the response to a previously subthreshold stimulus.

Our concepts of the electrophysiological actions of neuropeptides mainly derive from a combination of precise studies on simple systems and suggestive evidence from investigations on CNS circuits. The range of possible peptide effects has been clearly demonstrated in cultured spinal neurons. Using intracellular recording, Barker *et al.*[2] have shown that *Leu-ENK* can act as an inhibitory or excitatory neurotransmitter. Independent of these actions, modulation of the effects of the excitatory agent glutamate and, by a separate mechanism, of the effects of the inhibitory agents GABA and glycine were observed. A fifth action, alteration of the spike threshold, was also seen at higher *Leu-*

ENK concentrations. Individual cells exhibited more than one type of response, and in some, all five types were observed. In other systems, presynaptic modulatory effects of *ENK* have been reported,[71] notably decreased *SP* release in the trigeminal nucleus and cultured ganglion cells, decreased NE release in rat cortical slices, and increased release of excitatory transmitters acting on cortical pyramidal cells.

The studies on *SRIH* indicate a multipotency analogous to that described for *ENK*. In cultured cortical cells, *SRIH* increases the release of excitatory and inhibitory neurotransmitters, and the response to excitatory amino acids was additionally enhanced by postsynaptic modulation.[72] *Somatostatin* may also apparently act as an excitatory and inhibitory neurotransmitter.[37] The characteristics of the depolarization and increased firing frequency observed in cortical and hippocampal pyramidal cells and striatal neurons are suggestive of excitatory neurotransmitter action.[73] The inhibition of certain hypothalamic, cerebrocortical, and cerebellar neurons *in vivo* and inhibitory actions in hippocampal slices are consistent with inhibitory neurotransmitter action.[37,74] Similar actions may be involved in the depressant actions of *TRH* and *GnRH* observed in several areas, although the evidence is currently less convincing.[25] *Thyrotropin-releasing hormone* has also been shown to act as a subthreshold excitatory transmitter in the spinal cord[75] and inhibits cortical neuronal activity by a specific modulation of the response to certain (glutamate and aspartate) but not other (ACh) excitatory neurotransmitters.[74] *Substance P* also acts in a variety of independent ways involving pre- and postsynaptic actions.

In several instances, a role for voltage-sensitive Ca^{2+}-conductance channels has been proposed in peptide action. Modification of these channels would clearly have effects on transmitter release and has been suggested in the presynaptic effects of *SRIH*[76] and *ENK*.[77] A similar action could also explain the effects of *ENK* on spike threshold in cultured spinal neurons.[2]

The rapid time course and magnitude of the excitatory responses to *CCK* and *VIP* in cortical neurons and of *ATII* in the SON led authors to suggest a neurotransmitter role, although real evidence is lacking. In the hippocampus, a postsynaptic site of action of *CCK* indicating transmitter function has been shown.[78] The mechanism of the inhibitory actions of *GnRH* in various areas and of *NT* in the locus coeruleus and nucleus accumbens has yet to be determined.

A rather unusual type of peptide response has been described in the giant *DA* neuron of the snail.[79] *Vasopressin* decreased and $ACTH_{4-10}$ increased the membrane conductance, modifying spontaneous activity and the response to electrical signals but not the resting potential. Such an effect could be mechanistically transmittory by altering the membrane conductance to ion(s) whose equilibrium potential is close to the resting potential or neuromodulatory. The functional consequence is a depression ($ACTH_{4-10}$) or facilitation (*VP*) of the cell response to other inputs.

When *VP* or *OT* was applied iontophoretically to neurons in the lateral septum or dorsal hippocampus of the rat, an increase in spontaneous activity of the cells could be measured, resembling that evoked by glutamate. In addition, the response to glutamate was increased by application of the peptides, but apparently via a separate action.[80] Extracellular recordings obtained from nonpyramidal cells (interneurons) in hippocampal slices showed an excitatory effect on spontaneous activity by *OT* and *VP in vitro*.[81] The potencies of *OT*, *VP*, and other structural analogues are suggestive of the existence of receptors in the hippocampus similar to those of uterine smooth muscle cells: no correlation with vasopressor or antidiuretic activity was found.[81] Oxytocin and *VP* could modulate transynaptically the activity of a great number of pyramidal neurons in the hippocampus by a mechanism similar to that proposed for the effects of opiatelike peptides.

Neurohormonal effects include the actions of hypophysiotropic peptides on pituicytes and of other neuropeptides that reach their target cells via the blood or CSF. The potent excitatory effects of *ATII* in the SFO might be an example (see Section 3.3.6). The possibility that neurons may regulate other neurons by the diffusion of released peptides across the extracellular space is largely unexplored and will be very difficult to investigate. They have been described in the mollusc[2] but not as yet in mammalian systems. Nicoll et al.[75] have tentatively suggested that diffusion of *ENK* from the CA2 region of the hippocampus may mediate the effects of this peptide on opiate receptors on GABA interneurons in the CA1 region, where no *ENK* terminals have yet been found.[75] From the evidence that has accumulated to date, it seems likely that most neuropeptides will be discovered to have a repertoire of effects on target cells.

5. INTERACTIONS BETWEEN PEPTIDERGIC AND AMINERGIC SYSTEMS

Neuroanatomic studies using histo- and immunochemical techniques have shown a close association between peptidergic and aminergic systems. Since both monoamines and neuropeptides are involved in stress responses, learning, and the processing of acquired information, the question arises as to whether these two classes of neuroactive substances act in concert. At the moment, the majority of available data on the interaction between amines and peptides in the central nervous system is restricted to a simple assessment of an effect of the one on the level, turnover, or secretion of the other in a certain brain area.[82] We focus our attention on those cases in which the defined topographical distribution of amine and/or neuropeptide enables an approach to the problem to be made by localized microinjections and *in vitro* techniques using discrete groups of cells.

5.1. Dopaminergic Systems

5.1.1. The Tuberoinfundibular System

The DAergic pathways in the hypothalamus play an important role in regulation of pituitary hormone release. The predominant DAergic pathway in the hypothalamus originates in the ARC and PVN perikarya and projects to the

ME. Fibers also pass through the ME to the posterior and intermediate lobes of the pituitary. These pathways are frequently referred to as the tuberoinfundibular and tuberohypophyseal systems, respectively.[83] True DA synapses have not been described; the terminals are located in close proximity to pituicyte processes, neurosecretory terminals, axons, and capillaries. A neuromodulatory role of a paracrine or hormonal nature is therefore likely, as is also suggested by the absence of DA uptake systems.[83] The best-understood role of this DAergic system is the inhibition of *PRL* release from the anterior pituitary via specific DA receptors on the lactotropes. There is also considerable evidence that DA terminals in the ME modulate the release of hypothalamic releasing factors (notably *TRH, GnRH*, and *SRIH*) and thus indirectly alter pituitary secretion. Dopamine terminals in this region are in close association with peptidergic fibers. Furthermore, the ability of DA to stimulate the release of *GnRH*[25] and *TRH*[25] from hypothalamic synaptosomes *in vitro* indicates a direct interaction.

Dopamine terminals in the intermediate lobe of the pituitary are found in close association with POMC-containing cells[11] and are likely to be responsible for the tonic inhibition of intermediate lobe α-*MSH* release by DA.

Iontophoresis of DA excites approximately 50% of neurosecretory cells in the PVN, and there are strong indications that DA fibers from periventricular areas within the hypothalamus and terminating in the PVN play an active part in the *OT* response to suckling.[52,84]

Regulation by peptides (e.g., *PRL*, *END*s, *ENK*s, and α-*MSH*) of the release of pituitary hormones (including feedback control) seems in many cases to be mediated by effects on the hypothalamic DA systems.[82,84] In some instances (*END*s and *ENK*s), an action at DA terminals is likely.

5.1.2. The Nigrostriatal System

Besides the hypothalamic–hypophyseal system described above, the only interaction between peptides and a well-defined DAergic pathway is in the nigrostriatal system. *Enkephalins* and *END*s decrease DAergic transmission in the pathway by decreasing the release of DA from terminals in the corpus striatum, an action that is mediated by presynaptic opiate receptors.[82] As a consequence of this effect, DA synthesis increases as a result of diminished feedback inhibition via DA autoreceptors.[82] Inhibition of DA release from neurons that inhibit striatal cholinergic interneurons may be the mechanism whereby *ENK* stimulates ACh release in the striatum.[82]

The nigrostriatal DAergic system may be involved in the effects of *VP* and *OT* on memory retrieval systems, since these peptides have opposite effects on both striatal DA levels and the behavioral parameter.[85] These effects of *VP* and *OT* do not seem to take place at the cell bodies in the SN, and an effect at the terminals similar to that described for opioid peptides has been suggested.[86]

The behaviorally active peptides related to *ACTH*[10] may have a different site of action on the same system. Studies on the effects of *ACTH* on DA concentration in the corpus striatum have yielded conflicting results.[82] Activation of DAergic nerve cell bodies in the SN by both α-*MSH* and an analogue of *ACTH*$_{4-9}$ has been described.[82] Interestingly, microinjections of *ACTH* into the SN elicited excessive grooming in the rat.[67]

Substance P also exerts a facilitatory influence on nigrostriatal DAergic neurons. It has recently been shown that *SP* is synthesized in the striatum and transported along the striatonigral pathway to terminals on DAergic neurons in the SN,[87] where it is released. Microiontophoresis increases the firing of some nigral neurons that contain a *SP*-sensitive adenylate cyclase.[88]

The presence of *NT* receptors on DAergic cells in the SN has been demonstrated. *Neurotensin* selectively excites DAergic neurons in the zona compacta of the SN. Moreover, *NT* can release DA from a variety of brain areas both *in vivo* and *in vitro*.[53]

5.2. Noradrenergic Systems

Adrenergic mechanisms have been implicated in feedback regulation of pituitary hormone release (e.g., *PRL* and *LH*[89]) and control of episodic neuroendocrine events, especially in circadian rhythms in the pineal.[90] Although the pathways involved are as yet unknown, direct effects of NE on release from hypothalamic synaptosomes (*TRH*),[84] complex multisynaptic pathways (anterior pituitary *ACTH*),[84] and peripherally released epinephrine (intermediate lobe *ACTH* in response to neurogenic stress)[91] have all been implicated. Both α- and β-adrenergic mechanisms can be involved and often exert opposing actions. Direct excitation of *OT* neurons via α-noradrenergic pathways is involved in the suckling reflex, whereas β-noradrenergic mechanisms suppress *OT* release from the terminals in the neurohypophysis.[52] *Vasopressin* release related to cardiovascular reflexes is probably mediated via α-noradrenergic input arising from the dorsal vagal complex and the locus coeruleus on *VP* neurons.[52]

The effective sites at which *VP* affects consolidation and retrieval processes coincide with brain regions in which fibers of the dorsal noradrenergic bundle terminate. The dorsal noradrenergic bundle originates in the locus coeruleus and projects to a large number of limbic and other areas including the hypothalamus. Studies on the effects of selective lesions of these projection areas and on Brattleboro rats lacking endogenous *VP* suggest a modulation by *VP* of NE transmission, either at the level of the presynaptic terminals[85] or at the postsynaptic receptors.[33] Facilitation of learning processes by *ACTH* analogues also involves changes in NE concentrations in this system, including the cell bodies in the locus coeruleus.[82] The extent to which the effects of peptides on NE are causative in the behavioral action cannot always be assessed. In the case of *VP*, some of the changes in NE turnover in response to peptide are dependent on the state (naive or trained) of the animals, indicating a consequential rather than causative role.[85]

Iontophoresis of *SP* into the locus coeruleus alters the firing (mostly excitatory) of neurons in this region, suggesting an excitatory role for this peptide on the limbic noradrenergic pathways. Moreover, changes in synthesis, utilization, and uptake systems of monoamines have been reported to be caused by *SP*.[92,93] However, a correlation of these changes with *SP* action on pain perception could not unequivocally be made.

5.3. Serotonergic Systems

Serotonergic (5-HT) systems have frequently been mentioned in relation to circadian, ultradian, and seasonal rhythms in peptide levels, turnover, and release. The SCN[94] and the pineal[90] play a crucial role in the generation and synchronization of biological rhythms, and both regions serve neuroendocrine transducer functions. The SCN is the main projection area for 5-HT fibers from the raphe nuclei, and cell bodies containing *GnRH*, *VP*, and *SRIH* are present. However, the ability of ACh rather than 5-HT to excite 80% of SCN neurons indicates that ACh is a more likely candidate as the transmitter involved in the generation of rhythms in neuropeptidergic activity.[94] In the case of *TRH* release in the ME, a direct inhibitory effect by 5-HT on the terminals has been demonstrated.[25] Peptidergic terminals in the raphe nucleus appeared to be involved in the behavioral effects of *OT* and *VP* on retrieval, as has been indicated by the effects of local injections of the peptides and of anti-*VP* antiserum.[33,85] Moreover, 5-HT pathways originating in the raphe nucleus are essential for the expression of the behavioral effects of *VP*.[33,85] However, there is evidence that the effects of *VP* on 5-HT neurons in the raphe nucleus are indirect, mediated via noradrenergic mechanisms.

Microinjections of *VP* into the dentate gyrus effectively improve memory consolidation and retrieval.[33] *Vasopressin* also selectively increases the synthesis and release of 5-HT in the dentate gyrus of hippocampal slices.[95] Since no 5-HT cell bodies are present in this preparation, a modulatory role affecting release from the terminals has been proposed. The *ACTH* analogues [L-Phe^7]- and [D-Phe^7]ACTH$_{4-10}$ also affect hippocampal 5-HT levels. These peptides exert opposite effects on both 5-HT levels and avoidance behavior,[10,82] suggesting a functional correlation, but direct interactions between the peptides and the 5-HT pathway have not been demonstrated.

6. INTERACTIONS BETWEEN PEPTIDERGIC AND NONMONOAMINERGIC SYSTEMS

6.1. Cholinergic Systems

Acetylcholine plays an important role in regulating the release of pituitary hormones. It is present in high concentrations in the ME and can reach the anterior pituitary in effective amounts via the portal vessels. Muscarinic receptors are present in the pituitary, and ACh can increase *GH* and depress *PRL* release from pituitary cells *in vitro*.[96] Acetylcholine also stimulates the release of *ACTH* from the intermediate lobe *in vitro*.[11] In this case, ACh probably reaches the pituicytes via nerve terminals in the intermediate lobe rather than via the portal vessels.

The septohippocampal cholinergic pathway appears to be a target site for several neuropeptides (β-*END*, *ENK*, *ACTH*$_{1-24}$, α-*MSH*, *SP*).[82,97] The action of β-*END* has been explained in terms of an interaction with cholinergic neurons at the level of the septum,[82] although it might also be mediated by GABA interneurons (see Section 6.2). Direct effects of $ACTH_{1-24}$ and α-*MSH* on hippocampal receptors[82] have also been proposed. The presence within the septum of peptidergic terminals (*TRH, SRIH, VP, SP, ATII,* and *VIP*), cell bodies (*GnRH, SRIH, SP,* and *VIP*), specific binding sites (*TRH, ATII*), and peptide-sensitive neurons (*GnRH, ATII*) raises the possibility of complex peptide/ACh interactions.

A high proportion of neurons that are excited by *SP* are also excited by ACh, although pharmacological studies indicate that different receptors are involved.[98] Both synergism and antagonism between *SP* and ACh have been reported. In cat Renshaw cells, *SP* action appears to involve a specific depression of the postsynaptic response to ACh.[99] Antagonism between *SRIH* and ACh has been reported in the periphery (the guinea pig ileum) to involve presynaptic inhibition of ACh release.[76]

6.2. Amino Acid Systems

Interactions between peptides and amino acid transmitters have been best described in the spinal cord. The presynaptic modulation of the release of and the postsynaptic modulation of the response to amino acid transmitters in cultured cells (from spinal cord and cerebral cortex) by *ENK*s and *SRIH* have been discussed in Section 4. *Substance P* shows comparable actions.[2,100] A role for GABA interneurons has been suggested for the effects of opiates on sensory afferents in the spinal cord[75] and for opiate stimulation of neurons in the CNS (hippocampal pyramidal cells and olfactory bulb mitral cells). Similarly, the stimulation by β-*END* of septohippocampal cholinergic pathways (Section 6.1) is probably exerted via an opiate-receptor-mediated inhibition of GABAergic interneuron activity in the septum.[97] *Thyrotropin-releasing hormone* has also been reported to alter GABA-mediated neurotransmission.[101]

There are very little data at present on the effect of amino acid transmitters on peptidergic systems. The observations that GABA inhibits *SP* in the SN[98] and other brain tissue[93] are consistent with the normal role of this amino acid as an inhibitory transmitter. GABA also inhibits *PRL* release via receptors in the anterior pituitary.[96]

6.3. Peptide–Peptide Interactions

The action of hypophysiotropic peptides on the pituitary is the best-described example of the action of a peptide via a peptidergic receptor on peptidergic cell bodies. The original concept that one hypothalamic peptide stimulates one type of pituicyte has had to be modified: *TRH*, for example, stimulates both *TSH* and *PRL* release.[25] The same pituitary cell may also have receptors for more than one peptide. For instance, both *CRF* and *VP* in pituitary blood stimulate *ACTH* (and β-*END*) release.[18] Some pituicytes are regulated by pairs of peptidergic factors having opposing actions, e.g., α-*MSH* in the intermediate lobe by *MIF* and *MRF*, and *GH* by *SRIH* and *GRF*, but direct interactions with separate pituicyte receptors are less certain.

Although currently less clearly understood, there is growing evidence of peptide–peptide interactions outside the pituitary. The ME is the most prominent site for such interactions. *Vasointestinal polypeptide*, which is present in terminals in the ME, inhibits the release of *SRIH* and *GnRH* from hypothalamic tissue *in vitro*. Thus, it is likely that the stimulation by *VIP* of the secretion of *GH* and several other pituitary hormones[51] is secondary to modulation of the release of hypophysiotrophic hormones into portal vessels. *Neurotensin* and *SP* have actions opposing *VIP*; that is, they stimulate *SRIH* release.[102]

At least two peptides, *ATII* and *ENK*, modulate *VP* release. *Angiotensin II* appears to act as an excitatory agent at the level of the nerve cell bodies. Iontophoretic application of this peptide to magnocellular neurons in the SON increases neuronal firing and *VP* release both *in vivo* and *in vitro*.[52] Intracerebral *ATII* alters the firing of both *OT* and *VP* neurons. In contrast, *ENK* has no effect on magnocellular firing rate but has been reported both to inhibit and stimulate *VP* release.[47] At the level of the pars nervosa, stereospecific opiate receptors have been described with properties very similar to those of brain. The location of the opiate receptors has been presumed to be preterminal on the neurosecretory fibers. However, evidence for the existence of opiate receptors on the pituicytes themselves has also been put forward.[47] A presynaptic modulation may also be the functional significance of the frequent proximity of *ENK* and *SP* terminals at several stations along the sensory pathways. It has been shown that opiate peptides decrease depolarization-induced release of *SP* from the spinal cord, trigeminal nucleus, and cultured dorsal root ganglia.[71,98]

7. COEXISTENCE OF PEPTIDES AND OTHER AGENTS

The coexistence of peptides and "classical" transmitters within a single neuron has been demonstrated or suggested in many regions. In most types of nerve endings, small and large granular vesicles can be distinguished. Peptides are usually present only in the large vesicle, whereas amines occur in both, so mixed peptide/amine large vesicles are possible.[70]

Coexistence of *SP* (several areas) and *TRH* (descending medullary raphe neurons with 5-HT and sometimes also with each other) has been described.[103,106] There is some indication that peptides and amines are stored in the same vesicles and can be released together[103,106] and have effects on postsynaptic cells of a synergistic[103] or antagonistic[104] nature. At first sight, this seems counterproductive but may permit fine regulation of target cell activity. In adrenal cells, *Met-ENK* and *Leu-ENK* exist together with NE in the same intracellular storage vesicles and are cosecreted.[105] Similar phenomena may occur in neural cells (e.g., the ganglion cells) in which *ENK* and NE are found in the same terminals. There is also evidence for the coexistence of *SRIH* and NE in peripheral autonomic ganglion cells and for *VIP* and ACh in peripheral nerves innervating exocrine glands.[70,106] In the latter case, *VIP* mainly causes vasodilatation and ACh mainly secretion, but each agent potentiates the action of the other.[106]

A *CCK*-like peptide occurs in a subpopulation of DA neurons in the SN and in the ventral tegmental area.[62,106] Corresponding terminals containing both agents are found, for example, in the nucleus accumbens and olfactory tubercle. Strong excitatory effects of *CCK* on neuronal firing rate have also been reported.[106] *Cholecystokinin* fragments increase DA release from nerve endings, most probably from the *CCK*/DA-containing neurons. The mechanism and function of this action of *CCK* are not yet known.[106] A subpopulation of neurons in the periaqueductal central gray contains both a *SP*- and a *CCK*-like peptide, and these neurons seem to project to the spinal cord.[106]

As well as the existence of peptide plus classical neurotransmitter combinations described above, there are several instances of peptide–peptide combinations. Both *CCK*[103] and *ENK*s[107] have been reported to be present in the *OT* and *VP* neurons projecting from the SON and PVN to the posterior pituitary. *Leu-Enkephalin* seems to be predominantly associated with *VP*, and *Met-ENK* with *OT*.[107,108] The *ENK*s appeared to inhibit the release of peptide hormones without altering neuronal firing. The *ENK*s were localized in synaptoid elements on glial cells (pituicytes) in the neural lobe,[108] whereas the stereospecific opiate receptors are present on the pituicytes rather than the neurosecretory fibers.[47]

8. BIOCHEMICAL EFFECTS

8.1. Receptors and Second Messengers

Peptide hormones, in common with catecholamine hormones and neurotransmitters, bind to receptors within the membrane. There may be exceptions to this general rule, as $ACTH_{1-24}$ can interact with lipid membranes in such a way that its N-terminal message enters the bilayer and adopts a helical structure.[109] Smaller sequences, $ACTH_{4-10}$ and $ACTH_{4-9}$, can enter cells.[110] However, cytosolic receptors for these peptides have not been described. The initial membrane response may lead to changes in a cytosolic constituent (the second messenger) that mediates the further consequences of ligand binding to the receptor.

Cyclic nucleotides and calcium have been indicated to be second messengers in many systems, and their actions frequently involve modulation of the activity of protein kinases that catalyze the phosphorylation of protein substrates. The last step in the chain, the alteration of cell function as a result of changes in the degree of phosphorylation of specific proteins, is, particularly in the case of nerve cells, the least understood. Although cyclic AMP and calcium can alter the degree of phosphorylation of some synaptic proteins, direct evidence that this is a mechanism of regulation of transmitter release or of membrane permeability is currently lacking.[111] Neurotransmitter metabolism may also be modified by altering phosphorylation. For example, the activity of tyrosine hydroxylase seems to be regulated by a cyclic-AMP-dependent phosphorylation.

Changes in cyclic AMP concentration do not necessarily indicate that this substance is the second messenger for a particular peptide effect. They may be parallel or consequent to the initial biochemical effect. For example, the depression of *GH* release by *SRIH* is accompanied by a decrease in cyclic AMP[37]; adenylate cyclase activity is stimulated by *TRH* as a result of an initial effect on Ca^{2+} influx[101]; *ACTH*-induced increased adenylate cyclase activity is probably the result of changes in the state of phosphorylation of membrane proteins.[112] Cyclic AMP concentrations may also be modulated by regulation of its degradation by the cyclic phosphodiesterase enzyme. Several peptides ($ACTH_{1-24}$, *SRIH*, *SP*, and *VP*, in descending order of potency) appear to inhibit this enzyme activity in rat brain *in vitro*. Since this is a soluble enzyme, inhibition by peptides can only occur subsequent to some membrane-mediated effect or internalization of the peptide.

Besides the second-messenger route for interaction with cytosolic components, ligand–receptor complexes may modify intracellular function directly following internalization of the complex. Internalization has been reported for several peptides,[113] but the physiological significance of this phenomenon is uncertain, and it may be related to ligand degradation and regulation of receptor availability.

The information currently available on neuropeptides is insufficient to allow identification of functional receptors. However, several pieces of the jigsaw—specific binding sites, effects on classical second messengers, and pharmacological data—are beginning to accumulate. The opioid peptides are exceptional in that putative receptors were identified long before the natural ligands were discovered. In contrast, research on receptors for other neuropeptides is hampered by the lack of specific antagonists, although there have been some recent advances in analogues of *VP* and *OT*.[81,114]

High-affinity binding sites have been described for all (except some *POMC*-related) neuropeptides. The density of binding is nonuniform across different brain regions, and, in general, the distribution correlates with that of high peptide concentration and presence in terminals. However, binding could represent sites other than the receptor, for example, vesicular storage mechanisms. This problem is exemplified by *SP*, the high-affinity binding of which is almost exclusively associated with vesicles.[115] Structure–activity studies on the effects of analogues of *VP*,[13,81] *SRIH*,[116] *TRH*,[25] and *CCK*[60] indicate that central receptors may differ from those in the periphery; moreover, more than one class of brain receptor may exist.[116] Central receptors for some peptides (*ATII*, insulin, *VP*,[81] and *VIP*) show characteristics very similar to those in the periphery.

Three types of opiate receptors (μ, δ, and κ), mainly characterized in peripheral tissues, may also be distinguished in the brain by their differing relative affinities for agonists and antagonists.[111] The μ-type receptors, which are the predominant form in the striatum, midbrain, and hypothalamus, appear to be located in discrete regions postsynaptically to peptide-containing terminals. Their localization is strikingly associated with primary afferent input zones of sensory neurons.[118] The δ receptors, in contrast, appear to be more diffuse, possibly indicating a presynaptic or extrasynaptic[118] localization, and are found predominantly in the frontal cortex.[118]

Interactions between neuropeptides may be important, since both agonist and antagonist activity at opiate receptors[119] and cooperative effects between β-*END* and α-*MSH* on melanosome dispersion have been described.[120] *Somatostatin* also interacts with opiate receptors and appears to be a partial agonist–antagonist *in vitro*.[119] Several authors have reported an interaction between *ACTH*-like peptides and opiate receptors. $ACTH_{1-39}$, $ACTH_{1-24}$, and γ-*MSH* (in concentrations 10–100 times higher than *ENK*) can displace β-*END*, Met-*ENK*, and naloxone.[121-124] Smaller sequences ($ACTH_{1-10}$, $ACTH_{4-10}$, and α-*MSH*) were inactive.[121] This indicates that at least some of the behavioral and neurochemical effects are mediated by nonopiate receptors.

Specific binding of *ACTH* peptides to CNS membranes has not yet been demonstrated. This may be because there is a very low concentration of high-affinity sites,[7] as has been indicated in adrenocortical cells, where similar problems were encountered.[125] The $ACTH_{1-24}$ molecule seems to contain at least three distinct portions with behavioral activity.[126] The recently reported changes in ternary structure of $ACTH_{1-24}$ following interaction with lipid bilayers[109] may also contribute to the diversity of its actions. This might imply the existence of different central receptors. In the adrenal cells, there are indications that separate sequences in the molecule are responsible for binding and for coupling to adenylate cyclase.[127] Thus, the different behavioral effects may also represent different second messenger mechanisms evoked by binding of the peptides to a common receptor.

Second messengers in the peripheral actions of several neuropeptides have been clearly indicated (e.g., Ca^{2+} in the release of *TSH* by *TRH*; Ca^{2+} and cyclic *GMP* in *bombesin* actions; Ca^{2+} in the steroidogenic[127] and lipolytic[128] actions of *ACTH* and in melanophore dispersion by *MSH*[129]; cyclic AMP in the antidiuretic effect of *VP*[130]). However, the extent to which these findings can be extrapolated to the brain is uncertain, particularly in view of the fact that the receptor characteristics are apparently different (see above).

Direct evidence that cyclic AMP is involved in neuropeptide actions in the brain is limited. *Substance P* increases adenylate cyclase activity in the hypothalamus, pineal gland, SN, and cultured neuroblastoma cells.[131] However, evidence linking the *SP*-sensitive adenylate cyclase present in DAergic neurons in the SN with the excitatory peptide effect is presently lacking. For instance, potentiation of *SP* effects by phosphodiesterase inhibitors and excitation by cyclic AMP analogues have not been shown.[88] *Somatostatin* decreases cyclic AMP concentrations in the pituitary and increases cyclic AMP levels in the hippocampus and caudate–putamen.[32] Changes in calcium uptake and release in hippocampal synaptosomes have also been described.[76] However, in the pituitary, neither calcium nor cyclic AMP seems to mediate *SRIH*-induced depression of *GH* release.[37] *Thyrotropin-releasing hormone* can also modify cyclic AMP accumulation in cortical brain tissue *in vitro*, but this activity only becomes evident when normal cyclic AMP concentrations have been depressed.[132] The high-affinity binding of *VIP* to synaptosomal preparations from different brain regions correlates with the stimulation of adenylate cyclase.[61] *Insulin*, at a 100-fold lower concentration than DA, elicited a comparable rise in cyclic AMP levels in slices from olfactory bulb.[133]

The behaviorally active analogue of *VP*, desglycyl-8-arginine *VP*, which has little peripheral action, potentiates placebo-induced increases in brain cyclic AMP concentrations.[134] The most notable responses were observed in the septum (1 h after dosing) and in the hippocampus (24 h after dosing), areas that have been suggested to be involved in the long-term effects of *VP* on consolidation (see Section 3.3.2). In the same brain areas, septum and hippocampus, electrophysiological effects[80,81] and specific binding[36] of *VP* to synaptosomal plasma membranes have been shown. So far, however, the indications in favor of a second-messenger role for cyclic AMP in *VP* actions are only indirect.

A number of neuropeptides (*SP*, *SRIH*, *VIP*, and *ENK*) modulate the β-adrenergic response of purified astrocytes *in vitro*.[135] The increase in intracellular cyclic AMP induced by NE is markedly enhanced by *SP* and *SRIH* and is inhibited by *ENK*; *VIP* also increases cyclic AMP in the absence of NE. The increases in cyclic AMP are mediated by a β-adrenergic receptor.[135] Both NE and *VIP* have been shown to stimulate glycogenolysis. There are indications that the cyclic AMP and glycogenolytic responses of brain tissue are primarily mediated by glial cells.[136]

Clonal cell lines of neural origin have contributed much to our understanding of opiate receptor functioning. These cells possess a single class of opiate receptors that is coupled to an adenylate cyclase system. Receptor occupancy by agonists, including *ENK*s and *END*s, causes a rapid inhibition of adenylate cyclase activity and a concomitant decrease in cyclic AMP concentration.[111] Interestingly, the receptor–agonist interaction also leads to delayed changes in cyclic AMP metabolism and sensitivity to opiates that may be a model for the development of tolerance and dependence.

In brain, the effects of endogenous opiates on cyclic nucleotides is more complex. β-*Endorphin* inhibits adenylate cyclase activity in membrane preparations from rat cortex and brainstem, whereas *Met-ENK* stimulates the enzyme in the brainstem and inhibits it in the cortex.[111] In striatal slices, cyclic AMP levels were slightly lowered by *ENK*s, but a massive increase in cyclic GMP formation was observed.[111] The phosphorylation of proteins in the neostratum is modulated by β-*END* by at least two mechanisms. Stimulation of the phosphorylation of selective synaptic plasma membrane proteins is blocked by naloxone.[137] Inhibition of the phosphorylation of two proteins was also observed, and this effect was naloxone insensitive. The inhibitory effect could be mimicked by *Met-ENK* at very much higher concentrations.

The effects of *ACTH* peptides may also involve opiate and nonopiate receptor mechanisms (see above). Both cyclic AMP and Ca^{2+} may be influenced by *ACTH* peptides[111] and may be important mediators of some of the effects of these peptides, but a role for these agents as initial second messengers is highly questionable. The biphasic modulation by $ACTH_{1-24}$ of both brain adenylate cyclase activity and phosphorylation of synaptic plasma membrane proteins[111] and other evidence[112] suggest a link between these two effects.

Structure–activity studies have shown a striking correlation between activity in eliciting excessive grooming behavior and the degree of phosphorylation of specific membrane proteins.[111] Changes in the turnover of brain synaptosomal polyphosphoinositides also occur in response to *ACTH* peptides. These phospholipids have been purported to play a central role in cellular Ca^{2+} fluxes.[67] $ACTH_{5-18}$ also alters lipid metabolism in rat brain synaptosomes *in vitro* by stimulation of an ester hydrolase.[138]

Besides classical second-messenger mechanisms, other modes of action of peptides may be envisaged. *Insulin*, for example, has been reported to activate cerebral Na^+/K^+-ATPase activity.[139]

8.2. Trophic Functions of Peptides

We have so far considered neuropeptide actions in relation to the activity of neuronal circuits. There are indications that peptides also play a role in the development, maintenance, or alteration of these circuits.[140] Both circulating hormones and locally released factors may have trophic functions. The best-described instances of effects of circulating hormones are the nonpeptide hormones thyroxin and steroids. So far, our knowledge of specific peptide trophic factors comes mainly from the studies on *NGF*.[141] This peptide is essential for the normal development of sympathetic and peripheral sensory nerves and for the maintenance of these cells in culture. In the mature animal, *NGF* is not essential for the integrity of the nerve but does alter neurotransmitter metabolism.[91]

The discovery of a new target for *NGF*, the pheochromocytoma cell, has been the basis for new insights into the molecular changes induced by this factor.[142] The clonal cell line PC12 displays in the absence of *NGF* the features of adrenal chromaffin cells; *NGF* is able to bring about a conversion into cells that are strikingly similar to sympathetic neurons. These cells respond to *NGF* in a dual way: (1) by a transcription-dependent priming that endows the potential for forming neuritelike structures and (2) by transcription-independent outgrowth on regeneration of neurites. The role of cyclic AMP in the mechanism involved in the second response is under debate.[142]

During development, maintenance of nerve cells appears also to depend on the establishment of functional synaptic contacts.[128] In peripheral target tissues, many of the neuropeptides have both rapid endocrine effects and long-term trophic influences. Within the nervous system, release of these peptides at synapses could provide highly localized trophic influences that may be the basis of neuronal plasticity and memory consolidation. Studies on the growth of cells in culture may be vital to the furtherance of our understanding of this field, as was the case for *NGF*. In such cells, a relationship between cyclic AMP and differentiation processes, including neurite extension, has been reported. There is evidence that *SP* stimulation of neurite extension in cultured neuroblastoma cells is related to increases in cyclic AMP levels.[143] As described above, many neuropeptides alter brain cyclic AMP concentrations, but, as yet, the link to trophic functions cannot be made. Interestingly, *insulin*, which has a structure and metabolic actions similar to *NGF*,[144] is required for the growth of a large number of nerve-tissue-derived cultures and has been called "a nerve survival factor."

Antibodies against α-*MSH* injected into rat fetuses resulted in impaired brain maturation,[24] supporting the idea that intermediate lobe peptides, which are present only in the fetal stages in primates, play a role in brain development.

$ACTH_{1-24}$ (10^{-9}–10^{-7} M) was found to be able to partially replace serum in primary cultures of chick embryo CNS neurons.[145] The effects of *ACTH* addition to serum-free medium were seen in increased cell survival, increased neurite formation, and alterations in various metabolic activities. Trophic influences of *ACTH* peptides (including $ACTH_{4-10}$) have also been reported in adult rats suffering peripheral nerve injury.[146,147] This effect was seen as a threefold increase in the number of outgrowing axons.[148] The number of both myelinated and unmyelinated axons was increased.[149] The diameter of the axons was decreased, so that the content of axoplasm seemed not to be changed.[149] In this instance, cyclic AMP cannot be the mediator, since $ACTH_{4-10}$ does not alter adenylate cyclase activity.[111] *Growth hormone* may exert more general trophic influences on the brain in common with other organs. This peptide can partially correct the effects of neonatal thyroidectomy.[25]

8.3. RNA and Protein Metabolism

Long-term changes in cells of the nervous system in relation to trophic influences imply changes in structural and/or functional proteins. An example is the priming of PC12 cells by *NGF* (see above).[148] Such changes may result from altered gene expression, involving altered protein and possibly also RNA synthesis. Data related to neuropeptide influences are rather scarce and mainly limited to *ACTH* peptides. In general, the metabolic changes observed cannot be related to defined peptidergic pathways, even in the pineal, which is sensitive to neuropeptides *in vitro*, and the regulation of which is mediated predominantly via noradrenergic input.[151] In general, the studies have revealed the capability of nervous tissue to respond metabolically to neuropeptides, and circumstantial evidence (e.g., correlated time courses or structure–activity dependence) suggests a functional relationship with behavioral effects.

8.3.1. RNA Content and Labeling

Caution is necessary in interpreting the results of RNA-labeling studies because the predominant species of pulse-labeled RNA is heterogeneous nuclear RNA, of which only part is a precursor for mRNA. The data on the influence of *ACTH* on brain RNA content and turnover are mainly derived from the effects of altered hormonal states as a result of hypophysectomy or massive doses of $ACTH_{1-24}$, a peptide with corticotrophic activity. Interestingly, the effects of $ACTH_{1-24}$ on pulse labeling of brainstem RNA were in opposite directions in intact and adrenalectomized rats.[152] This suggests that *ACTH* and corticosterone have opposing influences on subcortical RNA metabolism, analogous to their opposite effects on avoidance behavior.[153] Thus, whereas hypophysectomy decreases the RNA content of several brain regions, there are indications that *ACTH, GH*, and adrenal steroids are all involved in cerebral RNA turnover.[110] From autoradiographic data, it is suggested that *ACTH* can change the transport of RNA from the nucleus to the cytoplasm in spinal motoneurons.[154]

The requirement for relatively high doses of $ACTH_{1-24}$ and the ineffectiveness of $ACTH_{4-10}$ are characteristics shared by responses in RNA synthesis, membrane phosphorylation, and cyclic nucleotide concentration.

8.3.2. Protein Synthesis

As with RNA studies, although for different reasons, interpretation of *in vivo* labeling in terms of cellular rates of protein synthesis is difficult because the specific activity of the available precursor is hard to estimate (see also Chapter 2, Vol. 5). Manipulation of behavioral and hormonal states has been reported, although not consistently, to alter cerebral blood flow and deoxyglucose uptake[110] and thus may alter protein labeling by effects unrelated to the protein synthetic apparatus *per se*. However, in spite of these difficulties, replacement of individual peptides in hypophysectomized rats has indicated correlations between behavioral effects and brain protein synthesis.[110] Hypophysectomy causes a decrease in the pulse labeling of brainstem proteins *in vivo*. Chronic treatment with $ACTH_{1-10}$ or $ACTH_{4-10}$, peptides with no corticotrophic activity, reversed this effect, whereas the peptide [D-Phe^7]$ACTH_{1-10}$ decreased incorporation into protein.[155] These peptide treatments have been shown to partially reverse or exacerbate, respectively, the deficiency in acquiring a behavioral task caused by hypophysectomy.[156]

Incubation of subcortical brain slices from hypophysectomized rats with low concentrations (5×10^{-7} M) of $ACTH_{1-24}$ or $ACTH_{1-10}$ resulted in increases in protein labeling comparable to the *in vivo* effects.[155] Brains from hypophysectomized rats have a lower polysome content, compatible with decreased protein synthetic activity.[110,157,160] Reverse changes in polysome profiles and activity in cell-free protein synthesis by *ACTH* have been reported by two independent groups. Treatment of hypophysectomized rats with $ACTH_{1-10}$ or des-glycyl-8-lysine-*VP* during avoidance behavior experiments restored polysome profiles.[110] In this study, peptide treatment alone did not alter the profile, nor did unsuccessful training in the absence of peptide. It was therefore concluded that the peptides had a permissive effect on both the behavioral acquisition and the restoration of polysome profiles.[110]

In the other study,[157] treatment with *ACTH* in the absence of training was found to be effective. The noncorticotrophic peptide $ACTH_{1-10}$ restored polysome profiles and cell-free protein synthesis activity towards normal values, whereas corticotrophic peptides (for example, $ACTH_{1-24}$, $ACTH_{1-23}$, and $ACTH_{1-17}$) or glucocorticoids themselves increased these parameters to supranormal levels. Thus, it seems that *ACTH* peptides can exert a dual effect by direct action and indirectly by mediation of glucocorticoids. The same authors have shown[158] that similar effects can also be found in the liver, so the trophic actions are not restricted to nervous tissue.

Stimulation by *ACTH* peptides of protein labeling in the brain has been reported by a number of workers using different amino acid precursors and various techniques.[110] The effect of $ACTH_{4-10}$ appears to be specific to nerve tissue. *ACTH* also increases protein labeling in the spinal cord as shown by *in vivo* labeling, and, autoradiographically, an increase in labeling of spinal motorneurons was shown.[154] In a cell-free protein synthesis assay following *in vivo* peptide treatment, $ACTH_{4-10}$ was also active in stimulating protein synthesis at doses effective in accelerating peripheral nerve regeneration.[147,148,150] A different structure–activity relationship was observed for the increase in protein synthesis in cell-free extracts of the SN and excessive grooming behavior following i.c.v. injection of peptides.[110] Because the biochemical change was studied after the peptide-evoked behavioral response, it is not clear whether the increase in protein synthesis reflects a direct peptide action or is the result of an increased activity of the SN, a brain region essential for expression of the behavioral response to the peptide.[67]

Rather surprisingly, *ACTH* peptides also seem to be active when added *in vitro* to the cell-free protein synthesis assay system. A biphasic modulation was observed, low concentrations (10^{-8} to 10^{-7} M) of $ACTH_{1-24}$ being stimulatory and high concentrations being inhibitory. $ACTH_{4-10}$ gives a similar biphasic pattern, whereas the D-isomer [7-D-phe] $ACTH_{4-10}$ has a mirror-image concentration–effect curve.[159] The stimulatory influences were also observed with $ACTH_{1-10}$ and $ACTH_{1-16}$.[160] The inhibitory effects are exerted by a number of peptides that contain the basic region (15–18). The inhibitory action can be observed in systems in which only chain elongation is occurring and shows a resemblance to an interaction with Mg^{2+}-binding sites, where polycations such as spermine have been shown to act.[161] Concomitantly with the inhibition of cell-free protein synthesis, changes in phosphorylation of a polyribosomal protein (pp30) has been observed.[162]

Enzyme activities in the pineal gland show pronounced circadian rhythms, which appear to be predominantly the result of NE regulation, although peptide influences have also been proposed.[90,151] In *in vitro* whole-gland preparations, both the NE terminals from the innervating ganglion and the circadian rhythmicity in levels of protein synthesis are preserved.[156] In this preparation, stimulation of protein synthesis by the peptides $ACTH_{1-24}$, $ACTH_{1-16}$, and $ACTH_{5-18}$ was observed with concentrations as low as 10^{-11} M.[151] Des-glycyl-8-lysine-*VP* was also active, although higher concentrations were required. At high concentrations of *ACTH* peptides, inhibition of protein synthesis was observed. The sensitivity to peptides varied in parallel with base-line activity during the day. This observation and the sensitivity of pineal protein synthesis to β-NE agonists and antagonists, calcium, and cyclic AMP suggest that *ACTH* peptides exert their effects either pre- or postsynaptically by interaction with β-noradrenergic mechanisms.[151] This is another example of a modulation of the β-adrenergic response of glial-type cells by neuropeptides (cf. the response of astrocytes in culture[135]).

In summary, three types of effect may be identified among the *ACTH* peptides. First, small N-terminal sequences are able to modify protein synthesis directly by intracellular actions. Entire small fragments, $ACTH_{4-10}$, $ACTH_{4-9}$, and $ACTH_{1-10}$, can cross the cell membrane, whereas this N-terminal region of larger peptides may have access to the cytosol by transmembrane positioning or by internalization (see Section 8.1). The influence of these peptides on protein synthesis *in vivo*, in slices, and in cell-free systems shows stereospecificity at Phe^7. Similar specificity is observed for the effect of these peptides on avoidance behavior and peripheral nerve regeneration. Larger sequences, $ACTH_{1-16}$ or $ACTH_{5-18}$, are needed to show effects on membrane phosphorylation, cyclic AMP production, RNA metabolism, and effects on protein synthesis probably mediated by synaptic mechanisms such as in the pineal and SN. These effects show similar structure–activity relationships to those for the activity of *ACTH* on excessive grooming and for affinity for the opiate receptor.[67] A third site within the *ACTH* molecule may be formed by the peptides that share the basic sequence (15–18) and are either N-terminally or C-terminally elongated ($ACTH_{5-18}$ or $ACTH_{15-24}$, respectively). The physiological significance of this site in brain function is unknown. The notion of two or three sites of neurotrophic activity within *ACTH* corroborates with comparable activities on peripheral targets.[127]

Few data are available on peptide influences on specific proteins. Changes in nerve-specific enolase activity and protein S-100 have been reported for *ACTH* peptides.[163,164] Both proteins have been extensively mentioned in relation to learning situations and neuronal plasticity,[165] but their role remains obscure.

The priming of PC12 cells by *NGF* is accompanied by the transcription-dependent induction of three proteins, one of them a surface glycoprotein.[142]

9. FINAL COMMENTS

We have described peptidergic systems in terms of their constituent elements, thus emphasizing the basic similarities, which outweigh the differences between individual peptidergic systems. In a few cases, the complete system can, more or less, be visualized. *Substance P*, for example, appears to act as a neurotransmitter in certain defined neuronal pathways and also has trophic influences mediated by cyclic AMP. The modulatory role of *ENK*s in interneurons at relay stations on sensory pathways is compatible with the behavioral activities of these peptides. However, in most instances, the current state of knowledge is insufficient to allow description of a complete peptidergic system. The *POMC* family of peptides, for example, have been localized in defined pathways, but the significance of the presence within the same nerve terminal of several active principles with various and even opposing behavioral actions and the relative importance of actions at the opiate (e.g., β-*END*-evoked catatonia) or other, undefined, receptors are poorly understood.

The uncertainty as to the mode of action of peptides on their target cells reflects a general lack of understanding of the function of the nervous system as a whole. Neuropeptide research has opened new horizons in the field of neurobiology. Not only have these studies introduced the concept of subtle control mechanisms by the corelease of more than one active principle, but the analogy between CNS and peripheral hormonal peptidergic systems may provide an insight into central adaptional processes and neuronal plasticity.

ACKNOWLEDGMENTS. The authors have benefited greatly from a close scientific association with Professor David de Wied, who founded and continues to inspire neuropeptide research in the Netherlands.

The authors are supported by FUNGO Grant no.13-31-43 from the Netherlands Organization for the Advancement of Pure Research.

REFERENCES

1. Pearse, A. G. E., 1976, *Nature* **262**:92-94.
2. Barker, K. L., and Smith, T. G., Jr., 1980, *Prog. Brain Res.* **53**:169-192.
3. Elliott, G. R., and Barchas, J. D., 1980, *Hormones and the Brain* (D. De Wied and P. A. Van Keep, eds.), MTP Press, Lancaster, pp. 43-52.
4. Wiegant, V. M., and Gispen, W. H., 1976, *Molecular and Functional Neurobiology* (W. H. Gispen, ed.), Elsevier Biomedical Press, Amsterdam, pp. 235-254.
5. Kordon, C., 1981, *Adv. Physiol. Sci.* **14**:265-266.
6. Krieger, D. T., 1980, *Adv. Physiol. Sci.* **13**:151-175.
7. Witter, A., and De Wied, D., 1980, *Handbook of the Hypothalamus*, Volume 2 (P. J. Morgane and J. Panksepp, eds.), Marcel Dekker, New York, pp. 307-451.
8. Verhoef, J., Prins, A., Veldhuis, H. D., and Witter, A., 1982, *J. Neurochem.* **38**:1135-1138.
9. Rapoport, S. I., Klee, W. A., Pettigrew, K. D., and Ohno, K., 1980, *Science* **207**:84-86.
10. De Wied, D., 1981, *Adv. Physiol. Sci.* **13**:23-38.
11. De Kloet, E. R., Palkovits, M., and Mezey, E., 1981, *Pharmacol. Ther.* **12**:321-351.
12. Herbert, E., Birnberg, N., Civelli, O., Lissitzky, J.-C., Uhler, M., and Durring, L., 1982, *Adv. Biochem. Psychopharmacol.* **33**:9-18.
13. Smyth, D. G., Zakarian, S., Deakin, J. W. F., and Massey, D. E., 1981, *Peptides of the Pars Intermedia* (D. Evered and G. Lawrenson, eds.), Pitman Medical, London, pp. 79-96.
14. Jackson, S., Hope, J., Estivariz, F., and Lowry, P. J., 1981, *Peptides of the Pars Intermedia* (D. Evered and G. Lawrenson, eds.), Pitman Medical, London, pp. 141-162.
15. Bennett, H. P. J., Brown, C. A., and Solomon, S., 1982, *Brain Res.* **231**:454-460.
16. Parish, D. C., Smyth, D. G., Normanton, J. R., and Wolstancroft, J. H., 1983, *Nature* **306**:267-269.
17. Seizinger, B. R., and Höllt, V., 1982, *Adv. Biochem. Psychopharmacol.* **33**:27-34.
18. Gillies, G., and Lowry, P. J., 1982, *Frontiers in Neuroendocrinology*, Volume 7 (W. F. Ganong and L. Martini, eds.), Raven Press, New York, pp. 45-75.
19. Vale, W., Spiess, J., Rivier, C., and Rivier, J., 1981, *Science* **213**:1394-1397.
20. Selye, H., 1936, *Br. J. Exp. Pathol.* **17**:234-248.
21. Pedersen, R. C., and Brownie, A. C., 1980, *Proc. Natl. Acad. Sci. U.S.A.* **77**:2239-2243.
22. Berkenbosch, F., Tilders, F. J. H., and Vermes, I., 1983, *Nature* **305**:237-239.
23. Van Ree, J. M., Bohus, B., Csontos, K. M., Gispen, W. H., Greven, H. M., Nijkamp, F. P., Opmeer, F. A., De Rotte, G. A., Van Wimersma Greidanus, T. B., Witter, A., and De Wied, D., 1981, *Life Sci.* **28**:2875-2888.
24. Swaab, D. F., and Martin, J. R., 1981, *Peptides in the Pars Intermedia* (D. Evered and G. Lawrenson, eds.), Pitman Medical, London, pp. 196-217.
25. Prange, A. J., Nemeroff, C. B., Lipton, M. A., Breese, G. R., and Wilson, I. C., 1978, *Handbook of Psychopharmacology*, Volume 13 (L. L. Iversen, S. D. Iversen, and S. H. Snyder, eds.), Plenum Press, New York, pp. 1-107.
26. De Wied, D., 1969, *Frontiers in Neuroendocrinology*, Volume 3 (W. F. Ganong and L. Martini, eds.), Oxford University Press, New York, pp. 97-140.
27. Swaab, D. F., 1980, *Hormones and the Brain* (D. De Wied and P. A. van Keep, eds.), MTP Press, Lancaster, pp. 87-100.

28. Rivier, C., and Vale, W., 1983, *Nature* **305**:325–327.
29. Sterba, G., Naumann, W., and Hoheisel, G., 1980, *Prog. Brain Res.* **53**:141–158.
30. Swanson, L. W., and Sawchenko, P. E., 1983, *Annu. Rev. Neurosci.* **6**:269–324.
31. Gash, D. M., and Thomas, G. J., 1983, *Trends Neurosci.* **6**:197–198.
32. Bohus, B., 1981, *Int. J. Ment. Health* **9**:6–44.
33. Bohus, B., Conti, L., Kovács, G. L., and Versteeg, D. H. G., 1982, *Neuronal Plasticity and Memory Formation* (C. A. Marsan and H. Matthies, eds.), Raven Press, New York, pp. 75–87.
34. Laczi, F., Gaffori, O., De Kloet, E. R., and De Wied, D., 1983, *Brain Res.* **260**:342–346.
35. Burbach, J. P. H., Kovács, G. L., De Wied, D., Van Nispen, J. W., and Greven, H. M., 1983, *Science* **22**:1310–1312.
36. Pearlmutter, A. F., Costatini, M. G., and Loeser, B., 1983, *Peptides* **4**:335–341.
37. Rorstad, O. P., Martin, J. B., and Terry, L. C., 1980, *The Role of Peptides in Neuronal Function* (J. L. Barker and T. G. Smith, Jr., eds.), Marcel Dekker, New York, pp. 573–614.
38. Knigge, K. M., Hoffman, G. E., Joseph, S. E., Scott, D. E., Sladek, C. D., and Sladek, J. R., Jr., 1980, *Handbook of the Hypothalamus* (P. J. Morgane and J. Panksepp, eds.), Marcel Dekker, New York, pp. 63–164.
39. Moss, R. L., and Dudley, C. A., 1980, *The Role of Peptides in Neuronal Function* (J. L. Barker and T. G. Smith, Jr., eds.), Marcel Dekker, New York, pp. 455–478.
40. Spiess, J., Rivier, J., and Vale, W., 1983, *Nature* **303**:532–535.
41. Watson, S.J., and Akil, H., 1981, *Adv. Biochem. Psychopharmacol.* **28**:77–86.
42. Verhoef, J., Wiegant, V. M., and De Wied, D., 1982, *Brain Res.* **231**:454–460.
43. Wang, X-C., Burbach, J. P. H., Verhoef, C. J., and De Wied, D., 1983, *J. Biol. Chem.* **258**:7942–7947.
44. Hughes, J., Beaumont, A., Fuentes, J. A., Malfroy, B., and Unsworth, C., 1980, *J. Exp. Biol.* **89**:239–255.
45. Rossier, J., 1982, *Trends Neurosci.* **5**:179–180.
46. Stengaard-Pedersen, K., and Larsson, L.-I., 1981, *Peptides* **2**(Suppl. 1):3–19.
47. Lightman, S. L., Ninkovic, M., Hunt, S. P., and Iversen, L. L., 1983, *Nature* **305**:235–239.
48. Weber, E., Roth, K. A., and Barchas, J. D., 1981, *Biochem. Biophys. Res. Commun.* **103**:951–958.
49. Watson, S. J., and Akil, H., 1982, *Adv. Biochem. Psychopharmacol.* **33**:35–42.
50. Snyder, S. H., Innis, R. B., and Corêa, F. M. A., 1980, *The Role of Peptides in Neuronal Function* J. L. Barker and T. G. Smith, Jr., eds.), Marcel Dekker, New York, pp. 375–389.
51. Cooper, P. E., and Martin, J. B., 1980, *Ann. Neurol.* **8**:551–557.
52. Poulain, D. A., and Wakerley, J. B., 1982, *Neuroscience* **7**:773–808.
53. Goedert, M., 1984, *Trends Neurosci.* **7**:3–5.
54. Uhl, G. R., and Snyder, S. H., 1981, *Adv. Biochem. Psychopharmacol.* **28**:87–106.
55. Fernstrom, M. H., and Leeman, S. E., 1981, *Adv. Physiol. Sci.* **13**:155–165.
56. Dao, W. P. C., Yajima, H., Uitagawa, K., and Walker, R. J., 1981, *Adv. Physiol. Sci.* **14**:249–254.
57. Taylor, R. L., and Burt, D. R., 1982, *J. Neurochem.* **38**:1649–1656.
58. Kalivas, P. W., and Horita, A., 1980, *J. Pharmacol. Exp. Ther.* **212**:203–210.
59. Quirion, R., Schults, C. W., Moody, T. W., Pert, C. B., Chase, T. N., and Donohue, T. L., 1983, *Nature* **303**:714–716.
60. Snyder, S. H., 1980, *Science* **209**:976–983.
61. Said, S. I., 1980, *The Role of Peptides in Neuronal Function* (J. L. Barker and T. G. Smith, Jr., eds.), Marcel Dekker, New York, pp. 351–374.
62. Bunney, B. S., Grace, A. A., Homner, D. W., and Skirboll, L. R., 1982, *Adv. Biochem. Psychopharmacol.* **33**:429–436.
63. Rosenzweig, J. L., Havrankova, J., Brownstein, M., and Roth, J., 1981, *Adv. Physiol. Sci.* **13**:175–186.
64. Oomura, Y., and Kita, H., 1981, *Diabetologica* **20**:290–298.
65. Epstein, A. N., 1981, *Adv. Biochem. Psychopharmacol.* **28**:373–387.
66. Phillips, M. I., 1980, *The Role of Peptides in Neuronal Function* (J. L. Barker and T. G. Smith, Jr., eds.), Marcel Dekker, New York, pp. 389–430.

67. Gispen, W. H., Van Someren, H., and Schotman, P., 1981, *Adv. Physiol. Sci.* **13**:223–231.
68. Donohue, T. L. O., and Jacobowitz, D. M., 1980, *Polypeptide Hormones* (R. F. Beers, Jr., ed.), Raven Press, New York, pp. 203–222.
69. Reppert, S. M., Artman, H. G., Swaminathan, S., and Fisher, D. A., 1981, *Science* **213**:1256–1257.
70. Hökfelt, T., Lundberg, J. M., Schultzberg, M., Johansson, O., Skirboll, L., Änggård, A., Fredholm, B., Hamberger, B., Pernow, B., Rehfeld, J., and Goldstein, M., 1980, *Proc. R. Soc. (Lond.) [Biol.]* **210**:63–77.
71. Jessell, T. M., 1981, *Adv. Biochem. Psychopharmacol.* **28**:189–198.
72. Dichter, M. A., and Delfs, J. R., 1981, *Adv. Biochem. Psychopharmacol.* **28**:145–157.
73. Olpe, H.-R., Balcar, V. J., Bittinger, H., Rink, H., and Sieber, P., 1980, *Eur. J. Pharmacol.* **63**:127–133.
74. Pittman, A. J., 1981, *Adv. Physiol. Sci.* **14**:231–241.
75. Nicoll, R. A., Alger, B. E., and Jahr, C. E., 1980, *Proc. R. Soc. (Lond.) [Biol.]* **210**:133–149.
76. Macdonald, R. L., and Nowak, L. M., 1981, *Adv. Biochem. Psychopharmacol.* **28**:159–173.
77. Fischbach, G. D., Dunlap, K., Mudge, A., and Leeman, S., 1981, *Adv. Biochem. Psychopharmacol.* **28**:175–188.
78. Kelly, J. S., and Dodd, J., 1981, *Adv. Biochem. Psychopharmacol.* **28**:133–144.
79. Lichtensteiger, W., and Felix, D., 1980, *Neuropeptides and Neural Transmission*, IBRO Monograph Series, Volume 7 (C. A. Marsan and W. Z. Traczyk, eds.), Raven Press, New York, pp. 333–338.
80. Joëls, M., and Urban, I. J. A., 1982, *Neurosci. Lett.* **33**:79–84.
81. Mühlethaler, M., Sawyer, W. H., Manning, M. M., and Dreifuss, J. J., 1983, *Proc. Natl. Acad. Sci. U.S.A.* **80**:6713–6717.
82. Versteeg, D. H. G., 1980, *Pharmacol. Ther.* **11**:535–557.
83. Moore, K. E., and Demarest, K. K., 1982, *Frontiers in Neuroendocrinology*, Volume 7 (W. F. Ganong and L. Martini, eds.) Raven Press, New York, pp. 161–190.
84. Kordon, C., Enjalbert, A., Hery, M., Joseph-Bravo, P. I., Rotsztejn, W., and Ruberg, M., 1980, *Handbook of the Hypothalamus*, Volume 2 (P. J. Morgane and J. Panksepp, eds.), Marcel Dekker, New York, pp. 253–306.
85. Kovàcs, G. L., Bohus, B., and Versteeg, D. H. G., 1980, *Prog. Brain Res.* **53**:123–139.
86. Schulz, H., Kovàcs, G. L., and Telegdy, G., 1980, *Neuropeptides and Neural Transmission* (C. A. Marsan and W. Z. Traczyk, eds.), Raven Press, New York, pp. 343–349.
87. Torrens, Y., Michelot, R., Beaujouan, J. C., Glowinski, J., and Bockaert, J., 1982, *J. Neurochem.* **38**:1728–1734.
88. Nathanson, J. A., 1981, *Adv. Biochem. Psychopharmacol.* **28**:599–608.
89. Crowley, W. R., 1982, *Neuroendocrinology* **34**:381–386.
90. Cardinali, D. P., 1979, *Trends Neurosci.* **2**:250–253.
91. Smelik, P., 1981, *Front. Horm. Res.* **8**:1–11.
92. Oehme, P., Hecht, K., Piesche, L., Hiese, H., Morgenstern, E., and Popper, M., 1980, *Neuropeptides and Neural Transmission*, IBRO Monograph Series, Volume 7 (C. A. Marsan and W. Z. Traczyk, eds.), Raven Press, New York, pp. 73–84.
93. Traczyk, W. Z., and Luczynska, M., 1980, *Neuropeptides and Neural Transmission*, IBRO Monograph Series, Volume 7 (C. A. Marsan and W. Z. Traczyk, eds.), Raven Press, New York, pp. 165–180.
94. Zucker, I., and Carmichael, M. S., 1981, *Adv. Biochem. Psychopharmacol.* **28**:459–473.
95. Auerbach, S., and Lipton, P., 1982, *J. Neurosci.* **2**:477–482.
96. McCann, S. M., 1981, *Adv. Physiol. Sci.* **14**:121–129.
97. Wood, P. L., Cheney, D. L., and Costa, E., 1979, *Neuroscience* **4**:1479–1484.
98. Phillis, J. W., 1980, *The Role of Peptides in Neuronal Function* (J. L. Barker and T. G. Smith, Jr., eds.) Marcel Dekker, New York, pp. 615–652.
99. Belcher, G., and Ryall, R. W., 1977, *J. Physiol. (Lond.)* **272**:105–119.
100. Barker, J. L., Vincent, J.-D., and MacDonald, J. F., 1980, *Neuropeptides and Neural Transmission*, IBRO Monograph Series, Volume 7 (C. A. Marsan and W. Z. Traczyk, eds.), Raven Press, New York, pp. 93–103.

101. Grimm-Jørgensen, Y., 1980, *The Role of Peptides in Neuronal Function* (J. L. Barker and T. G. Smith, Jr., eds.), Marcel Dekker, New York, pp. 479–507.
102. Reichlin, S., 1981, *Adv. Biochem. Psychopharmacol.* **28**:573–579.
103. Emson, P. C., and Hunt, S. P., 1982, *Molecular Approaches to Neurobiology* (I. R. Brown, ed.), Academic Press, New York, pp. 255–283.
104. Pernow, B., 1980, *Neuropeptides and Neural Transmission*, IBRO Monograph Series, Volume 7 (C. A. Marsan and W. Z. Traczyk, eds.), Raven Press, New York, pp. 5–17.
105. Iverson, L. L., Lee, C. M., Gilbert, R. F., Hunt, S., and Emson, P. C., 1980, *Proc. R. Soc. (Lond.) [Biol.]* **210**:91–111.
106. Hökfelt, T., Lundberg, J. M., Skirboll, L., Johanssons, O., Schultzberg, M., and Vincent, S. R. 1982, *Co-transmission, Proceedings of a Symposium, 50th Annual Meeting of the British Pharmacological Society, Oxford* (A. C. Cuello, ed.), Macmillan, London, pp. 77–126.
107. Martin, R., and Voigt, K. H., 1981, *Nature* **289**:502–504.
108. Martin, R., Geis, R., Holl, R., Schäfer, M., and Voigt, K. H., 1983, *Neuroscience* **8**:213–227.
109. Gremlich, H. H., Fringeli, U. P., and Schwyzer, R., 1983, *Biochemistry* **22**:4257–4264.
110. Dunn, A. J., and Schotman, P., 1981, *Pharmacol. Ther.* **12**:353–372.
111. Wiegant, V. M., Zwiers, H., and Gispen, W. H., 1981, *Pharmacol. Ther.* **12**:463–491.
112. Wiegant, V. M., Reul, J. M. H. M., and Gispen, W. H., 1982, *Prog. Brain Res.* **56**:403–410.
113. Kolata, C. B., 1978, *Science* **201**:895–897.
114. Manning, M., and Sawyer, W. H., 1984, *Trends Neurosci.* **7**:6–9.
115. Mayer, N., Saria, A., and Lembeck, F., 1980, *Neuropeptides and Neural Transmission*, IBRO Monograph Series, Volume 7 (C. A. Marsan and W. Z. Traczyk, eds.), Raven Press, New York, pp. 19–29.
116. Srikant, C. B., and Patel, Y. C., 1981, *Endocrinology* **108**:341–343.
117. Terenius, L., 1980, *Receptors for Neurotransmitters and Peptide Hormones* (C. Pepeu, M. J. Kuhar, and S. J. Enna, eds.), Raven Press, New York, pp. 321–328.
118. Pert, C. A., 1981, *Adv. Biochem. Psychopharmacol.* **28**:117–131.
119. Terenius, L., 1976, *Eur. J. Pharmacol.* **38**:211–213.
120. Carter, R. J., Schuster, S., and Morley, J. S., 1979, *Nature* **279**:74–75.
121. Stengaard-Pedersen, K., and Larsson, L. I., 1981, *Acta Pharmacol. Toxicol. (Kbh.)* **48**:39–46.
122. Akil, H., Hewlett, W. A., Barchas, J. D., and Li, C. H., 1980, *Eur. J. Pharmacol.* **64**:1–8.
123. Oki, S., Nakao, K., Nakai, Y., Ling, N., and Imura, H., 1980, *Eur. J. Pharmacol.* **64**:161–164.
124. Gispen, W. H., and Isaacson, R. L., 1981, *Pharmacol. Ther.* **12**:209–246.
125. Ramachandran, J., Lee, C. Y., Keri, G., and Kenez-Keri, M., 1980, *Polypeptide Hormones* (R. F. Beers, Jr., ed.), Raven Press, New York, pp. 295–308.
126. Greven H. M., and De Wied, D., 1980, *Hormones and the Brain* (D. de Wied and P. A. van Keep, eds.), MTP Press, Lancaster, pp. 115–127.
127. Saez, J. M., Morera, A. M., and Dazord, A., 1981, *Adv. Cyclic Nucleotide Res.* **14**:563–579.
128. Schwandt, P., 1981, *Front. Horm. Res.* **8**:107–121.
129. Eberle, A. N., 1980, *Cellular Receptors for Hormones and Neurotransmitters* (D. Schulster and A. Levitzki, eds.), John Wiley & Sons, Chichester, New York, pp. 219–231.
130. Schwartz, I. L., Huang, C.-J., Fischmann, A. J., Masar, S. K., and Wyssbrod, H. R., 1981, *Dev. Endocrinol.* **13**:101–135.
131. Teichberg, V. I., and Blumberg, S., 1980, *Cellular Receptors for Hormones and Neurotransmitters* (D. Schulster and A. Levitzki, eds.), John Wiley & Sons, Chichester, New York, pp. 397–403.
132. Smith, J. R., 1981, *Life Sci* **28**:2065–2069.
133. Barbaccia, M. L., Chuang, D. M., and Costa, E., 1982, *Adv. Biochem. Psychopharmacol.* **33**:511–518.
134. Schneider, D. R., Felt, B. T., and Goldman, H., 1982, *Pharmacol. Biochem. Behav.* **16**:139–143.
135. Royon, G., Noble, M., and Mudge, A. W., 1983, *Nature* **305**:715–717.
136. Von Calker, D., Löffler, F., and Hamprecht, B., 1983, *J. Neurochem.* **40**:418–427.

137. Ehrlich, Y. H., Davis, L. G., Keen, P., and Brunngraber, E. G., 1980, *Life Sci.* **26**:1765–1772.
138. Arnaud, J., Nobili, O., and Boyer, J., 1981, *Biochem. Biophys. Res. Commun.* **103**:1167–1172.
139. Bernstein, H. G., Poeggel, G., Dorn, A., Luppa, H., and Ziegler, M., 1981, *Experientia* **37**:434–435.
140. Nelson, P. G., and Brenneman, D. E., 1982, *Trends Neurosci.* **5**:229–232.
141. Thoenen, H., Barde, Y.-A., and Edgar, D., 1981, *Adv. Biochem. Psychopharmacol.* **28**:263–273.
142. Burstein, D. E., and Greene, L. A., 1982, *Molecular Approaches to Neurobiology* (I. R. Brown, ed.), Academic Press, New York, pp. 159–177.
143. Narumi, S., and Maki, Y., 1978, *J. Neurochem.* **30**:1321–1326.
144. Martin, J. B., and Landis, D. M. D., 1981, *Adv. Biochem. Psychopharmacol.* **28**:673–689.
145. Daval, J. L., Louis, J. C., Gerard, M. J., and Vincendon, G., 1983, *Neurosci. Lett.* **36**:299–304.
146. Strand, F. L., and Smith, L. M., 1980, *Pharmacol. Ther.* **11**:509–533.
147. Bijlsma, W. A., Jennekens, F. G. I., Schotman, P., and Gispen, W. H., 1981, *Eur. J. Pharmacol.* **76**:73–79.
148. Bijlsma, W. A., Jennekens, F. G. I., Schotman, P., and Gispen, W. H., 1983, *Nerve Muscle* **6**:104–112.
149. Bijlsma, W. A., van Asselt, E., Veldman, H., Jennekens, F. G. I., Schotman, P., and Gispen, W. H., 1983, *Acta Neuropathol. (Berl.)* **62**:24–30.
150. Bijlsma, W. A., Schotman, P., Jennekens, F. G. I., and Gispen, W. H., 1983, *Neurosci. Lett.* **38**:297–302.
151. Schotman, P., Allaart, J., and Gispen, W. H., 1981, *Brain Res.* **219**:121–135.
152. Gispen, W. H., and Schotman, P., 1976, *Neuroendocrinology* **21**:97–110.
153. De Wied, D., 1977, *Acta Endocrinol.* **85**:9–19.
154. Jakoubek, B., Burešová, M., Hájek, I., Etrychova, J., Pavlik, A., and Radičová, A., 1972, *Brain Res.* **43**:417–428.
155. Reith, M. E. A., Schotman, P., and Gispen, W. H., 1977, *Mechanisms, Regulation and Special Functions of Protein Synthesis in the Brain* (S. Roberts, A. Lajtha, and W. H. Gispen, eds.), Elsevier Biomedical Press, Amsterdam, pp. 383–398.
156. Fekete, M., Bohus, B., and de Wied, D., 1983, *Neuroendocrinology* **36**:112–118.
157. Lando, D., and Raynaud, J. P., 1980, *Endocrinology* **107**:2063–2068.
158. Lando, D., Secchi, J., Roche, J., and Raynaud, J. P., 1980, *Endocrinology* **107**:2055–2062.
159. Schotman, P., and Allaart, J., 1981, *J. Neurochem.* **37**:1349–1352.
160. Schotman, P., van Heuven-Nolsen, D., and Gispen, W. H., 1980, *J. Neurochem.* **34**:1661–1670.
161. Schrama, L. H., Frankena, H., Edwards, P. M., and Schotman, P., 1984, *J. Neurosci. Res.* **11**:67–77.
162. Schrama, L. H., Weeda, E., Frankena, H., Edwards, P. M., and Schotman, P., 1983, *15th FEBS Meeting*, Elsevier Sci. Publ., Amsterdam, p. 284.
163. Zomzely-Neurath, C. E., and Walker, W. A., 1980, *Proteins of the Nervous System* (R. A. Bradshaw and D. M. Schneider, eds.), Raven Press, New York, pp. 1–57.
164. Hydén, H., 1976, *Prog. Brain Res.* **45**:83–100.

12

Neurosecretion

S. Arch and H. Gainer

1. INTRODUCTION

It is customary, and often useful, to begin a topical review of this sort with a general definition of the field to be covered. In the case of "neurosecretion," or the study of the "neurosecretory cell," the task of designing a definition with satisfactory precision is challenging. After all, all neurons participate in some sort of secretory activity. Indeed, expulsion of waste products and some form of membrane recycling are necessary and fundamental properties of all cells. Moreover, all neurons, with the exception of those linked by low-resistance electrical junctions, communicate with one another by the secretion of messenger compounds. Since "neurosecretion" is not conventionally used to encompass all secretory activities of all nerve cells, it is evident that a more restrictive definition is required.

The use of neurosecretion to define a particular class of neuronal activities began about 50 years ago, when histologists noted that some groups of neurons could be stained selectively by procedures that had been used to identify glandular and endocrine tissues.[1,2] In the vertebrate central nervous system, these neuronal groups were most readily seen in the hypothalamo–neurohypophyseal system. Since this system is invested with capillary beds, and "neurosecretory material" could be seen in accumulations adjacent to these beds, the inference that the neurosecretory cells released hormones into the circulation was quickly drawn. Further work showing the accumulation or depletion of neurosecretory material in concert with major changes in physiological status, e.g., seasonal reproduction or experimentally induced changes in water and electrolyte balance, strengthened the assignment of an endocrine role to these cells. With the application of increasingly refined physiological and biochemical methodologies, the neuroendocrine status of the hypothalamo–neurohypophyseal system can now be considered fully demonstrated.

A similar progression from histologically designated neurosecretory structures to physiologically defined neuroendocrine cells has occurred in studies

---

S. Arch • Biological Laboratories, Reed College, Portland, Oregon 97202. *H. Gainer* • Laboratory of Developmental Neurobiology, National Institutes of Health, Bethesda, Maryland 20205.

on a variety of neuronal preparations from phylogenetically diverse organisms. As a consequence, definitions of neurosecretion typically embody the assumption of endocrine function. Whether this essential equation of neurosecretory with neuroendocrine functions is a satisfactory strategy or not depends on a judgment regarding what constitutes an identifiably separate field of study. Stated as a question, one can ask what value the term neurosecretory serves when it is defined as a virtual synonym for neuroendocrine. Since neurosecretion originated as the study of those groups of neurons that appeared to be chemically distinct (by virtue of their staining properties) from conventional neurons, it would appear to describe a field concerned with a class of neurons sharing some form of chemical uniqueness. Equation with neuroendocrine function subverts this inference.

A more coherent approach to definition of neurosecretion would be to center attention on what, in fact, constitutes the chemical uniqueness of neurosecretory cells. The reason that hypothalamic and other neurosecretory cells stain differently is that they contain abundant large secretory vesicles, which, in turn, contain polypeptides. The chemistry that sets neurosecretory cells apart is the chemistry of peptide secretion. This observation is not new to definitions of neurosecretion. The "neurosecretory process," a description of the unique chemical physiology of the neurosecretory cell, has been stated[3] to consist of the synthesis and vesicular enclosure of the secretory product in the perikaryon followed by vesicle transport to the secretory terminal, where a period of storage may precede secretion. In short, the description of the physiology of a peptidergic neuron.

Nonetheless, to adopt the characteristic biochemistry and cellular physiology of peptide secretion as the defining features of the neurosecretory cell extends the field beyond its traditional compass. Until recently, this extension would not have been obvious; however, the application of immunologic techniques to neuroanatomic investigations has revealed a surprisingly broad distribution of putative peptidergic neurons in CNS preparations.[4] Coincident with these observations, pharmacological and electrophysiological study has disclosed peptide activity in a range of neuronal interactions extending from synaptic transmission through neuromodulation and paracrine effects to the more familiar endocrine roles. Hence, peptidergic neurons are not restricted to any special class of functional interactions with other neurons or effectors.

Early in the development of study of neurosecretory cells, it was possible to maintain that they represented a unique neuronal cell type with a restricted set of functions. The primary chemical characteristic that set them apart was the avidity with which they took up stains for proteins and sugars. Since the most obvious concentrations of stained cells were in areas of neurohemal contact, presumed neuroendocrine function became a secondary defining characteristic. Additional features, such as typical secretory vesicle diameter and extensive neuritic branching, are essentially derivative of the initial characterizations. As we have noted, consideration of contemporary information undermines attempts to define the neurosecretory cell in a way that retains the limited field of these traditional conceptions. We are left, therefore, with two options. Either we continue with the historically determined limitation of the purview of neurosecretion to those instances of peptidergic systems that function as neuroendocrine effectors or we can attempt to define the study of neurosecretion in more coherent, if necessarily broader, terms.

It is our view, at least from the perspective of neurochemistry, that limiting the concept of neurosecretion to endocrine function is an arbitrary circumscription of perspective. What sets neurosecretory cells apart is precisely what sets peptidergic neurons apart from neurons that exclusively secrete conventional neurotransmitters. There is no evidence that the morphological distance separating the secreting cell from its follower, or target, is a significant determinant of the secreting cell's physiology or chemistry. Thus, in our view, the neurosecretory cell is a peptide-secreting neuron, and the neurosecretory process describes the chemistry and physiology of synthesis, packaging, transport, storage, and exocytotic release of a peptide effector substance without regard for the particular morphological associations of the secretory neurons.

Although this definition represents a move toward a more inclusive perspective, our review leans heavily on results derived from neurosecretory systems in which secretion is endocrine. This is a practical decision based on the fact that such systems have proven more tractable to experimental investigation and therefore are better understood than neurosecretory systems acting in synaptic pathways. Even this restriction of focus does not reduce the relevant literature to manageable dimensions; hence, the extensive reviews by Gabe,[5] Sachs,[6] Cross et al.,[7] Mason and Bern,[3] Berlind,[9] and the heuristic overview by Maddrell and Nordmann[8] should be consulted for more inclusive treatments of neurosecretion as it is traditionally defined.

2. MORPHOLOGY

Since structure reflects function, the morphological features of neurosecretory cells that characterize them will be found in those aspects that serve unique functional roles. By taking the position that neurosecretory cells are the class of neurons that secretes peptides, it is unlikely that investigation of whole-cell morphology will reveal unique features of neurosecretory cell structure. Maddrell and Nordmann[8] propose, on the basis of a definition that limits neurosecretion to neuroendocrine cells, that neurosecretory cells are characterized by much more profuse axonal branching than is found in other neuron types. In their view, the extensive arborization serves to multiply greatly the number of secretory terminals and, in this way, to permit adequate quantities of peptide to be released into the large circulatory volume that must be affected.

If our broader definition is adopted, the convenience of this structural differentiation is lost. Physiological and anatomic evidence strongly suggests conventional synaptic roles for some peptides.[10-13] In such cases, extensive axonal branching, if it exists, would be in service of a broadly distributed synaptic field and the result of circuit specifications like those met by other synaptically active neurons. Thus, in the context of this chapter, neurosecretory cells can be considered to represent the full range of morphologies from the relatively simple synaptic model to the highly branched instances encountered in neuroendocrine cells. The key consideration at the whole-cell level is that of functional role, not some intrinsic property of the condition of being peptidergic. This point is made by consideration of the general morphologies of the neuroendocrine models on which we are concentrating our attention.

The bag cell neurons of the mollusk *Aplysia californica* are clustered in two masses of 250–400 cells each on the ventral surface of the pleurovisceral connective nerves as they emerge from the parietovisceral (abdominal) ganglion.[14] The individual cells are as large as 75 μm in diameter in mature animals and are multipolar.[15,16] Since most molluscan neurons are monopolar, the appearance of several neuritic projections from the bag cells represents an unusual situation. However, along with further branching of the neurites, bag cell multipolarity appears to represent a specialization to insure substantial investment of the diffusely organized hemal contact zone with large numbers of secretory terminals. Coggeshall,[14] Kaczmarek *et al.*,[15] and Haskins *et al.*,[16] have described a region of close intertwining of bag cell neurites just rostral to the clusters that may represent the region of electrical synaptic contact known to exist from electrophysiologic study.[17,18]

The neurohypophyseal neurons of mammals display a quite different functional morphology. Cell somata are grouped in the paraventricular and supraoptic areas. Each cell appears to bear several dendrites, and these, as well as the soma, are decorated with smaller projections (spines). A single axon emerges from each cell and courses with others through the median eminence as a tract that then terminates in the neurohypophysis. These axons throw off numerous branches as they enter the neurohypophysis. The extent of this branching can be appreciated from the estimate that approximately 2000 terminals are formed from each axon.[8]

An interesting morphological feature associated with neurohypophyseal neurons is periodic dilations along the axons, especially near the terminals. In their most developed form, these swellings appear as extended diverticula.[19,20] They therefore, differ from the *en passant* varicosities noted in a variety of aminergic and peptidergic neurons. At the ultrastructural level, the swellings (Herring bodies) in neurohypophyseal neurons are seen to be filled with characteristic neurosecretory granules and often to contain multivesicular bodies as well. Consequently, these swellings are thought to be storage and reclaimation areas for vesicles and their contents.[21-23] Similar putative turnover sites have been observed in caudal dorsal cell axons in *Lymnaea*[24] and the cells of the caudal neurosecretory system in teleosts.[25] Thus, they may represent a common means for segretating the readily secretable from the "older" neurosecretory substances. Whether this segregation is in the service of recycling or of warehousing against the advent of conditions of extreme secretory demand is not clear. Both ends could be served. Aside from this physiological question, the presence of these granule-filled swellings raises the more fundamental issue of organellar sequestration. It would be extremely interesting to learn what changes occur in the membranes of granules that result in their reassignment to storage, or degradation, sites separate from the active secretory zone.

It is at the subcellular level that neurosecretory cells become readily distinguishable from conventional transmitter-secreting neurons. Neurosecretory cells are specialized to produce a peptide secretory product. For unknown reasons, but perhaps because of regulatory requirements for relative intimacy between the nucleus and sites of protein synthesis, free ribosomes, rough endoplasmic reticula, and Golgi structures are essentially confined to perikaryal regions in neurons. Hence, the provisioning of the remote secretory terminals with secretory product in peptidergic cells is sustained by activity in the perikaryal cytoplasm and membrane systems.

Although the same might be said for cells secreting conventional transmitters, there is surely a quantitative difference of note. Whereas the perikaryal protein synthetic and processing machinery is essential for the provisioning of synthetic enzymes and new membrane to the secretory terminals in conventional neurons, neurotransmitter synthesis, vesicular enclosure, and membrane recycling operate semiautonomously in the terminal specializations.[26] Neurosecretory cells thus appear much more active in terms of protein synthesis: much of the perikaryal cytoplasm is filled with rough endoplasmic reticulum; Golgi structures are well-developed and numerous; and it is characteristic for the perikaryon to be populated with electron-opaque neurosecretory granules. These features are consistent with the general model described by Palade,[27] Steiner et al.,[28] and Blobel[29] for the production of proteins destined for export from cells. Such proteins are translated from mRNA associated with the ribosomes of the rough endoplasmic reticulum. Coincident with their synthesis, they pass into the intralamellar cisternae through which they move to the Golgi stacks. Passage through the Golgi results in their segregation, chemical modification, and enclosure in vesicles.[30] In the case of neurosecretory cells, the vesicles then enter axoplasmic transport.

The abundance of 150- to 300-nm vesicles that typify neurosecretory cells at the ultrastructural level is consistent with the blue–white reflectance seen with the light microscope on epiillumination. The vesicle diameter is sufficiently large to reflect light at the blue end of the visible spectrum.[31] Despite the predominance of a particular size class of vesicles in neurosecretory cells, heterogeneity of vesicles is common.[32,33] Both larger and smaller size classes have been reported to coexist with typical neurosecretory granules. Heterogeneity with respect to electron opacity has also been noted.[34] The physiological significance of these variations remains unclear.

Nonetheless, it can be suggested that variations in opacity and apparent structuring within vesicle matrices may be largely a result of variation in the conditions and success of fixation and staining procedures employed.[35] Variations in size, on the other hand, could reflect different functional states of the typical granule or may represent altogether different organelles. For example, vesicles larger than the typical neurosecretory granule may be primary lysosomes, which play a role in granule breakdown. Multivesicular bodies are seen in both the perikaryom and neurites of the bag cells and may represent the degradative phase of vesicle turnover. Smaller vesicles, which seem more characteristic of terminal regions than perikarya, present other possibilities. One of these, consistent with observations in other neurons,[36] is that the smaller vesicles represent the means of membrane recycling after exocytotic episodes.[37,38] Evidence has also been developed indicating that "microvesicles" are sites of Ca^{2+} sequestration.[39]

Still another possibility of considerable physiological interest has been supported by the finding that several populations of peptide-containing neurons also contain conventional neurotransmitters.[11,40] Although it is too early to tell how general such situations may be, it could be that peptidergic neurons routinely secrete, in addition to their nominal product, a conventional transmitter. Needless to say, besides accounting in part for vesicle heterogeneity in secretory terminals, such findings open up a wider range of regulatory possibilities while further blurring functional distinctions among neuron types.

3. ELECTROPHYSIOLOGY

Because of the "glandular" character originally ascribed to neurosecretory cells, there was uncertainty about the extent to which they resembled other neurons electrophysiologically. In retrospect, there is an irony in this concern, since, with the advent of microelectrode technology, not only have numerous neurosecretory cells been confirmed to produce regenerative action potentials in essentially the same fashion as other neurons,[3] but several nonneural glandular cells are known to display neuronlike electrophysiology as well.[41] Thus, it appears that specialization for secretion entails differentiation of the membrane mechanisms that underlie regenerative potential shifts irrespective of the developmental history and anatomic fate of the cell.

Over the years since the first intracellular recordings from neurosecretory cells,[42,43] attention has been drawn to the difference in duration between action potentials in neurosecretory cells and those in other neuron types.[3] Typical durations in nonneurosecretory cells are in the range of 1–5 ms, but it is not uncommon to find durations greater than 10 ms in neurosecretory cells.[44–47] Although this was puzzling for some time, it now appears that the likely cause of this discrepancy is the existence of a significant, late developing increase in Ca^{2+} conductance during the action potential in neurosecretory cells. For example, during a burst of impulses in the bag cells, there is a progressive broadening of the individual spikes and the development of a prominent inflection on the falling phase of the action potential wave form. Treatment with tetrodotoxin (TTX, a Na^+-channel blocker) abolishes the earliest and fastest component of the depolarization but has no effect on the significant depolarization underlying the inflection on the falling phase. This more slowly developing and prolonged depolarization is sensitive to blockade with Co^{2+} and is therefore considered to be a function of a Ca^{2+} inward current.[48]

The significance of mixed Na^+ and Ca^{2+} inward currents and the progressive enhancement of the Ca^{2+}-associated component during repetitive firing of neurosecretory cells is not known. It is possible that these phenomena represent features of membrane dynamics peculiar to the physiological roles of the systems that have been studied most extensively. These systems are neuroendocrine and characteristically secrete phasically, with variable and sometimes long intervals between secretory episodes. It may be that enhanced Ca^{2+} entry during a phasic discharge is necessary to insure that an adequately large amount of secretory product is released into the circulatory volume. Electrophysiological studies of molluscan synaptic transmission have shown that the duration of terminal depolarization is positively correlated with the amount of transmitter released.[49] A related observation pertinent to neuroendocrine function may be the enhancement of oxytocin secretion from rat neurohypophysis *in vitro* when stimulation frequency is increased while holding stimulus number constant.[50] Higher frequencies of impulse arrival at secretory terminals may facilitate Ca^{2+} entry and thereby hormone secretion. The situation may be analogous to that studied in the crustacean X organ–sinus gland neuroendocrine system. Propagation of impulses in the secretory axons appears to be strictly Na^+ dependent, yet the terminals display progressive impulse broadening that is sensitive to treatments that affect Ca^{2+} permeability.[51]

Although Ca^{2+} is clearly a requirement for secretion,[52-56] the role it is playing in these instances may be regulatory with respect to other membrane conductances as well. Prolongation of repolarization (i.e., spike broadening) may be a result of both enhanced Ca^{2+} conductance[57-59] and Ca^{2+}-mediated inactivation of the repolarizing, voltage-dependent K^+ current.[60,61] Calcium would, in this case, be acting as part of a positive feedback loop. Since the Ca^{2+} channel is voltage dependent, Ca^{2+} conductance will be sustained while the cell is depolarized. If a Ca^{2+}-associated process results in partial inactivation of the rectifying K^+ outward current, depolarization will be prolonged, i.e., spike broadening will occur, and more Ca^{2+} can enter the terminal cytoplasm. Moreover, repolarization may not be complete before the rapid depolization mechanisms once again can be activated. The result will be earlier firing of another action potential and, by iteration, the generation of a burst of impulses.

Calcium may also play a role in restablizing the membrane. Intracellular injection of Ca^{2+} has been shown to produce a hyperpolarization that is carried by K^+.[62] Thus, in addition to Ca^{2+}-inactivated K^+ channels, neurons in a wide variety of organisms appear to possess Ca^{2+}-activated K^+ channels.[63] It is probable, therefore, that the deep and long-lasting hyperpolarization that terminates bursts in some neurosecretory cells[64] is brought about by the Ca^{2+}-mediated activation of a K^+ conductance mechanism with a high threshold for Ca^{2+} concentration. Calcium, in this way, could be both the crucial trigger to exocytosis and the regulatory signal that insures that the neuron or terminal fires repetitively over a confined period of time, thereby producing a rapid build-up of secretory product in the extracellular fluids.

The behavior of individual neurons can be amplified by ensemble activity as well. In both the bag cells and the neurohypophyseal neurons, large numbers of cells are active simultaneously. In the case of the neurohypophyseal cells, this ensemble activity may be the product of common presynaptic input to the activated group.[65] For the bag cells, by contrast, brief synaptic input leads to a recruitment of impulse production in the cell population through mutual interaction at electrical junctions.[17,66]

It remains to be determined if neurosecretory cells acting in conventional synaptic pathways display prominent Ca^{2+} currents and the associated tendency to enter bursting modes of firing. If they do, an association between enhanced Ca^{2+} conductance and secretion of peptides specifically will have to be sought. Alternatively, the evidence from other neuron types[67] suggests that what differentiates neurosecretory cell membrane properties is not the uniqueness of the Ca^{2+} contribution to the impulse waveform but rather its more pronounced development.

4. SYNTHESIS OF NEUROSECRETORY PRODUCTS

Most of the detailed information about hormone biosynthesis in neurosecretory cells comes from studies on the bag cells in *Aplysia* and the magnocellular neurons in the hypothalamus of mammals. Aside from the fact that neurosecretory cells possess axons, there is little difference in principle between the biosynthetic mechanisms of these cells and those of other nonneural secretory cells.[27–29] In general, secreted peptides and proteins are synthesized on ribosomes associated with rough endoplasmic reticulum (RER) as larger precursor forms known as preproproteins, polyproteins, or, more specifically, preprohormones. These preproproteins contain amino acid sequences at their N-termini that are used as "signals" to direct them into the cisternae of the RER, where the "signal sequences" are then removed by enzymatic cleavage. The resulting proprotein (or prohormone) is then routed through the cell's organelle systems (i.e., smooth ER, transition vesicles, and Golgi apparatus) for appropriate assortment and posttranslational modification. The ultimate intracellular destination is one of the vesicle types that subserve the functions of intracellular metabolism, membrane reconstitution, and secretion.[78–81] It is now generally agreed that in addition to transcription and translation processes, various signal, or topogenic,[29] sequences and posttranslational processing events are further involved in determining the particular fates of peptides within the cell. Nonetheless, the specific nature of this molecular information, its generality, and the means by which it is decoded remain unknown.

Several sophisticated experimental paradigms have been developed to identify and characterize preproproteins in addition to the simple and fruitful pulse-chase procedures originally applied to the study of proinsulin.[82] The approaches include: (1) demonstration in classical pulse-chase experiments in intact cellular systems (*in vivo* or *in situ*) that the putative precursor and products are directly related at the molecular level by conservation of radioactivity, quantitative immunoprecipitation, and peptide mapping procedures; (2) demonstration by *in vitro*, cell-free translation of mRNA purified from the synthesizing tissue and incubated in the presence of heterologous translation systems (e.g., wheat germ and reticulocyte lysates), that a precursor form of the peptide is synthesized. These experiments can also provide information about the existence of signal sequences of prohormones and, when performed in the presence of microsomal membranes, may permit demonstration of several posttranslational processes (e.g., signal sequence cleavage, glycosylation, acetylation, etc.). (3) Examination with recombinant DNA techniques of the sequence order of the preprohormone as well as nucleotide sequence of the transcriptional unit. Complementary DNA (cDNA) can be produced from a purified mRNA template, cloned, and then sequenced or used as a probe for

Table I
Properties of Oxytocin and Vasopressin Precursors

Precursor	Molecular weight (by SDS gel analysis)	Constituents		
		Nonapeptide	Neurophysin	Carbohydrate
Rat (in vivo)				
Prohormone				
Vasopressin	19,500	+	+	+
Oxytocin	15,000	+	+	−
Bovine (in vitro)				
Preprohormone				
Vasopressin	21,000	+	+	−
(+ membranes and tunicamycin)	19,000	+	+	−
(+ membranes only)	23,000	+	+	+
Oxytocin	16,500	+	+	−
(+ membranes and ± tunicamycin)	15,500	+	+	−

identification of the preprohormone gene. Each of these experimental approaches has been applied to the bag cell and magnocellular neurosecretory systems.

Experiments pertaining to the biosynthesis of neuropeptide hormones were initiated some 18 years ago by Howard Sachs and his colleagues, who concluded that vasopressin and its associated carrier protein, neurophysin, were synthesized together as part of a large precursor protein and then cleaved enzymatically to the final peptide products.[83-85] These studies preceded the discovery of proinsulin by 3 years and can be considered the intellectual precursors of modern prohormone research. Because of the technological limitations of the time, Sachs and his associates were unable to identify the hypothesized vasopressin precursors. However, recent work has confirmed precursor existence, and the full amino acid sequence of the bovine vasopressin precursor is now known.[86]

Application of modern techniques to in vivo studies has allowed the identification of separate precursors (ca. 20,000 daltons) for vasopressin- and oxytocin-associated neurophysins synthesized by the rat supraoptic nucleus.[87-91] The use of the Brattleboro rat (hereditary diabetes insipidus) model system for in vivo biosynthesis and axonal transport studies, pioneered by Pickering and his colleagues,[92-94] was invaluable in establishing which of these precursors was related to vasopressin. Based on these data, and subsequent work on the neurophysin precursors synthesized in vivo,[95-97] it has been possible to provide evidence that these precursors (two forms for vasopressin, pI 5.6 and 6.1, and two forms for oxytocin, pI 5.1 and 5.4) were indeed common precursors for the neurophysins and nonapeptides as had been hypothesized by Sachs' group.

Some of the properties of these precursors are summarized in Table I. From these data, it can be concluded that the vasopressin precursor contains three peptide moieties: the vasopressin sequence, the vasopressin-associated neurophysin sequence, and a glycopeptide of less than 10,000 molecular weight, in this order from amino to carboxyl terminus.[97]

An independent line of evidence based on *in vitro* translation techniques has led to similar conclusions (Table I). Several laboratories[98-100] have shown that mRNA isolated from bovine, mouse, or rat hypothalami can serve as templates in cell-free translation systems to yield 20- to 25,000-dalton proteins that are precipitable by antibodies raised against neurophysins. In the most extensive series of cell-free translation experiments,[100-106] it has been shown (1) that bovine hypothalamic mRNA codes for separate common precursors for vasopressin and neurophysin II and for oxytocin and neurophysin I; (2) that inclusion of dog liver microsomes in the reticulocyte translation cocktail results in the conversion of both precursors from preprohormones to prohormones, although only the vasopressin prohormone becomes glycosylated; (3) that the *in vitro* translated vasopressin preprohormone has a molecular weight (on SDS gels) of 21,000, is cleaved to a 19,000-dalton prohormone when translation is performed in the presence of microsomes and tunicamycin (a drug that prevents glycosylation), but is glycosylated to a molecular weight of 23,000 in the presence of microsomes without tunicamycin; (4) that the oxytoxin preprohormone is about 16,500 daltons and is reduced to a 15,500-molecular-weight prohormone species by microsomes with or without tunicamycin; and (5) that tryptic mapping of the vasopressin preprohormone and prohormone indicates that the vasopressin follows the signal sequence and precedes the neurophysin II in the preprohormone.

The entire nucleotide sequence of the cloned cDNA encoding the bovine vasopressin neurophysin II preprohormone has recently been reported.[86] The corresponding amino acid sequence contains 166 amino acids, of which 19 appear to belong to the signal sequence. The order of the peptide components in the prohormone is the same as that which was predicted from the *in vivo* and *in vitro* experiments described above. The signal sequence, which begins with Met_{-19} at the N-terminus, has a central hydrophobic region and terminates with Ala_{-1}, which is immediately followed by the arginine vasopressin sequence. The nonapeptide sequence is separated from the neurophysin II sequence by Gly_{10}-Lys_{11}-Arg_{12}, which in turn is separated from the 39-amino-acid C-terminal peptide by a single Arg_{108}. The terminal 39-residue peptide is believed to be the glycopeptide moiety discussed earlier, since it contains a characteristic sequence, Asn_{114}-Ala_{115}-Thr_{116}, typical of N-asparagine-linked glycopeptides.

It was expected that the peptide components would be separated by basic amino acid residues from what has been found in other prohormones and from the results of previous tryptic mapping studies. However, although the Gly_{10}-Lys_{11}-Arg_{12} sequence interposed between vasopressin and neurophysin II is typical of prohormone cleavage sites where amidation also occurs, the single basic residue, Arg_{108}, between neurophysin II and the glycopeptide is not characteristic of other prohormone cleavage sites. These usually contain pairs of basic amino acids. Curiously, the Arg-Arg site in positions 105–106, which would be an expected cleavage site, appears not to be recognized by the con-

```
              1                  5                   10
         NH₂-Ala-Asn-Asp-Arg-Ser-Asn-Ala-Thr-Leu-Leu-
            (109)            (113)               (118)


              11                 15                  20
             Asp-Gly-Pro-Ser-Gly-Ala-Leu-Leu-Leu-Arg-
            (119)            (123)               (128)


              21                 25                  30
             Leu-Val-Gln-Leu-Ala-Gly-Ala-Pro-Glu-Pro-
            (129)            (133)               (138)


              31                 35                  39
             Ala-Glu-Pro-Ala-Gln-Pro-Gly-Val-Tyr-COOH
            (139)            (143)               (147)
```

Fig. 1. Amino acid sequence of the glycopeptide in the bovine AVP-neurophysin II precursor. The amino acid order of the naturally found bovine glycopeptide[108] is shown numbered 1-39; the numbers in parentheses show the corresponding positions in the prohormone.[86]

verting enzyme. The peptide bond between these residues is intact in bovine neurophysin II *in situ*[107]

The amino acid sequence of the glycopeptide is shown in Fig. 1. A glycopeptide with the identical amino acid sequence had previously been extracted from bovine pituitary, isolated, and sequenced by Smyth and Massey.[108] Similar glycopeptides with remarkable conservation of sequences were also found by these authors in pig and sheep pituitaries. A smaller, 17-amino-acid glycopeptide with exact correspondence to the 1-17 sequence in Fig. 2. (except for Ser at position 2) was isolated earlier from pig posterior pituitary by Holwerda.[109] In addition to the 1-17 and 1-39 glycopeptide sequences, several other fragments have been found. These correspond to sequences 1-10, 1-19, 13-39, and 26-39. Although these fragments have been interpreted to be "naturally occurring" peptides,[108] it remains to be determined whether they are formed during the biosynthetic process *in vivo*, represent natural degradative products, or are produced by proteolytic attack during the isolation procedures.

Although much attention, with respect to neurohormone biosynthesis, has been focused on oxytocin and vasopressin, a parallel development has occurred in the study of biosynthesis of the egg-laying hormone (ELH) in the bag cell neurons of *Aplysia*. In 1972, Arch[110] reported that the appearance of ELH in pulse-chase studies using isolated bag cell clusters was preceded by the biosynthesis of a 29,000-dalton precursor protein. Over the past 10 years, extensive biosynthesis studies have been made using this neurosecretory system.[56,110-116] The amino acid sequence of the ELH has been reported,[117] and a molecular weight of 4385 has been calculated from this sequence. It is therefore apparent

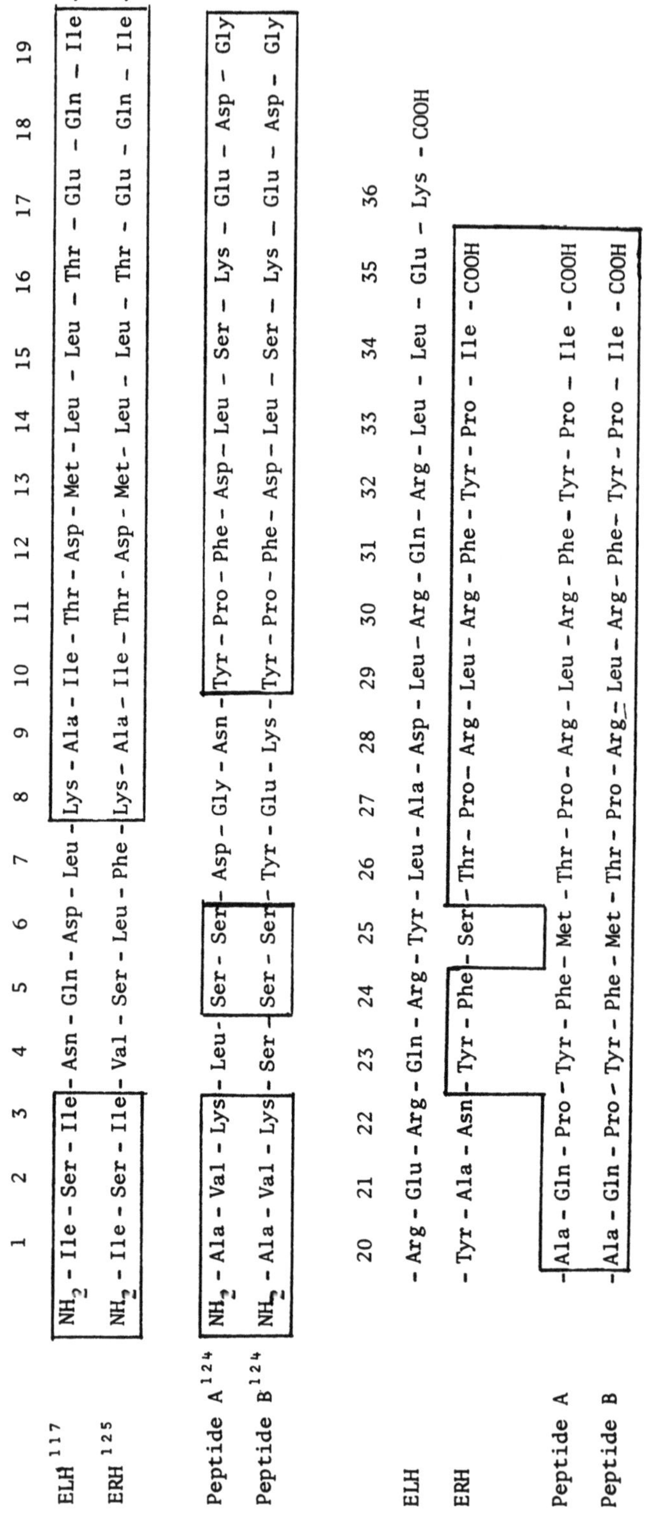

	1	2	3	4	5	6	7	8	9	10	11	12	13	14	15	16	17	18	19
ELH[117]	NH$_2$– Ile	– Ser	– Ile	– Asn	– Gln	– Asp	– Leu	– Lys	– Ala	– Ile	– Thr	– Asp	– Met	– Leu	– Thr	– Glu	– Gln	– Ile –
ERH[125]	NH$_2$– Ile	– Ser	– Ile	– Val	– Ser	– Leu	– Phe	– Lys	– Ala	– Ile	– Thr	– Asp	– Met	– Leu	– Thr	– Glu	– Gln	– Ile –
Peptide A[124]	NH$_2$– Ala	– Val	– Lys	– Leu	– Ser	– Ser	– Asp	– Gly	– Asn	– Tyr	– Pro	– Phe	– Asp	– Leu	– Ser	– Lys	– Glu	– Asp	– Gly
Peptide B[124]	NH$_2$– Ala	– Val	– Lys	– Ser	– Ser	– Ser	– Tyr	– Glu	– Lys	– Tyr	– Pro	– Phe	– Asp	– Leu	– Ser	– Lys	– Glu	– Asp	– Gly

	20	21	22	23	24	25	26	27	28	29	30	31	32	33	34	35	36
ELH	– Arg	– Glu	– Arg	– Gln	– Arg	– Tyr	– Leu	– Ala	– Asp	– Leu	– Arg	– Gln	– Arg	– Leu	– Leu	– Glu	– Lys – COOH
ERH	– Tyr	– Ala	– Asn	– Tyr	– Phe	– Ser	– Thr	– Pro	– Arg	– Leu	– Arg	– Phe	– Tyr	– Pro	– Ile – COOH		
Peptide A	– Ala	– Gln	– Pro	– Tyr	– Phe	– Met	– Thr	– Pro	– Arg	– Leu	– Arg	– Phe	– Tyr	– Pro	– Ile – COOH		
Peptide B	– Ala	– Gln	– Pro	– Tyr	– Phe	– Met	– Thr	– Pro	– Arg	– Leu	– Arg	– Phe	– Tyr	– Pro	– Ile – COOH		

Fig. 2. Amino acid sequences of ELH-related peptides. Boxes indicate homologies.

that the ELH forms only a small fraction of the 29,000-dalton prohormone. Evidence for the existence of additional peptides characteristic of the bag cells is available. Earlier studies[110-112] led to the inference that an acidic peptide (AP; pI 4.8) of approximately the same size as ELH (pI 9.3) is produced from the same prohormone. This peptide is transported and secreted from the bag cells.[118,119] At least three more peptides are typically recovered from medium into which the bag cells have been caused to secrete.[118] It is tempting to speculate that each of these is derived from the prohormone, but neither the characterization of the prohormone species nor that of the secreted species (except ELH and AP) is adequate to support such a speculation.

Recent *in vitro* translation and recombinant DNA studies by Scheller *et al.*[120] have confirmed the identity of an ELH precursor in the molecular weight range reported from *in situ* studies. Moreover, the cloned coding sequences provide the possibility for generating amino acid sequence information about candidate peptides in addition to ELH and AP in the prohormone.

Two other points of interest derive from these studies. First, ELH and several other peptides in the ELH precursor are bounded by pairs of basic amino acids, suggesting a specificity, similar to the case of vertebrate prohormones, for the prohormone-converting enzymes involved in posttranslational processing. Second, the *Aplysia* genome appears to contain at least five similar but distinct coding regions that contain the ELH sequence. The latter observation suggests that there is a gene family containing the ELH sequence. Since restriction mapping has shown that each of the putative genes differs outside the ELH sequence, translation and posttranslational processing should give rise to different but overlapping sets of peptides from each of these genes if they are expressed. The possibility thus exists, as Scheller and his associates have noted,[120] that differing physiological conditions may lead to alterations of the peptide output of the bag cells. It is also possible that various members of the ELH gene family are expressed in other tissues with dissimilar functional roles.

Figure 2 shows the amino acid sequences of several peptides in *Aplysia* that are found to occur naturally and to produce egg laying when injected into mature bioassay animals. Only ELH has been identified in the bag cells. The other three peptides have been isolated from the atrial gland. This glandular epithelial tissue is associated with the reproductive tract, but its function appears to be exocrine rather than endocrine.[121-123] Atrial gland peptides A and B cause egg laying, when bioassayed, by activating the bag cells to release ELH.[124] The peptide designated ERH can also act in this manner, although it shares with ELH the capacity to stimulate egg laying in animals from which the bag cells have been removed.[125] The sequence homologies between ELH from the bag cells and ERH from the atrial gland, on the one hand, and between ERH and A and B peptides within the atrial gland, on the other hand, suggest that the peptide complement of the atrial gland may represent a case of selective expression of one or more members of the ELH gene family. Since Scheller *et al.*[120] reported more than one *in vitro* translation product from bag cell mRNA that was reactive to an ELH antiserum, it may be possible as well to identify atrial gland peptide sequences in the bag cells.

The significant sequence homology between vasopressin and oxytocin (Table II) of the vertebrate neurohypophyseal system likewise indicates the existence of a gene family. This inference is further supported by the apparent sequence conservation in the neurophysins.[107] Cross-phyletic examinations of nonapeptide moieties (Table II) have prompted efforts toward a reconstruction of the evolutionary history of these sequences.[126] It is now generally believed that arginine vasotocin was the first vertebrate nonapeptide and that subsequent molecular events led to the evolution of separate vasopressinlike and oxytocinlike lines of nonapeptides. Similarly, detailed examination of neurophysin sequences would afford a still deeper insight into the history of the genes responsible for expression of the functional nonapeptides.

The emergence of a biologically active peptide in a biosynthetic process requires more than the elaboration and translation of a distinct mRNA. Several posttranslational modifications are necessary to fashion the final peptide products before secretion. Excellent sources of general information about these processes are available,[79-81] and only a brief commentary is offered here. The first event following translation of the preproprotein on the RER involves a signal protease ("signalase"), which cleaves off the signal peptide. The resultant proprotein can then be subjected to a variety of other processes within the RER. These include disulfide bond formation (of considerable importance in the vasopressin and oxytocin precursors) and the initial stages of N-asparagine-linked glycosylation, in which a core of sugars (glucosamine and mannose) is put on the precursor at Asp-X-Ser or Asp-X-Thr residues. The later stages of glycosylation, including trimming the core sugars and adding others, occur in the Golgi. Other posttranslational modifications include further proteolysis, phosphorylation, sulfation, methylation, amidation (at C-terminals), and acetylation (at N-terminals). Little is known, especially in neurosecretory cells, about the cell biology and enzymology of these processes, and these issues remain important subjects for future study.

What is clear is that the amidation and acetylation steps must occur following enzymatic cleavage of the precursor into its peptide products. The structural information available from a wide variety of proproteins indicates that the first cleavage enzyme is "trypsinlike," in that it has a specificity for basic amino acids (i.e., Lys, Arg). However, the enzyme need not be otherwise similar to pancreatic trypsin. Another enzyme implicated in proteolytic processing is a carboxypeptidase-B-like enzyme, which trims the remaining basic amino acid residue at the C-terminus.

Little is known about these converting enzymes in neurosecretory cells, and none has been purified to homogeneity. However, some insight about the location of converting enzyme activity can be gathered from axonal transport studies in the neurohypophyseal system. These studies show that most of the precursor is transported into the axon, where it is converted to the final peptides during axonal transport.[87,88,91] Since the neurosecretory vesicles are the principal vehicles for the axonal transport of the neurophysins, vasopressin, and oxytocin, it would follow logically that these organelles are the sites of prohormone conversion. A similar proposal based on other data was advanced earlier by Sachs's group.[85] Localization of the conversion process in secretory

Table II
Amino Acid Sequences of the Neurohypophyseal Hormone Family

Hormone	Sequence									Phylogenetic occurrence
	NH$_2$-1 Cys	2 Tyr	3 Ile	4 Gln	5 Asn	6 Cys	7 Pro	8 Arg	9 Gly—NH$_2$	
Arginine vasotocin (AVT)	—	—	—	—	—	⌐	—	—	—	All lower vertebrates (including cyclostomes), mammals (?)
Mesotocin (MT)	—	—	—	—	—	—	—	Ile	—	All nonmammalian tetrapods and lungfish.
Arginine vasopressin (AVP)	—	—	Phe	—	—	—	—	—	—	All mammals except pig family
Oxytocin (OT)	—	—	—	—	—	—	—	Leu	—	All mammals
Valitocin (VT)	—	—	—	—	—	—	—	Val	—	Sharks
Lysine vasopressin (LVP)	—	—	Phe	—	—	—	—	Lys	—	Pig family
Isotocin (IT)	—	—	—	Ser	—	—	—	Ile	—	Teleosts
Glumitocin (GT)	—	—	—	Ser	—	—	—	Gln	—	Rays
Aspartocin (AT)	—	—	—	Asn	—	—	—	Leu	—	Sharks

vesicles has also been proposed in the biosynthesis of insulin, glucagon, and somatostatin in islet cells of the pancreas.[127,128]

Given the above hypothesis that the secretory vesicle is the major site for enzymatic cleavage of the prohormone, one would expect that appropriate converting enzymes should be present in the vesicles. Recent experiments indicate that this is the case for a variety of tissues. Converting enzyme activity has been detected and partially characterized in secretory vesicles isolated from anglerfish pancreas islet cells,[127,128] neural and intermediate lobes of the rat pituitary,[129,130] and bovine posterior pituitary.[131] Analysis of the converting enzyme activities found in all of these secretory vesicles indicates that they are acid, thiol proteases with specificities of cleavage at pairs of basic amino acids. Thus, they differ from trypsin in their pH ranges of activity, their specificities of cleavage, and their enzyme inhibitor profiles (i.e., they are thiol proteases and not serine proteases) and are more like but still distinct from cathepsin B.

Since molecular biological study is beginning to reveal the kinds of enzymatic steps that are required to reduce a preprohormone to its functional components, specific enzymatic activities can be sought with more precision. It is already evident that more than one enzyme species is necessary to achieve primary cleavage, amino- or carboxy-terminal trimming, and acetylation, amidation, and phosphorylation. If all of these processes are taking place within the secretory vesicles, it is clearly necessary to begin thinking of these organelles as topologically differentiated structures and not simply as relatively inert packages for high concentrations of secretory product.

The peptidergic neuron is unique among the cell types specialized to secrete proteins in that the release site is generally a considerable distance from the site of secretory product synthesis. Moreover, the transport mechanism for delivery of the secretory product appears to operate independently of other processes such as electrical activity and secretory output in the cell. Thus, at least in principle, the existence of regulatory processes to adjust hormone synthesis to secretory demand might be expected. This is, however, an empirical issue, since the possibility that biosynthesis and secretion are not coupled cannot be excluded *a priori*. Evidence consistent with the occurrence of increased synthesis of hormonal product following increased secretory output has been obtained in both the neurohypophyseal and bag cell systems.

Rats maintained in a dehydrated state for 4–5 days exhibit nearly complete (90%) depletion of vasopressin and oxytocin.[132] Under comparable conditions, neurohypophyseal prohormone synthesis and processing have been reported to increase by a factor of 3 to 5.[84,87,88,133] Experiments performed on isolated bag cell organs induced to large-scale secretion by medium containing an elevated potassium concentration have revealed a 25–30% increase in prohormone synthesis that persists for several hours. Stimulation of synthesis appears to be selective for the ELH prohormone and to occur only under conditions in which secretion is induced.[134] The homeostatic function of this process would seem best served by a mechanism coupling hormone secretion to hormone synthesis; however, in neither experimental system has it been possible to separate hormone secretion from enhanced presynaptic transmitter secretion

5. AXONAL TRANSPORT

The phenomenon of axonal transport provides a unique tool with which the neurobiologist can interpret the cell biological organization of neurons. Since virtually all macromolecules in the axon and nerve terminals are derived from biosynthesis in the cell body and axonal transport, analysis of the molecular compositions of the various rate components of transport can be used for structure–function correlations. Since excellent reviews are available[68-70] on both the mechanisms and interpretations of axonal transport, only a brief statement about these issues is offered here. Orthograde transport can be divided conveniently into two general classes, the transport of (1) membrane-bounded organelles and (2) the cytoskeletal matrix. These classes, which can be further subdivided into various rate components, have been classified traditionally as fast and slow transport, respectively. The membrane-bounded organelle class moves at much more rapid rates (from 40 to 400 mm/day in mammals) than does the cytoskeletal protein (e.g., tubulin, intermediate filament, and actin) complex (0.1–6 mm/day). It has been argued that all fast transported proteins pass through the Golgi apparatus,[71] a view consistent with the fact that most organelles contain proteins that are glycosylated in the Golgi.

In addition to the orthograde transport mechanism, there is also a retrograde (toward the cell body) transport process. Retrograde transport (which moves at 50–70% the rate of orthograde fast transport) is usually measured and visualized by the uptake of exogenous radioiodinated tetanus toxin or nerve growth factor. This approach consequently biases towards the identification of membrane-bound vehicles (multivesicular bodies, tubular elements, etc.) in retrograde transport. Recent work using an alternative approach[72-74] has suggested that there may also be a slow component of retrograde transport, but this view is not widely accepted.

Evidence that neurosecretory proteins and peptides are synthesized in the magnocellular neuron perikarya and are intraaxonally transported to the terminals in the posterior pituitary within neurosecretory vesicles has been well documented by a variety of autoradiographic and biochemical techniques.[75,76] By making use of a micropunch technique for the dissection of small areas of rat brain, it has been possible to analyze some of the rate components of transport in the hypothalamo–neurohypophysial system of the rat.[77] These studies showed that the transport components behave similarly in neurosecretory cells as in other types of neurons, the only distinction being a relatively larger mass of protein, representing the neurosecretory vesicles, transported in the fast component (ca. 140 mm/day).

Morphological constraints have made the acquisition of reliable transport rate information more difficult in other neurosecretory systems. In any case, the available data do not point to correlations between transport rate and function of the secreted product. It appears unlikely as well that transport rate is linked to either secretion rate or membrane electrical activity.

6. SECRETION

The dominant morphological feature of neurosecretory cells is the abundance, on ultrastructural investigation, of 150- to 300-nm electron-opaque granules. These neurosecretory granules (NSG) are typically clustered in neuritic terminals either adjacent to other neurons or effectors or to hemal spaces. Biochemical, immunocytochemical, and autoradiographic studies have led to the generally held position that these granules are the repositories for the peptide products characteristic of this cell type. This conclusion is entirely consistent with that from studies of other neural and nonneural cells actively involved in secretion. Thus, although the possibility of release of neuropeptides by mechanisms other than vesicle-mediated exocytosis cannot be excluded logically, it seems most productive to proceed with the position that neurosecretory products are released into the extracellular space by fusion of the membrane of the NSG with the axolemma of the terminal.[135]

The most compelling indication of NSG–axolemma fusion is the "omega" profile seen in electron micrographs of secretory sites. Indeed, in some instances, electron-dense materials can be seen extending from the open vesicle matrix into the extracellular space. Despite the necessary caveat that such views are static and therefore not unambiguous, the most parsimonious interpretation is that the process of secretion of NSG contents has been visualized. The apparent simplicity of this solution to the problem of conveying a quantity of concentrated product from the inside to the outside of a cell masks important issues of mechanism. Secretory vesicles are thought to be negatively charged on their surfaces,[136–137] as, in all likelihood, is the inner face of the cell membrane. Hence, electrostatic repulsion must be overcome if the two membranes are to approach closely enough (0.5–1.0 nm) to permit fusion.[138] It is probable as well that waters of hydration bound both vesicle and plasma membranes, thereby imposing a further energetic barrier to apposition.[139–141] Recognition of these impediments to close membrane proximity must constrain models for the mechanics and chemistry of membrane fusion.

The ubiquity of the requirement for an elevation of cytoplasmic free Ca^{2+} has made Ca^{2+} dependence one of the defining criteria for a secretory event. Under physiological conditions, intracellular free Ca^{2+} is typically buffered at concentrations below 10^{-6} M. Results from studies employing direct intracellular injection of Ca^{2+} and of the Ca^{2+}-binding photoprotein aequorin[52,142,143] have permitted the estimate that transmitter secretion from the presynaptic element of the squid giant synapse depends on intraterminal free Ca^{2+} concentrations of 10^{-5} M or higher.[144,145] Since voltage-clamp studies on the same preparation have shown the existence of a voltage-dependent Ca^{2+} conductance,[146] the reasonable inference is that depolarization leads to Ca^{2+} channel activation and thus to the elevation of internal Ca^{2+} concentration.[147,148]

How the increase in free Ca^{2+} promotes secretion is not yet clear. It does seem likely, in view of the very rapid kinetics of secretory onset following depolarization, that Ca^{2+} action is confined to the region within a few hundred nanometers of the terminal membrane.[144] However, since the molecular machinery serving the process of secretion is likely to be concentrated in this area, this diffusional constraint may not limit the range of possible Ca^{2+} actions.[149] Among these, the reduction of electrostatic repulsion by charge screening[137] is appealing because of its simplicity but somewhat difficult to reconcile with the evidence for differences in divalent cation effectiveness in promoting secretion.[150]

More selective roles for Ca^{2+} have also been proposed. Typically, these involve either Ca^{2+} activation of binding sites on plasma and/or vesicle membranes or Ca^{2+}-dependent enzymatic activity. The Ca^{2+}-dependent aggregation of isolated chromaffin granules with one another and with plasma membrane in the presence of the protein synexin is an example of the former.[149,151] Although it remains to be determined if this aggregation phenomenon can be directly related to the physiological situation in intact tissues, there is evidence from structural studies of secretion that granule–granule fusion can occur before or just at the instant of secretory discharge.[123,152] Moreover, the Ca^{2+} requirement (200 μM) for *in vitro* aggregation is not inconsistent with levels that might be expected in the near vicinity of the secretory membrane.

With respect to Ca^{2+}-dependent enzymatic processes leading to membrane fusion, it has been argued that unusually high turnover times would be required to achieve the shortest-latency secretions.[138] However, this constraint could be relaxed in cases of prolonged bouts of secretion such as those encountered in neuroendocrine cells. In any case, it seems likely that, aside from charge screening, Ca^{2+} is probably involved in one or more selective interactions associated with exocytosis. Other divalent cations differ in the efficacy with which they can replace extracellular Ca^{2+}.[59] Barium ion and Sr^{2+} are effective Ca^{2+} substitutes, although they may exhibit different kinetics of action, whereas Mn^{2+}, and Co^{2+} have been shown to antagonize Ca^{2+}-mediated processes.[150,153,154] This sort of ion selectivity suggests the existence of specific binding sites. Nonetheless, despite the abundance of evidence for its crucial role in secretion, it is still necessary to state that we do not know what or how many things Ca^{2+} does.

The content of a NSG is not a uniform solution of a single peptide species. Inevitably, because the primary translation product is a precursor molecule, the nominal neuropeptide in any particular neurosecretory cell is probably accompanied by other peptides produced during precursor processing. Hence, the secretory output of neurosecretory cells is heterogeneous. Certainly, the most thoroughly documented case of this phenomenon is the cosecretion by neurohypophyseal terminals of oxytocin and vasopressin along with their respective neurophysins. Indeed, the fixed stoichiometry between the nonapeptides and their neurophysins over a range of secretion rates[55,155,156] constitutes a powerful confirmation of the exocytotic mechanism for secretion. It is now clear that the neurophysins constitute portions of the precursor molecules that are degraded to yield the nonapeptides (see previous section). Evidence that the neurophysins can bind the nonapeptides under appropriate ionic and pH conditions, conditions that are believed to exist within vesicles,[157] suggests that they may be of functional significance in osmotic stabilization of vesicles. It is not evident why this role would be essential for the NSG of the hypothalamo–neurohypophyseal system and not elsewhere as well. However, there is, at present, little evidence in support of the generality of carrier, or binding, proteins as constituents of NSGs in other neurosecretory systems.

Although other neurosecretory systems have been less thoroughly studied, there is evidence of cosecretion of more than one peptide. In cells producing proopiomelanocrotin (POMC), appropriate enzymatic cleavage could yield ACTH, MSH, β-endorphin, and Met-enkephalin.[158,159] Immunocytochemical study has revealed single hypothalamic neurons that contain the antigenic determinants for more than one of the possible cleavage products.[160] Because of difficulties imposed by heterogeneity in hypothalamic preparations and poor selectivity of stimulation pathways, it is not yet possible to assert that single neurons secrete more than one of these biologically active peptides. However, given the weight of the indirect data currently available, it would be surprising if they did not.

The bag cell neurons are known to secrete at least five separable peptide species.[118] The relationships among these are not known; however, there is evidence that at least ELH and an acidic peptide (AP) are derived from a common precursor.[112,115,116] The possibility exists that each secreted peptide is related to ELH but not directly to any of the others. This seemingly peculiar situation may arise from the existence of several separate coding sequences for ELH.[120] If each unique ELH-containing gene is expressed as a precursor that is then enzymatically processed, the net effect would be the production of ELH with differing ancillary products. The prominence of any one of these ancillary species might differ among animals and through time as a result of differential induction of the various mRNAs. A potentially useful approach to evaluating this hypothesis would be to examine peptide synthesis in single bag cell neurons to assess the degree to which there is idiotypic variation in the stoichiometry of secretory peptides.

Inclusion of the products of precursor breakdown in secretory granules suggests a degree of profligacy uncharacteristic of biological systems. However, this perception may be more apparent than real. If the POMC-producing neurons are capable of secreting more than one biologically active peptide, a new degree of regulatory control becomes possible. As physiological conditions change, different sets of processing enzymes might be activated to adjust the peptide content and stiochiometry of secretory output. Such a possibility is made obvious in this case because several of the potential products are already known to have physiological actions. In systems in which only one product has a known role, the appeal of this hypothesis for secretory plasticity will depend on the discovery of target actions for the cosecreted peptides. It would seem that secretion of ancillary peptides represents an ideal substrate for the action of natural selection in the development of chemical messengers. We might anticipate that continuing study will reveal hitherto undetected neuromodulatory and hormonal roles for the ancillary peptides secreted in nominally characterized neurosecretory systems.

The issue of multiple-effector secretion may already be resolved in one respect for several neurosecretory systems in which classical neurotransmitters coexist with identified peptides. Serotonin, epinephrine, norepinephrine, dopamine, and acetylcholine have each been found in association with a putative neuropeptide.[11,161] It does not appear, however, that any particular association is obligatory, since, for example, norepinephrine has been found in cells containing somatostatin, enkephalin, or neurotensin. Moreover, the distribution of strictly aminergic and cholinergic neurons is much more widespread than the distribution of cells containing both a peptide and a transmitter. The coexistence must therefore be important to a specific regulatory function and not be a general state of neurons. Caution is required in this respect, however, since it remains to be shown in all but a few cases that both peptide and transmitter are secreted and, if they are, what functions they serve. Indirect evidence for secretion is available from investigation of neurons in the raphe nuclei and dorsal horns, where immunoreactive substance P and serotonin have been shown to reside in the same dense-cored granules.[162] Proportional secretion of opioid peptides and catecholamines from chromaffin cells in culture has been reported.[163] Functional roles for presumably cosecreted vasoactive intestinal peptide and acetylcholine have been proposed for secretomotor cells of the autonomic system. Activity in such neurons is followed by atropine-sensitive exocrine secretion and atropine-resistant vasodilation.[11,164]

The probability of secretion of the contents of a NSG appears to change through time after its synthesis. In the hypothalamo–neurohypophyseal system, maximal secretion triggered *in vitro* can result in depletion of only about one-third the contents of the neural lobe.[38] Thus, there appears to be a limitation on the extent to which the pool of secretory product can be accessed. Both morphological and biochemical studies indicate that the releasable product is that most recently arrived from the perikaryon.[165] After a few days, this material enters a condition of much lower probability for release. This transition from high to low probability of release is most likely associated with the cycling of NSG from the active secretory terminals into the preterminal swellings, where they may remain for extended periods of time. Granules in the two areas are known to differ in buoyant density, average diameter, and osmotic lability.[166]

The relationship between these differences and the processes that route the aging NSG from secretory sites to swellings is not known. Since similarly detailed information is not available for other neurosecretory systems, the generality of the correlation between time of arrival and secretability cannot be determined. It is not evident, *prima facie,* that this is the only relationship between granule age and secretory access that can exist. A simple alternative would be some form of queuing in which newly arrived NSG take positions at the end of the line and become secretable only after older NSG have discharged. Studies of secretion in glandular tissues have not as yet disclosed any general relationship between granule "birthdate" and availability for secretion.

The final phase of the exocytotic secretory cycle is the reclamation of granule membrane from the axolemma. The necessity for such a process is dramatized by the estimate that NSG membrane would enlarge terminal plasma membrane area by 16% an hour under conditions of basal secretion alone in the neurohypophysis.[8] Clearly, much more rapid membrane expansion would occur with stimulated release. Presynaptic terminals at the vertebrate neuromuscular junction show pronounced folding after prolonged high-frequency stimulation.[36] It can be expected that high-intensity secretory activity in neurosecretory cells would result in similarly folded membrane profiles as the added membrane forced deformation of the terminal.

The mechanism for retrieval of membrane has been indicated in experiments employing superfusion of preparations with solutions containing markers for electron microscopic detection. Horseradish peroxidase (HRP), an impermeant enzyme, has been used frequently for this purpose, since its reaction products precipitate in place and are electron opaque. After bouts of secretion, HRP reaction products are typically found over electron-lucent membrane profiles in the terminal.[75] The most direct interpretation of such findings is that the HRP has been internalized from the medium by an endocytotic process. In some cases, endocytotic vesicles can be seen to be coated with a poorly resolved "fuzz."[167] This coating is probably the protein clathrin,[168] but, although its presence is strongly associated with the endocytotic process, its role in membrane retrieval is still a matter for speculation.

Once recovered, the fate of the membrane must be different in neurosecretory cells from the situation in nonpeptidergic neurons.[11] The enzymatic activities necessary for conventional neurotransmitter synthesis and vesicle enclosure are present in the presynaptic endings. Thus, membrane can be recycled locally. The requirement that secretory product synthesis and packaging take place in the perikaryon of peptidergic neurons places different demands on these cells. Either the reclaimed membrane must be degraded in the terminal, or it must be returned to the perikaryon by retrograde axonal transport, whereupon it can be recycled through the Golgi or broken down to provide substrates for new membrane assembly. There is evidence for the retrograde transport of HRP in the magnocellular neurons,[75] indicating that membrane is returned to the perikaryon. However, it has also been shown that membrane constituents may reside in secretory endings for long periods without appreciable retrograde transport.[169] It is conceivable that recovered membrane is dealt with differently depending on the economy of the cell at any particular time.

A more complex situation may exist in some neurosecretory cells secreting both a small transmitter species and a neuropeptide. Two distinct modes of membrane disposition may be in operation: local recycling for transmitter enclosure and long-distance recycling for peptide granule assembly. On the assumption that the chemistry of the membranes for the two types of vesicles is different in some respect, processes capable of recognizing unique membrane markers may be necessary to insure appropriate routing of the reclaimed structures.

7. CONCLUDING REMARKS

Understanding of the chemistry and physiology of the neurosecretory cell has advanced dramatically in the past decade. The greater attention such neurons are receiving is largely the result of the development of sophisticated and highly sensitive new analytical procedures for detection and characterization of proteins and peptides. Immunologic, high-performance chromatographic, and molecular biological techniques afford levels of specificity in localization, isolation, and characterization of proteinaceous species that earlier procedures simply could not approach. Perhaps not surprisingly, this refinement of empirical insight has not simplified our conception of the peptidergic neuron. It has, in fact, added new dimensions of complexity with which attempts at review and summary must deal.

The development of greater insight into the functional complexity of these cells is coincident with an appreciation of the possibilities for regulatory processes not previously expected. It is evident, for example, that NSGs contain more than one peptide species. Thus, neurosecretory cells secrete more than one peptide. What roles, if any, do the "extra" secreted peptides play in neuronal or metabolic physiology? Indeed, the existence of gene families suggests the possibility that the secreted peptide "profile" for a cell may change under changing conditions. This, in turn, would imply refined mechanisms for regulation of gene expression and posttranslational processing. Or, taking a different tack from the recognition that prohormones may contain a variety of potential individual peptide sequences, the possibility of more than one NSG type can be considered. As a result of selective assortment and concentration processes in the Golgi apparatus, NSG with different peptide contents may be produced. These, in turn, could have differing release site destinations (e.g., synaptic endings *vs.* neurohemal terminals) or different probabilities of exocytosis depending on stimulus conditions.

Although highly speculative, these sorts of possibilities are much less farfetched than they would have appeared only a few years ago. Moreover, they illustrate the unique and uniquely interesting promise of the study of neurosecretory cells. Although such cells have been demonstrated to share the defining biochemical characteristics of nonneural protein–secreting cells, they promise, by virtue of the fact that they are nerve cells, to disclose novel regulatory capabilities of greater subtlety and specificity. Thus, the continued biochemical exploration of the neurosecretory cell will make important contributions not only to our understanding of neural regulation but as well to our fundamental knowledge of cell biology. We share with Pickering[170] the view that the study of neurosecretion has come of age in the process of broadening its perspective. We can look forward to a challenging and exciting future for this crucial area of neurobiology.

REFERENCES

1. Speidel, C. G., 1919, *Carnegie Inst. Wash. Publ.* **13**:1–31.
2. Scharrer, E., 1928, *Z Vergl. Physiol.* **7**:1–38.
3. Mason, C. A., and Bern, H. A., 1977, *The Handbook of Physiology*, Volume 1 (E. R. Kandel, ed.), American Physiological Society, Bethesda, pp. 651–689.
4. Brownstein, M. J., 1980, *Proc. R. Soc. Lond.* [*Biol.*] **210**:79–90.
5. Gabe, M., 1966, *Neurosecretion*, Pergammon Press, New York.

6. Sachs, H., 1970, *The Handbook of Neurochemistry*, Volume 4, First Edition (A. Lajtha, ed.), Plenum Press, New York, pp. 373–428.
7. Cross, B. A., Dyball, R. E. J., Dyer, R. G., Jones, C. W., Lincoln, D. W., Morris, J. F., and Pickering, B. T., 1975, *Recent Prog. Horm. Res.* **31**:243–294.
8. Maddrell, S. H. P., and Nordmann, J. J., 1979, *Neurosecretion*, John Wiley & Sons, New York.
9. Berlind, A., 1977, *Int. Rev. Cytol.* **49**:171–251.
10. Snyder, S. H., 1980, *Science* **209**:976–983.
11. Hokfelt, T., Johansson, O., Ljungdahl, A., Lundberg, J. M., and Schultzberg, M., 1980, *Nature* **284**:515–521.
12. Kessler, J. A., Adler, J. E., Bohn, M. C., and Black, I. B., 1981, *Science* **214**:335–336.
13. Bitar, K. N., and Makhlouf, G. M., 1982, *Science* **216**:531–533.
14. Coggeshall, R. E., 1967, *J. Neurophysiol.* **30**:1263–1287.
15. Kaczmarek, L. K., Finbow, M., Revel, J.-P., and Strumwasser, F., 1979, *J. Neurobiol.* **10**:535–550.
16. Haskins, J. T., Price, C. H., and Blankenship, J. E., 1981, *J. Neurocytol.* **10**:729–747.
17. Kupfermann, I., and Kandel, E., 1970, *J. Neurophysiol.* **33**:865–876.
18. Dudek, F. E., and Blankenship, J. E., 1977, *J. Neurophysiol.* **40**:1301–1311.
19. Dellmann, H.-D., and Rodriguez, E. M., 1970, *Z. Zellforsch. Mikrosk. Anat.* **111**:293–315.
20. Morris, J. F., Tilly, G., and Hamilton, G., 1973, *Proc. Eur. Anat. Congr.* **3**:68–70.
21. Livingston, A., 1973, *Z. Zellforsch. Mikrosk. Anat.* **137**:361–374.
22. Rufener, C., 1974, *Neuroendocrinology* **13**:314–320.
23. Douglas, W. W., Nagasawa, J., and Schulz, R. A., 1971, *Nature* **232**:340–341.
24. Roubos, E. W., Schmidt, E. D., and Moorer-van Delft, C. M., 1981, *Cell Tissue Res.* **215**:63–78.
25. Bern, H. A., Yagi, K., and Nisioka, R. S., 1965, *Arch. Anat. Microsc. Morphol. Exp.* **54**:217–238.
26. Iverson, L. L., 1970, *The Neurosciences: Second Study Program* (F. O. Schmidt, ed.), Rockefeller University Press, New York, pp. 768–781.
27. Palade, G., 1975, *Science* **189**:347–358.
28. Steiner, D. F., Patzelt, C., Chan, S. J., Quinn, P. S., Tager, H. S., Nielsen, D., Lernmark, A., Noyes, B. E., Agarwal, K. L., Gabbay, K. H., and Rubenstein, A. H., 1980, *Proc. R. Soc. Lond. [Biol.]* **210**:45–59.
29. Blobel, G., 1980, *Proc. Natl. Acad. Sci. U.S.A.* **77**:1496–1500.
30. Rothman, J. E., 1981, *Science* **213**:1212–1219.
31. Thomsen, E., 1952, *J. Exp. Biol.* **29**:137–172.
32. Ishii, S., 1972, *Brain–Endocrine Interaction. Median Eminence: Structure and Function* (K. M. Knigge, D. E. Scott, and A. Weindl, eds.), Karger, Basel, pp. 119–141.
33. Stoeckart, R., Kreike, A. J., and Jansen, H. G., 1973, *Z. Zellforsch. Mikrosk. Anat.* **146**:501–515.
34. Vitry, G., Fondari, J., and Cougard, A., 1971, *C. R. Soc. Biol. (Paris)* **165**:361–364.
35. Morris, J. F., and Cannata, M. A., 1973, *J. Endocrinol.* **57**:517–529.
36. Heuser, J. E., and Reese, T. S., 1977, *The Handbook of Physiology*, Volume 1 (E. R. Kandel, ed.), American Physiological Society, Bethesda, pp. 261–294.
37. Bunt, A. H., 1969, *J. Ultrastruct. Res.* **28**:411–421.
38. Nordmann, J. J., and Morris, J. F., 1976, *Nature* **261**:723–725.
39. Blitz, A. L., Fine, R. E., and Toselli, P. A., 1977, *J. Cell Biol.* **75**:135–147.
40. Burnstock, G., 1976, *Neuroscience* **1**:239–248.
41. Petersen, O., 1976, *Physiol. Rev.* **56**:535–577.
42. Kandel, E. R., 1964, *J. Gen. Physiol.* **47**:691–717.
43. Morita, H., Ishibashi, T., and Yamashita, S., 1961, *Nature* **191**:183.
44. Frazier, W. T., Kandel, E. R., Kupfermann, I., Waziri, R., and Coggeshall, R. E., 1967, *J. Neurophysiol.* **30**:1288–1351.
45. Iwasaki, S., and Satow, Y., 1973, *Neuroendocrine Control* (K. Yagi and S. Yoshida, eds.), John Wiley & Sons, New York, pp. 85–110.
46. Bennett, M. V. L., Gimenez, M., and Ravitz, M. J., 1968, *Anat. Rec.* **160**:313–314.

47. Wilkens, J. L., and Mote, M. I., 1970, *Experientia* **26**:275–276.
48. Strumwasser, F., Kaczmarek, L. K., Jennings, K. R., and Chiu, A. Y., 1981, *Neurosecretion: Molecules, Cells, Systems* (D. Farner and K. Lederis, eds.), Plenum Press, New York, pp. 249–268.
49. Katz, B., and Miledi, R., 1967, *J. Physiol. (Lond.)* **192**:407–436.
50. Nordmann, J. J., and Dreifuss, J. J., 1972, *Brain Res.* **45**:604–607.
51. Cooke, I. M., 1977, *Peptides in Neurobiology* (H. Gainer, ed.), Plenum Press, New York, pp. 345–374.
52. Miledi, R., 1973, *P. R. Soc. (Lond.) [Biol.]* **183**:421–425.
53. Douglas, W. W., and Poisner, A. M., 1964, *J. Physiol. (Lond.* **172**:19–30.
54. Berlind, A., and Cooke, I. M., 1968, *Gen. Comp. Endocrinol.* **11**:458–463.
55. Uttenthal, L. O., Livett, B. G., and Hope, D. B., 1971, *Phil. Trans. R. Soc. Lond. [Biol.]* **261**:379–380.
56. Arch, S., 1972, *J. Gen. Physiol.* **59**:47–59.
57. Eckert, R., Tillotson, D., and Ridgeway, E. B., 1977, *Proc. Natl. Acad. Sci. U.S.A.* **74**:1748–1752.
58. Lux, H. D., and Heyer, C. B., 1977, *Neuroscience* **2**:585–592.
59. Hagiwara, S., and Byerly, L., 1981, *Annu. Rev. Neurosci.* **4**:69–125.
60. Eckert, R., and Lux, H. D., 1977, *Science* **197**:472–475.
61. Heyer, C. B., and Lux, H. D., 1976, *J. Physiol. (Lond.)* **262**:349–382.
62. Meech, R. W., 1974, *J. Physiol. (Lond.)* **237**:259–277.
63. Meech, R. W., 1978, *Annu. Rev. Biophys. Bioeng.* **7**:1–18.
64. Strumwasser, F., 1973, *Physiologist* **16**:9–42.
65. Poulain, D. A., and Wakerly, J. B., 1982, *Neuroscience* **7**:773–808.
66. Haskins, J. T., and Blankenship, J. E., 1979, *J. Neurophysiol.* **42**:356–367.
67. McAfee, D. A., and Yarowsky, P. J., 1979, *J. Physiol. (Lond.)* **290**:507–523.
68. Schwartz, J. H., 1979, *Annu. Rev. Neurosci.* **2**:467–504.
69. Grafstein, B., and Forman, D. S., 1980, *Physiol. Rev.* **60**:1167–1283.
70. Lasek, R. J., and Shelanski, M. L., 1981, *Neurosci. Res. Prog. Bull.* **19**:1–153.
71. Hammerschlag, R., Stone, G. C., Bolen, F. A., Lindsey, J. D., and Ellisman, M. H., 1982, *J. Cell Biol.* **93**:568–575.
72. Fink, D. J., and Gainer, H., 1980, *Science* **208**:303–305.
73. Fink, D. J., and Gainer, H., 1980, *J. Cell Biol.* **85**:175–186.
74. Gainer, H., and Fink, D. J., 1982, *Brain Res.* **233**:404–408.
75. Morris, J. F., Nordmann, J. J., and Dyball, R. E. J., 1978, *Int. Rev. Exp. Pathol.* **18**:1–95.
76. Pickering, B. T., 1978, *Essays Biochem.* **14**:45–81.
77. Fink, D. J., Russell, J. T., Brownstein, M. J., Baumgold, J., and Gainer, H., 1981, *J. Neurobiol.* **12**:487–503.
78. Wickner, W., 1980, *Science* **210**:861–868.
79. Freedman, R. B., and Hawkins, H. C. (eds.), 1980, *The Enzymology of Post-translational Modification of Proteins*, Academic Press, New York.
80. Koch, G., and Richter, D. (eds.), 1980, *Biosynthesis, Modification, and Processing of Cellular and Viral Polyproteins*, Academic Press, New York.
81. Zimmerman, M., Mumford, R. A., and Steiner, D. F. (eds.), 1980, *Precursor Processing in the Biosynthesis of Proteins*, New York Academy of Sciences, New York.
82. Steiner, D. F., Clark, J. L., Nolan, C., Rubenstein, A. H., Margoliash, E., Aten, B., and Oyer, P. E., 1969, *Recent Prog. Horm. Res.* **25**:207–282.
83. Sachs, H., and Takabatake, Y., 1964, *Endocrinology* **75**:943–948.
84. Takabatake, Y., and Sachs, H., 1964, *Endocrinology* **75**:934–942.
85. Sachs, H., Fawcett, P., Takabatake, Y., and Portanova, R., 1969, *Recent Prog. Horm. Res.* **25**:447–491.
86. Land, H., Schutz, G., Schmale, H., and Richter, D., 1982, *Nature* **295**:299–303.
87. Gainer, H., Sarne, Y., and Brownstein, M. J., 1977, *Science* **195**:1354–1356.
88. Gainer, H., Sarne, Y., and Brownstein, M. J., 1977, *J. Cell Biol.* **73**:366–381.
89. Brownstein, M. J., and Gainer, H., 1977, *Proc. Natl. Acad. Sci. U.S.A.* **74**:4046–4049.
90. Brownstein, M. J., Robinson, A. G., and Gainer, H., 1977, *Nature* **269**:259–261.

91. Gainer, H., and Brownstein, M. J., 1978, *Cell Biology of Hypothalamic Neurosecretion* (J. D. Vincent and C. Kordon, eds.), CNRS, Paris, pp. 525–542.
92. Burford, G. D., and Pickering, B. T., 1973, *Biochem. J.* **136**:1047–1052.
93. Jones, C. W., and Pickering, B. T., 1972, *J. Physiol. (Lond.)* **227**:553–564.
94. Pickering, B. T., Jones, C. W., Burford, G. D., McPherson, M., Swann, R. W., Heap, P. F., and Morris, J. F., 1975, *Ann. N.Y. Acad. Sci.* **248**:15–35.
95. Russell, J. T., Brownstein, M. J., and Gainer, H., 1979, *Proc. Natl. Acad. Sci. U.S.A.* **76**:6086–6090.
96. Russell, J. T., Brownstein, M. J., and Gainer, H., 1980, *Endocrinology* **107**:1880–1891.
97. Russell, J. T., Brownstein, M. J., and Gainer, H., 1981, *Neuropeptides* **2**:59–65.
98. Guidice, L. M., and Chaiken, I. M., 1979, *J. Biol. Chem.* **254**:11767–11770.
99. Lin, C., Joseph-Bravo, P., Sherman, T., Chen, L., and McKelvey, J. F., 1979, *Biochem. Biophys. Res. Commun.* **89**:943–950.
100. Schmale, H., Leipold, B., and Richter, D., 1979, *FEBS Lett.* **108**:311–316.
101. Richter, D., Schmale, H., Ivell, R., and Schmidt, C., 1980, *Biosynthesis, Modification, and Processing of Cellular and Viral Polyproteins* (G. Koch and D. Richter, eds.), Academic Press, New York, pp. 43–66.
102. Ivell, R., Schmale, H., and Richter, D., 1981, *Biochem. Biophys. Res. Commun.* **102**:1230–1236.
103. Schmale, H., and Richter, D., 1980, *FEBS Lett.* **121**:358–362.
104. Schmale, H., and Richter, D., 1981, *Proc. Natl. Acad. Sci. U.S.A.* **78**:766–769.
105. Schmale, H., and Richter, D., 1981, *Neuropeptides* **2**:151–156.
106. Schmale, H., and Richter, D., 1981, *Neuropeptides* **2**:47–52.
107. Pickering, B. T., and Jones, C. W., 1978, *Horm. Proteins Peptides* **5**:103–158.
108. Smyth, D. G., and Massey, D. E., 1979, *Biochem. Biophys. Res. Commun.* **87**:1006–1010.
109. Holwerda, D. A., 1972, *Eur. J. Biochem.* **28**:340–346.
110. Arch, S., 1972, *J. Gen. Physiol.* **60**:102–119.
111. Arch, S., Earley, P., and Smock, T., 1976, *J. Gen. Physiol.* **68**:197–210.
112. Arch, S., Smock, T., and Earley, P., 1976, *J. Gen. Physiol.* **68**:211–225.
113. Gainer, H., and Wollberg, Z., 1974, *J. Neurobiol.* **5**:243–261.
114. Loh, Y. P., Sarne, Y., Daniels, M., and Gainer, H., 1977, *J. Neurochem.* **29**:135–139.
115. Berry, R. W., 1981, *Biochemistry* **21**:6200–6205.
116. Berry, R. W., Trump, M. J., and Baylen, J. T., 1981, *Biochemistry* **21**:6206–6211.
117. Chiu, A. Y., Hunkapiller, M. W., Heller, E., Stuart, D. K., Hood, L. E., and Strumwasser, F., 1979, *Proc. Natl. Acad. Sci. U.S.A.* **76**:6656–6660.
118. Stuart, D. K., Chiu, A. Y., and Strumwasser, F., 1980, *J. Neurophysiology* **43**:488–498.
119. Arch, S., 1981, *Neurosecretion: Molecules, Cells, Systems* (D. Farner and K. Lederis, eds.), Plenum Press, New York, pp. 129–137.
120. Scheller, R. H., Jackson, J. F., McAllister, L. B., Schwartz, J. H., Kandel, E. R., and Axel, R., 1982, *Cell* **28**:707–719.
121. Arch, S., Smock, T., Gurvis, R., and McCarthy, C., 1978, *J. Comp. Physiol.* **128**:67–70.
122. Arch, S., Lupatkin, J., Smock, T., and Beard, M., 1980, *J. Comp. Physiol.* **141**:131–137.
123. Beard, M., Millecchia, L., Masuoka, C., and Arch, S., 1982, *Tissue Cell* **14**:297–308.
124. Heller, E., Kaczmarek, L., Hunkapiller, M., Hood, L., and Strumwasser, F., 1980, *Proc. Natl. Acad. Sci. U.S.A.* **77**:2328–2332.
125. Schlesinger, D. H., Babirak, S. P., and Blankenship, J. E., 1981, *Symposium on Neurohypophyseal Hormones and Other Biologically Active Peptides* (D. H. Schlesinger, ed.), Elsevier/North-Holland, Amsterdam, pp. 137–150.
126. Heller, H., and Pickering, B. T., 1970, *International Encyclopedia of Pharmacology and Therapeutics*, Section 41 (H. Heller and B. T. Pickering, eds.), Pergamon Press, Oxford, pp. 59–79.
127. Fletcher, D. J., Noe, B. D., Bauer, G. E., and Quigley, J. P., 1980, *Diabetes* **29**:593–599.
128. Fletcher, D. J., Quigley, J. P., Bauer, G. E., and Noe, B. D., 1981, *J. Cell Biol.* **90**:312–322.
129. Loh, Y. P., and Gainer, H., 1982, *Proc. Natl. Acad. Sci. U.S.A.* **79**:108–112.
130. Loh, Y. P., and Chang, T.-L., 1982, *FEBS Lett.* **127**:57–62.
131. Chang, T.-L., Gainer, H., Russell, J. T., and Loh, Y. P., 1982, *Endocrinology* **111**:1607–1614.

132. Jones, C. W., and Pickering, B. T., 1969, *J. Physiol. (Lond.)* **203:**499–558.
133. Valtin, H., Stewart, J., and Sokol, H. W., 1974, *The Handbook of Physiology,* Volume 7, American Physiological Society, Bethesda, pp. 131–171.
134. Berry, R. W., and Arch, S., 1981, *Brain Res.* **215:**115–123.
135. Normann, T. C., 1976, *Int. Rev. Cytol.* **46:**1–77.
136. Van Der Kloot, W., and Kita, H., 1973, *J. Membr. Biol.* **14:**365–382.
137. Dean, P. M., 1975, *J. Theor. Biol.* **54:**289–308.
138. Kelly, R. B., Deutsch, J. W., Carlson, S. S., and Wagner, J. A., 1979, *Annu. Rev. Neurosci.* **2:**399–446.
139. Bass, L., and Moore, W. J., 1966, *Proc. Natl. Acad. Sci. U.S.A.* **55:**1214–1217.
140. Remler, M. P., 1973, *Biophys. J.* **13:**104–117.
141. Cowley, A. C., Fuller, N., Rand, R. P., and Parsegian, V. A., 1977, *Biophys. J.* **17:**85A.
142. Llinas, R., Blinks, J. R., and Nicholson, C., 1972, *Science* **176:**1127–1129.
143. Llinas, R., and Nicholson, C., 1975, *Proc. Natl. Acad. Sci. U.S.A.* **72:**187–190.
144. Parsegian, V. A., 1977, *Society for Neuroscience Symposia,* Volume 2 (W. M. Cowan and J. A. Ferrendelli, eds.), Society for Neuroscience, Bethesda, pp. 161–171.
145. Llinas, R., and Heuser, J. E., 1977, *Neurosci. Res. Prog. Bull.* **15:**557–687.
146. Llinas, R., 1977, *Society for Neuroscience Symposia,* Volume 2 (W. M. Cowan and J. A. Ferrendelli, eds.), Society for Neuroscience, Bethesda, pp. 139–160.
147. Hubbard, J. I., Jones, S. F., and Landau, E. M., 1968, *J. Physiol. (Lond.)* **194:**355–380.
148. Hubbard, J. I., Jones, S. F., and Landau, E. M., 1968, *J. Physiol. (Lond.)* **196:**75–86.
149. Pollard, H. B., Creutz, C. E., Fowler, V., Scott, J., and Pazoles, C. J., 1982, *Cold Spring Harbor Symp. Quant. Biol.* **46:**819–834.
150. Katz, B., and Miledi, R., 1969, *J. Physiol. (Lond.)* **203:**459–487.
151. Creutz, C. E., Pazoles, C. J., and Pollard, H. B., 1978, *J. Biol. Chem.* **253:**2858–2866.
152. Orci, L., and Malaisse, W., 1980, *Diabetes* **29:**943–950.
153. Brandt, B. L., Hagiwara, S., Kidokoro, Y., and Miyazaki, S., 1976, *J. Physiol. (Lond.)* **263:**417–439.
154. Eckert, R., and Brehm, P., 1979, *Annu. Rev. Biophys. Bioeng.* **8:**353–383.
155. Fawcett, C. P., Powell, A. E., and Sachs, H, 1968, *Endocrinology* **83:**1299–1310.
156. Nordmann, J. J., Dreifuss, J. J., and Legros, J. J., 1971, *Experientia* **27:**1344–1345.
157. Gainer, H., 1981, *Neurosecretion and Brain Peptides* (J. B. Martin, S. Reichlin, and K. L. Bick, eds.), Raven Press, New York, pp. 5–20.
158. Herbert, E., Roberts, J., Phillips, M., Allen, R., Hinman, M., Budarf, M., Policastro, P., and Rosa, P., 1980, *Frontiers in Neuroendocrinology,* Volume 6 (L. Martini and W. F. Ganong, eds.), Raven Press, New York, pp. 67–101.
159. Herbert, E., Birnberg, N., Lissitsky, J.-C., Civelli, O., and Uhler, M., 1981, *Neurosci. Comment.* **1:**16–27.
160. Nilaver, G., Zimmerman, E. A., Defendini, R., Liotta, A. J., Krieger, D. T., and Brownstein, M. J., 1979, *J. Cell Biol.* **81:**50–58.
161. Hokfelt, T., Lundberg, J. M., Schultzberg, M., Johansson, O., Skirboll, L., Anggard, A., Fredholm, B., Hamberger, B., Pernow, B., Rehfeld, J., and Goldstein, M., 1980, *Proc. R. Soc. Lond. [Biol.]* **210:**63–77.
162. Pelletier, G., Steinbusch, H. W. M., and Verhofstad, A. A. J., 1981, *Nature* **293:**71–72.
163. Wilson, S. P., Chang, K.-J., and Viveros, O. H., 1982, *J. Neurosci.* **2:**1150–1156.
164. Lundberg, J. M., Anggard, A., Fahrenkrug, J., Hokfelt, T., and Mutt, Y., 1980, *Proc. Natl. Acad. Sci. U.S.A.* **77:**1651–1655.
165. Heap, P. F., Jones, C. W., Morris, J. F., and Pickering, B. T., 1975, *Cell Tissue Res.* **156:**483–497.
166. Nordmann, J. J., Louis, F., and Morris, S. J., 1979, *Neuroscience* **4:**1367–1379.
167. Heuser, J. E., and Reese, T. S., 1973, *J. Cell Biol.* **57:**315–344.
168. Pearse, B. M. F., 1975, *J. Mol. Biol.* **97:**93–98.
169. Swann, R. W., and Pickering, B. T., 1976, *J. Endocrinol.* **68:**95–108.
170. Pickering, B. T., 1981, *Neurosecretion: Molecules, Cells, Systems* (D. Farner and K. Lederis, eds.), Plenum Press, New York, pp. 415–430.

13

Control of Monoamine Synthesis by Precursor Availability

Candace J. Gibson

1. INTRODUCTION

Several important CNS monoamine neurotransmitters have as their precursors simple aromatic amino acids present in dietary protein. For instance, serotonin (5-hydroxytryptamine; 5-HT), the indoleamine neurotransmitter, is formed from the essential amino acid tryptophan, and the catecholamines, dopamine (DA), norepinephrine (NE), and epinephrine (E), are formed from the amino acid tyrosine (Figs. 1, 2). The circulating level of these amino acids changes with meal consumption, dependent on the dietary content of protein, carbohydrate, and fat.[1–4] The carrier system that transports these precursor amino acids across the blood–brain barrier is normally unsaturated, and changes in their circulating levels can raise or lower brain tryptophan and tyrosine concentration.[5,6] Once in the brain, and in the particular subset of amine-containing neurons, tryptophan and tyrosine can influence the synthesis of their respective neurotransmitters, since the rate-limiting synthetic enzymes are normally unsaturated with respect to substrate or, under certain conditions, may become precursor responsive.[7]

This chapter discusses factors that affect the circulating levels of these amino acid precursors and their uptake into brain; how brain precursor level influences synthesis and release of the monoamines; and what significance or function this ability to regulate CNS neurotransmission may serve.

2. PLASMA AMINO ACID COMPOSITION AND BRAIN PRECURSOR CONCENTRATION

2.1. Sources of Circulating Tryptophan and Tyrosine

The aromatic amino acids tryptophan and tyrosine are present in dietary vegetable and animal proteins. Tryptophan, an essential amino acid, is obtained

---

Candace J. Gibson • Department of Pathology, University of Western Ontario, London, Ontario N6A 5C1, Canada.

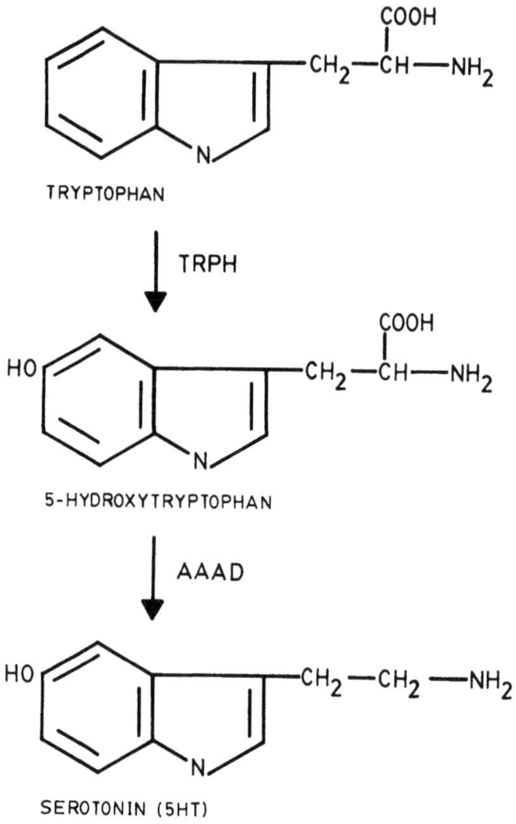

Fig. 1. Serotonin biosynthesis. The amino acid *l*-tryptophan is converted to 5-hydroxytryptophan (5-HTP) by the action of tryptophan hydroxylase (TRPH); 5-HTP is converted to serotonin (5-hydroxytryptamine; 5-HT) by the nonspecific enzyme, aromatic *l*-amino acid decarboxylase (AAAD).

solely by our dietary protein intake. It is usually the most limiting amino acid present in protein (from 0.5 to 1.6% of protein content). Several vegetable proteins (e.g., zein in corn) are deficient in tryptophan. Prolonged intake of these proteins can result in severe tryptophan deficiencies and reduced brain tryptophan and serotonin synthesis.[8]

Tyrosine can be derived in the body from the essential amino acid phenylalanine. The specific hydroxylating enzyme, phenylalanine hydroxylase, is present in the liver,[9] and following meal consumption and entry of amino acids into the portal circulation, about half of the ingested phenylalanine is converted into tyrosine. Thus, the circulating pool of free tyrosine comes directly from ingested tyrosine and indirectly from phenylanine. Tyrosine is plentiful in both animal and vegetable proteins.

As far as the CNS supply of tyrosine is concerned, the majority of the precursor for catecholamine synthesis comes directly as tyrosine. There is no phenylalanine hydroxylase in brain. There is some evidence that brain tyrosine hydroxylase can act on phenylalanine, converting it to tyrosine, which may then be available for catecholamine synthesis.[10,11] The extent to which this occurs *in vivo* is not known. At equimolar concentrations, tyrosine is a far

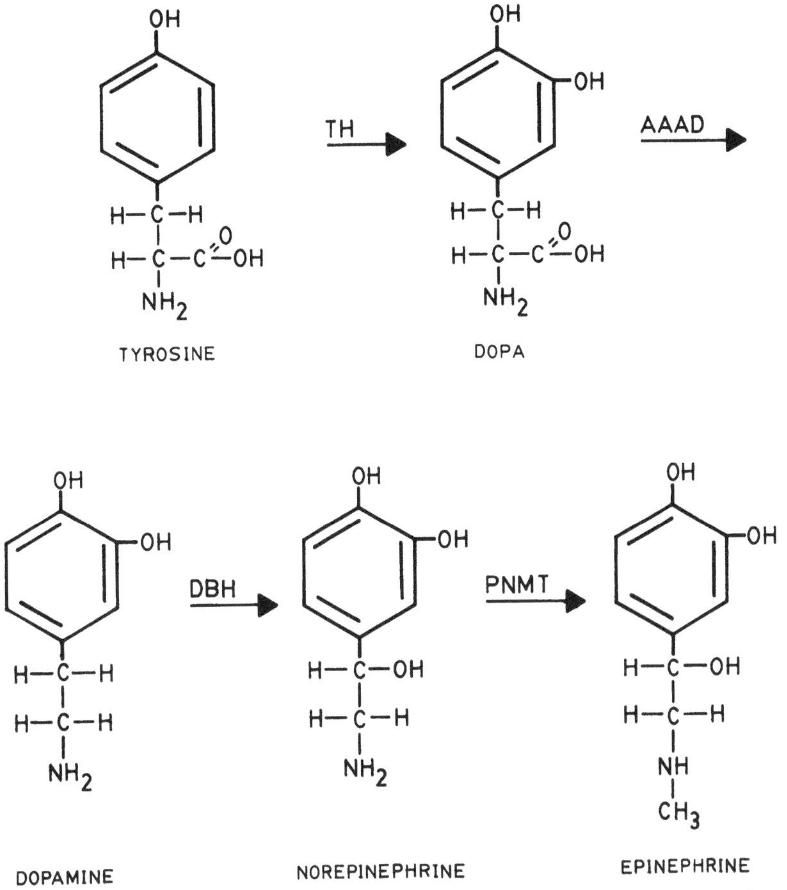

Fig. 2. Catecholamine biosynthesis. The amino acid *l*-tyrosine is converted to dihydroxyphenylalanine (DOPA) through the action of tyrosine hydroxylase (TH); DOPA is subsequently converted to the catecholamine dopamine (DA) by the enzyme aromatic *l*-amino acid decarboxylase (AAAD). Within norepinephrine neurons, dopamine is converted to norepinephrine (NE) by the enzyme dopamine-β-hydroxylase (DBH). In those neurons containing phenylethanolamine-N-methyltransferase (PNMT), epinephrine (E) is formed.

greater inhibitor of catecholamine synthesis from phenylalanine than *vice versa*, and catecholamine production from labeled phenylalanine is one-tenth that produced from labeled tyrosine precursor.[10] Certainly, when high doses of phenylalanine are given endogenously, or when plasma phenylalanine levels are abnormally high, as in phenylketonuria (PKU), the CNS production of the monoamines is greatly reduced.[12–14] This fact may be of importance in the administration of parenteral amino acid mixtures, many of which contain phenylalanine and the branched-chain amino acids but no tyrosine on the supposition that sufficient tyrosine will be produced in the liver.[15] Tyrosine may be produced in the periphery, but sufficient amounts may not enter the CNS because of the competitive nature of the other amino acids (Section 2.2).

2.2. Brain Uptake of Tryptophan and Tyrosine

The plasma amino acids enter the brain via specific carriers at the blood–brain barrier.[5,6] A large neutral amino acid (LNAA) carrier facilitates the brain uptake of the aromatic amino acids, tryptophan (TRP), tyrosine (TYR), phenylalanine (P), the branched-chain amino acids leucine (L), isoleucine (I), and valine (V), and others (methionine and histidine, for example).[5] The carrier system is stereospecific (*l*-amino acids), relatively unsaturated (K_m or association constant close to the plasma amino acid concentration), and competitive—an increased plasma concentration of one of the large neutral amino acids can decrease uptake of the others.[6]

The majority of amino acids circulate free in the blood; tryptophan, however, is an exception. Approximately 80% of tryptophan circulates bound to albumin,[16] with only 20% circulating in the free form. Both free and bound tryptophan are available for uptake into the brain; the affinity of tryptophan for the amino acid carrier is greater than its affinity for albumin.[17] Thus, free plasma tryptophan and a variable amount of the bound form are transported into brain.[18]

The competitive nature of the LNAA carrier means that brain neutral amino acid concentration will be determined not only by the plasma concentration of that particular amino acid but also by the plasma concentration of the competing neutral amino acids.[1,3] For all neutral amino acids, their brain concentration can be predicted on the basis of the plasma ratio of that particular amino acid of interest to the sum of the major competing neutral amino acids.[3] For instance, the plasma tryptophan ratio is the ratio of total plasma tryptophan concentration to the sum of the plasma concentrations of phenylalanine, tyrosine, and the three branched-chain amino acids; i.e., plasma tryptophan ratio = [TRP]/{[P] + [I] + [V] + [L] + [TYR]} (Fig. 3).

Plasma tryptophan ratio is a better predictor of brain tryptophan concentration than plasma tryptophan alone.[1] This is very clearly demonstrated following food consumption. Protein ingestion rapidly raises plasma tryptophan level and, if this were the only determinant of brain tryptophan uptake, might be expected to raise brain tryptophan levels. Exactly the opposite is seen (Fig. 3). Protein ingestion raises not only plasma tryptophan level but also the level of all circulating large neutral amino acids, and, hence, brain tryptophan level actually falls (Fig. 3). Consumption of a carbohydrate-containing meal, resulting in increased insulin secretion, lowers branched-chain amino acid concentration (as they are taken up into muscle to be catabolized). In this case, the decrease in circulating competitors results in an increased plasma tryptophan ratio, and brain tryptophan level rises (Fig. 3).[1,3] Thus, there is an inverse relationship between dietary protein content and brain tryptophan level; as dietary protein content is increased, the brain tryptophan level is reduced.

In the case of tyrosine, protein consumption raises plasma tyrosine to a greater extent than the increase in LNAAs (since plasma tyrosine is increased directly from the tyrosine in protein and indirectly from phenylalanine), and brain tyrosine level increases (Fig. 3). Following carbohydrate consumption, the secretion of insulin decreases plasma tyrosine, but not to the same extent

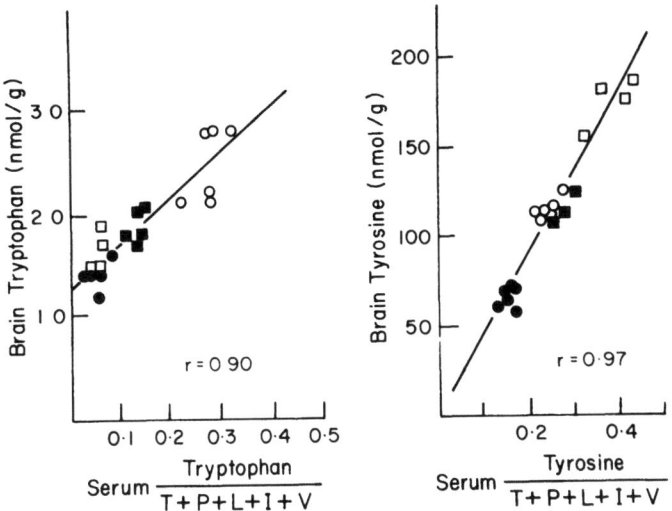

Fig. 3. Relationship between brain tryptophan level and the serum tryptophan ratio (left) and between brain tyrosine level and serum tyrosine ratio (right). Rats fasted overnight were killed 2 h after ingesting a single meal containing 0% casein (carbohydrate) (open circles, ○), 18% casein (closed squares, ■), or 40% casein (open squares, □). Control animals continued to fast during the period of food ingestion (closed circles, ●). Data were analyzed by linear regression; the values of the correlation coefficient (r) are significantly different from 0, $P < 0.01$ (Student's t-test). (Modified from Fernstrom and Faller.[3])

as the large neutral amino acids, and brain tyrosine concentration stays the same or rises slightly (Fig. 3).[2,3] Thus, there is a direct relationship between dietary protein content and brain tyrosine level; as dietary protein content is increased, the brain tyrosine level is increased.

Of course, plasma and brain tryptophan and tyrosine concentrations can be raised directly by their peripheral injection or by consumption of the pure compound.[19–22] Much higher concentrations of the brain amino acids result because of the lack of the dampening effect exerted by the other amino acids when meals are consumed. Tryptophan administration combined with carbohydrate results in a greater brain tryptophan concentration than tryptophan administration alone and may be useful therapeutically.[18,23]

2.3. Hormonal Influences on Plasma Amino Acid Pattern

Plasma amino acid pattern may also be influenced by certain hormones. As we have seen, insulin secretion after meal consumption has a profound effect on plasma amino acid concentrations.[1,24] Disturbances in insulin secretion, such as occur in diabetes, may be expected to affect plasma amino acid pattern and possibly brain tyrosine and tryptophan levels. This does seem to be true. In diabetic animals, plasma neutral amino acids are elevated (particularly the branched-chain amino acids), and, hence, brain tryptophan level is reduced.[25–27] Brain tyrosine may also be affected.[26] These changes in brain precursor availability ultimately affect brain monoamine synthesis.[25–27]

Corticosteroids have inductive effects on the major peripheral degradative enzymes for tryptophan and tyrosine, tryptophan pyrrolase and tyrosine transaminase, respectively.[28,29] The increased activity of the catabolic enzymes reduces their levels peripherally. Administration of large doses of corticosteroids *in vivo* does reduce plasma and brain levels of tyrosine and tryptophan and their monoamine transmitter products.[29,30] Very high doses of corticosteroids may induce CNS tyrosine transaminase.[30] These are really pharmacological experiments, and it is not clear that the plasma increases in corticosteroids following daily stresses, increases that are of much smaller magnitude, will have the same effects on these two amino acids.[31]

The sex steroids, estrogen and progesterone, may also affect plasma tyrosine and tryptophan level. Long-term use of oral contraceptive agents containing these steroids is associated with reduced plasma levels of the two amino acids and their plasma ratios.[32,33]

3. MONOAMINE BIOSYNTHETIC PATHWAYS

Once tryptophan and tyrosine have been taken up into the nerve terminal, the initial and rate-limiting step in their biosynthetic pathway is the hydroxylation of the indole ring of tryptophan by tryptophan hydroxylase (TRPH)[34,35] and of the catechol moiety of tyrosine by tyrosine hydroxylase (TH)[36] (Figs. 1, 2). These enzymes are specifically localized to those neurons that produce serotonin (TRPH)[37] and the catecholamines (TH).[38] The two hydroxylase enzymes require oxygen and tetrahydrobiopterin as cofactors. The resulting amino acids, 5-hydroxytryptophan (5-HTP) and dihydroxyphenylalanine (DOPA), are rapidly decarboxylated to form the neurotransmitters serotonin (5-HT) and dopamine (DA), respectively.[39] Dopamine is further hydroxylated in neurons that contain the copper-containing enzyme dopamine-β-hydroxylase (DBH) to form norepinephrine.[40] The enzyme phenylethanolamine-N-methyltransferase (PNMT) exists in specific brain regions (e.g., olfactory tubercle, brainstem nuclei) and here catalyzes the formation of epinephrine from norepinephrine (Figs. 1, 2).[41]

The initial hydroxylation step is rate limiting in both systems; the intermediates once formed are rapidly decarboxylated. Inhibition of these enzymes leads to a rapid decline in monoamine neurotransmitter level.[42,43] Thus, rates of monoamine synthesis are affected by treatments or conditions that influence this first step. This can occur in several different ways:

1. By changes in enzyme activity, both short-term conformational or kinetic changes[44-47] or long-term changes in actual amount of enzyme.[48]
2. By changes in precursor availability (tyrosine,[2,7,20] tryptophan,[1,7,19] or oxygen[49,50]).
3. By changes in cofactor availability (tetrahydrobiopterin[51]).

Brain monoamine vesicles may normally not be filled to maximum storage capacity with their transmitter, and pharmacological manipulations can increase vesicular monoamine content.[52] Since, in general, it is also the newly

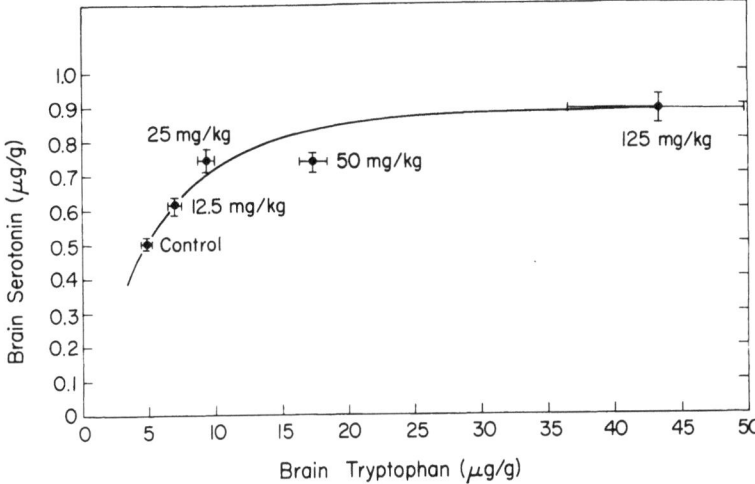

Fig. 4. Dose–response curve relating brain tryptophan and brain serotonin. Groups of ten rats received *l*-tryptophan (12.5, 25, 50, or 125 mg/kg, intraperitoneally) at noon and were killed 1 h later. Horizontal bars represent standard errors of the mean for brain tryptophan; vertical bars for brain serotonin. All brain tryptophan levels were significantly higher than control values ($P < 0.001$). All brain serotonin levels were significantly higher than control values ($P < 0.01$). (Reprinted from Fernstrom and Wurtman.[19])

synthesized amine that is most readily released,[53] changes in synthesis may result in changes in release and, hence, result in altered function or behavior.

3.1. Tryptophan Availability and Serotonin Synthesis

There is little evidence *in vivo* or *in vitro* for end-product inhibition of tryptophan hydroxylase. Thus, rates of reaction are determined by the enzyme's affinity for its substrate and by the availability of that substrate. This synthetic enzyme is not normally fully saturated with its tryptophan substrate. The affinity, expressed as an association constant or K_m, for tryptophan in *in vitro* preparations is approximately 30 to 50 μM.[34,35] *In vivo* calculations using 5-HTP accumulation following decarboxylase inhibition[54,55] are similar (about 25 μM) and are above the normal brain tryptophan concentration (20 μM).[1,54] Under normal conditions, the rate of 5-HT synthesis will be influenced by the availability of the substrate: increases in brain tryptophan concentration should increase brain 5-HT synthesis.

Administration of tryptophan peripherally increases rat brain tryptophan, serotonin, and its major acid metabolite, 5-hydroxyindoleacetic acid (5-HIAA).[19] The increases in 5-HT synthesis occur in a dose-dependent manner, small doses of tryptophan producing roughly linear increases in brain tryptophan and brain 5-HT concentration (Fig. 4). Larger doses (above 100 mg/kg), although producing further increases in brain tryptophan level, produce no greater increase in 5-HT, suggesting that the enzyme is now saturated with its substrate. Increased doses prolong the rise in brain 5-HT and 5-HIAA.[56,57]

Raising brain tryptophan level by consumption of a single carbohydrate meal also raises brain serotonin synthesis (5-HT level) and release (increased 5-HIAA).[1] Protein consumption, as expected because of the decrease in brain tryptophan, results in decreased serotonin synthesis.[1]

3.2. Tyrosine Availability and Catecholamine Synthesis

It is beyond the scope of this chapter to discuss all of the literature dealing with the short-term regulation of tyrosine hydroxylase activity. Since this chapter deals primarily with precursor control of monoamine synthesis, I now discuss those aspects of enzyme activity that allow precursor dependence to be exhibited. Over the past few years our understanding of TH regulation has evolved from simple end-product inhibition to a complex conformational change involving enzyme phosphorylation that results in altered affinities for cofactor and end product.[44–47]

Partially purified TH preparations exhibit end-product inhibition *in vitro*.[36] Synthetic activity decreases in the presence of the catechols DOPA, DA, or NE. It was postulated that a small pool of catecholamines existed free in the cytosol in contact with the enzyme and controlled its activity during nerve stimulation.[58] For example, when firing rates increased, this pool was diminished, and synthetic activity increased.

Certain factors *in vivo* could not be reconciled with this hypothesis. Administration of DA receptor blockers or lesioning of the nigrostriatal tract leads to increased DOPA synthesis acutely with no change in DA level.[59] It soon became apparent, as purification techniques improved, that TH could be phosphorylated *in vitro*[47,60,61] (and probably is *in vivo* during increased neuronal firing)[62,63] and exists in two forms—one a nonphosphorylated basal state, and the other a phosphorylated "activated" state. The activated form of the enzyme exhibits a much reduced affinity for its catechol end products (10–76 mM DA or NE in the active form *versus* 0.1 mM in the basal form) and a much increased affinity for its tetrahydrobiopterin cofactor (10–20 μM in the active form *versus* 500–900 μM in the basal form).[44–47,60,62,63] Since biopterin levels are relatively low in brain (100 μM, well below the K_m for biopterin in the basal state), cofactor concentration could be more important in regulating TH activity than the level of the end-product. Tetrahydrobiopterin is heterogeneously distributed in brain, coincident with tyrosine and tryptophan hydroxylase activities.[64] Its administration intraventricularly increased the incorporation of labeled tyrosine into striatal catechols.[51]

When the nerve is stimulated, and firing rates increased (via direct electrical stimulation or following lesions or administration of receptor blockers), tyrosine hydroxylase becomes phosphorylated, affinity for its end products is reduced, and that for the cofactor is increased, and synthesis is enhanced.[44–47,62,63] The exact mechanism coupling increased neuronal firing rate to enzyme phosphorylation is not known but may involve a multistep process mediated via receptors, changes in ion flux (particularly calcium), and protein kinase activity.

Table I
Relationship between Brain Tyrosine Level and Catecholamine Synthesis

Treatment		Brain tyrosine level (change from control, %)	CA or metabolite level (change from control, %)	Reference
Basal conditions				
Catecholamine systems: DOPA accumulation				
Leucine	100 mg/kg	−22	−25	20
Tyrosine	20 mg/kg	+29	+7	20
	50 mg/kg	+77	+12	
	100 mg/kg	+116	+15	
Tyrosine	20 mg/kg	+69	+25	65
(fasted rats)	50 mg/kg	+114	+44	
	100 mg/kg	+187	+32	
Norepinephrine systems: MHPG accumulation				
Tyrosine 100 mg/kg		+68	+16	66
Dopamine systems: striatal HVA accumulation				
Tyrosine 100 mg/kg		+202	+12	67
Activated conditions				
Norepinephrine systems: MHPG accumulation				
Cold stress + tyrosine, 100 mg/kg		+68	+60	66
Hypertensive rat + tyrosine, 235 mg/kg[a]		+187	+45	91
Dopamine systems: striatal DOPA, DA, or HVA accumulation				
γ-Butyrolactone + tyrosine, 200 mg/kg		+46 +48	+18 (DA) +26 (DOPA)	70
Haloperidol + tyrosine, 100 mg/kg		+120	+69	67
Nigrostriatal lesions + tyrosine, 250 mg/kg[a]		+185	+150	68

[a] Tyrosine was given as the methylester form. In all other experiments *l*-tyrosine was used and injected intraperitoneally 1–2 h prior to sacrifice.

In vitro determinations of tyrosine's affinity for tyrosine hydroxylase are extremely variable (K_m = 20 to 100 μM) and depend on the cofactor used, the enzyme preparation, and the method of assay.[36,44–48] Differing results may be partially explained by differing proportions of activated and unactivated enzyme that are present in the given conditions. *In vivo* determinations of tyrosine saturation (using DOPA accumulation following decarboxylase inhibition) suggest that the enzyme is 75 to 80% saturated with its substrate.[2,55] There is some leeway for affecting brain catechol synthesis normally, in particular, when brain tyrosine level is lowered (for example, in fasted animals). The peripheral administration of tyrosine raises brain tyrosine level and brain DOPA accumulation by 15 to 25%,[20,55] whereas treatments that lower brain tyrosine level (i.e., administration of competing large neutral amino acids) reduce brain DOPA accumulation[2,20,55] (Table I). In rats fasted overnight, the brain tyrosine level falls to values below the K_m for tyrosine (0.05 mM), and the same doses of tyrosine result in a greater accumulation of DOPA (Table

I).[65] Consumption of protein meals raises plasma and brain tyrosine and, hence, brain DOPA accumulation.[2]

These changes in DOPA accumulation following decarboxylase inhibition may not occur when the catecholomine (CA) systems are intact. In regions such as striatum, dopamine level is maintained by intraneuronal (inhibitor DA autoreceptors) and interneuronal elements (postsynaptic receptors; multisynaptic feedback loops), and changes in synthesis and release are rapidly compensated.[59] This may be true only in dopaminergic systems as opposed to noradrenergic systems (which do not exhibit the same type of tight anatomic connections and negative feedback loops) or even only in striatal DAergic systems. DOPA accumulation following tyrosine administration does vary regionally: the greatest increases occur in the cerebral hemispheres, predominantly a noradrenergic region.[55]

In noradrenergic systems, tyrosine may be more effective in influencing rates of synthesis. Following tyrosine administration to probenecid-treated rats, there is an increased accumulation of the major NE metabolite, 3-methoxy-4-hydroxyphenylglycol sulfate (MHPG-SO$_4$).[66] In contrast, striatal homovanillic acid content (HVA, the major DA metabolite) does not increase significantly following tyrosine administration to rats also treated with probenecid.[67] When the rat is pretreated with the DA receptor blocker haloperidol, tyrosine administration does increase striatal HVA accumulation (Table I).[67]

Following lesions of the nigrostriatal tract, remaining DA neurons increase their firing rates, and DA turnover is accelerated. Tyrosine administration further increases DA metabolite accumulation (both HVA and dihydroxyphenylacetic acid, DOPAC) in the remaining neurons.[68] Thus, DA neurons will also respond to increased precursor availability in situations in which neuronal firing rates and tyrosine hydroxylase activity has been increased. Other examples of this tyrosine responsiveness include increased HVA following chronic reserpine treatment,[69] increased striatal DA after γ-butyrolactone (GBL) administration,[70] increased MHPG accumulation in cold-stressed rats,[66] and increased HVA after DA receptor blockade and treatment with a DA reuptake blocker, amfenolic acid.[71] In all cases, increased CNS tyrosine is associated with an increase in catecholamine synthesis (Table I).

The specificity of tyrosine for "activated" catecholamine systems is also demonstrated in the retina. The dopaminergic amacrine cells of the retina contain TH that is activated by the physiological stimulus of light.[72,73] Peripheral tyrosine administration raises retinal tyrosine level in both light- and dark-exposed retina, but DA turnover is accelerated by tyrosine only in light-exposed eyes.[74]

All of these situations of tyrosine responsiveness are associated with activation of the synthetic enzyme tyrosine hydroxylase (in some situations following an increase in neuronal activity).[44-46] The exact mechanism coupling tyrosine hydroxylase activation to increased dependence on tyrosine availability is not known. The accelerated synthesis rates may simply use up tyrosine in the vicinity of the enzyme, or there may be some synergistic effect of increased biopterin affinity on the enzyme's subsequent interaction with tyrosine.

3.3. Requirements for Precursor Control of Monoamine Synthesis

In summary, the circumstances that allow precursor control to occur include (1) that the level of circulating precursor change in response to the ingestion of food or the pure amino acid,[1,3,4,21,22] (2) that the carrier mechanism that transports the precursor neutral amino acids into brain not be fully saturated so that changes in the plasma level will result in changes in the brain level of the precursor (what is more important is that this occurs in a highly predictable fashion based on the ratio of the precursor to the other competing large neutral amino acids present in the blood) (Fig. 3),[1,3,5,6] and (3) that the rate-limiting enzyme in monoamine synthesis be unsaturated and free from significant feedback control by end product. In serotonergic systems, the rate-limiting enzyme is not fully saturated with its amino acid substrate and not subject to feedback control (either by end product or by multisynaptic feedback loops). This allows changes in the brain precursor level to influence the rate at which 5-HT is synthesized and released (Fig. 4).[19,21,55–57] In catecholaminergic systems, under normal conditions the enzyme may not be fully saturated, and small increases in CA synthesis can occur in systems in which there is no feedback. For instance, small increments in NE synthesis occur in CNS neurons following tyrosine treatment of probenecid-treated rats[66] and in peripheral NE systems, as seen in increased urinary NE and metabolite excretion following tyrosine ingestion (24-h urinary CAs are increased by 25% after tyrosine consumption).[75] When catecholaminergic neurons are made to fire more frequently and/or tyrosine hydroxylase activity is increased, increased brain tyrosine will lead to increased CA synthesis (Table I). Activation of TH results in a much reduced affinity for end product and an increased affinity for cofactor.[44–47]

4. CONSEQUENCES OF PRECURSOR CONTROL OF MONOAMINE SYNTHESIS

Increased synthesis following precursor administration leads to increased release of the monoamines (as measured by changes in metabolites) and to behavioral changes. The changes in monoamine metabolite level, although indicative of release of the parent neurotransmitter, are not, by themselves, proof of increased release (they could simply reflect increased metabolism). Tryptophan injection does lead to a moderate elevation of serotonin release into the perfused ventricles of the rat[76] coincident with the maximum elevation of 5-HT level in brain tissue. Similar studies have not been conducted on the catecholamine systems. There are numerous reports of the physiological, endocrinological, and behavioral changes concomitant with tryptophan or tyrosine administration (for reviews, see refs. 7,77–79). Data on behavioral effects are often conflicting and may reflect differences in precursor dosage and time course, time of day effects, and the route of administration (i.e., whether the precursor is given as the pure compound or with food). A few examples are discussed below to illustrate the general requirements for precursor control, e.g., the importance of the amino acid ratio and the influence of increased neuronal activity.

Tryptophan has been extensively studied, with reports of its effects on sleep,[80] locomotor activity,[81-83] pain sensitivity,[84-86] aggression,[87] and eating behavior.[88] In general, tryptophan has a calming or behavioral depressant action, inducing sleep in low doses,[80] reducing locomotor activity,[81] and decreasing muricidal behavior in rats.[87] Tryptophan reduces food intake, reducing the total amount of food that animals consumed during the following 24 h and reducing meal size.[88]

The effect on food intake may be specific for a particular dietary component. Treatments that enhanced central serotonergic transmission selectively diminished carbohydrate consumption.[89] In long-term experiments in which weanling rats were allowed to choose between diets differing in protein and carbohydrate content, the plasma tryptophan ratio was inversely correlated to the proportion of protein consumed.[90]

Dietary protein and carbohydrate content can specifically influence brain tryptophan and serotonin level by the effects on plasma amino acid pattern.[1] Dietary carbohydrate increases the concentration of plasma tryptophan relative to the concentration of the competing neutral amino acids, i.e., the plasma tryptophan ratio, increasing brain tryptophan and 5-HT content; on the other hand, dietary protein decreases the plasma tryptophan ratio and hence brain tryptophan and serotonin. The changes in brain serotonin synthesis and release may influence subsequent eating behavior, thus completing a behavior feedback loop. As plasma amino acid pattern and brain tryptophan content vary with the amount of protein and carbohydrate in the diet, serotonin-producing neurons will exhibit inverse responses to the proportion of protein and carbohydrate in each meal. Thus, serotonergic neurons may act as sensors of food-induced changes in plasma and brain amino acid composition.

The behavioral effects of tyrosine have been less extensively studied. Tyrosine has been reported to increase locomotor activity in mice[12] and aggressive behavior in mice when given as a dietary supplement.[82,83] Tyrosine administration reduces the hyperprolactinemia associated with chronic reserpine treatment (presumably by increasing dopamine synthesis).[69]

The specificity of tyrosine for activated catecholamine systems is demonstrated by its effect on blood pressure under two different physiological conditions. Blood pressure may be controlled by central mechanisms that, in general, will tend to lower pressure and by peripheral mechanisms that will tend to raise it. In spontaneously hypertensive rats, brainstem MHPG levels rise rapidly coincident with a blood-pressure-lowering effect of peripheral tyrosine administration.[91] Addition of tyrosine to the diet or consumption of a high-protein meal will also lower blood pressure in these rats.[92] Intraventricular administration of tyrosine lowers blood pressure in SH rats, confirming the central action of the peripherally administered tyrosine.[93] Hypotension produced by profound blood loss is accompanied by an increased synthesis and release of peripheral catecholamines from the sympathetic nerves and the adrenals in an attempt to raise lowered blood pressure. In this case, peripheral tyrosine administration raises blood pressure via increased synthesis of peripheral adrenal catecholamines.[94]

In spontaneously hypertensive rats, tyrosine administration had its greatest effect on central NE systems, whose turnover was increased in an attempt to lower blood pressure. Conversely, when blood pressure was lowered, tyrosine had its greatest effect on the accelerated systems, in this case, the peripheral nerves and adrenals, and the increased circulating catecholamines raised blood pressure. In normotensive rats, tyrosine administration had slight but insignificant effects on blood pressure (the small increases in peripheral CA synthesis were probably insufficient to cause a change in blood pressure).

Many scientists have been reluctant to believe that the brain should be susceptible to the vagaries of changing diet and that food consumption could influence CNS neurotransmission. On the contrary, it may make some sense that an animal is able to measure and respond to changes in environmental stimuli, i.e., to reduce carbohydrate and caloric consumption when presented with sufficient food stores, to change locomotor activity to make the finding of food more likely, or to regulate the diurnal production of hormones, sleep, etc. An animal that is able to respond to subtle environmental changes may have an enormous adaptive advantage.

5. FUTURE RESEARCH

Future studies in this area will concentrate on two areas: one, the use of the individual purified nutrients as "drugs" to be used in certain human diseases in which deficiencies of the particular neurotransmitter exist; the other, an investigation of the behavioral changes that occur as a result of our normal consumption of meals of varying composition (i.e., how much of our behavior is actually determined by the foods we eat).

The basic biochemistry of precursor control suggests that these compounds may be of use clinically in diseases associated with alterations or deficiencies in monoamine neurotransmission. For instance, where there is a loss of MA nerve terminals (as a result of a pathological lesion as in Parkinson's disease or as a result of pharmacological manipulations), remaining nerve terminals may compensate by increasing their firing rates, and precursor therapy may be beneficial. Tryptophan has been tested in many conditions including depression, myoclonus, migraine headaches, sleep disorders, Parkinson's and Alzheimer's diseases, pain syndromes, and in appetite control (for reviews see refs. 7,77–69). Tyrosine is currently being investigated as an antidepressant[7,78,95] and in parkinsonian patients.[7,78,96] Based on the animal studies, tyrosine may also have utility in the regulation of blood pressure (as pure tyrosine or through regulation of the protein intake in the diet).

The use of the amino acids as therapeutic agents offers several advantages. The responses to increased precursor availability are usually self-limited; e.g., serotonin synthesis usually only increases to the point of full enzyme saturation (Fig. 4). In fact, at much higher doses, serotonin neurons may actually decrease their firing rates to compensate for increased transmitter synthesis and release.[97] Excess amino acid substrate is metabolized along normal metabolic routes and is relatively quickly eliminated. Increases in neurotransmission brought about by precursor may be more specific, since they are dependent on the location of the synthetic enzymes, tryptophan and tyrosine hydroxylase (not like L-DOPA, for instance, which can be taken up and decarboxylated by non-catecholamine-containing neurons or even nonneuronal elements[98]). Further specificity is conferred in that neurons that are firing more frequently exhibit the greatest degree of precursor responsiveness. Most other monoaminergic drugs act equally at all synapses and hence can produce unwanted side effects.

Few studies have looked at the behavioral effects of meals by varying composition, and this represents the newest and perhaps the most difficult area of research. Care must be taken to control all other aspects of the environment to insure that it is only the diet that is having an influence on behavior, and diets must be properly designed so that they are not investigating the effects of unphysiological proportions of nutrients but of changes in food consumption that are likely to occur in the normal diet. In preliminary studies, locomotor activity has proved to vary with dietary protein intakes.[99] Over a range of differing protein intakes [from no protein (0%) through low protein (6 and 12%) to normal protein (18%)], rats showed a spectrum of changes in activity that were correlated to protein intake; with decreasing protein, locomotor activity decreased.

Future research may reveal that subtle changes in mood and daily performance are attributable to our dietary intake and reinforce the old maxim that "we are what we eat."

REFERENCES

1. Fernstrom, J. D., and Wurtman, R. J., 1972, *Science* **178**:414–416.
2. Gibson, C. J., and Wurtman, R. J., 1977, *Biochem. Pharmacol.* **26**:1137–1142.
3. Fernstrom, J. D., and Faller, D. V., 1978, *J. Neurochem.* **30**:1531–1538.
4. Fernstrom, J. D., Wurtman, R. J., Hammarstrom-Wiklund, B., Rand, W. M., Munro, H. N., and Davidson, C. S., 1979, *Am. J. Clin. Nutr.* **32**:1912–1922.
5. Blasberg, R., and Lajtha, A., 1966, *Brain Res.* **1**:86–104.
6. Pardridge, W. M., and Oldendorf, W. H., 1977, *J. Neurochem.* **28**:5–12.
7. Wurtman, R. J., Hefti, F., and Melamed, E., 1981, *Pharmacol. Rev.* **32**:315–335.
8. Fernstrom, J. D., and Wurtman, R. J., 1971, *Nature* **234**:62–64.
9. Kaufman, S., 1974, *Aromatic Amino Acids in the Brain* (G. E. W. Wolstenholme and D. W. Fitzsimons, eds.), Elsevier, Amsterdam, pp. 85–115.
10. Karobath, M., and Baldessarini, R. J., 1972, *Nature* **236**:206–208.
11. Bagchi, S. P., and Smith, T. M., 1979, *Res. Commun. Chem. Pathol. Pharmacol.* **26**:447–458.
12. Gibson, C. J., Deikel, S. M., Young, S. N., and Binik, Y. M., 1982, *Psychopharmacology* **76**:118–121.
13. McKean, C. M., 1972, *Brain Res.* **47**:469–476.
14. Yuwiler, A., and Louttit, R. T., 1961, *Science* **134**:831–832.
15. Thom, J. C., Victor, T., Pichanick, A. M. E., Hunter, J. C., and Pretorius, L., 1981, *South Afr. Med. J.* **20**:946–949.
16. McMenamy, R. H., and Oncley, J. L., 1958, *J. Biol. Chem.* **233**:1436–1447.
17. Yuwiler, A., Oldendorf, W. H., Geller, E., and Braun, L., 1977, *J. Neurochem.* **28**:1015–1023.
18. Wurtman, R. J., and Pardridge, W. M., 1979, *J. Neural Transm.* [*Suppl.*] **15**:227–236.
19. Fernstrom, J. D., and Wurtman, R. J., 1971, *Science* **173**:149–152.

20. Wurtman, R. J., Larin, F., Mostafapour, S., and Fernstrom, J. D., 1974, *Science* **185**:183–184.
21. Eccleston, D., Ashcroft, G. W., Crawford, T. B. B., Stanton, J. B., Wood, D., and McTurk, P. H., 1970, *J. Neurol. Neurosurg. Psychiatry* **33**:269–272.
22. Melamed, E., Glaeser, B., Growdon, J. H., and Wurtman, R. J., 1980, *J. Neural Transm.* **47**:299–306.
23. Fernstrom, J. D., and Wurtman, R. J., 1971, *Science* **174**:1023–1025.
24. Fernstrom, J. D., and Wurtman, R. J., 1972, *Metabolism* **21**:337–342.
25. MacKenzie, R. G., and Trulson, M. E., 1978, *J. Neurochem.* **30**:205–211.
26. Crandall, E. A., and Fernstrom, J. D., 1980, *Diabetes* **29**:460–466.
27. Crandall, E. A., Gillis, M. A, and Fernstrom, J. D., 1981, *Endocrinology* **109**:310–312.
28. Litwack, G., and Diamondstone, T. I., 1962, *J. Biol. Chem.* **237**:469–472.
29. Green, A. R., Sourkes, T. L., and Young, S. N., 1975, *Br. J. Pharmacol.* **53**:287–292.
30. Laborit, H., and Thuret, F., 1977, *Agressologie* **18**:83–88.
31. Curzon, G., 1972, *J. Psychiatr. Res.* **9**:243–252.
32. Møller, S. E., 1979, *Lancet* **2**:472.
33. Møller, S. E., 1981, *Neuropsychobiology* **7**:192–200.
34. Tong, J. H., and Kaufman, S., 1975, *J. Biol. Chem.* **250**:4152–4158.
35. Ichiyama, A., Nakamura, S., Nishizuka, Y., and Hayaishi, O., 1970, *J. Biol. Chem.* **245**:1699–1709.
36. Nagatsu, T., Levitt, M., and Udenfriend, S., 1964, *J. Biol. Chem.* **239**:2910–2917.
37. Joh, T. H., Shikimi, T., Pickel, V. M., and Reis, D. J., 1975, *Proc. Natl. Acad. Sci. U.S.A.* **72**:3575–3579.
38. Pickel, V. M., Joh, T. H., and Reis, D. J., 1975, *Brain Res.* **85**:295–300.
39. Lovenberg, W., Weissbach, H., and Udenfriend, S., 1962, *J. Biol. Chem.* **237**:89–93.
40. Axelrod, J., 1972, *Pharm. Rev.* **24**:233–243.
41. Hokfelt, T., Fuxe, K., Goldstein, M., and Johansson, O., 1974, *Brain Res.* **66**:235–251.
42. Koe, B. K., and Weissman, A., 1966, *J. Pharmacol. Exp. Ther.* **154**:499–516.
43. Spector, S., Sjoerdsma, A., and Udenfriend, S., 1965, *J. Pharmacol. Exp. Ther.* **147**:86–95.
44. Zivkovic, B., Guidotti, A., and Costa, E., 1974, *Mol. Pharmacol.* **10**:727–735.
45. Roth, R. H., Morgenroth, V. H., and Salzman, P. M., 1975, *Naunyn Schmeidebergs Arch Pharmacol* **289**:327–343.
46. Morgenroth, V. H., Walters, J. R., and Roth, R. H., 1976, *Biochem. Pharmacol.* **25**:655–661.
47. Lovenberg, W., Bruckwick, E. A., and Hanbauer, I., 1975, *Proc. Natl. Acad. Sci. U.S.A.* **72**:2955–2958.
48. Black, I. B., and Reis, D. J., 1975, *Brain Res.* **84**:269–278.
49. Davis, J. N., and Carlsson, A., 1973, *J. Neurochem.* **20**:913–915.
50. Davis, J. N., Carlsson, A., MacMillan, V., and Siesjo, B. K., 1973, *Science* **182**:72–74.
51. Kettler, R., Bartholini, G., and Pletscher, A., 1974, *Nature* **249**:476–478.
52. West, D. P., and Fillenz, M., 1980, *J. Neurochem.* **35**:1323–1328.
53. Glowinski, J., Besson, M. J., Cheramy, A., and Thierry, A. M., 1972, *Adv. Biochem. Psychopharmacol.* **6**:93–109.
54. Carlsson, A., and Lindqvist, M., 1973, *J. Neural Transm.* **34**:79–91.
55. Carlsson, A., and Lindqvist, M., 1978, *Naunyn Schmiedebergs Arch. Pharmacol.* **303**:157–164.
56. Eccleston, D., Ashcroft, G. W., and Crawford, T. B. B., 1965, *J. Neurochem.* **12**:493–503.
57. Moir, A. T. B., and Eccleston, D., 1968, *J. Neurochem.* **15**:1093–1108.
58. Alousi, A., and Weiner, N., 1966, *Proc. Natl. Acad. Sci. U.S.A.* **56**:1491–1496.
59. Carlsson, A., Kehr, W., and Lindqvist, M., 1974, *Adv. Biochem. Psychopharmacol.* **12**:135–143.
60. Joh, T. H., Park, D. H., and Reis, D. J., 1978, *Proc. Natl. Acad. Sci. U.S.A.* **75**:4744–4748.
61. Letendre, C. H., MacDonnell, P. C., and Guroff, G., 1977, *Biochem. Biophys. Res. Commun.* **74**:891–897.
62. Weiner, W., Lee, F.-L., Dryer, E., and Barnes, E., 1978, *Life Sci.* **22**:1197–1216.
63. Masserano, J. M., and Weiner, N., 1979, *Mol. Pharmacol.* **16**:513–528.
64. Levine, R. A., Kuhn, D. M., and Lovenberg, W., 1979, *J. Neurochem.* **32**:1575–1578.

65. Gibson, C. J., 1977, *Factors Controlling Brain Catecholamine Biosynthesis: Effect of Brain Tyrosine*, Ph.D. Thesis, M.I.T., Cambridge.
66. Gibson, C. J., and Wurtman, R. J., 1978, *Life Sci.* **22**:1399–1406.
67. Scally, M. C., Ulus, I. H., and Wurtman, R. J., 1977, *J. Neural Transm.* **41**:1–6.
68. Melamed, E., Hefti, F., and Wurtman, R. J., 1980, *Proc. Natl. Acad. Sci. U.S.A.* **77**:4305–4309.
69. Sved, A. F., Fernstrom, J. D., and Wurtman, R. J., 1979, *Life Sci.* **25**:1293–1300.
70. Sved, A. F., and Fernstrom, J. D., 1981, *Life Sci.* **29**:743–748.
71. Fuller, R. W., and Snoddy, H. D., 1982, *J. Pharm. Pharmacol.* **34**:117–118.
72. Iuvone, P. M., Galli, C. L., Garrison-Gund, C. K., and Neff, N. H., 1978, *Science* **202**:901–902.
73. Iuvone, P. M., Galli, C. L., and Neff, N. H., 1978, *Mol. Pharmacol.* **14**:1212–1219.
74. Gibson, C. J., Watkins, C. J., and Wurtman, R. J., 1982, *Retina* **2**:332–340.
75. Alonso, R., Agharanya, J. C., and Wurtman, R. J., 1980, *J. Neural Transm.* **49**:31–43.
76. Ternaux, J. P., Boireau, A., Bourgoin, S., Hamon, M., Hery, F., and Glowinski, J., 1976, *Brain Res.* **101**:533–548.
77. Growdon, J. H., 1979, *Nutrition and the Brain*, Volume 3 (R. J. Wurtman and J. J. Wurtman, eds.), Raven Press, New York, pp. 117–181.
78. Growdon, J. H., and Gibson, C. J., 1982, *Current Neurology*, Volume 4 (S. H. Appel, ed.), John Wiley & Sons, New York, pp. 117–144.
79. Young, S. N., and Sourkes, T. L., 1975, *Advances in Neurochemistry*, Volume 2 (B. W. Agranoff and M. H. Aprison, eds.), Plenum Press, New York, pp. 133–191.
80. Hartmann, E., 1977, *Am. J. Psychiatry* **134**:366–370.
81. Modigh, K., 1973, *Psychopharmacologia* **30**:123–134.
82. Thurmond, J. B., Lasley, S. M., Corking, A. L., and Brown, J. W., 1977, *Pharmacol. Biochem. Behav.* **6**:475–478.
83. Thurmond, J. B., Kramarcy, N. R., Lasley, S. M., and Brown, J. W., 1980, *Pharmacol. Biochem. Behav.* **12**:525–532.
84. Lytle, L. D., Messing, R. B., Fisher, L., and Phebus, L., 1975, *Science* **190**:692–694.
85. King, R. B., 1980, *J. Neurosurg.* **53**:44–52.
86. Hosobuchi, Y., Lamb, S., and Bascom, D., 1980, *Pain* **9**:161–169.
87. Gibbons, J. L., Burr, G. A., Bridger, W. H., and Leibowitz, S. F., 1979, *Brain Res.* **169**:139–153.
88. Latham, C. J., and Blundell, J. E., 1979, *Life Sci.* **24**:1971–1978.
89. Wurtman, J. J., and Wurtman, R. J., 1979, *Life Sci.* **24**:895–904.
90. Ashley, D. V. M., and Anderson, G. H., 1975, *J. Nutr.* **105**:1412–1421.
91. Sved, A. F., Fernstrom, J. D., and Wurtman, R. J., 1979, *Proc. Natl. Acad. Sci. U.S.A.* **76**:3511–3514.
92. Osumi, Y., Tanaka, C., and Takaori, S., 1974, *Jpn. J. Pharmacol.* **24**:715–720.
93. Yamori, Y., Fujiwara, M., Horie, R., and Lovenberg, W., 1980, *Eur. J. Pharmacol.* **68**:201–204.
94. Conlay, L. A., Maher, T. J., and Wurtman, R. J., 1981, *Science* **212**:559–560.
95. Gelenberg, A. J., Wojcik, J. D., Growdon, J. H., Sved, A. F., and Wurtman, R. J., 1980, *Am. J. Psychiatry* **137**:622–623.
96. Growdon, J. H., and Melamed, E., 1980, *Neurology (N.Y.)* **30**:396.
97. Gallager, W., and Aghajanian, G. K., 1976, *Neuropharmacology* **15**:149–158.
98. Karobath, M., Diaz, J. L., and Huttunen, M. O., 1971, *Eur. J. Pharmacol.* **14**:393–396.
99. Chiel, H. J., and Wurtman, R. J., 1981, *Science* **213**:676–678.

14

Cerebral Subsystems and Isolated Tissues

Henry McIlwain

1. INTRODUCTION

Parts of the brain, as of other organs of the animal body, can be remarkably autonomous. Survival of much of their functioning for a few hours can require only a minimal supply of materials. When relevant investigations began in the author's laboratory in the 1940s, this was readily accepted with respect to cold-blooded animals such as frogs and small fish, of which the almost intact, isolated brain remained electrically active when supplied with simple bathing solutions. Comparable success was not obtained with preparations from mammalian cerebral systems. However, parts of the brain from laboratory animals and from man had been employed for many years as tissue slices in studying other aspects of cerebral functioning, especially respiration and metabolite interconversion. This was extended first to the metabolic maintenance in tissue slices of their content of labile materials including ATP and phosphocreatine and second to the metabolic maintenance of gradients in Na^+ and K^+ ions between the tissues and bathing solutions. It was then shown that application to the tissues of fluctuating electrical potentials caused changes in the tissue content of the substances just described.[1,2] The electrical stimuli led to breakdown of the labile compounds, to diminution of the concentration gradients, and to increase in the energy-yielding processes of respiration and glycolysis.

Such metabolic responses to stimulation were found in human cerebral tissues[3] obtained at neurosurgery as well as from a variety of other mammalian species, and they prompted further attempts to obtain electrical responses from excised cerebral tissues. These successively revealed[4,5] resting cell potentials and their displacement by chemical means and by electrical stimulation, immediate impulse generation in response to stimulation, and also the ability of some regions to give long-term modification of cell firing as a result of brief stimuli.[6,7] In the 18 years since these results were obtained, investigations using

---

Henry McIlwain • Department of Biochemistry, St. Thomas's Hospital Medical School, London SE1 7EH, England.

isolated tissues and subsystems from the brain have greatly proliferated. They have been applied in metabolic and neurotransmitter studies and in electrophysiological and histological studies, with findings collated in reviews and collective volumes.[8-10]

The special advantages of using isolated cerebral tissues and subsystems include the following:

1. The traumatic process of isolating a chosen subsystem of the brain is not carried out immediately before analysis or fixation for histological examination; by contrast, the measurements follow a period of reestablishing tissue composition and functioning under chosen, optimal, conditions.
2. Defined bathing fluids replace the cerebrospinal fluid and interstitial fluids of the brain *in situ*. The bathing fluids are readily accessible for analysis or for making additions of potentially neuroactive substances. They can be supplied by superfusion at varying speeds and for chosen purposes and can either imitate or can greatly extend the interstitial fluids normal to the tissue.
3. Defined electrical conditions can be applied to the isolated tissues as field potentials from localized extracellular stimulating electrodes or from intracellular electrodes. Arrays of recording electrodes can be employed and rapidly maneuvered for multiple observations of electrical output. Temporally patterned stimuli can be applied and accurately timed in relation to imposed or observed changes in chemical conditions.
4. Multiple tissue samples are obtainable from a chosen animal; by choice of the areas sampled, interaction between cerebral regions can be examined without involvement of the rest of the brain or of the animal body. Concomitant observation, microscopically, can be made of the placement of stimulating and recording electrodes and, when necessary, of localized superfusion.

2. PREPARATIVE METHODS

Collected information, pictorial descriptions, and diagrams are available.[2,4,8,9,11] Desirable objectives, which are discussed separately below, are as follows:

1. To minimize the time and the handling of the tissue that intervene between stopping blood supply to the brain or to the part of the brain concerned and having the chosen tissue under incubation conditions *in vitro*.
2. To prepare the tissue isolate in reproducible and minimally damaged fashion with appropriate criteria of reproducibility.
3. To choose incubation conditions capable of maintaining the tissue property which it is desired to investigate: incubating solutions, gas phase, and apparatus come under consideration.

2.1 Obtaining the Isolate Minimally Altered

In planning day-to-day experiments, all preparations that can be made ahead of tissue removal, such as those of solutions and apparatus, should be carried out first. In handling the biological material rapidly, it is usually desirable for two or three workers to collaborate in sharing responsibility for apparatus and supplies, preparing tissues, and their initial weighing and mounting in any tissue-holding arrangements that are adopted.

The brain from small laboratory animals is removed post-mortem; from others, portions may be taken under anesthesia, and in appropriate instances human biopsy samples may be obtained with neurosurgical collaboration.[3,5] In the latter instance, and also when slaughterhouse material is transported, excised blocks of tissue are placed in small, otherwise empty, watertight containers, which are carried in ice at 0°C. In all other instances, it is recommended that the brain or excised portion immediately be placed on a cutting table e.g., an 8-cm square of plastic that is horizontal and cemented to a heavy base that elevates it 6 cm above the bench and on which a filter paper moist with the incubating fluid has been freshly placed. At room temperature, 12–18°C, the chosen tissue blocks are dissected free from the remainder of the brain; methods for obtaining these blocks are described in Sections 4–8 below. Tissue slices are then cut to their chosen tissue thickness with a bow cutter spaced with a tissue guide with coverslip insert (Section 2.2). When tissue weight is required, the coverslip plus tissue are rapidly weighed, e.g., by torsion balance, at this stage. The coverslip plus tissue is then transferred to the fluid of a slice chamber (Section 3.1) or of a mounting bath,[4] and the slice floated free from the coverslip. The chamber or bath contains the buffered incubation fluid already oxygenated and at 37–39°C.

The time between cessation of blood supply and the incubation of tissues in oxygenated fluids at 37–39°C should be noted with a view to minimizing it or appraising its effects.

2.2. Choice and Criteria of Preparative Methods

Unless the tissue function being observed proves sensitive to oxygen, there is marked advantage in using a gas phase of 90–100% oxygen. With tissue of typical respiratory rate, the oxygen that it consumes in respiration can be replaced by diffusion to depths of 0.15–0.2 mm from an outer surface, allowing a slice exposed on two sides to be 0.3–0.4 mm thick and still adequately oxygenated. Thicker slices show increased production of lactic acid from glucose, low values for potassium content, and fewer or modified electrical responses; these measures can be used as criteria for adequacy of tissue oxygenation under otherwise normal conditions.

To obtain tissue sections of a chosen thickness reproducibly, different forms of templates or machines were used. Recommended[4] is a glass guide, which consists of a microscope slide about 4 × 8 cm on one face of which, near its long sides, have been cemented strips of coverslip 5 × 65 mm, leaving a central recess of appropriate depth (0.2–0.4 mm; see below). When this guide is in use, a coverslip (No. 0, 22 × 32 mm) made lightly adherent with moisture from condensed breath or a few microliters of water is placed in this recess. The guide with adherent coverslip is inverted over the tissue block that is to be sliced, and the chosen section cut with a bow cutter.

The bow cutter consists of a narrow, 1 mm, section of blade edge 75 mm long taken from razor blading and held in a small frame the shape of a bow saw. The total length of the frame is 12 cm, and it exerts a tension of 200 g on the blade, keeping it sufficiently rigid. In use, the central coverslip adhering to the cutting guide rests on the tissue block to be sliced, and the blade of the bow cutter must touch the lateral cemented coverslip sections at each side of the guide. The bow cutter is then moved forward into the tissue with a cutting motion. The tissue slice so produced adheres to the central coverslip of the guide, which is then detached, leaving the coverslip plus slice. Handled by the coverslip, the pair are transferred to a balance if slice weight is needed and then to incubating fluid in a mounting bath or in the apparatus in which tissue properties are to be observed. The same guide then receives another coverslip of known weight for cutting a further slice; as successive slices are cut, a colleague weighs them and transfers them to incubating solutions, where they are floated free from the coverslip and placed as necessary into transfer holders or into electrode systems or simple incubating vessels.

The thickness of slices cut in this fashion is approximately the depth of the guide recess plus half the thickness of the cutting blade and minus the thickness of the central coverslip. A recess of 0.38–0.4 mm, measured by engineer's micrometer, gives a slice about 0.35 mm thick. The thickness of slices cut using a particular guide can be calculated from their weight and area: slices, e.g., of neocortex and of 1 to 2 cm^2 are weighed and then are floated over millimeter-ruled paper placed below a Petri dish of incubating fluid.

2.3. Alternative Methods

Procedures in which tissue is cooled, e.g., to 0°C, before cutting are not recommended for general use, although a special occasion for their use was noted in Section 2.1. Cooling dissociates microtubules and diminishes energy-yielding reactions. Procedures in which tissues, before or after cutting, come into contact with solutions other than oxygenated incubation fluids containing glucose or a surrogate are also to be avoided for general use. Again, there are exceptions when nonaqueous fluids or deliberately depleted or enriched media have been used.[4,11] Placing large blocks of tissue, e.g., a rat hemisphere or a 1-mm-thick slice, in warm oxygenated media is also undesirable, as the preparations are oxygenated to a limited depth only, and the outer portions appear to be subjected to deleterious influences from the remainder. Thus, guinea pig piriform lobe so treated did not afford the electrical responses of Section 5. Cutting procedures have been widely recommended in which the tissues to be used are compressed between a broad cutting blade with wedge-shaped cutting edges and a screwed-together template; these are undesirable, as the full width of the blade abrades two tissue surfaces in the course of producing a slice.

Sections 4–8 contain descriptions of the initial preparation of tissue blocks separated from the brain by an initial dissection. As an alternative to the manual slicing of Section 2.2, such blocks may be cut mechanically. A *vibratome*[9,12,13] enables blocks of tissue a few millimeters in size to be sliced under fluid by a fine blade that is given a rapid cutting motion electrically; the progress of cutting can be observed microscopically. Chopped tissues[4,14] have a distinct sphere of application and have occasionally been used for electrophysiological studies.[15] It is to be noted, however, that even when made to similar dimensions, chopped tissues are not equivalent to tissues sectioned with a cutting motion. More damage was evident histologically in chopped cerebellar tissue than in slice; the chopped tissue also gave smaller responses to kainic acid and to glutamate when measured in terms of cyclic GMP formation.[16] The degree of damage depends on the sharpness of the blade.[15]

2.4. Incubation and Superfusion Techniques

Incubation fluids normally used[4,11] resemble extracellular fluids with which the tissues are in contact *in vivo*, that is, interstitial fluids and cerebrospinal fluid. They thus resemble the Krebs–Ringer fluids (K–Rs) adopted earlier for metabolic studies. Data are available[4,9,17] on preferred and alternative buffers, oxidizable substrates, and other constituents. Thus, for specific purposes, there have been added to such salines Mn^{2+}, Co^{2+}, amino acids, glutathione, fumarate, ascorbate, creatine, adenosine, guanosine, norepinephrine, histamine, gangliosides, cerebrospinal fluid, and fractions from plasma proteins. Trial of many such additions may be advisable before it is concluded that isolated tissues do not reproduce a particular phenomenon observable *in situ*.

In relation to a particular tissue sample, the chosen incubation fluid(s) may be employed as a single batch, as a succession of smaller amounts, or by superfusion. In each case, the volume of fluid used should be considered in relation to the volume of blood or cerebrospinal fluid available to or passing through or over the tissue *in situ* during the experimental period chosen. Thus, for an experiment of 1 h, 3 ml of fluid are commonly provided for 30–100 mg of tissue, and this is approximately the volume of blood flowing through the tissue *in situ*. It would, however, correspond more closely to *in situ* conditions to supply a series of much smaller volumes batchwise or by superfusion, as the larger volume can at the beginning of such an experiment leach materials from the tissue or provide a reservoir in which they accumulate.

As an alternative to the tissue being immersed in fluid, it may be supported in such a way that fluid flows over it and constitutes only a thin film above it and in some cases also below it.[4,18,19] In many experiments, tissues have been completely immersed in K–Rs during a preincubation period of some minutes, established as necessary for resynthesis of labile constituents modified during tissue preparation, and have subsequently had the fluid withdrawn to leave the tissue resting on a fiber grid with minimal fluid on its upper surface, which received stimulating and recording electrodes and/or chemical additions.

3. TYPES OF OBSERVATIONS FEASIBLE

Cerebral tissue slices or isolated subsystems are associations of some 10^5–10^6 cells retaining most of the relationships to each other that arose during their growth *in situ*. They are surviving tissues, understandably different from cultured tissues and represent perhaps 10^{-3} to 10^{-4} of a typical mammalian brain. Despite the increase in surface caused by their preparation, this additional surface is still only a minute fraction of the outer cell surface of the component tissue cells. The isolates thus offer good access to interstitial fluids but not to intracellular components. Because cell structure is largely intact, the isolates are suitable for examining phenomena of cellular control by metabolic and other techniques. Intercellular relationships are also largely intact, especially synaptic phenomena, for synaptic structures are usually among the neural components of the brain most resistant to physical distortion. The neurotransmitter and electrophysiological components of synaptic transmission are thus among the phenomena most extensively studied in isolated cerebral subsystems and tissues. Sections 3.1–3.4 give some general examples illustrating these observations, and Sections 4–8 give details relating to individual cerebral regions.

3.1. Metabolic Responses to Applied Agents

Among agents evoking characteristic responses from the isolates, electrical techniques most impressively demonstrate their continued functioning. The metabolic responses observed include those of Section 1, several of which have been measured in relation to the frequency, duration, amplitude, and temporal patterning of applied stimuli.[4,9,11] The stimuli were first applied as field potentials from electrodes placed so as to enclose most of the tissue samples within the potential gradients generated. Specific effects, however, were seen when stimulating pulses were applied to localized electrodes placed in relation to structural elements of chosen tissues. This gave indirect evidence for impulse conduction within isolates.

Metabolic responses of isolates to electrical stimulation required normal cation balance and substrates in bathing media, and the substances in which changes were induced included energy-rich phosphates, cations, and the turnover of phosphoproteins and of certain phospholipids. A number of substances inhibited responses to electrical stimulation, including general depressants, analgesics, anticonvulsants, phenothiazines, and tetrodotoxin.[4,5,11] Protoveratrine or 30 mM potassium salts, which like electrical potential gradients caused depolarization, paralleled only some of the resultant changes in the tissues.[4]

3.2. Neurotransmitter Synthesis, Metabolism, and Output

Isolated tissues and cerebral subsystems are suitable for such studies, and the present notes are brief to avoid overlap in content with Volumes 4, 5, and 6. The most valuable data are obtained from apparatus that allows the isolates to be simultaneously superfused and electrically stimulated.[4,8,9,19,20] It is then possible to measure precursor uptake, basal output of neurotransmitter and related compounds, and their output during and after various forms of stimulation. Such data can, further, be compared with parallel results obtained with the additional presence of chosen inhibitory or other modulating agents. The methods are accordingly much used in study of transmitter antagonists and synergists and in analysis of drug action.

Several stages subsequent to neurotransmitter release are also susceptible to study in cerebral isolates, including especially the resultant changes in cyclic nucleotide content and protein phosphorylation.[21-23] Apart from the investigation of already established neurotransmitters, evidence for adenosine as neurohumoral agent in cerebral systems was largely obtained by examination of isolated neocortical tissues.[19,21,23,24] Observations regarding acetylcholine, catecholamines, serotonin, and excitatory and inhibitory amino acids are made in Sections 4 to 8, below.

3.3. Electrical Responses to Stimulating Agents

The isolates *in vitro* do not usually exhibit spontaneous cell firing; "injury discharges" are not in evidence. Resting membrane potentials observed by intracellular microelectrodes have been found to be within ranges observed in cerebral cells *in vivo*.[6,18] In the isolated tissue, their modulation with change in bathing solutions is readily observed; diffusion of depolarizing substances to various depths within tissue slices has been measured.[18,25] The size of regions of negative potential has also been measured and correlated with cellular dimensions in the tissues examined. Such observations employed a solidly mounted tissue chamber[18] with rigidly clamped micrometers to carry the glass microelectrodes. These, stimulating electrodes, and micropipettes for addition of solutions could be maneuvered through a central aperture above a tissue slice that was in or at the surface of incubating fluids and supplied with moist O_2–CO_2 to the fluid and above its surface.

This apparatus also served for observations of action potentials by extracellularly placed electrodes either at the tissue surface or penetrating it.[7,26] Cell firing was readily initiated in a wide variety of isolates by stimuli similar to those that modified cell firing in the brain *in situ*.

3.4. Histological Examination of Isolates

Interpretation of chemical or electrical findings with tissues or isolated subsystems usually needs knowledge of the cellular and subcellular structures carried by the specimen being examined; instances are quoted in Sections 3.3, 4, and 6. The structure of rat brain tissue slices in terms of nerve tracts and nuclei is described in excellent detail in a 60-page atlas.[54] The extent to which cell structure and organelles are preserved in variously treated tissues has also been made the subject of specific investigations. Well-oxygenated tissues in normal, good media were well preserved, and effects of hypoxia in causing tissue damage could be recognized.[27]

An examination of microtubules within axons, dendrites, and synaptic boutons of neocortical slices[28] showed them to be largely stable to the temperature changes involved in tissue preparation and incubation. Several of these categories of structures have been enumerated quantitatively in incubated preparations of rat hippocampus using electron microscopy[15]; this is described in more detail in Section 6.4, for morphological changes have been reported to follow electrical stimulation of the isolates during incubation. If due measures are taken for preservation of labile constituents, subcellular fractionation of incubated tissues can be carried out to localize metabolites, including cyclic AMP[29]

4. NEOCORTEX

Neocortical preparations from the rat, guinea pig, and man were the first mammalian cerebral tissues that afforded metabolic responses to electrical excitation in isolation.[1-3] As the basis for this, the tissues were shown to exhibit differential ion concentrations, resting membrane potentials, and spike discharges.[4-6,11,18] The phenomena were exhibited in tissue sections 0.35 mm in thickness, and in the guinea pig the size and incidence of regions of negative potential were correlated with the cell structure of the cortical layers.[25] The rate and extent of depolarization induced by specified electrical conditions were modified by endogenous agents and by a number of added reagents including centrally acting drugs.[4,11,30] These observations were made with intracellular micropipette electrodes, which also showed diminution in membrane potentials to be caused by increased extracellular [K^+], by glutamate and other excitatory amino acids, and by hypoxia.[18,31]

With extracellular stimulating and recording electrodes, tissues 0.2 mm thick from the outer layer of guinea pig neocortex responded to individual stimuli applied to the former outer surface by producing two successive negative waves.[32] Both components of this direct cortical response were concluded to be derived from the distal portion of apical dendrites; histological examination of such slices showed them to consist only of the molecular layer of the cortex. The neocortex of the rat was included with the striatum in an *in vitro* preparation[33] that showed spontaneous and evoked activity in a glucose bicarbonate Krebs solution. Stimulation from points at the cortex caused spikes in a defined radial region of the striatum with a 3- to 5-ms latency. The response was concluded to be mediated by a monosynaptic input that required the presence of calcium salts in bathing solutions. Chloride ions are also required for normal cell firing in isolated neocortical preparations,[34] and the mechanism of the depressant action of adenosine on cell firing has been examined in isolated neocortex.[35] A β-adrenoreceptor-activated adenylate cyclase is concluded to be involved in linkage between electrically induced depolarization and energy-metabolic responses of neocortical tissues.[36]

5. PIRIFORM LOBE: OLFACTORY CORTEX AND LATERAL OLFACTORY TRACT

The piriform lobe of rats or guinea pigs has provided a valuable system in which a simply prepared section can include a clearly demarcated input, the lateral olfactory tract, to innervate major cellular components of the same section.[7,26] In small rodents, the lateral olfactory tract forms part of the surface of the piriform lobe and is visible as a white band leading from the olfactory bulb to its connections in the cortex.[7,9,19,26] After the brain of a guinea pig or rat is taken out, the rhinal fissure can be used as a landmark to separate from the neocortex of each hemisphere a block of tissue that includes the piriform cortex and the tract. Such a block with its original outer surface uppermost is placed on a cutting table; a bow cutter and guide (Section 2.1) now yield a slice 0.35 mm thick and weighing about 12–15 mg from each hemisphere.

Such slices carry up to 1 cm of myelinated tract and so enable satisfactory separation of stimulating and recording electrodes. Typically, slices have been preincubated immersed in Krebs–Ringer solutions, which subsequently are withdrawn to leave the tissue with an upper surface drained but in a moist O_2–CO_2 atmosphere; the solutions are then supplied by flow to the lower tissue surface, or the tissue is superfused.

In initial investigations,[7] fine ball-tipped silver wires were used for stimulation near the cut end of the tract, and a similar electrode for recording at various points on the tract and at the cortical surface. Greatest response was shown in the prepiriform area and, to a single stimulus, consisted of a conducted impulse followed by a negative wave 15 to 30 ms in duration. Responses to variously grouped stimuli gave evidence for the negative wave being composed of postsynaptic potentials of cortical neurons; spike potentials of such neurons were also seen.

Thick slices of the piriform lobe (600 μm) have also been examined, but necessarily at a lower temperature of 25°C.[37,38] These afforded intracellularly recorded potentials of some -75 mV in cellular elements whose membrane resistance was also measured; ketamine and other anesthetics increased the resistance. Up to 3 h sojourn *in vitro* was needed for responses to excitation to return to the tissues, possibly on account of the subnormal temperature employed. The cellular origin of the responses was investigated by recording at different depths in the tissues, which were subsequently examined histologically. Spike generation and its dependence on ionic environment were measured, including observations with tetrodotoxin and tetraethylammonium salts. Interactions of excitatory and inhibitory amino acids were displayed.

6. HIPPOCAMPUS

The hippocampus is intriguing structurally and functionally; behavioral studies and lesioning techniques have demonstrated roles in relation to emotional and drive mechanisms and involvement in cognition and memory fixation. During 15 years of *in vitro* investigation, many of its notable features have been investigated in isolated preparations; the lamellar structure of particular hippocampal regions has made tissue slices, cut in appropriate orientations, natural units for study. In laboratory animals, the hippocampus is readily recognized and dissected; when sliced, its structural features can be seen directly or at small magnification and used to guide the placement of pipettes, electrodes, and lesions. Major fiber systems of the hippocampus travel perpendicularly to its longitudinal axis and are so organized as to afford distinct cellular and dendritic layers.

6.1. Dentate Gyrus

The isolated preparations that were initially used[39,40] consisted only of parts of the hippocampus and were obtained from the brain of a guinea pig after removing the brainstem and exposing the regio inferior of the hippocampus with the dentate gyrus. A slice 0.25 mm thick × 3 × 5 mm was taken from this region with a narrow blade and guide and incubated in a glucose bicarbonate Krebs–Ringer in apparatus (Section 3.3) that allowed manipulation of surface electrodes while maintaining an O_2–CO_2 atmosphere. Silver ball electrodes were placed in defined relationships to the granule cell layer; stimulation gave one or two negative spikes superposed on a slow negative wave lasting 20–30 ms, phenomena closely similar to those obtained by comparable stimulation *in vivo* and concluded to indicate an excitatory postsynaptic potential plus the synchronous discharge of granule cells that it induced. Acetylcholine at 0.1 μM or greater in the presence of eserine suppressed the spike and wave obtained on stimulation while also generating spontaneous discharge. From these experiments and others with intracellular recording and in chloride-free media, acetylcholine was concluded to act at presynaptic terminals on the granule cell bodies and proximal dendrites; the cholinergic tracts came from the septal region.[41]

Slices of rat dentate gyrus cut at 200- to 350-μm thickness by vibratome (Section 2.3) at chosen orientation to the hippocampal axis were examined[41] during superfusion at a few milliliters per minute. This, in a chamber with a thin fluid phase only over the tissue and a moist oxygen-rich gas phase above (Section 3.3), allowed stable intracellular resting potentials of 50 to 70 mV to be recorded for up to 5 h from granule cells. The granule cell potentials were found to be sensitive to electrophoretic application of serotonin, γ-aminobutyrate, and glutamate. Evoked field potentials near the granule cell body layer were also sensitive to lowering glucose concentrations to 2 mM and below.[42,42a] On intracellularly measured potentials, γ-aminobutyrate and serotonin had similar effects: both caused depolarization and inhibition of spontaneous action potentials.[41] Evaluation by current injection showed similar reversal potentials for the two actions, giving evidence for mediation by alteration in membrane permeability to a common ion, probably chloride. It is suggested, however, that the origins of the fibers acting by serotonin and by γ-aminobutyrate differ from one another and also from those acting by glutamate (Table I).

6.2. The CA1 Region

Many of the experimental opportunities, advantages, disadvantages, and requirements for *in vitro* studies of the present type have been examined in relation to the pyramidal cells of the CA1 hippocampal region in the guinea pig and rat.[8,43,44] These studies showed close resemblance to *in vivo* observations

Table I
Connections in Hippocampal Regions Involved in Studies with In Vitro Preparations[a]

Region	Receives input from	Other observations
Dentate gyrus, granule cells	Septal area; cholinergic	Also releases adenosine
	Entorrhinal cortex; glutaminergic	The perforant pathway to granule cell dendrites
	Locus coerulus; noradrenergic	—
	Medial raphe; serotonergic	Serotonin application to granule cells inhibits
CA1, pyramidal cells	i. Schaffer commissural afferents, stratum radiatum	Activation via i to iv monosynaptic
	ii. Alveus and striatum oriens via fimbria	Efferent fibers also go to fimbria
	iii. Alvear pathways from entorhinal region	—
	iv. Perforant pathway from stratum radiatum	—
	v. Mossy fibres from dentate, hilar region	Polysynaptic
CA3, pyramidal cells	Mossy fibers, axons of granule cells after passing dentate hilus	—

[a] Data from refs. 8,9,40,42.

in resting potentials, spike characteristics, and excitatory and inhibitory postsynaptic potentials under the two conditions. Dendritic spiking was shown to be calcium dependent. For examination of the rat brain *in vitro*, it was removed under anesthesia and cooled in incubation fluid to 4 to 6°C, and a block carrying the hippocampus was dissected free. This was chopped to yield slices of 300–400 μm, which were placed in baths at 33–37° in oxygenated fluids. Tissues remained active for 8–12 h, and single-cell penetrations with microelectrodes could be maintained for up to 4 h with flow rates of a few milliliters of fluid per minute. Resting membrane potentials of 66 mV were observed from pyramidal cells of the region. These could be excited monosynaptically to give spikes by stimulating via the routes i to iv of Table I; in all of these instances, latencies were 6–9 ms, and the pathways were concluded to be monosynaptic. Excitatory and inhibitory postsynaptic potentials were also observed. Antidromic stimulation was carried out and gave evidence for projection from the CA1 cells through the fimbria.

Examined in preparations of guinea pig hippocampus,[44] pyramidal cells of the CA1 region gave resting membrane potentials of 63 mV and a spike amplitude of 71 mV. Excitatory postsynaptic potentials were given by stimulation from most of the stratum radiatum, leading to action potentials in 3–10 ms. Pulses arriving antidromically from the alveus could also stimulate, but substimulatory pulses gave inhibitory postsynaptic potentials. Severing the tissue with small knives to allow only limited input to the dendritic field of CA1 pyramidal cells showed that only some 5% of the incoming fibers were sufficient to excite the cells. By this technique, synapses on the dendrites proximal to the cell body were shown to be functionally different from more distal synapses. An excitatory postsynaptic potential recorded from the CA1 dendritic region in response to stimulation of Shaffer commissural projections was diminished by 5 μM adenosine, whereas theophylline facilitated the response, actions favoring adenosine as extracellular modulator.[8]

γ-Aminobutyrate applied iontophoretically to CA1 pyramidal cells reduced the amplitude of their population spike, blocking spontaneous and induced unit activity.[44] Dendritic responses could be blocked by local application of γ-aminobutyrate while still leaving the cell body excitable. The effects of γ-aminobutyrate were blocked by bicuculline or picrotoxin and were concluded to be mediated by movements of chloride ions. Glutamate applied electrophoretically caused depolarization and a burst of spike discharges; about 1 s was needed for repolarization. Localized sensitive areas were observed on the dendritic tree, and glutamate was concluded to be an excitatory transmitter at CA1 excitatory synapses.[44]

Confirmatory experiments with rat hippocampal preparations showed release of glutamic acid, measured isotopically, from sections that had been preincubated for 10 min in media containing labeled glutamic and aspartic acids.[45] After the sections had been washed free from excess radioactivity, continued incubation with superfusion showed diminishing output of label until the tissue was stimulated at the Schaffer collateral commissural fibers entering the region superior. Recording at the pyramidal cell layer showed electrical responses, and the fluids collected from the dendritic zone at the stratum radiatum now showed a sharply increased quantity of isotopically labeled glutamic acid. Liberation of aspartate was not increased by the stimulation, nor was that of glutamate when the superfusing fluids contained no added calcium salts.

6.3. The CA3 Region

Transverse sections 0.3 mm thick of the rat or guinea pig hippocampus, including the dentate gyrus, subiculum, and presubiculum, have been examined[39,40] after preincubation for 30 min at 37°C in bicarbonate–glucose Krebs–Ringer solutions. Neurons of the CA3 region were activated by stimuli applied to silver-ball electrodes in the granular or subgranular layer of the dentate gyrus. A negative field potential developed, which increased in amplitude and complexity when the chlorides of bathing solutions were largely replaced by propionate ions and was then followed by afterdischarges. Intracellular recordings from the CA3 neurons showed a small excitatory postsynaptic potential and a long-lasting inhibitory potential after granule layer stimulation. The latency of the action potential triggered by the EPSP indicated the occurrence of chemical transmission; the IPSP could be inhibited by 1 μM bicuculline.[46]

6.4. Long-Lasting Potentiation of Synaptic Transmission in Hippocampal Preparations

Brief bursts of high-frequency stimuli applied to hippocampal regions in rats and guinea pigs can give an unusual long-lasting facilitation of synaptic transmission, which persists for some weeks. Comparable phenomena have been reproduced in isolated hippocampal preparations maintained in vitro.[47,48] In preparations from rat brains that had been cooled to 4°C in bicarbonate–glucose incubation media, slices were made by chopping at 500-μm intervals and examined at 32–34°C in the oxygenated media in a slice chamber admitting electrodes and microscopic observation.[15] Stimulating electrodes about 3 mm apart were used on the stratum radiatum in the CA1 region and its junction with CA3. Occasional stimuli produced in the dendritic zone a potential measured extracellularly to be about 2 mV. High-frequency bursts of stimuli (three of 200/s for 0.5 s) increased two- to threefold the rate of rise in potential when this was measured again at subsequent periods some 10 min or several hours afterwards.

Examined electron microscopically,[15] incubated unstimulated slices (except near their cut surfaces) were found to be relatively unchanged when compared with tissue similarly fixed without incubation; portions of tissue near the surfaces newly made by cutting were, however, found to contain swollen elements. Quantitative appraisal of the length of postsynaptic densities, of the areas of dendritic spines, and of the width of spine processes showed that the stimuli that induced potentiation brought about measurable cytological changes. These consisted of (1) an increase in the number of synapses on dendritic shafts and (2) a change in shape of dendritic spines, which became less elongated and more rounded. The increase in shaft synapses was about 50% ($P < 0.02$), as was that of the ratio of shaft to spine synapses ($P < 0.005$). Shaft synapses represented only 2 to 3% of all synapses in the apical dendrite zone in which they were enumerated; thus, the change caused by the stimulation was in a minority of the synapses of the tissue; whether it had occurred on the dendritic shafts of pyramidal cells or of interneurons was unspecified.[15]

These phenomena probably have metabolic bases. The long-term potentiation was found not to occur in calcium-free media, suggesting the change to be at the level of synaptic transmission.[8] Potentiated slices were subjected to subcellular fractionation to yield synaptosomal membrane preparations, and following a clue from other studies,[22] the degree of protein phosphorylation in the membranes was appraised. The stimulation yielding long-term potentiation was found to have induced greater phosphorylation in a particular protein species of molecular weight about 40,000. This did not occur in Ca^{2+}-free media and was attributed to the potentiating stimuli causing Ca^{2+} accumulation intracellularly in a fashion that did not occur when the same number of stimuli were spread over a longer time. The alterations in membrane composition and area of synaptic contact were considered responsible for the potentiation.

7. OPTIC TRACT, LATERAL GENICULATE, AND SUPERIOR COLLICULUS

This system is similar to that of Section 5, the olfactory, in including a powerful sensory input by a well-defined tract to regions rich in synaptic interconnections. In the guinea pig, an optic tract carrying at one end the optic chiasma and at the other the superior colliculus can be dissected free from the remainder of the brain in 2–3 min and maintained at 37°C in a chamber with Krebs–Ringer bicarbonate–glucose medium.[19,49] A number of metabolic measurements have been made with such superfused isolates: on stimulation of the optic tract, output of lactic acid and of [^{3}H]adenine derivatives increased following their labeling during a preincubation period, as did that of [^{3}H]compounds following uptake of [^{3}H]serotonin. The output of serotonin caused by electrical stimulation did not occur in the presence of 1 μM lysergic acid diethylamide.

Results of stimulating the tract electrically from the tips of extracellular silver wires have been observed with intracellularly and extracellularly placed micropipette electrodes.[13] These showed excitatory and inhibitory postsynaptic potentials in the lateral geniculate nucleus that were blocked by low concentrations of Ca^{2+} salts or by raising Mg^{2+} to 6.3 mM. Observed at extracellular electrodes, 1 μM bicuculline blocked the inhibitory potentials, and the excitatory potentials were seen as small positive waves, the S-potentials, with superposed spikes. Serotonin blocked these potentials and was concluded to do so by suppressing transmitter release from the optic nerve terminals in the lateral geniculate nucleus. An analysis such as this is much more readily made in the isolated system than *in vivo*, when bicuculline is convulsant.

Analogous preparations from the rat and cat, some made by cutting with a vibratome (Section 2.3), have also been examined *in vitro*.[50]

8. OTHER SYSTEMS

Some further neural systems can be mentioned only briefly despite their importance. Detailed investigations of the cerebellum and spinal cord carried out *in vivo* make them attractive subjects for study as isolates. Guinea pig cerebellar preparations were made by cutting 0.1-mm sections in a plane perpendicular to the long axis of a folium; following preincubation and perfusion, spontaneous cell firing was detected at the dendrites of Purkinje cells with electrodes in the molecular and Purkinje cell layers. Such firing could also be accentuated by 9 mM KCl and by weak electrical stimulation.[12] Stronger electrical stimulation at the formerly superficial part of the molecular layer gave complex spike trains closely similar to those of the climbing fiber response seen *in vivo*. This latter response remained in the presence of γ-aminobutyrate, which inhibited the spontaneous discharges.

The spinal cord of the mouse, hemisected longitudinally and superfused with solutions equilibrated with 95% O_2, was stimulated at the spinal roots or at the dorsal tracts, and responses measured as field potentials and also from single cells.[9] The firing patterns recognized included those seen in Renshaw cells. Intracellular recordings were also obtained from spinal motoneurons of kittens and young rats; the preparations, in isolation, showed excitatory postsynaptic potentials and spontaneous activity.

The retina can be regarded as the prototype of tissue slices, for it already exists as a sheet of neural tissue sufficiently thin to be oxygenated by diffusion from its outer surfaces and is attached only at its rim and at the optic nerve. It was examined metabolically in the 1920s[51] in parallel with thinly sliced tissues whose status *in vitro* was being appraised. The studies at that time included events initiated by incident light on the photoreceptor systems, and more recently a variety of electrophysiological events have been examined, in part to act as a model of phenomena in central neural systems.[17]

9. OUTLOOK

Cerebral subsystems and isolated tissues have shown their value at several stages of neurochemical development. Initially, their main uses concerned energy metabolism and cell firing; more recently, they were applied to chemical and electrical aspects of neurotransmission via neurohumoral systems and especially at synapses. The isolates have contributed to displaying the great chemical and regional diversity of cerebral makeup. With present measurements an isolate no longer represents, e.g., the neocortex or brainstem but a specific part of it. Future developments, it has been suggested,[52,53] may include a concern with particular intracellular regions, especially with the elongated arrays of postsynaptic structures exposed to materials undergoing cytoplasmic transport in the long dendritic branches of large neurons. Any part of the brain, sectioned with due regard to its structural elements, can be expected to yield an isolate that displays important aspects of its role *in situ* or, if not, to generate soluble problems in a search to find why not.

Use of thin tissue isolates of 10–100 mg was initially regarded as a necessity dictated by the need to provide material supplies by diffusion; now their use can be seen also to be a contribution to obtaining simpler and more readily investigated subsystems. Isolates of this size carry (Section 3) some 10^5–10^6 cells, but as a large neuron may synapse with some 10^3–10^5 others, such cell groupings may indeed be needed to display a major role played by neurons, that is their specialization for cell interactions. By their elongated shape and arborizations, many neurons occupy domains much greater than their cell volumes. Attempts to develop functioning systems based on "isolated neurons" have not met with much success because of the structural strength and functional importance of synaptic contacts. If nondamaging cleavage at the synaptic clefts succeeded, there would be left, except in favorable cases, a major problem of disentangling cell processes to yield an "isolated neuron." At present, it remains that the minimum environment of a large, intact cerebral neuron includes its contacts with some 10^3 to 10^5 other cells, and thus isolates of 1–10 mg such as have been used in some investigations recounted above are not greatly above this minimum. Miniaturized electronic components and developed microspectrometry[36] enable more chemical and electrical events to be seen within such minimal isolates, whose translucency already invites combined studies of these types.

REFERENCES

1. McIlwain, H., 1956, *Physiol. Rev.* 36:355–375.
2. McIlwain, H., 1961, *Methods Med. Res.* 9:230–236.

3. McIlwain, H., Ayres, P. J. W., and Forda, O., 1952, *J. Ment. Sci.* **98**:265–272.
4. McIlwain, H., 1975, *Practical Neurochemistry* (H. McIlwain, ed.), Churchill Livingstone, Edinburgh, pp. 105–207, 275–292.
5. McIlwain, H., 1975, *The Nervous System*, Volume 1 (D. B. Tower, ed.), Raven Press, New York, pp. 535–539.
6. Li, C.-L., and McIlwain, H., 1957, *J. Physiol. (Lond.)* **139**:178–190.
7. Yamamoto, C., and McIlwain, H., 1966, *J. Neurochem.* **13**:1333–1343.
8. Lynch, G., and Schubert, P., 1980, *Annu. Rev. Neurosci.* **3**:1–22.
9. Kerkut, G. A., and Wheal, H. V. (eds.), 1981, *Electrophysiology of Isolated Mammalian CNS Preparations*, Academic Press, London.
10. Dingledine, R. (ed.), 1984, *Brain Slices* Plenum Press, New York.
11. McIlwain, H., and Bachelard, H. S., 1984, *Biochemistry and the Central Nervous System*, 5th edition, Churchill Livingstone, Edinburgh (in press).
12. Hounsgaard, J., and Yamamoto, C., 1979, *Exp. Brain Res.* **37**:387–398.
13. Yamamoto, C., and Sawada, S., 1981, *Electrophysiology of Isolated Mammalian CNS Preparations* (G. A. Kerkut and H. V. Wheal, eds.), Academic Press, London, pp. 233–255.
14. McIlwain, H., and Buddle, H. L., 1953, *Biochem. J.* **53**:412–420.
15. Lee, K., Oliver, M., Schottler, F., and Lynch, G., 1981, *Electrophysiology of Isolated Mammalian CNS Preparations* (G. A. Kerkut and H. V. Wheal, eds.), Academic Press, London, pp. 189–211.
16. Garthwaite, J., Woodhouse, P. L., Collins, M. J., and Balazs, R., 1979, *Brain Res.* **173**:373–377.
17. Ames, A. III., 1981, *J. Neurochem.* **37**:867–877.
18. Gibson, I. M., and McIlwain, H., 1965, *J. Physiol. (Lond.)* **176**:261–283.
19. Heller, I. H., and McIlwain, H., 1973, *Brain Res.* **53**:105–116.
20. McIlwain, H., and Snyder, S. H., 1970, *J. Neurochem.* **17**:521–530.
21. Kakiuchi, S., Rall, T. W., and McIlwain, H., 1969, *J. Neurochem.* **16**:485–491.
22. Williams, M., and Rodnight, R., 1977, *Prog. Neurobiol.* **8**:183–250.
23. Daly, J. W., 1977, *Cyclic Nucleotides in the Nervous System*, Plenum Press, New York.
24. McIlwain, H., 1979, *Physiological and Regulatory Functions of Adenosine and Adenine Nucleotides* (H. P. Baer and G. I. Drummond, eds.), Raven Press, New York, pp. 361–367.
25. Hillman, H. H., and McIlwain, H., 1961, *J. Physiol. (Lond.)* **157**:263–278.
26. Campbell, W. J., McIlwain, H., Richards, C. D., and Somerville, A. R., 1967, *J. Neurochem.* **14**:937–938.
27. Cohen, M. M., 1973, *Biochemistry, Ultrastructure and Physiology of Cerebral Anoxia, Hypoxia and Ischemia*, Karger, Basel.
28. Jones, D. H., Gray, E. G., and Barron, J., 1980, *J. Neurocytol.* **9**:493–504.
29. Newman, M., and McIlwain, H., 1978, *Biochem. J.* **170**:73–79.
30. Hillman, H. H., Campbell, W. J., and McIlwain, H., 1963, *J. Neurochem.* **10**:325–339.
31. Bradford, H. F., and McIlwain, H., 1966, *J. Neurochem.* **13**:1163–1177.
32. Yamamoto, C., and Kawai, N., 1967. *Exp. Neurol.* **19**:176–187.
33. Miller, J. J., 1981, *Electrophysiology of Isolated Mammalian CNS Preparations* (G. A. Kerkut and H. V. Wheal, eds.), Academic Press, London, pp. 309–336.
34. Richards, C. D., and McIlwain, H., 1967, *Nature* **215**:704–707.
35. McIlwain, H., 1981, *Purinergic Receptors* (G. Burnstock, ed.), Chapman & Hall, London, pp. 163–198.
36. Keller, E., and Cummins, J. T., 1980, *J. Neurochem.* **35**:1329–1334.
37. Harvey, J. A., Scholefield, C. N., and Brown, D. A., 1974, *Brain Res.* **76**:235–245.
38. Scholefield, C. N., 1978, *J. Physiol. (Lond.)* 275:547–557, 559–566.
39. Yamamoto, C., 1972, *Exp. Brain Res.* **14**:423–435.
40. Yamamoto, C., 1972, *Exp. Neurol.* **35**:154–164.
41. Assaf, S. Y., Crunelli, V., and Kelly, J. S., 1981, *Electrophysiology of Isolated Mammalian CNS Preparations* (G. A. Kerkut and H. V. Wheal, eds.), Academic Press, London, pp. 153–188.
42. Bachelard, H. S., and Cox, D. W. G., 1981, *J. Physiol. (Lond.)* **317**:62–64P;
42a. Cox, D. W. G., and Bachelard, H. S., 1982, *Brain Res.* **239**:527–534.

43. Schwartzkroin, P. A., 1981, *Electrophysiology of Isolated Mammalian CNS Preparations* (G. A. Kerkut and H. V. Wheal, eds.), Academic Press, London, pp. 15–50.
44. Langmoen, I. A., and Andersen, P., 1981, *Electrophysiology of Isolated Mammalian CNS Preparations* (G. A. Kerkut and H. V. Wheal, eds.), Academic Press, London, pp. 51–105.
45. Wieraszko, A., and Lynch, G., 1979, *Brain Res.* **160**:372–376.
46. Yamamoto, C., Matsumoto, K., and Takagi, M., 1980, *Exp. Brain Res.* **38**:469–477.
47. Schwartzkroin, P., and Wesker, K., 1975, *Brain Res.* **89**:107–119.
48. Andersen, P., Sundberg, S. H., Sveen, O., and Wigstrom, H., 1977, *Nature* **266**:736–737.
49. Kawai, N., and Yamamoto, C., 1969, *Int. J. Neuropharmacol.* **8**:437–449.
50. Godfraind, J. M., and Kelly, J. S., 1981, *Electrophysiology of Isolated Mammalian CNS Preparations* (G. A. Kerkut and H. V. Wheal, eds.), Academic Press, London, pp. 257–284.
51. Krebs, H. A., 1981, *Otto Warburg: Cell Physiologist, Biochemist, Eccentric*, Clarendon, Oxford.
52. McIlwain, H., 1978, *Prog. Neurobiol.* **11**:189–203.
53. McIlwain, H., 1980, *Neuroscience* **5**:1393–1411.
54. Cuello, A. C., and Carson, S., 1983, *Brain Microdissection Techniques, IBRO Handbook Series*, Volume 2, Wiley, Chidester.

15

Memory

Bernard W. Agranoff

1. INTRODUCTION

For present purposes, learning is defined as the tendency of an organism to increase its probability of responding to a stimulus in a prescribed fashion, whereas memory of the learned task refers to performance of the response when the stimulus is presented at some later time. Entire fields of endeavor within the discipline of psychology deal with the nature and magnitude of stimuli, the responses evoked, and their temporal interdependence. The nature of the inferred physicochemical alterations in the brain that underlie behavioral change or of their electrophysiological or structural concomitants remains largely unknown, but not for a want of interest or effort. Our lack of understanding is all the more striking in view of the significant strides made during the past 30 years in the elucidation of the molecular bases of biological processes, and especially over the past decade in the neurosciences. Progress in learning and memory research prior to 1972 was summarized in the first edition of the *Handbook*.[1] The present chapter emphasizes developments in the interim. The references include a number of reviews and monographs that may be of additional value to the reader[2-9] (see Y. Tsukada, Chapter 16, this volume).

1.1. The Role of Macromolecular Synthesis in Behavior— Interventive and Correlative Approaches

Consideration of the above definitions may raise the question of how learning and memory can be considered separately, since we cannot demonstrate one without the other. To do so, one must further distinguish between short-term memory (STM) and long-term memory (LTM). Short-term memory is acquired at the time of learning. It mediates, for example, the improvement in performance of a task by the subject within a single multitrial training session. Long-term memory mediates retention of the learned response, tested in a

---

Bernard W. Agranoff • Neuroscience Laboratory Building and Mental Health Research Institute, University of Michigan, Ann Arbor, Michigan 48109.

subsequent session a few hours or even years later. The distinction is an important one, since STM and LTM can be separated experimentally: inhibition of protein or of RNA synthesis in the brain has little effect on the ability of an animal to acquire a new task (STM formation), but formation of LTM of the new task is blocked.[2,4,5] It can be shown in some cases that LTM does not begin to form until after the training session is completed. The approach that uses agents that block memory formation is generally classified as interventive (or disruptive). The other major category of experiments applicable to learning and memory studies in whole animals is designated as correlative. With the correlative approach, alterations in the brain resulting from environmental input(s) are sought. For example, increases in synaptic complexity have been reported following environmental enrichment of animals for several weeks, a result that indicates the importance of new synapse formation.[10] Increased brain RNA diversity is also reported.[11] Correlative neurochemical–behavioral experiments often employ radioisotopic tracers. This approach has not generally been as successful as interventive strategies, in part because of the difficulty in distinguishing purported alterations arising from stress, exercise, etc. from those related to cognitive aspects of a training session. Also, alterations may be anticipated to be minute and thus difficult to distinguish from the background of ongoing metabolic events.

1.2. Further Progress in the Role of Macromolecular Synthesis in Mediation of LTM Formation

Although the interventive approach strongly implicates macromolecular synthesis in the preservation of newly learned behavior, the nature of the requisite process(es) is unclear. Regional inhibition studies in the mouse[12-14] and in the rat[15,16] suggest that the amygdala and hippocampus are important sites. Although such experiments might localize a site of action of an inhibitory agent, they tell us little about the specific population of cells that may be involved and even less regarding which proteins must be synthesized in order for LTM to form. Such indications might be derived from correlative studies (see Section 2.6). It is relevant to recall at this point that neuronal macromolecular synthesis is largely if not entirely confined to the cell body. Newly synthesized protein reaches presynaptic and postsynaptic sites via anterograde transport "down" the axon and "up" the dendrite. The relatively slow rates of eucaryotic transcription and translation in addition to those for axonal transport[17,18] would appear to predict what has been inferred experimentally: learning, which can occur in milliseconds, does not depend on macromolecular synthesis. As has been suggested elsewhere,[2,19] the acquisition process (STM formation) may be mediated by posttranslational activities at the synapse. Thus, the known rapid formation of STM and the delayed formation of LTM may reflect the rapidity of posttranslational modification of proteins relative to that of synthesis of the peptide sequence.

How synaptic relationships might be altered in behavioral contexts may be governed by the same rules that determine how they are formed initially and how they are reformed in response to injury.[20] Knowledge of the details of developmental and regenerative growth may thus reveal useful information regarding behavioral change. Of particular interest is the appearance of growth-associated proteins (GAPs) in neurons whose axons have been injured and in which regrowth has been initiated.[21,22] Might the cell body not respond similarly when more limited synaptic "growth" occurs related to behavioral change?

2. RECENT PROGRESS IN BIOCHEMICAL APPROACHES TO BEHAVIOR

2.1. A Phyletic Sampling

Relevance to biochemical mechanisms of behavior is cited in experiments dealing with species ranging from bacteria to primates.[2,6,7] Whether the proposals and claims are justified will be judged by evidence not yet at hand. The unraveling of biochemical mechanisms that mediate chemotaxis in bacteria has been largely the result of effective use of genetic mutants. By this means, the molecular basis of the function and regulation of receptors, of locomotion, and of the intervening primitive logic that directs bacterial movement in a chemosensory gradient has been elucidated.[23] Habituation and sensitization phenomena have been demonstrated in aneural cells. In such experiments, light, heat, mechanical, or electrical stimuli are applied to protozoa or coelenterates, and a stimulus-specific under- or overresponse is demonstrated. Such behavioral modifications are of the nonassociative variety. The altered state, which persists for minutes or hours, has been attributed to alterations in Ca^{2+} fluxes.[24] The claims are much less compelling than those for sensitization, habituation, and more recently of associative learning in the mollusk (see Section 2.2.1 below). Mollusks are of particular interest because their nervous system contains a relatively small number of large, identifiable neurons.

Invertebrate species with rapid life cycles are potentially attractive for genetic studies on behavior.[25] Long-term associative learning has been documented in social insects such as bees,[26] but such species are not readily amenable to biochemical or genetic manipulation. A large number of behavioral mutants have been identified in *Drosophila melanogaster*,[27] a species whose suitability for genetic studies is well known. Cold narcosis is reported to produce a time-dependent block of LTM formation, although in contrast with results in higher species, it is reportedly insensitive to cycloheximide, a blocker of protein synthesis. Retention of active avoidance learning in the praying mantis is reportedly blocked by this agent.[28] "Bait shyness," food aversion following coupling of ingestion with a noxious stimulus, is spread widely through the animal kingdom, including invertebrates[29,30] (see also Section 2.2.1). The genetic approach has also been used extensively in nematodes, and a number of behavioral mutants have been identified.[31]

Both interventive and correlative studies on memory have tended to use small vertebrates, in particular the goldfish, the chick, and the mouse.[2] This probably reflects the low cost of these animals, the availability of suitable training paradigms that produce quantifiable evidence of learning and memory, and the benefits derived from using small animals with favorable brain/body weight ratios (1% or more) for the efficient use of radioisotopes following systemic injection.

The nonhuman primate has the obvious advantage of close anatomic homology of brain structures with human brain, but its expense greatly limits possibilities for biochemical experiments, which generally require large numbers of animals for tissue analysis. Autoradiographic experiments permit a large number of measurements to be made on a single animal and therefore efficiently approaches such problems as localization of brain regions that might be metabolically activated during learning or memory formation. Even so, subjects must be sacrificed for each time point, each drug dose, etc. Autoradiographic experiments related to behavior in primates[32] may eventually be extended to noninvasive studies in humans (see Section 2.7.1).

2.2. Reduced Systems

If one invasively identifies brain circuitry in which observed alterations correlate predictably with gross behavior, the search for underlying altered morphology and biochemistry is greatly facilitated. One would also hope that such demonstrated localized changes could be shown to mediate the behavioral change, although this is not an easy task. In the past, behavioral scientists were quite discouraged by attempts to localize the "engram," a hypothesized physical manifestation of associative learning, by means of brain lesions. It now appears that a number of preparations are available in which a discrete lesion indeed specifically blocks the formation of memory. An example is a recent report[33] that unilateral cerebellar lesions in the rabbit prevent learning of a classically conditioned eyelid and nictitating membrane response without an effect on training of the contralateral eye or on the unconditioned response on the lesioned side.

Alternatively, isolated brain tissue may be shown to respond *in vitro* to electrical stimuli with resultant long-lasting physiological and biochemical changes that can be inferred to be related to alterations underlying behavioral change in the intact organism (see Section 2.2.2).

2.2.1. Molluscan Preparations

Kandel and associates have demonstrated habituation and sensitization of a gill withdrawal reflex in *Aplysia* to be attributable to presynaptic modifications in identified neurons and that Ca^{2+} flux directly or indirectly plays a role.[34] Although habituation appears to be mediated by homosynaptic mechanisms, sensitization is believed to be mediated via a serotonergic presynaptic excitatory input and, further, to involve a cyclic-AMP-mediated step. The cyclic-AMP-regulated phosphorylation of an M_r 160 K protein in the membranes of sensory cells has been proposed.[35] Sensitization can be elicited by a single noxious stimulus and persists 20–60 min, whereas repetitive stimulation can lead to sensitization of days' or even weeks' duration.[36] Athough the terms "short-" and "long-term" memory have been used to describe these phenomena, caution must be taken not to confuse this terminology with STM and LTM as used in studies in higher animals to denote two qualitatively different kinds of memory produced as a result of a single training session and separable by the use of agents that block consolidation, e.g., inhibitors of protein synthesis. Whether LTM as defined by studies in higher animals can be demonstrated in *Aplysia* is not known. At present, biochemical correlates of molluscan sensitization, such as elevated cyclic AMP, do not persist sufficiently long to explain altered behavior over a period of days. Other possible mechanisms such as the induction of new regulatory subunits of a cyclic-AMP-dependent protein kinase must be invoked.[37] Evidence of associative conditioning in mollusks, i.e., CS–US coupling that observes the requisite temporal contingencies, has in fact been reported in *Aplysia*[38] as well as in other mollusks.[39] In *Hermissenda*, a reduced K^+ current has been demonstrated in an identified neuron in the visual system following pairing of light and of rotation but not in randomized controls.[40] An associated increase in phosphorylation of an M_r 20 K protein has also been reported.[41]

Given the degree of restraint and dissection necessary for such studies, the question arises whether behavior can be measured in isolated tissues. In fact, a preparation of the nervous system and lips of the mollusk *Limax maximus* is reported to mediate a feeding response as well as training and retention of an aversive conditioning paradigm.[42]

2.2.2. The Mammalian Brain Slice

Among *in vitro* preparations in which brain tissue can be used for electrophysiological models of conditioning, much emphasis has been placed on the hippocampal slice.[43,44] Heterosynaptic excitation produces long-lasting electrical changes, and altered metabolic changes are reported as described in Section 2.6.2 below.

Following conditions of long-term potentiation (LTP) in the hippocampal slice, there is decreased subsequent phosphorylation of an M_r 40 K protein by [^{32}P]ATP.[44,45] This is interpreted to indicate an increase in the steady-state phosphorylated level of the protein, leaving less apoprotein for the *post-hoc* labeling. The protein has been identified as the α subunit of a mitochondrial enzyme, pyruvate dehydrogenase (PDH), and it has been proposed that the LTP leads to altered mitochondrial metabolism.[46] It has been suggested, for example, that the phosphorylated (inhibited) form of the enzyme leads to lowered acetyl CoA levels and therefore decreased rates of acetylcholine synthesis. It has alternatively been proposed that PDH regulation is linked to Ca^{2+} sequestration by mitochondria. Development of LTP is dependent on the presence of Ca^{2+} and is accompanied by increased cellular Ca^{2+} levels. More recently, glutamate has been implicated as a neurotransmitter in this preparation, and a Ca^{2+}-activated protease has been reported to increase the number of synaptic glutamate receptors.[47]

Synaptic membranes isolated from stimulated hippocampal slices are reported to support an increase in *in vitro* phosphorylation of an M_r 52 K protein,[48] and it has also been reported that the stimulated hippocampal slices incorporate increased radioactivity from added labeled valine into proteins that subsequently appear in the extracellular medium.[49]

2.3. Imprinting

Imprinting, a tendency seen in chicks shortly after hatching to follow a visual object to which they are exposed, is reported to be accompanied by increased incorporation of labeled uracil into the region of the anterior forebrain roof.[50] Various behavioral controls indicate that the increase is specific for the establishment of the imprinted behavior. In split-brain preparations in which only one side of the brain appears to be imprinted on the basis of behavioral criteria, the increased incorporation is reported to occur only on the trained side. Autoradiographic and lesion studies indicate that the effect is specific to the medial hyperstriatum ventrale.[51] Lesion studies indicate that loss of this group of cells blocks retention of the imprinted response.[52]

Other possible biochemical alterations, including changes in biochemical markers of cyclic nucleotide and acetylcholine metabolism, have been suggested, but it is less clear whether they are also specific to imprinting or are confined to a specific anatomic region.[50] Although there are changes in brain amino acid pools during training, it is reported that there is additionally a specific increase in labeling of an acidic protein fraction presumed to be enriched in tubulin in the anterior dorsal forebrain as a function of successful imprinting. There is also increased quinuclidinyl benzilate (QNB) binding in the posterior dorsal forebrain and midbrain.[53] Studies in imprinted chicks injected with [^{14}C]-labeled 2-deoxyglucose indicate increased glucose utilization in the anterior forebrain as a function of imprinting.[54] The possible involvement of the pituitary–adrenal cortical axis in imprinting of ducklings has been reported. Injection of corticosterone reduces the approach response, whereas ACH_{1-10} augments the behavior.[55]

2.4. Kindling

Repeated administration of a subconvulsive electrical stimulus to a specific brain region such as the amygdala over a period of days results in progressively increasing discharge activity, leading eventually to a seizure. There is no evidence for pathological change, and the phenomenon has been proposed to constitute an experimental model for epilepsy.[56] Because of its long-lasting nature, kindling has also been proposed as a useful model of memory formation. The administration of inhibitors of protein synthesis block formation of the kindled response,[57] and it has been proposed that the mechanism of the block may be to prevent formation of altered synaptic relationships.[58] The possible involvement of specific neurotransmitter systems has been proposed. Muscarinic receptors, measured by QNB binding, are reduced in the kindled hippocampus, a fact that appears to be independent of acetylcholine production.[59] Alterations in dopaminergic function are also reported.[60]

2.5. Interventive Agents

The use of blockers of macromolecular synthesis to separate STM from LTM formation has been discussed elsewhere.[1-9] Extensive parametric investigations in the mouse have established details of the depth and duration of block in protein synthesis necessary to prevent memory formation of a passive-avoidance task.[61] In the chick, evidence is reported for multiple stages of memory formation on the basis of sequential consolidation curves generated by agents that affect ion transport, calcium availability, and protein synthesis.[62]

It has been proposed that the various blockers of protein synthesis may selectively affect metabolism of a neurotransmitter or its receptor, and amnestic effects can be demonstrated with agents that block neurotransmitter synthesis as well as with agonists and antagonists of their receptors. Evidence implicating various neuropharmacological agents has been reviewed extensively.[4,63]

2.5.1. Microtubule-Binding Agents

A variety of drugs, including colchicine, vinblastine, and vincristine have long been known as antimitotic agents. They do not readily cross the blood–brain barrier,[64] although behavioral effects are reported in the mouse when they are combined with injection of a dose of anisomycin (a protein synthesis inhibitor) insufficient to be amnestic itself.[65] When the various microtubule-binding agents are injected directly into the brain, they block axonal transport and hence effectively break the flow of chemical information between the cell body and its processes. Intracerebrally injected colchicine is reported to block formation of memory[66] as well as imprinting[67] in the chick. It has also been reported to block memory formation in the goldfish.[68] In each of these experiments, it is important to distinguish specific effects on memory from performance decrements resulting from possible toxic effects of these agents unrelated to the block of LTM formation.

2.5.2. Antibodies

The identification of increasing numbers of brain-specific proteins and recent progress in techniques for the preparation and isolation of both mono- and polyclonal antibodies have led to renewed interest in the use of antibodies as potential disruptive agents for the study of behavior. A series of studies on the effects of antibodies to gangliosides by Rapport and co-workers in the rat indicate the anti-GM_1 antiserum selectively blocks memory consolidation.[69] Antisera to galactocerebroside were without effect. Immunoglobulin G from an antiserum prepared against a mixture of human gangliosides injected i.p. into rats produced an unexpected facilitatory effect.[70] Whether antisera can cross the blood–brain barrier in sufficient amounts to produce behavioral effects is unproven. Antisera injected into pregnant rats reportedly affect behavior in the progeny.[69,70] Antibodies to the S-100 protein have also been reported to have behavioral effects, but antibodies to 14-3-2 do not.[69]

2.5.3. Peptides

Both anterior and posterior pituitary factors have been reported to play a role in behavior.[71,72] The posterior pituitary hormone vasopressin appears to facilitate performance in animals and in humans, although the mechanisms by which they act may be quite different. In animals, a lack of vasopressin is reported to affect acquisition of one-trial passive avoidance but not of multitrial active avoidance. Studies with vasopressin-blocking drugs suggest that there may be a link between the behavioral effect and the pressor activity of this peptide.[73] Fragments and synthetic analogues of ACTH, an anterior pituitary hormone, have been implicated in behavioral mechanisms. A number of analogues, including Organon 2766, a slowly metabolized peptide based on the structure of $ACTH_{4-10}$, have potencies in behavioral assays several magnitudes greater than ACTH or the structurally related hormone, α-MSH. As is the case with vasopressin, the action of ACTH and related peptides appears to be complex. Studies in animals and humans indicate that its major effect is to increase attention rather than to affect cognitive processes.[74] The relationship of effects of peptides on learning and memory, on a grooming response in rodents, and on biochemical actions has been discussed by Gispen and associates.[75,76]

Behaviorally active peptides are reported to enhance phosphorylation of B-50, a protein kinase that modulates enzyme activity that in turn phosphorylates the plasma membrane phospholipid, phosphatidylinositol phosphate, to its bisphosphate form (see also ref. 48). The polyphosphoinositides have been implicated in receptor activation. In the CNS, they have been shown to be highly correlated with muscarinic receptor activation.[77] A role for opiate peptides in learning and memory, particularly in fear conditioning, has been proposed.[78,79] Of possible relevance, the recently described dynorphins are highly active in inhibiting phosphorylation of B-50 kinase.[80]

2.6. Correlative Changes

2.6.1. Protein Synthesis

Although blockers of RNA and protein synthesis can be demonstrated to interfere with consolidation of memory, there is little evidence for altered RNA or protein labeling from injected precursors that can be attributed specifically to LTM formation.[2,4,5] Examples are the reported selective synthesis of M_r 26 K and a 32 K protein and their secretion into the spinal fluid of goldfish being trained in a balancing task, reported by Shashoua[81] (but see also comments in ref. 4). Also, Hyden and co-workers[82] have observed increases in labeling of the brain-specific proteins S-100 and 14-3-2 in rats being trained in reversal of handedness.

2.6.2. Protein Phosphorylation

Since posttranslational modifications could occur quite rapidly in the synaptic region, they have been discussed as candidates for the biochemical basis of STM formation.[2,19] Thus, phosphorylation has been proposed to constitute the molecular basis of long-term potentiation in hippocampal slices (see Section 2.2.2). A similar alteration in the phosphorylated state of the α subunit of PDH following passive-avoidance training in rats has been reported[83] (see also Section 2.2.2.).

2.6.3. Structural Correlates

It has been postulated and is generally accepted that behavioral change is accompanied by altered synaptic connections. Whether such alterations can be expected to be expressed in measurable altered morphology is, however, questionable. A single neuron can have 10^4–10^5 synaptic connections, and even a simple behavior is believed to be mediated by networks of neurons in which only a small fraction of the terminals of any given neuron may be altered. Indirect evidence in support of the hypothesis may be derived by comparison of neuronal branching in animals that have had relatively little behavioral stimulation during development with those that have been exposed to a wide variety of experiences. Altered dendritic branching in experimental animals resulting from extensive behavioral experience has been reported in fish,[84] rats,[85] and monkeys.[86] Most extensive studies have been performed in the rat.[87] Comparison of morphology and biochemical parameters of rats raised in environmentally complex conditions compared to those raised under environmentally improverished conditions reveals a wide variety of alterations in neurotransmitter enzyme levels,[87,88] tubulin synthesis,[89] and in unique-sequence RNA.[11]

2.7. Neurochemical Studies on Memory in Humans

Human behavior, including ideational and linguistic capabilities, clearly separates the human from all other species. The highly developed hemispheric lateralization of the human brain has been correlated with alterations in unique-sequence RNA in right and left temporal cortex specimens obtained on autopsy.[90] Electrophysiological studies in living subjects such as averaged evoked potentials and complex analyses of electroencephalograms have proven useful as behavioral correlates, but only recently have noninvasive biochemical approaches been developed.

2.7.1. Noninvasive Neurochemical Studies

Gamma-emitting inert gases such as $^{77}K_r$ and $^{133}X_e$ have been used for determining regional cerebral blood flow. One can demonstrate, for example, altered regional circulation in an appropriate cortical region following sensory stimulation or a motor response.[91] Association cortex shows increased circulation in relation to speech as anticipated from inferences from the neurological literature. The development of positron-labeled nuclides has greatly increased the possibilities of this approach. These short-lived cyclotron-produced radioisotopes, including ^{11}C, ^{13}N, ^{15}O, and ^{18}F, have already proven useful for the study of brain metabolism. For example, [^{11}C]2-deoxyglucose or [^{18}F]fluorodeoxyglucose can be injected intravenously, and the isotopic distribution pattern in brain can be detected subsequently by a positron emission tomographic (PET) scanner.[92] Such techniques have already successfully shown correlates of visual and auditory stimulation. In the latter case, distinction between left and right temporal cortex glucose utilization can be demonstrated, depending on how auditory information is processed, e.g., language vs. music. It may be anticipated that changes in glucose utilization as a result of learning may be detected in the future in analogy to findings with [^{14}C]deoxyglucose in the bird[54] and monkey.[32]

When labeled deoxyglucose or its fluorinated derivatives are administered by bolus injection, conversion to the trapped phosphate is mostly complete by 10 to 15 min. Regional distribution is not measured until 45–60 min, however, the additional time is required for free deoxyglucose to clear the brain. When the deoxyglucose method is used to study behavioral correlates, the inferred metabolic alteration must then be sustained for a reasonable fraction of the initial 15-min interval. If not, the change may go undetected, since it will be averaged prior to measurement. To observe more transient correlates, the use of cerebral blood flow techniques may be preferable. It is of interest that Le Doux et al. have recently demonstrated altered cerebral blood flow in the rat brain following an auditory stimulus as a result of prior aversive conditioning.[92]

Nuclear magnetic resonance can be used to produce high-resolution brain images based on relaxation times of protons in body water. Although this use of NMR as a structural instrument has obvious important applications, it must be remembered that nonimaging NMR spectrometry is also a powerful tool for the measurement of a number of atoms present in living tissues and of their chemical environments via chemical shift data. Thus, [^{31}P]-containing intracellular intermediates such as nucleotides, sugar phosphates, and inorganic phosphate can be measured noninvasively in a volume of underlying tissue by means of surface coils[94] as a function of ischemia, seizure, etc.

2.7.2. Neuropharmacological Approaches

The administration of various drugs in attempts to reverse human memory deficits, particularly those seen in aging, may eventually furnish clues regarding the physiological mechanisms of memory. For example, the use of cholinergic[95,96] or catecholaminergic[8] agents suggests that specific neurotransmitter pathways may be physiologically involved and affected in the disease. The use of ACTH-like peptides[97] or of vasopressin[98] in humans has been reported to facilitate performance, although it is questionable whether cognitive processes *per se* are improved by these agents. Interpretation of the value of drugs to treat memory deficits in aging populations is complicated by the high probability that what is described in a patient as a memory deficit may contain elements of mental depression and loss of attentiveness, which often accompany the aging process and affect performance of psychological tests.

ACKNOWLEDGMENT. Supported in part by the Wiley Buchanan Fund.

REFERENCES

1. Agranoff, B. W., 1971, *Handbook of Neurochemistry*, Volume 6 (A. Lajtha, ed.), Plenum Press, New York, pp. 203–219.
2. Agranoff, B. W., 1981, *Basic Neurochemistry*, 3rd ed. (G. J. Siegel, R. W. Albers, B. W. Agranoff, and R. Katzman, eds.), Little, Brown, Boston, pp. 801–820.
3. Agranoff, B. W., 1982, *Changing Concepts of the Nervous System* (A. R. Morrison and P. L. Strick, eds.), Academic Press, New York, pp. 717–728.
4. Dunn, A. J., 1980, *Annu. Rev. Psychol.* **31**:343–390.
5. Rainbow, T. C., 1979, *Neurochem. Res.* **4**:297–312.
6. Rosenzweig, M. R., and Bennett, E. L. (eds.), 1976, *Neural Mechanisms of Learning and Memory*, MIT Press, Cambridge.
7. Tsukada, Y., and Agranoff, B. W. (eds.), 1980, *Neurobiological Basis of Learning and Memory*, John Wiley & Sons, New York.
8. Squire, L. R., and Davis, H. P., 1981, *Annu. Rev. Pharmacol. Toxicol.* **21**:323–356.
9. Rose, S. P. R., 1981, *Neuroscience* **6**:811–821.
10. Greenough, W. T., 1976, *Neural Mechanisms of Learning and Memory* (M. R. Rosenzweig and E. L. Bennett, eds.), MIT Press, Cambridge, pp. 255–278.
11. Grouse, L. D., Schrier, B. K., Bennett, E. L., Rosenzweig, M R., and Nelson, P. G., 1978, *J. Neurochem.* **30**:191–203.
12. Eichenbaum, H., Quenon, B. A., Heacock, A. M., and Agranoff, B. W., 1976, *Brain Res.* **101**:171–176.
13. Boast, C. A., and Agranoff, B. W., 1978, *Society for Neuroscience Abstracts*, Vol. 4, p. 255.
14. Flood, J. F., Smith, G. E., and Jarvik, M. E., 1980, *Brain Res.* **197**:153–165.
15. Kesner, R. P., Partlow, L. M., Bush, L. G., and Berman, R. F., *Brain Res.* **209**:159–176.
16. Berman, R. F., Kesner, R. P., and Partlow, L. M., 1978, *Brain Res.* **158**:171–188.
17. Schwartz, J., 1979, *Annu. Rev. Neurosci.* **2**:467–504.
18. Ochs, S., 1983, *Handbook of Neurochemistry*, 2nd ed., Volume 5 (A. Lajtha, ed.), Plenum Press, New York, pp. 355–379.
19. Agranoff, B. W., 1980, *Neurobiological Basis of Learning and Memory* (Y. Tsukada and B. W. Agranoff, eds.), John Wiley & Sons, New York, pp. 135–147.
20. Cotman, C. W. (ed.), 1978, *Neuronal Plasticity*, Raven Press, New York.
21. Skene, J. H. P., and Willard, M., 1981, *J. Cell Biol.* **89**:86–95, 96–103.
22. Skene, J. H. P., and Willard, M., 1981, *J. Neurosci.* **1**:419–426.
23. Koshland, D. E., Jr., 1980, *Annu. Rev. Neurosci.* **3**:43–75.
24. Eisenstein, E. M., Brunder, D. G., and Blair, H. J., 1982, *Neurosci. Biobehav. Rev.* **6**:183–194.
25. Quinn, W. G., and Gould, J. L., 1979, *Nature* **278**:19–23.
26. Menzel, R., 1968, *Z. Vergl. Physiol.* **60**:82–102.
27. Dudai, Y., 1977, *J. Comp. Physiol.* **114**:69–89.
28. Jaffe, K., 1980, *Physiol. Behav.* **25**:367–371.
29. Garcia, J., Hankins, W., and Rusiniak, K., 1974, *Science* **185**:824–831.
30. Gelperin, A., 1975, *Science* **189**:567–570.
31. Grigliatti, T. A., Hall, L. M., Rosenbluth, R., and Suzuki, D. T., 1973, *Mol. Gen. Genet.* **120**:107–114.
32. Bugbee, N. M., and Goldman-Rakic, P. S., 1981, *Soc. Neurosci. Abstr.* **7**:239.
33. Lincoln, S. S., McCormick, D. A., and Thompson, R. F., 1982, *Brain Res.* **242**:190–193.
34. Klein, M., Shapiro, E., and Kandel, E. R., 1980, *J. Exp. Biol.* **89**:117–157.
35. Kandel, E. R., 1981, *Nature* **293**:697–700.
36. Pinsker, H. M., Hening, W. A., Carew, T. J., and Kandel, E. R., 1973, *Science* **182**:1039–1042.
37. Eppler, C. M., Palazzolo, M. J., and Schwartz, J. H., 1982, *J. Neurosci.* **2**:1692–1704.
38. Walters, E. T., Carew, T. J., and Kandel, E. R., 1981, *Science* **211**:504–506.
39. Farley, J., and Alkon, D. L., 1980, *Science* **210**:1373–1375.
40. Alkon, D. L., Lederhendler, I., and Shoukimas, J. J., 1982, *Science* **215**:693–695.

41. Neary, J. T., Crow, T., and Alkon, D. L., 1981, *Nature* **293**:658–660.
42. Chang, J. J., and Gelperin, A., 1980, *Proc. Natl. Acad. Sci. U.S.A.* **77**:6204–6206.
43. Andersen, P., Sundberg, S. H., Sveen, O., and Wigstrom, H., 1977, *Nature* **266**:736–737.
44. Lynch, G., and Schubert, P., 1980, *Annu. Rev. Neurosci.* **3**:1–22.
45. Lynch, G., Browning, M., and Bennett, W. F., 1979, *Fed. Proc.* **38**:2117–2122.
46. Browning, M., Baudry, M., Bennett, W. F., and Lynch, G., 1981, *J. Neurochem.* **36**:1932–1940.
47. Siman, R., Baudry, M., and Lynch, G., 1982, *Soc. Neurosci. Abstr.* **8**:883.
48. Bar, P. R., Tielen, A. M., Lopes da Silva, F. H., Zwiers, H., and Gispen, W. H., 1982, *Brain Res.* **245**:69–79.
49. Duffy, C., Teyler, T. J., and Shashoua, V. E., 1981, *Science* **212**:1148–1151.
50. Rose, S. P. R., 1980, *Neurobiological Basis of Learning and Memory* (Y. Tsukada and B. W. Agranoff, eds.), John Wiley & Sons, New York, p. 179.
51. Horn, G., McCabe, B. J., and Bateson, P. P. G., 1979, *Brain Res.* **168**:361–373.
52. Bateson, P. P. G., Horn, G., and McCabe, B. J., 1978, *J. Physiol. (Lond.)* **275**:70P.
53. Longstaff, A., and Rose, S. P. R., 1981, *J. Neurochem.* **37**:1089–1098.
54. Kohsaka, S., Takamatsu, K., Aoki, E., and Tsukada, Y., *Brain Res.* **172**:539–544.
55. Martin, J., 1978, *Science* **200**:565–566.
56. Goddard, G. V., and Douglas, R. M., 1975, *Can. J. Neurol. Sci.* **2**:385–394.
57. Jonec, V., and Wasterlain, C. G., 1979, *Exp. Neurol.* **66**:524–532.
58. Cain, D. P., Corcoran, M. E., and Staines, W. A., 1980, *Exp. Neurol.* **68**:409–419.
59. Dasheiff, R. M., and McNamara, J. O., 1980, *Brain Res.* **195**:345–353.
60. Engel, J., Jr., and Sharpless, N. S., 1977, *Brain Res.* **136**:381–386.
61. Flood, J. F., and Jarvik, M. E., 1976, *Neural Mechanisms of Learning and Memory* (M. R. Rosenzweig and E. L. Bennett, eds.), MIT Press, Cambridge, pp. 483–507.
62. Gibbs, M. E., and Ng, K. T., 1977, *Biobehav. Rev.* **1**:113–136.
63. Hunter, B., Zornetzer, S. F., Jarvik, M. E., and McGaugh, J. L., 1977, *Handbook of Psychopharmacology*, Volume 8: *Drugs, Neurotransmitters and Behavior* (L. L. Iversen, S. D. Iversen, and S. H. Snyder, eds.), Plenum Press, New York, pp. 531–577.
64. Bennett, E. L., Alberti, M. H., and Flood, J. F., 1981, *Pharmacol. Biochem. Behav.* **14**:863–869.
65. Flood, J. F., Landry, D. W., Bennett, E. L., and Jarvik, M. E., 1981, *Pharmacol. Biochem. Behav.* **15**:289–296.
66. Bell, G. A., and Morgan, I. A., 1981, *Behav. Brain Res.* **2**:301–322.
67. Cherfas, J. J., and Bateson, P., 1978, *Behav. Biol.* **23**:27–37.
68. Cronly-Dillon, J., Carden, D., and Birks, C., 1974, *J. Exp. Biol.* **61**:443–454.
69. Rapport, M. M., Karpiak, S. E., and Mahadik, S. P., 1979, *Fed. Proc.* **38**:2391–2396.
70. Rick, J. T., Gregson, A. N., Leibowitz, S., and Adinolfi, M., 1980, *Dev. Med. Child Neurol.* **22**:719–724.
71. Rigter, H., and Crabbe, J. C., 1979, *Vitam. Horm.* **37**:153–241.
72. de Wied, D., and Versteeg, D. M. G., 1979, *Fed. Proc.* **38**:2348–2354.
73. LeMoal, M., Koob, G. F., Koda, L. Y., Bloom, F. E., Manning, M., Sawyer, W. H., and Rivier, J., 1981, *Nature* **291**:491–493.
74. de Wied, D., 1977, *Ann. N.Y. Acad. Sci.* **297**:263–274.
75. Jolles, J., Zwiers, H., Dekker, A., Wirtz, K. W. A., and Gispen, W. H., 1981, *Biochem. J.* **194**:283–291.
76. Jolles, J., Aloyo, V. J., and Gispen, W. M., 1982, *Molecular Approaches to Neurobiology* (I. R. Brown, ed.), Academic Press, New York, pp. 285–316.
77. Fisher, S. K., and Agranoff, B. W., 1982, *Phospholipids in the Nervous System* Volume 1: *Metabolism* (L. A. Horrocks, G. B. Ansell, and G. Porcellati, eds.), Raven Press, New York, pp. 301–313.
78. Rigter, H., Jensen, R. A., Martinez, J. L., Jr., Messing, R. B., Vasquez, B. J., Liang, K. C., and McGaugh, J. L., 1980, *Proc. Natl. Acad. Sci. U.S.A.* **77**:3729–3732.
79. Gallagher, M., and Kapp, B. S., 1978, *Life Sci.* **23**:1973–1978.
80. Zwiers, H., Aloyo, V. J., and Gispen, W. H., 1981, *Life Sci.* **28**:2545–2551.
81. Shashoua, V. E., 1981, *Neurochem. Res.* **6**:1129–1147.

82. Hyden, H., and Ronnback, L., 1979, *Behav. Neural Biol.* **27**:294–301.
83. Morgan, D. G., and Routtenberg, A., 1981, *Science* **214**:460–471.
84. Coss, R. G., and Globus, A., 1978, *Science* **200**:787–790.
85. Wesa, J. M., Chang, F.-L. F., Greenough, W. T., and West, R. W., 1982, *Dev. Brain Res.* **4**:253–257.
86. Pysh, J. J., and Weiss, G. M., 1979, *Science* **206**:227–229.
87. Walsh, R. N., 1980, *Int. J. Neurosci.* **11**:77–89.
88. Bennett, E. L., 1976, *Neural Mechanisms of Learning and Memory* (M. R. Rosenzweig and E. L. Bennett, eds.), MIT Press, Cambridge, pp. 279–287.
89. Jorgensen, O. S., and Meier, E., 1979, *J. Neurochem.* **33**:381–382.
90. Grouse, L., Omenn, G. A., and McCarthy, B. J., 1973, *J. Neurochem.* **20**:1063–1073.
91. Lassen, N. A., 1980, *Cerebral Metabolism and Neural Function* (J. V. Passonneau, R. A. Hawkins, W. D. Lust, and F. A. Welsh, eds.), Williams & Wilkins, Baltimore, pp. 144–160.
92. Le Doux, J. E., Thompson, M. E., Iadecola, C., Tucker, L. W., and Reis, D. J., 1983, *Science* **221**:576–577.
93. Phelps, M. E., Mazziotta, J. C., and Huang, S. C., 1982, *J. Cereb. Blood Flow Metab.* **2**:113–162.
94. Brownell, G. L., Budinger, T. F., Lauterber, P. C., and McGeer, P. L., 1982, *Science* **215**:619–626.
95. Drachman, D. A., 1977, *Neurology (Minneap.)* **27**:783–790.
96. Bartus, R. T., Dean, R. L. III, Beer, B., and Lippa, A. S., 1982, *Science* **217**:408–417.
97. Sandman, C. A., George, J., McCann, T. R., Nolan, J. D., Kaswan, J., and Kastin, A. J., 1977, *J. Clin. Endocrinol. Metab.* **44**:884–891.
98. Gold, P. W., Weingartner, H., Ballenger, J. C., Goodwin, F. K., and Post, R. M., 1979, *Lancet* **2**:992–994.

16

Neurochemical Correlates of Learning Impairment

Yasuzo Tsukada

1. INTRODUCTION

Since neurobiological approaches to the elucidation of the mechanisms of learning and memory usually require the use of animal behaviors (conditioned reflexes, operant conditioning, etc.) as an essential tool for experimentation, studies in this field have been confronted with great difficulties, especially in quantitative analysis. When the historical breakthrough in the field of genetics revealed the role of DNA as the bearer of genetic information, neurobiologists also began to search for similar mechanisms of information processing involving base sequences of RNA and protein as memory-bearing substances in nerve cells. Others tried to analyze metabolic changes, especially of proteins and amino acids, in the brain of animals after learning.

These approaches attracted attention as novel efforts toward a molecular explanation for memory. Later, "memory molecules" became a target of investigation, as exemplified by memory transfer experiments. More recently, the finding of psychotropic agents that act as mood modulators has given further indirect evidence for the hypothesis that memory has a physicochemical basis. It has been very difficult, however, to identify specific changes related to the process of learning or memory, because a stimulus to the brain may cause various secondary changes through nonspecific alteration of physiological activities, including cerebral circulation and hormone secretion.

In the search for experimental models of learning and memory, posttetanic potentiation and rigidity following spinal damage are sometimes used as simple models. More frequently used models are learned behavior of higher mammals, including avoidance conditioning reflex, funambulation (tight-rope walking), transfer of handedness, and operant discrimination learning. Imprinting, a peculiar persistent learning process seen mainly in birds during early brain development, can also be used as a model of plasticity of the nervous system.

Yasuzo Tsukada • Department of Physiology, Keio University School of Medicine, Tokyo, Japan.

Many investigators have made a distinction between short-term memory processes with transient events and long-term processes with highly stable storage of information. The former may be caused by some reversible molecular modification in the synaptic membrane, and the latter may be accompanied by the formation of new molecules and produced by intracerebral memory fixation within about 1 h after learning; it is not known whether the long-term process requires the short-term process as an essential prerequisite (see Chapter 15, Memory).

2. STRATEGIES FOR NEUROCHEMICAL STUDIES

Neurochemical approaches to the study of learning and memory can be classified as follows: (1) analysis of neurochemical changes accompanying a particular learning process (correlative study); (2) investigation of the changes of learned behavior caused by administration of metabolic inhibitors or experimental brain damage (interventive study); (3) isolation and identification of "memory substances" and transfer of a particular memory to untrained animals by administrating them (memory transfer); (4) analysis of neurochemical correlates of learning impairment[1,2] (correlative study of developmental learning disability).

2.1. Correlative Studies

In these kinds of studies, differences in the metabolism of the brains of trained and untrained animals are examined. Any difference found may be directly or indirectly involved in the chemical process of learning. The neurochemical correlates identified in this way include alterations in ammonia,[3,4] amines,[5,6] glycolipids,[7] nucleotides,[8] RNA,[9–13] proteins,[13–17] and glycoproteins.[18] Among these, RNA and proteins have been intensively studied with respect to their metabolic changes with learning.

Hydén et al.[9,10] suggested the formation of new mRNA based on the finding that funambulation training caused a change in RNA base composition in the nuclei of Deiter's neurons in rats. They also reported an increase of RNA and a change in its base composition in the neurons of sensory cortex along with increased production of S-100 protein in the hippocampus[15] in the training of handedness transfer in rats. The latter experiment employs the contralateral hemisphere as a control, making it possible to exclude nonspecific factors including changes in cerebral circulation.

Glassman et al.[11,12,16,17] used radioisotopes to study changes in intracerebral syntheses of RNA and protein accompanying electric-shock avoidance learning in mice. They tried to define the effect of learning by the combined use of quiet and yoked controls and found increased synthesis of RNAs of all sizes and of protein in the trained animals. Studies of protein synthesis using [³H]lysine have shown that avoidance learning causes an increase of synthesis not only in the brain but in the liver as well. These phenomena may be explained by increases in precursor pool size caused by an increase in blood circulation and acceleration of intracellular metabolism. There remains doubt about the specificity of the reported metabolic changes in RNA and protein. Also, the finding[19,20] that administration of ACTH or its fragments[4-10] increased protein synthesis in the brain and liver suggests the possibility of hormonal involvement. De Wied et al.[21] have shown that ACTH is behaviorally active in rats in recovering impaired behavior.

Tsukada et al.[3,4] reported a change in ammonia metabolism in the rat brain accompanying the conditioning of an avoidance reflex induced by using light as the conditioned stimulus. The stimulus, presented for 5 s, caused a transient increase of brain ammonia, which rapidly combined with glutamic acid to form glutamine. However, when the stimulus was given repeatedly over a prolonged time, the increase in ammonia was not observed. They postulated that the ammonia response in the brain following the stimulus could be altered by conditioning and presented the change of amino acid metabolism as a chemical correlate of learning.

The imprinting[22] formed in birds immediately after hatching is thought to represent a kind of plasticity in the brain to sensory stimuli, giving a model of functional change in the brain accompanying the learning process. Bateson et al.[23-30] studied the metabolic changes in RNA and protein in the chick brain after imprinting with a colored flashing light. Increased synthesis of RNA and protein was found in the forebrain roof but not in the basal forebrain or midbrain, suggesting a localized biochemical event.

A similar experiment was performed in our laboratory.[31-33] Within a few hours after hatching in a dark room, the chick was given an imprinting stimulus with a moving red ball for 45 min and then returned to darkness for 2 days. The same stimulus was then given, and the following behavior of the chick was recorded. Animals showing this behavior for up to 10 min were put into a learning group, whereas those that did not (about half of the all chicks) were used as controls. On the next day, chicks of both groups were injected intraperitoneally with [^{14}C]2-deoxyglucose (2-DG, 10 µCi) and were given the same stimulus for 45 min. During the exposure of the imprinting stimulus, each chick was placed in a plastic cylinder to limit movement. After 45 min, the brain was frozen and sectioned into frozen slices, which were rapidly dried and placed in contact with X-ray film for autoradiography. When the incorporation of 2-DG into the brain tissue was compared between the imprinted and nonimprinted groups, significantly increased incorporation was found in the medial hyperstriatum ventrale (MHV) and lateral neostriatum (LN) in the imprinted chick (Fig. 1). This finding of localized increase of glucose metabolism with imprinting behavior corresponds well with the gross biochemical findings by Bateson et al.

2.2. Changes of Learned Behavior Caused by Administration of Metabolic Inhibitors or Experimental Brain Damage

Correlative studies as described above have strongly suggested that the process of learning is accompanied by *de novo* synthesis of RNA and protein as well as increased metabolism of glucose in the brain. On the basis of these

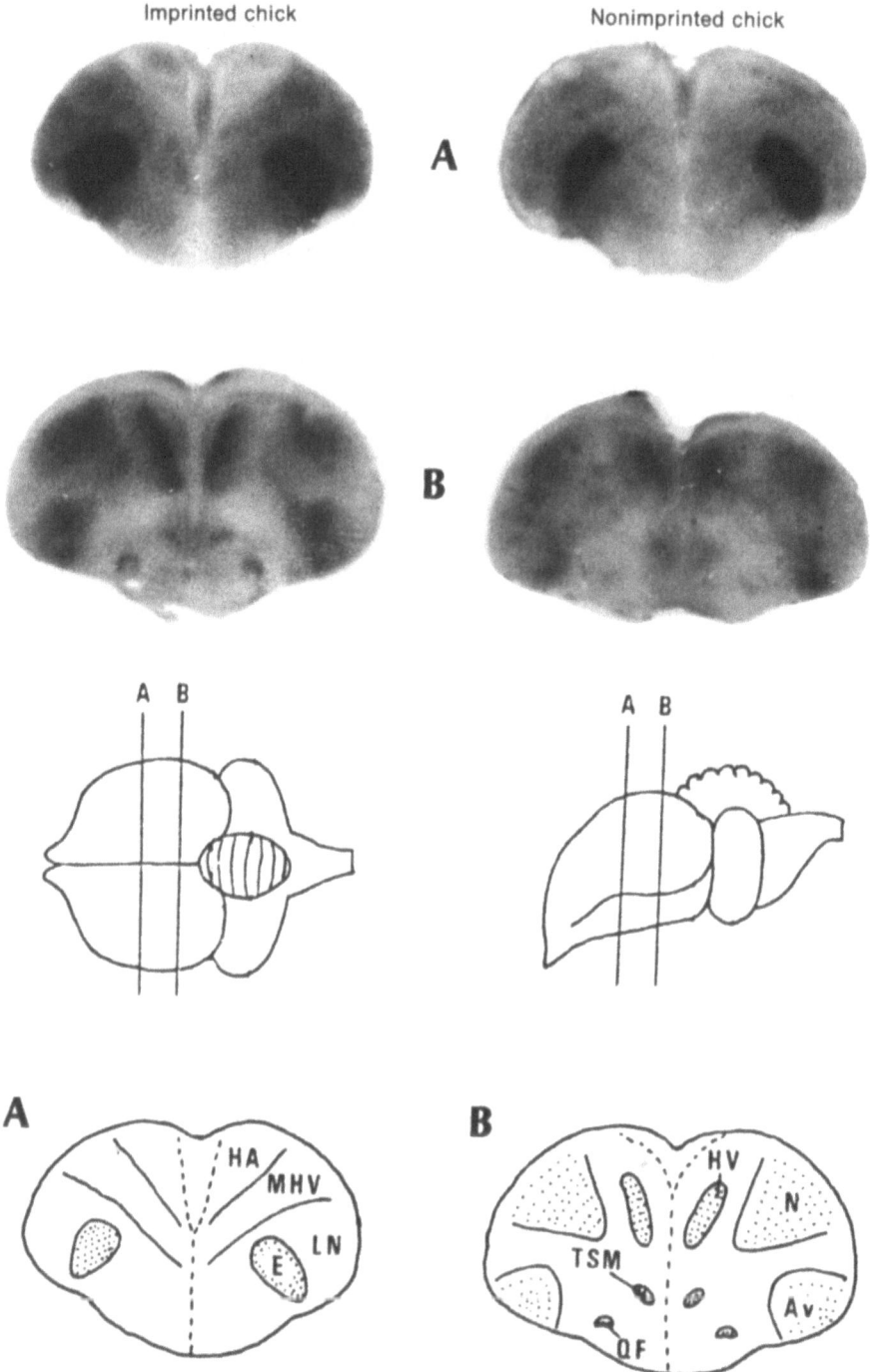

Fig. 1. (a) Autoradiographs of the brains of imprinted and nonimprinted chicks.[31] The sites of the sections (A, B) are shown in (b). (b) Diagram of autoradiographic sites in part a. Abbreviations: Av, archistriatum pars ventralis; E, ectostriatum; HA, hyperstriatum accesorium; HV, hyperstriatum ventrale; LN, lateral neostriatum; MHV, medial hyperstriatum ventrale; N, neostriatum; QF, tractus quintofrontalis; TSM, tractus septomesencephalicus.

findings, another kind of study was performed in which a potent inhibitor of RNA or protein synthesis was injected into animals, or some particular region of the brain was destroyed or removed, in order to analyze effects on the learning process. Severe impairment of learning ability was reportedly caused by injection of 8-azaguanine[34] or actinomycin D,[35,36] potent inhibitors of RNA synthesis, in goldfish and mice; however, these agents are so toxic, particularly in mammals, that it seems difficult to reach a definite conclusion. It was reported that injection of cytosine arabinoside,[37] a potent inhibitor of DNA synthesis, into the brain of goldfish did not affect learning ability, suggesting no direct involvement of DNA synthesis in the process of learning.

Agranoff et al.[38,39] performed an important study regarding the effect of protein synthesis inhibitors on learning in the goldfish, in which puromycin or cycloheximide could be injected intracranially. They observed a rapid increase of correct responses of avoidance of electric shock when light was used as a conditioned stimulus in an experimental tank (short-term memory). The trained fish were returned to their home tanks, and when the same conditioned stimulus was applied 3–30 days later, they showed the same high ratio of correct avoidance response, suggesting consolidation of memory or formation of long-term memory.

They then injected protein inhibitors intracranially at various intervals after learning. By this experiment, it was found that the inhibitor did not affect the process of short-term memory formation and that consolidation of long-term memory could be completed within 1 h after learning under conditions of normal protein synthesis in the brain. On the other hand, when trained fish were detained in the experimental tank, consolidation of memory did not begin even 1 h after learning, and susceptibility to puromycin persisted beyond 1 h. Thus, it was shown that fixation of memory occurred only if the fish were in the home tank. Details of this experiment have been reported elsewhere.[39]

Other studies[40,41] using mice also showed that electric shock immediately after each learning session completely inhibited memory formation and that injection of amphetamine, ACTH, or norepinephrine greatly delayed memory fixation, producing an effect antagonistic to that of protein synthesis inhibitors. These observations underscore the importance of identifying conditions that might trigger protein synthesis following learning.

The implication of protein synthesis in memory does not necessarily indicate the formation of a "memory substance." It would be more relevant to propose that such protein synthesis may be related to chemical and structural changes in nerve cells or synaptic regions required to establish a new functional network of neural circuits.

In our laboratory,[32,33] the effect of the removal of medial hyperstriatum ventrale (MHV) or lateral neostriatum (LN) by electrocautery in ducklings on a following behavior formed by imprinting was studied on the basis of a previously described finding that imprinting in the chicks produces an elevation of metabolism in these brain regions. After establishment of imprinting behavior, bilateral removal of MHV caused loss of following behavior. But removal of LN itself had no effect (Table I). Salzen et al.,[42,43] however, reported an effect of LN destruction on a stimulus discrimination in imprinting. The dif-

Table I
Effect of Brain Lesions on Imprinting Behavior in Ducklings[a]

	Sham operated	LN Lesion		MHV Lesion	
		Unilateral	Bilateral	Unilateral	Bilateral
Preoperated	9:30	10:00	—	10:00	—
Postoperated					
30 min	1:57	0:15	0:21	0:43	0:15
1 h	4:07	1:05	1:29	1:11	0:10
2 h	9:17	1:57	3:33	1:09	0:07
24 h	9:02	8:48	7:31	9:07	0:09

[a] Values indicate the time for following behavior and are expressed as means (min:s) of three experiments. Data from refs. 32,33.

ferent imprinting method employed may be the cause of this variation. When cycloheximide (20 μg), a potent protein synthesis inhibitor, was injected into the MHV of duckling bilaterally immediately after hatching, and the imprinting stimulus was then given, the following behavior to an imprinting stimulus did not appear (Fig. 2). Again, injection of a protein synthesis inhibitor into LN had no effect. Discrimination learning ability, which was tested by pecking activity under different brightness as conditioned stimulus and diurnal rhythm of locomotor activity, was shown to be unaffected in the ducklings treated with bilateral removal of MHV.

These findings suggest that MHV may be essential for the establishment of imprinting, a specialized form of learning, and that the formation of a new neural network may follow the induction of protein synthesis in this region after an imprinting stimulus. This newly formed network may then mediate following behavior.

In regard to avoidance learning in mice, it has been reported[44–47] that bilateral injection of cycloheximide into the hippocampus, striatum, or amygdala will cause memory loss, whereas that injection into the mesencephalic reticular formation, thalamus, or cerebrum has no effect. These observations may provide a clue to study the localization of memory storage (see also Chapter 15).

A finding[48] that injection of inhibitors of axonal transport such as colchicine and vinblastine impairs electric-shock avoidance learning in goldfish and mice suggests that transport of newly formed protein may be necessary for memory fixation. It has also been reported[49] that ouabain administration to chick brain impairs learning of shock avoidance.

Hydén et al.[15] reported that previous injection of antibody to S-100 protein, a brain-specific protein, into the rat brain impaired handedness transfer learning, suggesting an as yet unknown role of this protein in the learning process.

Studies[50,51] have been performed on the relationship between the process of learning and intracerebral levels of monoamines such as dopamine (DA), norepinephrine (NE), and serotonin (5-HT), revealing an amnestic effect of elevated 5-HT level. In our laboratory,[52,53] the ability to acquire operant bright-

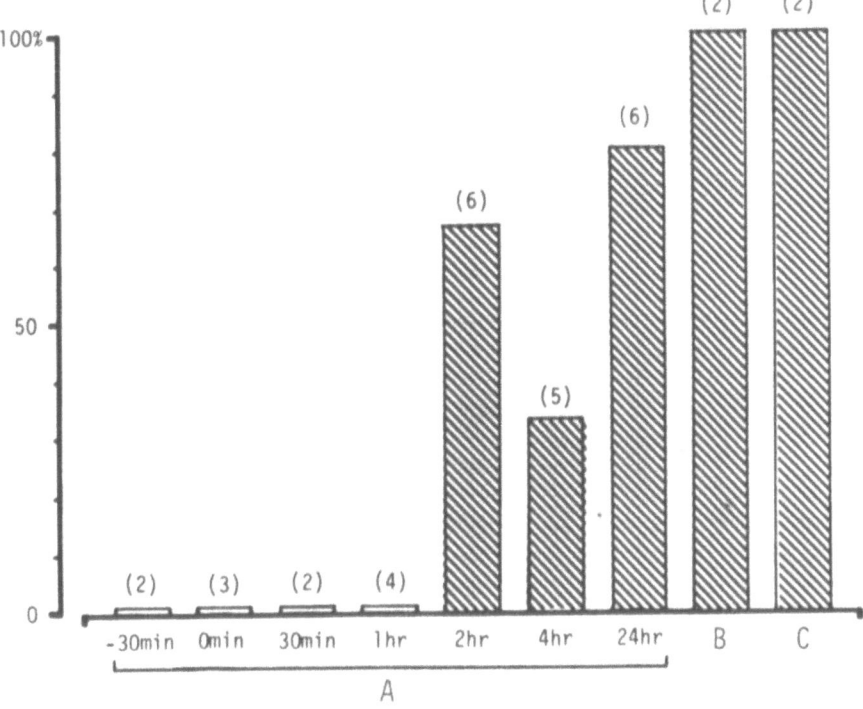

Fig. 2. Imprinted response (duration of following behavior) as percent of those of the controls.[33] (A) Cycloheximide (CHX) (20 μg) was injected into MHV bilaterally before (30 min before) or after (immediately after, 30 min, 1 h, 2 h, 4 h, or 24 h after) the first exposure of imprinting stimulus. (B) Cycloheximide (20 μg) was injected into LN bilaterally immediately after the exposure. (C) Cycloheximide (1 μg) was injected into MHV bilaterally after the exposure. Number of ducklings in each experiment is shown in parentheses.

ness discrimination learning was analyzed in rats injected with monoamine precursors (L-DOPA and L-5-HTP), but elevation of intracerebral monoamines measured in various parts of the brain had no effect on the observed learned response. A similar result was obtained in monkeys.[75] Only the injection of L-5-HTP at doses higher than 100 mg/kg caused a marked reduction of behavioral activity in rats.

2.3. Isolation and Identification of Memory Substances

McConell[54] proposed that specific RNA molecules act as memory substances on the basis of findings obtained in their digestion experiments with RNase and injection of extracted RNA in *Planaria*. On the other hand, Ungar et al.[55,56] reported successful memory transfer in rats injected intraperitoneally with a peptide obtained from the brain of rats trained to prefer a well-lighted condition to darkness. According to their contention, this memory peptide (scotophobin) is made up of 15 amino acids, and memory transfer was obtained with synthetic peptide. It has not been proved, however, that the injected peptide penetrated the blood–brain barrier.

Table II
Biochemical Markers for Brain Development[a]

Brain weight
DNA content
RNA content
Protein content
Contents of brain-specific proteins (S-100, MBP, Enolase isozyme, Thy-1, GFA, P-400)
Contents of neurotransmitters (5-HT, NE, DA, ACh, GABA, Glu, neuropeptides) and activities of related enzymes
2′,3′-Cyclic nucleotide 3′-phosphodiesterase (CNPase)
Lipid content

[a] These markers may indicate number of cells, cell size, ratio of neurons and glia, synaptogenesis, glial process, and myelinogenesis.

Considering the fact that injection of some neuropeptides[57] including ACTH, MSH, vasopressin (and their active fragments), and GH reportedly enhances learning ability through their effect on nonspecified nuclei in the thalamus, one can speculate that behavioral activities of several peptides may explain the so-called transfer effects.

2.4. Neurochemical Correlates of Learning Impairment in Animals with Maldeveloped Brain

Since brain development is controlled by environmental factors (nutrition, hormones, sensory stimuli, etc.) within a genetically determined framework, it can be significantly altered with various interventions in the prenatal or neonatal period. Using animal models of brain maldevelopment produced in this way, investigators are exploring neurochemical correlates of learning impairment in order to elucidate the basic chemical process of learning.

In our laboratory,[2,53,58–62] several experimental models have been produced by applying various chemical agents affecting rat brain development in the perinatal period (vulnerable period). In the neonatal rat brain, proliferation of nerve cells has been almost completed, and glial proliferation in the cerebral cortex and differentiation of neurons and glia in the cerebellum are still occurring. Therefore, intervention in this period may selectively disturb myelination or synaptogenesis, which proceed postnatally. Chemical markers and parameters measured in our studies are listed in Table II.

Experimental models of cretinism, phenylketonuria (PKU), hydrocortisone intoxication (HC), and malnutrition utilized rats (Table III). Experimental microencephaly was induced in the offspring by injecting 20 mg/kg of methylazoxymethanol (MAM) into pregnant rats on day 15 of gestation. Methylazoxymethanol, the aglycone of cycasin, is known to inhibit DNA synthesis and to intervene in the prenatal division of cerebral neurons.[62–64]

As shown in Table IV, DNA content per cerebrum was reduced in HC and MAM groups, indicating loss of cells. In the other models, hypotrophy of

Table III
Experimental Models of Brain Damage

Abnormality	Age
Hypothyroid rat (cretinism) (100 μCi ^{131}I injection)	Neonatal immediately after birth
Experimental phenylketonuric rat (PKU) (L-phenylalanine loading)	Neonatal 0 day to 8th month
Hydrocortisone-intoxicated rat (40 mg/kg injection)	Neonatal 0 day to 3rd day
Undernourished rat (12% casein diet)	Neonatal 1st day to 8th month
MAM-treated rat (20 mg/kg injection)	Prenatal 15th gestational day

individual cells was suggested by reduced brain weight and normal DNA content per cerebrum. Significant changes in monoamine levels were found only in the MAM-treated group, which showed roughly doubled levels in the cerebrum (Table V). Observations with fluorescence microscopy confirmed the chemical analyses. It was assumed that aminergic neurons were concentrated in MAM-treated rat on the basis of the finding that MAM treatment on day 15 of gestation affects largely the neurons other than aminergic ones, leaving the latter, which develop earlier. These findings have been confirmed by other investigators.[65,66] In the case of PKU, serotonin levels were reduced significantly but were restored quickly when phenylalanine loading ceased.

Table IV
Cerebral Weight and DNA Content in Rat Cerebrum Following Altered Development[2,53,58,59,62]

Condition	Cerebral wt. (mg)[a]	DNA(μg)/cerebrum[a]
Control (4 months)	1224 ± 32 (47)	929 ± 53 (47)
Cretin (4 months)	*1063 ± 59* (18)	907 ± 73 (18)
Control (7 months)	1332 ± 65 (4)	1160 ± 60 (4)
PKU (7 months)	*1207 ± 58* (4)	1130 ± 40 (4)
Control (20 days)	796 ± 73 (30)	1442 ± 82 (10)[b]
Hydrocortisone (20 days)	*675 ± 51* (30)	*1169 ± 23* (10)[b]
Control (4 months)	1223 ± 33 (3)	
Hydrocortisone (4 months)	*1004 ± 41* (3)	
Control (8 months)	1810 ± 30 (5)	1190 ± 80 (5)
Undernourished (8 months) (12% casein diet)	*1590 ± 40* (5)	1110 ± 160 (5)
Control (12 months)	1310 ± 50 (3)	960 (2)
MAM (12 months)	*670 ± 30* (3)	*580* (2)

[a] Values are expressed as mean ± S.D. with number of rats shown in parentheses. Values in *italics* are significantly different from corresponding controls by Student's *t*-test with two-tailed distributions ($P < 0.01$). DNA content was measured according to the method of Burton[72] after fractionation by the method of Schmidt and Thannhauser.[73]
[b] DNA content was determined by fluorometry using diaminobenzoic acid.[74]

Table V
Monoamine Contents of Rat Cerebral Hemisphere[2,53,58,59,62]

	Serotonin[a]	Dopamine[a]	Norepinephrine[a]
Control	1.11 ± 0.09 (4)	0.40 ± 0.11 (4)	0.30 ± 0.12 (4)
Hypothyroid	1.33 ± 0.12 (4)	0.80 ± 0.12 (4)	0.47 ± 0.05 (4)
PKU	*0.54 ± 0.07* (3)	0.86 ± 0.07 (3)	0.24 ± 0.06 (3)
Undernourished	1.04 ± 0.09 (5)	0.40 ± 0.15 (5)	0.39 ± 0.05 (5)
MAM	*1.83 ± 0.53* (7)	*1.25 ± 0.25* (5)	*0.52 ± 0.05* (5)

[a] Values are expressed as mean (μg/g.w.w.) ± S.D. with number of rats (aged 4 months) shown in parentheses. Values in *italics* are significantly different from controls by Student's *t*-test with two-tailed distributions ($P < 0.01$).

2′,3′-Cyclic nucleotide 3′-phosphodiesterase (CNPase) has been used as a marker enzyme of the CNS myelin sheath.[67-70] In order to define the profile of myelination in the CNS, CNPase activity in the brain of model animals was measured. The results revealed significant reductions of the enzyme activity in animals with cretinism, PKU, hydrocortisone intoxication, and malnutrition. No significant change was found in rats with MAM-induced microencephaly (Table VI). This suggested that these neonatal interventions inhibited the functional differentiation of oligodendroglia, resulting in poor myelination. In the case of the HC group, it was found by measuring [^{3}H]leucine incorporation into protein that protein synthesis in the brain (including cerebellum) was significantly inhibited. Also, the incorporation of labeled thymidine into DNA was clearly suppressed, resulting in inhibition of weight gain of the brain.

Because lowered pituitary growth hormone (GH) secretion from the hypophysis was found in the HC rats,[61] an experiment was performed in which bovine pituitary GH (10 mg/kg per day) was injected into these animals for 20

Table VI
CNPase Activity in Rat Cerebrum[2,53,58,59,62]

	Units/mg protein[a]	Change (%)
Control (1 month)	1.50 ± 0.06 (24)	
Cretin (1 month)	1.09 ± 0.10 (31)	−27
Control (6 months)	2.61 ± 0.16 (5)	
Cretin (6 months)	1.80 ± 0.31 (5)	−31
Control (7 months)	3.16 ± 0.10 (5)	
PKU (7 months)	2.76 ± 0.13 (5)	−13
Control (4 months)	3.09 ± 0.07 (3)	
Hydrocortisone (4 months)	2.34 ± 0.05 (3)	−24
Hydrocortisone + bovine GH (4 months)	2.79 ± 0.02 (3)	−9
Control (4 months)	3.63 ± 0.24 (5)	
Undernourished (4 months)	2.91 ± 0.17 (5)	−20
Control (3 months)	3.58 ± 0.09 (4)	
MAM (3 months)	3.66 ± 0.29 (4)	0

[a] Values are expressed as mean ± S.D. with number of rats shown in parentheses.

successive days immediately after birth. This treatment resulted in a recovery of intracerebral DNA and protein syntheses and CNPase activity, suggesting restoration of brain development, particularly a myelination-stimulating effect of pituitary GH.

Operant brightness discrimination learning ability of these animal models was examined in 4-month-old male rats. Bright light (8000 lx) and dim light (80 lx) were used as conditioned stimuli, and reinforcement was performed by a variable interval (VI) schedule in which bar pressing on presentation of bright light gave a food pellet once every 15 s on the average, whereas bar pressing on presentation of dim light provided no food. Each of the conditioned stimuli was given for 30 s and presented 20 times at random in one session. Training was carried out with one session per day. Learning performance was evaluated by the correct response ratio (percentage of the number of the correct responses in the total number of responses), and when this ratio was more than 85% for three successive days, the learning criterion in the control group was considered to have been attained. It is assumed that in this kind of discrimination learning, emotional excitement was much less than in avoidance learning.

Control rats attained the learning criterion within 25 sessions, but none of the animals with cretinism, PKU, or hydrocortisone intoxication attained the criterion even after 35 sessions, indicating marked impairment of discrimination learning ability (Fig. 3). Cretinous rats showed almost complete loss of learning ability. In hydrocortisone-intoxicated animals, injection of bovine pituitary GH recovered almost normal learning ability, suggesting a strong correlation between myelination in the cerebrum and learning ability (Fig. 4). Malnourished rats also had impairment of learning ability, which could not be significantly improved by injection of porcine pituitary GH. The defect may have been caused by lack of nutrients in this case.

These results indicate an important role of pituitary GH in cerebral development, suggesting an effect on myelination through glial differentiation and maturation.

Another study[60,71] showed that Snell dwarf mice, a mutant mouse (dw/dw), which genetically lack pituitary GH and TSH, had lowered CNPase activity

Table VII
CNPase Activity of the Cerebrum on Snell Dwarf Mice (dw/dw)[60,61]

Treatments	CNPase activity[a]	
	+/?	dw/dw
None	1.53 ± 0.07 (12)	0.90 ± 0.05 (12)
bGH	1.50 ± 0.13 (7)	1.15 ± 0.06 (5)
bGH + T$_4$	1.70 ± 0.03 (13)	1.56 ± 0.04* (15)
TSH	1.33 ± 0.08 (3)	1.02 ± 0.04 (7)

[a] Values are expressed as mean (units/mg protein) ± S.E.M. with number of animals shown in parentheses. Values with asterisks are significantly different from non-treated dwarf mice by Student's t-test with two-tailed distributions (*$P < 0.01$).

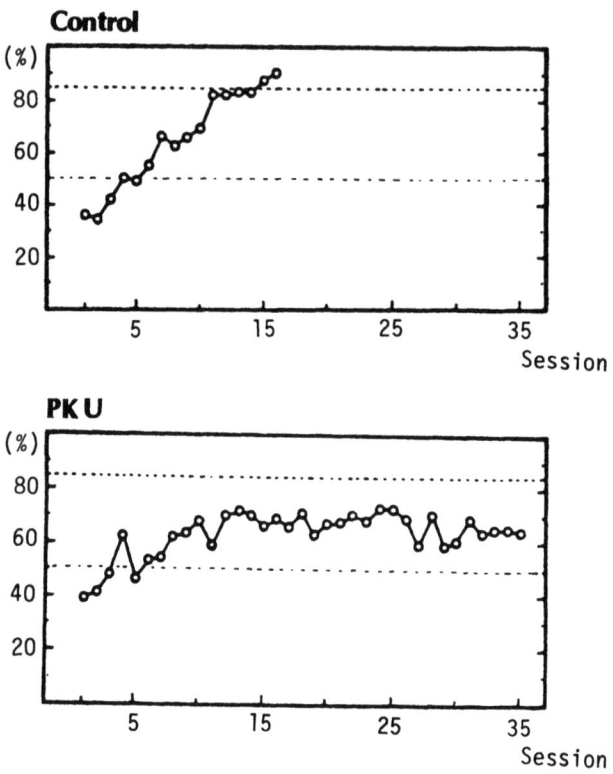

Fig. 3a. Correct response ratio of brightness discrimination learning test in control and PKU rats.[2,58] Each point represents the mean value from 5 rats.

Table VIII
Relationship between Cerebral Components and Learning Ability in Experimental Models of Brain Impairment[a]

	Body weight	Cerebral weight	Cerebral DNA	Cerebral CNPase	Learning ability
Hypothyroid (cretinism)	↓↓	↓	~	↓	↓↓
Experimental phenylketonuric rat (PKU)	↓	↓	~	↓	↓
Hydrocortisone-intoxicated rat	↓↓	↓	↓	↓	↓
Undernourished rat	↓↓	↓↓	~	↓	↓↓
MAM-treated rat	~	↓↓	↓↓	~	↓

[a] The determinations were carried out on 4-month-old rats. CNPase, 2′,3′-cyclic nucleotide 3′-phosphodiesterase; learning ability, brightness discrimination learning. ↓, slightly decreased; ↓, moderately decreased; ↓↓, markedly decreased; ~, unchanged.

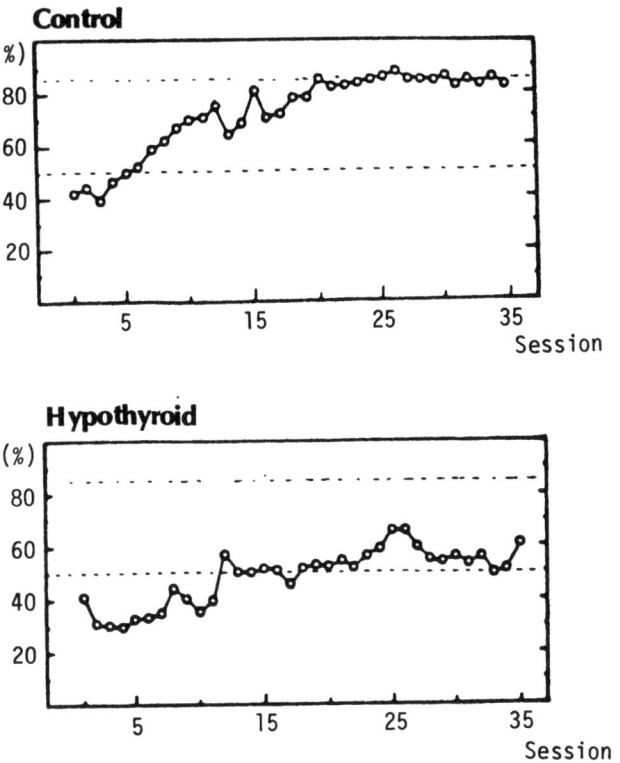

Fig. 3b. Correct response ratio of brightness discrimination learning test in control and hypothyroid rats.[2] Each point represents the mean value of 5 rats.

in the cerebrum, indicating impaired myelination there (Table VII). Behavioral activity of dwarf mice was reduced, and diurnal rhythm of behavior was also lost. Daily injection of bovine GH combined with T_4 into these dwarf mice elevated cerebral CNPase activity, increased locomotor activity, and recovered diurnal rhythm of behavior. However, injection of TSH or T_4 alone had no effect on either myelination or behavior. This also indicated an important role of pituitary GH in myelination in the cerebrum in the presence of thyroid hormone and suggested that myelination might be the key factor of behavior control.

In contrast to the other models, rats with MAM-induced microencephaly could eventually attain criterion after 35 sessions in spite of the marked loss of cerebral cells and weight (both decreased to about 60% of the control value), indicating a comparatively well-preserved learning ability.

As summarized in Table VIII, our experimental models showed a strong correlation between cerebral CNPase activity and discrimination learning ability, suggesting a serious effect of poor cerebral myelination on learning. Poor myelination may perhaps result in jamming of information because of defective insulation of nerve fibers, functionally abnormal synaptogenesis, and disturbance of the triggering of protein synthesis necessary for memory fixation.

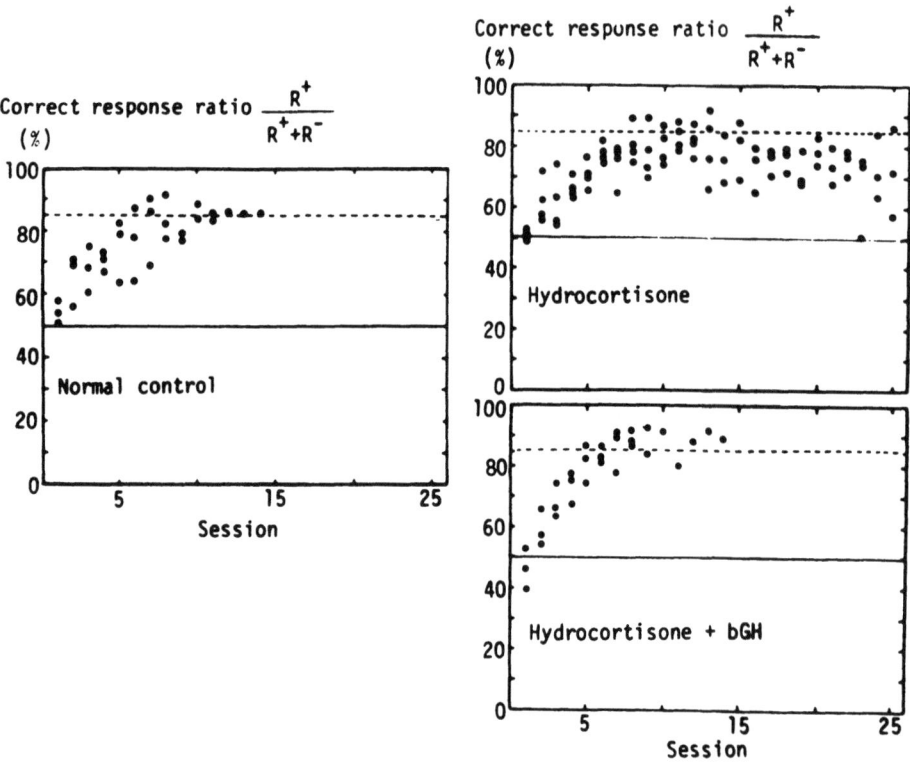

Fig. 4. Correct response ratio of brightness discrimination test in hydrocortisone-intoxicated and rehabilitated rats. Each point represents the correct response ratio indicating of each experimental rat.[61]

It could be concluded from these findings that postnatal maturation of the oligodendroglia plays quite an important role in the development of higher brain functions. Our ongoing studies using neurological mutant animals such as reeler, shiverer, quaking, weaver, etc. with genetically defective brain development may further contribute to the elucidation of the neurochemical correlates of abnormal learning ability.

3. CONCLUSION

Neurochemical studies of learning and memory have been steadily advancing, and it appears to be established that both the short-term and long-term processes of memory are accompanied by chemical changes in the brain. Intervention in short-term memory with electric shock or ouabain administration may involve a change of ion permeability on the synaptic membrane or interference with some transient chemical modification, e.g., glycosylation or phosphorylation of protein. As for long-term memory, intracerebral protein synthesis may be necessary for its consolidation, which may involve functionalization of previous inoperative synapses or neogenesis of particular synapses. The newly formed protein may not necessarily be specific; it may well be a protein ordinarily contained in the synaptic structures. The site of memory fixation in the brain may be identified with localized administration of inhibitors of protein synthesis, and it would be much less difficult to isolate the kind of protein produced there with current biochemical techniques.

Isolation of mRNA mediating the production of relevant proteins in the process of memory fixation would finally be possible, resulting in elucidation of the genetic code for the structure of such proteins. At the current stage of exploration, an important question is when and how the signal for protein synthesis in the process of memory fixation is produced. The studies in this field will always require great efforts in meticulously discriminating relatively slight specific responses directly connected to the process of memory or learning from nonspecific metabolic changes as long as learned behavior, difficult to quantify, is used as an index.

REFERENCES

1. Tsukada, Y., 1978, *Shinkei Shinpo* 22:1027–1042.
2. Tsukada, Y., Nomura, M., Nagai, K., Kohsaka, S., Kawahara, H., Ito, M., and Matsutani, T., 1977, *Behavioral Neurochemistry* (J. M. R. Delgado and F. V. Defeudis, eds.), Spectrum, New York, pp. 63–84.
3. Tsukada, Y., Takagaki, G., Sugimoto, S., and Hirano, S., 1958, *J. Neurochem.* 21:295–303.
4. Tsukada, Y., 1966, *Prog. Brain Res.* 21A:268–291.
5. Seiden, L. S., Brown, R. M., and Lewy, A. J., 1973, *Chemical Modulation of Brain Functions*, (H. C. Sabelli, ed.), Raven Press, New York, pp. 261–275.
6. Deutsch, J. A., 1973, *The Physiological Basis of Memory* (J. A. Deutsch, ed.), Academic Press, New York, pp. 59–76.
7. Dunn, A. J., Entingh, D., Entingh, T., Gispen, W. H., Machlus, B., Perumal, R., Rees, H. D., and Brogan, L., 1974, *The Neurosciences, Third Study Program*, (F. O. Schmitt and F. G. Worden, eds.), MIT Press, Cambridge, pp. 679–684.
8. Etingh, D., Damstra-Entingh, T., Dunn, A., Wilson, J. E., and Glassman, E., 1974, *Brain Res.* 70:131–138.
9. Hydén, H., and Egyházi, E., 1962, *Proc. Natl. Acad. Sci. U.S.A.* 48:1366–1373.
10. Hydén, H., and Egyházi, E., 1964, *Proc. Natl. Acad. Sci. U.S.A.* 52:1030–1035.
11. Glassman, E., 1969, *Annu. Rev. Biochem.* 38:605–646.
12. Glassman, E., and Wilson, J. E., 1970, *Brain Res.* 21:157–168.
13. Bateson, P. P. G., Horn, G., and Rose, S. P. R., 1972, *Brain Res.* 39:449–465.
14. Beach, G., Emmens, M., Kimble, D. P., and Lickey, M., 1969, *Proc. Natl. Acad. Sci. U.S.A.* 62:692–696.
15. Hydén, H., and Lange, P. W., 1970, *Exp. Cell Res.* 62:125–132.
16. Rees, H. D., Brogan, L. L., Entingh, D. J., Dunn, A., Shinkman, P. G., Damstra-Entingh, T., Wilson, J. E., and Glassman, E., 1974, *Brain Res.* 68:143–156.
17. Hershkowitz, M., Wilson, J. E., and Glassman, E., 1975, *J. Neurochem.* 25:687–694.
18. Bogoch, S., 1968, *The Biochemistry of Memory*, Oxford University Press, London, New York.
19. Semigonovsky, B., and Jakoubek, B., 1971, *Brain Res.* 35:319–323.
20. Dunn, A. J., 1976, *Neural Mechanisms of Learning and Memory* (M. R. Rosenzweig and E. L. Bennett, eds.), MIT Press, Cambridge, pp. 311–320.
21. De Wied, D., and Gispen, W. H., 1977, *Peptides in Neurobiology*, (H. Gainer, ed.), Plenum Press, New York, pp. 397–448.
22. Hess, E. H., 1974, *Imprinting*, Van Nostrand Reinhold, New York.
23. Bateson, P. P. G., Horn, G., and Rose, S. P. R., 1975, *Brain Res.* 84:207–220.

24. Bateson, P. P. G., Horn, G., and Rose, S. P. R., 1972, *Brain Res.* **39**:449–465.
25. Bateson, P. P. G., Rose, S. P. R., and Horn, G., 1973, *Science* **181**:576–578.
26. Horn, G., McCabe, B. J., and Bateson, P. P. G., 1979, *Brain Res.* **168**:361–373.
27. Horn, G., Rose, S. P. R., and Bateson, P. P. G., 1973, *Brain Res.* **56**:227–237.
28. Rose, S. P. R., Bateson, P. P. G., Horn, A. L. D., and Horn, G., 1970, *Nature* **225**:650–651.
29. Rose, S. P. R., and Haywood, J., 1977, *Biochemical Correlates of Brain Structure and Function* (A. N. Davison, ed.), Academic Press, New York, pp. 249–292.
30. Rose, S. P. R., 1980, *Neurobiological Basis of Learning and Memory* (Y. Tsukada and B. W. Agranoff, eds.), John Wiley & Sons, New York, pp. 179–191.
31. Kohsaka, S., Takamatsu, K., Aoki, E., and Tsukada, Y., 1979, *Brain Res.* **172**:539–544.
32. Tsukada, Y., and Takamatsu, K., 1982, *Excerpta Medica International Congress Series 620 (Proceedings of the Symposium on the Pharmacology of Learning and Memory)*, pp. 59–69.
33. Takamatsu, K., 1982, *Keio Igaku* **59**:221–240.
34. Dingman, W., and Sporn, M. B., 1961, *J. Psychiatr. Res.* **1**:1–11.
35. Cohen, H. D., and Barondes, S. H., 1966, *J. Neurochem.* **13**:207–211.
36. Agranoff, B. W., Davis, R. E., Casola, L., and Lim, R., 1967, *Science* **158**:1600–1601.
37. Casola, L., Lem, R., Davis, R. E., and Agranoff, B. W., 1968, *Proc. Natl. Acad. Sci. U.S.A.* **60**:1389–1395.
38. Agranoff, B. W., Davis, R. E., and Brink, J. J., 1966, *Brain Res.* **1**:303–309.
39. Agranoff, B. W., 1980, *Neurobiological Basis of Learning and Memory* (Y. Tsukada and B. W. Agranoff, eds.), John Wiley & Sons, New York, pp. 135–147.
40. Flexner, J. B., and Flexner, L. B., 1971, *Proc. Natl. Acad. Sci. U.S.A.* **68**:2519–2521.
41. Gibbs, M. E., 1976, *Pharmacol. Biochem. Behav.* **4**:305–309.
42. Salzen, E. A., Parker, D. M., and Williamson, A. J., 1975, *Brain Res.* **24**:145–157.
43. Salzen, E. A., Parker, D. M., and Williamson, A. J., 1978, *Exp. Brain Res.* **31**:107–116.
44. Flexner, J. B., Flexner, L. B., and Stellar, E., 1963, *Science* **141**:57–59.
45. Flexner, L. B., Flexner, J. B., De La Haba, G., and Roberts, R. B., 1965, *J. Neurochem.* **12**:535–541.
46. Barondes, S. H., 1970, *Int. Rev. Neurobiol.* **12**:177–205.
47. Eichenbaum, H., Quenon, B. A., Heacock, A. M., and Agranoff, B. W., 1976, *Brain Res.* **101**:171–176.
48. Murakami, T. H., 1980, *Neurobiological Basis of Learning and Memory* (Y. Tsukada and B. W. Agranoff, eds.), John Wiley & Sons, New York, pp. 165–178.
49. Watts, M. E., and Mark, R. F., 1971, *Proc. R. Soc. Lond. [Biol.]* **178**:454–464.
50. Murphy, D. L., and Redmond, D. E., Jr., 1975, *Catecholamines and Behavior*, Volume 2, Plenum Press, New York, pp. 73–104.
51. Essman, W. B., 1973, *Current Biochemical Approaches to Learning and Memory*, Spectrum, New York, pp. 159–188.
52. Kohsaka, S., Nagai, K., and Tsukada, Y., 1977, *Bull. Jpn. Neurochem. Soc.* **16**:29–32.
53. Kohsaka, S., and Tsukada, Y., 1980, *Neurobiological Basis of Learning and Memory*, John Wiley & Sons, New York, pp. 149–164.
54. McConnell, J. V., 1966, *Annu. Rev. Physiol.* **28**:107–136.
55. Ungar, G., Desiderio, D. M., and Parr, W., 1972, *Nature* **238**:198–202.
56. Ungar, G., 1974, *Life Sci.* **14**:595–604.
57. Jolles, J., Aloyo, V. J., and Gispen, W. H., 1982, *Molecular Approaches to Neurobiology* (I. R. Brown, ed.), Academic Press, New York, pp. 285–316.
58. Kohsaka, S., and Tsukada, Y., 1979, *Keio J. Med.* **28**:97–108.
59. Tsukada, Y., Kohsaka, S., and Nagai, K., 1979, *Proceedings International Nutrition Conference. Behavioral Effects of Energy and Protein Deficits*, NIH Publication No. 79 1906, pp. 12–21.
60. Noguchi, T., Sugisaki, T., Watanabe, M., Tsukada, Y., and Tanabe, M., 1980, *Growth and Growth Factors* (K. Shizume and K. Takano, eds.), University of Tokyo Press, Tokyo, pp. 203–230.
61. Noguchi, T., Sugisaki, T., Watanabe, M., Kohsaka, S., and Tsukada, Y., 1982, *J. Neurochem.* **38**:246–256.
62. Matsutani, T., Nagayoshi, M., Tamaru, M., and Tsukada, Y., 1980, *J. Neurochem.* **34**:950–956.

63. Spatz, M., and Laqueur, G. L., 1968, *Proc. Soc. Exp. Biol. Med.* **129**:705–710.
64. Matsumoto, H., Spatz, M., and Laqueur, G. L., 1972, *J. Neurochem.* **19**:297–306.
65. Johnston, M. V., and Coyle, J. T., 1980, *J. Neurochem.* **34**:1429–1441.
66. Johnston, M. V., Carman, A. B., and Coyle, J. T., 1981, *J. Neurochem.* **36**:124–128.
67. Kurihara, T., and Tsukada, Y., 1967, *J. Neurochem.* **14**:1167–1174.
68. Tsukada, Y., Nagai, K., and Suda, H., 1980, *J. Neurochem.* **34**:1014–1017.
69. Tsukada, Y., and Suda, H., 1980, *Cell. Mol. Biol.* **26**:493–504.
70. Sims, N. R., and Carnegie, P. R., 1978, *Adv. Neurochem.* **3**:1–41.
71. Noguchi, T., Sugisaki, T., and Tsukada, Y., 1982, *J. Neurochem.* **38**:257–263.
72. Burton, K., 1956, *Biochem. J.* **62**:315–323.
73. Schmidt, G., and Thannhauser, S. J., 1945, *J. Biol. Chem.* **161**:83–89.
74. Kissane, J. M., and Robins, E., 1958, *J. Biol. Chem.* **233**:184–188.
75. Tsukada, Y., Kishimoto, H., and Nagai, K., 1975, *Contemporary Primatology*, S. Karger, Basel, pp. 56–66.

17

Behavioral and Neurochemical Effects of ACTH

Willem Hendrik Gispen and Henk Zwiers

1. INTRODUCTION

The role that peptide hormones play in the regulation of target cell metabolism is well documented. Numerous results suggest that these hormones are in a similar way important to brain cell metabolism. Focusing on corticotropin (ACTH), it seems reasonable to expect that its neurochemical action somehow underlies its behavioral effects. Thus, a great deal of recent research has dealt with the interrelationship of ACTH, brain metabolism, and behavior. The present chapter aims to summarize some of the results obtained in studies on ACTH, brain, and behavior. In particular, recent progress in studies of healthy, aged, or mentally disabled human beings is discussed to illustrate the potential clinical importance of the peptides derived from ACTH. In the last part of the chapter we review the evidence that ACTH, by affecting events associated with the function of synaptic membranes, may influence ongoing transmission in certain synapses, presumably contributing to its modulatory action on behavior in animal and man.

2. ACTH

2.1. ACTH: Peptide Hormone and Neuropeptide

The classical endocrine view is that the cells of the pituitary are the only source of the peptide hormone ACTH. When it became apparent that ACTH and/or melanotropin (MSH) were capable of influencing the nervous system and behavior in experimental animals and man (for a recent review see De Wied and Jolles[1]), interest became focused on the presence and possible origin of ACTH and congeners in the nervous system. It was shown that immuno-

---

Willem Hendrik Gispen and Henk Zwiers • Division of Molecular Neurobiology, Rudolf Magnus Institute for Pharmacology, and Institute of Molecular Biology, State University of Utrecht, 3584 CH Utrecht, The Netherlands.

reactive ACTH was present in the cerebrospinal fluid (CSF).[2] Although the major part of the blood in the portal vessels flows from the hypothalamus to the pituitary,[3] the existence of a retrograde blood flow towards the brain has been identified and thus can explain the presence of ACTH in brain.[4] Peptides in general have difficulty in passing the brain barrier systems, but it has been demonstrated that ACTH can reach central nervous structures via the vascular system of the pituitary stalk or the liquor space around the gland.[5-8] Moldow and Yalow[9] maintain that the pituitary could be the sole source of ACTH and related peptides in the body. However, numerous reports in recent literature based on refined immunochemical, morphological, and biochemical techniques suggest that, in addition to the pituitary, the nervous system is also capable of producing ACTH and congeners.[10-12] Immunocytochemical observations are consistent with the notion that opiomelanocortin peptides, including ACTH-like peptides, are present in neurons of the arcuate nucleus and in distinct pathways running from this nucleus to the limbic system, midbrain, and hindbrain. There is also evidence for additional, extrahypothalamic neuronal networks as well.[10,13] ACTH and other opiomelanocortin peptides are present in the development of the fetal brain.[10,14] Cells containing ACTH-related antigenic determinants in pituitary gland and brain develop independently of one another,[15] again pointing to at least two ACTH-containing systems in the body.

It has been suggested that the posttranslational processing of the pro-opiomelanocortin precursor[16] is similar for the intermediate lobe of the pituitary and brain.[17] This implies that in brain the production of $ACTH_{1-16}$, which is capable of all currently known central nervous system (CNS) actions of ACTH,[1] together with corticotropinlike intermediate lobe peptide (CLIP; $ACTH_{18-39}$) is feasible.[18] Indeed, preliminary studies revealed the generation of $ACTH_{1-16}$ and $ACTH_{18-39}$ from $ACTH_{1-39}$ in rat brain synaptic membranes *in vitro* (X.-C. Wang, J. H. P. Burbach, and D. De Wied, unpublished observations).

In summary, ACTH and some of its fragments are present in nervous tissue and may derive either from the pituitary via a specific transport system or via the circulation or from peptidergic neurons via local synthesis.[4]

2.2. ACTH as a Messenger

Irrespective of its role as hormone in the periphery or as neuropeptide in the brain, the peptide ACTH is the bearer of multiple active sequences and may activate different receptors involved in the same or different biological response(s).[19-21] In view of the fact that ACTH from five different sources is invariably an open-chain peptide (in contrast, e.g., to insulin), Schwyzer[20] suggested a linear or sychnologic organization of the information encoded in this peptide. This means that discrete (sometimes overlapping) sequences of adjacent amino acids are responsible for the different components of the total biological action of ACTH.[22] In addition to the active sites (messages) in the molecule, Schwyzer[20] distinguished auxiliary sites, which he described in terms of their function as address and potentiator. The address site adds receptor-specific affinity, and the potentiator enhances potency and intrinsic activity.

Effects of ACTH

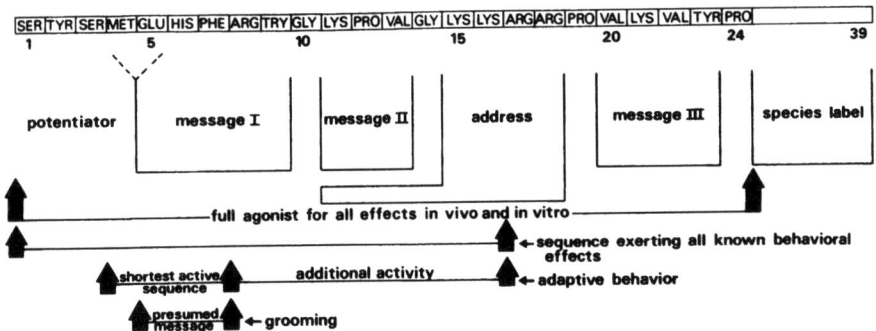

Fig. 1. Primary structure of the N-terminus of ACTH. The assumed functions of the various sequences are indicated according to Schwyzer,[20] illustrating the multiplicity of messages encoded in the peptide. The sequence 1–16 exerts all known behavioral effects; the sequence 4–7 is the shortest active fragment both in adaptive behaviors and in excessive grooming. (Revised after Schwyzer,[20] with permission.)

In aqueous solutions, the peptide is a flexible molecule and, in binding to the more rigid receptor, might adapt itself to the recognition part of the receptor to insure optimal interaction with a given message.[19,23] The three messages in the biologically active core of the whole $ACTH_{1-39}$ molecule are the sequences 4/5–9, 11–13, and 15–18, activating different receptors (Fig. 1).[20] As discussed elsewhere,[24–26] posttranslational processing of ACTH or other open-chain neuropeptides may result in the simultaneous release of the different message sequences, which may act synergistically or antagonistically in the modulation of a nervous system response. It has been proposed that disturbances in the control of local neuropeptide supply may underlie certain psychopathological states.[1,25]

In view of the multiple message sites and the many different effects of ACTH on the central and peripheral nervous systems, it is assumed that the effects of ACTH on the nervous system are the result of interaction with multiple receptors.[24] Peptide recognition is thought to be mediated by specific receptors located at the outside of the target cell membrane. Unfortunately, despite efforts of several independent groups of investigators using classical radioligand-binding assays, no specific, high-affinity, and low-capacity binding sites for ACTH could be identified in brain membranes.[27] The same author proposed that the high potency of ACTH neuropeptides in eliciting behavioral responses is indicative of the presence of a receptor system with the extremely high-affinity and low-capacity characteristics that interfere with detection in binding assays. However, in high concentrations, ACTH and congeners have selective affinity for brain opiate receptors.[28–32] It is still under debate whether some of the behavioral effects of ACTH result from ACTH–opiate receptor activation.[1,33–38]

Recently, evidence has accumulated pointing to complex interactions between ACTH and (neuronal) cell membranes not necessarily involving peptide-membrane receptor interactions (see Section 4.5).

3. ACTH AND BEHAVIOR IN EXPERIMENTAL ANIMALS AND MAN

In this section, we review some of the literature on the role that ACTH and congeners play in the modulation and control of certain behaviors. We choose to cover the literature on peptide effects in active and passive avoidance paradigms, as most of the studies on the neuropharmacology, neural substrate, and critical peptide structure have been performed in relation to these behavioral actions. Furthermore, two compulsive behaviors seen after central administration of ACTH are discussed, namely, the stretching and yawning syndrome and the induction of excessive grooming. The effects of the peptide on the latter behavior are treated in great detail, as they illustrate the complexity of the peptide–brain and –behavior interaction. The potential clinical relevance of neuropeptides renders it necessary to take into account their behavioral profile in social situations. Hence, the effects on sociosexual behavior in animals are discussed. Finally, the major findings in human studies to date are reviewed to underscore the importance of ACTH and congeners for brain and behavior in the old or diseased human. For extensive recent reviews on the effects of ACTH/MSH-like peptides on animal and human behavior, the reader is referred to several papers or monographs.[1,24,39–45]

3.1. Conditioned Avoidance Behavior

Conditioned avoidance behavior has been used extensively to assess the role that ACTH plays in the modulation of adaptive behavior in the rat. However, a variety of nonaversely motivated behaviors are influenced by ACTH/MSH-like peptides as well.[1,47] It seems that ACTH and related peptides from both pituitary and brain are of importance for the control of active avoidance learning.[48] However, what the relative contributions of peptides derived from these two sources are remain to be determined.

Although under certain conditions ACTH has been shown to improve acquisition of shock-motivated active avoidance behavior in rats, in general, extinction of the conditioned avoidance response seems to be more sensitive to the peptide treatment.[39] Mirsky et al.[49] were among the first to demonstrate effects of ACTH on the acquisition of an escape response in the rat. In later years, it became apparent that variables such as circadian rhythm, shock intensity, etc. were extremely important in the expression of modulatory influences of ACTH on acquisition of avoidance.[50–56] In adenohypophysectomized, hypophysectomized, or amygdalectomized rats, which all show a slow rate of avoidance acquisition, the effect of ACTH on avoidance acquisition can be demonstrated more readily.[48,57–59] In contrast to that of vasopressin and congeners, the behavioral influence of ACTH-like peptides is of short duration.[59]

Murphy and Miller[60] observed that administration of ACTH during acquisition resulted in a delay of extinction. More pronounced effects of the peptide treatment on extinction were obtained when the peptide was given during the extinction phase. Again, it was shown that the effect of the peptide was of relatively short duration.[61] Since it has been demonstrated that ACTH-like peptides retard extinction of a pole-jumping avoidance response both after systemic and after intracerebroventricular (i.c.v.) administration in a dose-dependent manner, this behavioral paradigm has been used to identify the active site in the ACTH molecule and the neural substrate sensitive to ACTH. It appeared that midbrain–limbic structures are essential to the expression of this behavioral activity of the peptide (see for review ref. 62).

Ever since it was observed that systemic administration of ACTH influences behavior, one has wondered whether corticosteroids mediate the behavioral effects of ACTH. At present, there is no doubt that both ACTH and steroids can influence behavior—sometimes in opposite directions—independently of each other.[63] Thus, although ACTH and steroids may affect similar behaviors by interaction with the same neural substrate,[64] the behavioral effects of ACTH do not depend on the release or presence of adrenal steroids.[45] Thus, there is a direct ACTH–brain interaction in the peptide modulation of avoidance behavior, and hence De Wied[40,58] formulated a neuropeptide concept long before the identification of peptidergic networks.

A number of recent reviews have dealt with structure–activity relationships between peptides derived from proopiomelanocortin and adaptive behaviors.[1,65–69] At first it was assumed that $ACTH_{4-10}$ was the shortest sequence with the full behavioral activity of $ACTH_{1-39}$.[65] Later studies revealed that, in fact, $ACTH_{4-7}$ is the smallest sequence (Fig. 1). The phenylalanine (Phe) in position 7 plays a key role in the mediation of the behavioral effects of ACTH. Substitution of the D-enantiomer of Phe at position 7 appeared to facilitate rather than delay extinction.[70,71] Other D-substitutions, which in general improve metabolic stability among other effects, were without this opposing shift in the direction of the behavioral effect. Although it was originally believed that [D-Phe^7]$ACTH_{4-10}$ contained a new intrinsic behavioral activity, it is now assumed that the D-peptide may counteract some of the effects of endogenous ACTH-like peptides on brain cell firing, possibly leading to opposite effects in active conditioned avoidance behavior.[1]

Structure–activity studies not only revealed a dissociation of structural requirements necessary to activate adrenal or brain receptors, they also pointed to differences in melanotropic and CNS requirements. For instance, the replacement in $ACTH_{1-17}$ of arginine (Arg) by lysine (Lys) at position 8, which is accompanied by loss of steroidogenic and melanotropic activity, did not reduce the behavioral activity.[1] Tryptophan (Trp) in position 9 is essential for melanotropic activity but not for the behavioral activity of the sequence 4–10.[1,65] Testing numerous structural modifications, Greven and De Wied[65] synthesized an $ACTH_{4-9}$ analogue (Org 2766: H-Met(O)-Glu-His-Phe-D-Lys-Phe-OH) that appeared to be 1000-fold more potent in inhibiting extinction than the sequence $ACTH_{4-10}$.

As has been discussed in Section 2.2, ACTH is considered to have multiple message sites, which can activate different receptors, sometimes leading to the same response. Greven and De Wied[66] observed residual behavioral activity in the sequence 7–16, which could be increased to the same level as that seen after $ACTH_{4-10}$ and $ACTH_{4-7}$ by C-terminal elongation to 7–16. Specific modifications of 7–16 led to a millionfold increase of activity.[67] Thus, with respect to avoidance extinction as well, there is a redundancy encoded in the ACTH molecule. The principle of potentiating dormant activity by appropriate N- or C-terminal chain elongation is also observed in other ACTH–CNS structure–activity studies.[24] Sandman and co-workers[41,72] assume that the multiplicity of information in the N-terminal part of the ACTH molecule, effective in adaptive behavior, is not so much a reflection of message repetition but rather that different parts of the molecule code for different behavioral processes.

Little is known about the crucial steric conformation of the peptide molecule at the presumed CNS receptor site.[66] In their review on structure–activity relationships of ACTH-, β-LPH-, and MSH-derived peptides, Van Nispen and Greven[69] proposed that $ACTH_{4-10}$ forms an α-helix, bringing the Phe and Met residues in close proximity and allowing a peptide receptor interaction of residues 4 (Met) and 8 (Arg) (see also ref. 66).

In addition to the studies using active avoidance paradigms, ACTH effects on passive avoidance behavior have been widely studied.[1,73] Using a one-trial passive avoidance task,[74] one can, in principle, separately study the effects of the peptide on acquisition/consolidation (prior to or immediately after the learning trial) and on retrieval processes (prior to the retention test).

With the whole ACTH molecule and postlearning trial treatment, it was observed that the peptide may facilitate later retention performance.[75,76] These authors reported an inverted U-shaped dose–response curve, a phenomenon characteristic of a variety of central effects of ACTH. Whether this inverted U is a feature of receptor or substrate or is the result of differential interaction with more than one peptide-sensitive system/receptor is not clear. Although there is some confusion as to the effectiveness of short N-terminal fragments under these conditions, the prevailing view is that $ACTH_{4-10}$ is only effective when given prior to the retention test.[1,45] Fekete and De Wied[77] reasoned that the use of high doses of the peptides, as done in the studies of Flood *et al.*[78] and Martinez *et al.*,[79] may have obscured proper analysis of when and at what level $ACTH_{4-10}$ or Org 2766 facilitate the passive avoidance performance. Furthermore, they determined that the effect of $ACTH_{4-10}$ lasted 3 to 6 h, whereas that of the potentiated Org 2766 amounted to at least 24 h. Interestingly, [D-Phe⁷]$ACTH_{4-10}$ also facilitates passive avoidance behavior with strength comparable to that of $ACTH_{4-10}$ and duration comparable to that of Org 2766.[65,77]

An extremely important observation is that Org 2766, after subcutaneous injection of low doses (nanogram quantities), facilitated and after high doses (microgram quantities) inhibited passive avoidance retention.[80] For the first time this differential effect on one given behavior could be ascribed to the involvement of different sites in the short molecule. The inhibitory effect was apparently located in the C-terminal part of the molecule. In further studies, these authors demonstrated that the facilitation residing in the N-terminal part of the molecule was not related to a neural substrate containing naloxone-sensitive opiate receptors, whereas the C-terminal inhibitory sequence expressed its activity through interaction with a neural substrate containing opiate receptor.[37] Org 2766 itself has no affinity for naloxone or dihydromorphine receptors in brain.[30]

Although the studies of Gold[73] using the whole ACTH molecule could not exclude a permissive role of adrenal steroid hormones, the effectiveness of Org

2766 again suggests that a direct peptide–brain interaction is responsible for the modulatory role of ACTH in passive avoidance behavior.

3.2. The Stretching and Yawning Syndrome

The first behavioral stereotypy reported to occur after intracranial injection of ACTH into mammals was the stretching and yawning syndrome (SYS).[81,82] In rodents, the onset of this syndrome is preceded by bouts of excessive grooming (see Section 3.3). The SYS was first described in dogs but could be equally elicited by intracisternal administration of ACTH in monkeys, rabbits, cats, guinea pigs, mice, and rats.[83–87] The SYS is seen after central and not after systemic treatment with the peptide. There is one report, however, indicating that chronic treatment of mice with a high dose of α-MSH resulted in enhanced yawning activity.[88] Intraarachnoidal administration of a crude ACTH preparation in doses up to 0.02 μg/kg in man caused nausea and vomiting but no stretching.[89] Gessa et al.[90] have shown that the act of stretching is accompanied by EEG and behavioral arousal. The most sensitive brain regions studied were the hypothalamic areas lining the third ventricle. Direct injection of ACTH into these regions also produced behavioral arousal, which was interpreted as reflecting sexual excitement. In later years, the research of these investigators has focused on sexual arousal induced by central application of ACTH and congeners (see Section 3.4).

Structure–activity studies pointed to a message site in $ACTH_{5-10}$, although the activity of this sequence *per se* was very reduced.[83,91] It was noted that the SYS could be antagonized by chlorpromazine, atropine, and morphine.[83] Urbá-Holmgren et al.[92] reported that in the rat the act of yawning involves a central cholinergic, muscarinic mechanism. Wood et al.[93] reported that the SYS elicited in rats by i.c.v. injection of α-MSH or $ACTH_{1-24}$ was paralleled by a twofold elevation in the hippocampal turnover rate of acetylcholine (ACh). Later, these authors suggested that the peptides activate hippocampal receptors and that the septal cholinergic input into the hippocampus is necessary to elicit the SYS.[94] At variance with a specific cholinergic mechanism seems to be the report by Mogilnicka and Klimek,[95] who found that drugs stimulating central dopaminergic transmission induce yawning in rats. Huston[96] has shown that yawning and penile erection in rats can be brought about by spreading depression, which has been reported to result from microinjection of $ACTH_{1-24}$ into neocortex and hippocampus.[97] Whatever the neural substrate underlying the SYS is, evidence has been presented that the substrate responsible for SYS is different from those involved in the effect of ACTH on adaptive behaviors and for the display of excessive grooming.[98]

3.3. Excessive Grooming

Grooming or maintenance behavior is a common species-characteristic movement pattern with readily definable components. The "care of the body surface" behavior is an important behavior as it removes ectoparasites, dirt, and other foreign substances and may also aid in wound healing.[99] However, it is assumed that grooming behavior in the rat may subserve other functions as well, e.g., thermoregulation, regulation of social behavior, and deactivation or displacement activity (see Spruijt and Gispen[38]). The latter function is thought to play a role in the grooming seen in novel or potentially stressful situations. Novel as well as stressful stimuli are among the most potent activators of the hypothalamo–pituitary–adrenal system.[100] However, a number of studies have pointed out that the grooming seen after exposure to novel or stressful stimuli is not mediated through the pituitary–adrenal system.[101–103] Yet, the pituitary hormone ACTH seems to be crucial in the initiation of this stress-related grooming. Dunn et al.[104] injected specific antibodies against ACTH into the cerebroventricular system of the rat and found that this treatment specifically inhibited the stress-related grooming response. Long before it was recognized that central peptidergic systems could play a modulatory role in the regulation of grooming behavior in the rat,[34,104–106] it was reported that intracranial or intraventricular injection of ACTH into rats resulted in the display of excessive grooming in a dose-dependent manner.[81,83,85] In rodents, the excessive grooming response is interrupted by episodes of stretching and yawning (see Section 3.2).

Although in the induction of SYS, peripheral gonadal steroids seem to play a permissive role,[87,91,107] the induction of the excessive grooming response is independent of the endocrine system. The response can be elicited in castrated, hypophysectomized, or adrenalectomized rats.[85,102] As noted by Haun and Haltmeyer,[108] i.c.v. injection of ACTH elevates the level of circulating corticosterone. Subsequently, Wiegant et al.[102] documented that this was not the result of ACTH escaping from the CSF into the circulation but that ACTH may in fact activate a neural system involved in the regulation of the pituitary–adrenal system. In addition, these authors showed that this neural substrate was dissociated from that underlying the grooming response. Also the facts that only central, and not peripheral, injection of ACTH elicits grooming and that ACTH fragments with no known classical endocrine effect produce the response make it almost certain that ACTH-induced excessive grooming is the result of a direct effect of the peptide on brain structures.

3.3.1. Pattern of ACTH-Induced Grooming

Recently, Spruijt and Gispen[38] studied the pattern of grooming behavior in ACTH- and saline-treated rats. ACTH-treated rats groomed for a much longer time than saline-treated rats did after the i.c.v. injection and placement in a novel observation box. However, if the data are converted to proportion of time spent per grooming element as a percentage of the total time of grooming, no major differences between the two groups were apparent. In Fig. 2, the relative element duration and relative transition frequency within the grooming repertoire is given. Clearly, the major pathways are similar for both groups of rats. Grooming often is initiated by vibration and follows a cephalo-caudal progression.[38,109] In saline-treated or untreated rats, the scratching/licking paw element normally concludes a grooming episode. However, in ACTH-treated rats, there are important new transitions from scratching to body

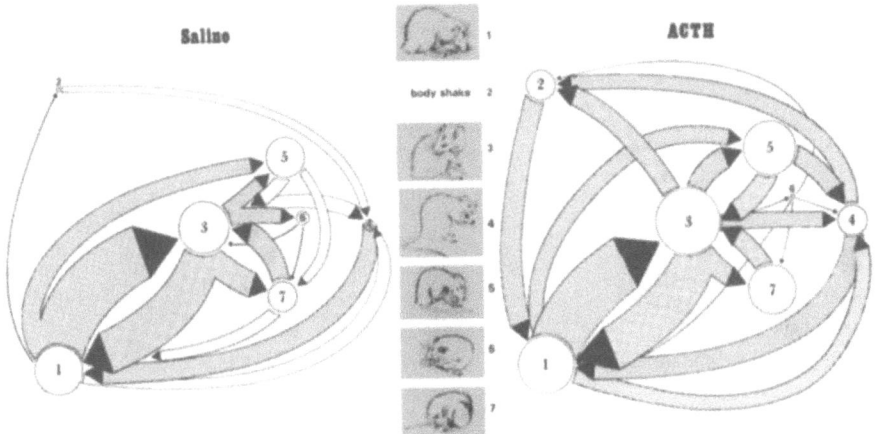

Fig. 2. Structure of novelty/saline- and ACTH-induced grooming behavior in the rat. The circles represent the log of the relative duration in seconds spent per element. The arrows indicate transition direction, and their width represents the square root of the number of transitions. Elements: 1, head washing; 2, body shake; 3, body grooming; 4, vibration; 5, scratching/licking of paw; 6, tail sniffing; 7, anogenital grooming. $ACTH_{1-24}$ 3 µg/3 µl i.c.v. or 3 µl saline. (From Spruijt and Gispen.[38])

grooming and vibration, leading via body shakes to face washing. Thus, ACTH treatment did not result in an increase in the number of bouts of grooming but in a prolongation of bout duration, as via face washing the rats resume their grooming behavior.[34,38]

3.3.2. Is Grooming Peptide Specific?

Although ACTH was the first peptide reported to induce excessive grooming, presently a great variety of peptides have been reported to enhance the display of excessive grooming (Table I). One should bear in mind that not all publications have reported on the response in sufficient detail to allow proper comparison. From those that have, it emerges that one should be cautious in generalizing the behavioral effects across peptides. To date, only ACTH/MSH and low doses of $dynorphin_{1-13}$ produce a grooming response that does not differ in structure from that seen in untreated rats when placed in a novel observation box. Thus, if central peptidergic networks or neurohumoral supply is of physiological importance in the initiation of grooming behavior, it is assumed that the endogenous active principle is a peptide with features common to the abovementioned peptides.[38]

3.3.3. Structure–Activity Studies

The grooming response to i.c.v. injection of ACTH has been very useful in demonstrating the complexity with which information is encoded in peptides (see Section 2.2; Fig. 1). $ACTH_{1-24}$ but neither its composing sequences 1–10 or 11–24 nor their equimolar combination will elicit excessive grooming. The

Table I
Peptides and Excessive Grooming in Rodents

Peptide	Activity[a]	References
ACTH	+	81
ACTH$_{1-24}$	+	85
ACTH$_{4-10}$	−	83,85,86,112
[D-Phe^7]ACTH$_{4-10}$	+	85,86,112
Org 2766	−	85
α-MSH	+	81,85,115
β-MSH	+	85
γ-MSH	−	110,111
Somatostatin	+/−	353,354
TRH	±	147
Vasopressin	±	355
Oxytocin	±	355
Prolactin	+	356
β-LPH	?	84
β-endorphin	+	357
γ-endorphin	−	357
α-endorphin	±	357
Enkephalins	−	357
Dynorphin$_{1-23}$	+	358,359
[dT]β-endorphin	+	38,135
[dT]dynorphin$_{1-13}$	+	38,135
Substance P	+	360,361
Eledoisin	+	362
Bombesin	+	362
Neurotensin	+	363
Angiotensin	−	11

[a] +, activity; ±, minor activity; −, no activity; ?, questionable.

fact that ACTH$_{1-24}$, α-MSH, and β-MSH were effective was ascribed to the common presence of the 4–10 sequence; γ-MSH, which contains the sequence Met-Gly-His-Phe-Arg, however, did not induce the response.[110,111] However, administration of the 4–10 sequence *per se* was without effect.[85,112] Gradual shortening at the C-terminus or at the N-terminus revealed that the message site is somewhere in the region 5–13,[85] possibly the sequence 5–7 (Fig. 1).[112] The inactivity of 7–10 but the activity of 5–14 and 4–7 are in support of this view. Interestingly, like the effect of ACTH on the retardation of extinction of active conditioned avoidance behavior (see Section 3.1), the [D-Phe^7]ACTH$_{4-10}$ analogue is unique among D-enantiomer substitutions in that it is active in inducing excessive grooming.[85,86,113] Other D-enantiomer substitutions such as the ACTH$_{4-9}$ analogue Org 2766 (see Gispen *et al.*[85]) are without effect on the activity of 4–10 and make it unlikely that improved metabolic stability can account for the expression of the behavioral effect.[114]

Besides specific proteolysis, —NH$_2$ acetylation and —COOH amidation or glycosylation are known examples of posttranslational modifications crucial to the biological activity of a given peptide. With respect to grooming behavior, the full activity of α-MSH [(Ac-Ser^1)—ACTH$_{1-13}$—NH$_2$] and the low or absent activity of $ACTH_{1-13}$—NH_2 provide a striking example of the effect of a single-step posttranslational activation.[85,115]

The structure–activity relationship can at best be understood if one assumes (1) that the stereoconformation of the peptide at its receptor site is crucial to proper receptor activation and (2) that expression of biological activity of a given message sequence depends on proper —COOH or —NH_2 terminus elongation.[20,22,67] The latter principle of expression of dormant activity can also be found for other ACTH–CNS relationships.[24]

3.3.4. Neural Substrate

The studies of the neural substrate underlying ACTH-induced excessive grooming involve both experiments with rats bearing lesions in certain brain regions and experiments employing neuropharmacological manipulation. The studies using the former approach revealed differences between ACTH effects on extinction of conditioned avoidance behavior, sexual excitement, and SYS on the one hand and grooming behavior on the other hand.[98] Electrolytic destruction of the parafascicular area, large parts of the septal complex, or small parts of ventral or dorsal hippocampus, regions known to be crucial to the effect of ACTH on avoidance extinction,[62,116,117] did not suppress ACTH-induced excessive grooming. In addition, lesioning of the medial preoptic area, known to be of importance to the effect of ACTH on sexual excitement,[90] was without effect on grooming. Lesions in the amygdala, mammillary bodies, and the dorsal and ventral hippocampus enhanced ACTH-induced SYS but were without effect on excessive grooming. Only large lesions of the hippocampal complex interfered with the ACTH-induced grooming. Evidence was obtained to suggest that hippocampectomy influenced the sensitivity of another, yet unknown brain structure important to ACTH-induced excessive grooming and that higher amounts of ACTH were required to initiate grooming.[98,118] There are suggestions that hippocampectomy may affect neurotransmission in the nucleus accumbens and neostriatum.[119–121]

Subsequently, it was found that the dopamine receptor blocker haloperidol was more effective in reducing ACTH-induced excessive grooming in hippocampectomized than in sham-operated control rats.[122] This finding was interpreted to reflect altered dopaminergic sensitivity in hippocampectomized rats and to confirm the involvement of dopaminergic systems in ACTH-induced grooming (see below). As far as we know, the only brain lesion to irreversibly suppress ACTH-induced excessive grooming is the destruction of large parts of the zona compacta and zona reticulata of the substantia nigra.[34] Although such rats are capable of normal grooming after water immersion, even high doses of ACTH have only a minor effect. The effect of the lesion is to eliminate grooming bouts of long duration without affecting bout structure *per se*.[34] Interestingly, local bilateral injection of small quantities of ACTH into the substantia nigra induces some but not maximal excessive grooming in the rat.

It has been shown that peripheral administration of a variety of presumed dopamine (DA) receptor blockers suppresses ACTH-induced grooming.[123–125] Cools[126] proposed that there are at least two functionally different DA receptors in the brain, one mediating excitation, the other inhibition. These two systems originate in the substantia nigra and the ventral tegmental area and project asymmetrically in the nucleus accumbens and neostriatum. It was concluded that ACTH-induced excessive grooming is very sensitive to a shift in the balance of these two systems in nucleus accumbens and neostriatum.[123,127,128]

A great number of studies have indicated that central effects of ACTH may involve a neural substrate containing opiate receptors.[28–31,33,35,83,129–134] Indeed, peripheral administration of naloxone, naltrexone, or naloxazone inhibits ACTH-induced grooming.[33,133] As discussed above, low doses of β-endorphin or of dynorphin$_{1-13}$ also initiate excessive grooming. Hence, it is conceivable that the grooming induced by ACTH is the result of a direct activation of opiate receptors by the peptide. Yet, evidence has been obtained to suggest that such a mechanism is unlikely. First, none of the peripherally administered drugs were proven to act at the site at which ACTH initiated the behavior. Secondly, dynorphin$_{1-13}$ and dynorphin$_{2-13}$ are nearly equally potent in inducing the grooming response and the behavior even after both peptides are suppressed by naloxone.[135,136] As the presence of the N-terminal tyrosine residue is of crucial importance to opiate receptor activation,[137,138] it is unlikely that the dynorphin$_{2-13}$ activates brain opiate receptors. Thirdly, ACTH$_{4-10}$ and [D-Phe^7]ACTH$_{4-10}$ both have similar affinity for brain opiate receptors,[30] but only the latter peptide induces excessive grooming. Plomp and Van Ree[139] suggested that the active site for opiate receptor activity resided in the ACTH sequence 7–9, whereas the active site for grooming is assumed to reside in 5–7.

Therefore, after reviewing all the current evidence, it seems most appropriate to consider the involvement of an opiate receptor in ACTH-induced grooming to be at a site other than the primary peptide–brain interaction.[38] Such a view is supported by the careful evaluation of the time course of grooming observed after various manipulations.[140,141] These authors suggest that novelty/stress-related and ACTH-induced grooming depend on a primary opiate-insensitive and a secondary opiate-sensitive neural substrate. Interestingly, the effects of Org 2766 on passive avoidance behavior also indicate an opiate/nonopiate system.[37]

As more centrally acting drugs with more or less defined mechanisms of action are being tested, it becomes apparent that the notion of dopaminergic substrate-containing opiate receptors underlying ACTH-induced grooming may be too simple.[124] Benzodiazepines decrease grooming in response to novelty but not to ACTH or β-endorphin.[142] The neuroleptics haloperidol and clozapine, but not sulpiride, decrease the ACTH-induced grooming. The antidepressants amitryptiline, nomifensine, and mianserine all suppress the response. Evidence was presented showing that the suppression by these drugs was not the result of competing behavioral activity elicited by the drug itself.[124] Furthermore, novelty/stress-related grooming is assumed to involve the histamine H$_1$ receptor[143] and α-adrenergic receptors.[34] Clearly there is a need for a more defined description of the neural substrate underlying ACTH-induced grooming. Nonetheless, at present it seems justified to emphasize the role of the substantia nigra as a structure and dopamine as transmitter being affected by ACTH.

3.3.5. Single-Dose Tolerance

An interesting feature of ACTH-induced grooming, namely, the development of short-term tolerance, has been extremely useful in the study of similarities and dissimilarities in neural substrates underlying peptide-induced grooming.[118,144–146] Recently, the development of long-term tolerance was demonstrated after i.c.v. infusion of $ACTH_{1-24}$ over days by means of a minipump,[147] but there are no signs of tolerance when $ACTH_{1-24}$ injections are given repeatedly 24 h apart.[144,148] When the interval between two injections is shorter than 12 h, a reduced amount of grooming after the second injection is observed. This reduction can be overcome by increasing the dose in the second injection. Furthermore, a cross tolerance among β-endorphin, [D-Phe^7]ACTH$_{4-10}$ and $ACTH_{1-24}$ has been demonstrated, suggesting similarities in neural substrate.[144] No such cross tolerance exists for prolactin and $ACTH_{1-24}$, indicating differences in neural substrate.[146] The tolerance is not dependent on endocrine mechanisms but can be prevented by neuropharmacological manipulation (haloperidol or naloxone) prior to the first injection of $ACTH_{1-24}$.[128,144] Interestingly, the single-dose tolerance is reminiscent of what is seen in catecholaminergic systems involved in thermoregulation.[149,150] Recently, Spruijt et al.[151] demonstrated that prior injection of $ACTH_{1-24}$ into the cerebroventricular system prevented the effect of a second injection 4 h later into the substantia nigra. The reversed order, first bilateral injection into the substantia nigra followed by a second injection into the ventricles, did not result in the development of tolerance. Such data again underscore the suggestion that there must be an additional neural site involved in ACTH-induced excessive grooming.

3.3.6. Biological Significance

Originally, lack of insight into the biological significance of the grooming response to i.c.v. $ACTH_{1-24}$ led to the view that the grooming test was an excellent but artificial way to study the complex ACTH–brain interaction and to guide neurochemists to brain structures sensitive to ACTH. Such a view is no longer tenable. In various recent reviews one may find the notion that this grooming response may very well relate to other known effects of ACTH on adaptive behaviors, and hence its usefulness in the study of the mechanism of action of ACTH has gained tremendously.[1,34,38,141,152]

Although several interpretations seem tenable, at present we prefer to consider ACTH-induced grooming in the context of a behavioral deactivation mechanism operative after the arousing influence of ACTH.[38,106] In an attempt to describe the effect of ACTH on adaptive behaviors in a unifying mechanism, Bohus and De Wied[45] assumed that ACTH may augment the state of arousal in certain brain structures and so increase the probability of a given behavioral performance. Thus, in line with this hypothesis, we assume that i.c.v. administration of ACTH and congeners increases the state of activation of the rat. The rat tries to lower this enhanced level of arousal to a new level of internal homeostasis. Probably, grooming is selected as the best strategy to adopt, as, in the absence of other urging stimuli, this behavior is not dependent on time and place.[38,106] It may be that the grooming related to novelty or stress resulting from conflict situations or other stressful stimuli is initiated by the release of ACTH and congeners from peptidergic pathways.[104,153] Once we accept that the grooming behavior is a strategy selected by the rat in aroused conditions, it is not surprising that dopaminergic transmission in the neostriatum is crucial to the display of this response, since this structure is involved in the selection mechanism whereby the optimal behavioral strategy is determined.[154]

3.4. Sexual Behavior

Several authors have reported that ACTH and related peptides may influence sexual behavior of both male and female mammals, notwithstanding that the results are sometimes contradictory.[1,91] There is evidence that gonadal[91,155] and adrenal[156-158] steroid hormones may play a permissive role in these peptide actions.

Bertolini et al.[91,155,159] reported that i.c.v. injection of ACTH and related peptides induces sexual arousal in cats, rabbits, and rats. This is characterized by recurrent episodes of penile erections and ejaculations accompanied by copulatory movements even in the absence of a receptive female partner. In fact, the male rat does not seek to copulate with a male or receptive female partner.[91] Implantation of $ACTH_{1-24}$ in the septal–preoptic region results in penile erection in the squirrel monkey.[160] Systemic administration of ACTH suppresses copulatory behavior of male rabbits in a novel environment.[161] Repeated subcutaneous administration of $ACTH_{4-10}$ increases intromission and ejaculation latency in castrated male rats maintained on a threshold dose of testosterone.[162]

Intraventricular infusion of $ACTH_{1-24}$, but not of $ACTH_{4-10}$, in female rabbits induces a sexual posture similar to lordosis.[107,108] Systemic treatment with $ACTH_{1-24}$ but not with $ACTH_{4-10}$ of estrogen-primed ovariectomized rats also initiates a lordotic response.[156,163] Other authors showed that α-MSH and $ACTH_{4-10}$ but not $ACTH_{1-39}$ or $ACTH_{1-16}$ biphasically affect lordotic behavior of ovariectomized rats receiving submaximal doses of progesterone.[164,165]

In analyzing the permissive role of adrenal steroids in ACTH-induced lordotic behavior, De Cantazaro and associates[157,158] observed that i.c.v. administration of ACTH in estrogen-primed ovariectomized and ovariectomized–adrenalectomized female rats induces lordotic behavior, whereas peripheral administration is only effective in ovariectomized females. Therefore, they suggest a difference in mechanism between the two routes of administration, the peripheral effect being mediated by adrenal steroids.[158]

Using sexual reward as a positive reinforcement, Bohus et al.[162] showed that systemic treatment of $ACTH_{4-10}$ delayed extinction of a runway response in male rats that were allowed to copulate in the goal box during the training period. Furthermore, with an increasing barrier technique, it was observed that ovariectomized rats that received $ACTH_{4-10}$ endured more aversive treatment to reach a partner male rat.[163] Such experiments do not primarily assess the effect of the peptide on sexual behavior itself but rather reveal effects of the peptide on motivational and/or attentional processes as studied in other adaptive experimental designs (see Section 3.1).

In summary, there is no doubt that under certain conditions ACTH and congeners can influence male and female sexual behavior. However, more research is necessary to understand the nature of the permissive role that adrenal and/or gonadal steroids play and to determine whether the enhanced display of copulatory elements indeed leads to increased sexual activity with receptive partners.

3.5. Social Behavior

Extremely relevant to the potential clinical use of ACTH and congeners is the study of their behavioral profile in social situations in rats. For agonistic behavior, that is, the behavior observed during interaction of at least two conspecifics in which propensities to aggression and flight are both activated,[166] Leshner[167] proposed a hormone–behavior relationship in which the hormonal state determines the behavior and, conversely, in which the behavior affects the hormonal state. By use of a trained fighter mouse as the aversive stimulus in a standard passive avoidance paradigm, it was found that the pituitary-adrenal system hormones affected social avoidance differently than avoidance in shock-mediated aversion. Specifically, it was found that corticosterone rather than ACTH is the primary hormone in control of such behavior (see for review Leshner et al.[168]). A similar conclusion was reached based on studies in which a series of repeated tests of submissiveness was employed and the influence of pituitary–adrenocortical hormone levels was determined.[168–170] Scott and Frederickson[171] had already demonstrated that ACTH may affect certain forms of agonistic behavior in mice and rats. They argued that changes in corticosterone levels interact with the experience of defeat to affect submissiveness. Their studies with respect to submission suggested that the most likely sequence of events is that defeat leads to an increase in corticosterone levels, which than facilitates the occurrence of submission. Also, Schuurman[172] demonstrated a correlation between defeat and an exaggerated pituitary–adrenal response.

In a review on hormonal influences on attack and defense in the mouse, Brain[173] concludes that acute treatment with ACTH may enhance aggression independently from the adrenal or gonadal steroids but that chronic treatment results in the opposite effect.

Studies by File[174] revealed that under standardized conditions manipulation of the test environment markedly influenced the duration of active social interaction of pairs of rats. She proposed that the anxiety-producing influences elicited by environmental manipulation reduced the social interaction. Since anxiety is likely to activate the pituitary–adrenal system,[100] the roles of ACTH and steroids in the control of social behavior seen under these circumstances were explored. It was found that corticosterone increased,[175] whereas $ACTH_{1-24}$, independently of its steroidogenic effect, and $ACTH_{4-10}$ decreased social activity.[176,177] Recently, Niessink and Van Ree[178] confirmed the findings of File with respect to $ACTH_{1-24}$ and $ACTH_{4-10}$. Both groups of researchers reported that in contrast to $ACTH_{4-10}$, Org 2766 in certain test and treatment conditions increased the social interaction between pairs of rats.[134,178,179] There is little doubt that the effects of $ACTH_{4-10}$ and Org 2766 are the result of a direct effect on brain structures, as was previously documented for other behavioral effects of these peptides.[1]

However, the abovementioned studies of File and Niessink are—in contrast to the studies of Leshner—restricted to direct effects of the peptides on social behavior. They do not take long-term behavioral adaptations as a consequence of the interaction between "experience and treatment" into account. With respect to a possible clinical use, behavioral adaptations are extremely important.

3.6. Behavioral and Clinical Studies in the Human

Based on the numerous data from animal studies suggesting that ACTH and congeners, by improving motivation and attention, facilitate behavioral adaptation, a variety of studies have been performed investigating the behavioral profile in human volunteers and patients (see recent reviews[41,43,44,180-182]).

The first human data on ACTH and $ACTH_{1-10}$ reported by Endröczi et al.[183] concerned changes in EEGs of young normal subjects that were interpreted to reflect peptide effects on attentional processes. Reviewing the available literature, Branconnier[181] concludes that despite the relatively small number of studies, ACTH and its small N-terminal fragments affect EEG activity recorded during stimulation or information processing.

The literature on the effects of acute treatment with $ACTH_{4-10}$ or Org 2766 on psychomotor performance and attention in the human is confusing. Gaillard[180] devotes a special section to methodological issues that make interpretation and comparison of the literature so difficult. The behavioral effects of ACTH on human performance are not easily explained by an increase in general arousal, as autonomic and cortical measures are not affected during rest conditions.[180,181] Gaillard[180] ascribes the improvement in performance during a monotonous serial test situation to an effect of the peptide on task-oriented motivation or sustained attention.

Sustained attention is deduced from the observation that subjects are more able to maintain alertness in tasks that normally show a decline in performance as a function of time at work.[180] Acute treatment with the ACTH peptides seems not to facilitate selective attention. Hence, there is a remarkable resemblance to the interpretation of the animal data in terms of task-oriented motivation or attention.[41,45] Although the effects of the ACTH-like peptides are far from dramatic, they seem to have one advantage over presently available stimulants in that they affect vigilance without adversive side effects.[180,181] Although some studies suggest that the peptide facilitates perception,[184,185] the majority of the data suggest no direct effects on a perceptual process.[180] Likewise, there is little or no convincing evidence for improvement of cognitive performance, learning, or memory processing in the human being.[181] Org 2766, for example, has no effect on immediate or delayed memory, on verbal or visual memory, or on memory retrieval.[182]

Thus, the small but significant influence of acute treatment of healthy humans with ACTH-like peptides is on sustained attention. At present, the most pronounced effect of Org 2766, however, is mood improvement in elderly people after subchronic oral administration (see Section 3.7).

Of special interest are the studies by Sandman and Kastin on the effect of $ACTH_{4-10}$ and Org 2766 on mentally retarded patients.[41] To date, three studies have been completed,[186-188] and the overall picture is that retarded subjects evidenced enhanced orienting responses to novel stimuli and also retained the ability to discriminate relevant from irrelevant information after treatment with the ACTH-like peptides. As in the studies on elderly people, peptide treatment improved the sociability of the patients as evidenced by increased patient–patient and patient–supervisor contact.[41]

3.7. Aging

Increased imbalance in the hormonal influence on brain homeostasis[189] and the decreased bioavailability of neuropeptides in brain peptidergic networks[43] have been postulated to underlie in part the deteriorated mental performance evidenced in old age.

Evidence is rapidly accumulating that neuropeptides may be of beneficial effect on the mood and mental performance of the aged human.[43] Notably, the $ACTH_{4-9}$ analogue Org 2766 has profound effects in the aged animal and man.[190] Considering the mechanism of action of this peptide, it is noteworthy that ACTH and some of its N-terminal fragments have been shown to exert trophic effects on nervous tissue.[11,191-193] The trophic effect is evidenced by a stimulation of RNA and protein synthesis[11] and of glucose utilization.[194,195] The potential clinical importance of this feature of ACTH is best illustrated by (1) the marked facilitation of functional recovery and repair of damaged peripheral nerve in rats after subchronic treatment with fragments of ACTH[196-201] and (2) the beneficial effect of peptide treatment in patients with various neuromuscular diseases.[197] From animal studies, it appeared that the peptides affected the function of the central motor neuron rather than the postsynaptic, muscle component of the neuromuscular junction.[202] Furthermore, Landfield et al.[189] have shown that both hippocampal morphological correlates of brain aging and age-dependent behavioral impairments in the rat are reduced by long-term treatment with Org 2766. Such long-term treatment may not only restore a disturbed peptide supply of the brain but may also counteract the known loss of neuronal plasticity in old age.[203] This notion is also compatible with the recent findings that ACTH increases the fluidity of synaptic plasma membranes prepared from the brain of aged rats[204-206] (see also Section 4.5).

Recently, a number of excellent reviews have addressed the animal and human studies on the role of ACTH-like peptides in old age.[43,44,190,207,208] Long treatment of rats with Org 2766 did not affect the life-span of rats.[208] Behaviorally, old rats seem easily distracted while traversing a learned route through a maze. Org 2766 protected performance from this distraction.[208] Similarly, old-age-impaired reversal learning in rats is also counteracted by long-term treatment with this peptide.[189] In old-aged humans, chronic treatment with Org

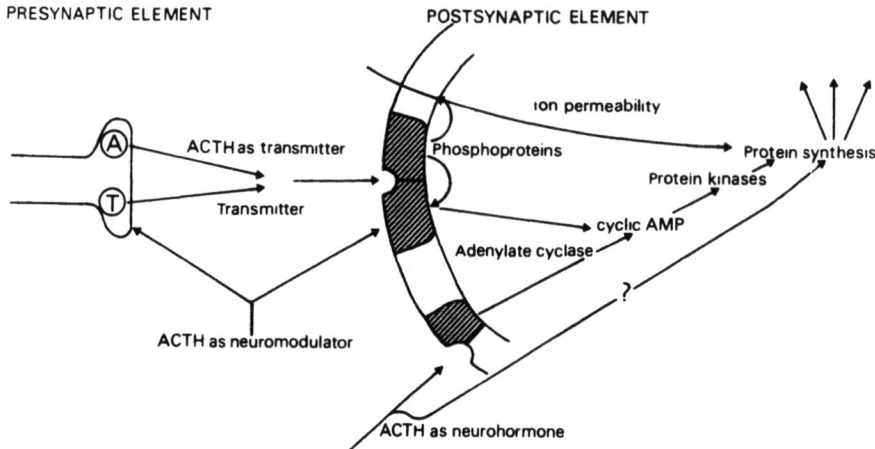

Fig. 3. Modes of action of ACTH in the brain. This figure illustrates the three currently explored modes of action of ACTH. (1) ACTH as putative transmitter. On depolarization of a peptidergic terminal containing ACTH (A), the peptide is released into the synaptic cleft and diffuses to the postsynaptic side of the synapse. (2) ACTH as neurohormone. Circulating ACTH may affect brain tissue as well as any other of its target cells and may exert its trophic effect on protein synthesis via a receptor coupled to adenylate cyclase. (3) ACTH as neuromodulator. The peptide may derive from the same or an adjacent terminal as the transmitter (T) and modulates either the synthesis or release of the transmitter presynaptically or the response of the cell to transmitter receptor activation postsynaptically, presumably involving changes in ion permeability of the cell membrane. (From Gispen et al.[300])

2766 primarily seems to improve mood.[44,207] Branconnier et al.,[209] employing a serial reaction time test in mildly senile subjects, found that $ACTH_{4-10}$ caused some suppression of increased reaction times resulting from continuous performance over 15 min. Furthermore, the subjects displayed reduced depression and confusion. Org 2766 reduced anxiety and depression and enhanced feelings of competence.[210,211] Also, in mildly to moderately demented patients, the peptide treatment seemed to make the patient more sociable.[212]

Irrespective of whether ACTH and congeners amend brain pathology or counteract physiological aging of the brain, it is clear that the peptide ACTH has great potential to improve brain function in old age.

4. NEUROCHEMICAL MECHANISM OF ACTION OF ACTH

In recent years, many reviews have been written on the mechanism of action of neuropeptides. Currently, three modes of action are considered: as neurotransmitters, as neurohormones, and as neuromodulators[11,192,213] (see Fig. 3). With respect to ACTH and its shorter fragments, there is still insufficient evidence to document a role as neurotransmitter.[192] Most of the literature on the neurochemical mechanism of action of ACTH deals with trophic or modulatory effects. Recently, the neurohormonal trophic influence, i.e., the stimulation of brain RNA and protein synthesis, has been reviewed extensively.[11,24,191,193,214] In this chapter, we concentrate on what is presently known about the modulatory effects of ACTH on synaptic function, especially on synaptic membrane processes.

4.1. Neurotransmitters

4.1.1. Catecholamines

Over the past 10 years, numerous groups of investigators have attempted to obtain information concerning a possible involvement of brain catecholamines in the effects of ACTH and its congeners on animal behavior (see for review ref. 215). Turnover studies based on catecholamine disappearance revealed that systemic treatment with $ACTH_{4-10}$ enhanced brain norepinephrine (NE).[216,217] In studies using an assay in which the capacity of the brain to convert [^{3}H]tyrosine into [^{3}H]catecholamine was measured, systemic administration of $ACTH_{4-10}$ was found to accelerate brain catecholamine synthesis.[218] In similar experiments, Dunn et al.[220] observed that $ACTH_{1-24}$ and $ACTH_{4-10}$ caused an increase in the conversion of [^{3}H]tyrosine into [^{3}H]dopamine but not into [^{3}H]NE.[219] Whether or not the peptides directly influence regional tyrosine hydroxylase activity in the brain is as yet unclear, as the data are hampered by apparent variability.[220–222]

Further evidence for modulatory influences of ACTH on brain catecholamines was provided by Endröczi.[223–225] Administration i.c.v. or implantation into the locus coeruleus of ACTH fragments caused a regional increase in the disappearance of i.c.v. administered [^{3}H]NE. The work of Lichtensteiger and colleagues also indicates that peptides sharing the N-terminal part with ACTH (e.g., α-MSH) have a direct effect on catecholamine neurons.[226,227] Telegdy and Kovács[228,229] observed regional changes in DA and NE concentrations after systemic treatment of rats with ACTH. Delanoy et al.[230] reported that i.c.v. $ACTH_{1-24}$ increased the accumulation of [^{3}H]DA from [^{3}H]Tyr in a subsequent in vitro assay only in slices of frontal cortex and not of hippocampus or striatum. There was no effect on the synthesis of NE. As several of the rats receiving $ACTH_{1-24}$ did not display excessive grooming, there was no correlation with the behavioral response.[230] $ACTH_{1-24}$ increased stimulation-evoked NE release from sympathetic nerves by acting on presynaptic ACTH receptors.[231]

Although certainly not complete, the literature cited above indicates that brain catecholamine systems can be affected by ACTH and congeners. Versteeg,[215] however, concluded that the nature of this influence of the ACTH peptides on brain catecholamines (direct or indirect) was still unresolved. In part, the seemingly confused information concerning the effects of ACTH on catecholamine neurotransmission may have arisen from differential effects of pre- or postsynaptic influences of the peptides depending on the conditions used.

4.1.2. Serotonin, GABA, and Acetylcholine

In contrast to the many papers on brain catecholamines, little is known about ACTH and brain serotonin (5-HT). Leonard[217,232] reported a decrease in brain level after chronic $ACTH_{4-10}$ treatment and an enhanced conversion of Trp to 5-HT only in rat cortex after chronic treatment with α-MSH. Telegdy and Kovács[228,229] demonstrated acute effects of ACTH on the 5-HT concentration of hypothalamus, mesencephalon, and hippocampus. Ramaekers et al.[233] found that changes in hippocampal 5-HT levels but not those in other brain parts were correlated with the changes in passive avoidance behavior after administration of $ACTH_{4-10}$.

Recently, Kendall et al.[234] reported that γ-aminobutyric acid (GABA) receptor binding in rat brain can be modified by changes in the circulating levels of ACTH. Systemic administration of ACTH and $ACTH_{4-10}$ caused an increase in midbrain and striatal GABA receptor binding. The increase seems to be associated with the appearance of low-affinity, high-capacity binding sites not observed in untreated animals.[234]

Early studies by Torda and Wolff[235,236] indicated that the synthesis of ACh by brain slices is increased by ACTH and diminished by hypophysectomy. Wood et al.[93,94] found a 50–100% increase in hippocampal ACh turnover rate following i.c.v. administration of $ACTH_{1-24}$. Based on several findings, these authors suggest that ACTH exerts this stimulatory effect in the hippocampus by activating receptors localized in the hippocampus. In contrast, Botticelli and Wurtman[237] concluded that the effect was on septohippocampal cholinergic neurons rather than on hippocampal receptors. Strand and colleagues[197,238] showed that ACTH not only stimulated the accumulation of [^{14}C]choline by whole-brain synaptosomes in vitro but also accelerated choline release. Interestingly, in synaptosomes from different brain regions, both stimulatory (hippocampus, hypothalamus) and inhibitory (thalamus, cerebellum) effects of ACTH on choline uptake were found.

4.2. Cyclic AMP

Within the framework of the classical second-messenger concept,[239] all or part of the central effects of ACTH could result from activation of an adenylate-cyclase-coupled membrane receptor. In fact, the available literature on ACTH and brain cyclic AMP is confused.

In broken-cell preparations, some authors showed no effect,[240,241] whereas others found a biphasic effect of ACTH on adenylate cyclase activity.[242] The latter authors demonstrated that the effects of ACTH very much depended on the concentrations of Ca^{2+}, Mg^{2+}, and ATP.[242,243] Maximal stimulation of adenylate cyclase occurred at 0.1 μM $ACTH_{1-24}$, which is within the range of that necessary for full activation of adenylate cyclase in peripheral target cells.[244] At high concentrations, $ACTH_{1-24}$ inhibits brain adenylate cyclase, possibly by lowering the degree of phosphorylation of certain membrane proteins essential in adenylate cyclase regulation.[242,243,245-247] Structure–activity studies revealed that the sequences 1–24, 1–16, and 4–7 elicited this inhibitory response, whereas the sequences 11–24, 1–10, and 4–10 were without effect.[242]

In slices of rat cerebral cortex, no effect of ACTH on the accumulation of cyclic AMP was observed,[248] whereas $ACTH_{1-24}$ enhanced the level of cyclic AMP but not of cyclic GMP in slices from rat posterior thalamus and neostriatum.[242,249] In regions obtained from rabbit brain, it was shown that ACTH$_{1-24}$ at micromolar concentrations stimulates the *in vitro* accumulation of cyclic AMP only in a number of circumventricular organs, namely, the choroid plexi of the lateral, third, and fourth ventricles, the pineal gland, the subcommissural organ, and the area postrema.[250]

Intrathecal administration of microgram quantities of ACTH in rabbits increased the cyclic AMP but not the cyclic GMP concentration in the cerebrospinal fluid, indicating that ACTH may also affect brain cyclic AMP production *in vivo*.[251,252] Likewise, i.c.v. administration of 1 μg ACTH$_{1-16}$ into rats resulted in a significant increase of septal cyclic AMP levels measured after microwave fixation of the brain tissue.[242] As the magnitude of the change (+27%) was of the order of that seen by others after electrical stimulation of nerve cell bodies,[253] and the septum is important for the expression of the behavioral activity of ACTH (see Section 3.2), a direct effect of ACTH on septal cells seems possible, although other explanations cannot be ruled out.[244]

In summary, N-terminal fragments of ACTH can modulate the activity of brain adenylate cyclase under certain experimental conditions. However, there is a variety of evidence that numerous CNS effects of ACTH are not mediated by cyclic AMP as second messenger (see Section 4.3).

4.3. Protein Phosphorylation

Phosphorylation is one of the most intensively studied posttranslational, covalent modifications of proteins.[254] A classical example illustrating the importance of protein phosphorylation for the cell is its regulatory role in carbohydrate metabolism.[255] It appears that protein kinases catalyzing the transfer of phosphate from ATP of proteins, together with protein phosphatases catalyzing the removal of the phosphate group from the protein, were directly and indirectly subject to regulation by a number of cellular constituents such as cyclic nucleotides, ions, hormones, and neurotransmitters.[255] The mechanism of action of the hormone ACTH has been studied extensively in two of its target tissues, adrenal cortex, where ACTH induces the synthesis and release of steroids, and brain, where ACTH acts as a neuroactive peptide. Briefly, some literature is reviewed here on the adrenal effects of ACTH on protein phosphorylation; a more detailed survey of the effects of ACTH on brain membrane phosphorylation is given. For effects of ACTH on cytosolic protein phosphorylation in the brain, the reader is referred to Schotman *et al.*[256]

4.3.1. Adrenal Cortex

In 1958, Haynes[257] reported that slices of beef adrenal cortex tissue showed enhanced activity of the enzyme phosphorylase on incubation with ACTH. It was realized[258] that this activation was caused by a cyclic-AMP-mediated effect on the phosphorylation state of phosphorylase. Murakami and Ichii[259] reported that subcutaneously injected ACTH enhanced *in vivo* [^{32}P]phosphate incorporation into rat adrenal proteins. The effect was only observed in the microsomal fraction and not in the nuclear, mitochondrial, or cytosolic fractions.

Using intact adrenocortical cells, Podesta et al.[260] found that on incubation with [^{32}P]phosphate, a great number of proteins incorporated radioactivity. It appeared that ACTH$_{1-24}$ (and cyclic AMP) induced a rapid and transient increase in the phosphorylation of a M_r 150 K cytosolic protein.

In recent studies, Koroscil and Gallant[261,262] demonstrated that ACTH$_{1-24}$ induced changes in the phosphorylation of several adrenocortical proteins located in various subcellular fractions. Using a photoaffinity-radiolabeled cyclic AMP derivative, these authors showed that ACTH$_{1-24}$ selectively activated the type I cyclic-AMP-dependent protein kinase within the microsomal fraction. Bristow et al.[263] reported that subcellular fractions enriched in plasma membranes from adrenal cortex retain the capacity to phosphorylate endogenous membrane proteins and, furthermore, that the phosphorylation of these proteins is markedly enhanced by the addition of cyclic AMP. The action of ACTH on the adrenal cortex seems to involve the following train of events: ACTH binds to its receptor on the surface of the plasma membrane of the adrenocortical cell, where it stimulates adenylate cyclase. Enhanced levels of cyclic AMP lead to enhanced activity of cyclic-AMP-dependent protein kinase, resulting in phosphorylation of some, as yet poorly characterized, proteins. It is hypothesized that these proteins are involved in steroid biosynthesis and release.[262]

4.3.2. Function in Brain

Brain tissue is the richest source of protein-phosphorylating activity known in mammalian tissue.[264,265] This may suggest that protein phosphorylation is involved in one of brain tissue's characteristic functions, namely, chemical neurotransmission.[266] Indeed, evidence has been obtained for a modulatory role of protein phosphorylation in neurotransmitter biosynthesis[267,268] and release.[269,270]

A number of functional correlates between phosphorylation of total protein[271,272] and of specific proteins[273-276] on the one hand and electrical events at the synapse on the other hand have been described. It has been suggested that the process of opening or closing ion channels in the synaptic membrane is mediated by phosphorylation of membrane proteins.[277,278] Evidence to support such a notion has been obtained using toad bladder,[279] avian erythrocytes,[280-282] *Aplysia* neurons,[283,284] and membranes obtained from *Torpedo californica* that are highly enriched in ACh receptors.[285-287]

In line with the presumed involvement of protein phosphorylation in the regulation of synaptic ion channels are the findings that various neurotransmitters, directly or via cyclic AMP, affect phosphate incorporation into synaptic proteins,[288-292] including their own receptor complexes.[287] However, there is other evidence to suggest that the changes in synaptic membrane protein phosphorylation are modulating the activity of various membrane-associated enzymes.[247,293,294]

4.3.3. Synaptic Plasma Membrane

In view of a possible identical mechanism of action of ACTH in brain and in adrenal cortex (receptor–adenylate cyclase–cyclic AMP), we studied the influence of behaviorally active fragments of ACTH on synaptic plasma membrane (SPM) phosphorylation.

When SPM is incubated in the presence of [γ-^{32}P]ATP, phosphorylation of a great number of proteins is observed. ACTH$_{1-24}$ in concentrations of 10^{-5}–10^{-4} M inhibits the phosphorylation of some relatively low-molecular-weight proteins (M$_r$ 17–48 K),[295,296] whereas cyclic AMP enhances the phosphorylation of certain proteins with a higher molecular weight (M$_r$ 50–80 K).[295] The involvement of different phosphoprotein bands and the opposite direction of the peptide and the cyclic nucleotide effects make it highly unlikely that the peptide effect could have been mediated by cyclic AMP.[295] No effect of the peptide was observed on membrane ATPase activity, and for at least one protein (B-50, M$_r$ 48 K), the evidence indicated that the reduced phosphorylation in the presence of this peptide did not reflect enhanced membrane-bound protein phosphatase activity but inhibited kinase activity.[297]

As peptide-induced release of the kinase from the membrane might underlie the observed inhibition of B-50 phosphorylation by ACTH, the supernatant from SPM incubated with the peptide was inspected. Only one major protein (M$_r$ 41 K) was detected.[298] However, based on the discrepancies between the structural requirement for ACTH-like peptides to inhibit B-50 phosphorylation in SPM and to release the 41 K protein from SPM, it was concluded that these two peptide effects are only remotely related.[299]

The phosphorylation of B-50 was dependent on the presence of Mg^{2+} and Ca^{2+}.[300] Based on its relative migration on sodium dodecylsulfate (SDS) polyacrylamide gels and its Ca^{2+}-dependent and cyclic-AMP-independent phosphorylation, the phosphoprotein B-50 may be identical to band γ-5,[301] F$_1$,[302] or protein 47K.[204] By two-dimensional analysis and peptide mapping, B-50 has been proven to be identical to protein p54p (Ca^{2+}).[303] It appeared that addition of calmodulin did not markedly stimulate the phosphorylation of B-50.[276,294,303] This may suggest that calmodulin does not serve as transducer of the Ca^{2+} to stimulate the B-50 kinase; however, one cannot exclude the possibility that membrane-bound calmodulin serves as the intermediate.

Structure–activity studies showed that besides ACTH$_{1-24}$, the shorter sequences 5–18 and 1–16 were fully active, ACTH$_{1-13}$ has reduced activity, and ACTH$_{1-10}$, ACTH$_{11-24}$, and their combination were inactive in inhibiting the B-50 phosphorylation in SPM.[297] Interestingly, the effect of ACTH fragments on the phosphorylation of B-50 is very similar to that found for the induction of excessive grooming.[34,85] Therefore, the effect of i.c.v. administration of a behaviorally active ACTH fragment on endogenous SPM phosphorylation in vitro was investigated. Administration of microgram quantities of ACTH$_{1-24}$ i.c.v. in rat and subsequent preparation of SPM after 30 min resulted in an increased amount of in vitro incorporated [^{32}P]phosphate into the same five phosphoprotein bands[304] that also responded after in vitro administration of ACTH. There appears to be a U-shaped dose–response curve of phosphate incorporation into SPM protein bands between 30 and 3000 ng of injected ACTH$_{1-24}$. This effect of in vivo ACTH treatment on in vitro endogenous phosphorylation was also time dependent, with maximal effect 30 min after the peptide injection.[304] As observed in other in vivo/in vitro approaches, changes induced in brain phosphorylation systems under *in vivo* conditions may persist in a *post hoc in vitro* assay.[274,305,306] If ACTH also inhibited SPM protein kinase(s) *in vivo*, the resulting SPM preparation of an animal so treated would have a higher percentage of unphosphorylated proteins in the ACTH-affected bands. Since ACTH is washed out during the preparation of SPM (H. Zwiers, unpublished observations), the subsequent *in vitro* SPM phosphorylation assay will then result in higher phosphate incorporation in ACTH-sensitive bands than in the saline-treated controls (see Weller and Rodnight[307]).

4.3.4. Phosphoprotein B-50 and B-50 Protein Kinase

Treatment of SPM with 0.5% Triton X-100 in 75 mM KCl solubilized 15% of the total B-50 protein kinase activity and preserved the sensitivity of the enzyme to $ACTH_{1-24}$. Column chromatography of the solubilized material over diethylaminoethylcellulose (DEAE-cellulose) pointed to the presence of multiple protein kinase activities, one of which was the ACTH-sensitive B-50 protein kinase.[308] The column fractions containing the B-50 protein kinase were subjected to ammonium sulfate precipitation, and a protein fraction (55–80% precipitate, further referred to as ASP55-80) enriched in both kinase and B-50 was prepared. This fraction showed a linear incorporation of [^{32}P]phosphate into endogenous B-50 with time for at least 10 min, and a maximal incorporation of 0.65 mol phosphate/mol B-50 was reached after 60 min. The inhibition by $ACTH_{1-24}$ of the protein kinase was concentration dependent; the half-maximal effective concentration was 5×10^{-6} M, 10–50 times lower than that obtained using intact SPM. The B-50 protein kinase required both Mg^{2+} and Ca^{2+} for optimal activity.[300,309] After two-dimensional electrophoresis on polyacrylamide slab gels, the B-50 protein kinase and the B-50 protein could be further identified, purified, and characterized. The isoelectric point (IEP) of the kinase is 5.5, and the M_r 70 K, whereas the IEP of the substrate protein B-50 is 4.5, and the M_r 48 K. Amino acid analysis on microgram quantities of purified kinase and B-50 protein revealed basic/acid amino acid ratios in agreement with the respective IEPs.[309]

In recent years there have been several reports on the purification of cyclic-nucleotide-independent protein kinases from rat brain.[310–312] Because several similarities were observed between B-50 protein kinase and another Ca^{2+}-sensitive protein kinase (kinase C) obtained from rat brain,[311] we have undertaken a comparison of the properties of these brain protein kinases and reached the conclusion that the B-50 kinase and kinase C are very similar, if not identical, as judged by their sensitivity to $ACTH_{1-24}$, $ACTH_{1-10}$, phospholipids, chlorpromazine, and proteolytic activation.[313] Both purified proteins were able to phosphorylate B-50. The observations that phospholipids stimulated the phosphorylation state of B-50[313] and that the phosphorylation state of B-50 affects the metabolism of polyphosphoinositides[293,314] implies an intimate relationship between proteins and lipids in the synaptic membrane. A similar conclusion was reached by Nishizuka's group.[315]

4.3.5. Localization of B-50

Subcellular fractionation of brain tissue followed by endogenous phosphorylation of various subfractions *in vitro* clearly showed that ACTH-sensitive B-50 phosphorylation was highly enriched in the light SPM fraction,[265,276] and Sörensen et al.,[316] using a variety of isolation techniques, concluded that B-50 phosphorylation was enriched in presynaptic membranes of low buoyant density. Both *in vitro* and after *in vivo* labeling, B-50 appeared to be one of the major phosphorylated proteins in SPM.[317]

Recently we reported on the distribution of immunoreactive B-50 protein in rat brain. An antiserum was used that was raised against the phosphoprotein band B-50 isolated from rat SPM by SDS slab gel electrophoresis. In agreement with the presumed presynaptic localization of B-50, the antiserum reacted with tissue components in brain regions rich in synaptic contacts. In contrast, white matter and cell perikarya were virtually without immunostaining.[318] The data suggest that at the cellular level there is a restricted localization of the B-50 protein (i.e., synaptic region) but that at the brain regional level the protein seems ubiquitous. In order to determine whether B-50 was specific for nervous tissue, we monitored the endogenous phosphorylation of proteins in the particulate and the cytosolic subcellular fractions of homogenates from a variety of organs. Phosphorylation of B-50 occurred only in particulate fractions from brain. Both two-dimensional analysis and immunochemical detection revealed the presence of B-50 in particulate material from brain and not in that from other tissues. Therefore, it was concluded that the localization of B-50 seems to be confined to the nervous tissue.[265]

The regional distribution of endogenous B-50 phosphorylation was studied using SPM obtained from individual brain regions. The highest value was found in SPM from septal origin, the lowest in SPM from the medulla spinalis.[265] Interestingly, $ACTH_{1-24}$ is at least ten times more effective in inhibiting B-50 phosphorylation in septal SPM than in SPM from whole brain.[319] Under optimal conditions, the IC_{50} for $ACTH_{1-24}$-induced inhibition of B-50 phosphorylation in septal SPM is in the order of $3-5 \times 10^{-7}$ M. The septal area has previously been reported to be essential for the expression of some of the behavioral effects of ACTH-like peptides.[320] Other studies also point to a special relationship between ACTH and the septum: stereospecific uptake[321,322] and specific transport of pituitary ACTH to the septal complex,[8] ACTH-containing nerve fiber projections to the septum,[323] and an ACTH-induced increase in septal cyclic AMP *in vivo*.[242]

4.4. Lipid Phosphorylation

Not only phosphoproteins but also phospholipids are involved in processes that are crucial for synaptic membrane function. The inositol phospholipids are implicated in receptor activation, Ca^{2+} entry through the membrane, and membrane (de)polarization.[324–326] The original interest in phosphatidylinositol (PI) arose from the observation that ACh-treated pancreas showed a large increase in the incorporation of radioactive phosphate into phosphatidic acid (PA) and

PI.[324,327] It is not clear what the initial event in the so-called "PI response" is. It has been proposed that triphosphoinositide (TPI) rather than PI has a primary role in the responses of cells to external stimuli such as hormones and neurotransmitters.[326,328–330] Further work is in progress to characterize the role that these changes in polyPI play in the influx of Ca^{2+}.

4.4.1. Adrenal Cortex

In a series of papers, Farese et al. reported on the effect of $ACTH_{1-24}$ on the levels of phospholipids in the adrenal. They found that $ACTH_{1-24}$ increased adrenal concentrations, of PA, PI, DPI, and TPI severalfold[331] and that the observed increases were dependent on the presence of Ca^{2+}.[332] It appeared that the effect could also be elicited by cyclic AMP.[333] Besides the increase in the levels of phospholipids belonging to the phosphatidate–inositide pathway, $ACTH_{1-24}$ also increased levels of phosphatidylcholine and phosphatidylethanolamine.[332] There is some evidence for a direct effect of $ACTH_{1-24}$ on one essential enzyme, glycerol-3-PO_4-acyltransferase.[332] It may be that activation by $ACTH_{1-24}$ of the transferase is caused by enhanced phosphorylation of this enzyme, which would be in agreement with the reported effects of cyclic AMP on adrenal cortical cell membrane protein phosphorylation.[263]

4.4.2. Brain

Jolles et al.[334] incubated synaptosomal fractions of rat brain prelabeled with inorganic orthophosphate in the presence of $ACTH_{1-24}$ and measured the amount of label recovered in PI, DPI, TPI, and PA. In a time- and dose-dependent manner, the peptide decreased the amount of label in the polyPI and PA. Most sensitive to the addition of the peptides were DPI and TPI. Recently, it was shown in a lysed synaptosomal fraction that a stimulating effect of ACTH can be seen in the absence of Ca^{2+}.[314] The potency of the various ACTH-like peptides to enhance TPI synthesis in this lysed synaptosomal fraction decreased in the order $ACTH_{1-24}$ > $ACTH_{5-18}$ > $ACTH_{1-16}$ > $ACTH_{1-13}$. $ACTH_{4-10}$ and $ACTH_{11-24}$ were ineffective.[326]

In an effort to relate these peptide-induced changes in polyPI metabolism to the previously reported effect of the peptide on membrane phosphoproteins (see above), DPI and [γ-^{32}P]ATP were added to the ASP55-80 membrane protein fraction containing the B-50 protein kinase/B-50 substrate protein complex.[293] The only two labeled compounds appeared to be the B-50 protein and TPI. Apparently, this fraction, in addition to protein kinase activity, also contains DPI-kinase activity. An inverse relationship exists between the degree of phosphorylation of B-50 and the amount of TPI produced. These and other findings strongly suggest that B-50 protein phosphorylation and membrane polyPI metabolism are correlated events.[326] Such a notion was underscored by data on the simultaneous effect of $ACTH_{1-24}$ on B-50 protein kinase and DPI-kinase activity in the same ASP55-80 membrane protein fraction. A dose-dependent inhibition of B-50 phosphorylation was observed, again accompanied by a stimulation of TPI production.[293]

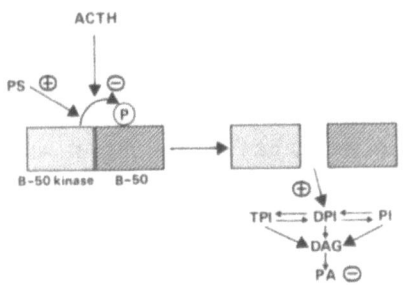

Fig. 4. Relationship between ACTH-sensitive phosphorylation of protein B-50 and the metabolism of phosphoinositides. In this working model, B-50 in its phosphorylated form is bound to the B-50 kinase in the presynaptic membrane. By inhibiting the phosphorylation of B-50, ACTH induces the dissociation of the complex. As the presence of other lipids such as phosphatidylserine stimulates B-50 protein kinase, and ACTH affects membrane lipid fluidity, it is not clear what the exact ACTH-sensitive step in the inhibition of the B-50 protein kinase is. Whatever the mechanism, the now dissociated free B-50 protein acts as a modulator of the system that controls the phosphorylation state of the phosphoinositides, presumably at the level of the conversion of DPI to TPI by DPI kinase (see also Zwiers et al.[319]).

Recently, more evidence was collected pointing to protein B-50's involvement in TPI metabolism. By using immunoglobulins (IgG) directed specifically against B-50, we were able to block the phosphorylation of this protein in the synaptic membrane. In parallel samples, we again observed an enhanced TPI labeling. Thus, with both ACTH and IgG-B-50, we were able to affect the phosphorylation state of B-50, and both independent treatments led to enhanced synthesis of TPI.[335] By analogy to the regulation of the activity of cyclic-AMP-dependent protein kinases, where the phosphorylation state of the inhibitory regulatory unit is a determining factor in the release of the free, active catalytic unit,[336,337] the hypothesis was introduced that B-50 could be a regulatory protein of DPI-kinase and that the phosphorylation state of the regulatory protein B-50, which can be modulated by ACTH, determines the level of TPI in the membrane[293] (Fig. 4).

In view of the high affinity of TPI for Ca^{2+} (dissociation constant TPI–Ca^{2+}, 2.7×10^{-6} M[338]), which is of the same order of magnitude as calmodulin–Ca^{2+} (4–18 μM),[339] the physiological role for membrane TPI levels may be connected with the fine regulation of intrasynaptosomal Ca^{2+} levels.[340] Thus, it was proposed that ACTH, via this mechanism, could alter the influx of Ca^{2+} into the presynaptic terminal and hence modulate neurotransmitter exocytosis.[319]

4.5. Synaptic Plasma Membrane Fluidity

In the preceding paragraphs we discussed the effects of ACTH on various processes occurring in the synaptic plasma membrane, namely, adenylate cyclase activity, protein phosphorylation, and polyPI metabolism. No high-affinity receptors for ACTH in rat brain have as yet been demonstrated,[27] opening the possibility that at least part of the membrane-associated effects of ACTH are mediated by a more general mechanism. Schwyzer[20] discussed the possibility that ACTH may interact with the target cell membrane independent of binding to a specific membrane receptor. Based on experiments using artificial lipid bilayer membranes, he proposes that ACTH might span the membrane, exposing its N-terminus to the inside. Alternatively, ACTH might exert a direct effect on membrane fluidity, such as has been described for other hormones and neurotransmitters.[341] Furthermore, the lipid fluidity of membranes seems to affect the activity of the various membrane proteins, such as carriers,[342] receptors,[343,344] phosphoproteins,[204] and enzymes.[345] By use of the fluorescence depolarization method[346] as an indicator of membrane fluidity, it was found that $ACTH_{1-24}$ fluidized SPM in a dose-dependent way, the lowest effective concentration being 10^{-5} M. As ATP was not present in the test system, it is highly unlikely that phosphorylation of proteins and/or lipids is required for the peptide-induced change in SPM microviscosity. In the adrenal cortex, numerous studies pointed to the implication of long-chain polyunsaturated fatty acids in the action of ACTH.[347-352]

In a series of experiments the structure–activity relationship for the effect on SPM lipid fluidity was assessed. Besides $ACTH_{1-24}$, the sequences 1–39, 1–16, and 5–18 were also active. The smaller sequences 1–10, 4–10, and 11–24 and the analogue Org 2766 were inactive. As there is a striking similarity between this structure–activity relationship and that for the inhibition of B-50 phosphorylation,[300] it is tempting to speculate that these ACTH-induced events are related. Further work is in progress to assess the ACTH-induced changes in membrane fluidity in more detail in order for us to understand more fully the significance of these findings.

5. CONCLUDING REMARKS

In the foregoing paragraphs we have reviewed some but certainly not all of the presently known modulatory influences of ACTH and congeners on animal and human behavior. It is clear that these peptides have profound effects on central nervous structures involved in the expression of a variety of behaviors. There is reason to believe that both pituitary- and brain-derived ACTH molecules are of importance in this respect. Further research is necessary to enable a fuller comprehension of the interrelationship of the two ACTH-producing systems.

As cell metabolism underlies cell functioning, it is of great interest to identify neurochemical events sensitive to ACTH-like peptides that exert behavioral effects. The present chapter reveals the possibility that ACTH, by interfering with a multifunctional protein–phosphoinositide complex in the presynaptic membrane, may influence ongoing transmission in certain synapses (see Fig. 4). Further studies on the temporal and topographical distribution of the neurochemical actions of ACTH are required before the link with synaptic and behavioral effects can be established. It seems probable that administered peptides act on neurons that normally receive peptidergic projections. If and to what extent general influences on membrane fluidity are responsible for ACTH-induced inhibition/stimulation of membrane-bound enzymes needs further clarification.

The mechanisms of action of neuropeptides that modulate or control the behavior may have characteristics in common. Hence, it is hoped that the studies on ACTH are of relevance in the search for the neurochemical mechanism of action of behaviorally active neuropeptides in general.

ACKNOWLEDGMENTS. The authors are greatly indebted to Drs. David De Wied and Philippa Edwards for helpful discussions and to Mrs. Lia Claessens, Mrs. Greet Hoekstra, and Mr. Ed Kluis for their aid in preparing the manuscript.

REFERENCES

1. De Wied, D., and Jolles, J., 1982, *Physiol. Rev.* **62**:976–1059.
2. Allen, J. P., Kendall, J. W., McGilvra, R., and Vancura, C., 1974, *J. Clin. Endocrinol.* **38**:586–593.
3. Harris, G. W., 1955, *Neural Control of the Pituitary Gland*, Edward Arnold, London.
4. De Kloet, E. R., Palkovits, M., and Mezey, E., 1981, *Pharmacol. Ther.* **12**:321–351.
5. Page, R. B., Munger, B. L., and Bergland, R. M., 1970, *Am. J. Anat.* **146**:273–301.
6. Bergland, R. M., and Page, R. B., 1978, *Endocrinology* **102**:1325–1338.
7. Bergland, R. M., Blume, H., Hamilton, A., Monica, P., and Paterson, R., 1980, *Science* **210**:541–543.
8. Mezey, E., Palkovits, M., De Kloet, E. R., Verhoef, J., and De Wied, D., 1978, *Life Sci.* **22**:831–838.
9. Moldow, R., and Yalow, R. S., 1978, *Proc. Natl. Acad. Sci. U.S.A.* **75**:994–998.
10. Swaab, D. F., Achterberg, P. W., Boer, G. J., Dogterom, J., and van Leeuwen, F. W., 1981, *Endogenous Peptides and Learning and Memory Processes* (J. L. Martinez, Jr., R. A. Jensen, R. B. Messing, H. Rigter, and J. L. McGaugh, eds.), Academic Press, New York, pp. 7–36.
11. Tonnaer, J. A. D. M., Wiegant, V. M., De Jong, W., and De Wied, D., 1982, *Brain Res.* **236**:417–428.
12. Krieger, D. T., Liotta, A. S., Brownstein, M. J., and Zimmerman, E. A., 1980, *Recent Prog. Horm. Res.* **36**:277–344.
13. Watson, S. J., Richard, C. W., and Barchas, J. D., 1978, *Science* **200**:1180–1182.
14. Chatelain, A., Dubois, M. P., and Dupouy, J. P., 1976, *Cell Tissue Res.* **169**:335–344.
15. Schwartzberg, D. G., and Nakane, P. K., 1982, *Endocrinology* **110**:855–864.
16. Chrétien, M., Benjannet, S., Gossard, F., Gianoulakis, C., Crine, P., Lis, M., and Seidah, N. G., 1979, *Can. J. Biochem.* **57**:1111–1121.
17. Watson, S. J., and Akil, H., 1980, *Hormones and the Brain* (D. de Wied and P. A. van Keep, eds.), MTP Press, Lancaster, pp. 73–86.
18. Scott, A. P., Bloomfield, G. A., Lowry, G. A., Gilke, P. J., London, T. J., and Rees, L. H., 1976, *Peptide Hormones* (J. A. Parson, ed.), University Park Press, Baltimore, pp. 247–271.
19. Schwyzer, R., 1977, *Ann. N.Y. Acad. Sci.* **297**:3–24.
20. Schwyzer, R., 1980, *Trends Pharmacol. Sci.* **3**:327–331.
21. Schwyzer, R., 1980, *Proc. R. Soc. (Lond.) [Biol.]* **210**:5–20.
22. Schulster, D., and Schwyzer, R., 1980, *Cellular Receptors for Hormones and Neurotransmitters* (D. Schulster and A. Levitzki, eds.), John Wiley & Sons, New York, pp. 197–217.
23. Schwyzer, R., 1963, *Pure Appl. Chem.* **6**:265–295.
24. Gispen, W. H., van Ree, J. M., and de Wied, D., 1977, *Int. Rev. Neurobiol.* **20**:209–250.
25. De Wied, D., 1978, *Dev. Neurosci.* **4**:113–122.
26. Burbach, J. H. P., and De Wied, D., 1982, *Neuronal Plasticity and Memory Formation* (C. Ajmone Marsan and H. Matthies, eds.), Raven Press, New York, pp. 153–158.
27. Witter, A., 1980, *Receptors for Neurotransmitters and Peptide Hormones* (G. C. Pepeu, M. Kuhar, and L. Enna, eds.), Raven Press, New York, pp. 407–414.
28. Terenius, L., 1975, *J. Pharm. Pharmacol.* **27**:450–452.
29. Terenius, L., 1976, *Eur. J. Pharmacol.* **38**:211–213.
30. Terenius, L., Gispen, W. H., and De Wied, D., 1975, *Eur. J. Pharmacol.* **33**:395–399.
31. Gispen, W. H., Buitelaar, J., Wiegant, V. M., Terenius, L., and De Wied, D., 1976, *Eur. J. Pharmacol.* **39**:393–397.

32. Akil, H., Hewlett, W. A., Barchas, J. D., and Li, C. H., 1980, *Eur. J. Pharmacol.* **64**:69–77.
33. Wiegant, V. M., Gispen, W. H., Terenius, L., and De Wied, D., 1977, *Psychoendocrinology* **2**:63–69.
34. Gispen, W. H., and Isaacson, R. L., 1981, *Pharmacol. Ther.* **12**:209–246.
35. Jacquet, Y. F., 1978, *Science* **201**:1032–1034.
36. Amir, S., Blair, R., Shizgal, P., and Amit, Z., 1979, *Science* **205**:424–425.
37. Fekete, M., and De Wied, D., 1982, *Eur. J. Pharmacol.* **81**:441–448.
38. Spruijt, B., and Gispen, W. H., 1983, *Hormones and Behaviour in Higher Vertebrates* (Y. Balthazart, E. Pröve, and R. Gilles, eds.), Springer-Verlag, Berlin, pp. 118–136.
39. De Wied, D., and Gispen, W. H., 1977, *Peptides in Neurobiology* (H. Gainer, ed.), Plenum Press, New York, pp. 397–448.
40. De Wied, D., 1977, *Life Sci.* **20**:195–204.
41. Sandman, C. A., and Kastin, A. J., 1981, *Pharmacol. Ther.* **13**:39–60.
42. Martinez, J. L., Jr., Jensen, R. A., Messing, R. B., Rigter, H., and McGaugh, J. L., 1981, *Endogenous Peptides and Learning and Memory Processes*, Academic Press, New York.
43. De Wied, D., and Van Ree, J. M., 1982, *Life Sci.* **31**:709–719.
44. Pigache, R. M., 1984, *Clinical Pharmacology and Psychiatry* (L. Gram, E. Usdin, S. Dahl, and P. Kragh-Sørensen, eds.), Macmillan, London, (in press).
45. Bohus, B., and De Wied, D., 1980, *General, Comparative and Clinical Endocrinology of the Adrenal Cortex* (I. Chester Jones and W. H. Henderson, eds.), Academic Press, London, pp. 265–347.
46. O'Donohue, T. L., and Dorsa, D. M., 1982, *Peptides* **3**:353–395.
47. Kastin, A. J., Olson, R. D., Sandman, C. A., and Coy, D. H., 1981, *Endogenous Peptides and Learning and Memory Processes* (J. L. Martinez, Jr., R. A. Jensen, R. B. Messing, H. Rigter, and J. L. McGaugh, eds.), Academic Press, New York, pp. 563–577.
48. Fekete, M., Bohus, B., and De Wied, D., 1983, *Neuroendocrinology* **36**:112–118.
49. Mirsky, I. A., Miller, R., and Stein, M., 1953, *Psychosom. Med.* **15**:574–588.
50. Bohus, B., and Endröczi, E., 1965, *Acta Physiol. Acad. Sci. Hung.* **26**:183–189.
51. Beatty, D. A., Beatty, W. A., Bowman, R. E., and Gilchrist, J. C., 1970, *Physiol. Behav.* **5**:939–944.
52. Ley, K. F., and Corson, J. A., 1971, *Experientia* **27**:958–959.
53. Ley, K. F., and Corson, J. A., 1972, *Int. J. Psychobiol.* **2**:265–271.
54. Pagano, R. R., and Lovely, R. H., 1972, *Physiol. Behav.* **24**:759–764.
55. Stratton, L. O., and Kastin, A. J., 1974, *Horm. Behav.* **5**:149–155.
56. Bohus, B., and De Wied, D., 1981, *Endogenous Peptides and Learning and Memory Processes* (J. L. Martinez, Jr., R. A. Jensen, R. B. Messing, H. Rigter, and J. L. McGaugh, eds.), Academic Press, New York, pp. 59–77.
57. De Wied, D., 1964, *Am. J. Physiol.* **207**:255–259.
58. De Wied, D., 1969, *Frontiers in Neuroendocrinology* (W. F. Ganong and L. Martini, eds.), Oxford University Press, New York, pp. 97–140.
59. Bohus, B., Gispen, W. H., and De Wied, D., 1973, *Neuroendocrinology* **11**:137–143.
60. Murphy, J. V., and Miller, R. E., 1955, *J. Comp. Physiol. Psychol.* **48**:47–49.
61. Bohus, B., Nyakas, C., and Endröczi, E., 1968, *J. Neuropharmacol.* **7**:307–314.
62. Van Wimersma Greidanus, T. B., 1977, *Front. Horm. Res.* **4**:129–139.
63. De Wied, D., Van Delft, A. M. L., Gispen, W. H., Weijnen, J. A. W. M., and Van Wimersma Greidanus, T. B., 1972, *Hormones and Behavior* (S. Levine, ed.), Academic Press, New York, pp. 135–171.
64. Veldhuis, H. D., and De Kloet, E. R., 1982, *Neuroendocrinology* **34**:374–380.
65. Greven, H. M., and De Wied, D., 1973, *Prog. Brain Res.* **39**:429–442.
66. Greven, H. M., and De Wied, D., 1977, *Front. Horm. Res.* **4**:140–152.
67. Greven, H. M., and De Wied, D., 1980, *Hormones and the Brain* (D. De Wied and P. A. Van Keep, eds.), MTP Press, Lancaster, pp. 115–127.
68. Coy, D. H., and Kastin, A. J., 1980, *Pharmacol. Ther.* **10**:657–668.
69. Van Nispen, J. W., and Greven, H. M., 1982, *Pharmacol. Ther.* **16**:67–102.
70. Bohus, B., and De Wied, D., 1966, *Science* **153**:318–320.

71. De Wied, D., 1969, *Proc. Int. Cong. Endocrinol.* **184**:310-316.
72. Sandman, C. A., Beckwith, B. G., and Kastin, A. J., 1980, *Peptides* **1**:277-280.
73. Gold, P. E., and Delanoy, R. L., 1981, *Endogenous Peptides and Learning and Memory Processes* (J. L. Martinez, Jr., R. A. Jensen, R. B. Messing, H. Rigter, and J. L. McGaugh, eds.), Academic Press, New York, pp. 79-98.
74. Ader, R., Weijnen, J. A. W. M., and Moleman, P., 1972, *Psychon. Sci.* **26**:125-128.
75. Gold, P. E., and Van Buskirk, R., 1976, *Horm. Behav.* **7**:509-517.
76. Gold, P. E., and Van Buskirk, R., 1976, *Behav. Biol.* **16**:387-400.
77. Fekete, M., and De Wied, D., 1982, *Pharmacol. Biochem. Behav.* **16**:387-392.
78. Flood, J. F., Jarvik, M. E., Bennett, E. L., and Orme, A. E., 1976, *Pharmacol. Biochem. Behav.* **5**:(Suppl. 1):41-51.
79. Martinez, J. L., Jr., Vasquez, B. J., Jensen, R. A., Soumireu-Mourat, B., and McGaugh, J. L., 1979, *Pharmacol. Biochem. Behav.* **10**:145-147.
80. Fekete, M., and De Wied, D., 1983, *Pharmacol. Biochem. Behav.* **17**:177-182.
81. Ferrari, W., 1958, *Nature* **181**:925-926.
82. Ferrari, W., 1958, *Arch. Ital. Sci. Farmacol.* **3**:133.
83. Ferrari, W., Gessa, G. L., and Vargiu, L., 1963, *Ann. N.Y. Acad. Sci.* **104**:330-345.
84. Izumi, K., Donaldson, J., and Barbeau, A., 1973, *Life Sci.* **12**:203-210.
85. Gispen, W. H., Wiegant, V. M., Greven, H. M., and De Wied, D., 1975, *Life Sci.* **17**:645-652.
86. Rees, H. D., Dunn, A. J., and Iuvone, P. M., 1976, *Life Sci.* **18**:1333-1340.
87. Rodriquez-Sierra, J. F., Terasawa, E., Goldfoot, D. A., and De Wied, D., 1981, *Horm. Behav.* **15**:77-85.
88. Segawa, T., Kawakasi, K., and Yajima, H., 1973, *Jpn. J. Pharmacol.* **23**:121-123.
89. Floris, E., 1963, *Boll. Soc. Ital. Biol. Sper.* **39**:558-560.
90. Gessa, G. L., Pisano, M., Vargiu, L., Crabai, F., and Ferrari, W., 1967, *Rev. Can. Biol.* **26**:229-236.
91. Bertolini, A., Gessa, G. L., and Ferrari, W., 1975, *Sexual Behavior: Pharmacology and Biochemistry* (M. Sandler and G. L. Gessa, eds.), Raven Press, New York, pp. 247-257.
92. Urbá-Holmgren, R., González, R. M., and Holmgren, B., 1977, *Nature* **267**:261-262.
93. Wood, P. L., Malthe-Sörensen, D., Cheney, D. L., and Costa, E., 1978, *Life Sci.* **22**:673-678.
94. Wood, P. L., Cheney, D. L., and Costa, E., 1979, *J. Pharmacol. Exp. Ther.* **209**:97-103.
95. Mogilnicka, E., and Klimek, V., 1977, *Pharmacol. Biochem. Behav.* **7**:303-305.
96. Huston, J., 1971, *Nature* **232**:274-275.
97. Jakobartl, L., and Huston, J. P., 1977, *Neurosci. Lett.* **5**:189-192.
98. Colbern, D., Isaacson, R. L., Bohus, B., and Gispen, W. H., 1977, *Life Sci.* **21**:393-402.
99. Borchelt, P. L., 1980, *Comparative Psychology, An Evolutionary Analysis of Animal Behavior* (M. Ray Denny, ed.), John Wiley & Sons, New York, pp. 362-384.
100. Mason, J. W., 1968, *Psychosom. Med.* **30**:576-607.
101. Jolles, J., Rompa-Barendregt, J., and Gispen, W. H., 1979, *Behav. Neural Biol.* **25**:563-572.
102. Wiegant, V. M., Jolles, J., Colbern, D., Zimmerman, E., and Gispen, W. H., 1979, *Life Sci.* **25**:1791-1796.
103. Gispen, W. H., Brakkee, J. H., and Isaacson, R. L., 1980, *Behav. Biol.* **29**:481-486.
104. Dunn, A. J., Green, E. J., and Isaacson, R. L., 1979, *Science* **203**:281-283.
105. Katz, R. J., and Roth, K. A., 1979, *Neurosci. Lett.* **13**:209-212.
106. Jolles, J., Rompa-Barendregt, J., and Gispen, W. H., 1979, *Horm. Behav.* **12**:60-72.
107. Baldwin, D. M., Haun, C. K., and Sawyer, C. H., 1974, *Brain Res.* **80**:291-301.
108. Haun, C. K., and Haltmeyer, G. C., 1975, *Neuroendocrinology* **19**:201-213.
109. Richmond, G., and Sachs, B. D., 1980, *Behaviour* **75**:82-95.
110. Van Ree, J. M., Bohus, B., Csontos, K. M., Gispen, W. H., Greven, H. M., Nijkamp, F. P., Opmeer, F. A., De Rotte, A. A., Van Wimersma Greidanus, Tj. B., Witter, A., and De Wied, D., 1981, *Life Sci.* **28**:2875-2888.
111. O'Donohue, T. L., Handelmann, G. E., Loh, Y. P., Olton, D. S., Leibowitz, J., and Jacobowitz, D. M., 1981, *Peptides* **2**:101-104.
112. Wiegant, V. M., and Gispen, W. H., 1977, *Behav. Biol.* **19**:554-558.

113. Wiegant, V. M., Colbern, D., Van Wimersma Greidanus, T. B., and Gispen, W. H., 1978, *Brain Res. Bull.* **3**:167-170.
114. Witter, A., Greven, H. M., and De Wied, D., 1975, *J. Pharmacol. Exp. Ther.* **193**:853-860.
115. O'Donohue, T. L., Handelmann, G. E., Chaconas, T., Miller, R. L., and Jacobowitz, D. M., 1981, *Peptides* **2**:333-344.
116. Bohus, B., and De Wied, D., 1967, *Physiol. Behav.* **2**:221-223.
117. Van Wimersma Greidanus, T. B., Bohus, B., and De Wied, D., 1974, *Neuroendocrinology* **14**:280-288.
118. Elstein, K., Hannigan, J. H., Jr., and Isaacson, R. L., 1981, *Behav. Neural Biol.* **32**:248-254.
119. Bär, P. R., Gispen, W. H., and Isaacson, R. L., 1981, *Pharmacol. Biochem. Behav.* **14**:305-312.
120. Isaacson, R. L., 1981, *Adv. Physiol. Sci.* **17**:485-500.
121. Isaacson, R. L., Hannigan, J. H., Jr., Springer, J., Ryan, J., and Poplawsky, A., 1983, *Integrative Neurohumoral Mechanisms: Physiological and Clinical Aspects* (E. Endröczi, L. Angelucci, and D. De Wied, eds.), Elsevier Biomedical Press, Amsterdam, pp. 23-34.
122. Isaacson, R. L., and Colbern, D., 1981, *Physiol. Psychol.* **9**:260-262.
123. Wiegant, V. M., Cools, A. R., and Gispen, W. H., 1977, *Eur. J. Pharmacol.* **41**:343-345.
124. Traber, J., Klein, H. R., and Gispen, W. H., 1982, *Eur. J. Pharmacol.* **80**:407-414.
125. Guild, A. L., and Dunn, A. J., 1982, *Pharmacol. Biochem. Behav.* **17**:31-36.
126. Cools, A. R., 1977, *Adv. Biochem. Psychopharmacol.* **16**:215-225.
127. Cools, A. R., Wiegant, V. M., and Gispen, W. H., 1978, *Eur. J. Pharmacol.* **50**:265-268.
128. Gispen, W. H., Ormond, D., ten Haaf, J., and De Wied, D., 1980, *Eur. J. Pharmacol.* **63**:203-207.
129. Zimmermann, E., and Krivoy, W., 1973, *Prog. Brain Res.* **39**:383-394.
130. Gispen, W. H., and Wiegant, V. M., 1976, *Neurosci. Lett.* **2**:159-164.
131. Jacquet, Y. F., 1979, *Trends Neurosci.* **2**:140-143.
132. Jacquet, Y. F., 1979, *Science* **205**:425.
133. Dunn, A. J., Childers, S. R., Kramacy, N. R., and Villiger, J. W., 1981, *Behav. Neural Biol.* **31**:105-109.
134. File, S. E., 1981, *Peptides* **2**:255-260.
135. Aloyo, V. J., Spruijt, B., Zwiers, H., and Gispen, W. H., 1982, *Soc. Neurosci. Abstr.* **8**:371.
136. Aloyo, V. J., Spruijt, B., Zwiers, H., and Gispen, W. H., 1983, *Peptides* **4**:833-836.
137. Frederickson, R. C. A., 1977, *Life Sci.* **21**:23-41.
138. Chavkin, C., and Goldstein, A., 1981, *Proc. Natl. Acad. Sci. U.S.A.* **78**:6543-6547.
139. Plomp, G. J. J., and Van Ree, J. M., 1978, *Br. J. Pharmacol.* **64**:223-227.
140. Reinstein, D. K., Hannigan, J. H., Jr., and Isaacson, R. L., 1981, *Behav. Neural Biol.* **32**:248-254.
141. Isaacson, R. L., Hannigan, J. H., Jr., Brakkee, J. H., and Gispen, W. H., 1983, *Brain Res. Bull.* **11**:289-293.
142. Dunn, A. J., Guild, A. L., Kramarcy, N. R., and Ware, M. D., 1981, *Pharmacol. Biochem. Behav.* **15**:605-608.
143. O'Callaghan, J., Horowitz, G. P., and Isaacson, R. L., 1982, *Behav. Neural Biol.* **53**:368-374.
144. Jolles, J., Wiegant, V. M., and Gispen, W. H., 1978, *Neurosci. Lett.* **9**:261-266.
145. Wiegant, V. M., Jolles, J., and Gispen, W. H., 1978, *Dev. Neurosci.* **4**:447-450.
146. Drago, F., Bohus, B., Gispen, W. H., Scapagnini, U., and De Wied, D., 1983, *Brain Res.* **263**:277-282.
147. Gmerek, D. E., and Cowan, A., 1983, *Eur. J. Pharmacol.* **88**:339-346.
148. Colbern, D. L., Isaacson, R. L., Green, E. J., and Gispen, W. H., 1978, *Behav. Biol.* **23**:381-387.
149. Rosenfeld, G. C., and Burks, T. R., 1977, *J. Pharmacol. Exp. Ther.* **202**:654-659.
150. Schwartz, J. C., Costentin, J., Matres, M. P., Protais, P., and Baudry, M., 1978, *Neuropharmacology* **17**:665-685.
151. Spruijt, B., Brakkee, J., and Gispen, W. H., 1984, *Eur. J. Pharmacol.* (in press).
152. Gispen, W. H., 1982, *Scand. J. Psychol.* [*Suppl.*] **1**:16-25.

153. Krieger, D. T., Liotta, A., and Brownstein, M. J., 1977, *Brain Res.* **128**:575-579.
154. Cools, A. R., 1980, *Behav. Brain. Res.* **1**:361-378.
155. Bertolini, A., Vergoni, W., Gessa, G. L., and Ferrari, W., 1968, *Life Sci.* **7**(part II):1203-1206.
156. Feder, H. H., and Ruf, K. B., 1969, *Endocrinology* **84**:171-174.
157. De Cantazaro, D., and Gorzalka, B. B., 1980, *Pharmacol. Biochem. Behav.* **12**:201-206.
158. De Cantazaro, D., Gray, D. S., and Gorzalka, B. B., 1981, *Physiol. Behav.* **26**:207-213.
159. Bertolini, A., Vergoni, W., Gessa, G. L., and Ferrari, W., 1969, *Nature* **221**:667-669.
160. MacLean, P. D., 1973, *Hormones and Brain Function* (K. Lissàk, ed.), Plenum Press, New York, pp. 379-389.
161. Korányi, L., Endröczi, E., and Tárnok, F., 1965/66, *Neuroendocrinology* **1**:144-157.
162. Bohus, B., Hendrickx, H. H. L., Van Kolfschoten, A. A., and Krediet, T. G., 1975, *Sexual Behavior: Pharmacology and Biochemistry* (M. Sandler and G. L. Gessa), Raven Press, New York, pp. 269-275.
163. Meyerson, B. J., and Bohus, B., 1976, *Pharmacol. Biochem. Behav.* **5**:539-545.
164. Thody, A. J., De Rotte, A. A., and Van Wimersma Greidanus, T. B., 1979, *Brain Res. Bull.* **4**:213-216.
165. Wilson, C. A., Thody, A. J., and Everard, D., 1979, *Behaviour* **13**:293-300.
166. Schuurman, T., 1981, Thesis, University of Groningen, Groningen.
167. Leshner, A. I., 1978, *An Introduction of Behavioral Endocrinology*, Oxford University Press, New York.
168. Leshner, A. I., Merkle, D. A., and Mixon, J. F., 1981, *Exogenous Peptides Learning and Memory Processes* (J. L. Martinez, Jr., R. A. Jensen, R. B. Messing, H. Rigter, and J. L. McGaugh, eds.), Academic Press, New York, pp. 159-179.
169. Roche, K. E., and Leshner, A. I., 1979, *Science* **204**:1343-1344.
170. Leshner, A. I., 1980, *Prog. Brain Res.* **53**:427-438.
171. Scott, J. P., and Frederickson, E., 1951, *Physiol. Zool.* **24**:273-309.
172. Schuurman, T., 1980, *Prog. Brain Res.* **53**:415-420.
173. Brain, P. F., 1980, *Prog. Brain Res.* **53**:391-413.
174. File, S. E., 1980, *J. Neurosci. Methods* **2**:219-238.
175. File, S. E., Vellucci, S. V., and Wendtlandt, S., 1979, *J. Pharm. Pharmacol.* **31**:300-305.
176. File, S. E., Vellucci, S. V., and Wendtlandt, S., 1978, *Exp. Brain Res.* **32**:R53.
177. File, S. E., 1979, *Brain Res.* **171**:157-160.
178. Niessink, R. J. M., and Van Ree, J. M., 1984, *Life Sci.* **34**:961-970.
179. Clarke, A., and File, S. E., 1981, *Brain Res. Bull.* **6**:39-46.
180. Gaillard, A. W. K., 1981, *Endogenous Peptides and Learning and Memory Processes* (J. L. Martinez, Jr., R. A. Jensen, R. B. Messing, H. Rigter, and J. L. McGaugh, eds.), Academic Press, New York, pp. 181-196.
181. Branconnier, R. J., 1981, *Pharmacol. Ther.* **14**:161-175.
182. Pigache, R. M., and Rigter, H., 1981, *Front. Horm. Res.* **8**:193-207.
183. Endröczi, E., Lissàk, K., Fekete, T., and De Wied, D., 1970, *Prog. Brain Res.* **32**:254-262.
184. Sandman, C. A., McCanne, G. J., Nolan, T. R., Kaswan, J. D., and Kastin, A. J., 1977, *J. Clin. Endocrinol. Metab.* **44**:884-890.
185. O'Hanlon, J. F., Fussler, C., Sancin, E., and Grandjean, E. P., 1978, *Report of the Swiss Federal Institute of Technology*, Zurich.
186. Sandman, C. A., George, J., Walker, B., Nolan, J. D., and Kastin, A. J., 1976, *Pharm. Biochem. Behav.* **5**(Suppl. 1):23-28.
187. Walker, B. B., and Sandman, C. A., 1979, *Am. J. Ment. Def.* **83**:346-352.
188. Sandman, C. A., Walker, B. B., and Lawton, C. A., 1980, *Peptides* **1**:109-114.
189. Landfield, P. N., Sundberg, D. K., Smith, M. S., Eldridge, J. C., and Morris, M., 1980, *Peptides* **1**:(Suppl. 1):185-196.
190. Pigache, R. M., 1982, *Neuropeptides and Hormone Modulation of Brain Function and Homeostasis* (J. M. Ordy, J. R. Sladek, and B. Reisberg, eds.), Raven Press, New York, pp. 88-123.
191. Dunn, A. J., and Gispen, W. H., 1977, *Biobehav. Rev.* **1**:15-23.
192. Gispen, W. H., 1980, *Prog. Brain Res.* **53**:193-206.

193. Dunn, A. J., and Schotman, P., 1981, *Pharmacol. Ther.* **12**:353–372.
194. Delanoy, R. D., and Dunn, A. J., 1978, *Pharmacol. Biochem. Behav.* **9**:21–26.
195. McCulloch, J., Kelly, P. A. T., and Van Delft, A. M. L., 1982, *Eur. J. Pharmacol.* **78**:151–158.
196. Strand, F. L., and Kung, T. T., 1980, *Peptides* **1**:135–138.
197. Strand, F. L., and Smith, C. M., 1980, *Pharmacol. Ther.* **11**:509–534.
198. Bijlsma, W. A., Jennekens, F. G. I., Schotman, P., and Gispen, W. H., 1981, *Eur. J. Pharmacol.* **76**:73–79.
199. Bijlsma, W. A., Jennekens, F. G. I., Schotman, P., and Gispen, W. H., 1981, *Dev. Neurosci.* **13**:411–416.
200. Bijlsma, W. A., Jennekens, F. G. I., Schotman, P., and Gispen, W. H., 1983, *Muscle Nerve* **6**:104–112.
201. Bijlsma, W. A., Jennekens, F. G. I., Schotman, P., and Gispen, W. H., 1983, *Acta Neuropathol.* (Berl.) **62**:24–30.
202. Gonzales, E., and Strand, F. L., 1979, *Physiologist* **22**:46.
203. Lynch, G., 1982, *Aging*, Raven Press, New York, pp. 201–206.
204. Hershkowitz, M., Heron, D., Samuel, D., and Shinitzky, M., 1982, *Prog. Brain Res.* **56**:419–434.
205. Hershkowitz, M., Zwiers, H., and Gispen, W. H., 1982, *Biochim. Biophys. Acta* **692**:495–497.
206. Van Dongen, C., Hershkowitz, M., Zwiers, H., De Laat, S., and Gispen, W. H., 1984, *Aging of the Brain* (W. H. Gispen and Y. Traber, eds.), Elsevier Biomedical Press, Amsterdam, pp. 101–114.
207. Rigter, H., Van Delft, A. M. L., and Pigache, R. M., 1983, *Integrative Neurohumoral Mechanisms: Physiological and Clinical Aspects* (E. Endröczi, L. Angelucci, D. De Wied, and U. Scapagnini, eds.), Elsevier Biomedical Press, Amsterdam, pp. 449–456.
208. Rigter, H., 1982, *The Psychopharmacology of Aging* (D. Wheatley, ed.), Oxford University Press, Oxford, pp. 97–112.
209. Branconnier, R. J., Cole, J. O., and Cardos, G., 1979, *Psychopharmacology* **61**:161–165.
210. Ferris, S. H., and Reisberg, B., 1981, *Paper presented at the III World Congress of Biological Psychiatry*, Stockholm.
211. Wilner, A. E., 1981, *Paper presented at III World Congress of Biological Psychiatry*, Stockholm.
212. Braverman, A., Hamdy, R., Meisner, P., and Perera, N., 1981, *Paper presented at III World Congress of Biological Psychiatry*, Stockholm.
213. Barker, J. L., and Smith, T. G., Jr., 1979, *Adv. Exp. Med. Biol.* **116**:3–25.
214. Schotman, P., Reith, M. E. A., Van Wimersma Greidanus, T. B., Gispen, W. H., and De Wied, D., 1976, *Molecular and Functional Neurobiology* (W. H. Gispen, ed.), Elsevier, Amsterdam, pp. 309–344.
215. Versteeg, D. H. G., 1980, *Pharmacol. Ther.* **11**:535–557.
216. Versteeg, D. H. G., 1973, *Brain Res.* **49**:483–485.
217. Leonard, B. E., 1974, *Arch. Int. Pharmacodyn. Ther.* **207**:242–253.
218. Versteeg, D. H. G., and Wurtman, R. J., 1975, *Brain Res.* **93**:552–557.
219. Iuvone, P. M., Morasco, J., Delanoy, R. L., and Dunn, A. J., 1978, *Brain Res.* **139**:131–139.
220. Dunn, A. J., Iuvone, P. M., and Rees, H. D., 1976, *Pharmacol. Biochem. Behav.* **5**(Suppl. 1):139–145.
221. Dunn, A. J., Gildersleeve, N. B., and Gray, H. E., 1978, *J. Neurochem.* **31**:977–982.
222. Van Loon, G. R., Sole, M. J., Kamble, A., Kim, C., and Green, S., 1978, *Ann. N.Y. Acad. Sci.* **297**:284–293.
223. Endröczi, E., Hraschek, A., Nyakas, C., and Szabo, G., 1975, *Cellular and Molecular Base of Neuroendocrine Processes* (E. Endröczi, ed.), Akademiai Kiado, Budapest, pp. 607–618.
224. Endröczi, E., 1976, *Acta Physiol. Acad. Sci. Hung.* **48**:59–64.
225. Endröczi, E., 1977, *Neuropeptide Influences on the Brain and Behavior* (L. H. Miller, C. A. Sandman, and A. J. Kastin, eds.), Raven Press, New York, pp. 179–187.
226. Lichtensteiger, W., and Lienhart, R., 1977, *Nature* **266**:635–637.
227. Lichtensteiger, W., and Monnet, F., 1979, *Life Sci.* **25**:2079–2087.

228. Telegdy, G., and Kovács, G. L., 1979, *Central Nervous System Effects of Hypothalamic Hormones and Other Peptides* (R. Collu, A. Barbeau, J. R. Ducharme, and J. G. Rochefort, eds.), Raven Press, New York, pp. 189–205.
229. Telegdy, G., and Kovács, G. L., 1979, *Brain Mechanisms in Memory and Learning: From Single Neuron to Man* (M. A. B. Brazier, ed.), Raven Press, New York, pp. 249–268.
230. Delanoy, R. L., Kramarcy, N. R., and Dunn, A. J., 1982, *Brain Res.* **231**:117–129.
231. Göthert, M., 1981, *Eur. J. Pharmacol.* **76**:295–296.
232. Leonard, B. E., Kafoe, W. F., Thody, A. J., and Shuster, S., 1976, *J. Neurosci. Res.* **2**:39–45.
233. Ramaekers, F., Rigter, H., and Leonard, B. E., 1978, *Pharmacol. Biochem. Behav.* **8**:547–551.
234. Kendall, D. A., McEwen, B. S., and Enna, S. J., 1982, *Brain Res.* **236**:365–374.
235. Torda, C., and Wolf, H. G., 1952, *Am. J. Physiol.* **169**:140–149.
236. Torda, C., 1953, *Am. J. Physiol.* **173**:176–178.
237. Botticelli, L. J., and Wurtman, R. J., 1981, *Nature* **289**:75–76.
238. Veals, J., 1979, Ph.D. Thesis, New York University, New York.
239. Sutherland, E. W., 1972, *Science* **177**:401–408.
240. Burkhard, W. P., and Gey, K. F., 1968, *Helv. Physiol. Pharmacol. Acta* **26**:197–198.
241. Von Hungen, K., and Roberts, S., 1973, *Eur. J. Biochem.* **36**:391–401.
242. Wiegant, V. M., Dunn, A. J., Schotman, P., and Gispen, W. H., 1979, *Brain Res.* **168**:565–584.
243. Wiegant, V. M., Reul, J. M. H. M., and Gispen, W. H., 1982, *Prog. Brain Res.* **56**:397–404.
244. Wiegant, V. M., Zwiers, H., and Gispen, W. H., 1981, *Pharmacol. Ther.* **12**:463–490.
245. Richards, J. M., Tierney, J. H., and Sevisloclei, H., 1981, *J. Biol. Chem.* **256**:8889–8891.
246. Whittemore, S. R., Lenox, R. H., Hendley, E. D., and Ehrlich, Y. H., 1981, *Neurochem. Res.* **6**:775–785.
247. Ehrlich, Y. H., Whittemore, S. R., Garfield, M. K., Graber, S. G., and Lenox, R. H., 1982, *Prog. Brain Res.* **56**:375–396.
248. Forn, J., and Krishna, 1971, *Pharmacology* **5**:193–204.
249. Wiegant, V. M., and Gispen, W. H., 1975, *Exp. Brain Res.* [suppl.] **23**:219.
250. Rudman, D., 1978, *Endocrinology* **103**:1556–1561.
251. Rudman, D., and Isaacs, J. W., 1975, *Endocrinology* **97**:1476–1480.
252. Rudman, D., 1976, *Neuroendocrinology* **20**:235–242.
253. Korf, J., and Sebens, J. B., 1978, *J. Neurochem.* **32**:463–468.
254. Weller, M., 1979, *Protein Phosphorylation: The Nature, Function, and Metabolism of Proteins, which Contain Covalently Bound Phosphorus*, PION, London.
255. Krebs, E. G., and Beavo, J. E., 1979, *Annu. Rev. Biochem.* **48**:923–959.
256. Schotman, P., Frankena, H., Schrama, L. H., and Edwards, P. M., 1982, *Prog. Brain Res.* **56**:213–235.
257. Haynes, R. C., 1958, *J. Biol. Chem.* **233**:1220–1222.
258. Riley, G. A., and Haynes, R. C., 1963, *J. Biol. Chem.* **238**:1563–1570.
259. Murakami, N., and Ichii, S., 1973, *Endocrinol. Jpn.* **20**:421–424.
260. Podesta, E. J., Milani, A., Steffen, M., and Neher, R., 1979, *Biochem. J.* **180**:355–363.
261. Koroscil, T. M., and Gallant, S., 1980, *J. Biol. Chem.* **255**:6276–6283.
262. Koroscil, T. M., and Gallant, S., 1981, *J. Biol. Chem.* **256**:6700–6707.
263. Bristow, A. F., Schulster, D., and Rodnight, 1981, *Biochim. Biophys. Acta* **675**:24–28.
264. Carstens, M., and Weller, M., 1979, *Biochim. Biophys. Acta* **551**:420–431.
265. Kristjansson, G. I., Zwiers, H., Oestreicher, A. B., and Gispen, W. H., 1982, *J. Neurochem.* **39**:371–378.
266. Gispen, W. H., and Routtenberg, A., (eds.), 1982, *Prog. Brain Res.* **56**:
267. Costa, E., Kurosawa, A., and Guidotti, A., 1976, *Proc. Natl. Acad. Sci. U.S.A.* **73**:1058–1062.
268. Joh, T. H., Park, D. H., and Reis, D. J., 1978, *Proc. Natl. Acad. Sci. U.S.A.* **75**:4744–4748.
269. Brennan, M. J. W., and Cantrill, R. C., 1980, *J. Neurochem.* **35**:506–508.
270. Sieghart, W., Theoharides, T. C., Alper, S. L., Douglas, W. W., and Greengard, P., 1978, *Nature* **279**:329–331.

271. Heald, P. J., 1957, *Biochem. J.* **66**:659–663.
272. Trevor, A. J., and Rodnight, R., 1965, *Biochem. J.* **95**:889–896.
273. Browning, M., Dunwiddie, T., Bennett, W., Gispen, W. H., and Lynch, G., 1979, *Science* **203**:60–62.
274. Browning, M., Baudry, M., and Lynch, G., 1982, *Prog. Brain Res.* **56**:317–338.
275. Bär, P. R., Schotman, P., Gispen, W. H., Lopes da Silva, F. H., and Tielen, A. M., 1980, *Brain Res.* **198**:478–484.
276. Bär, P. R., Tielen, A. M., Lopes da Silva, F. H., Zwiers, H., and Gispen, W. H., 1982, *Brain Res.* **245**:69–79.
277. Rodnight, R., 1975, *Metabolic Compartmentation and Neurotransmission Relation to Brain Structure and Function* (S. Berl, D. D. Clark, and D. Schneider, eds.), Plenum Press, New York, pp. 205–208.
278. Greengard, P., 1976, *Nature* **260**:101–108.
279. Dousa, T. P., 1973, *Life Sci.* **13**:1033–1040.
280. Rudolph, S. A., and Greengard, P., 1974, *J. Biol. Chem.* **249**:5684–5687.
281. Alper, S. L., Beam, K. G., and Greengard, P., 1980, *J. Biol. Chem.* **255**:4864–4871.
282. Palfrey, H. C., Alper, S. L., and Greengard, P., 1980, *J. Exp. Biol.* **39**:103–155.
283. Drummond, A. H., Benson, J. A., and Levitan, I. B., 1980, *Proc. Natl. Acad. Sci. U.S.A.* **77**:5013–5017.
284. Kaczmarek, L. K., Jennings, K. R., Strumwasser, F., Nairn, A. C., Walter, U., Wilson, F. D., and Greengard, P., 1980, *Proc. Natl. Acad. Sci. U.S.A.* **77**:7487–7491.
285. Gordon, A. S., Davis, L. G., and Diamond, I., 1977, *Proc. Natl. Acad. Sci. U.S.A.* **74**:263–267.
286. Gordon, A. S., Davis, L. G., Milfray, D., and Diamond, I., 1977, *Nature* **267**:539–540.
287. Gordon, A. S., and Diamond, I., 1980, *J. Supramol. Struct.* **14**:163–174.
288. Williams, M., 1976, *Brain Res.* **109**:190–195.
289. Forn, J., and Greengard, P., 1978, *Proc. Natl. Acad. Sci. U.S.A.* **75**:5195–5199.
290. Hullihan, J. P., Wilson, J. E., and Williams, M., 1979, *Biochim. Biophys. Acta* **583**:232–240.
291. Dolphin, A. C., and Greengard, P., 1981, *Nature* **289**:76–79.
292. Lemos, J. R., Novak-Hofer, I., and Levitan, I. B., 1982, *Nature* **298**:64–65.
293. Jolles, J., Zwiers, H., Van Dongen, C., Schotman, P., Wirtz, K. W. A., and Gispen, W. H., 1980, *Nature* **286**:623–625.
294. Rodnight, R., 1982, *Prog. Brain Res.* **56**:1–25.
295. Zwiers, H., Veldhuis, D., Schotman, P., and Gispen, W. H., 1976, *Neurochem. Res.* **1**:669–677.
296. Mahler, H. R., Kleine, L. P., and Sörensen, R. G., 1981, *Regulatory Peptides: Functional and Pharmacological Aspects*, Raven Press, New York.
297. Zwiers, H., Wiegant, V. M., Schotman, P., and Gispen, W. H., 1978, *Neurochem. Res.* **3**:455–463.
298. Aloyo, V. J., Zwiers, H., and Gispen, W. H., 1982, *J. Neurochem.* **38**:871–875.
299. Aloyo, V. J., Zwiers, H., and Gispen, W. H., 1982, *Integrative Neurohumoral Mechanisms: Neuropeptides and Psychosomatic Processes* (E. Endröczi, D. De Wied, L. Angelucci, and U. Scapagnini, eds.), Elsevier Biomedical Press, Amsterdam, pp. 489–497.
300. Gispen, W. H., Zwiers, H., Wiegant, V. M., Schotman, P., and Wilson, J. E., 1979, *Adv. Exp. Med. Biol.* **116**:199–224.
301. Gower, H., and Rodnight, E., 1982, *Biochim. Biophys. Acta* **716**:45–52.
302. Routtenberg, A., 1982, *Prog. Brain Res.* **56**:349–374.
303. Mahler, H. R., Kleine, L. P., Ratner, N., and Sörensen, R. G., 1982, *Prog. Brain Res.* **56**:27–48.
304. Zwiers, H., Wiegant, V. M., Schotman, P., and Gispen, W. H., 1977, *Mechanism, Regulation and Special Functions of Protein Synthesis in the Brain* (S. Roberts, A. Lajtha, and W. H. Gispen, eds.), Elsevier/North Holland Biomedical Press, Amsterdam, pp. 267–272.
305. Ehrlich, Y. H., Rabjohns, R. R., and Routtenberg, A., 1977, *Pharmacol. Biochem. Behav.* **6**:169–175.
306. Holmes, H., Rodnight, R., and Kapoor, R., 1977, *Pharmacol. Biochem. Behav.* **6**:415–420.
307. Weller, M., and Rodnight, R., 1973, *Biochem. J.* **133**:387–389.

308. Zwiers, H., Tonnaer, J., Wiegant, V. M., Schotman, P., and Gispen, W. H., 1979, *J. Neurochem.* **33:**247–256.
309. Zwiers, H., Schotman, P., and Gispen, W. H., 1980, *J. Neurochem.* **34:**1689–1699.
310. Greengard, P., 1979, *Trends Pharmacol. Sci.* **1:**27–29.
311. Inoue, M., Kishimoto, A., Takai, Y., and Nishizuka, Y., 1977, *J. Biol. Chem.* **252:**7610–7616.
312. Miyamoto, E., Fukunaga, K., Matsui, K., and Iwasa, Y., 1981, *J. Neurochem.* **37:**1324–1330.
313. Aloyo, V. J., Zwiers, H., and Gispen, W. H., 1982, *Prog. Brain Res.* **56:**303–315.
314. Jolles, J., Zwiers, H., Dekker, A., Wirtz, K. W. A., and Gispen, W. H., 1981, *Biochem. J.* **194:**283–291.
315. Takai, Y., Minakuchi, R., Kikkawa, U., Sano, K., Kaibuchi, K., Yu, B., Matsubara, T., and Nishizuka, Y., 1982, *Prog. Brain Res.* **56:**293–307.
316. Sörensen, R. G., Kleine, L. P., and Mahler, H. R., 1981, *Brain Res. Bull.* **7:**57–61.
317. Oestreicher, A. B., Zwiers, H., Gispen, W. H., and Roberts, S., 1982, *J. Neurochem.* **39:**683–692.
318. Oestreicher, A. B., Zwiers, H., Schotman, P., and Gispen, W. H., 1981, *Brain Res. Bull.* **6:**145–153.
319. Zwiers, H., Jolles, J., Aloyo, V. J., Oestreicher, A. B., and Gispen, W. H., 1982, *Prog. Brain Res.* **56:**405–417.
320. Van Wimersma Greidanus, T. B., Bohus, B., and De Wied, D., 1975, *Anatomical Neuroendocrinology* (W. E. Stumpf and L. D. Grant, eds.), S. Karger, Basel, pp. 284–289.
321. Verhoef, J., Palkovits, M., and Witter, A., 1977, *Brain Res.* **126:**89–104.
322. Verhoef, J., Witter, A., and De Wied, D., 1977, *Brain Res.* **131:**117–128.
323. Pelletier, G., and Leclerc, R., 1979, *Endocrinology* **104:**1426–1433.
324. Michell, R. H., 1975, *Biochim. Biophys. Acta* **415:**81–148.
325. Michell, R. H., 1979, *Trends Biochem. Sci.* **4:**128–131.
326. Jolles, J., Aloyo, V. J., and Gispen, W. H., 1982, *Molecular Approaches to Neurobiology* (I. R. Brown, ed.), Academic Press, New York, pp. 285–316.
327. Hokin, L. E., and Hokin, M. R., 1959, *J. Biol. Chem.* **234:**1387–1390.
328. Michell, R. H., Kirk, C. J., Jones, L. M., Downes, C. P., and Creba, J. A., 1981, *Phil. Trans. R. Soc. (Lond.) [Biol.]* **296:**123–137.
329. Abdel-Latif, A. A., Green, K., Smith, J. P., McPherson, J. C., and Matheny, J. L., 1978, *J. Neurochem.* **30:**517–525.
330. Hawthorne, J. N., and Pickard, M. R., 1979, *J. Neurochem.* **32:**5–14.
331. Farese, R. V., Sabir, M. A., and Vandor, S. L., 1979, *J. Biol. Chem.* **254:**6842–6844.
332. Farese, R. V., Sabir, M. A., and Larson, R. E., 1981, *Endocrinology* **109:**1895–1901.
333. Farese, A. V., and Sabir, M. A., 1979, *Biochim. Biophys. Acta* **575:**299–304.
334. Jolles, J., Wirtz, K. W. A., Schotman, P., and Gispen, W. H., 1981, *FEBS Lett.* **105:**110–114.
335. Oestreicher, A. B., Van Dongen, C., Zwiers, H., and Gispen, W. H., 1983, *J. Neurochem.* **41:**331–340.
336. Ehrlichman, J., Rosenfeld, R., and Rosen, O. M., 1974, *J. Biol. Chem.* **249:**5000–5003.
337. Granot, J., Mildvan, S. A., and Kaiser, E. T., 1980, *Arch. Biochem. Biophys.* **205:**1–17.
338. Cho, T. M., Cho, J. S., and Loh, H. H., 1978, *Proc. Natl. Acad. Sci. U.S.A.* **75:**784–788.
339. Lin, Y. M., Liu, Y. P., and Cheung, W. Y., 1974, *J. Biol. Chem.* **249:**4943–4954.
340. Buckley, J. T., and Hawthorne, J. N., 1972, *J. Biol. Chem.* **247:**7218–7223.
341. Luly, P., and Shinitzky, M., 1979, *Biochemistry* **18:**445–450.
342. Yuli, I., Wilbrandt, W., and Shinitzky, M., 1981, *Biochemistry* **20:**4250–4256.
343. Heron, D., Shinitzky, M., Hershkowitz, M., and Samuel, D., 1980, *Proc. Natl. Acad. Sci. U.S.A.* **77:**7463–7467.
344. Heron, D., Israeli, M., Hershkowitz, M., and Samuel, D., 1981, *Eur. J. Pharmacol.* **72:**361–364.
345. Sanderman, H., Jr., 1978, *Biochim. Biophys. Acta* **515:**209–237.
346. Shinitzky, M., and Barenholz, Y., 1978, *Biochim. Biophys. Acta* **515:**367–394.
347. Schrey, R. P., and Rubin, R. P., 1979, *J. Biol. Chem.* **254:**11234–11241.
348. Rubin, R. P., Sink, L. E., Schrey, M. P., Day, A. R., Kao, C. S., and Freer, R. J., 1979, *Biochem. Biophys. Res. Commun.* **90:**1364–1370.

349. Rubin, R. P., 1982, *Fed. Proc.* **41**:2181–2187.
350. Arnaud, J., Nobili, O., and Boyer, J., 1981, *Biochim. Biophys. Acta* **663**:401–407.
351. Arnaud, J., Nobili, O., and Boyer, J., 1981, *Biochem. Biophys. Res. Commun.* **100**:1167–1172.
352. Arnaud, J., Nobili, O., and Boyer, J., 1980, *Biochim. Biophys. Acta* **617**:524–528.
353. Havleck, V., Rezek, M., and Friesen, H., 1976, *Pharmacol. Biochem. Behav.* **4**:455–465.
354. Dokas, L. A., Zwiers, H., Coy, D. M., and Gispen, W. H., 1982, *Eur. J. Pharmacol.* **88**:185–193.
355. Delanoy, R. L., Dunn, A. J., and Tintner, R., 1978, *Horm. Behav.* **11**:348–362.
356. Drago, F., Canonico, P. L., Bitetti, R., and Scapagnini, U., 1980, *Eur. J. Pharmacol.* **65**:457–458.
357. Gispen, W. H., Wiegant, V. M., Bradbury, A. F., Hulme, E. C., Smyth, D. G., Snell, C. R., and De Wied, D., 1976, *Nature* **264**:794–795.
358. Katz, R. J., 1980, *Neuropharmacology* **19**:801–803.
359. Zwiers, H., Aloyo, V. J., and Gispen, W. H., 1981, *Life Sci.* **28**:2545–2551.
360. Katz, R. J., 1979, *Neurosci. Lett.* **12**:393–401.
361. Katz, R. L., 1980, *Int. J. Neurosci.* **10**:187–189.
362. Katz, R. L., 1980, *Neuropharmacology* **19**:143–146.
363. Van Wimersma Greidanus, T. B., and Rinkel, G. J. E., 1983, *Eur. J. Pharmacol.* **88**:117–120.

18

Pain Transmission

Tony L. Yaksh

1. PAIN-TRANSMITTING SYSTEM

Unconditioned stimuli applied to peripheral tissues including skin, viscera, and muscle at intensities sufficient to distort or damage tissue will evoke verbal reports of pain in humans and coordinated efforts to escape (i.e., pain behavior) in animals. The classic description by Beecher[1] of wounded men in battle not reporting pain in spite of extensive tissue damage is a commonly cited example indicating, however, that there is not an invariant relationship between the physical stimulus and the behavioral sequelae. As is discussed below, the rostrad transmission of afferent information at every synaptic link is subject to a considerable and variable degree of inhibition and modulation by intrinsic neural systems. In the following discussion, I briefly review (1) the complex substrates whereby neural activity generated by high-intensity mechanical, thermal, or chemical stimuli gain access to higher brain centers and (2) systems that regulate transmission through these systems.

1.1. Peripheral Afferents

Somatic and visceral information relating the organism to its environment gains access to the central nervous system by myelinated (Group II/Aβ*: diameter, 6–12 μm; conduction velocity, 30–100 m/sec) and small, lightly myelinated and unmyelinated primary afferents (Group III/Aδ: diameter, 1–6 μm; conduction velocity, 4–30 m/sec. Group IV/C: diameter, <1 μm; conduction velocity, <2.5 m/sec). These afferent fiber systems have their cell bodies in the dorsal root ganglia. It should be noted that even "unmyelinated" fibers commonly have a Schwann cell investment, although unlike the case for myelinated fibers, this covering is irregular, and a single Schwann cell may envelop several small-diameter axons,[4] a unit referred to as a Remak bundle.

---

* Aβ–Aδ–C refers to the fiber designation based on conduction velocities and is referred to as the St. Louis classification.[2] Groups II, III, and IV represent the corresponding fiber populations described in muscle on the basis of relative diameter.[3]

Tony L. Yaksh • Department of Neurosurgical Research, Mayo Clinic, Rochester, Minnesota 55905.

Characteristically, two classes of neurons are found within the dorsal root ganglia: large-diameter type A cells giving rise to large-diameter myelinated fibers and small-diameter type B cells giving rise to smaller-diameter myelinated/unmyelinated fibers.[5] Aside from the relative density of intracellular elements (see Lieberman[5a]) and the difference with regard to the state of myelination of the axon, histochemical analysis has revealed the afferents and their associated cell bodies to have a differential chemistry. Thus, extralysosomal acid phosphatase and cholinesterase have been shown to be preferentially associated with type B cell bodies.[6-9] Recent advances in immunohistochemistry have also revealed differential content with regard to peptides (see below).

It is important to note that considerable evidence indicates that the axon and distal terminal are the site of considerable metabolic activity. Electron microscopic examination of axon profiles have typically demonstrated the presence of agranular vesicles and mitochondria in the free nerve endings.[10] As the terminals do not possess ribosomes with which to manufacture peptides and proteins, these materials are obtained following synthesis in the cell body by an energy-dependent transport system mediated by microtubules or neurofilaments.[11,12]

In the periphery, the terminals of primary afferents (1) ramify extensively, (2) lose their perineural sheath, (3) show increasingly irregular Schwann cell covering, and (4) either develop specialized endings (such as the pacinian corpuscle) or lose their Schwann cell and, for example, in the case of cutaneous afferents, become contiguous with the ependymal layer.[13-16]

Centrally, the dorsal root ganglian send their axons through the dorsal root, where the myelin sheath changes at a point just proximal to the ganglion from Schwann cells to glia. The majority of these sensory fibers enter the dorsal horn via the dorsal root entry zone, as indicated by the law of Bell and Magendie.[17] Current evidence indicates, however, that a small proportion of the afferent fibers enters the spinal gray through the ventral root.[18,19] The distribution and ramification of the dorsal root as a function of fiber size and myelination within the dorsal gray are discussed further below. Information from the head and upper neck derives from a similar system, which has its cells of origin in the trigeminal ganglia located in the floor of the cranium.

1.1.1. Physiological Response Properties of Primary Afferents

Recording from single afferent fibers in peripheral nerve fascicles reveals discriminable populations of afferents that respond with a continuous discharge to high-intensity thermal (>45°C) and mechanical stimuli (Von Frey hair > 2 g)[20-24] and to the topical application of certain chemicals known to evoke pain behavior (see below). A significant proportion of the unmyelinated afferents have also been demonstrated to respond to a combination of stimuli such as thermal, mechanical, and/or chemical; i.e., they are polymodal.[22,23,25,26]

That these afferent fiber populations, which respond preferentially to high-intensity stimulation, are relevant to pain behavior is supported by several observations.

First, the direct stimulation of axons reveals that the activation of certain populations of small, slowly conducting axons is a sufficient and apparently necessary prerequisite for the elicitation of pain behavior in a variety of experimental animal preparations[27,28] and in human.[29,30-32,36]

Second, *in situ* recording from single fibers with microelectrodes in humans has demonstrated that peripheral thermal, mechanical, or chemical stimuli giving rise to verbal reports of pain invariably give rise to activity in fibers that conduct at velocities commonly associated with lightly myelinated and unmyelinated fibers.[22,23,25,33] Phenomenologically, it is known that the report of pain consists of an initial "sharp" followed by a dull throbbing sensation. These are referred to as first and second pain, respectively. Human studies have demonstrated that the activation of fibers conducting at Aδ velocity elicits verbal descriptions that resemble first pain, whereas C fiber activity is associated with the production of dull throbbing sensations commonly associated with so-called second pain (see Vallbo *et al.*[35]).

Third, the relevance of small fiber activity to the pain event is substantiated by the observation that selective blockade of large fibers by a local compression, leaving small fiber transmission intact, has no effect on reported pain. In contrast, selective blockade of C-fiber activity with local anesthetics, leaving A-fiber volleys intact, resulted in a blockade of the pain report, rendering a strong stimulus to be reported as a sharp sensation.[32,34]

In spite of these observations and the clear role of small fibers in pain transmission, we cannot exclude the likelihood that large fibers may play a significant role in conveying information regarding the nature of the stimulus.[37,38] A further caveat to the above comments on the role of small fibers is that although the evocation of pain behavior under normal circumstances is associated with activity in small afferent fiber populations, not all small afferent fibers are uniquely associated with the transmission of information generated by tissue-damaging stimuli. Thus, populations of C fibers associated with muscles are activated by mild mechanical distortion,[39,40] and mild temperature changes ($<40°C$) may activate populations of lightly myelinated cutaneous afferent systems.[33,41] That such systems may still contribute to the overall pain syndrome, particularly under other conditions such as inflammation, is of considerable interest and is discussed below.

Anatomically, the peripheral terminals of the majority of the small primary afferents whose activity appears to be related to pain behavior possess anatomically undifferentiated endings. Several organ systems, notably the cornea, are known to uniformly possess unmyelinated terminals, and yet all sensations, light touch, temperature, chemical sensitivity, and pain, can be sensed in the cornea.[42] The apparent lack of readily differentiable nerve endings suggests that the transduction mechanism leading to a generator potential and nerve impulse must reside within the fabric of the terminal membranes. With regard to mechanical stimulation, parallels with studies made on the pacinian corpuscle suggest that local changes in membrane ionic permeability may be brought about by mechanical distortion.[43,44] With regard to thermal stimulation, it appears reasonable to presume that changes in protein configuration secondary to temperature may produce changes in the configurations of the ion channels resulting in altered permeability prior to depolarization.[45]

Following the local application of a high-intensity mechanical stimulus, a red flush appears at the site of the stimulus surrounded by a bright red flare and a local edema.[46] This "Triple" response results in part from an extravassation of plasma as a result of increased permeability of the capillary epithelium and a widespread arterial dilatation. This inflammatory response is commonly associated with a local reduction in the magnitude of the stimulus that otherwise elicits a pain response. Thus, mild heat damage applied to peripheral receptive fields of an afferent results in a significant increase in the magnitude of the discharge evoked by a given stimulus in polymodal nociceptors[25,47,48] and high-threshold mechanoreceptors[49] and a paradoxical discharge of cold receptors in the presence of noxious heat.[50] This enhancement of sensitivity is also accompanied by a prolongation in the discharge otherwise evoked by the stimulus as compared to that produced by the application of the same stimulus to the nonsensitized nerve ending. A chemical intermediary with a prolonged half-life that alters the environs of the nerve terminal in the area of injury to alter its activity is an intuitively inviting hypothesis.

On the basis of the physiological observations, it is apparent that such intermediaries may be associated with three functions: (1) the activation of small afferents and the subsequent evocation of a pain event; (2) the facilitation of the ongoing discharge evoked by a given stimulus; and (3) changes in the vascular permeability that would lead to extravassation, reddening, and edema.

Considerable effort has been extended to determine what chemical intermediaries, if any, might mediate these observed affects. A number of endogenous agents deriving from tissue (potassium, prostanoids), plasma (amines such as histamine, serotonin, and kinins), or nerve terminals (substance P) have been proposed.

Serotonin is found in mast cells and platelets,[51] and histamine in granules of mast cells in basophils and platelets.[52] The release of these amines from storage may be evoked by mechanical trauma heat, or by-products of tissue damage, most notably neutrophils, lysosomal materials, anaphylactic toxins secondary to complement activation, thrombin, collagen, as well as certain lipidic acids of the arachidonic cascade.[53-56]

The presence of prostanoids, a large family of structurally diverse C_{20} lipid acids, in tissues reflects their synthesis by membrane-bound lipoxygenase or cyclooxygenase enzymes secondary to the activation of phospholipase and the subsequent local liberation of arachidonic acid. There is not space in the present discussion to examine this complex family of agents, many of which have physiologically opposing functions (see refs. 57–59). Nevertheless, it is clear that during inflammation there are significant increases of many of the lipidic acids including PGE_2 and $PGF_{2\alpha}$.[60] Such agents have been shown to possess significant physiological effects on vascular permeability and neuronal function. It is important to note that the so-called nonsteroidal antiinflammatory agents such as acetylsalicylic acid and indomethacin are known to be potent inhibitors of cyclooxygenase, and their application, which results in a diminution of hypersensitivity and swelling, is accompanied by a significant reduction in the levels of the prostanoids generated by that enzyme.[61]

Bradykinin is one of a series of kinins that are generated by the cascade of chemical events that occur secondary to the activation by a variety of stimuli of factor XII of the Hageman clotting system.[62] This agent has been reported to be released from skin by noxious stimulation.[63] Extracellular potassium is known to accumulate secondary to muscle activity. Isometric stretching of muscle in cat resulted in an approximate doubling of the levels of extracellular potassium.[64]

Several agents that are known to be contained in nerve terminals, for example, substance P, vasoactive intestinal polypeptide, and cholecystokinin, are commonly thought to be released from the central terminals of the primary afferent axons, where they may play a neurotransmitter role (see below). Considerable evidence clearly suggests, however, that these agents are also transported distally to the nerve terminals,[65] where, as noted above, profiles of synaptic vesicles and mitochondria have been reported. That some of these peptides, notably substance P, exist within a releasable pool at the peripheral nerve terminal is suggested by experiments carried out with the tooth pulp, where it has been shown that the release of substance P immunoreactivity is remarkably enhanced by stimulation of the innervating sensory nerve.[66] Although it may be argued that this is not a physiological preparation and that under most conditions neuronal activity travels centrally and not peripherally, the presence of antidromic activity in the sensory nerve can originate from depolarization of central afferent terminals leading to antidromically conducted spikes.[67] Alternatively, as noted, peripheral terminals of afferents are known to ramify extensively. Stimulation of one of the branches could lead to an orthodromic spike that could antidromically invade an adjacent ramification. Either likelihood would provide a mechanism for such a peripheral release under the appropriate circumstances. Notably, such a release of vasoactive agents would in fact account for the surrounding flare (vasodilatation) that occurs in an area surrounding a local injury.[68,69]

If agents that are present in tissue and respond to injury are intermediaries in the pain response to an injury, then it is reasonable to question whether they can, in fact, activate small-diameter afferent fibers and produce pain when applied locally. Histamine, serotonin, acetylcholine, bradykinin, and potassium have all been shown to excite primate afferents when injected by a close intraarterial injection into their receptive fields.[20,70,71] The direct application of histamine and bradykinin onto the skin under an exposed blister base or the direct subcutaneous injection of these agents has been reported to evoke a pain response in man.[72-74] In animals, the intraarterial administration of serotonin, histamine, or bradykinin will evoke species-specific signs of stress such as considerable agitation, vocalization, efforts to escape, pupillary dilatation, and changes in heart rate and blood pressure.[70,75] In other words, certain agents that are released secondary to tissue damage in fact result in the activation of small afferents and in the evocation of a pain event. Two considerations should, however, be made in interpreting this literature: (1) the concentrations of drug required to produce these effects are commonly several orders of magnitude in excess of that which may be found in tissue extracts, and (2) these agents rarely if ever activate one population of fibers. Commonly, such application may alter activity in large-diameter, rapidly conducting fibers as well as those afferent fiber populations that are more commonly associated with the pain event.

In contrast to the ability of the above agents to directly activate nerve terminals and produce pain, there appear to be classes of agents that serve not to produce pain but to facilitate the response to a given stimulus. Thus, Handwerker[76] noted that perfusion of the receptive field of thermal nociceptive afferents with PGE_2 had no effect on the resting activity of this fiber but did produce a dose-dependent increase in the discharge rate of the fiber response to a given stimulus. Though these prostanoids have little effect on pain behavior when applied intraarterially or directly to blister bases at very high doses, they do appear to facilitate the behavioral response to other algogenic stimuli.[77,78] Thus, intraarterial or intradermal injections of PGE_1 will augment the response produced by injections of bradykinin.[79] Significantly, studies on angina pectorlis have indicated that temporary occlusion of the coronary artery will sensitize the heart to the algogenic effects of intraarterial bradykinin. This has led to the suggestion that angina may result from the neurohumoral activation of afferent terminals by the joint release of both bradykinin and prostaglandins. Importantly, this sensitization is inhibited by prior treatment with prostaglandin synthesis inhibitors.[80]

The above description does not give credit to the complexity of the mechanisms whereby endogenous material released in the vicinity of the peripheral afferent terminal secondary to high-intensity stimulation may participate in the transduction of sensory information. A general outline, however, may be perceived. First, there are two classes of chemical agents, those that clearly activate high threshold peripheral afferents at relatively low concentrations and result in pain following a local application (e.g., histamine, bradykinin, serotonin) and those that facilitate the pain evoked by a variety of chemical and physical stimuli (e.g., the prostanoids and substance P) but are relatively ineffective in evoking pain themselves. Tissue damage will be accompanied by the generation of orthodromic activity. The subsequent generation of antidromic activity would give rise to the release of agents contained in the nerve terminals, notably substance P. Such a peripheral release of substance P would likely result in an increase in capillary permeability (and subsequent extravasation), resulting in the movement into tissue of large protein complexes as well as agents known to evoke pain and increase small fiber activity. The release of such agents might be coincident with the local synthesis of prostanoids; e.g., bradykinin is known to evoke prostaglandin synthesis,[81] and their presence is known to facilitate the algogenic activity of agents that result in neuronal discharge.

Thus, thermal stimulation will sensitize high-threshold afferents,[82] and the infusion of prostaglandins into the receptive field will facilitate the response of a single afferent to a given thermal stimulus. It should be noted that it is not likely that a single essential chemical mediator exists. Thus, the sensitizing effects of temperature on polymodal C fibers in the rabbit ear were examined, and a variety of antagonists were administered intraarterially. Carboxypeptidase B (a bradykinin deactivator), indomethacin (a cyclooxygenase inhibitor), mepyramine (a prostaglandins antagonist), and methysergide (a putative serotonin antagonist) each resulted in a decrement in the sensitization of the afferent. The net result was, however, that the maximum antagonism occurred when all of these agents were coadministered, suggesting that a number of algogenic systems may be contributing to the final response.[47]

1.1.2. Dorsal Horn Systems Related to the Transmission of Information Generated by High-Intensity Stimulation of Somatic and Visceral Afferents

The gray matter of the spinal cord is commonly separated into laminae according to the original cytoarchitectual descriptions of Rexed.[83] Briefly, the outermost layer of cells is referred to as lamina I. Laminae II and III refer to the next two layers of smaller cells, which lie in a region referred to according to its gross anatomic appearance as the substantia gelatinosa (gelatinous substance in unfixed tissue). Laminae IV and V are referred to jointly as the nucleus proprius and form the neck of the dorsal gray matter. Deeper laminae (VI, VII, and VIII) fall within the intermediate horn of the gray matter; whereas the deepest laminae (IX, X, and XI) are generally referred to as those portions constituting the ventral horn of the spinal gray.

The majority of afferent fibers enter the dorsal gray matter through the dorsal root entry zone, which lies adjacent to the dorsal columns, with larger myelinated fibers tending to aggregate medially and smaller myelinated and unmyelinated fibers distributing laterally.[84,85] Small fibers enter the dorsal gray from above, whereas large myelinated fibers tend to follow a ballistic trajectory into the deeper portions of the dorsal horn before making a turnaround to terminate in the more dorsal laminae (III, IV, and V) of the gray matter.[86] Although the distribution of primary afferents is commonly drawn as being limited to the segment of entry, in fact, virtually all primary afferents bifurcate and distribute rostrally and caudally along a longitudinal axis before entering into the dorsal gray. Large fibers may project rostrally in the dorsal column as far as the dorsal column nuclei (although the majority of them terminate at the segmental levels prior to that rostral extent).[87] In contrast, the rostrocaudal distribution of small afferent collaterals traveling in the medial aspect of the tract of Lissauer appears to be limited to only several millimeters.[85,88]

Although the majority of afferents clearly enter the dorsal gray through the dorsal root entry zone, considerable evidence (anatomic and electrophysiological) has begun to accrue that significant populations of unmyelinated primary afferents with their cell bodies in the dorsal root ganglia may enter the spinal gray through the ventral root.[18,19] The presence of these afferents may provide an explanation for the phenomenon that intense pain can be generated by synchronous activation of peripheral afferents following dorsal rhizotomy and for the disappointing therapeutic outcome of such interventions in chronic pain.[91,92]

The dorsal horn of the spinal cord is an extremely complex region and features interplays between virtually all categories of afferents. Nevertheless, with regard to nociceptive transmission, electrophysiological experiments combined with localization studies have revealed the existence of three functionally definable classes of neurons that may receive information generated by high-intensity stimulation.

1.1.3. Marginal Cells

Marginal cells, designated by virtue of their distribution in the most superficial lamina of the dorsal horn (marginal zone), are large cells with dendrites oriented transversely across the extent of the dorsal horn. Morphological studies carried out following rhizotomy have indicated that the terminals of primary afferents are largely limited to the distal dendrites of the marginal cells.[85]

Retrograde transport studies and antidromic stimulation of the thalamus have indicated that a proportion of these neurons in fact project via the ventrolateral tract to supraspinal centers. It should be noted, however, that a significant proportion of lamina I neurons have projections limited to an intersegmental distribution.[87,93–95]

Electrophysiological studies have indicated that there are at least three distinguishable categories of neurons that exist within Lamina I of the spinal cord: (1) cells that are excited by afferents having conduction velocities in the Aδ/C fiber range and that respond to intense mechanical distortion of the skin; (2) cells that are activated by polymodal afferents and conduct in the range of C fibers; and (3) cells that are excited by innocuous cooling of the skin by the activation of afferents having conduction velocities of an Aδ fiber.[96] As expected, such observations are in accord with the tract-tracing studies that indicate the projection of fine-caliber afferent fibers into the most superficial layer of the dorsal gray matter (see below). These cells receive input from group III and group IV muscle afferents as well, suggesting the convergence of muscle and cutaneous input.[97] Such convergence between different organ systems on a single neuron appears to be a common occurrence for second-order neurons in the dorsal horn of the spinal cord.

That marginal neurons may play a significant role in transmitting information related to the pain message is suggested by the early observations of Kuru,[98] who noted that following successful ventrolateral cordotomies (see below), which destroy spinopetal pathways, that a significant proportion of marginal neurons showed retrograde chromatolysis in those patients whose cordotomies had proven successful in aleviating their pain.

1.1.4. Gelatinosa Neurons

The substantia gelatinosa consists of a complex neuropil, which may be subdivided into lamina II and lamina III on the bases of cellular density and differences in the state of myelination of innervating efferents.* Morphologically, lamina II may be further divided into an outer and inner layer (see Cervero and Iggo[99]).

Afferent innervation as examined by electron microscopy indicates that the outer portion of lamina II, containing small cells, receives only unmyelinated terminals, whereas the inner layer receives a mixture of both small mye-

---

* There is considerable controversy over whether the nomenclature "substantia gelatinosa" should be associated with the Rexed classification lamina III or with II and III.[99] In case of doubt, the particular laminar designation is preferred in place of the general term. Clearly, the two regions can be anatomically and physiologically distinguished.

linated and unmyelinated endings. Lamina III receives a preponderance of myelinated terminals associated with Aβ and Aδ primary afferents.[100-103]

The distribution of axons of gelatinosa neurons is complex and at present only partially understood. It is clear that a class of these neurons project axons within the lateral tract of Lissauer.[85,104] Others of these neurons also are likely to send axons in a longitudinal orientation intersegmentally and likely constitute the cells of origin for the so-called dorsal intercornual tract, which is known to run longitudinally in the gray matter.[85] Finally, neurons that lie within the substantia gelatinosa are known to contribute axons that travel in the contralateral ventrolateral pathway and, according to retrograde transport and antidromic activation studies, reach as far as the thalamus.[105,106]

Electrophysiologically, this region has proven to be extremely complex. It is clear that this region possesses cells that receive mixed input from Aβ, Aδ, and C fibers. Single-unit recordings from cells located within laminae II and III have been shown to possess a number of novel properties. In general, cells in these regions have been shown to be activated by Aδ/C fiber stimulation. Functionally, a significant proportion of cells appear to respond to light brush and touch.[107-110] Perhaps most intriguing regarding these cells, at the present time, is their peculiar response properties. Several investigators have observed that brief stimulation will often evoke prolonged discharges in a large population of the cells (see above). Cervero and colleagues,[111] in an extensive series of investigations, have suggested that gelatinosa neurons may be placed in a number of categories according to their response properties in the presence of noxious and innocuous afferent input. Thus, in these studies, those neurons that were inhibited by innocuous input were excited by noxious input, and *vice versa*.

1.1.5. Lamina V Neurons

The anatomic organization of lamina V cells was observed to be such that the cell bodies were in Lamina V with their dendrites extending dorsally to arborize within the substantia gelatinosa. As indicated previously, Aβ fibers are known to course deeply within the gray matter and ascend in a radial fashion to terminate in lamina III. These observations clearly suggest that lamina V neurons are spatially organized to receive input from both Aβ and Aδ/C terminals on their dendrites. These neurons were observed to project to the ventrobasal thalamus and into the mesencephalon as well as to provide certain propriospinal projections within the spinal cord.[90,93,112]

Wall,[113,114] in a series of investigations, observed that neurons that lay in the vicinity of Rexed lamina V responded with increasing frequency as the intensity of stimulation of the peripheral field was increased. This property led to their being referred to as "wide dynamic range" neurons (see ref. 115). Consistent with this property, these neurons showed clear signs of convergence between Aβ- and Aδ/C-fiber-mediated functions. These neurons demonstrate considerable modality as well as organ convergence. Thus, neurons within the nucleus proprius (Lamina V) were excited by activation of sympathetic afferents, coronary artery occlusion, as well as by noxious cutaneous input applied within the skin dermatomes that corresponded to the segmental localization of the neurons in question.[116] In addition, these cells showed a clear property of temporal facilitation wherein repeated stimuli applied at short intervals would produce a self-sustaining, high-frequency discharge referred to as "windup."[117]

Though the terms "wide dynamic range neuron" and "lamina V neuron" are commonly used interchangeably, cells with such physiological properties are not uniquely limited to the fifth lamina of the spinal cord. Physiologically defined, wide dynamic range neurons have been described as widely distributed in laminae IV, VI, and VII.[118]

This class of neurons (wide dynamic range) has received the most attention with regard to their serving as a second-order link in the ascending system by which information regarding the properties of high-intensity stimulation gains access to rostral structures, and a likely role in the pathway that mediates the rostral transmission of nociceptive information appears likely. Psychophysical studies correlating the reported pain response in humans and in animals are in close accord with the response properties of these wide dynamic range neurons.[119,120]

Thus, within the dorsal horn of the spinal cord, one is able to discriminate three anatomically distinct populations of neurons that may act as links whereby afferent input generated by high-intensity stimulation and transmitted by Aδ/C fibers gains access to higher centers. It should be noted that these neurons functionally fall into two major classes: those that appear to respond largely to high-intensity input generated by Aδ and/or C fibers activated by tissue-damaging heat or mechanical stimuli (i.e., nociceptive specific) and those that respond in an increasing fashion with increasing intensity in response to a wide category of afferent input [light touch versus heavy mechanical, thermal stimuli ranging from 30°C (innocuous) to 45°C (painful), etc.].

The relative contribution of the two types of neurons is not clear. The theoretical significance of such findings bears on the question of whether the "pain event" is coded by a pathway that is uniquely activated only by high-intensity stimuli; i.e., is there a unique pain pathway? The presence of nociceptive specific neurons would permit such a possibility. Alternatively, the output of the wide dynamic range neuron may be driven by noxious and innocuous stimuli. The pain event then might depend not on a unique, specific pathway but on the frequency coding of activity in a pathway that carries information relevant to the perception of a wide range of stimulus intensities. Practically, the relevance of these theoretical considerations is that in the latter case, surgical intervention can never accomplish a selective blockade of the transmission evoked by high-intensity stimulation.

A caveat that should be made that these studies on the electrophysiology of dorsal horn neurons have been carried out in animals that are anesthetized or in animals which were rendered decerebrate and/or decerebrate/spinal. Clearly, the presence of anesthesia will itself serve to attenuate the transmission of certain types of stimuli and to alter the apparent response properties of the neurons.[121,122] We know that such drug treatments in fact may block "pain" by blocking sensory transmission at the level of the spinal cord. The use of spinal or decerebrate/spinal animals alters completely the complexion of the descending facility and inhibitory tone, which is known to govern sensory transmission through the spinal cord. Further discussion of this point appears below.

1.2. Ascending Pathways by Which High-Intensity Stimulation Gains Access to Brain Centers

Classic observations indicated that unilateral transection of the ventrolateral tract would produce an impressive contralateral pain deficit in patients suffering from otherwise intractable pain.[123,124] This observation lent early support to the notion that fibers traveling in this quadrant represented an essential link in the "pain pathway."

Although the ventrolateral quadrant appears to contain a significant proportion of fibers that are relevant to the transmission of noxious information, and, indeed, the prominent effects of lesions in this region on pain perception argue for its relevance, several lines of investigation clearly indicate that the ventrolateral quadrant is not alone in its role in the rostral transmission of nociceptive information. Three lines of evidence suggest alternate pathways: (1) it is not unusual for the pain sensation to reoccur 3 months to a year following an otherwise successful cordotomy; (2) contralateral pain sensation will often be observed following a unilateral lesion, suggestive of projections that may travel bilaterally; and (3) high-intensity stimulation applied to an area rendered analgesic by a cordotomy will often evoke sensations of diffuse burning pain commonly associated with the activation of C fibers,[124] suggesting that alternate pathways not traveling through the cordotomized system are being brought into play by a particular mode of stimulation. As is pointed out below, considerable evidence exists to support the idea that there are no fewer than four potential classes of pathways by which high-intensity stimulation may travel from the first-order afferent to higher centers. These are (1) the ventrolateral funiculus (consisting of fibers of the spinoreticular, spinomesencephalic, spinothalamic pathways), (2) the ventral funicular pathways, (3) the dorsal funicular pathways (including the spinocervical tract), and (4) the intersegmental spinal pathways (including the tract of Lissauer, the dorsolateral propriospinal system, and the dorsal intracornual tract).

1.2.1. Ventral Funicular Pathways

Cells that contribute their projecting axons in this system have been extensively studied. As noted above, Kuru[98] observed retrograde chromatolytic reactions in neurons of the marginal zone and of the deeper laminae (IV and V) in patients who had undergone clinically effective cordotomies. Significantly, more advanced physiological and anatomic techniques such as the use of antidromic stimulation of various brainstem and diencephalic centers as well as the retrograde transport of label such a horseradish peroxidase have largely confirmed these early findings, suggesting that neurons that project into the thalamus and to the brainstem originate in the marginal zone, substantia gelatinosa, and in the neurons whose cell bodies lie within the nucleus proprius (laminae IV, V, and VI) (see above). Tract-tracing studies have indicated that a large proportion of the fibers that travel in the ventrolateral tract appear to originate in the dorsal and intermediate gray matter and cross in the dorsal commissure at levels one to two segments rostral from their point of origin.[125] These cells arrange themselves somatotopically such that axons are displaced laterally by the input from the more rostral segments as they descend rostrally.[126]

Although the ventrolateral aspect of the cord is commonly associated with the primary ascending system in the ventral quadrant, there is considerable evidence that an equivalent set of pathways lies adjacent to the ventral sulcus. Notably, this system is seldom lesioned during surgical intervention for pain because of its proximity to the anterior spinal artery. Nevertheless, Kerr and colleagues[85] have demonstrated that axons in this system arise from cells that lie in the marginal layer and in the deeper laminae. Systematic investigations of the physiological properties of these axons have not been carried out.

Although the term ventrolateral quadrant is often times taken to be synonymous with the spinothalamic system, it is clear that a large proportion of the ascending fibers that lie in the ventrolateral quadrant do not reach the thalamus. Thus, in man, on the basis of degeneration studies, it has been suggested that almost half of the fibers in the ventrolateral quadrant will terminate at a level caudal to the rostral aspect of the inferior olive.[127] Examination by retrograde transport and antidromic activation clearly indicates that ventrolateral fibers may be categorized into three populations: those that project to the thalamus (spinothalamic), those that project to the mesencephalon (spinomesencephalic), and those that project to the brainstem reticular formation (spinoreticular) fibers. These projection systems are discussed below.

1.2.2. Dorsofunicular Pathways

The most obvious dorsofunicular pathways are those that travel in the dorsal columns, although lesions of this system have largely failed to alter the pain threshold,[128,129] and stimulation of this system appears to give rise to tactile sensations.[130] A second pathway that travels in the dorsal funiculus is the spinocervical tract, which travels ipsilaterally in the dorsolateral quadrant of the spinal cord. Although the spinocervical tract is clearly present in cat, its presence is reduced as one ascends the phylogenetic scale such that it is practically nonexistent in humans.[131–133] Nevertheless, physiological studies have clearly indicated in cat and in primate that several types of spinocervical neurons can be identified that may be activated not only by tactile but by noxious thermal and mechanical stimuli as well.[134–136]

1.2.3. Intersegmental Links

Although the long ascending pathways are clearly relevant to the perception of pain, several observations suggest the relevance of nonfunicular pathways. First, there is commonly a recovery of pain sensation even following bilateral tractotomies.[125] Systematic studies in primates using well-defined stimuli have clearly confirmed the reocurrence of pain reactivity following spinal cordotomies.[237] Second, intense visceral pains are commonly not abolished even by bilateral ventral cordotomies.[125] Third, systematic animal studies have indicated that multiple hemisections at different cord levels will have surprisingly little effect on either the behavioral or the autonomic responses otherwise evoked by strong stimuli.[137] Fourth, destruction directed at the dorsal gray matter, e.g., spinonucleolysis, has been shown to produce prolonged pain relief associated with a number of intractable pain syndromes.[138] Such procedures leave the long funicular pathways intact and, as a result, offer strong support for the importance of local systems in the dorsal gray that may be relevant to the transmission of nociceptive information. Together, these observations suggest that short projecting systems may travel ipsilaterally and contribute to the rostral transmission of pain information. Segmental pathways that may contribute to these phenomena are the following:

1. Lissauer tract. The medial portion of this system appears to consist largely of collaterals of the small primary afferents that travel rostrally and caudally following bifurcation prior to entering the dorsal horn.[85,139,140] The lateral portion of this tract lies in the dorsolateral funiculus just lateral to the dorsal root entry zone and consists of small fibers deriving from neurons in the gelatinosa and the marginal layers as well as some collaterals of primary afferents.[141] Lesions of the dorsolateral quadrant that involve the tract of Lissauer have been observed to elevate the nociceptive threshold.[142]
2. Dorsolateral propriospinal system. This ipsilateral projection system travels lateral to the axis of the nucleus proprius and appears to contain axons originating from the neurons of the gelatinosa and the marginal zone.[84] As noted above, several classes of marginal cells have been demonstrated to be activated by high-intensity $A\delta$ and C fiber input, and it is possible that such neurons may project in this pathway.
3. Dorsal intracornual tract. This system consists of small-diameter myelinated fibers that travel longitudinally through the medial aspects of the nucleus proprius. As rhizotomies have failed to significantly alter the distribution of these fibers, it seems likely that they arise from neurons intrinsic to the spinal cord.[85]

It should be noted that although ample evidence suggests that alternate systems may play a role in pain transmission, the role of the intrinsic segmental pathways outlined above is clearly not proven. One must not forget the significant contralateral pain relief that is accorded by unilateral ventrolateral cordotomy, indicating the significance of long pathways that are crossed. As indicated, the segmental pathways outlined above are largely organized in an ipsilateral fashion, although intersegmental crossing fibers are clearly present.

1.3. Supraspinal Systems in Pain Transmission

Association of specific brainstem sites with the throughput of information generated by peripheral tissue-damaging stimuli has been carried out on the basis of anatomic connectivity as well as by electrophysiological investigations of excitatory inputs. On the basis of terminal projection fields, three principal targets have been identified for fibers that project in this ventrolateral system: the medulla, the mesencephalon, and the diencephalon.

1.3.1. Medulla

Unilateral ventrolateral tractotomy has been shown to produce a significant ipsilateral and contralateral degeneration in the medial and paralateral medullary reticular formation.[143,144] The majority of the projections to the medulla are predominantly of an uncrossed nature. Thus, extensive degeneration is not observed following midline myelotomies.[145] On the other hand, the injection of HRP in the medial medulla has given evidence of ipsilateral and contralateral labeled cell bodies in laminae IV, V, and VIII.[146] It should be noted in this regard that degeneration studies have consistently pointed out that crossing fibers can be seen following unilateral cervical cord section at many levels rostral to the caudal medulla.[144] Such crossing could in part account for the presence of large and often bilateral receptive fields for these neurons (see below). Significantly, only a few signs of marginal cell innervation were observed.[90,146]

A variety of studies have suggested that medullary neurons will project to the intralaminar and ventrobasal nuclear complex of the thalamus.[147-149] As such, this represents part of the classical spinoreticulothalamic pathway, which was suggested by Bowsher and colleagues.[148] This sytem is also referred to as the paleospinothalamic system.

Neurons in these regions are reported to be activated by application of a wide variety of stimuli to their receptive fields. Of the nuclei in the medullary region, those neurons that lie within the nucleus gigantocellularis appear to be most effectively activated by input deriving from somatic stimulation of peripheral afferents conducting at the velocity of Aδ and C fibers.[150-152] Receptive fields for these neurons generally exist both ipsi- and contralaterally and will frequently encompass an entire quadrant if not the entire body,[153,154,275] suggesting the presence of extensive supraspinal convergence of the spinofugal systems.

The electrophysiological characteristics of neurons within the lateral medullary reticular formation has prompted the suggestion that this system may play an important role in mediating pain behavior in unanesthetized animals. Thus, Casey[155-157] and others,[158,159] in an extensive series of investigations, demonstrated that (1) the frequency of discharge of single neurons in the nucleus gigantocellularis is directly proportional to the intensity of the electrical stimulus applied to a peripheral nerve, (2) the intensity of stimulation that evokes escape behavior in the unanesthetized cat is similar to the intensity that results in the maximum discharge rate of nucleus gigantocellularis neurons, (3) latency of escape behavior is closely correlated with the frequency of discharge of the nucleus gigantocellularis neuron, (4) electrical stimulation of the nucleus gigantocellularis will evoke escape behavior in rats and cats and serve as an unconditioned stimulus in Pavlovian conditioning paradigms, and (5) lesions of the nucleus gigantocellularis have been shown to attenuate the response of the organism to a stimulus that would otherwise produce an escape response.

1.3.2. Mesencephalon

The mesencephalon may for our present purposes be divided into the mesencephalic reticular formation and the medial central gray matter. These regions receive crossed projections from ascending spinopetal neurons, and these mostly terminate along the lateral aspects of the central gray and to a lesser degree into the subjacent reticular formation.[147] Of particular importance with regard to the projections of the mesencephalic reticular formation and central gray are their rostral projections to the midline and intralaminar nucleus of the thalamus and caudal hypothalamus. Strong reciprocal connections exist between the central gray and the subjacent tegmentum as well as caudally back to the medullary level.[160]

Electrophysiological recording from cells within the mesencephalic central gray and subadjacent reticular formation has indicated the existence of cells that are differentially responsive to innocuous and noxious cutaneous stimulation and appear to be broadly classified as wide dynamic range neurons.[161-163] Electrical stimulation of peripheral afferents giving rise to activity in afferents that conduct at C fiber velocity produces a high-frequency discharge in neurons within this mesencephalic region.[161-163] As in the medullary reticular formation, there are few if any cells that can be deemed uniquely nociceptive, i.e., activated only by high-intensity mechanical or thermal input. Similarly, neurons in the mesencephalic reticular formation display a significant degree of convergence, and bilateral receptive fields including the whole body are not uncommon.[161-163]

Studies to determine the relevance of this sytem to the pain event have been carried out. Classical work with electrical stimulation has indicated that focal activation of the central gray and, to a limited extent, the adjacent mesencephalic reticular formation is associated with reports of dysphoria in man.[166] In cats and dogs, such stimulation serves to evoke so-called sham rage (i.e., flattening of the ears, distress vocalization, pupil dilatation, and vigorous efforts to escape).[164,165] These results are in accord with the observation that fibers of the ascending spinothalamic system terminate in this region. Electrical stimulation, as in the medulla, presumably mimics activity in such fibers. We must remember that it is not possible to exclude the possibility that the terminals that are found within the mesencephalon deriving from spinal neurons may in fact represent collaterals of spinothalamic systems that also gain access to diencephalic centers. In such cases, focal stimulation would serve to activate such fibers antidromically and mimic the ascending message generated by a peripheral stimulus in the spinothalamic system. In this sense, it is important to note that extensive lesions of the mesencephalic central gray or subadjacent reticular formation have largely failed to produce any significant change in the pain response,[167,168] suggesting that this region, though capable of evoking pain behavior when terminals are locally activated by focal electrical stimulation, are not essential for the manifestation of an animal's response to an otherwise aversive stimulus.

1.3.3. Diencephalon

To date, three nuclear groupings in the thalamus have been demonstrated to receive significant information from systems known to be associated with activity generated by high-intensity peripheral stimuli: the posterior nuclear complex, the ventrobasal complex, and the medial intralaminar nuclear complex.

1.3.3a. Posterior Nuclear Complex. The posterior nuclear complex consists of the caudal pole of the ventromedial thalamus lying rostrally to the medial geniculate.[169] Anatomic studies have indicated that input into this region derives to a great extent from the ventrolateral tract in the thalamus as well as from lemniscal input from the dorsal column nuclei and the spinocervical tract.[170-172] This nucleus has been demonstrated to project to the posterior portion of the somatosensory area (I and II) of the cortex.[169,173]

Electrophysiological studies carried out in the posterior nuclear complex have yielded variable results. In the early studies by Poggio and Mountcastle,[174] over 60% of the neurons examined had large bilateral receptive fields and displayed a high-frequency steady discharge in response to high-intensity peripheral stimulation. Later studies failed to find this proportion of nocisponsive neurons.[175,176]

Lesions of the posterior nuclear complex in primate have been shown to reduce the responsiveness of animals to high-intensity mechanical stimulation,[177] though thalamic lesions directed at this region in humans have been controversial, with the effects at best being transient.[178]

1.3.3b. Ventrobasal Complex. The ventrobasal complex is the primary somatosensory nucleus of the thalamus and consists of the nucleus ventralis posterior and nucleus ventralis lateralis. These nuclear complexes project in a somatotopically organized fashion to the somatosensory cortex (SI and SII).[179-181] Consistent with the observation that the majority of the neurons in the ventrobasal thalamus are responsive to innocuous tactile or thermal stimulus,[182-184] the ventrobasal nuclei have been thought to receive primarily lemniscal input from the dorsal column nuclei.[185] On the other hand, it has been clearly demonstrated that the spinothalamic tract does provide an input into the ventrobasal complex,[172] and this is in accord with the electrophysiological finding that a small population of neurons in this region do show a discriminable response to noxious stimulation.[184,186] Retrograde transport studies with HRP have indicated that spinal cells projecting to the lateral aspects are found predominantly in the marginal layer and in the n. proprius (regions showing a preponderance of nociceptive neurons). Those cells projecting medially were most commonly found in the intermediate region and ventral horn.[187]

Lesions of the ventrobasal complex have been shown to alter somatosensory discrimination and produce transient analgesia in cats[188] and humans.[125]

Electrical stimulation of this region generally produces nonnoxious paresthesias,[189] although reports of highly localized pain after low-intensity focal stimulation of the ventrobasal nucleus has been reported in humans.[190]

1.3.3c. Medial Intralaminar Nuclear Complex. The intralaminar nuclei consisting of n. paracentralis, n. centralis medialis, n. centromedian, n. centralis lateralis, and n. parafascicularis form a thin layer around the lateral aspect of the nucleus medialis dorsalis. Input into these nuclei is thought largely to be contributed by fibers traveling in the ventrolateral quadrant (i.e., spinothalamic).[172] A second source of input into the central median parafascicular region is the projections that are thought to originate in the medial medulla and mesencephalon.[191,192] Recent studies with retrograde transport have pointed to surprisingly specific projections from subpopulations of marginal neurons to the dorsal portion of the n. submedius of the medial thalamus.[236] Projections from the intralaminar nucleus distribute diffusely to wide areas of the cortex including the frontal parietal limbic regions and form the anatomic basis for the physiological defined diffuse reticular activating system.[194]

Electrophysiological studies of the medial and intralaminar nuclear complexes have demonstrated the presence of cells responsive to noxious stimuli that appear to encode stimulus frequency in terms of the duration of frequency discharge.[195] Stimulation in large bilateral receptive fields at intensities that drive activity in Aδ and C fibers will produce significant activation of neurons in the medial intralaminar nuclei.[196] Such observations have been made in detail in animals and have also been largely confirmed in human.[197]

Lesion studies carried out in rat, cat, and primate have produced discrepant results, with significant increases in the nociceptive threshold being observed by some[198,199] but not by other[200,201] investigators. In man, relief of intractable pain associated with metastatic cancer has been obtained following lesions of the medial thalamus, particularly those directed at the centromedian/parafascicular/medialus dorsalis complex.[202]

1.3.4. Cortex

Thalamocortical projections originating in the ventrobasal nucleus and in the posterior thalamic complex are known to project, respectively, to the anterior and posterior regions of the second somatosensory area of the cortex.[179,204–206] Consistent with the response properties of neurons in those thalamic groups, the majority but not all of the cells in the anterior region are activated by light tactile input, whereas those in the posterior portion of the somatosensory area tend to be polysensory, and a significant proportion have been observed to respond to high-intensity stimulation in a fashion that resembles neurons observed earlier in the posterior thalamic group.[207,208] Although emphasis has been placed on the role of the second somatosensory area, Kenshalo and colleagues[209] noted that a preponderance of ventrobasal neurons that responded to noxious stimuli could be antidromically driven by focal electrical stimulation of SI.

The interpretation of such observations is made difficult by the associational properties of the cortex and in part reveals the difficulties in philosophically defining the meaning of the word "pain." Early work clearly demonstrated that the severing of thalamofugal fibers to the frontal cortex would result in an abolition of the pain report. Classically, such individuals would report that they "felt the pain" but that it did not bother them.[210] It should be noted that this is fundamentally different from the types of reports that are commonly obtained when lesions are directed at portions of the ascending sensory system and made at the level of the diencephalon or below. Thus, following ventrolateral tractotomy, there is a significant attenuation in the magnitude of the pain reported. Similarly, lesions in the medial thalamus are associated with a loss of the pain associated with metastatic cancer. In the case of the cortex, however, the reported clinical finding is one of disregard. This suggests that the question of "pain" is a multipropertied one in which the sensory message that gives rise to the input generating a perceptual event may be separated from the meaning of the perceptual event itself.

1.4. Summary of the Rostral Projection System

High-intensity stimulation applied to peripheral organs such as skin and viscera gives rise to activity in small myelinated and lightly myelinated fiber systems that enter the spinal cord to make synapses in the dorsal horn with second-order neurons of three classes: marginal cells, gelatinosa neurons, and so-called lamina V or wide dynamic range cells. These cells project rostrally, both ipsilaterally and contralaterally, in the ventral and ventrolateral quadrants to higher brainstem centers.

In addition, there are clearly multisynaptic pathways that travel within the gray matter, of which there are at least three, i.e., the dorsal intercornual tract, the dorsolateral fasiculus proprius, and the tract of Lissauer. The major ascending pathways give rise to projections that reach medullary (medial and paramedial regions), mesencephalic (central gray and subjacent reticular formation), and diencephalic (ventrobasal complex, posterior nuclear group, and medial and intralaminar) nuclear groups. Importantly, although it is known that ascending systems that project to the brainstem may in fact terminate there, it is likely that rostrally projecting fibers may give rise *en passant* to significant collateral projections that terminate in these more caudal regions. Nevertheless, classically, one can separate such projections that enter the medulla and then give rise to second-order connections going into the thalamus, i.e., the spinoreticulothalamic tracts (paleospinothalamic tract), from those projections to the thalamus that are essentially direct (neospinothalamic tract). A similar organization can be described for spinomesencephalic projections, where in fact significant rostral connections with the midline diencephalon are known to originate in the medial mesencephalon. The most rostral point of mapping of such systems reponding to high-intensity stimulation is the thalamocortical projections from the nuclear groups that are known to receive spinothalamic and bulbothalamic input.

Several caveats should be added to the above comments. First, it should be noted that the study of the supraspinal elements of the pain pathway is rendered difficult for several reasons. First, lesion studies that remove certain portions of the system do not tell us what the functions are of that structure that we have removed; rather, they tell us what the nervous system does in the absence of that structure. It is thus possible that a given region might indeed play a role in the pain mediating the animal's response to an otherwise noxious stimulus but that its lesion fails to produce any observable effects on the animal's behavior because it represents one of several substrates that are associated with the processing of sensory information.

Secondly, many investigators seek to determine what role a system has by focally activating a region by locally depolarizing the region with a stereotaxtically directed electrode. Such stimulation will, however, activate not only local dendrites and cell bodies but also, in both orthodromic and antidromic fashion, all fibers of passage. Thus, if a fiber system sends a collateral into a given region and one electrically stimulates in that region, the net result may well be the antidromic activation of the main fiber trunk via the axon collateral. The antidromic spike will then invade the main system to proceed both orthodromically and antidromically with predictable consequences. Such a phenomenon may in fact occur following stimulation in an area such as the mesencephalic central gray, where it is likely that many of the spinal mesencephalic terminals are in fact collaterals of spinothalamic fibers that terminate in the diencephalon. In such cases, stimulation of the mesencephalic central gray would produce its effects not necessarily by the activation of a local circuit but by the antidromic activation of the spinothalamic system, which would in effect serve to mimic the results of high-intensity peripheral stimulation reaching rostral and caudal systems.

The third problem with the study of supraspinal units is the necessity in most cases of carrying out the experiments under some kind of anesthetic. Both practical (e.g., holding of the cell, maintenance of stable blood pressure) and humane considerations dictate the use of an adequate anesthetic. Such a drug overlay will clearly serve to alter the transsynaptic movement of information.

It should be noted that one of the essential difficulties in interpreting whether or not a brainstem neural substrate plays a role in the perception of "pain" arises because of the complexity of the "pain event." Thus, following high-intensity stimulation, one observes a variety of behavioral responses: changes in sympathetic outflow (i.e., hypertension, vasoconstriction), voluntary motor responses (i.e., escape behavior, lever pressing, running), and reflex activity (i.e., ventral root reflexes). The mechanisms underlying each of these aspects of the pain event may be separated. Clearly, one can evoke many of the changes in blood pressure and the reflex response in animals that conventionally are not thought to perceive pain, i.e., decerebrate or decerebrate–spinal animals.

It is thus quite possible that examining the discharge of neurons within certain systems may in fact reflect their role in the mediation of specific aspects of the pain event. Thus, a lesion that destroys the ability of a high-intensity stimulus to evoke a change in blood pressure or a spinal reflex does not necessarily indicate that the manipulation has blocked "pain" or, in fact, that that system plays any role as a fundamental link between the peripheral stimulus and the pain report. In the most trivial case, a simple blockade of efferent activity would in fact produce a complete attenuation of the pain response as measured by a behavioral task (such as bar pressing or escape) but in fact would clearly not be defined as producing analgesia. Similarly, sympathetic blocking agents would attenuate the effects on blood pressure produced by high-intensity stimulation but again would not reflect a blockade of nociceptive transmission. As an example, one might consider the role of the medullary nuclei (or any neural system). Though its activity may covary with the overall behavioral response of the animal to a high-intensity stimulus, the "meaning" of the neural activity may be related to arousal, changes in autonomic outflow, but not in fact be fundamentally involved in the substrate underlying the pain event. The complexity of the role a system plays is progressively compounded at the higher levels, where there is a progressive convergence of the many anatomically diverse substrates.

2. MODULATORY SYSTEMS THAT CONTROL THE PROCESSING OF SENSORY-EVOKED ACTIVITY

Though it is clear in the above brief discussion that an ordering of neural substrates through which high-intensity stimulation may pass can be made, the presence of such a system does not indicate the essential complexity of the organization of the transmission system through which this sort of information may pass. A general pattern that has begun to emerge in virtually all sensory circuits is the presence of powerful modulatory influences, which interact at every synaptic link. Detailed discussions of such modulatory circuitry have been given elsewhere,[211-213] but, in brief, the following discussion summarizes a number of the circuits that have thus far been examined.

2.1. Spinal Cord

Primary afferent input into the dorsal horn of the spinal cord appears to activate a number of modulatory circuits. Dorsal root potentials associated with the activation of primary afferents are known to alter the amount of afferent transmitter release per impulse and serve to reduce the efficiency of transmission through this first synaptic link.[67] Although the precise mechanism underlying this system is not currently understood, significant evidence suggests that neurons within the substantia gelatinosa that are activated by afferent input may terminate presynaptically on primary afferents and exert their inhibitory effects through the release of GABA.[214] It should be noted that stimulation of the dorsal column, which presumably activates collaterals of large primary afferents, has been shown to inhibit dorsal horn interneurons[215] and produce the clinical relief of pain[216] (but see ref. 217). This particular observation fulfilled a prediction of the gate control theory proposed by Melzack and Wall.[218]

In addition to the presynaptic inhibition presumably mediated by GABA, other systems are also known to modulate dorsal horn function. Glycine has been shown to produce a powerful postsynaptic inhibition and is found within the dorsal horn.[215] The discovery of endogenous opioids, e.g., the endorphins, in neurons intrinsic to the substantia gelatinosa[219] in conjunction with the ability of such agents to inhibit the discharge of dorsal horn neurons by a receptor-mediated mechanism, which is presumably both pre- and postsynaptic,[220] has led to the concept of an intrinsic opiate modulatory circuit within the gelatinosa.

2.2. Descending Modulation

In addition to those systems that are intrinsic to the spinal cord, it has long been known that descending pathways would modulate activity in flexor reflex afferents.[221,222] Stimulation of reticulospinal and raphe spinal systems have been shown to modulate nociceptive information transfer in virtually every spinopetal pathway examined including the spinoreticular and the spinothalamic systems.[223-225] Significantly, stimulation of many of these brainstem systems tends to have a preferential (although not absolutely exclusive) effect on the activity evoked in these neurons by the activation of Aδ/C fiber primary afferents. Significantly, the origins of these descending systems are not uncommonly found within regions that receive input from spinopetal systems. Thus, stimulation of neurons within the mesencephalic central gray is able to inhibit discharge of neurons in the dorsal horn of the spinal cord (see above).

Considerable work has been carried out on the pharmacology of the descending systems that mediate this spinopetal modulation of spinal cord function. Ample evidence has pointed to the significance of descending monoamine pathways (serotonin, norepinephrine) that originate in a number of pontine and medullary nuclei (see ref. 212). The iontophoretic application of these monoamines in the dorsal cord has been demonstrated to exert a powerful modulatory influence over sensory processing, particularly that evoked by high-intensity stimulation of peripheral afferents.[226]

The control of sensory-evoked transmission is not limited to the synapses that lie within the dorsal horn of the spinal cord but may occur at supraspinal sites as well. It has been demonstrated, for example, that stimulation within the mesencephalic central gray will inhibit the discharge of neurons in the nucleus gigantocellularis in response to peripheral stimuli.[222] Electrical activation of the SII somatic sensory cortex will inhibit the discharge of neurons in the posterior thalamic complex.[228] Similarly, cortical control over activity in the medial and intralaminar nuclear groups has also been suggested.[229] In this regard, it is significant that electrical stimulation in the caudate will inhibit activity in the medial and intralaminar nuclei[230] and in primates has been shown to reduce the affective component of the behavior evoked by strong cutaneous electrical stimulation.[231]

In short, at every link in the pathway it is clear that there are mechanisms that can alter the efficiency of transmission through the synapse. Current investigations carried out in a number of laboratories including our own have in fact suggested that many of these systems may be activated in a reflex fashion by the sensory input that they serve to modulate.

For example, it has been demonstrated that high-intensity peripheral stimulation is able to increase the release of descending 5-HT and norepinephrine originating from descending monoamine pathways.[232] Similarly, such stimulation is also able to increase the levels of endorphinlike activity released from the spinal cord.[233] In view of the modulatory effects produced by the focal application of these agents onto dorsal horn neurons (see above), it appears reasonable to presume that such observations imply that certain types of peripheral input are able to activate the release of agents that, via a local receptor system, exert a modulatory influence over transmission at terminals that presumably serve to regulate the rostral transmission of nociceptive information.

The role of these modulatory systems has been much debated. Clearly, the intrathecal application of opiates and α agonists as well as the stimulation of descending pathways have been observed to alter the animal's response to otherwise aversive stimulus, suggesting their role in producing "analgesia." On the other hand, the bias of the experimental design normally precludes seeing other sensory effects of these descending systems. It is known, for example, that stimulation of the sensory motor cortex will depress the activation of spinothalamic neurons evoked by Aβ input and have little effect on the discharge evoked by continued high-threshold stimulation.[234]

It thus appears possible that modulatory circuits may serve other regulatory purposes than simply to suppress the input generated by painful stimulus and may therefore have consequences other than to produce analgesia. In light of the clear relevance of these systems to the control of nociceptive input, it behooves us to consider the possible significance of these systems *vis-à-vis* sensory processing.

2.2.1. Gain Reduction

Modulatory circuits activated in a reflex fashion by systematic input may serve to reduce the gain and therefore the amount of information transferred through a system (the spinofugal tracts) that has limited channel capacity. It should be noted that this does not necessarily imply that we would "perceive less pain." If we listen to the radio with the volume turned very low, we realize from experience that a kettle drum is nevertheless very loud, and we perceive it as such because we know that the gain on the radio has been reduced. Thus, although the physical stimulus coming out of the radio is reduced, we are aware of the intensity of that stimulus from prior experience and from our knowledge of the role of the volume control on the radio. This allows us to tune the radio for maximum sensitivity (when the physical input is near threshold) and to decrease the sensitivity when the stimulus would otherwise exceed the dynamic range of the system.

2.2.2. Response Specificity

These modulatory systems may serve to control the response specificity of the neuron. As noted previously, one of the difficulties in studying the response properties of neurons in the spinal cord and brainstem is the need to surgically alter the animal or suppress it by means of anesthetics. It is not uncommon, in decerebrate–spinal animals, to observe that the dorsal horn of the spinal cord has a large number of "wide dynamic range neurons," i.e., neurons that respond not only to Aβ but to Aδ and C fiber input as well. Moreover, many of these neurons appear to display a considerable degree of convergence among various organ systems, i.e., cutaneous, visceral, and muscle. The observations that stimulation of the input from the brainstem, i.e., raphe, mesencephalic central gray, and so forth, produces a selective suppression of the discharge evoked by Aδ/C fibers whereas stimulation of somatosensory cortex preferentially inhibits the discharge evoked by Aβ fibers clearly suggest that under certain conditions the activation of certain descending pathways may serve to inhibit selectively the activity evoked by one or another of the inputs normally impinging on a cell; i.e., such descending pathways would serve to render a neuron modality specific.

Thus, as to whether or not there is a specific pain pathway in the spinal cord, it has often been pointed out that many of the neurons in the spinal cord respond promiscuously to many modalities. Such an organization would clearly suggest that no afferent system could be "pain specific" under any circumstances. In fact, the present considerations regarding the dynamics of an intact neuraxis make it probable that the spinal cord has many modality-specific neurons, but they are rendered so by programming that is in fact controlled by mechanisms that lie within the supraspinal systems.

In light of the above comments, it is clear that the existence of substrates that respond to high-intensity stimulation provides a potential mechanism for the rostrad transmission of information generated by such stimuli. It is also clear that the modulatory circuits constituting an integral part of this particular substrate render it subject to modification by a variety of mechanisms. These systems may be reflexly activated, i.e., as in the inhibition generated by large primary afferents. Alternatively, the regulation may occur through descending pathways, the activity of which can be regulated by supraspinal systems known to be subject to modification by complex processes such as classical and operant conditions. These interactive substrates, therefore, provide some insight as to why the behavioral phenomenon of "pain" has been so difficult to investigate and why, for example, the psychophysics of pain has been a perplexing subject to study.

As noted in the very beginning, there are those situations in which the effective stimulus (extensive tissue damage) does not lead to behavioral sequelae that are in accord with the descriptive word "pain." Far from being disheartening, however, this complexity offers the possibility of pharmacological and physiological interventions that may specifically alter various aspects of this transmission process. At present, our limited knowledge of the pharmacology of these modulatory circuits as well as of the pharmacology of the ascending pain system has only revealed the first blurry outlines of the complex physiological system whose interplay is governed on a moment-to-moment basis by a wide variety of environmental and organismic influences.

ACKNOWLEDGMENTS. I would like to thank Ms. Ann Rockafellow for the preparation of this manuscript. This paper was written while the author had support from the Mayo Foundation and NS 16541.

REFERENCES

1. Beecher, H. K., 1957, *Pharmacol. Rev.* **9**:59–290.
2. Gasser, H. S., 1950, *J. Gen. Physiol.* **33**:651–690.
3. Lloyd, D. P. D., 1943, *J. Neurophysiol.* **6**:293–315.
4. Ochoa, J., 1976, *The Peripheral Nerve*, Chapman and Hall, London, pp. 106–158.
5. Andres, K. H., 1961, *Z. Zellforsch. Mikrosk. Anat.* **55**:1–48.
5a. Lieberman, A. R., 1976, *The Peripherase Nerve*, Chapman and Hall, London, pp. 188–278.
6. Csillik, B., and Knyihar, E., 1978, *Progress in Neurobiology*, Volume 10, Pergamon Press, Oxford, pp. 203–230.
7. Knyihar, E., and Csillik, B., 1976, *Exp. Brain Res.* **26**:73–87.
8. Kalina, M., and Bubis, J. J., 1976, *Experientia* **25**:386–387.
9. Kalina, M., and Wolman, M., 1970, *Histochemie* **22**:100–108.
10. Tervo, T., and Palkama, A., 1978, *Acta Anat.* **102**:164–175.
11. Droz, B., 1973, *Brain Res.* **62**:383–394.
12. Bisby, M. A., 1976, *Gen. Pharmacol.* **7**:387–393.
13. Cauna, N., 1969, *J. Comp. Neurol.* **136**:81–98.
14. Boivie, J., and Perl, E. R., 1975, *MTP International Review of Science, Physiology Series One*, Volume 3 (C. C. Hung, ed.), University Press, Baltimore, pp. 303–411.
15. Chouchkov, C. N., 1972, *Z. Mikrosc. Anat. Forsch.* **86**:273–288.
16. Cauna, N., 1966, *Touch, Heat and Pain: CIBA Foundation Symposium* (A. V. de Reuck and J. Knight, eds.), Churchill, London, pp. 117–127.
17. Cranefield, P. E., 1979, *The Way In and The Way Out. Francois Magendie, Charles Bell and the Roots of the Spinal Nerves*, Futura Publishing, New York.
18. Coggeshall, R. E., Coulter, J. D., and Willis, W. D., Jr., 1974, *J. Comp. Neurol.* **153**:39–58.
19. Yamamoto, T., Takahashi, K., Satomi, H., and Ise, H., 1977, *Brain Res.* **126**:350–354.
20. Burgess, P. R., and Perl, E. R., 1973, *Handbook of Sensory Physiology*, Volume II (A. Iggo, ed.), Springer-Verlag, Berlin, Heidelberg, New York, pp. 29–78.
21. Georgopoulos, A. P., 1976, *J. Neurophysiol.* **39**:71–83.
22. Bessou, P., and Perl, E. R., 1969, *J. Neurophysiol.* **32**:1025–1043.
23. Beck, P. W., Handwerker, H. O., and Zimmerman, M., 1974, *Brain Res.* **67**:373–386.
24. Dubner, R., and Beitel, R. E., 1976, *Adv. Pain Res. Ther.* **1**:155–160.
25. Beitel, R. E., and Dubner, R., 1976, *J. Neurophysiol.* **39**:1160–1175.
26. Kumazawa, T., and Perl, E. R., 1977, *J. Neurophysiol.* **40**:1325–1338.
27. Zotterman, Y., 1939, *J. Physiol. (Lond.)* **95**:1–28.
28. Vyklicky, L., 1979, *Adv. Pain Res. Ther.* **3**:727–746.
29. Van Hees, J., and Gybels, J. M., 1972, *Brain Res.* **48**:397–400.
30. Torebjork, H. E., and Hallin, R. G., 1974, *J. Neurol. Neurosurg. Psychiatry* **37**:653–664.
31. Torebjork, H. E., and Hallin, R. G., 1974, *Brain Res.* **67**:387–403.
32. Torebjork, H. E., and Hallin, R. G., 1976, *Sensory Function of the Skin in Primates* (Y. Zotterman, ed.), Pergamon Press, Oxford, pp. 475–487.
33. LaMotte, R. H., and Campbell, J. N., 1978, *J. Neurophysiol.* **41**:509–528.
34. Torebjork, H. E., and Hallin, R. G., 1973, *Exp. Brain Res.* **16**:321–332.
35. Vallbo, A. D., Hagbarth, K.-E., Torebjork, H. E., and Wallen, B. G., 1979, *Physiol. Rev.* **59**:917–957.
36. Konietzky, F., Perl, E. R., Trevino, D., Light, A., and Hensel, H., 1981, *Exp. Brain Res.* **42**:219–222.
37. Willer, J. C., Boureau, F., and Albe-Fessard, D., 1978, *Brain Res.* **152**:358–364.
38. Willer, J. C., Boureau, F., and Albe-Fessard, D., 1980, *Brain Res.* **201**:465–470.
39. Iggo, A., 1961, *J. Physiol. (Lond.)* **155**:52P–53P.

40. Kumazawa, T., and Mizumura, K., 1977, *J. Physiol. (Lond.)* **273**:179–194.
41. Meyer, R. A., and Campbell, J. W., 1981, *Brain Res.* **224**:149–152.
42. Lele, P. P., and Weddell, G., 1956, *Brain* **79**:119–154.
43. Loewenstein, W. R., 1971, *Handbook of Sensory Physiology*, Volume 1, Springer-Verlag, Berlin, Heidelberg, New York, pp. 267–290.
44. Catton, W. T., 1970, *Physiol. Rev.* **50**:297–318.
45. Bannister, L. H., 1976, *The Peripheral Nerve* (D. N. Landon, ed.), Chapman and Hall, London, pp. 396–463.
46. Lewis, T., 1942, *Pain*, Macmillian, New York.
47. Perl, E. R., 1976, *Adv. Pain Res. Ther.* **1**:17–28.
48. Torebjork, H. E., and Hallin, R. G., 1979, *Adv. Pain Res. Ther.* **3**:121–131.
49. Fitzgerald, M., and Lynn, B., 1977, *J. Physiol. (Lond.)* **265**:549–563.
50. Dubner, R., Sumino, R., and Wood, W. I., 1975, *J. Neurophysiol.* **38**:1373–1389.
51. Essman, W. B., 1978, *Availability, Localization and Disposition*, Volume 1. *Serotonin in Health and Disease* (W. B. Essman, ed.), Spectrum Publications, New York, pp. 15–178.
52. Green, J. P., Johnson, C. L., and Weinstein, H., 1978, *Psychopharmacology: A Generation of Progress* (M. A. Lipton, A. DiMascio, and K. F. Killam, eds.), Raven Press, New York, pp. 319–332.
53. Kaliner, M., and Austen, K. F., 1975, *Annu. Rev. Pharmacol.* **15**:177–189.
54. Uvnas, B., 1978, *Handbuch der Experimentallelen Pharmacologie*, Volume 18 (Part 2) (M. Rocha e Silva, ed.), Springer-Verlag, New York, pp. 75–92.
55. Sullivan, T. J., and Parker, C. W., 1979, *J. Immunol.* **122**:431–436.
56. Morrison, D. C., and Henson, P. M., 1978, *Immediate Hypersensitivity: Modern Concepts and Developments* (M. K. Bach, ed.), Marcel Dekker, New York, pp. 431–502.
57. Harris, R. H., and Ramwell, P. W., 1979, *Annu. Rev. Physiol.* **41**:653–668.
58. Moncada, S., and Vane, J. R., 1979, *Pharmacol. Rev.* **30**:293–331.
59. McGiff, J. C., 1981, *Annu. Rev. Pharmacol. Toxicol.* **21**:479–509.
60. Hamberg, M., and Jonsson, C.-E., 1973, *Acta Physiol. Scand.* **87**:240–245.
61. Moncada, S., and Vane, J. R., 1979, *Adv. Intern. Med.* **24**:1–22.
62. Chan, J. V. C., Burrowes, C. E., and Movat, H. Z., 1978, *Agents Actions* **8**:65–72.
63. Winkelmann, R. K., 1968, *The Skin Senses* (D. R. Kenshalo, ed.), Charles C Thomas, Springfield, Illinois, pp. 499–511.
64. Hnik, P., Holas, M., Krekvle, I., Křiž, N., Mejsnar, J., Smielsko, V., Ujec, E., and Vyskčil, F., 1976, *Pfluegers Arch.* **362**:85–94.
65. Hokfelt, T., Johansson, O., Ljungdahl, A., Lundberg, J., and Schultzberg, M., 1980, *Nature* **284**:515–521.
66. Olgart, L., Gazelius, B., Brodin, E., and Nilsson, G., 1977, *Acta Physiol. Scand.* **101**:510–512.
67. Schmid, R. F., 1971, *Ergeb. Physiol.* **63**:21–101.
68. Gamse, R., Holzer, P., and Lembeck, F., *Br. J. Pharmacol.* **68**:207–213.
69. Hagermark, O., Hokfelt, T., and Pernow, B., 1978, *J. Invest. Dermatol.* **71**:233–235.
70. Juan, H., and Lembeck, F., 1974, *Naunyn Schmiedbergs Arch. Pharmacol.* **283**:151–164.
71. Fock, S., and Mense, S., 1976, *Brain Res.* **105**:459–469.
72. Armstrong, D., Dry, R., Keele, C. A., and Markham, J. W., 1953, *J. Physiol. (Lond.)* **120**:326–351.
73. Elliot, D. E., Horton, E. W., and Lewis, G. M., 1960, *J. Physiol. (Lond.)* **153**:473–480.
74. Keele, C. A., and Armstrong, D., 1964, *Monographs of the Physiological Society*, Volume 12 (H. Barcroft, H. Davson, and W. D. M. Paton, eds.), Edward Arnold, London, pp. 1–374.
75. Guzman, F., Braun, C., and Lim, R. K. S., 1962, *Arch. Int. Pharmacodyn.* **136**:353–384.
76. Handwerker, H. O., 1976, *Adv. Pain Res. Ther.* **1**:41–45.
77. Solomon, L. M., Juhlin, L., and Kirschbaum, M. D., 1968, *J. Invest. Dermatol.* **51**:280–282.
78. Tyers, M. B., and Haywood, H., 1979, *Agents Action* **6**:65–78.
79. Chahl, L. A., and Iggo, A., 1977, *Br. J. Pharmacol.* **59**:343–374.
80. Staszewska-Barczak, J., Ferreira, S. H., and Vane, J. R., 1976, *Cardiovasc. Res.* **10**:314–327.

81. Juan, H., and Lembeck, F., 1976, *Agents Actions* **6**:642–645.
82. Fitzgerald, M., and Lynn, B., 1977, *J. Physiol. (Lond.)* **265**:549–563.
83. Rexed, B., 1954, *J. Comp. Neurol.* **100**:297–380.
84. Cajal, S. Ramon y, 1909, *Histologie du Systeme Nerveux de l' Hommes et des Vertebres*, 1952 reprint, Instituto Ramon y Cajal, Madrid.
85. Kerr, F. W. L., 1975, *Pain* **1**:325–356.
86. Light, A. R., and Perl, E. R., 1979, *J. Comp. Neurol.* **186**:117–132.
87. Sterling, P., and Kuypers, H. G. J. M., 1967, *Brain Res.* **4**:1–15.
88. Wall, P. D., and Yaksh, T. L., 1978, *Exp. Neurol.* **60**:570–583.
89. Craig, A. D., and Burton, H., 1981, *J. Neurophysiol.* **45**:443–466.
90. Kevetter, G. A., Haber, L. H., Vezierski, R. P., Chung, J. M., Martin, R. F., and Willis, W. D., 1982, *J. Comp. Neurol.* **207**:61–74.
91. Coggeshall, R. E., Applebaum, M. L., Fazen, M., Stubbs, T. B., and Sykes, M. T., 1975, *Brain* **98**:157–166.
92. Onofrio, B. M., and Campa, H. K., 1972, *J. Neurosurg.* **36**:751–755.
93. Trevino, D. L., Maunz, R. A., Bryan, R. N., and Willis, W. D., 1972, *Exp. Neurol.* **34**:64–77.
94. Price, D. D., and Mayer, D. J., 1975, *Pain* **1**:59–72.
95. Willis, W. D., Kenshalo, D. R., Jr., and Leonard, R. B., 1979, *J. Comp. Neurol.* **188**:543–574.
96. Kumazawa, T., and Perl, E. R., 1978, *J. Comp. Neurol.* **177**:417–434.
97. Cervero, F., Iggo, A., and Ogawa, H., 1976, *Pain* **2**:5–24.
98. Kuru, M., 1949, *Sensory Paths in the Spinal Cord and Brain Stem of Man*, Sogensya, Tokyo.
99. Cervero, F., and Iggo, A., 1980, *Brain* **103**:717–772.
100. Gobel, S., 1978, *J. Comp. Neurol.* **180**:375–393.
101. Gobel, S., 1978, *J. Comp. Neurol.* **180**:395–414.
102. Ralston, H. J., 1979, *J. Comp. Neurol.* **184**:619–642.
103. Light, A. R., and Perl, E. R., 1979, *J. Comp. Neurol.* **186**:117–132.
104. LaMotte, C., 1977, *J. Comp. Neurol.* **172**:529–562.
105. Willis, W. D., Leonard, R. B., and Kenshalo, D. R., Jr., 1978, *Science* **202**:986–988.
106. Fukushima, T., and Kerr, F. W. L., 1979, *J. Comp. Neurol.* **183**:169–184.
107. Kumazawa, T., and Perl, E. R., 1978, *J. Comp. Neurol.* **177**:417–434.
108. Cervero, F., Molony, V., and Iggo, A., 1977, *Brain Res.* **136**:545–569.
109. Wall, P. D., Merrill, E. G., and Yaksh, T. L., 1979, *Brain Res.* **160**:245–261.
110. Light, A. R., Trevino, D. L., and Perl, E. R., 1979, *J. Comp. Neurol.* **186**:151–172.
111. Cervero, F., Molony, V., and Iggo, A., 1979, *Brain Res.* **175**:351–355.
112. Giesler, G. J., Cannon, J. T., Urca, J. C., and Liebeskind, J. C., 1978, *Science* **202**: 984–986.
113. Wall, P. D., 1960, *J. Neurophysiol.* **23**:197–210.
114. Wall, P. D., 1967, *J. Physiol. (Lond.)* **188**:403–423.
115. Price, D. D., and Dubner, R., 1977, *Pain* **3**:307–338.
116. Foreman, R. D., Ohata, C. A., and Gerhart, K. D., 1978, *Neural Mechanisms in Cardiac Arrhythmias* (P. J. Schwartz, A. M. Brown, A. Malliani, and A. Zanchetti, eds.), Raven Press, New York, pp. 191–207.
117. Mendell, L. M., 1966, *Exp. Neurol.* **16**:316–332.
118. Brown, P. B., Fuchs, J. L., and Tapper, D. N., 1975, *J. Neurophysiol.* **38**:19–25.
119. Price, D. D., Dubner, R., and Hu, J. W., 1976, *J. Neurophysiol.* **39**:936–953.
120. Price, D. D., and Mayer, D. J., 1975, *Pain* **1**:59–72.
121. DeJong, R. H., Robles, R., and Heavner, J. E., 1970, *Anesthesiology* **32**:440.
122. Heavner, J. E., 1975, *Pain* **1**:239.
123. Spiller, W. G., and Martin, E., 1912, *J.A.M.A.* **58**:1489–1490.
124. White, J. C., and Sweet, W. H., 1955, *Pain, Its Mechanisms and Neurosurgical Control*, Charles C Thomas, Springfield, Illinois.
125. White, J. C., and Sweet, W. H., 1969, *Pain and the Neurosurgeon*, Charles C Thomas, Springfield, Illinois.
126. Hyndman, O. R., and Van Epps, C., 1939, *Arch. Surg.* **38**:1036–1053.

127. Bowsher, D., 1976, *Pain* **2**:361–378.
128. Cook, A. W., and Browder, E., 1965, *Arch. Neurol.* **12**:72–79.
129. Vierck, C. J., Jr., Hamilton, D. M., and Thornby, J. I., 1971, *Exp. Brain Res.* **13**:140–158.
130. Nashold, B. S., Jr., and Friedman, N., 1972, *J. Neurosurg.* **36**:590–597.
131. Morin, F., 1955, *Am. J. Physiol.* **183**:245–252.
132. Nijensohn, D. E., and Kerr, F. W. L., 1975, *J. Comp. Neurol.* **161**:459–470.
133. Truex, R. C., Taylor, M. J., Smythe, M. Q., and Gildenberg, P. L., 1970, *J. Comp. Neurol.* **139**:93–104.
134. Brown, A. G., and Franz, D. N., 1969, *Exp. Brain Res.* **7**:231–249.
135. Cervero, F., Iggo, A., and Molony, V., 1977, *J. Physiol. (Lond.)* **267**:537–538.
136. Kniffki, K. D., Mense, S., and Schmidt, R. F., 1977, *Proc. Int. U. Physiol. Sci.* **13**:393.
137. Basbaum, A. I., 1973, *Exp. Neurol.* **40**:699–716.
138. Nashold, B., Urban, B., and Zorub, D. S., 1976, *Adv. Pain Res. Ther.* **1**:959–963.
139. Ranson, S. W., 1913, *J. Comp. Neurol.* **23**:259–281.
140. Ranson, S. W., 1914, *J. Comp. Neurol.* **24**:531–545.
141. Szentagothai, J., 1964, *J. Comp. Neurol.* **122**:219–239.
142. Kennard, M. A., 1954, *J. Comp. Neurol.* **100**:511–524.
143. Bowsher, D., 1957, *Brain* **80**:606–622.
144. Mehler, W. R., Feferman, M. E., and Nauta, W. J. H., 1960, *Brain* **83**:718–750.
145. Kerr, F. W. L., and Lippman, H. H., 1974, *Adv. Neurol.* **4**:147–156.
146. Kerr, F. W. L., and Fukushima, T. F., 1980, *Pain* (J. J. Bonica, ed.), Raven Press, New York, pp. 47–61.
147. Bowsher, D., 1975, *Brain Res.* **95**:211–220.
148. Bowsher, D., 1976, *Pain* **2**:361–378.
149. Mancia, M., Marginelli, M., Mariotti, M., Spreafico, R., and Broggi, G., 1974, *Brain Res.* **69**:297–314.
150. Casey, K. L., 1969, *Exp. Neurol.* **25**:35–56.
151. Goldman, P. L., Collins, W. F., Taub, A., and Fitzmartin, J., 1972, *Exp. Neurol.* **37**:597–606.
152. Guilbaud, G., Besson, J. M., Oliveras, J. L., and Wyon-Maillard, M. C., 1973, *Brain Res.* **63**:131–140.
153. Anderson, S. D., Basbaum, A. I., and Fields, H. L., 1977, *Brain Res.* **123**:363–368.
154. Cedarbaum, J. M., and Aghajanian, G. K., 1978, *Life Sci.* **23**:1383–1392.
155. Casey, K. L., 1971, *Int. J. Neurosci.* **2**:15–28.
156. Casey, K. L., and Keene, J. J., 1973, *Brain Unit Activity during Behavior* (M. I. Phillips, ed.), Charles C Thomas, Springfield, Illinois, pp. 115–129.
157. Keene, J. J., and Casey, K. L., 1973, *Physiol. Behav.* **10**:283–287.
158. Halpern, B. P., and Halverson, J. D., 1974, *Behav. Biol.* **11**:215–229.
159. Pearl, G. S., and Anderson, K. V., 1976, *Adv. Pain Res. Ther.* **1**:259–265.
160. Chi, C. C., 1970, *J. Comp. Neurol.* **139**:259–272.
161. Barnes, K. L., 1976, *Exp. Neurol.* **50**:180–193.
162. Young, D. W., and Gottschaldt, R. M., 1976, *Exp. Neurol.* **51**:628–636.
163. Eickhoff, R., Handwerker, H. O., McQueen, D. S., and Schick, E., 1978, *Pain* **5**:99–113.
164. Spiegel, E. A., Keltzkin, M., and Szekely, E. G., 1954, *J. Neuropathol. Exp. Neurol.* **13**:212–220.
165. Skultety, F. M., 1963, *Arch. Neurol.* **8**:608–620.
166. Nashold, B. S., Jr., Wilson, W. P., and Slaughter, D., 1969, *J. Neurosurg.* **30**:14–24.
167. Liebman, J. M., Mayer, D. J., and Liebeskind, J. C., 1973, *Behav. Biol.* **9**:299–306.
168. Deakin, J. F. W., and Dostrovsky, J. O., 1978, *Br. J. Pharmacol.* **63**:159–165.
169. Burton, H., and Jones, E. G., 1976, *J. Comp. Neurol.* **168**:249–302.
170. Kerr, F. W. L., 1975, *J. Comp. Neurol.* **159**:335–356.
171. Boivie, J., 1971, *Brain Res.* **28**:459–490.
172. Boivie, J., 1979, *J. Comp. Neurol.* **186**:343–370.
173. Jones, E. G., and Powell, T. P. S., 1971, *J. Comp. Neurol.* **143**:185–216.
174. Poggio, G. F., and Mountcastle, V. B., 1960, *Bull. Johns Hopkins Hosp.* **106**:266–316.
175. Curry, M. J., 1972, *Brain Res.* **44**:439–462.

176. Berkley, K. J. 1973, *J. Neurophysiol.* **36**:940–952.
177. Schwartzman, R. J., 1970, *Arch. Neurol.* **23**:419–429.
178. Hassler, R., 1960, *Acta Neurochir.* **8**:353–423.
179. Jones, E. G., and Powell, T. P. S., 1969, *Brain Res.* **13**:298–318.
180. Sugita, K., Mutsuga, N., Takaoka, Y., and Doi, T., 1972, *Confin. Neurol.* **34**:265–274.
181. Friedman, D. P., and Jones, E. G., 1980, *Brain Res.* **191**:249–252.
182. Poggio, G. F., and Mountcastle, V. B., 1963, *J. Neurophysiol.* **26**:775–806.
183. Burton, H., Forbes, D. J., and Benjamin, R. M., 1970, *Brain Res.* **24**:179–190.
184. Harris, F. A., 1980, *Exp. Neurol.* **68**:27–49.
185. Boivie, J., 1978, *J. Comp. Neurol.* **178**:17–48.
186. Nyquist, J. K., 1975, *Exp. Neurol.* **48**:123–135.
187. Willis, W. D., Kenshalo, D. R., Jr., and Leonard, R. B., 1979, *J. Comp. Neurol.* **188**:543–574.
188. Glassman, R. D., Forgus, M. W., Goodman, J. E., and Glassman, H. N., 1975, *Exp. Neurol.* **48**:460–492.
189. Talairach, J., Hecaen, M., David, M., Monnier, M., and de Ajurioguerra, J., 1949, *Rev. Neurol.* **81**:4–24.
190. Halliday, A. M., and Logue, V., 1972, *Neurophysiology Studied in Man* (G. G. Somjen, ed.), Excerpta Media, Amsterdam, pp. 221–230.
191. Bowsher, D., Mallart, A., Petit, D., and Albe-Fessard, D., 1968, *J. Neurophysiol.* **31**:288–300.
192. Pearl, G. S., and Anderson, K. V., 1980, *Brain Res. Bull.* **5**:203–206.
193. Mancia, M., Broggi, G., and Margnelli, M., 1971, *Brain Res.* **25**:638–641.
194. Jones, E. G., and Leavitt, R. Y., 1974, *J. Comp. Neurol.* **154**:349–378.
195. Emmers, R., 1981, *Pain: A Spike Coded Message in Brain*, Raven Press, New York.
196. Dong, W. K., Ryu, H., and Wagman, I. H., 1978, *J. Neurophysiol.* **41**:1592–1613.
197. Ishijima, B., Yoshimasu, N., Fukushima, T., Hori, T., Sekino, H., and Sano, K., 1975, *Confin. Neurol.* **37**:99–106.
198. Kaelber, W. W., Mitchell, C. L., Yarmat, A. J., Afifi, A. K., and Lorens, S. A., 1975, *Exp. Neurol.* **46**:282–290.
199. Marburg, D. J., 1973, *Int. J. Neurosci.* **5**:153–158.
200. Delacour, J., and Borst, A., 1972, *J. Comp. Physiol. Psychol.* **80**:458–468.
201. Yaksh, T. L., Yeung, J. C., and Rudy, T. A., 1977, *Neuropharmacology* **16**:107–114.
202. Sugita, K., Mutsuga, N., Takaoka, Y., and Doi, T., 1972, *Confin. Neurol.* **34**:265–274.
203. Whitsel, B. L., Petrucelli, L. M., and Werner, G., 1969, *J. Neurophysiol.* **32**:170–183.
204. Jones, E. G., and Powell, T. P. S., 1970, *Brain* **93**:37–56.
205. Saporta, S., and Kruger, L., 1979, *Brain Res.* **178**:275–295.
206. Burton, H., and Jones, E. G., 1976, *J. Comp. Neurol.* **168**:249–302.
207. Carreras, M., and Andersson, S. A., 1963, *J. Neurophysiol.* **26**:100–126.
208. Whitsel, B. L., Petrucelli, L. M., and Werner, G., 1969, *J. Neurophysiol.* **32**:170–183.
209. Kenshalo, D. R., Giesler, G. J., Leonard, R. B., and Willis, W. D., 1979, *J. Neurophysiol.* **43**:1594–1614.
210. Elithorn, A., Piercy, M., and Crosskey, M. A., 1955, *J. Neurol. Neurosurg. Psychiatry* **18**:34–43.
211. Fields, H. L., and Basbaum, A. I., 1978, *Annu. Rev. Physiol.* **40**:217–248.
212. Yaksh, T. L., and Hammond, D. L., 1982, *Brain Stem Control of Spinal Mechanisms* (B. Sjolund and A. Bjorklund, eds.), Oxford University Press, New York, pp. 470–472.
213. Yaksh, T. L., Hammond, D. L., and Tyce, G. M., 1981, *Fed. Proc.* **40**:2786–2794.
214. Game, C. J. A., and Lodge, D., 1975, *Exp. Brain Res.* **23**:75–84.
215. Hillman, P., and Vall, P. D., 1969, *Exp. Brain Res.* **9**:284–306.
216. Shealy, C. N., Mortimer, J. T., and Hagfors, N. R., 1970, *J. Neurosurg.* **32**:560–564.
217. Fox, J. L., 1974, *Surg. Neurol.* **2**:59–64.
218. Melzack, R., and Wall, P. D., 1965, *Science* **150**:971–980.
219. Hokfelt, T., Ljungdahl, A., Elde, R., Nilsson, G., and Terenius, L., 1977, *Proc. Natl. Acad. Sci.* **74**:3081–3085.
220. Yaksh, T. L., 1981, *Pain* **11**:293–346.

221. Engberg, I., Lundberg, A., and Ryall, R. W., 1968, *J. Physiol. (Lond.)* **194**:201–223.
222. Engberg, I., Lundberg, A., and Ryall, R. W., 1968, *J. Physiol. (Lond.)* **194**:225–236.
223. Willis, W. D., Haber, L. H., and Martin, R. F., 1977, *J. Neurophysiol.* **40**:968–981.
224. Fields, H. L., and Anderson, S. D., 1978, *Pain* **5**:333–349.
225. Lovick, T. A., and Wolstencroft, J. H., 1979, *Pain* **7**:135–145.
226. Headley, P. M., Duggan, A. W., and Griersmith, B. T., 1978, *Brain Res.* **145**:185–189.
227. Morrow, T. J., and Casey, K. L., 1976, *Adv. Pain Res. Ther.* **1**:503–510.
228. Curry, M. J., 1972, *Brain Res.* **44**:439–462.
229. Bowsher, D., 1966, *The Thalamus* (D. P. Purpura and M. D. Yahr, eds.), Columbia University Press, New York, pp. 99–108.
230. Kuromi, H., Satoh, M., and Takagi, H., 1973, *Eur. J. Pharmacol.* **24**:317–320.
231. Lineberry, C., and Vierck, C., 1975, *Brain Res.* **98**:110–134.
232. Tyce, G. M., and Yaksh, T. L., 1981, *J. Physiol. (Lond.)* **314**:513–529.
233. Yaksh, T. L., and Elde, R. P., 1981, *J. Neurophysiol.* **46**:1056–1075.
234. Coulter, J. D., Foreman, R. D., Beall, J. E., and Willis, W. D., 1976, *Adv. Pain Res. Ther.* **1**:271–277.
235. Haber, L. H., Moore, B. D., and Willis, W. D., 1982, *J. Comp. Neurol.* **207**:75–84.
236. Craig, A. D., and Burton, H., 1981, *J. Neurophysiol.* **45**:443–466.
237. Vierck, C. J., and Luck, M. M., 1979, *Brain* **102**:233–248.

19

The Neurochemical Study of Sleep

Antonio Giuditta, Carla Perrone Capano, and Gigliola Grassi Zucconi

> *We are such stuff as dreams are made on; and our little life is rounded with a sleep.*
>
> Shakespeare, *The Tempest*, IV, I, 156

1. INTRODUCTION

Knowledge of the complex relationship existing between sleep and brain biochemistry has been accumulating at a relatively slow pace. Progress in this field was thoroughly reviewed approximately 6 years ago,[1,2] and there is little evidence that the rate of this progress has accelerated at all since then. Yet, such a goal should be considered an essential endeavor of the neurosciences. Although neurochemical studies cannot be taken as the only methodology leading to the goal, it is becoming increasingly evident that the final sentence of the chapter describing the role of sleep will not be written until we will know the type of operations a sleeping brain is performing at a molecular level. In turn, it is just as obvious that to design rewarding experiments at a molecular level, a general understanding of the biology of sleep is required, and questions should be asked within the framework of available hypotheses on the role of sleep. One should also remember that brain is not the only target organ for the functions of sleep. Other bodily activities are influenced by sleep, as is mental performance.

This type of reasoning has set the main guidelines in the writing of this chapter.

Antonio Giuditta and Carla Perrone Capano • Institute of General Physiology, Faculty of Sciences, University of Naples, and International Institute of Genetics and Biophysics, Naples, Italy. *Gigliola Grassi Zucconi* • Institute of Cell Biology, Faculty of Sciences, University of Perugia, Perugia, Italy.

2. A BRIEF OUTLINE OF THE BIOLOGY OF SLEEP

From the behavioral point of view, sleep may be defined as a state of decreased responsiveness to external stimuli, provided these remain below a given intensity. The lack of oriented behavior does not imply a lack of movements. Gross body movements and contractions of certain muscles occur during sleep. Considerably more information may be gathered from analysis of brain electrical activity. In man, rhythmic oscillations at a frequency of about 10 Hz (α waves) occur during relaxed wakefulness, particularly from the occipital cortex. They are superseded by waves of lower amplitude and higher frequencies (30–40 Hz, β waves) during active wakefulness.

More complex patterns are seen during sleep. They have been instrumental in separating the state of quiet or synchronized sleep (SS) from that of active or paradoxical sleep (PS). The latter phase has been the last to be discovered but has since attracted a lot of experimental and theoretical interest. It is characterized by an electroencephalographic (EEG) activity resembling active wakefulness that occurs while the organism is deeply asleep. These surprising features explain the origin of the name.

The EEG patterns of SS are completely different. They have been used to distinguish four stages of SS that may be considered to differ from each other chiefly in a progressive slowing down of wave frequency accompanied by a corresponding increase in wave amplitude. Thus, the slowest waves (0.5–2 Hz, δ waves) prevail during stages 4 (more than 50% of the recorded time) and 3 (from 20% to 50%). They are believed to arise from the synchronous activity of large populations of neurons. The name SS originated from this behavior. Stages 1 and 2 are the first to appear normally and are characterized by a progressive slowing down of wave frequencies. Spindles and K complexes appear in stage 2. The former waveform corresponds to oscillations of 8–12 Hz whose amplitude declines after an initial increase. The K complexes are identified as a fast shift followed by a slower oscillation.

Simpler EEG patterns prevail in other mammals and in lower vertebrates.[3] Definitive proof of the presence of SS and PS exists only for mammals and birds, i.e., for homeothermic animals. However, EEG activity recorded from some reptilian species has been attributed to a primitive form of PS. This claim has not been generally accepted. Only a primitive form of SS appears to be present in lower vertebrates. Next to nothing is known with regard to the occurrence of sleep in invertebrates, at least from the point of view of electrical activity. From behavioral observations, invertebrates are known to undergo cyclic variations in activity (rest–activity cycle), which may represent the early phylogenetic origin of the more complex sleep–wakefulness cycle present in higher vertebrates and in man.

The late evolutionary appearance of PS stands in striking contrast with its early appearance and prevalence during the ontogenetic cycle.[4] In man as well as in other mammals, PS or a more primitive form of PS often designated as active sleep (AS) markedly prevails during late fetal life and early postnatal development. It declines progressively thereafter to adult values. In animal species that are immature at birth, SS is clearly of later occurrence. It then becomes progressively more structured as it attains its adult features.[5] The apparent discrepancy between ontogenetic and phylogenetic data may be resolved if it is accepted that a primitive form of PS appears early in evolution, perhaps earlier than SS.[3,6] Ontogenetic development is further characterized by a progressive structuring of the ultradian sleep–wakefulness rhythmicity according to a circadian modulation[7,8] and by the sequential organization of SS and PS episodes. In the adult man, a period of sleep normally begins with SS (stages 1 to 4 in successive steps). A PS episode follows and is in turn followed by another period of SS terminating with another PS episode. This cycling continues throughout the night with a progressive lengthening of PS episodes and shortening of stages 3 and 4 of SS. A full cycle (from one PS episode to the next) lasts approximately 90 min. Shorter cycles may be present in early development and in different animal species.

The degree of tonic contraction of postural muscles, which is quite high and variable during wakefulness, decreases markedly during SS and is completely abolished during PS. Gross body movements occur during SS, whereas twitches of several muscle groups intervene during PS, particularly at limb extremities, vibrissae, and eyes. The latter movements frequently occur simultaneously on both eyes (rapid eye movements or REM) and are responsible for the additional name given to PS, i.e., REM sleep. Phasic movements are more extensive and conspicuous during AS in early developmental stages. Analogous movements have been traced back to even earlier stages (embryo), when AS may not be recognized on the basis of EEG and other neurophysiological criteria.[9] The overall reduction in muscular activity associated with sleep is presumably the cause of the slight but significant decrease in metabolic rate (*ca.* 10%) observed in comparison with the basal rate of relaxed wakefulness.[10]

In general, autonomic functions such as heart frequency, blood pressure, respiratory frequency, gastric secretion, etc. undergo a slight but noticeable dampening during SS. On the other hand, their intensity increases and may sharply oscillate during PS. As a gross but useful generalization, it may be said that sympathetic tone becomes slightly lower during SS, whereas it decreases further but becomes quite variable during PS. At the same time, the tone of the parasympathetic nervous system is increased during SS and may transiently deactivate during PS. It is believed that homeostatic mechanisms efficiently regulate autonomic functions at a somewhat reduced level during SS but become inactive during PS. The behavior of the thermoregulatory, cardiovascular, and respiratory apparatus supports this conclusion.[11]

Important hormonal changes occur during sleep.[12] The first to be observed regarded GH, which was found to increase sharply during stages 3 and 4 of the first SS episode in man. On the other hand, ACTH and related adrenal cortical hormones increase during the late part of the night in loose association with periods of PS. When sleep is displaced to the light period of the day, the peak of GH release follows SS, but no significant variation occurs in the ACTH rhythm. These observations have been interpreted to indicate the autonomous nature of the circadian ACTH oscillation and, conversely, the tight association of GH release with SS. More recently, however, the occurrence of SS has been dissociated from GH release.[13] No relationship between GH release and SS has been noted in nonprimate mammals. On the other hand, in man, hormones including prolactin, testosterone, LH, and FSH (the latter two chiefly during puberty) are likewise preferentially released during sleep. The anabolic nature of most of these hormones has provided support for the hypothesis that sleep subserves an anabolic role for the brain and for the organism as a whole.[14] The complex influences exerted by SS and PS on muscular activity, autonomic functions, and hormonal release illustrate quite well that sleep cannot be considered a state concerned merely with the brain. Rather, these observations emphasize the involvement of the whole organism.

Genetic studies carried out on inbred strains of mice have shown that some features of sleep are genetically determined. They include the daily amount of SS and PS, their circadian distribution, the occurrence of spindles, and other related traits. The daily amounts of SS and PS appear to depend on separate determinants.[15]

Sleep features of several mammalian species have been related to a number of constitutional and ecological variables. For instance, daily amounts of SS have been inversely related to body size, whereas PS has been inversely related to predatory danger.[16] More recent analyses have further refined and extended these relationships.[6]

An additional series of experiments has regarded the possible relationship of SS with torpor and hibernation. Development of mature SS in the opossum appears to be associated with development of thermoregulatory capacities. In addition, entrance into hibernating periods takes place from states of SS. This evidence and other observations have led to the conclusion that SS may be the first step in a continuum of states aimed at curtailing energy expenditures.[17]

In an analogous perspective, the amount of SS has been found to increase with physical excercise[18] and with starvation.[19] Visual load may be another determinant of SS duration.[20] More conspicuous effects are produced by lengthening the period of previous wakefulness. A circadian influence has also been noted.[8]

A somewhat different attitude has prevailed with regard to PS determinants. A vast body of literature has related PS time with previous PS deprivation (PSD),[21] with intellectual capacities,[22] with periods of recovery from brain damage,[23] and, more intriguingly, with the occurrence of a previous learning experience.[24-26] Apparent conflict on the latter relationship has been traced to the nature of the training task with regard to the innate learning capacity of an animal species. Only tasks that are difficult to learn induce an extra amount of PS in the following period. The Seligman distinction between prepared and unprepared learning has provided the basis for this conceptualization. On the other hand, discrepancies remain unresolved with regard to the period after acquisition during which the extra amount of PS becomes evident. According to some studies, this may be limited to the first hours after training, whereas other studies suggest a considerably more protracted increase. Additional evidence has shown that PSD inflicted in the posttraining period prevents later recall of the learned behavior. In studies of inbred strains of mice differing in their learning abilities, the effect of posttraining PSD has been related to genetic determinants.[27]

For additional reading on the biology of sleep, there are a number of books and articles.[28-34]

3. HYPOTHESES ON THE FUNCTIONS OF SLEEP

Only limited theoretical effort has been concerned with considering sleep a unitary phenomenon fully integrated in sleep–wakefulness cycling.[1] Sleep (or PS) has been associated with the category of instinctual behaviors and has been considered the consummatory phase of appetitive periods occurring during wakefulness (or SS). Little more than generic cues have emerged from these hypotheses[35,36] with regard to the organismal needs subserved by sleep. For instance, it has been suggested that sleep is required to bring the brain within homeostatic boundaries, but the nature of the physiological parameters to be controlled has been left unclear.[37] It is implicit in this view that sleep represents a period of recovery from the wear and tear of wakefulness. Such recovery might be concerned with the remaking of constituents depleted or simply modified during wakefulness or, alternatively, with the disposition of noxious material accumulated during wakefulness. The latter view was probably the first to appear in print in the guise of the hypnotoxin theory of Pieron.[38] In turn, this led to or became associated with the idea that sleep was little more than an intoxicated state of the brain (close to coma) involving or perhaps being induced by an ischemic and/or anoxic condition.

Little remains of this version of the recovery hypothesis today. Indeed, brain activity may be just as high and even higher during sleep (mainly PS) than during wakefulness, as indicated by measures of neuronal unit activity,[39] cerebral blood flow, and O_2 uptake,[12] and a number of neurochemical indices (*vide infra*). The idea of material compounds accumulating during wakefulness to be disposed of during sleep has led, however, to the recognition and identification of sleep-inducing factors. A different fate has blessed the former version of the recovery hypothesis. Anabolic processes have been assumed to prevail during sleep on the basis of much circumstantial although appealing evidence.[14] They have been thought largely to concern protein synthesis, although direct verification of the latter point has been limited. It seems likely nonetheless that the recovery processes that presumably occur during sleep aim at reconstituting long-term functions of the brain. They certainly do not concern such steady-state activities as, for instance, the uphill driving of ionic gradients, which is continuously operating on excitable membranes and does not require the setting aside of a special mode of operation of the brain. It is a moot point whether anabolic processes should prevail during SS or during PS. Different authors have made different suggestions.

Partly related to the recovery theory but quite distinct from it is the hypothesis that sleep represents a period of energy saving for the organism. Relationships of SS with metabolic rate[10,16] and with lethargy and thermoregulatory control[17] and of PS with loss of thermoregulation[12] have all lent support to this hypothesis.[40] Although relaxed wakefulness would provide almost as much sparing of energy as sleep (metabolic rate falls only about 10% in the latter condition), sleep has been considered a way of enforcing immobility on animal species confronted with limited food supply and a high metabolic rate.

A basically distinct theory only superficially amenable to the recovery concept is encompassed by the idea that sleep, particularly PS, is a period of reprogramming of brain circuitry during which irrelevant information may be cleared away while recent useful acquisitions may become permanently or more orderly associated with preexisting programs.[41] The intriguing possibility that mutual interactions of this kind may concern innate behaviors has been suggested by Jouvet.[42] Since the concept of recovery implies a return to an original state, whereas reprogramming subsumes the gaining of a different state, the two concepts are mutually exclusive on this account. Much of what has been found with regard to the relationships between PS and memory processes supports this hypothesis. In addition, the elevated amounts of PS normally present during early brain development[4,5] may also be considered in line with this hypothesis in view of the fact that most brain circuitry is laid down during such stages.

Earlier hypotheses concerning PS had suggested its involvement in brain development,[4] in the maintenance of the catecholamine system in the CNS,[43] in the development and maintenance of oculomotor pathways,[44] or its role as a "sentinel" state in the periodic monitoring of the environment.[45] The latter two theories appear to have mainly a historical interest.

4. POSSIBLE APPROACHES IN THE NEUROCHEMICAL STUDY OF SLEEP

The experimental designs that have been adopted in the neurochemical study of sleep may be conveniently divided into two main categories according to whether they deal (1) with the effects of sleep or PS deprivation (SD or PSD) or, alternatively, (2) with the biochemical behavior of brain tissue during sleep.

4.1. Deprivation Studies

Deprivation studies have been prompted by the need to circumvent the difficulty of investigating an elusive state such as normal sleep, which is not readily reproduced in a laboratory setting and at the investigator's will. Indeed, in this approach a comparison is made between two states of wakefulness, one of which is marked by the previous absence of sleep. Such strategy has been even more cogently required in the case of PS, whose nature is considerably more episodic and elusive. In the latter case, the whole structure of sleep remains unbalanced. The condition of PSD is generally induced by placing animals on a small platform surrounded by water. The method relies on the atonia of postural muscles that accompanies PS episodes as an automatic mean of aborting them. Under this condition, the animal would fall into the water and awaken, although it would still be able to obtain a large share of its SS time in view of the sufficient amount of muscle tone remaining during SS.

These theoretical expectations are fulfilled only under certain experimental conditions.[46] They are inevitably accompanied by a considerable amount of stress, which may presumably be mimicked by placing animals on a larger platform. In such a condition, animals may obtain their share of PS without falling in the water while still exposed to the same type of stressors, such as confinement to a small surface, dampness, and the like. Other types of stressors, i.e., immersing the animal in water at 19°C for 1 h, have been suggested, but it is hard to see how closely this condition or even the confinement on a large platform may reproduce the amount and the quality of stress experienced by animals placed on smaller platforms.

More recently, other methods of PSD have been proposed, but as yet they have not been adopted for neurochemical studies.[47] The above considerations and the fact that several PSD studies made with the platform technique have been carried out without resorting to a stress control group invite a cautious attitude in the interpretation of the results. Pharmacological means have also been used to induce PSD. Criticism of these attempts stems chiefly from consideration of the possible secondary effects of the drugs used. Since active compounds generally interfere with neurotransmitter turnover and mechanism of action, their side effects might plausibly include blockade of a given neuronal system at sites or at times not strictly related to PS regulation.

Similar considerations may apply to SD studies inasmuch as the conditions of SD have generally been induced by pharmacological means or by enforced physical exercise.

In addition to these methodological pitfalls, deprivation studies suffer from a more basic shortcoming inasmuch as they provide only second-hand knowledge of the biochemical behavior of brain during sleep. In principle, following a line of reasoning closely shaped in the recovery hypothesis, one would expect an accumulation (or a depletion) of materials whose rate of turnover has undergone some variation during the deprivation period. In addition, one would be able to detect such variation using appropriate radioactive precursors. However, even if the observed effects were to be attributed with confidence to the state of deprivation *per se*, they would still represent a sort of complementary replica of the biochemical changes supposedly taking place during sleep or PS, respectively. As result, they would be able to provide only suggestive evidence.

4.2. Studies during Sleep

Studies carried out during sleep do not suffer from the methodological and interpretative shortcomings outlined with regard to the deprivation studies. In general, they have yielded more compelling and direct data on the metabolism of brain during sleep. It should be noted that most studies of this kind have dealt with SS periods. Only very few investigations have been concerned with PS episodes, undoubtedly because of the methodological difficulties connected with studying periods of such short duration in laboratory animals.

Particularly in earlier studies, unreliable methods have been used to induce sleep and to assess its occurrence. For instance, sleep has been obtained by pharmacological means, by turning on a strong light, or by forcing animals into a previous period of physical exercise. In addition, assessment of the state of sleep has been made solely on the basis of behavioral observation. These methods may lead to error inasmuch as (1) induced sleep may not be comparable to spontaneous sleep; (2) the nature of the previous waking experience may influence brain metabolism during sleep; and (3) behavioral observation is inadequate to detect the possible presence of periods of wakefulness and to specify the type of sleep that has occurred.

With these words of caution, we may now pass to examine the main neurochemical findings. The data are presented in three sections dealing, respectively, with (1) energy and intermediary metabolism; (2) amino acids and proteins, and (3) nucleic acids.

5. SLEEP NEUROCHEMISTRY

5.1. Energy and Intermediary Metabolism

5.1.1. Paradoxical Sleep Deprivation Studies

Increased levels of brain lactate (29%), pyruvate (22%), and malate (18%) were found in adult rats after 4 days of PSD induced by the platform technique.[48] Similar but less pronounced effects were noted in rats kept on larger platforms or immersed in water at 19°C for 1 h daily. No change occurred in a number of other carbohydrate and energy metabolites. The results suggested that the rate of brain glycolysis was enhanced by stress and, presumably, by PSD *per se*.

Previous analyses of brain glycogen carried out under comparable conditions were in agreement with this conclusion.[49] In adult rats kept in the condition of PSD for 3 days, the concentration of glycogen decreased to a marked degree in several brain regions including hippocampus, caudate nucleus, and caudal brainstem but not in the frontal and occipital cortex. The effect chiefly concerned the fraction of bound glycogen and was also observed in animals kept on larger platforms. The glycogenolytic effect of PSD appeared to be under the control of adrenergic and cholinergic systems, since it was abolished by the previous administration of atropine, propranolol, or reserpine.[50] In a later paper,[51] it was reported that the concentration of brain glycogen in a teleost fish displayed a circadian fluctuation and attained its lowest value concomitant with the period of behavioral activity.

A related observation that might be taken to support the concept of an enchanced rate of brain metabolism during PSD concerned the brain levels of inorganic cations.[52] A lower concentration of K^+ (*ca.* 20%) was found in the telencephalon and in the remaining brain of rats after 10 days of PSD. A comparable decrease was present in plasma. On the other hand, the concentration of other cations, i.e., Na^+, Ca^{2+}, and Mg^{2+}, remained essentially unchanged in brain and in plasma. Unfortunately, in this experiment, the lack of a stress control group and the prolonged period of deprivation prevent assignment of the effects to PSD *per se*.

5.1.2. Studies during Sleep

A relevant series of experiments has been devoted to the study of brain lactate and other metabolites involved in glycolysis, the tricarboxylic acid cycle, and energy metabolism during sleep. The main outcome of these studies points to the conclusion that in the sleeping brain the glycolytic rate depends to a substantial degree on the nature of the previous waking experience. This was clearly suggested in an experiment carried out with rats (250–300 g) previously exposed for several days to the experimental cage and to the recording apparatus, whereas other animals were first exposed to the experimental conditions only few hours before analysis.[53] In the former group, the concentration of brain lactate and pyruvate decreased 21% and 33%, respectively, after a short period of sleep (*ca.* 20 min) in comparison with awake controls. Variations were in the opposite direction in nonhabituated animals. Lactate increased 64% during sleep, and pyruvate became 54% higher. The latter change was not significant. At the same time, ATP and creatine phosphate decreased 11% and 15%, respectively, whereas nonsignificant trends in the opposite direction were observed in habituated rats. A relevant finding concerned the concentration of brain metabolites in awake animals, which did not differ in the two experimental groups. Electroencephalographic recording did not show the presence of overt differences in the sleep patterns of habituated and nonhabituated animals, but rats of the former group were quiet and prone to sleep when placed in the experimental situation, whereas nonhabituated rats were restless and started to fall asleep only after a few hours.

It may be concluded that some form of brain activity related to the stress and novelty of being in a foreign environment increased the rate of glycolysis during the period of ensuing sleep but not during wakefulness, i.e., while the novel environment was being experienced. This behavior may be compared to the energetic debt contrived during physical exercise and paid in the ensuing period of rest. It is worth noting that in the case of the sleeping brain, the nature of the debt incurred and of the neural (perhaps also humoral) activity that represents it is completely unknown. Conversely, when the previous period of wakefulness remained relaxed and quiet, the brain glycolytic rate decreased markedly during sleep.

A comparable situation presumably prevailed in earlier experiments, in which the concentration of brain lactate decreased in the sleeping brain.[54,55] On the other hand, the contribution of stress and fatigue may explain the smaller decrease (10%)[56] or lack of change[57] noted in other work. Indeed, in the first of the latter two reports, rats (150 g) were kept 4 h in a rotating drum before being allowed 60 min sleep (their body weight decreased 7 g), whereas in the experiment reported by Shimizu *et al.*,[57] younger rats (30–45 g) were implanted with cortical and muscle electrodes only 3–5 days before the experiment. It should be noted that the fall in brain lactate occurring during sleep is less conspicuous (*ca.* 15%) in rats of this body weight (50 g) than in adult rats (28%).[55] It may be surmised, furthermore, that the stressful effects of surgery were still present in the young rats at the time of the experiment.

In the sleeping rats analyzed by Van den Noort and Brine,[56] additional changes occurred in several other brain metabolites. Whereas there was a fall in AMP (21%) and ADP (10%), an increase occurred in ATP (10%), creatine phosphate (20%), fructose disphosphate (16%), and glucose (28%). It is unclear to what degree these variations were induced by the forced physical exercise inflicted on these animals. Indeed, in this experiment, the concentration of brain lactate was considerably higher than in other reports irrespective of the state of sleep or wakefulness.[2]

It should be pointed out that the studies reviewed so far were concerned exclusively with determinations of brain metabolites and not with measures of metabolic fluxes. As a result, inferences about changes in metabolic rate occurring during sleep should be considered as reasonable but tentative interpretations.

Further evidence that the sleeping brain may utilize variable amounts of energy according to the nature of the previous waking experience was provided by a paper[58] dealing with the incorporation of [^{32}P]orthophosphate into different chemical fractions obtained from rat brain. As discussed in this interesting work and in a preceding paper,[59] the radioactivity of a phosphoprotein fraction was found to be markedly increased after 30 min of sleep. Whereas in the earlier experiment sleep was assessed by behavioral observation of 3-week-old rats and the radioactive precursor was injected systemically, in the later work, adult rats (250–300 g) were implanted with cortical and muscle electrodes and with an intraventricular cannula 3 weeks before the experiment. This procedure permitted sleep to be determined by EEG criteria and labeled phosphate to be injected directly into the brain. The main result revealed that increased labeling of the phosphoprotein fraction occurred during sleep only if the animals had not been accustomed to the experimental environment. No change was present in habituated rats. In the latter group, the only significant effect concerned the specific activity of the lipid fraction, which underwent a significant fall during sleep.

In an ingenious set of experimental procedures, the phosphoprotein involved in the sleep effect was purified to homogeneity and identified as glucose-6-phosphatase.[60] The enzyme is present throughout the brain and appears to be localized in a neuronal perikaryal fraction, bound to plasma and intracellular membranes. Apart from its hydrolytic activity, glucose-6-phosphatase catalyzes transfer of phosphate from a donor molecule to glucose.[61] In these reactions, a histidine residue becomes phosphorylated. The increased phosphorylation of the brain enzyme during sleep might be open to several interpretations, including the suggestion that it reflects an enhanced rate of hydrolysis of glucose-6-phosphate.

In vivo experiments carried out with rats systematically injected with [^{14}C]2-deoxyglucose supported the latter conclusion.[62] Forty-five minutes after the injection, when the cerebral concentration of the labeled sugar and of its phosphorylated derivative [^{14}C]2-deoxyglucose-6-phosphate had virtually reached a steady state, animals were allowed to sleep for 15 min or were kept awake for the same length of time by gentle manipulation. Brain analyses showed that the concentration of labeled 2-deoxyglucose-6-phosphate decreased 30% in the sleeping brain.

Previous experiments carried out *in vitro* with a variety of preparations including slices of rat brain had supported a hydrolytic role of the enzyme in the intracellular release of free sugar.[63] On the other hand, only a slight increment (15%) in the activity of brain glucose-6-phosphatase was observed in sleeping animals.[62] This apparent contradiction may be resolved by assuming that in brain cells the availability of glucose-6-phosphate to the enzyme site is regulated by a microsomal carrier protein, as it is in other tissues.[64]

It appears that glucose-6-phosphatase is on the luminal aspect of the endoplasmic reticulum, whereas the transport protein is located on its cytoplasmic side. If this condition applies to brain cells, it will be reasonable to conclude that the enhanced phosphorylation of the enzyme observed during sleep results from an increased availability of glucose-6-phosphate to the interior of the endoplasmic reticulum.[62] The ultimate physiological significance of this arrangement in brain remains unclear.

Recently, a striking demonstration that the brain of habituated animals uses up less glucose during SS has been given by Kennedy et al.[65] using the 2-deoxyglucose method devised by Sokoloff et al.[66] Adult rhesus monkeys implanted with electrodes several weeks earlier and perfectly accustomed to the experimental condition were injected intravenously with the labeled sugar a few minutes after the end of the first PS episode. The experiment was terminated 30 min later. In comparison with animals kept awake by occasional gentle jostling of the experimental chamber, sleeping monkeys presented lower mean rates of glucose utilization in each of 75 brain structures examined. Significance thresholds were reached in 44 structures including cerebral cortex, cerebellum, and thalamus, where the decrease was most marked. Variable rates of glucose uptake reflecting the columnar and stratified organization of the cerebral cortex were observed in awake animals. The variability disappeared almost completely in sleeping monkeys. Analysis of the sleep period by polygraphic recording revealed that the effect was to be attributed to SS. On the average, states 3 and 4 accounted for almost 80% of the experimental time, whereas short episodes of PS and wakefulness occurred occasionally and only during the later, less relevant part of the incorporation period.

These findings should be contrasted to the analogous data reported by Petitjean et al.[67] for the brain of sleeping cats. In this experiment, animals were implanted with electrodes only 3–4 days before administration of labeled 2-deoxyglucose and were left free to sleep or remain awake during the following 30-min period. In comparison with cats displaying an average of only 7 min of stage 1 SS (slow waves interrupted by cortical activation), sleeping cats (average of 20.6 min of stage 1 SS) presented a general tendency to a higher rate of glucose uptake in all regions of the gray matter examined. The increase was most pronounced in the medial habenula (46%), parietal cortex (33%), and nucleus raphe pontis (28%) but did not reach statistical significance in any structure. Glucose utilization remained essentially unchanged in the white matter. In a previous paper[68] concerning presumably the same animals, the rate of glucose utilization had been reported to increase markedly with sleep in the choroid plexuses. In the choroid plexuses of the fourth ventricle, the rate of glucose uptake was linearly related to the amount of SS. These findings were ascribed to a hypothetical secretion of sleep-inducing factors by the plexuses or to their more intense secretory activity during sleep. In the analogous experiment carried out with the rhesus monkey,[65] careful examination of the choroid plexuses did not reveal any significant differences between wakefulness and sleep.

Although these discrepancies may be attributed to a number of intervening factors,[67] a presumably relevant difference may regard the degree of stress experienced by the two animal groups during the wakeful period preceding the analyses. Indeed, the amount of stress was very low in the case of the monkeys (surgery completed several weeks earlier; previous repeated exposure to the experimental environment; analyses made during the second SS episode) but was presumably not negligible in the case of the cats (surgery completed only 3–4 days earlier; essentially no previous exposure to the experimental condition; analyses made during a period of alternating wakefulness and sleep). If further work will substantiate the relevance of this difference, the metabolic rate of the sleeping brain will become even more compellingly tied to the nature of the previous waking experience.

In a more recent report,[68a] [^{14}C]2-deoxyglucose autoradiography was used in the rat to determine the relative metabolic activity (RMA) of several brain regions during SS and, most notably, during PS. To increase the amount of the latter state in the experimental period, some of the rats were deprived of PS by the platform technique for 3 days. A general decrease in RMA occurred during SS in sensory and motor areas, including cerebellum and a large part of the thalamus, while an increase was observed only in some limbic regions, such as the hypothalamic core and the dentate fascia of the hippocampus. A different pattern prevailed during PS. In this state a significant increase in RMA was present in the reticular core, substantia nigra, dentate fascia and stratum moleculare of the hippocampus, and several regions of the extrapyramidal motor system. On the other hand, a decrease occurred in the cerebellum and in the ventral thalamic nucleus. In the attempt to relate the RMA of each brain region to the level of neuronal activity known to prevail in that region during SS and PS, some discrepancies were noted. They were attributed to the additional dependence of metabolic activity upon cellular functions other than those reflected in neuronal discharge. In view of the fact that the experiments were carried out with habituated animals, the general decrease in RMA observed during SS is in line with the interpretation that brain activity during SS is related to the nature of the previous waking experience.

Sensory stimulation has been shown to induce differential effects on the rate of glucose uptake of the activated brain regions according to the vigilance state of the animal.[69] In this interesting work, a parallel analysis of neuronal firing and glucose uptake (labeled-deoxyglucose method) was made in artificially prepared cats during the presentation of appropriate visual stimuli. The cats had previously been subjected to SD by having been kept in a slowly revolving drum overnight. In the lateral geniculate body, the marked increase in glucose uptake induced by the stimulus during wakefulness was almost completely abolished during SS. In the visual cortex, the effect was less marked and was essentially confined to the deeper layers of the cortex. The biochemical changes appeared to reflect corresponding variations in the rate and temporal pattern of neuronal firing.

Since glucose is the main substrate oxidized by brain under normal conditions, the variations in rate of glucose uptake observed during SS should be associated with corresponding changes in cerebral oxygen utilization. Only a limited number of observations have been made of this point. In an early paper[70] concerning sleep-deprived human subjects examined in the late part of the night, no significant difference was noted in the rate of cerebral oxygen utilization between sleep (assessed behaviorally) and wakefulness. Recently, this conclusion was traced[65] to the anomalous behavior of one of the six sleeping subjects, whose rate of oxygen uptake had actually increased 34%. After the subject was excluded as a statistical outlier, the average value of cerebral oxygen uptake recorded during sleep became significantly lower (11%) than during wakefulness. This result appeared to be in line, at least qualitatively, with the response of the monkey brain, whose overall rate of glucose uptake was calculated to decline approximately 30% during sleep.[65]

Besides the possibility of a species difference, the less marked decrease in oxygen uptake recorded in the human subjects might be ascribed to a circadian effect or, more reasonably, to the previous SD period they had experienced. An additional possible explanation is suggested by the work of Magnes et al.[71] on perfused cat brain utilizing uniformly labelled [^{14}C]glucose. With this artificial preparation, it was found that the specific activity of the CO_2 produced by the brain became significantly lower during periods of EEG synchronization induced by mesencephalic brainstem section than during EEG activation. In turn, this suggested that during periods of EEG synchronization the rate of glucose oxidation might decrease at the expense of other substrates. A comparable effect was not observed when cortical synchronization was induced by barbiturate treatment or by thalamic stimulation. This possibility remains to be verified during periods of natural sleep.

An interesting result has been obtained by Karnovsky et al.[62,62a] with regard to brain glycogen. The concentration of this compound was found to increase markedly during the first few minutes of SS and to decrease just as rapidly after the transition to wakefulness. As the rats were apparently not accustomed to the experimental situation, the experiment suggests that under these conditions glycogen may accumulate in brain during SS concomitantly with a faster rate of glycolysis[53] and a faster rate of hydrolysis of glucose-6-phosphate.[58,62] However, since similar analyses were not made with habituated animals, it is too early to conclude that the glycogen effect is likewise to be attributed to an influence of the novel environment. It is also possible that glycogenogenesis, increased glycolytic flux, and enhanced turnover of glucose-6-phosphatase may not occur in the same cellular compartment.

Analyses of brain glycogen have also concerned the period of sleep following 3 days of PSD induced by the platform technique.[49] Results were dramatically dependent on the amount of sleep allowed to the deprived rats and showed the existence of regional variations in response. In the frontal and occipital cortex, where glycogen levels remained unaffected by PSD, the first 3 h of sleep brought about a significant decrease chiefly in the fraction of bound glycogen. Only after 6 or 9 h of sleep did glycogen levels return to normal. In other brain regions, i.e., caudate nucleus, caudal brainstem, and hyppocampus, the concentration of glycogen was decreased by PSD, but it remained unchanged in the stress control group. In the same regions, the decrease persisted or became even more conspicuous during the first 3 h of sleep. Again, only after 6 or 9 h of sleep was the normal level regained. Since the period of sleep that follows PSD is particularly rich in PS episodes, the degradation of glycogen occurring during the initial part of sleep might be attributed to PS. Direct proof of this possibility is lacking.

This hypothesis might explain the apparent discrepancy existing between these data and data demonstrating an increase in brain glycogen during sleep.[62] Indeed, in the latter case, sleep was essentially devoid of PS. The glycogenolytic effect prevailing in most brain regions during the first hours of sleep[49] suggests again that the previous waking experience, in this case PSD, created some sort of debt, which was repaid at the expence of glycogen as soon as sleep was allowed. Only after the debt had been repaid could the glycogen level be restored. If PS is indeed involved in the glycogenolytic effect observed during the initial portion of sleep that follows a period of PSD, it would appear that PS is a state of the brain during which the debt accumulated during PSD may be dissipated. The nature of this debt may not be entirely the same as that created during wakefulness by a novel and stressful experience.

The need to take into adequate consideration the cellular basis of the metabolic changes of the sleeping brain was emphasized by the work carried out by Hydén and Lange[72] on the succinoxidase activity of single neurons and glia cells isolated from the caudal reticular formation of rabbits. The animals used in this experiment were habituated to sleep in individual boxes by a prolonged period of training. On the last day, they were killed after 90 min of sleep as assessed behaviorally. In comparison with animals kept awake for 1 h by gentle handling, neuronal perikarya dissected from the nucleus reticularis gigantocellularis displayed a level of succinoxidase activity that was almost threefold higher. At the same time, a significant decrease occurred in glial cells. A comparable but less conspicuous increase (60%) in enzyme activity was noted in the neurons of the nucleus reticularis pointis oralis. Again, no significant change was recorded in glia cells. The effects appeared to be specific for the reticular cells inasmuch as no changes were observed in trigeminal and hypoglossal neurons. Analogous changes were not observed under barbiturate anesthesia.[73] Under this condition, neuronal succinoxidase activity was markedly lower than during physiological sleep. The authors interpreted their data in terms of the hypothesis that neurons and nearby glia cells form a cooperating supracellular unit whose molecular components undergo relative shifts according to functional demands. The intensity of the latter factor may not be entirely represented by firing frequency. It is unknown whether the changing levels of succinoxidase reflect enzyme activation, enzyme synthesis, or transfer of active protein from intracellular or transcellular sites.

Intracellular levels of cyclic nucleotides change significantly in brain during sleep.[74] Adult rats (350 g) were habituated for 5–7 days to a special restraining apparatus and killed thereafter by microwave irradiation following 10 min of wakefulness or SS or 2 min of PS. In several brain regions, but not in the cerebellum, the concentration of cyclic AMP decreased during SS or PS. A similar decrease was observed with regard to cyclic GMP in hippocampus, midbrain, pons–medulla, and cerebellum. No change was noted in the cortex, but in the striatum, the level of cyclic GMP was higher during PS than in the other two states. In the brainstem regions, the concentration of cyclic GMP was more elevated during PS than during SS.

Other physiological parameters that may be taken to represent indices of metabolic rate are cerebral blood flow (CBF), brain temperature, and neuronal firing. They have all been measured during sleep. The literature accumulated on neuronal activity in different brain areas during wakefulness, SS, and PS is quite extensive.[39] In general, it may be said that sleep states bring about a redistribution of neuronal activity that, on the whole, tends to decrease somewhat during SS and to increase more markedly during PS. Measured values of CBF and brain temperature are in general accord with these observations.[12] However, some discrepancies remain. For instance, in human subjects in comparison with wakefulness, a period of SS was reported to induce either a moderate increase[70,75] or a moderate decrease[76,77] in CBF. In the cat, SS brought about an overall 15% increase in CBF. Only ten out of 25 brain regions displayed the effect, which was highest in the association cortex and in cerebellum (20–30% rise). In this experiment, however, cats were forced to walk on a treadmill for 1–2 days before analyses.[78] Likewise, in the Mangold et al.[70] study, subjects were kept awake until early in the morning before being analyzed. There is room, therefore, to suggest an influence of the previous waking experience on the rate of CBF during the subsequent period of sleep. A similar influence might be hypothesized with regard to neuronal activity.

It does not seem that a systematic investigation on these points has been undertaken. Yet, it would be highly desirable if these metabolic and physiological indices were analyzed in the same animal species and under comparable conditions. A concerted effort of this kind would definitely establish whether the response of any of the variables examined shows a consistent differential effect during sleep depending on the nature of the previous waking experience. Until homogeneous data are available, the observed discrepancies in the behavior of some of the metabolic and physiological parameters should be regarded as only apparent. In addition, it should be recalled that brain metabolism depends to a considerable extent on glial elements, whose activity cannot be monitored by conventional neurophysiological techniques.

More concordant results have been obtained with regard to PS. In animals as well as in human subjects, a marked and consistent increase in CBF has been described by several authors in comparison to SS and, less often, in comparison to wakefulness.[12] In the paper by Reivich et al.,[78] blood flow increased in all 25 brain regions examined. Increments over the state of wakefulness ranged from 62% in the cerebellar white matter to 173% in cochlear nuclei. It has already been mentioned that this study was carried out on cats forced to move on a treadmill for 1–2 days and therefore likely to be fatigued and stressed. The contribution of these factors in determining the extent of the observed variations is possible but unknown. Brain temperature rises occurring during PS have been taken as an additional indication of an enhanced tissue metabolism. Other interpretations are based on the inflow of warmer blood to the brain.[12] The two hypotheses may not be mutually exclusive. These observations should be viewed in the more general perspective of the loss of thermoregulation that characterizes PS.[11] At least on a qualitative basis, the CBF increase observed during PS is in accord with the higher level of neuronal activity recorded during this state.[39] A quantitative and regional analysis of this correlation is, however, lacking.

Before closing this section, we should alert the reader to a number of relevant observations that might have a bearing on our discussion of brain metabolic rate during sleep. These observations have led to the formulation of the ischymetric hypothesis of sleep regulation.[79] In brief, the hypothesis states that sleep is promoted by an increased rate of cellular metabolism and, conversely, antagonized by a reduced metabolic rate. In favor of the hypothesis are the correlations found between meal size and subsequent sleep, the sleep-promoting effects of infusions of nutrient, especially if associated with an infusion of insulin, the manipulation of the circadian distribution of sleep brought about by reversal of lipogenetic and lipolytic activity, and the sleep-suppressing action of starvation in lean but not obese rats. It is possible, however, that cellular sensing of nutrient availability is just one of several permissive mechanisms that monitor external and internal conditions, e.g., temperature, predators, and relate them to the appropriate transition between sleep and wakefulness. From this point of view, the ischymetric hypothesis would not necessarily be predictive of the rate of brain metabolism during sleep except for the cell receptors themselves, which are likely to be located in the hypothalamus.

5.2. Amino Acids and Proteins

5.2.1. Paradoxical Sleep Deprivation Studies

The concentrations of GABA, aspartate, glutamate, and several other amino acids were reported to undergo complex variations in six different regions of the brain of adult cats deprived of PS for 3 days.[80] For example, GABA increased in the mesencephalic reticular formation and in the frontal cortex but decreased in the caudate nucleus; aspartate increased in hippocampus; glutamate did not vary. At variance with an earlier paper describing approximately similar changes,[81] in this study possible stress effects were monitored, at least qualitatively, by extending the analyses to a group of animals deprived of PS for 3 days but allowed free access to sleep for 8 h a day during the period of deprivation. Somewhat different results were reported in rat brain after 3 days of PSD.[82] Glutamate, aspartate, and glutamine increased, but no change occurred to GABA. More significant and marked rises (two- to threefold) were noted for several other amino acids, including serine, glycine, alanine, phenylalanine, and isoleucine. The lack of a stress control group prevents assessment of the specificity of these effects.

Likewise, only presumptive data (for lack of a stress control group) concern the rate of synthesis of some brain amino acids from glucose in rats (*ca.* 100 g) subjected to 2 days of PSD.[83] In comparison to an untreated control group, the specific activity of glutamate, aspartate, and GABA was found to be higher in PS-deprived animals when referred to the specific activity of brain glucose. The effect was significant 15 min after the systemic injection of [^{14}C]glucose. No significant difference was observed with regard to brain glucose up to 60 min after the injection.

Similar considerations apply to the observations that in rats subjected to PSD the activity of brain tyrosine hydroxylase undergoes a significant increase[84] and that the level of brain ammonia may rise three- to fivefold.[85] Indeed, in neither experiment was a stress control group included. In the latter paper, a comparable increase was reported also to occur in the rate of ammonia production from endogenous brain substrates. In addition, significant increments were observed in the activity of glutamine synthetase (*ca.* 11%), glutaminase (53%), and AMP deaminase (159%).

The conclusion that PSD influences brain ammonia metabolism was supported by a more extensive series of experiments.[86] In rats subjected to PSD, the concentration of ammonia increased significantly in the soluble fraction obtained from each of three brain regions (brainstem, cerebellum, and cerebrum), whereas only a slight increase was observed in animals kept on larger platforms. Variable changes were observed in the mitochondrial and microsomal fractions. In the latter fractions, less intense variations prevailed in the stress control group. At the same time, the level of glutamine was found to be lower throughout the brain in both experimental conditions. In addition, PSD induced an increment in the activity of glutaminase (most conspicuous in cerebral mitochondria) and in the activity of glutamine synthetase (in cerebral microsomes). Increases of lower magnitude were observed in the stress control group. Comparable changes in the rate of ammonia formation and in glutamine concentration have been briefly reported to be reversible following a period of normal sleep.[87]

Cytochemical methods have been used to assess the content of proteins (and of basic proteins) in neurons and glia cells of different brainstem nuclei under conditions of PSD induced by the platform technique. In general, the effects (protein loss and decrease in cell volume) were more marked in neurons than in glia cells and tended to become less conspicuous after the first day of PSD. A subsequent period of sleep was sufficient to restore normal values. Regional differences in response were also found.[88] A slight increase in the activity of acid proteinases and a later slight decrease in the activity of neutral proteinases have been described in rat brain after PSD. In all of these studies, the lack of a stress control group makes interpretation of the results doubtful.

Attempts to follow the rate of amino acid uptake and incorporation into the brain using the PSD approach have yielded only partially clarifying insights. In the first experiments of this kind to be carried out, adult rats were deprived of PS for 2 days and then left in the same experimental condition or allowed to sleep for 6 additional hours. In the *in vivo* studies,[89] a mixture of [^{3}H]-labeled amino acids was given by intraperitoneal injection at the start of the last 6 h, whereas in the *in vitro* studies,[90] slices of telencephalon and brainstem were taken at the end of the experimental period and incubated for 45 min with a mixture of [^{3}H]amino acids. It should be noted that no stress control group was analyzed in either of the two experiments. In the former experiment, in comparison with cage controls, no variation was noted in either group in the specific activity of the TCA-soluble fraction of the telencephalon, cerebellum, and brainstem. On the other hand, a general decrease occurred in the specific activity of the protein fraction in the PSD group. The difference was significant in the telencephalon, in the cerebellum, and in the nuclear and soluble fractions of the brainstem. The decrease in relative specific activity was less pronounced in the group of rats allowed free sleep and remained significant only in the telencephalon. The latter observation may indicate that a longer period of recovery is required to normalize cortical rates of protein synthesis. The finding may be related to comparable observations made on cortical glycogen content.[49] It is of interest that in the Bobillier *et al.* study[89] no significant change occurred in the specific activity of the protein fraction in liver, although the uptake of [^{3}H]amino acids by this organ was significantly lower in the PSD group.

Determination of the rate of protein synthesis in brain slices[90] did not appear to be an appropriate method to reveal variations such as those reported in the *in vivo* studies. The specific activity of the protein fraction remained the same in all conditions, and the only significant change concerned the TCA-soluble fraction of the brainstem, whose radioactivity increased *ca.* 40% in the PSD group.

A more adequate design was adopted in a recent experiment[91] in which rats (250 g) were exposed to the condition of PSD or placed on larger platforms for 3 days. At this time, they received an intracisternal injection of [^{3}H]amino acids and were allowed to incorporate for 6 additional hours while being restored to their original condition or allowed free sleep. Inspection of the ratios of radioactivities calculated between the protein and the TCA-soluble fraction indicated that in comparison with normal animals cerebellar values were higher in the PSD condition and still higher in the stress condition. The period of free sleep did not modify the latter increment but significantly enhanced the former ratio. It appeared that the rate of protein synthesis in the cerebellum was increased by placement on a platform and that PS exerted a permissive effect on this influence.

Values of percent incorporation were, however, inconsistent with this view. In cerebral tissue, the protein–pool ratios were not statistically different except for a marked decrease induced by sleep in the value recorded for the animals kept on larger platforms. The latter group of animals had values of percent incorporation significantly higher than those of all other animal groups. In the brainstem, the only significant effect (a decrease) was observed in the protein–pool ratio of the rats allowed free sleep after PSD. A more general observation was the consistent increase in the protein–pool ratio in the stress condition as compared to PSD. The difference was most marked in cerebral tissue, where it almost reached significance. This result may be taken in further support of the concept that PS has a permissive role on the increase in brain protein synthesis induced by confinement of rats to a platform.

These conclusions are somewhat weakened by the consideration that the data have all been calculated with regard to the radioactivity of the acid-soluble pool. They might therefore reflect the behavior of the latter fraction, particularly after a pulse of 6 h, when most of the radioactive precursors would have left the tissue.

It is striking, nonetheless, that a 3-day period of confinement on small or larger platforms and the ensuing period of recovery sleep induce analogous relative variations in the concentration of acid mucopolysaccharides (mainly chondroitin sulfate) of rat brain.[92] In the latter experiment, it was observed that in the cerebral hemispheres and in the cerebellum the extent of the change was more pronounced in rats kept on larger platforms than in the PSD group. Following 6 h of sleep, the effect tended to wane in the former case but became even more conspicuous in the PSD group. The same pattern of changes was observed irrespective of the direction of the effect, which tended towards an increase in cerebral hemispheres and towards a decrease in cerebellum. Even more compellingly than the findings on protein synthesis,[91] these data appear to indicate that an unfamiliar and stressful condition, such as that of being confined on a platform, induces brain biochemical effects that require PS for their full expression. If PS is prevented, a debt is created to be repaid later by a further change in the same direction as soon as sleep (with its PS rebound) is allowed.

5.2.2. Sleep Deprivation Studies

Of several amino acids measured in the brain of rats (*ca.* 100 g) forced to walk on a treadmill for 24 h, only GABA increased (15–19%) in comparison with normally awake controls.[93] A suggestion of a more active synthesis of brain glutamine was obtained in rats of comparable body weight after 2 days of SD induced by a similar method.[83] In comparison with quiet controls, the specific radioactivity of glutamine was found to be significantly higher relative to that of brain glucose or glutamate up to 1 h after the subcutaneous injection of uniformly labeled [^{14}C]glucose. More conspicuous effects were noted in the medulla oblongata and in the pons 30 min after the injection. A comparable increase was present in the specific activity of glutamate, aspartate, and GABA, but only if the determinations were made 15 min after the injection.

Sleep deprivation induced by phenamine has been used to study the content of proteins (and of basic proteins) in single cells of different brainstem nuclei.[88,94] Marked losses of protein were found in the neuronal cytoplasm and in the glia cells of the supraoptic nucleus, particularly after 1 day of SD. Less conspicuous losses prevailed in the cells of the locus coeruleus, particularly with regard to the basic proteins. These effects were readily reversed after a period of normal sleep.

In perhaps the earliest paper on this subject to appear, Shapot[95] reported that in comparison with normal controls, rats subjected to SD for 1 day displayed a marked decrease in the incorporation of [^{35}S]methionine into brain protein. Analyses were made 30 min after the subarachnoidal injection of the radioactive amino acid. In animals similarly deprived but allowed to sleep during the period of incorporation, protein labeling was considerably higher. The possible contribution of the treatment adopted to induce SD (phenamine and teasing) is difficult to evaluate.

5.2.3. Studies during Sleep

Variations in the level of some brain amino acids have been reported to occur during sleep.[93] In comparison with awake rats (100 g), a period of sleep (20–30 min) induced by a strong light or allowed to take place spontaneously after 1 day of SD brought about an increase in the concentration of aspartate (29%) and GABA (15–19%).

In a more artificial preparation, i.e., a cat with surgical lesions at various levels in the brain, a synchronous EEG pattern was associated with a markedly higher release of GABA from the cortical surface. Conversely, the release of glutamate was higher during periods of EEG desynchronization.[96] In a comparable preparation (immobilized and artificially ventilated cat), the transition from an activated to a synchronous EEG pattern was accompanied by a 15–20% increase in the activity of cortical aspartate transaminase.[97] With regard to the latter two experiments, it is unknown whether the same behavior can be observed under more physiological conditions.

At a cellular level, a period of normal sleep was reported to increase the protein content of glia cells in the rat supraoptic nucleus. Essentially no change occurred in neurons. A more marked increase (60%) was observed in the content of glial basic proteins, but in neuronal perikarya the increase was more limited (20%).[88] These results cannot be generalized to other brain nuclei as, for instance, in the red nucleus sleep brought about little or no change in glia cells and even a moderate decrease (14%) in neurons.[94] In a previous study from the same laboratory, it had been noted that the activity of neutral proteinases increased slightly during S in the cerebral hemispheres and in the bulb but not in cerebellum.[94]

Only a few observations were concerned with the rate of cerebral protein synthesis during sleep. In young rats (26 days old) kept awake for 3 h by handling or allowed to sleep for the latter half of this period, it was found that in comparison with caged animals the uptake of [^{3}H]amino acids by brainstem slices was decreased 25% by wakefulness but was increased by sleep.[89] An even more pronounced decrease (almost 50%) was found in the specific activity of the protein fraction during wakefulness, but essentially no change occurred during sleep. Similar effects were observed in the subcellular fractions prepared from the slices, although statistical significance was attained only with regard to the microsomal and soluble fractions of waking rats. Comparable changes occurred in the telencephalon, but the only significant difference concerned the protein specific activity of sleeping rats. In considering these results, one should not forget the caveat that physiological effects may be damped or somehow modified during the preparation of the slices. Additional data concerned the uptake and the incorporation of [³H]amino acids by rat pups (18 g) under *in vivo* conditions. After a pulse of 2 h, the only significant effect was present in the TCA-soluble fraction of the telencephalon, which decreased about 10% when the rats were kept awake by gentle handling.

A previous paper by the same group[98] discussed similar analyses made on a pharmacologically manipulated preparation. An inhibitor of the synthesis of serotonin acting at the level of tryptophan hydroxylase (*p*-chlorophenylalanine) was given to adult cats to induce insomnia. Two days later, a group of animals received 5-hydroxytryptophan (5-HTP) while another group received saline; 5-HTP is a direct precursor of serotonin and was given to bypass the inhibited site. Following an additional period of 2 h, a mixture of [³H]amino acids was injected intraventricularly, and incorporation was allowed to proceed for 4 h. During the incorporation period, cats treated with saline remained awake, whereas animals treated with 5-HTP displayed almost normal amounts of SS and a markedly increased amount of PS (a rebound phenomenon). In comparison with the former group and with untreated cats, in the sleeping animals, the uptake of radioactive amino acids was considerably lower in the cervical spinal cord and in the caudal brainstem. On the other hand, the level of protein synthesis in the telencephalon was found to be directly related to the amount of PS (and SS) present in the first hour of incorporation. Fractionation of the soluble proteins from the telencephalon by polyacrylamide gel electrophoresis revealed that a high-molecular-weight band was considerably more labeled in the sleeping cats.

A direct analysis of the rates of brain protein synthesis during wakefulness, SS, and PS was made by Brodskii *et al.*[99] by measuring the incorporation of [³H]leucine in homogenates of biopsy samples of the cat parietal cortex. After animals were accustomed to the experimental condition for several days, a special needle was used to obtain about 50 mg of tissue from one side of the brain following 20 min of SS (or wakefulness) or 1.5 min PS. The next day, a comparable amount of tissue was taken from the contralateral side after 20 min of wakefulness. In this way, comparisons were made in the same cat, thus reducing the degree of variability. Similar rates of incorporation were measured when the initial biopsy sample was taken during wakefulness or during PS. On the other hand, the rate of incorporation was approximately 50% lower in the sample taken during SS.

The striking decrease in the rate of brain protein synthesis observed during SS does not support the anabolic role of sleep, at least with regard to this state. It is during SS, however, that most anabolic hormones are preferentially released.[12] Nonetheless, in the Brodskii *et al.*[99] experiment, cats were accustomed to the experimental condition for several days. It is not unreasonable to suppose that, as in the case of the metabolic processes linked to energy production, the rate of brain protein synthesis during SS may be sensitive to the nature of the previous wakeful period and may undergo an increase if the experimental animal is exposed to a novel environment or somehow stressed before sleep. If this view is correct, the anabolic hypothesis would remain valid only under certain conditions. In other words, SS may be an anabolic period for the brain.

With respect to PS, the available evidence supports or suggests the view that the rate of brain protein synthesis is comparable to that prevailing during wakefulness. The possibility of a modulatory influence exerted by the previous waking experience should also be verified with regard to PS. Evidence on this point will be definitely more difficult to obtain in view of the short episodic nature of PS and of its interspersion within longer periods of SS. As *in vivo* analyses of brain protein synthesis yield a far from adequate time resolution, experiments in which both states of sleep are present during the period of incorporation should be complemented by parallel analyses carried out during comparable periods of SS. Only tentative conclusions may be drawn otherwise.

5.3. Nucleic Acids

5.3.1. Paradoxical Sleep-Deprivation Studies

A considerable loss of brain RNA and DNA was reported to occur in adult rats after 6 or 10 days of PSD induced by the platform method.[85] In this study, chiefly designed to investigate brain ammonia metabolism, nucleic acids were considered one of the endogenous sources of brain ammonia whose concentration and rate of formation increased dramatically after PSD. In a later study, the results concerning brain nucleic acids could not be reproduced.[100] In the latter paper, the condition of PSD was reported to bring about significant reductions in the wet weight of some brain regions and of other organs.

Variable losses of RNA were reported to be induced by PSD in single cells of different brainstem nuclei.[88,94] In the supraoptic nucleus of adult rats, a progressive depletion of RNA was observed in the neuronal cytoplasm but not in glia cells. The effect was maximal after 24 h (*ca.* 40%), decreased with longer periods of PSD, and was readily reversed by sleep. RNA changes stemmed mainly from corresponding variations in the perikaryal volume. In the nucleus coeruleus, similar results were obtained, but the content of RNA decreased in glia cells as well. In neurons, the effect persisted after 2 days of PSD. In all of the experiments described above, the lack of a stress control group makes interpretation of the results uncertain.

5.3.2. Sleep Deprivation Studies

Similar cytochemical methods have been used to investigate the content of RNA in single cells of different brainstem nuclei after periods of SD induced by phenamine treatment.[88,94] In the supraoptic nucleus, a loss of RNA occurred concurrently in neurons and glia cells and persisted to about the same degree (25–30%) with periods of SD lasting a few days. Comparable effects were observed in the nucleus coeruleus except that in this region the content of neuronal RNA returned to normal after 2 days of SD and attained higher values after 4 days of SD. All of these changes were reversed by a short period of sleep.

5.3.3. Studies during Sleep

In the supraoptic nucleus of adult rats, periods of normal sleep produce an increase in the RNA content of glia cells (*ca.* 25%) but not of neurons.[88] A comparable increase was observed in the gliocytes of the nucleus coeruleus.[94] A slight but significant increase in nuclear diameter was observed in the neurons of the rabbit nucleus coeruleus concomitant with periods of EEG activation.[101]

The possible influence of sleep on cortical RNA synthesis has been considered in a series of papers. Adult rabbits implanted with dural electrodes were injected subarachnoidally with radioactive orotate. At the end of the experimental period, cortical RNA was extracted by a phenol method and size fractionated by centrifugation in a sucrose density gradient. Biochemical data were related to the amount of SS occurring during the period of incorporation, usually lasting 60 min.

From comparisons of the RNA radioactivity of a gradient region enriched in ribosomal RNA species (28 S to 50 S) with that present in a gradient region containing only heterogeneous RNA (hRNA, >50 S), it was found that the ratio increased markedly with decreasing amounts of SS. The effect became more evident when the period of incorporation was extended to 105 min. These results were interpreted to suggest that during SS the rate of cortical rRNA synthesis became less prominent in comparison with the rate of formation of hRNA.[102]

Additional experiments supported this view and indicated a selective involvement of neuronal cells in this behavior. In brief, from experiments using the same animal preparation but providing radioactive orotate by an intravetricular route, it was found that comparable changes in the sedimentation profile of radioactive RNA were observed only in a purified fraction of large nuclei. In particular, at high values of SS, the relative amount of the 45 S rRNA precursor and the ratios between 28 S rRNA and 32 S rRNA were significantly lower than at low percentage values of SS. These variations were clearly indicative of a decreased rate of synthesis and processing of rRNA during SS.[103] This conclusion is in line with reports of a lower rate of cerebral protein synthesis during SS.[99,104] Since the purified fraction of large nuclei contained mainly neuronal and astroglial nuclei, these results were ascribed to either one or both cell populations. No comparable variations were observed in a purified fraction of small nuclei, largely of oligodendroglial origin.

In the same experiment, it was observed that the nuclear content of radioactive RNA increased progressively with increasing amounts of SS in the fraction of large nuclei but not in the fraction of small nuclei. This finding could not be attributed to an accumulation of rRNA, since its rate of synthesis and processing appeared to decrease during SS, but presumably reflected the involvement of a different family of nuclear RNA, most likely hnRNA.[103]

In a third type of analysis, a purified fraction of neuronal perikarya was prepared from cortical tissue and further subdivided into its crude nuclear and cytoplasmic compartments. The main result of this approach was that the content of radioactive RNA increased progressively with increasing amounts of SS in the fraction of neuronal nuclei, whereas it slightly decreased in the neuronal cytoplasm. This result definitely involved neurons in the accumulation of radioactive RNA, which had previously been noted during SS in the fraction of large nuclei. It suggested, in addition, that the depressing effect of SS on the amount of radioactive RNA recovered in the neuronal cytoplasm could be attributed to a reduced transfer of newly synthesized nuclear RNA, presumably containing rRNA but also other RNA species.[105]

These findings indicated that brain transcription events are modulated by SS in a cell-specific fashion. Apart from the involvement of rRNA, the suggestion that newly synthesized hnRNA may accumulate in neuronal nuclei during SS appears most intriguing. It is known that hnRNA contains primary transcripts involved in the processing and control of mRNA synthesis. More interestingly, it is known that the sequence complexities of hnRNA and mRNA are severalfold higher in brain than in other organs.[106] The brain specificity of a large share of these transcripts suggests their involvement in specific brain functions, most likely those related to differentiation events and to the processing and maintenance of neural information. In view of the large body of literature relating sleep, and particularly PS, to memory processing,[25,26] it may not be surprising to find that sleep exerts an influence on the turnover of cerebral hnRNA. It is intriguing in this regard that inhibitors of RNA synthesis, such as actinomycin D and actinomycin S_3, injected intraventricularly in rats at the beginning of the light period have recently been found to increase the amount of SS and PS in the following dark period.[106a]

Very little work has concerned the behavior of cerebral DNA in relation to sleep. Apart from the report of a substantial loss of brain DNA following PSD,[85] which was, however, not confirmed in a later investigation,[100] the field has remained silent in this respect. Yet sleep, particularly PS, has been implicated in brain growth and development[4,9] and thus might be expected to exert an influence on the rate of proliferation of brain cells and on the associated process of DNA replication. Only a bare beginning has been made in this direction. Male rat pups kept in a condition of PSD by daily injections of chlorimipramine during the second and third postnatal weeks were found to contain less DNA in the cerebral cortex and more DNA in the hypothalamus at the age of 15 months. An increased DNA content was found in the hippocampus of female rats treated in the same way. Brain weight changes were also observed. Additional effects regarded the behavior of these animals.[107]

A similar experimental design has been used in a preliminary report describing a conspicuous increase in the incorporation of [^{3}H]thymidine in the DNA of fetal brain after the daily injection of chlorimipramine for 5 days in the gestating mothers.[108] Although a pharmacological treatment represents the only available method of depriving animals of PS at a perinatal stage, only an examination of the effects of additional drugs inducing PSD will establish the specificity of these findings.

Additional reasons to suggest the need for investigating the relationship between sleep and brain DNA in more detail relate to the presence of a DNA turnover in adult brain[109] and the involvement of cerebral DNA synthesis in learning processes.[110] It should also be mentioned that extensive work carried out on procaryotic and eucaryotic systems has drastically changed our previous concept of DNA as a metabolically inert molecule. A much more flexible attitude prevails now to account for features such as amplification, magnification, transcription, and genomic rearrangement. There is no reason to believe that brain cells may be prevented from adopting similar or analogous features if they were to be of some use in brain activity.

It is against this general background that one should examine the data of a preliminary report[111] describing the presence of a highly significant inverse correlation between concentration of rat brain radioactive DNA and a threshold amount of PS. In this experiment, rats were injected with [^{3}H]thymidine before being exposed to massive shuttle-box training and eventually allowed a period of rest. The basic idea was to assume that brain macromolecules synthesized during a waking experience might somehow be modified in the following sleep period. In particular, since neural information gathered during a training period may require an extra amount of PS for further processing,[25,26] it was expected that PS might exert a more incisive influence on the molecular counterparts of that experience. If additional work provides further support to this idea, neurochemical investigations aimed at elucidating the role of sleep might adopt an additional experimental design besides the deprivation approach and the studies made during sleep. This third modality might be appropriately identified as a "sequential" approach in view of the logic that has suggested it. The rationale for such an approach has previously been reported.[1]

6. SLEEP-INDUCING FACTORS

Despite its obvious relevance in a chapter dealing with the neurochemistry of sleep states, the literature that has accumulated on the involvement of neurotransmitters and other putative regulatory compounds in the transitions and maintenance of the vigilance states are not reviewed here. This choice is justified by the following reasons: (1) the very large size and intricacy of the available information, which would require a chapter of its own, (2) the availability of excellent reviews on this topic,[112,113] and, lastly but most cogently, (3) the belief that at the present time this type of neurochemistry, although appropriate in revealing and extending neurophysiological mechanisms, has only a superficial relationship to what the neurochemistry of sleep should be about, i.e., investigating the functional role of sleep. In brief, an adequate coverage of the neurotransmitter field would have substantially altered the emphasis of this chapter.

An analogous line of reasoning does not apply to the case of sleep-inducing factors, which appear to act at a higher hierarchical level and may lead us closer to the basic aspects of sleep. For these reasons, brief coverage of this topic will be given in the following pages.

Sleep-inducing factors have been obtained from two different types of preparation, i.e., from animals prevented from entering sleep and from animals spontaneously immersed in sleep. In the former approach, the underlying assumption relies on a time-dependent accumulation of the factor.

The active compounds have been assayed following their intraventricular or systemic administration into recipient animals by analysis of a number of relevant features associated with sleep such as posture, locomotor activity, and, most importantly, EEG activity.

The initial search for a sleep factor accumulating during wakefulness was started by Legendre and Pieron.[114] These authors obtained active extracts from blood or serum and, in later attempts, from the cerebrospinal fluid (CSF) of dogs kept awake for several days. Extracts were injected in large volumes into the circulation and, subsequently, into the fourth ventricle of recipient dogs. The experiments were carried out under inadequate conditions, but the results supported the notion that a sleep-inducing substance was accumulating in the brain of awake animals. The active compound appeared to be heat labile, nondialyzable, and insoluble in ethanol. Later work,[115] still concerned with dogs, was carried out with more refined techniques and gave partially positive results whose interpretation was complicated by the concomitant presence of undesired side effects. Still further technical refinements were introduced by Pappenheimer et al.[116] Their work was carried out with goats kept awake by an automatic device while the animals' CSF was slowly collected. The initial assay consisted in the intraventricular infusion of the extract into rats, whose nocturnal locomotor activity was significantly reduced. In additional experiments, the active extract was shown to induce other essential features of sleep, including an EEG pattern typical of SS.

A relevant observation was that the amplitude of the slow waves was higher than normal and similar to that observed after a period of SD.[117] The purification and chemical characterization of the active compound was finally accomplished using human urine as starting material. Factor S, as it came to be known, turned out to be a small glycopeptide containing glutamic acid, alanine, diaminopimelic acid, and muramic acid in molar ratios of 2:2:1:1. As the latter compound is a normal constituent of bacterial cell walls, the possibility of an adventitious contamination has been considered but is considered unlikely on a number of counts.[118] It is still possible that factor S may originate from the chemical modification of a hypothetical peptidoglycan absorbed in the intestine. Several muramylpeptides are known to antagonize sleep, and a muramyldipeptide induces sleep at a considerably higher dose than factor S. It is of interest that muramyldipeptides are known for their immunostimulant activity. Factor S is active at very low doses (pmol·kg^{-1}) when injected intraventricularly. It may act at the level of the brainstem.

A sleep-inducing factor has been extracted from the brainstem of sleep-deprived rats.[119] The latter compound appears to be similar if not identical to factor S.

Although these factors are concerned essentially with the induction of SS, some preliminary observations have been made with regard to the accumulation of a PS-inducing factor in the CSF of cats.[120] The ingenious assay method developed in this work was essentially based on the utilization of an insomniac cat. The animal was deprived of SS and PS by treating it 2 days earlier with p-chlorophenylalanine, an inhibitor of serotonin synthesis acting at the level of tryptophan hydroxylase. With this assay method, it was shown that CSF from PS-deprived cats contains a PS-inducing factor. This factor appears to be present in lower concentration in the CSF of normal cats. It is believed that serotonin may be mediating the release of such factor.

This work should be related to an independent series of experiments that indicated that the perfusate from the cat mesencephalic reticular formation obtained during periods of PS contains a PS-inducing peptide.[121] The most convincing demonstration was provided by the observation that antibodies raised against some of the perfusate proteins selectively suppressed PS when infused into the mesencephalic area of a recipient cat. Little or no effect was found on SS. Proteins with similar activity were present in lower concentration in the perfusate obtained from waking cats. Previous work from the same group had indicated that the amount of protein present in the perfusate varied in a cyclic manner concomitantly with the sleep–wake cycle and reached its highest level during periods rich in PS episodes. Further evidence relating the elevation in protein concentration of the perfusate with PS was obtained by selectively suppressing PS with chloramphenicol or with lesions of the preoptic area. It is unknown whether the PS-inducing factors identified by the two groups of investigators are identical.

The suppressive effect of chloramphenicol on PS just mentioned brings in an additional approach, which has been followed by several people in the effort to relate brain protein synthesis with sleep. In this type of work, inhibitors of protein synthesis were given to cats, rats, and mice with the aim of analyzing their effects on the following periods of sleep. On the whole, the data indicate that PS but not SS is selectively suppressed by these inhibitors. In turn, this observation has suggested that the onset of PS may require synthesis of a brain peptide. Cycloheximide, anisomycin, puromycin, chloramphenicol, and, to some extent, erythromycin have all been shown to decrease PS time.[122–124] In cats a similar effect also occurs with regard to the PS rebound that follows PSD.

The suppressive effect has been traced to an increased delay in PS onset and to a reduced frequency of PS episodes. Phasic aspects of PS appear to be most affected.[125] Some uncertainties remain, however. For instance, thiamphenicol, a close structural analogue of chloramphenicol, is unable to suppress PS in cats despite its known inhibitory effect on mitochondrial protein synthesis.[124] In the same animal species, an intraventricular injection of cycloheximide appears to have no immediate effect on PS and may actually exert a delayed and prolonged enhancement of PS time. A suppressive effect is produced, however, by an intraperitoneal injection of cycloheximide.[122] It is also unclear how similar effects on PS may be exerted by inhibitors known to act on cytoribosomes and mitoribosomes, respectively. On the other hand, in view of the lack of a suppressive effect exterted on PS by inhibitors of RNA synthesis, it appears likely that the peptide presumed to trigger PS may be synthesized on long lived mRNA.[106a] In general accord with the idea that peptides may exert a gating role on PS are the observations that GH is released during stages 3 and 4 of SS in man,[12] i.e., before the first PS episode, and that GH administration exerts an enhancing influence on PS time in cats[126] and in rats.[31]

Other work on sleep-inducing factors has concerned the identification of a compound released into the circulation during the hypnogenic stimulation of thalamic centers in the rabbit.[127] Such a factor, identified as δ-sleep-inducing peptide (DSIP), has also been obtained by chemical synthesis. It is a nonapeptide that induces SS at doses in the range of tens to hundreds of nanomoles per kilogram following its intraventricular or intravenous administration, respectively. Its effects are relatively short lasting in comparison to those exerted by factor S. Thus, DSIP has been suggested to exert its hypnogenic action by acting as an agonist at the level of the opiate receptors.[128]

Several other naturally occurring compounds have likewise been implicated in the humoral control of sleep. Among these, it is worth mentioning melatonin for reducing sleep latency and increasing sleep time,[129] vasointestinal peptide for increasing PS time,[130] and arginine vasotocin for its extremely potent effect (10^{-17} mol) on EEG patterns.[131]

The roles of several of these compounds and their interplay with the sleep factors described above are still to be elucidated. It may not be unreasonable to expect that sleep is regulated by a multitude of humoral factors and neuronal systems corresponding to the different, albeit interrelated, functions presumably subserved by sleep.

7. CONCLUSION

It is always tempting for an author to remind the reader of his previous insight on one or another aspect of a research topic; and we will not resist the temptation at this point. The following are excerpts taken from the concluding section of our previous review article on the biochemistry of sleep.[1]

> ... the glycolytic flux in brain may change during sleep depending on the previous waking experience of the organism. ... What is still obscure is the nature of the energy requiring reactions which appear to be activated when sleep follows a novel experience and, in addition, the types of cellular elements which are involved in the process. ...

The experimental work that has been added since then to the literature on the neurochemistry of sleep states appears to confirm the validity of our conclusions. A substantial but still incomplete body of evidence indicates that during SS the mammalian brain may require different levels of energy expenditure according to the nature of the previous waking periods. Novel and stressful experiences may induce a shift towards a faster metabolic rate, whereas, on the other hand, a period of relaxed wakefulness may favor the transition into an energy-saving attitude. The hypothesis suggests that some if not all of the discrepancies noted during SS in the behavior of several linked indices of metabolic and physiological activity may depend on the variable nature of the previous waking experience. These discrepancies might be resolved (or confirmed) by the simultaneous monitoring of two or more of these indices under the same or comparable conditions. If the hypothesis proves correct, it will become clear that during SS the brain switches into an energy-saving mode only if no preexisting debt has been contrived during wakefulness. Payment of the debt may be equated with tissue restoration. The neural nature of the debt is still unknown.

Much weaker evidence has been gathered with regard to the supposed anabolic action of SS, at least as far as brain protein synthesis is concerned. One of the few direct determinations made on this point rather indicates a substantial inhibition.[99] It is possible, however, that the nature of the previous waking experience may likewise exert a comparable modulatory influence on the synthesis, processing, and degradation of brain macromolecules during SS. No experimental verification of this possibility has yet been attempted, but the available evidence suggests that SS may modify cerebral RNA synthesis and processing. On more general terms, an involvement of brain macromolecules in sleep events is suggested by their role in learning and memory processes and by the contribution given by sleep, especially PS, to the further elaboration of the memory trace. It is actually possible that the involvement of brain macromolecules in sleep events may be related to the nature of the debt incurred during some types of wakeful periods.

A more homogeneous picture emerges with regard to PS. Determinations of a host of metabolic and physiological parameters agree in emphasizing the concept of PS as an activated state of the brain. Anabolic rates, e.g., protein synthesis, are also enhanced during PS. The latter observation indicates that the high energy fluxes connected to an elevated rate of neuronal firing may be simultaneously directed at supporting an increased rate of macromolecular synthesis. This conclusion weakens the line of reasoning according to which anabolic reactions are likely to prevail in the sleeping brain in view of the apparently discordant behavior of energy-requiring processes, which tend to decrease, and various metabolic indices such as CBF or oxygen utilization, which tend to remain relatively unchanged. According to our hypothesis, such discrepancies might rather find an adequate explanation in the nature of the previous waking experience.

The physiological debt created by PSD and repaid in the following PS rebound has already started to show some of its biochemical concomitants. This is true, for instance, of the protein and RNA content of some neurons and glia cells, whose loss during PSD is replenished in the period of free sleep. On the other hand, the behavior of other biochemical parameters such as the rate of protein synthesis or the content of acid mucopolysaccharides and glycogen may run in a different direction. Indeed, during the initial period of sleep that follows PSD, the latter indices change in the same direction as during PSD. In other words, the contrived debt, far from being repaid, is actually becoming worse. Together with the observation that the change is less intense during PSD than in the stress control group, these findings appear to support the concept of a permissive role of PS in the manifestation of some biochemical concomitants of a novel and stressful situation. The nature and mechanisms of such a role are fully unknown. Of course, these are only tentative suggestions, since in the period of sleep that follows a condition of PSD, the episodes of PS, although more frequent, are embedded in SS.

In conclusion, can we conceive of PS as a period of consistently elevated activity of the brain and of SS as a state ranging from low to high levels of activity according to the nature of the previous wakefulness? Most of what we presently know supports such a hypothesis, but of course we do not know much. And we certainly lack direct measures of biochemical and physiological parameters of PS following different types of wakefulness. To pose another question: does a PS episode remain the same irrespective of the level of activity the brain has manifested during the previous period of SS? At first sight, this would seem an unlikely possibility, which only those who consider PS and SS as completely independent states would be willing to support. Only further work in the field will provide us with an answer.

In any event, it is intriguing that the flexible behavior of SS with regard to energy fluxes and activity appears to leave a way to a more sustained level of operations, fully comparable to that of PS, the earlier we go back in ontogenetic development. In the near-term sheep fetus, CBF actually increases appreciably in the transition from an EEG pattern resembling PS to an EEG pattern resembling SS.[132] Is this merely a special feature of the sheep, or is it a more general property of early stages of brain (and sleep) development? In the latter case, it might be extremely revealing to clarify the nature of the factors that keep SS so extremely active at a perinatal age. By analogy, it may be presumed that at later stages of development they are substituted by some type of waking experience such as those associated with a novel and stressful environment.

Very little work has been done with regard to the involvement of brain macromolecules in the sequential relationship that is known to exist between learning and PS. Of special relevance in this regard might be the suggestion that the brain operates on the information gathered and already preliminarily processed during wakefulness by submitting it to different sets of sequential operations, which take place during SS and PS, respectively.[1] This "sequential" hypothesis appears to have received some preliminary support from a recent experiment dealing with cerebral DNA.[111]

Incidentally, the "sequential" hypothesis may find adequate support in the analogy between the workings of the brain and of the intestine. As the latter organ sequentially digests the food taken through the mouth, the sleeping brain may operate sequentially on the information gathered during wakefulness. Moreover, as secretory or motor activities in the intestine are scanty or absent if no food is available, likewise, the operations of the sleeping brain will not be detected and understood until proper food is provided, i.e., unless relevant information is acquired by the brain. The "sequential" approach may therefore prove to be a necessary avenue to follow in the attempts to elucidate the role of sleep at a molecular level.

If we may go back to our excerpts,[1]

> The processing of information carried out by mammalian brain continues throughout the sleep period with different operations from those prevailing during wakefulness. . . . We may hope to define the broad outlines of information processing by analogy with the operations of food uptake and assimilation.

The reader will admit that our insight was all inclusive.

We may foresee, however, a possible difficulty with our "sequential" hypothesis. As does the whole body of biology (but not of quantum mechanics), the hypothesis implies that the arrow of time runs in one direction only. In this respect, if some apparently well-controlled experiments performed by Dunne[133] are confirmed, we will be forced to conclude that the information found in our dreams, i.e., during PS, may come from future events as well as from the past. What type of "sequential" operations will account for such processing?

REFERENCES

1. Giuditta, A., 1977, *Biochemical Correlates of Brain Structure and Function* (A. N. Davison, ed.), Academic Press, New York, pp. 293–337.
2. Karnovsky, M. L., and Reich, P., 1977, *Advances in Neurochemistry*, Volume 2 (B. W. Agranoff and M. H. Aprison, eds.), Plenum Press, New York, pp. 213–275.
3. Karmanova, I. G., 1982, *Evolution of Sleep: Stages of the Formation of the Wakefulness-Sleep Cycle in Vertebrates*, S. Karger, Basel.
4. Roffwarg, H. P., Muzio, J. N., and Dement, W. C., 1966, *Science* **152**:604–619.
5. Navelet, Y., Benoit, O., and Bouard, G., 1982, *Electroencephalogr. Clin. Neurophysiol.* **54**:71–78.
6. Meddis, R., 1979, *Brain, Behaviour and Evolution* (D. A. Oakley and H. C. Plotkin, eds.), Methuen, London, pp. 99–125.
7. Groos, G., 1981, *Sleep 1980* (W. P. Koella, ed.), S. Karger, Basel, pp. 42–51.
8. Winfree, A. T., 1982, *Nature* **297**:23–27.
9. Corner, M. A., Mirmiran, M., and Bour, H., 1981, *Sleep 1980* (W. P. Koella, ed.), S. Karger, Basel, pp. 236–245.
10. Brebbia, D. R., and Altshuler, K. Z., 1965, *Science* **150**:1621–1623.
11. Parmeggiani, P. L., 1980, *Physiology in Sleep* (J. Orem and C. D. Barnes, eds.), Academic Press, New York, pp. 97–143.
12. Orem, J., and Barnes, C. D. (eds.), 1980, *Physiology in Sleep*, Academic Press, New York.
13. Mendelson, W. B., Jacobs, L. S., Gillin, J. C., and Wyatt, R. J., 1979, *Psychoneuroendocrinology* **4**:341–349.
14. Adam, K., and Oswald I., 1977, *J. R. Coll. Physicians* **11**:376–388.
15. Oliverio, A., 1980, *Waking Sleeping* **4**:155–166.
16. Allison, T., and Cicchetti, D. V., 1976, *Science* **194**:732–734.
17. Walker, J. M., and Berger, R. J., 1980, *Prog. Brain Res.* **53**:255–278.
18. Shapiro, C. M., Bortz, R., Mitchell, D., Bartel, P., and Jooste, P., *Science* **214**:1253–1254.
19. MacFadyen, U. M., Oswald, I., and Lewis, S. A., 1973, *J. Appl. Physiol.* **35**:391–394.
20. Horne, J. A., and Walmsley, B., 1976, *Phycophysiologia* **13**:115–120.
21. Parmeggiani, P. L., Cianci, T., Calasso, M., Zamboni, G., and Perez, E., 1980, *Electroencephalogr. Clin. Neurophysiol.* **50**:293–302.
22. Feinberg, I., 1968, *Science* **159**:1256.
23. Greenberg, R., and Dewan, E., 1969, *Nature* **223**:183–184.
24. Vogel, G. W., 1975, *Arch. Gen. Psychol.* **32**:749–761.
25. McGrath, M. J., and Cohen, D. B., 1978, *Psychol. Bull.* **85**:24–57.
26. Pearlman, C. A., 1979, *Neurosci. Biobehav. Rev.* **3**:57–68.
27. Kitahama, K., Valatx, J.-L., and Jouvet, M., 1981, *Physiol. Behav.* **27**:41–50.
28. Oswald, I., 1962, *Sleeping and Waking*, Elsevier, Amsterdam.
29. Kleitman, N., 1963, *Sleep and Wakefulness*, The University of Chicago Press, Chicago, London.
30. Hartmann, E. L., 1973, *The Functions of Sleep*, Yale University Press, New Haven.
31. Drucker-Colin, R. R., and McGaugh, J. L., 1977, *Neurobiology of Sleep and Memory*, Academic Press, New York.
32. Webb, W. B., and Cartwright, R. D., 1978, *Annu. Rev. Psychol.* **29**:223–252.
33. Weitzman, E. D., 1981, *Annu. Rev. Neurosci.* **4**:381–417.
34. Monnier, M., Bremer, F., Gaillard, J. M., Hediger, H., Horne, J. A., Parmeggiani, P. L., Passouant, P., and Rossi, G. F., 1980, *Experientia* **36**:1–142.
35. Moruzzi, G., 1969, *Arch. Ital. Biol.* **107**:175–216.
36. Mancia, M., 1975, *Sleep 1974* (P. Levin and W. P. Koella, eds.), S. Karger, Basel, pp. 5–13.
37. Moruzzi, G., 1972, *Ergeb. Physiol.* **64**:1–165.
38. Pieron, H., 1913, *Le Problème Physiologique du Sommeil*, Masson, Paris.
39. Steriade, M., and Hobson, J. A., 1976, *Prog. Neurobiol.* **6**:155–376.
40. Horne, J. A., 1977, *Physiol. Phychol.* **5**:403–408.

41. Dewan, E. M., 1970, *Sleep and Dreaming* (E. Hartmann, ed.), Little, Brown, Boston, pp. 295-307.
42. Jouvet, M., 1980, *Prog. Brain Res.* **53**:331-346.
43. Stern, W. C., and Morgane, P. J., 1974, *Behav. Biol.* **11**:1-32.
44. Berger, R., 1969, *Psychol. Rev.* **76**:144-164.
45. Snyder, F., 1966, *Am. J. Psychiatry* **123**:121-136.
46. Hicks, R. A., Okuda, A., and Thomsen, D., 1977, *Am. J. Psychol.* **90**:95-102.
47. Van Hulzen, Z. J. M., and Coenen, A. M. L., 1980, *Physiol. Behav.* **25**:807-811.
48. Mendelson, W., Guthrie, R. D., Guynn, R. D., Harris, R. L., and Wyatt, R. J., 1974, *J. Neurochem.* **22**:1157-1159.
49. Karadžić, V., and Mršulja, B., 1969, *J. Neurochem.* **16**:29-34.
50. Mršulja, E. B., and Rakič, L. M., 1970, *J. Neurochem.* **17**:455-456.
51. Mršulja, B. B., and Rakić, L. M., 1974, *J. Exp. Marine Biol. Ecol.* **15**:43-48.
52. Heiner, L., Godin, Y., Mark, J., and Mandel, P., 1968, *J. Neurochem.* **15**:150-151.
53. Reich, P., Geyer, S. J., and Karnovsky, M. L., 1972, *J. Neurochem.* **19**:487-497.
54. Richter, D., and Dawson, R. M. C., 1948, *Am. J. Physiol.* **154**:73-79.
55. Cocks, J. A., 1967, *Nature* **215**:1399-1400.
56. Van den Noort, S., and Brine, K., 1970, *Am. J. Physiol.* **218**:1434-1439.
57. Shimizu, H., Tabushi, K., Hishikawa, Y., Kakimoto, Y., and Kaneko, Z., 1966, *Nature* **212**:936-937.
58. Reich, P., Geyer, S. J., Steinbaum, L., Anchors, J. M., and Karnovsky, M. L., 1973, *J. Neurochem.* **20**:1195-1205.
59. Reich, P., Driver, J. K., and Karnovski, M. L., 1967, *Science* **157**:336-338.
60. Anchors, J. M., and Karnovsky, M. L., 1975, *J. Biol. Chem.* **250**:6408-6416.
61. Nordlie, R. C., 1974, *Current Topics in Cellular Regulation*, Volume 8 (B. L. Horecker and E. R. Stadtman, eds.), Academic Press, New York, pp. 33-117.
62. Karnovsky, M. L., Burrows, B. L., and Zoccoli, M. A., 1980, *Cerebral Metabolism and Neural Function* (J. V. Passonneau, R. A. Hawkins, W. D. Lust, and F. A. Welsh, eds.), Williams & Wilkins, Baltimore, pp. 359-366.
62a. Karnovsky, M. L., Reich, P., Anchors, J. M. and Burrows, B. L., 1983, *J. Neurochem.* **41**:1498-1501.
63. Anchors, J. M., Haggerty, D. F., and Karnovsky, M. L., 1977, *J. Biol. Chem.* **252**:7035-7041.
64. Nilsson, O. S., Arion, W. J., De Pierre, J. W., Dallner, G., and Ernster, L., 1978, *Eur. J. Biochem.* **82**:627-634.
65. Kennedy, C., Gillin, J. C., Mendelson, W., Suda, S., Miyaoka, M., Ito, M., Nakamura, R. K., Storch, F. I., Pettigrew, K., Mishkins, M., and Sokoloff, L., 1982, *Nature* **297**:325-327.
66. Sokoloff, L., Reivich, M., Kennedy, C., Des Rosiers, M. H., Patlak, C. S., Pettigrew, K. D., Sakuroda, O., and Shinohara, M., 1977, *J. Neurochem.* **28**:897-916.
67. Petitjean, F., Seguin, S., Des Rosiers, M. H., Salvert, D., Buda, C., Janin, M., Debilly, G., Jouvet, M., and Bobillier, P., 1981, *C.R. Acad. Sci. [Series III] (Paris)* **292**:211-214.
68. Bobillier, P., Seguin, S., Petitjean, F., Buda, C., Salvert, D., Janin, M., Chouvet, G., Souchier, C., and Jouvet, M., 1980, *C.R. Acad. Sci. [D] (Paris)* **291**:91-96.
68a. Ramm, P. and Frost, B. J., 1983, *Sleep*, **6**:196-216.
69. Livingstone, M. S., and Hubel, D. H., 1981, *Nature* **291**:554-561.
70. Mangold, R., Sokoloff, L., Conner, E., Kleinerman, J., Therman, P.-O. G., and Kety, S. S., 1955, *J. Clin. Invest.* **34**:1092-1100.
71. Magnes, J., Allweis, C., and Abeles, M., 1967, *J. Neurochem.* **14**:859-871.
72. Hydén, H., and Lange, P. W., 1965, *Science* **149**:654-656.
73. Hamberger, A., Hydén, H., and Lange, P. W., 1966, *Science* **151**:1394-1395.
74. Ogasahara, S., Taguchi, Y., and Wada, H., 1981, *Brain Res.* **213**:163-171.
75. Seylaz, J., Pinard, E., Mamo, H., Goas, J. Y., Luft, A., and Correze, J. L., 1975, *Brain Work: The Coupling of Function, Metabolism and Blood Flow in the Brain* (D. H. Ingvar and N. A. Lassen, eds.), Munksgaard, Copenaghen, pp. 235-252.
76. Townsend, R. E., Prinz, P. N., and Obrist, W. D., 1973, *J. Appl. Physiol.* **35**:620-625.
77. Sakai, F., Meyer, J. S., Karakan, I., Yamaguchi, F., and Yamamoto, M., 1979, *Neurology (N.Y.)* **29**:61-67.

78. Reivich, M., Isaacs, G., Evarts, E., and Kety, S. S., 1968, *J. Neurochem.* **15**:301–306.
79. Danguir, J., and Nicolaidis, S., 1980, *Am. J. Physiol.* **238**:E307–E312.
80. Karadžić, V., Mičić, D., and Rakić, L., 1971, *Experientia* **27**:509–511.
81. Mičić, D., Karadžić, V., and Rakić, L. M., 1967, *Nature* **215**:169–170.
82. Davis, J. M., Himwich, W. A., and Stout, M., 1969, *Biol. Psychiatry* **1**:387–390.
83. Mark, J., Godin, Y., and Mandel, P., 1969, *J. Neurochem.* **16**:1263–1272.
84. Sinha, A. K., Ciaranello, R. D., Dement, W. C., and Barchas, J. D., 1973, *J. Neurochem.* **20**:1289–1290.
85. Haulica, I., Ababei, L., Teodorescu, C., Rosca, V., Haulica, A., Moisiu, M., and Haller, C., 1970, *J. Neurochem.* **17**:823–826.
86. Levental, M., Rakić, L., and Rusić, N., 1972, *Arch. Int. Physiol. Biochem.* **80**:861–870.
87. Dostálová, K., Navrátil, J., and Skeelnovský, A., 1978, *Activ. Nerv. Super. (Praha)* **20**:57–58.
88. Demin, N. N., and Rubinskaja, N. L., 1974, *Dokl. Akad. Nauk. SSSR* **214**:940–942.
89. Bobillier, P., Sakai, F., Seguin, S., and Jouvet, M., 1974, *J. Neurochem.* **22**:23–31.
90. Bobillier, P., Sakai, F., Seguin, S., and Jouvet, M., 1971, *Life Sci.* **10**:1349–1357.
91. Shapiro, C., and Girdwood, P., 1981, *Neuropharmacologia* **20**:457–460.
92. Levental, M., Šušić, V., Rusić, M., and Rakić, L., 1975, *Arch. Intern. Physiol. Biochim.* **83**:221–232.
93. Godin, Y., and Mandel, P., 1965, *J. Neurochem.* **12**:455–460.
94. Demin, N. N., 1974, *Advances in Neurochemistry* (E. M. Kreps, et al., eds.), Nauka, Leningrad, pp. 29–39.
95. Shapot, V. S., 1957, *Metabolism of the Nervous System* (D. Richter, ed.), Pergamon Press, New York, pp. 257–262.
96. Jasper, H. H., Khan, R. T., and Elliot, K. A. C., 1965, *Science* **147**:1448–1449.
97. Steriade, M., Costantinescu, E., and Apostol, V., 1969, *Brain Res.* **13**:177–180.
98. Bobillier, P., Froment, J.-L., Seguin, S., and Jouvet, M., 1973, *Biochem. Pharmacol.* **22**:3077–3090.
99. Brodskii, V. Y., Gusatinskii, V. N., Kogan, A. B., and Nechaeva, N. V., 1974, *Dokl. Akad. Nauk SSSR* **215**:748–750.
100. Balestrieri, S., D'Onofrio, G., and Giuditta, A., 1980, *Neurochem. Res.* **5**:1251–1264.
101. Bubenik, G., and Monnier, M., 1972, *Exp. Neurol.* **35**:1–12.
102. Vitale-Neugebauer, A., Giuditta, A., Vitale, B., and Giaquinto, S., 1970, *J. Neurochem.* **17**:1263–1273.
103. Giuditta, A., Rutigliano, B., and Vitale-Neugebauer, A., 1980, *J. Neurochem.* **35**:1259–1266.
104. Feldman, G. L., Fedorenko, G. M., and Gusatinskii, V. N., 1979, *Tsitologiia* **21**:429–433.
105. Giuditta, A., Rutigliano, B., and Vitale-Neugebauer, A., 1980, *J. Neurochem.* **35**:1267–1272.
106. Kaplan, B. B., 1983, *Handbook of Neurochemistry*, 2nd ed., Volume 2 (A. Lajtha, ed.), Plenum Press, New York, pp. 1–26.
106a. Uezu, E. and Matsumoto, J., 1983, *Jpn. J. Physiol.* **33**:129–133.
107. Corner, M. A., Mirmiran, M., Bour, H. L. M. G., Boer, G. J., van de Poll, N. E., van Oyen, H. G., and Uylings, H. M. B., 1981, *Prog. Brain Res.* **53**:347–356.
108. Grassi-Zucconi, G., Belia, S., Franciolini, F., and Giuditta, A., 1982, Abstract 4th Meeting European Society for Neurochemistry, Abstract 343.
109. Perrone-Capano, C., D'Onofrio, G., and Giuditta, A., 1982, *J. Neurochem.* **38**:52–56.
110. Giuditta, A., 1983, *Handbook of Neurochemistry*, Volume 5 (A. Lajtha ed.), 2nd edition, Plenum Press, New York pp. 251–276.
111. Ambrosini, M. V., Scaroni, R., Chiurulla, C., and Giuditta, A., 1983, *Neurosci. Lett.* [*Suppl.*] **14**:58.
112. Jouvet, M., 1972, *Ergeb. Physiol.* **64**:166–307.
113. Gillin, J. C., Mendelson, W. B., Sitaram, N., and Wyatt, R. J., 1978, *Annu. Rev. Pharmacol. Toxicol.* **18**:563–579.
114. Legendre, R., and Pieron, H., 1913, *Z. Allg. Physiol.* **14**:235–262.
115. Schendorf, J. G., and Ivy, A. C., 1939, *Am. J. Physiol.* **125**:491–505.
116. Pappenheimer, J. R., Miller, T. B., and Goodrich, C. A., 1967, *Proc. Natl. Acad. Sci. U.S.A.* **58**:513–517.

117. Pappenheimer, J. R., 1979, *Johns Hopkins Med. J.* **145**:49–56.
118. Krueger, J. M., Pappenheimer, J. R., and Karnovsky, M. L., 1982, *J. Biol. Chem.* **257**:1664–1669.
119. Nagasaki, H., Iriki, M., Inoué, S., and Uchizono, K., 1974, *Proc. Jpn. Acad.* **50**:241–246.
120. Sallanon, M., Buda, C., Janin, M., and Jouvet, M., 1982, *Brain Res.* **251**:137–147.
121. Drucker-Colín, R. R., De Gómez-Puyon, M. T., Gutiérrez, M. C., and Dreyfus-Cortés, G., 1980, *Exp. Neurol.* **69**:563–575.
122. Stern, W. C., Morgane, P. J., Panksepp, J., Zolovick, A. J., and Jaliwiec, J. E., 1972, *Brain Res.* **47**:254–258.
123. Rojas-Ramírez, J. A., Aguilar-Jiménez, E., Posadas-Andrews, A., Bernal-Pedraza, J. G., and Drucker-Colín, R. R., 1977, *Psychopharmacologia* **53**:147–150.
124. Petitjean, F., Buda, C., Janin, M., David, M., and Jouvet, M., 1979, *Psychopharmacogia* **66**:147–153.
125. Drucker-Colín, R. R., Gutiérrez, M. C., and Bernal-Pedraza, J. G., 1980, *Front. Horm. Res.* **6**:138–155.
126. Stern, W. C., Jalowiec, J. E., Shabshelowitz, H., and Morgane, P. J., 1975, *Horm. Behav.* **6**:189–196.
127. Schoenenberger, G. A., Maier, P. F., Tobler, H. J., Wilson, K., and Monnier, M., 1978, *Pfluegers Arch.* **376**:119–129.
128. Tissot, R., 1981, *Neuropsychobiology* **7**:321–325.
129. Holmes, S. W., and Sugden, D., 1982, *Br. J. Pharmacol.* **76**:95–101.
130. Rion, F., Cespuglio, R., and Jouvet, M., 1981, *C.R. Acad. Sci. [Series III] Paris* **293**:679–682.
131. Borbély, A. A., and Tobler, I., 1980, *Trends Pharmacol. Sci.* **1**:356–358.
132. Mann, L. I., Duchin, S., and Weiss, R. R., 1974, *Am. J. Obstet. Gynecol.* **119**:533–538.
133. Dunne, J. W., 1981, *An Experiment with Time*, Papermac (Macmillan), New York.

20

Composition of Intraocular Fluids and the Microenvironment of the Retina

Laszlo Z. Bito

1. INTRODUCTION

On the basis of its embryonic derivation, morphological organization, and function, the retina must be regarded as part of the central nervous system (CNS). A unique aspect of the CNS is its ability to maintain local extracellular chemical environments that are vastly different from those of other organ systems. Our understanding of the chemical microenvironment of the brain and the mechanisms that maintain its homeostasis is largely based on studies on the composition of cerebrospinal fluid (CSF). Similar information about the retina can be gained from studies on the chemical composition of the intraocular fluids (IOFs), particularly the vitreus* adjacent to the retina, since there are no appreciable diffusion barriers between the vitreus and the extracellular fluid (ECF) of the retina.[1,2]

The normal functioning of neuronal tissues, including the retina, depends greatly on the chemical composition of their extracellular milieu. Clearly, the electrical activity of these tissues is determined equally by intracellular and extracellular concentrations of ions. Many normal constituents of the ECFs, such as amino acids and autacoids, which either modulate the release and action of neurotransmitters or are putative transmitters themselves. Furthermore, virtually all nutrients and metabolic precursors required by the neuronal and glial elements of the retina are derived from its ECF compartments, and virtually all of its metabolic by-products must be eliminated through these compartments. Thus, a detailed knowledge of the retinal chemical microenvironment is required in order to understand its normal functions and its pathologies as

---

* Use of the nouns "vitreus" and "aqueus" as synonyms for "vitreous humor" and "aqueous humor," respectively, is discussed in Section 4.

Laszlo Z. Bito • College of Physicians and Surgeons, Columbia University, New York, New York 10032.

well as to formulate physiological solutions or culture media suitable for maintaining the retina or its components during *in vitro* studies.

Maintenance of the well-controlled chemical environments of the brain and retina depends on the permeability properties and on the facilitated and active transport functions of the blood–brain and blood–ocular barriers, respectively.[1] In spite of steady efforts made by a handful of investigators, the complexity of these transport processes has only just begun to be understood. Consequently, our knowledge of the composition of the retinal ECF and its homeostasis is still rudimentary. However, current information indicates that although the chemical composition of the retinal microenvironment resembles that of the rest of the CNS, it differs greatly from that of the anterior segment of the eye as well as from those of other tissues of the body.[2]

2. RELEVANT LITERATURE AND THE SCOPE OF THIS CHAPTER

Several textbooks, monographs, and review articles dealing with specific aspects of the IOFs and retinal neurochemistry have been published in recent years. The topics most extensively covered include aqueus* composition and dynamics,[3–7] morphological[8] and physiological[9] aspects of aqueus outflow mechanisms, and the control of intraocular pressure.[4,5,9] A recent chapter by Balazs and Denlinger[10] provides a detailed review of the structure, macromolecular constituents, and physical properties of the vitreus. Evidence of the transport functions of the retinal choroid[11] and the role of the "retinal pigment epithelium" in maintaining the microenvironment and function of the retina under normal and pathological conditions have also been reviewed.[12] Species variations in retinal vascularization have been discussed in relation to limitations imposed on vascular and avascular retinas by oxygen and glucose availability.[13] In addition, specific aspects of retinal function and metabolism, including nutritional requirements, toxicology, and neurotransmitter systems, have been addressed in the proceedings of a recent symposium sponsored by the National Research Council.[14]

The goal of this chapter is to provide information on the microenvironment of the vertebrate retina by summarizing data on the concentration of solutes in the intraocular fluids, with emphasis on the concentration of solutes in the vitreus, a compartment that can be regarded as one of the great cisterns of the CNS.[2] The reader is referred to the publications mentioned above for more detailed discussions of retinal neurochemistry and other relevant aspects of ocular physiology.

3. THE INTRAOCULAR FLUID COMPARTMENTS

The vitreus of most mammals is a gel as a result of the presence of a network of collagen fibers.[10] Typically, these fibers are more densely packed

---

* Use of the nouns "vitreus" and "aqueus" as synonyms for "vitreous humor" and "aqueous humor," respectively, is discussed in Section 4.

at the periphery of the vitreous body. In fact, in many species the center of the vitreous contains an essentially collagen-free liquid pocket, and in some of these species this pocket enlarges with age.[15] However, even the vitreus gel contains less than 0.5% collagen, which can easily be separated from the viscous liquid by filtration or ultracentrifugation. The viscosity of the collagen-free liquid vitreus arises primarily from the presence of hyaluronic acid, which varies in concentration in different species, but the sum of collagen and hyaluronic acid in the vitreus rarely exceeds 1%.[10,16] Although the collagen fiber network and the hyaluronic acid contained between these fibers may hinder the diffusion of high-molecular-weight proteins and other macromolecules, they have little or no effect on the diffusion of electrolytes or other low-molecular-weight solutes,[17] including nutrients and metabolic by-products. Thus, although the vitreus is a gel in most species, it behaves as a stationary fluid compartment in regard to solute exchange with surrounding tissues.

4. DEFINITIONS AND METHODOLOGY

Morphologists have used the term "vitreous body" for the connective tissue compartment bordered by the retina, the posterior chamber, and the lens, whereas physiologists have generally referred to the liquid component contained within its collagen network or to the liquid obtained after removal of the collagen network as the "vitreous humor." According to terminology recently adopted by Balazs and Denlinger,[10] "vitreus" will be used as a noun referring to either the gel or the liquid within this compartment or to the fluid derived from it. The general term "intraocular fluids" (IOFs) will refer to both the aqueus and vitreus, irrespective of whether the vitreus is in the liquid or gel state in a given species.

In most cases, when a single value is given for the concentration of a constituent in the vitreus, it refers to a sample obtained by the perscleral introduction of a hypodermic needle into the liquid pocket of the vitreus. It should be noted, however, that if a large-gauge needle is used and considerable suction is applied, some of the gel vitreus may be liquified, yielding a larger liquid sample than that normally present. Because the liquid pocket is in the central region of the vitreus, its chemical composition is influenced by the composition of both the anterior and the posterior vitreus segments and hence can be regarded as representative of the vitreus as a whole.

When separate values are given for the anterior and posterior segments of the vitreus, these samples, unless otherwise indicated by a footnote, were obtained by bisection of the globe at or slightly behind the equator, using a microtome knife over a divided collection platform, followed by removal of the collagen fibers from each sample by filtration or centrifugation.[18] In some studies, isolated globes were frozen, and the vitreus dissected into two or more regions. Such procedures have been shown, however, to produce severe "freezing-out" artifacts, leading to an increased concentration of solutes toward the center of the vitreus. A release of some solutes into the IOFs from the surrounding tissues may also occur as a result of freezing. For these reasons, values obtained from frozen eyes are presented only if more reliable values are not available and these values are indicated by a footnote.

Because "aqueous humor," meaning "watery fluid," is redundant, many authors have simply used the word "aqueous" to refer to the fluid in the anterior segment of the eye. However, that word is an adjective, and its use as a noun is improper. For this reason, and to be consistent with the usage of "vitreus," the noun "aqueus" is used to refer to this fluid.

The plane of the iris divides the aqueus compartment into the anterior and the posterior chambers. Much of the aqueus from both chambers can be removed easily by paracentesis through the cornea. If a hypodermic needle is inserted into the posterior chamber, about 15–25% of the total volume of the aqueus can be withdrawn. Because reflux of fluid from the anterior chamber is prevented by the valvelike action of the iris as it falls against the lens, it can be assumed that fluid obtained in this manner is from the posterior chamber.[18,19] The anterior chamber can then be emptied by paracentesis through the cornea, using a second hypodermic needle. The fluids thus obtained from the posterior and anterior chambers are referred to as posterior and anterior aqueus, respectively.

In some species, such procedures, especially removal of the whole aqueus volume, can result in the breakdown of the blood–aqueus barrier[20] and/or the reflux of plasma or blood through the outflow channels of the chamber angle.[21] Thus, once aqueus is removed, the anterior chamber is refilled with a fluid, the so-called "secondary aqueus," which is vastly different from normal aqueus in its chemical composition. Although knowledge of the chemical composition of the secondary aqueus and the mechanism involved in its formation is important to an understanding of the pathophysiology of ocular irritative and inflammatory reactions, values of the chemical composition of the secondary aqueus are not included herein.

Apparently based on its embryonic derivation, the epithelial layer of the choroid, which is one of the most important sites for the homeostasis of the retinal microenvironment, is generally but incorrectly referred to as the "retinal pigment epithelium."[12,22] However, this epithelium is not attached to the retina and, in fact, has its basement lamina on the choroid. Hence, it is structurally and functionally an integral part of the choroid.[2] Furthermore, the name "choroid" implies a structure that resembles the chorion, the outer nutritive coat of the embryo, which has an epithelial lining. Thus, only if this epithelium is regarded as a part of the choroid does this tissue resemble the chorion or its namesake, the choroid plexus.[2] For these reasons, throughout this text "choroid" or "retinal choroid" is used to refer to the outer nutrient coat of the retina, including its epithelium, whereas its epithelial lining will be referred to as the "choroidal epithelium."

The two layers of epithelial cells covering the ciliary processes can be more accurately conceptualized as two separate, apposed epithelia, that of the ciliarly stroma and that of the posterior chamber, rather than as a single two-layered epithelium.[2] This concept explains the apex-to-apex organization of these cells which, if thought of as comprising a single epithelium, must be regarded as a histological oddity. Thus, the term "ciliary epithelia," rather than "epithelium," is used to refer to the two cell layers covering the ciliary processes.

5. CONCENTRATION GRADIENTS WITHIN THE INTRAOCULAR FLUID COMPARTMENTS

There are regional differences in the chemical composition of CSF and, presumably, in the composition of the ECFs of the brain that are caused by metabolic activities and differences between the transport activities of the choroidal and extrachoroidal regions of the blood–brain barrier systems.[23-26] For the most part, these gradients are small, providing a relatively homogeneous chemical environment suitable for central neuroectodermal tissues. However, in the eye, the ectodermal and mesodermal tissues of the anterior segment coexist with the neuroectodermal tissue of the retina without significant diffusional barriers. Therefore, the chemical composition of the IOFs has much greater diversity. Large concentration gradients are maintained between the IOFs of the anterior and posterior segments of the eye and even between the anterior and posterior segments of the vitreous by the absorptive and secretory transport functions of the blood–aqueous and blood–retinal barriers, and by the metabolic activities of the tissues surrounding these fluid compartments.[2] Although early studies concluded that the aqueous resembles the CSF in its chemical composition, it is now clear that the composition of aqueous humor resembles that of plasma ultrafiltrate; it is the vitreous adjacent to the retina that closely resembles CSF in its chemical composition.[2,7]

Interpretation of the steady-state gradients of solutes within the IOF compartments and of their relationship to plasma dialysate is a primary key to understanding the chemical microenvironment of the retina. Clearly, the concentration of a substance in the retinal microenvironment can be expected to be the same as in IOFs if its concentration in all IOF compartments is the same as that in plasma dialysate. On the other hand, if the concentration of a substance increases or decreases toward the retina within the IOFs, its concentration within the retinal ECF must correspondingly be even higher or even lower than that in vitreous. Thus, an understanding of such concentration gradients enables us to estimate the concentration of solutes within the microenvironment of the retina.

For these reasons, the values in the following tables were taken, whenever possible, from studies in which the concentration of a solute in the anterior and posterior aqueous and vitreous, as well as in blood, were reported. Whenever values for only one IOF compartment and plasma were determined in a study, simultaneously determined vitreous and plasma values are reported preferentially over simultaneously determined values for aqueous and plasma. In a few cases, however, values for each fluid listed in these tables had to be drawn from different sources for a particular species. Clearly, differences in the physiological and dietary state of animals, and differences in sample handling and analytical techniques, make it difficult to assign significance to small concentration gradients.

6. THE TRANSPORT FUNCTION OF THE BLOOD—OCULAR BARRIER SYSTEMS

Most mammalian retinas are well vascularized.[27] Substantial evidence indicates that the tight-junctional endothelium of the retinal vasculature, like that of the brain, has active and facilitated transport functions.[2,28] However, the retinas of some vertebrates have a vascular bed that is limited either to the vitreal surface and/or to small regions of the total retinal area, and most mammalian retinas have avascular regions. In such avascular retinas or retinal regions, the choroid must play a primary role in maintaining the chemical environment of the retinal ECF. However, structures that contribute to the chemical homeostasis of the fluid in the posterior chamber—i.e., the ciliary processes and, in birds, the pecten, an apparently secretory tissue that projects into the vitreus cavity from the optic nerve head—must be regarded as parts of the blood–retinal barrier system, since there are no diffusional barriers between the vitreus and the retinal ECF.[2]

Although elucidation of the transport functions of the various regions of these barrier systems is clearly a prerequisite for understanding the homeostasis of IOFs, the information presently available from *in vitro* studies is rather limited, frequently contradictory, and often beset with artifacts.[1,2,6,29] The large variations in the reported electrical resistance of *in vitro* preparations of retinal choroid or ciliary processes clearly reflect varying degrees of damage inflicted on these delicate tissues during their isolation or mounting in Ussing-type chambers. Furthermore, substantial evidence indicates that the properties of these epithelia, especially the ciliary epithelia, are affected by a variety of autacoids, including mediators known to be released in response to various stimuli, including manipulation of these tissues. Thus, it is most unlikely that such *in vitro* studies accurately reflect the normal *in vivo* direction and/or rate of net solute fluxes.

When small fluid volumes were introduced between the retina and the choroid to create an artificial detachment of the retina, the rate of resorbtion of this fluid was reduced by ouabain.[30] However, such detachment can be expected to affect the secretory functions of the choroidal epithelium, as well as the transmembrane transport functions of the glial and neuronal elements of the retina. Analysis of subretinal fluid obtained from artificial or spontaneous retinal detachments have also been performed but are most unlikely to reflect the normal chemical composition of retinal ECF.

In some studies, microelectrode techniques have been used to estimate the concentration of some electrolytes in the ECF of the retina.[11] However, such studies have required the prolonged, extensive manipulation of the animal and the eye, or have been performed on isolated retinas maintained without arterial perfusion in Ringer's solutions, which do not accurately reflect the normal chemical environment of the retina. Although such studies have proven to be most valuable in detecting rapid changes in electrolyte concentrations as a result of phototransduction or neuronal activity, their relevance to the *in vivo* steady-state concentrations of solutes in the microenvironment of the retina has not been established.

7. SPECIES DIFFERENCES AND THE RELIABILITY OF INFORMATION ON HUMAN INTRAOCULAR FLUIDS

Sampling of the CSF is a routine clinical procedure, and analysis of CSF has been used for decades as a primary tool for diagnosing neurological diseases. In contrast, paracentesis of the anterior chamber for analytical or diagnostic purposes is rarely performed on patients, and, in spite of recent improvements in vitrectomy techniques, methods for collecting diagnostic vitreus samples have not been developed. Since the transparency of the eye permits direct observation of the retina, and since intraocular pressure can be measured with noninvasive techniques, there has been little impetus to develop diagnostic techniques that would involve cannulation of the IOF compartments. Consequently, information on the chemical composition of the human vitreus is rudimentary. The few available values have generally been obtained either from diseased eyes that had been surgically removed or from normal eyes obtained several hours or even a day after death. Since the concentrations of many solutes in the IOFs change very rapidly after death,[31] such values cannot be assumed to represent normal streadystate concentrations.

In view of the limited body of reliable data available on the composition of human IOFs, an understanding of species variations in IOF composition is of paramount importance. Such understanding is also required for the interpretation of the limited information available on some other species. Clearly, such interpretations or extrapolations can only be based on an understanding of basic trends in species variations and their relationship to evolutionary, environmental, and behavioral factors.

8. THE CONCENTRATION OF MAJOR SOLUTES IN MAMMALIAN INTRAOCULAR FLUIDS AND IN THE RETINAL MICROENVIRONMENT

8.1. Sodium, Chloride, and Bicarbonate

In general, the concentration of sodium (Na^+) in the IOFs is slightly lower than or equal to that in blood plasma (Table I). To determine whether or not Na^+ and/or one of the major anions are transported actively from blood into the IOFs, their concentrations in the IOFs, which contain negligible levels of plasma proteins, have to be compared to their Donnan distribution values or to their concentrations in a plasma dialysate.

By dialyzing the aqueous humor of dogs and cats against their own plasma and determining the concentrations of Na^+ and Cl^- before and after establishment of equilibrium, Davson and Luck[42] showed that, relative to plasma, there is indeed an excess of Na^+ and Cl^- in the aqueous humor of these species. In the bovine and rhesus, the concentration of Na^+ in the IOFs was reported to be slightly greater than that in simultaneously obtained blood serum. These observations are consistent with the concept that the secretion of aqueous humor by the ciliary processes involves the active transport of Na^+ and/or

Table I
Sodium Concentration (mmol/kg H_2O ± 1 S.E.M.) in the Intraocular Fluids and Blood Plasma or Serum of Vertebrates[a]

Species	Aqueus Anterior	Aqueus Posterior	Vitreus	Plasma or [serum]
Rabbit	153 ± 4.4(32)	159 ± 1.2(32)		163 ± 1.2(32)
		142.5(3)	140.0(3)	
Woodchuck (active)		151 ± 3.4(33)	147 ± 1.4(33)	152 ± 1.9(33)
(hibernating)		149 ± 4.3(33)	148(33)	154 ± 2.7(33)
Sheep	145 ± 2.10(34)	151 ± 1.90(34)		144 ± 1.41(34)
Bovine		146 ± 1.3(35)	148 ± 2.2(35)	[143 ± 1.7] (35)
Horse		121(36)	119(36)	[146] (36)
Cat		159 ± 1.5(37)		[163 ± 1.6] (37)
Dog		146 ± 4.1(38)		[145 ± 3.1] (38)
Rhesus		155 ± 12.0(39)		[148 ± 3.6](39)
		153 ± 4.2(40)		[148 ± 1.5] (40)
Human			138 ± 12(41)[b,c]	[142 – 156] (41)[b]

[a] Numbers in parentheses indicate references in this and all succeeding tables.
[b] Samples obtained from infants and children only (≤13 yrs.). Vitreus and serum obtained from different patients.
[c] Vitreus collected >1 h after death caused by systemic diseases or accidents.

Cl^- into extracellular channels between adjacent nonpigmented epithelial cells of the ciliary processes, thereby creating high local osmolarity and, consequently, an osmotic water flow that results in the production of a slightly hypertonic fluid in the posterior chamber.[6]

The very low Na^+ concentration reported for the IOFs of the horse (Table I) is most unexpected since, in most ECFs, Na^+ accounts for almost one-half of the total osmolarity. Furthermore, the Cl^- concentration in the IOFs of the horse was reported to be 4 to 11% higher than that in serum (Table II). Thus, we cannot assume that in this species the IOFs are hypotonic to blood plasma. This would imply either that the Na^+ values given for the IOFs were erroneously low or that some other cations(s) must be present in the equine IOFs at unusually high concentrations.

In the aqueus of some species, such as the rabbit, there is a relative excess of bicarbonate and a relative deficiency of Cl^- as compared to those concentrations in blood plasma, whereas in other species this relationship is reversed[3] (Table II vs. Table III). Although some investigators have postulated that bicarbonate plays a primary role in the secretion of aqueus,[43] it has been shown that some species, including primates, do not have excess bicarbonate in their IOFs.[40,44] Thus, the relative roles of Na^+, Cl^-, and bicarbonate in aqueus production remain controversial. Although this controversy will ultimately have to be resolved by detailed studies on the transport functions of the isolated ciliary bodies of different species, the concentration of these solutes in the aqueus, especially the posterior aqueus, must be considered in the interpretation of *in vitro* findings, since the extensive manipulation of the tissue required for such studies may alter the transport functions of the ciliary epithelia.

Table II
Chloride Concentration (mmol/kg H_2O ± S.E.M.) in Intraocular Fluids and Blood Plasma or Serum of Vertebrates

Species	Aqueus Anterior	Aqueus Posterior	Vitreus	Plasma or serum
Rabbit	101.0(45)	96.5(45)	104.7(45)	108.8(45)
Wood chuck				
(active)		103 ± 3.4(33)	116.1 ± 1.4(33)	105 ± 1.2(33)
(hibernating)		112 ± 3.8(33)	109(33)	100 ± 3.2(33)
Sheep	113 ± 1.5(34)	117 ± 2.0(34)		104 ± 1.7(34)
Bovine		123 ± 1.4(35)	124 ± 1.7(35)	[103 ± 1.1] (35)
Horse		125.1(46)	118(36)	[113.6] (46)
Cat		131(42)		124(42)
Dog	128.1 ± 0.6(47)	132.7 ± 1.0(47)		117 ± 0.8(47)
Rhesus		139 ± 7.87(39)		[109 ± 4.7](39)
		125 ± 1.0(40)		[107 ± 1.7] (40)
Rhesus			117 ± 12(41)[a,b]	[109 ± 4.7](39)
		134 ± 5.3(48)[c]		109 ± 3.5(48)[c]
		126 ± 0.9(44)[d]		117 ± 0.8(44)

[a] Samples obtained from infants and children (≤13 yrs.) Vitreus and serum obtained from difffrent patiets.
[b] Vitreus collected >1 h after death caused by systemic diseases or accidents.
[c] Samples obtained from patients with various ocular diseases.
[d] Aqueus obtained from eyes of patients with senile cataracts.

Table III
Bicarbonate Concentration (mmol/kg H_2O ± 1 S.E.M.) in the Intraocular Fluids and Blood Plasma or Serum of Vertebrates

Species	Aqueus Anterior	Aqueus Posterior	Vitreus Anterior	Vitreus Posterior	Plasma or [serum]
Rabbit	27.7 ± 0.6(19)	34.1 ± 0.6(19)			24.0 ± 0.7(19)[a]
	30.1 ± 0.8(49)	36.0(50)	31.3 ± 0.8(50)[b]	23.9 ± 0.6(50)[b]	19.4 ± 1.1(49)[a]
		27.7(51)		19.5(51)	25.6(51)[c]
Sheep	23.5 ± 0.9(34)	18.0 ± 1.1(34)			28.3 ± 1.0(34)[c]
Bovine		36.0(52)			25.9 ± 1.4(53)[d]
Cat		30.4(42)			25.3(42)[d]
Rhesus		22.5 ± 0.9(40)			[18.8 ± 1.6] (40)[d]
Human		19.6 ± 0.3(48)[e]		15(54)[f]	26.5 ± 1.3(48)[d,e]
		21.5 ± 0.4(44)[g]			25.9 ± 0.3(44)[c]

[a] Blood collected by cardiac puncture.
[b] Vitreus dissected from frozen eyes from region adjacent to the ciliary body (anterior vitreus) and from region adjacent to the retina (posterior vitreus).
[c] Arterial.
[d] Venous.
[e] Samples obtained from patients with various ocular diseases.
[f] Mean postmortem interval 1¾.
[g] Aqueus obtained from eyes of patients with senile cataracts.

Although the concentration of either Cl^- or bicarbonate is typically greater in the aqueus than in plasma dialysate, the concentration of these ions in the vitreus approximates that in plasma dialysate (Tables II and III). These findings suggest that one or both of these major anions are actively transported by the ciliary epithelia in connection with aqueus production but that no physiologically significant net transport of Na^+, Cl^-, or bicarbonate occurs across the choroid or retinal vasculature into the vitreus. On the contrary, evidence suggests that active, ouabain-sensitive Na^+ transport across the choroid is responsible for removal of subretinal fluid after experimental induction of retinal detachment.[30] It remains to be seen whether or not similar choroidal absorptive Na^+ transport processes and consequent fluid transport are active when the retina is in normal proximity to the choroid. In any case, it is reasonable to assume on the basis of current information that in the retinal ECF, the concentrations of these three ions are virtually identical to the values given in Table I, II, and III for their respective concentrations in the vitreus.

8.2. Potassium

The concentrations of potassium (K^+) in the IOFs of the mammalian eye are generally highest in the posterior aqueus and/or in the anterior segment of the vitreus adjacent to the lens (Table IV). This can be attributed to active transport of K^+ across the ciliary epithelia into the posterior chamber, accumulation of K^+ by the lens from the posterior chamber across its anterior surface, and leakage of K^+ through the posterior surface of the lens into the anterior vitreus.[7,18] Indeed, the existence of active K^+ transport across the ciliary epithelia has been substantiated by the use of intravenous infusion of rubidium,[18] and a lenticular pump–leak mechanism for K^+ has been demonstrated *in vitro*.[60-62] Furthermore, the K^+ concentration in the vitreus of aphakic eyes is lower than that in the vitreus of phakic eyes and is well below that in the aqueus or blood plasma.[63] These findings suggest that in the typical mammalian eye, there is a net flux of K^+ from the blood into the posterior chamber, through the vitreus and ECF of the retina, across the blood–retinal barriers, and back into the blood. Thus, secretory K^+ transport across the ciliary processes and absorptive K^+ transport across the blood–retinal barriers maintain a high-K^+ environment for the mesodermal and ectodermal tissues of the anterior segment while providing the retina with a lower-K^+ environment possibly similar to that of the brain.[7]

8.3. Calcium and Magnesium

Because relatively high percentages of calcium (Ca^{2+}) and magnesium (Mg^{2+}) are bound to plasma proteins in blood, it is necessary to compare the concentrations of these divalent cations in IOFs to those in a plasma ultrafiltrate or in an *in vivo* dialysate[64] in order to elucidate their ocular site(s) of transport. In all species in which the Ca^{2+} concentrations in plasma dialysate and IOFs have been determined simultaneously, gradients of Ca^{2+} within the IOF compartments and between these compartments and plasma dialysate have gen-

Table IV
Potassium Concentration (mmol/kg H_2O ± S.E.M.) in the Intraocular Fluids and Blood Plasma or Serum of Vertebrates

Species	Aqueous		Vitreus		Plasma or [serum]
	Anterior	Posterior	Anterior	Posterior	
Rabbit	4.3 ± 0.17(55)	4.7 ± 0.09(55)	4.7 ± 0.12(55)	4.6 ± 0.13(55)	4.6 ± 0.22(55)
Woodchuck (active)	4.4 ± 0.1(33)		5.3 ± 0.1(33)		4.5 ± 0.14(33)
(hibernating)	4.8 ± 0.3(33)		5.8(33)		4.8 ± 0.2(33)
Sheep	4.9 ± 0.17	5.0 ± 0.20(34)	5.6(56)		4.5 ± 0.12(34)
Pig	4.9(56)		5.3(56)		6.1(56)
Bovine	4.9 ± 0.05(57)		4.9 ± 0.15(57)		[7.3 ± 0.08] (57)
Horse	4.86(46)		4.95(36)		[5.15(46)]
Cat	4.2 ± 0.09(55)	4.7 ± 0.08(55)	5.2 ± 0.13(55)	5.0 ± 0.12(55)	4.1 ± 0.24(55)
Dog	4.7 ± 0.06(55)	5.3 ± 0.05(55)	5.2 ± 0.05(55)	5.1 ± 0.05(55)	4.4 ± 0.19(55)
Rhesus	3.6 ± 0.18(55)	4.1 ± 0.16(55)	3.6 ± 0.19(55)		4.2 ± 0.20(55)
Human			5.5(58)[a]		[3.6–5.2] (59)

[a] Value obtained by extrapolating the regression of K^+ concentration in the vitreus versus time after death back to 0 time.

Table V
Calcium Concentration (mmol/kg H_2O ± 1 S.E.M.) in the Intraocular Fluids and Blood Plasma or Serum Vertebrates

Species	Aqueus Anterior	Aqueus Posterior	Vitreus Anterior	Vitreus Posterior	Plasma or [serum]	Plasma dialysate
Rabbit	1.7 ± 0.02(55)	1.7 ± 0.02(55)	2.0 ± 0.04(55)	2.0 ± 0.05(55)	3.1 ± 0.09(55)	2.1 ± 0.05(55)
Woodchuck (active)	1.3 ± 0.04(33)		1.5 ± 0.05(33)		2.7 ± 0.07(33)	
(hiberrating)	1.9 ± 0.10(33)		2.2(33)		2.9 ± 0.05(33)	
Bovine	1.6 ± 0.05(57)		1.8 ± 0.05(57)		[2.6 ± 0.03] (57)	
Horse	1.6(36)		1.7(36)		[2.5] (36)	
Cat	1.4 ± 0.04(55)	1.4 ± 0.04(55)	1.5 ± 0.05(55)	1.6 ± 0.04(55)	2.4 ± 0.05(55)	1.5 ± 0.05(55)
Dog	1.5 ± 0.03(55)	1.5 ± 0.04(55)	1.7 ± 0.04(55)	1.8 ± 0.05(55)	2.6 ± 0.06(55)	1.5(55)
Rhesus	1.3 ± 0.03(55)	1.3 ± 0.06(55)	1.3 ± 0.02(55)		2.4 ± 0.05(55)	1.4(55)
Human			1.6 ± 0.05(41)[a,b]		[1.8–3.0](41)[a]	

[a] Samples obtained from infants and children only (≤13 yrs.). Vitreus and serum obtained from different patients.
[b] Vitreus collected >1 h after death caused by systemic diseases or accidents.

erally been small (Table V). Although the transport of Ca^{2+} across the ciliary processes must be facilitated to maintain a Ca^{2+} concentration comparable to that in plasma dialysate in the freshly secreted fluid, there is no evidence of a large excess of Ca^{2+} in any IOF compartment and, hence, no evidence of active Ca^{2+} transport into the IOFs. It is reasonable to assume, therefore, that under steady-state conditions, the concentration of Ca^{2+} in the ECF of the mammalian retina is similar to that in the vitreus and in the CSF. Hence, the Ca^{2+} concentration in the ECFs of the brain and retina can be assumed to be comparable.

In contrast to Ca^{2+}, Mg^{2+} concentration gradients between IOF compartments are pronounced, as are species variations in these concentrations (Table VI). Among the three mammalian species in which the Mg^{2+} concentrations in the anterior and posterior aqueus and in the anterior and posterior segments of the vitreus have been determined simultaneously,[55] the highest Mg^{2+} concentrations have always been found in the posterior vitreus. It is noteworthy that in all three of these species, the Mg^{2+} concentration in the posterior vitreus was greater than that in the posterior aqueus, implying a net flux of Mg^{2+} from the blood into the retinal ECF across the blood–retinal barrier(s). Whether or not species variations in the concentration of Mg^{2+} in the vitreus reflect differences in some aspects of retinal function remains to be examined.

8.4. Inorganic Phosphate

Whenever IOFs and serum have both been analyzed for inorganic phosphate in the same study, the concentration of this ion in the aqueus has been much lower than that in serum (Table VII). In rabbits, the only species in which aqueus and vitreus values have been reported simultaneously, the concentration of phosphate in the vitreus was found to be even lower than that in the aqueus. Although a higher phosphate concentration in human vitreus was reported in one study, the vitreus was obtained an average of 9 h after death; thus, the value reported may be artifactually high. Although more studies on the concentrations of phosphate in IOFs are clearly needed, it is reasonable to conclude at this time that the phosphate concentration in IOFs, especially the vitreus, and hence in the retinal ECF is generally lower than that in blood plasma.

8.5. Glucose and Lactic Acid

In most species studied, the concentration of glucose in blood plasma or serum exceeds that in the aqueus, where, in turn, the concentration exceeds that in the vitreus (Table VIII). The fact that the glucose concentration in the posterior chamber of the rabbit is only slightly less than that in simultaneously obtained blood plasma suggests that glucose enters the posterior chamber by facilitated transport. In the absence of such transport, a significant deficiency of glucose in the newly secreted aqueus would be expected, especially since the lens must derive its glucose primarily from the posterior chamber.

Table VI
Magnesium Concentration (mmol/kg H_2O ± 1 S.E.M.) in the Intraocular Fluids and Blood Plasma or Serum of Vertebrates

Species	Aqueus		Vitreus		Plasma or [serum]	Plasma dialysate
	Anterior	Posterior	Anterior	Posterior		
Rabbit	0.71 ± 0.02(55)	0.75 ± 0.02(55)	1.00 ± 0.04(55)	1.08 ± 0.02(55)	0.94 ± 0.04(55)	0.8 ± 0.03(55)
Woodchuck (active)	0.96 ± 0.02(33)		1.1 ± 0.03(33)		0.79 ± 0.03(33)	
(hibernating)	1.08 ± 0.03(33)		1.15(33)		1.45 ± 0.04(33)	
Bovine	0.43 ± 0.01(35)		0.37 ± 0.03(35)		[0.61 ± 0.04](35)	
Horse	1.1(36)		0.83(36)		[1.2](36)	
Cat	0.50 ± 0.03(55)	0.45 ± 0.02(55)	0.51 ± 0.02(55)	0.54 ± 0.03(55)	0.98 ± 0.08(55)	0.7 ± 0.02(55)
Dog	0.59 ± 0.01(55)	0.53 ± 0.01(55)	0.64 ± 0.01(55)	0.69 ± 0.03(55)	0.78 ± 0.03(55)	0.59(55)
Rhesus	0.59 ± 0.02(55)	0.65 ± 0.02(55)	0.63 ± 0.02(55)		0.62 ± 0.04(55)	0.46(55)
Human			1.32 ± 0.03(41)[a,b]		[0.70–1.19](41)[a]	

[a] Samples obtained from infants and children only (≤13 yrs.). Vitreus and serum obtained from different patients.
[b] Vitreus collected >1 h after death caused by systemic diseases or accidents.

Table VII
Phosphate Concentration (mmol/kg H₂O) in the Intraocular Fluids and Blood Serum of Vertebrates

Species	Aqueus	Vitreus		Serum
		Anterior	Posterior	
Rabbit	0.34 ± 0.02(65)	0.10 ± 0.01(65)		0.61 ± 0.03(65)
	0.24 ± 0.01(66)	0.05 ± 0.003(66)[a]	0.04 ± 0.003(66)[a]	0.59 ± 0.02(66)
		0.06 ± 0.004(66)		
Bovine		0.1–0.03(67)		0.64 ± 0.11(68)
Cat	0.16(69)			0.61(69)
Dog	0.14(69)			0.44(69)
Rhesus	0.14 ± 0.01(40)			0.68 ± 0.04(40)
Human	0.20(69)	0.4(70)[b]		0.36(69)

[a] Vitreus dissected from frozen globes.
[b] Samples collected an average of 9 h after death.

The finding that the glucose concentration in the vitreus is generally lower than that in the aqueus or blood plasma cannot be attributed to use of glucose by the lens, because the steady-state concentration of glucose in the vitreus of aphakic eyes is also very low.[63] Rather, the low glucose concentration in the vitreus must be a result of its high rate of use by the retina and/or a result of the removal of glucose from the retinal ECF by the blood–retinal barriers.

Table VIII
Glucose Concentration (mmol/kg H₂O ± 1 S.E.M.) in the Intraocular Fluids and Blood Plasma or Serum of Vertebrates

Species	Aqueus		Vitreus		Plasma or [serum]
	Anterior	Posterior	Anterior	Posterior	
Rabbit	5.4(45)	5.6(45)	3.0(45)		5.7(45)[a]
	6.6 ± 0.04(71)		3.4 ± 0.03(71)		7.22(71)
Woodchuck (active)	6.8 ± 0.39(33)		4.8 ± 0.19(33)		10.8 ± 0.58(33)
(hibernating)	4.3 ± 0.50(33)		3.9(33)		5.1 ± 0.44(33)[a]
Bovine	3.3 ± 0.06(72)		2.8 ± 0.23(73)[b]	2.9 ± 0.08(73)[b]	[3.8 ± 0.14](73)
Horse	5.4(36)		5.4(36)		[5.1](36)[c]
			3.3 ± 0.02(73)[b]		[5.1](73)
Cat	4.6 ± 0.21(74)		3.2 ± 0.15(74)		6.0 ± 0.28(74)[d]
Dog	4.7 ± 0.93(38)				[5.7 ± 0.45](38)
Rhesus	1.7 ± 0.64(39)				[3.7 ± 0.78](39)
	3.0 ± 0.22(40)				[4.1 ± 0.58](40)[c]
Human	3.2 ± 0.1(75)[e]		3.44(70)[f]		5.0 ± 0.08(75)[e]

[a] Blood collected by cardiac puncture.
[b] Vitreus dissected from frozen globes.
[c] Venous.
[d] Arterial.
[e] Cataractous eyes; capillary blood obtained from fasting patients.
[f] Samples collected an average of 9 h post-mortem.

Table IX
Lactic Acid Concentration (mmol/kg H$_2$O ± 1 S.E.M.) in the Intraocular Fluids and Blood Plasma or Serum of Vertebrates

Species	Aqueus		Vitreus		Plasma or [serum]
	Anterior	Posterior	Anterior	Posterior	
Rabbit	9.2(45)	9.8(45)	12.1(45)		10.3(45)[a]
Woodchuck (active)	7.5 ± 0.4(33)		9.0 ± 0.5(33)		9.8 ± 1.3(33)
(hibernating)	3.8 ± 0.3(33)		3.9(33)		1.0 ± 0.2(33)[a]
Bovine			5.1 ± 0.4(73)[b]	5.3 ± 0.5(73)[b]	[4.2](73)
	2.9(31)[c]		2.6(31)[c]		
Human	4.3 ± 0.3(48)[d]		7.8(48)		1.8 ± 0.2(48)[d,e]

[a] Blood collected by cardiac puncture.
[b] Vitreus dissected from frozen globes.
[c] Samples collected 2-3 h post-mortem.
[d] Samples collected from patients with various ocular diseases.
[e] Venous.

Although it is generally assumed that the retina has a very high glucose requirement, this requirement may be overestimated, since the techniques used to measure it, such as superfusion or removal of the retina, may well stimulate or otherwise alter its metabolic activity.

Thus, the low glucose concentration in the retinal ECF most likely results, at least in part, from absorptive transport processes across the blood–retinal barriers. Facilitated entry of glucose at some regions of the blood–ocular barrier coupled with active removal at other regions could provide adequate glucose supply while still maintaining a low concentration in the retinal ECF.[2] Active transport of glucose from the CSF back into the blood has been shown to occur across the choroid plexus and is apparently responsible for maintaining a low glucose concentration in the ventricular system and in the ECF of adjacent regions of the brain.[25] Although the bulk of the glucose used by fully vascularized retinas must be derived from the chorodial and retinal vasculatures, the vitreus must serve as an effective source of glucose during short periods of ischemia and, hence, must be regarded as an emergency reservoir.[2] In turn, the ciliary processes must be important sources of glucose for the vitreus.

In several species studied, the concentration of lactic acid has been found to be higher in the vitreus than in either the aqueus or blood plasma (Table IX). This must result, at least in part, from the release of lactic acid from the retina, since the concentration of lactic acid in the vitreus of aphakic rabbit eyes is not significantly lower than that in normal phakic rabbit eyes.[63]

8.6. Ascorbic Acid

The concentration gradients of ascorbic acid between blood plasma and IOFs of some species are larger than those of any other major solute, indicating the existence of active processes capable of transporting this vitamin against up to 20- or 30-fold concentration gradients (Table X). However, the species

Table X
Ascorbic Acid Concentration (mmol/kg H_2O ± 1 S.E.M.) in the Intraocular Fluids and Blood Plasma or Serum of Vertebrates

Species	Aqueus		Vitreus		Plasma or [serum]
	Anterior	Posterior	Anterior	Posterior	
Rabbit	1.1(45)	1.3(45)	0.5(45)		0.1(45)
Guinea pig	0.9(76)		1.0(76)		0.05 ± 0.001(76)
Rat	0.03(77)		0.1(78)		[0.1](79)
Woodchuck (active)	0.5 ± 0.1(33)		0.6 ± 0.1(33)		0.1 ± 0.02(33)
(hibernating)	0.5 ± 0.1(33)		0.5 ± 0.1(33)		0.2 ± 0.02(33)
Sheep			0.5(78)		
Bovine	1.1(76)		0.8(76)		
Horse	1.2(80)		1.2(80)		0.04 ± 0.002(80)
Cat	0.1 ± 0.01[c]		0.2 ± 0.02	0.2 ± 0.02[c]	0.2 ± 0.02[c]
Rhesus	1.2 ± 0.04(81)	1.4 ± 0.04(81)			[0.2](40)
Human	1.1 ± 0.3(48)[a]		2.0(78)[b]		0.04 ± 0.02(48)[a]

[a] Samples collected from patients with various ocular diseases.
[b] Vitreus collected from fetal eyes.
[c] L. Z. Bito (unpublished data).

variations in the concentrations of this vitamin in the IOFs are also greater than those of all other major solutes. In cats, the ascorbic acid concentration in the aqueous humor is half that in blood plasma, whereas in the horse it is 30 times greater than that in plasma. In the rabbit, the ascorbic acid concentration in the vitreus is lower than that in the aqueus, whereas in the rat and some other species, the reverse is true. Thus, in some species the primary site of active transport of ascorbic acid into the IOFs must be the ciliary processes, whereas the blood–retinal barrier(s) may be the primary site in other species.

In all species that actively transport ascorbic acid into the eye, the ascorbic acid concentration in the retinal microenvironment is several times greater than that in blood plasma and is comparable to or even higher than that in most other regions of the brain. At this time, however, we cannot rule out the possibility that even in species that have a much lower concentration of ascorbic acid in their IOFs, i.e., one that does not significantly exceed that in their blood plasma, ascorbic acid is actively transported from the choroid into the retinal ECF. It is possible that in such species, continuous loss of ascorbic acid or its metabolites at other regions of the blood–ocular barrier system, perhaps including the retinal circulation, prevents the reflection of choroidal transport in an elevated ascorbate concentration in the vitreus. It should be noted, for example, that in the cat eye, the ascorbic acid concentration in the vitreus, especially the posterior vitreus, is slightly higher than that in the aqueus, raising the possibility that even in these species the ascorbate concentration in the retinal ECF may be significantly higher than that in blood plasma.

Based on an apparent correlation between species variations in the ascorbic acid concentration of the aqueous humor and the behavioral adaptations of different species, Ringvold[77] suggested that the ascorbic acid in the aqueus protects the anterior segment against photodynamic effects by filtering out ultraviolet light in diurnal species. The correlation between presumed exposure to light and ascorbic acid concentrations in the aqueus is indeed quite striking among the 19 species considered by this author. It should be pointed out, however, that there are clear exceptions to this relationship, since some diurnal birds have very low ascorbic acid concentrations in their IOFs[16] (L. Z. Bito, unpublished data). In addition, ascorbic acid concentrations are high in other species, not only in the eye, where photodynamic effects may be important, but also in the CSFs and the brain, areas of the CNS that are well protected from light. Thus, this vitamin is likely to have other important but as yet undetermined biochemical or physiological roles for the normal functioning of CNS tissues.

8.7. Amino Acids

The concentrations and concentration gradients of amino acids in the IOFs show considerable species variation (Table XI). For example, in the aqueus, especially the anterior aqueus of the sheep, the amino acid concentration is generally higher than that in plasma; in humans, this relationship is reversed. It should be noted, however, that the amino acid concentrations in blood plasma fluctuate considerably, so the values presented by some authors may not truly represent steady-state concentrations.[86] Furthermore, the samples from which values for the concentration of amino acid in the human aqueus have been obtained came from eyes affected by a variety of pathological conditions and hence may not represent the normal steady-state distribution of these amino acids.

Only in one species, the dog, have separate amino acid analyses been performed on the anterior and posterior segments of the vitreus with precautions taken to obtain blood samples that were representative of preexisting, normal amino acid levels.[86] The concentrations of most amino acids in this species were higher in the posterior chamber than in other IOF compartments. In addition, the amino acid concentrations in the posterior vitreus were generally lower than in the anterior vitreus and, with few exceptions, much lower than in blood plasma. The vitreus of rabbits also has lower concentrations of amino acids than the aqueus humor or plasma (Table XI).

These observations not only suggest that amino acids enter the posterior chamber by facilitated and/or active transport processes but that there is a continuous loss of amino acids through the vitreus to the retina. This loss to the retina may be assumed simply to represent the use of some amino acids. However, studies with a nonmetabolizable amino acid analogue, cycloleucine, clearly demonstrated that some amino acids are lost from the vitreus through absorptive active transport from the ECF of the retina back into the circulation.[84,87] Thus, the low amino acid concentration in the ECF of the mammalian retina is generally maintained, at least in part, by absorptive transport processes across the blood-retinal barrier(s).[2,7]

Taurine is actively accumulated in the retina, either from the vitreus or directly from the circulation, and, at least in cats, is essential for normal visual function.[88] However, this may simply reflect intracellular taurine accumulation

Table XI

Amino Acid Concentrations (μmol/kg H_2O) in Intraocular Fluids and Blood Plasma of Vertebrates

	Human[82a]		Human[83b]		Sheep[84c]			Rabbit[85]				Dog[86]			
					Aqueus			Aqueus				Aqueus		Vitreus	
	Aqueus	Plasma[d]	Aqueus	Plasma[d]	Ant.	Post.	Plasma[d]	Ant.	Post.	Vitreous	Plasma[c]	Ant.	Post.	Ant.	Post.	Plasma[d]
Alanine	307 ± 10	326 ± 15	255 ± 26	371 ± 23	600 ± 130	570 ± 150	230 ± 60	480	445	74	302	450	620	160	70	290
Arginine	105 ± 4	71 ± 3	102 ± 29	63 ± 12	240 ± 30	220 ± 20	120 ± 40	272	276	45	96	140	100	60	40	110
Aspartic acid					200 ± 20	180 ± 30	20 ± 0	55	87	19	29	280	210	350	0	20
Cystine	15 ± 2[f]	134 ± 5[f]	BD[g]	106 ± 12	260 ± 20	240 ± 40	140 ± 10	30	8	BD[g]	54	190	210	150	40	180
Glutamic acid	10 ± 1	58 ± 4	3 ± 1	33 ± 5	1090 ± 270	860 ± 150	100 ± 20	295	401	169	178	230	230	140	60	130
Glutamine			952 ± 42	713 ± 27								460	820	280	110	0
Glycine	38 ± 6	274 ± 23	13 ± 2	259 ± 21	960 ± 410	930 ± 440	780 ± 250	614	314	169	1190	110	200	130	50	290
Histidine	69 ± 2	78 ± 2	66 ± 18	65 ± 13	90 ± 20	80 ± 30	40 ± 20	210	181	20	116	70	60	20	20	70
Isoleucine	66 ± 3	53 ± 3	54 ± 9	48 ± 7	180 ± 30	180 ± 30	80 ± 20	116	121	10	112	100	290	50	30	100
Leucine	139 ± 5	104 ± 5	122 ± 15	97 ± 10	260 ± 50	280 ± 50	110 ± 20	174	191	18	163	190	230	60	60	120
Lysine	167 ± 7	254 ± 9	173 ± 44	241 ± 22	160 ± 30	180 ± 60	100 ± 30	423	397	70	212	180	180	100	90	230
Methionine	59 ± 12	24 ± 5	40 ± 5	17 ± 2	50 ± 10	60 ± 20	50 ± 20	23	26	BD[g]	14	60	60	40	20	30
Phenylalanine	95 ± 3	48 ± 1	100 ± 14	46 ± 6	110 ± 20	120 ± 20	40 ± 10	97	141	11	97	90	150	50	40	40
Proline	47 ± 4	231 ± 12	25 ± 4	244 ± 17				267	223	48	323					
Serine					600 ± 100	520 ± 100	250 ± 50	585	597	130	420	130	160	110	30	270
Taurine	59 ± 5	61 ± 2	42 ± 4	56 ± 5		170 ± 50		138								
Threonine	132 ± 6	114 ± 6	121 ± 14	112 ± 14	180 ± 80	180 ± 30	80 ± 20		149	25	165	80	110	0	10	130
Tryptophan				50 ± 7		430 ± 60		24	4	BD[g]	13	200				
Tyrosine	94 ± 5	53 ± 3	92 ± 13	215 ± 14	170 ± 50		50 ± 10	101	161	21	58	140	170	70	40	50
Valine	286 ± 12	216 ± 10	279 ± 30		410 ± 40		160 ± 20	230	214	45	195		200	100	60	40
Total	1696	2096	2431	2737	5770	5200	2350	4140	4482	972	3744	3094	4000	1870	770	2110

[a] From ten patients with retinitis pigmentosa (mean ± S.E.M.).
[b] From 41 patients with cataracts and/or a variety of other ocular disorders (mean ± S.E.M.).
[c] Mean ± S.D.
[d] Venous.
[e] Cardiac puncture.
[f] Half-cystine.
[g] Below detection limit.

Table XII
Osmolality (mOsm/kg H₂O) of Intraocular Fluids and Blood Plasma of Some Mammals

Species	Aqueus		Vitreus		Plasma or [serum]
	Anterior	Posterior	Anterior	Posterior	
Rabbit	302 ± 3.1(7)		296 ± 1.1(7)		299 ± 1.5(7)
Woodchuck (active)			311 ± 3.4(33)		310 ± 3.2(33)
(hibernating)			304 ± 1.7(33)		302 ± 2.0(33)
Cat	317 ± 1.6(7)	321 ± 2.2(7)	317 ± 1.4(7)		309 ± 2.6(7)
Dog	307 ± 1.5(7)	308 ± 1.8(7)	306 ± 3.0(7)	302 ± 2.8(7)	300 ± 1.7(7)
Rhesus		305 ± 0.5(40)			[306 ± 1.4](40)

and does not necessarily imply its net active transport across the blood–retinal barrier(s). In fact, it appears that the absorptive transport of taurine exceeds its secretory transport across the choroidal epithelium.[11] Furthermore, in spite of the high concentration of taurine in the retina, especially in the photoreceptors, its concentration in the vitreus of normal kittens is only about one-half of that in plasma.[88] Thus, the concentration of taurine in the retinal ECF, like that of other amino acids, is kept well below its plasma concentration.

We conclude, therefore, that the concentrations of amino acids in the IOFs of the anterior segment are comparable to or even higher than those in blood plasma, that their concentrations in the ECF of the retina are much lower than those in blood plasma, and that, for the most part, their concentrations in the retinal ECF are similar to those in the CSF and, presumably, to those in the ECFs of the brain.

8.8. Osmolality

In general, the aqueus obtained from the posterior chamber is slightly hyperosmotic to plasma (Table XII). This is expected, since the production of aqueus by the ciliary processes is best explained on the basis of a standing osmotic gradient within the intercellular spaces between nonpigmented cells of the ciliary epithelia.[6] According to this model, an isosmotic fluid could only be produced if the channels between the cells were infinitely long.[89]

In all species for which such information is available, both the anterior aqueus and the vitreus are slightly hyposmotic to the posterior aqueus. In dogs, the only species for which separate values for anterior and posterior vitreus are also available, it has been reported that the osmolarity of the posterior vitreus is below that of the anterior vitreus but still slightly hyperosmotic to plasma (Table XII). Although it should be remembered that metabolic activities such as the conversion of glucose to lactic acid or the production of H_2O from carbohydrates can alter local osmolarity, this decline in osmolarity away from the site of aqueus secretion suggests that there is a net flux of water across most regions of the blood–ocular barriers, including the blood–retinal barriers.

Table XIII
Total Protein Concentration (mg/100 ml ± 1 S.E.M.) in the Intraocular Fluids and Blood Plasma or Serum of Vertebrates

Species	Aqueus	Vitreus	Plasma or [serum]
Rabbit	39 ± 5.2(90)	18 ± 1.2(90)[a]	6000–7000(91)
Guinea pig		10 ± 0.8(90)[a]	4680(92)
Rat		16 ± 4.9(90)[a]	5700–5800(93–95)
Woodchuck (active)	124 ± 44(33)	105 ± 25(33)	6875 ± 497(33)
(hibernating)	52 ± 4.0(33)	85 ± 18(33)	8321 ± 240(33)
Sheep	45 ± 9.4(90)	43 ± 1.5(90)[b]	
Pig		51 ± 1.6(90)[b]	
Bovine	24 ± 2.6(90)	50 ± 2.6(90)[b]	5400–6700(96)
Horse	19 ± 1.6(90)	24 ± 1.9(90)[b]	
	20(36)	65(36)	[7369](36)
Cat	27 ± 2.0(90)	20 ± 1.4(90)[a]	7090 ± 140(97)
Dog	36 ± 0.9(90)	43 ± 2.4(90)[b]	5650 ± 260(97)
Human	30–60(98)		7000–7500(98)
	49 ± 6.1(90)[c]	46 ± 4.4(90)[a,c]	
		62 ± 1.7(41)[d,e]	[4200–8000](41)[d]

[a] Vitreus obtained from frozen globe.
[b] Vitreus homogenized followed by ultracentrifugation.
[c] Samples collected 3 h after death.
[d] Samples obtained from infants and children only (≤13 yrs.). Vitreus and serum obtained from different patients.
[e] Samples collected >1 h after death caused by systemic diseases or accidents.

At any rate, based on currently available information and theoretical considerations, we can conclude that the retinal ECF, like the ECF of the brain, is essentially isotonic with or possibly slightly hypertonic to blood plasma.

8.9. Soluble Proteins and Enzymes

The total concentration of soluble proteins in the IOFs (Table XIII) is generally very low—only about 0.5–1% of that in plasma. However, the aqueus contains all of the major plasma protein fractions, although in proportions different from those in plasma. For example, albumin, a low-molecular-weight protein, tends to be more abundant in the aqueus of most species than high-molecular-weight γ-globulins.[3,99] This is consistent with the concept that the presence of these proteins in IOFs results from leakage through the blood–ocular barriers. However, this relationship between molecular size and relative concentrations of soluble proteins in the IOFs is by no means universal,[3] and the possibility that some of these proteins enter the eye by specific mechanisms cannot be ruled out.

Several enzymes have also been demonstrated in IOFs (Table XIV). In general, there is greater species variation in enzyme activity than in the total protein concentrations in the IOFs. Enzyme activity is generally greater in the vitreus than in aqueus but tends to be lower in the IOF compartments than in plasma.

Table XIV
Enzyme Concentrations (U/liter Except Where Otherwise Noted) in the Intraocular Fluids and Blood Serum of Vertebrates

Enzymes, species	Aqueous	Vitreus	Serum
Lactic dehydrogenase			
Rabbit (90)	8.5 ± 1.3	28.3 ± 6.3	134.6 ± 23.1
Rat (100)	885 ± 304	738 ± 189	2837 ± 297
Sheep (90)	—	24.6 ± 3.4	491 ± 34
Pig (90)	115 ± 11	20.4 ± 1.8	1079 ± 115
Bovine (90)	7.9 ± 0.3	42.6 ± 1.4	1331 ± 63
Horse (90)	6.4 ± 0.8	14.5 ± 1.2	339 ± 33
Cat (90)	4.7 ± 0.6	27.9 ± 3.7	106 ± 14
Dog (90)	3.6 ± 0.5	10.2 ± 1.4	
Rhesus (100)	146 ± 4.7	234 ± 29	1666 ± 155
Aldolase (100)			
Rabbit	30.1[a]	20 ± 2.0[a]	179 ± 20[a]
Guinea pig	—	23 ± 4.8[a]	95 ± 13[a]
Rat	6.7 ± 1.7	210 ± 24	239 ± 19
Bovine	5.8 ± 0.3	8.0 ± 0.4	45.1 ± 5.9
Cat	0.9 ± 0.2	16.8 ± 3.23	16.7 ± 3.94
Rhesus		8.9 ± 1.17	94 ± 13
Phosphohexose isomerase			
Rabbit (90)	11.7 ± 1.9	26.9 ± 6.0	369 ± 47
Sheep (90)	—	13.4 ± 1.8	303 ± 38
Pig (90)	—	32.6 ± 1.6	303 ± 33
Bovine (90)	8.8 ± 1.7	28.0 ± 2.2	402 ± 19
Horse (90)	3.8 ± 0.5	8.4 ± 0.6	31.0 ± 4.3
Cat (90)	7.0 ± 0.9	32.7 ± 3.5	356 ± 58
Dog (90)	4.0 ± 0.5	17.4 ± 3.7	—
Rhesus (100)	—	90 ± 11[b]	443 ± 41[b]
Malate dehydrogenase (90)			
Rabbit	18.9 ± 1.1	32.2 ± 2.8	449 ± 51
Sheep	14.5 ± 3.1	39.8 ± 4.2	353 ± 79
Pig	—	63.2 ± 3.9	—
Bovine	9.3 ± 0.5	45.3 ± 1.6	353 ± 21
Horse	9.1 ± 0.4	19.0 ± 1.4	379 ± 54
Cat	7.9 ± 0.8	45.4 ± 5.0	59.0 ± 6.6
Dog	5.8 ± 0.5	30.0 ± 1.8	—

[a] Sibley–Lehninger units/ml.
[b] Bodansky units/ml.

In some species, most notably rabbits, the concentration of soluble proteins in the vitreus is even lower than that in the aqueous. However, this is not a general trend since, in many species, the protein concentrations in these fluid compartments are essentially identical, whereas in others, including the dog, horse, and bovine, the soluble protein concentration in the vitreus is greater than that in the aqueous. Thus, the normal concentrations of proteins in the IOFs and in the microenvironment of the retina are very low, about 1% of the concentration of proteins in plasma, and are similar to those in the CSF.

9. THE HOMEOSTASIS OF THE MICROENVIRONMENT OF THE MAMMALIAN RETINA

Although numerous studies have demonstrated the homeostatic control of the concentration of solutes in CSF,[25] similar studies on the IOFs are generally lacking, probably because the fluid compartments of the typical mammalian eye are not suited for repeated sampling. Some studies have reported the effect of altered plasma concentrations of solutes such as ascorbic acid,[101-103] potassium,[18] and amino acids[104] on the concentrations of these solutes in the aqueus. However, these studies have not included values for the vitreus and hence are not applicable to the evaluation of the homeostasis of the retinal microenvironments.

The remarkable sophistication of the homeostatic mechanisms that control the chemical composition of mammalian IOFs is probably best illustrated by the observation that the concentrations of most solutes in the aqueus and vitreus of normothermic and hibernating woodchucks are essentially identical.[33] The concentration of most solutes in the IOFs is maintained in spite of a difference of more than 30°C in the body temperature of this species during its normo- and hypothermic states and in spite of the fact that the concentrations of many solutes in the blood plasma of woodchucks are greatly affected by the thermoregulatory state of the animal (Tables I–VI, VIII–X).

10. THE MICROENVIRONMENT OF THE AVIAN RETINA AND ITS HOMEOSTASIS

Although this chapter deals primarily with the microenvironment of the mammalian retina, avian eyes offer some unique advantages for the study of the retinal microenvironment. Although the avian retina is extremely well developed, it is completely avascular and must depend on the vitreus for its metabolic exchanges to a much greater extent than mammalian retinas of comparable complexity.[2] Furthermore, the eyes of birds typically have a cortical layer of liquid vitreus that is separated from the retina only by a very thin gel layer.[105] Thus, a liquid vitreus sample that is in close equilibrium with the retina can be readily obtained from most birds.[106]

In one study on the chicken, the high Mg^{2+} concentration in this liquid vitreus layer was maintained at a remarkably stable level even when the Mg^{2+} concentration in blood plasma was either increased severalfold by intravenous infusions or reduced to less than half of its normal value by dietary depletion.[106] These results demonstrate not only that the Mg^{2+} concentration in the local environment of the avian retina is maintained at a very high value but that it is under effective homeostatic control within an extended range of plasma concentrations.[106] Homeostatic control of the mammalian retinal microenvironments should be investigated with regard to solutes such as Mg^{2+}, K^+, and amino acids, which can be expected to affect retinal function.

11. RECOMMENDATIONS FOR THE COMPOSITION OF ARTIFICIAL PHYSIOLOGICAL SOLUTIONS FOR RETINAL RESEARCH

Since the chemical environment of the retina differs from the environments of all other tissues of the body with the possible exception of the brain, typical Ringer's solutions or tissue culture media designed for the maintenance of peripheral tissues are not suitable for *in vivo* superfusion or *in vitro* studies on the retina or for the maintainance of cells derived from it. Table XV is meant to serve as a guide for the preparation of physiological solutions resembling the normal chemical environment of the retina of different species. It many cases, the concentration of a particular solute in the microenvironment of the retina cannot be effectively estimated from currently available values. Moreover, different studies require physiological solutions having different buffering capacities, or necessitate alterations in the concentrations of certain constituents. Therefore, this table presents ranges of acceptable concentrations for each solute for several species. It is expected that fine adjustments in the concentrations of some solutes can be best achieved by optimizing some relevant physiological parameter(s).

In Table XV, smaller ranges have been given for solutes whose concentration within the retinal ECF could be estimated with reasonable accuracy. However, a larger range is given when the concentrations of solutes in the vitreus were not available in the literature. In these instances, extrapolations had to be made based on the concentration of the solutes in the aqueus or plasma and the known relationship between these fluids and the vitreus in other species. The reader is also referred to the tables that have been presented for each solute (Tables I–XIV) and to the references indicated therein for further information.

The species listed in Table XV are those for which the best values are available and/or those that are commonly used in visual research. Although the woodchuck is not currently used in retinal research, recommendations for this species are included since this hibernator, like heterotherms in general, offers some unique advantages for many types of studies on the retina and visual process. For example, it is most likely that retinas of hibernating mammals can be maintained *in vitro* at a reduced temperature for much longer periods than retinas of other mammals. Furthermore, retinas of such heterotherms could be studied at two distinctly different temperatures, i.e., at 37°C and at 4°C, both *in vivo* and *in vitro*, allowing detailed study of some metabolic and electrophysiological processes, especially fast events.

Since values for all potential experimental animals could not be included, a more general recommendation applicable to other species is included in the last column of Table XV. Since the chemical composition of the blood plasma of most species is readily available from compendia such as Volume III of *The Biology Data Book*,[107] the information listed in the last column of Table XV should serve as a general guide for the preparation of a physiological solution suitable for retinal research on most species.

Table XV

Recommended Concentration Ranges (mmol/kg H_2O unless Noted Otherwise) for the Preparation of Physiological "Ringer's Solutions" or Tissue Culture Media Used for Retinal Superfusions, Maintenance of Retinas in Vitro, or for the Culture of Retinal Cells

	Rabbit	Woodchuck (active)	Cat	Dog	Primate (Human)	Other species
Sodium	140–145	147–152	155–162[a]	140–145[a]	135–150	0–5% below plasma [Na^+]
Chloride	102–106	110–116	120–132[a]	120–130[a]	110–120	Sum of [Cl^-] plus [HCO_3^-]
Bicarbonate	20–25	10–30[a]	20–30[a]	10–20[a]	15–25	5–10% below [Na^+]
Potassium	3–4.5	3–5	3–5	3–5	2.5–5	10–30% below plasma [K^+]
Calcium	1.5–2.0	1.3–1.6	1.5–1.8	1.7–1.9	1.3–1.6	0–10% above [Ca^{2+}] in plasma dialysate
Magnesium	1.0–1.2	1.0–1.3	0.5–0.7	0.6–0.8	0.6–1.3	0–50% above [Mg^{2+}] in plasma dialysate
Phosphate	0.04–0.1	<0.4[a]	<0.4[a]	<0.4[a]	<0.5	10–30% of blood serum [phosphate]
Glucose	2–3.5	3–5	2–3.5	2–4[a]	3–4	30–50% below plasma [glucose]
Lactic acid	6–12	6–10	6–10[a]	6–10[a]	4–9	5–10 mM
Ascorbic acid	0.3–0.5	0.4–0.6	0.1–0.2	0.1–0.5[a]	0.8–2	Up to 30-fold over plasma [ascorbate]
Amino acids[b]	0.7–1.5	0.7–1[a]	0.7–1[a]	0.7–1	0.7–1.5[a]	30–50% below plasma [amino acids][b]
Osmolality[c]	295–300	308–312	310–315	300–305	300–310[a]	Within 0–5 mOsm of plasma osmolality
Total soluble proteins[d]	10–20	80–110	10–20	30–50	30–60	<1% of total [protein] in plasma

[a] Based on aqueous values and/or on comparison with other species.
[b] The sum of all measured amino acids in μmol/kg H_2O; see Table XI for relative proportions of individual amino acids.
[c] mOsm/kg H_2O.
[d] mg/100 ml.

It should also be noted that because the composition of the retinal microenvironment is very similar to that of the CSF,[7] CSF itself can be regarded as a suitable medium for the maintainance of the retina. Since relatively large volumes of CSF can be obtained easily from the cisterna magna of most species, this may be a practical solution for use in some short-term experiments.

For maintainance of tissues *in vitro*, it is customary to use glucose concentrations higher than or equal to the normal blood plasma glucose levels. In the case of peripheral tissues, this may be acceptable, because the glucose concentration in the blood varies considerably under normal physiological conditions, and because these tissues have a well-established tolerance to high glucose levels. However, in central nervous tissues, the glucose concentration is kept well below that of blood plasma.[25] Glucose concentrations even approaching that of plasma must be regarded as unphysiological for the retina.

Concentrations of amino acids as high as those typically used in tissue culture media should also be considered as unphysiological for the retina. Such high amino acid concentrations may be especially unsuitable for use in neurobiological or electrophysiological studies on this organ because many amino acids are putative or demonstrated neurotransmitters, false transmitters, or inhibitors of neuronal transmission. Thus, when amino acids are used at concentrations higher than those generally existing in the retinal ECF, it can be expected that retinal function will be abnormal. The same can be said when Mg^{2+} of K^+ concentrations uncharacteristic of the retinal ECF are used in *in vitro* studies. By the same token, the use of low concentrations of ascorbic acid or its omission from solutions must be regarded as potentially detrimental to retinal function, especially in species that accumulate this vitamin in their IOFs against a 10- to 30-fold concentration gradient.

Although retinal explants or cells derived from the retina may be maintained in tissue culture media that were developed for maintenance of cell lines derived from nonneuroectodermal tissues or from tumors, this may simply reflect the adaptability of cells to abnormal environmental conditions. If the only goal of the experiment is the survival or propagation of cells, such media may be adequate. However, if maintenance of the unique characteristics of these cells is desired, the normal *in vivo* chemical environment of the retina should be reproduced as closely as possible.

12. CONCLUSIONS

Detailed studies on the concentration gradients within and between the IOF compartments and between these compartments and blood plasma allow the elucidation of the chemical microenvironment of the retina. Such gradients also indicate the existence and site(s) of inwardly or outwardly oriented active and facilitated transport processes across the blood–ocular barrier system. Although *in vitro* studies on the transport functions of the blood–ocular barriers will ultimately be required to document their existence and to elucidate their mechanisms, it cannot be automatically assumed that the rates of solute fluxes or even their direction, as measured *in vitro*, reflect the actual fluxes of these solutes *in vivo* under normal steady-state conditions. The isolation of such unusually delicate tissues as the choroid and the ciliary processes, and their maintenance in Ussing-type chambers can be expected and, in some cases have been shown, to alter their permeability and transport characteristics. Thus, the results of such *in vitro* experiments should always be discussed in light of the known distribution of the substance in question in the IOFs under normal steady-state conditions.

Analysis of the available data on the concentrations of various solutes in the IOFs and in blood plasma clearly indicates that the chemical composition of the microenvironment of the retina is distinctly different from that of the anterior segment of the eye and from the ECFs of most other tissues, although it appears to resemble closely the microenvironment of the brain. In general, as compared to blood plasma, this microenvironment is characterized by high magnesium and ascorbic acid concentrations, low concentrations of amino acids, glucose, potassium, and phosphate, and extremely low concentrations of soluble proteins. There is also evidence that the concentration of at least some of these solutes is maintained in the microenvironment of the retina within narrower ranges than the physiologically tolerable fluctuations in the concentrations of these solutes in blood plasma. This indicates the existence of local homeostatic mechanisms.

The retina, because of its easy accessibility and relatively simple organization, is an ideal region of the CNS for *in vivo* and *in vitro* studies. Because of its thinness, the retina can be expected to survive *in vitro* for prolonged periods if maintained in an appropriate solution in a relatively undamaged state. However, solutions that were originally designed for use in gut baths or for maintaining cell lines derived from peripheral tissues or tumors cannot be regarded as appropriate for the retina. An effort should therefore be made to develop optimal physiological solutions based on the chemical microenvironment of the normal *in situ* retina for the *in vivo* suprafusion or for the *in vitro* maintenance of this organ or cells derived from it. It must be remembered, however, that there are considerable species variations in the normal chemical environment of the retina, so that a physiological solution ideally suited for the maintainance of the retina of one species cannot be assumed to be equally suitable for other species.

ACKNOWLEDGMENTS. The preparation of this chapter and much of the work emanating from the author's laboratory and reported herein were supported by USPHS Research Grant EY 00333 from the National Eye Institute. The author wishes to thank R. A. Baroody, E. M. Klein, and C. Hoopes for their invaluable assistance with the preparation of this manuscript.

REFERENCES

1. Bito, L. Z., Davson, H., and Fenstermacher, J. D. (eds.), 1977, *The Ocular and Cerebrospinal Fluids*, Academic Press, London.
2. Bito, L. Z., and DeRousseau, C. J., 1980, *The Blood–Retinal Barriers* (J. G. Cunha-Vaz, ed.), Plenum Press, New York, pp. 133–163.

3. Davson, H., 1969, *The Eye*, 2nd ed., Volume 1 (H. Davson, ed.), Academic Press, London, pp. 67–186.
4. Davson, H., 1969, *The Eye*, 2nd ed., Volume 1 (H. Davson, ed.), Academic Press, London, pp. 187–272.
5. Bill, A., 1975, *Physiol. Rev.* **55**:383–417.
6. Cole, D. F., 1977, *Exp. Eye Res.* **25**(Suppl.):161–176.
7. Bito, L. Z., 1977, *Exp. Eye Res.* **25**(Suppl.):273–289.
8. Tripathi, R. C., 1977, *Exp. Eye Res.* **25**(Suppl.):403–407.
9. Bill, A., 1977, *Exp. Eye Res.* **25**(Suppl.):291–304.
10. Balazs, E. A., and Denlinger, J. L., 1983, *The Eye*, 3rd ed., Volume 1A (H. Davson, ed.), Academic Press, London, pp. 533–589.
11. Steinberg, R. H., and Miller, S. S., 1979, *The Retinal Pigment Epithelium* (K. M. Zinn and M. F. Marmor, eds.), Harvard University Press, Cambridge, pp. 205–225.
12. Zinn, K. M., and Marmor, M. F. (eds.), 1979, *The Retinal Pigment Epithelium*, Harvard University Press, Cambridge.
13. Chase, J., 1982, *Ophthalmology* **89**:1518–1525.
14. Dowling, J. E., Proenza, L. M., and Atwell, C. W. (eds.), 1982, *Retina* **2**:231–375.
15. Denlinger, J. L., Eisner, G., and Balazs, E. A., 1980, *Exp. Eye Res.* **31**:67–79.
16. Balazs, E. A., Laurent, T. C., Laurent, U. B. G., DeRoche, M. H., and Bunney, D. M., 1959, *Arch. Biochem. Biophys.* **81**:464–479.
17. Maurice, D. M., *The Blood–Retinal Barriers* (J. G. Cunha-Vaz, ed.), Plenum Press, New York, pp. 165–178.
18. Bito, L., and Davson, H., 1964, *Exp. Eye Res.* **3**:283–297.
19. Kinsey, V. E., 1953, *Arch. Ophthalmol.* **50**:401–417.
20. Eakins, K. E., 1977, *Exp. Eye Res.* **25**(Suppl.):483–498.
21. Raviola, G., 1974, *Invest. Ophthalmol.* **13**:828–858.
22. Zinn, K. M., and Benjamin-Henkind, J. V., 1979, *The Retinal Pigment Epithelium*, Harvard University Press, Cambridge, pp. 1–31.
23. Bito, L. Z., and Davson, H., 1966, *Exp. Neurol.* **14**:264–280.
24. Bito, L. Z., 1969, *Science* **165**:81–83.
25. Bradbury, M., 1979, *The Concept of a Blood–Brain Barrier*, John Wiley & Sons, New York.
26. Bradbury, M. W. B., and Sarna, G. S., 1977, *Exp. Eye Res.* **25**(Suppl.):249–257.
27. Wise, G. N., Dollery, C. T., and Henkind, P. (eds.), 1971, *The Retinal Circulation*, Harper & Row, New York.
28. Betz, A. L., and Goldstein, G. W., 1980, *Exp. Eye Res.* **30**:593–605.
29. Cuhna-Vaz, J. G. (ed.), 1980, *The Blood–Retinal Barriers*, Plenum Press, New York.
30. Marmor, M. F., Abdul-Rahim, A. S., and Cohen, D. S., 1980, *Invest. Ophthalmol. Vis. Sci.* **19**:893–903.
31. Bito, L. Z., and Salvador, E. V., 1970, *Exp. Eye Res.* **10**:273–287.
32. Cole, D. F., 1959, *Br. J. Ophthalmol.* **43**:268–287.
33. Bito, L. Z., and Roberts, J. C., 1974, *Comp. Biochem. Physiol.* **47A**:173–193.
34. Cole, D. F., 1972/73, *Ophthalmic Res.* **4**:1–7.
35. Tron, E., 1927, *Albrecht von Graefes Arch. Klin. Exp. Ophthalmol.* **118**:713–722.
36. Duke-Elder, W. S., 1930, *Br. J. Ophthalmol.* (Suppl.) **4**:1–72.
37. Davson, H., 1939, *J. Physiol.* (Lond.) **96**:194–201.
38. Heywood, R., and Street, A. E., 1974, *Res. Vet. Sci.* **17**:401–403.
39. Heywood, R., and Street, A. E., 1975, *J. Med. Primatol.* **4**:296–298.
40. Gaasterland, D. E., Pederson, J. E., MacLellan, H. M., and Reddy, V. N., 1979, *Invest. Ophthalmol. Vis. Sci.* **18**:1139–1150.
41. Blumenfeld, T. A., Mantell, C. H., Catherman, R. L., and Blanc, W. A., 1979, *Am. J. Clin. Pathol.* **71**:219–223.
42. Davson, H., and Luck, C. P., 1957, *J. Physiol.* (Lond.) **137**:279–293.
43. Maren, T. H., 1974, *Invest. Ophthalmol.* **13**:479–484.
44. Becker, B., 1957, *Arch. Ophthal.* **57**:793–800.
45. Reddy, D. V. N., and Kinsey, V. E., 1960, *Arch. Ophthalmol.* **63**:715–720.
46. Duke-Elder, W. S., 1927, *Biochem. J.* **21**:66–77.

47. Wistrand, P., Nechay, B. R., and Maren, T. H., 1961, *Acta Pharmacol. Toxicol.* **17**:315–336.
48. De Benardinis, E., Tieri, O., Polzella, A., and Iuglio, N., 1965, *Exp. Eye Res.* **4**:179–186.
49. Green, H., Bocher, C. A., and Leopold, I. H., 1955, *Arch. Ophthalmol.* **53**:472–477.
50. Green, H., Sawyer, J. L., and Leopold, I. H., 1957, *Arch. Ophthalmol.* **57**:85–89.
51. Aurricchio, G., and Ambrosio, A., 1953, *Boll. Soc. Ital. Biol. Sper.* **29**:1172–1175.
52. Davson, H., 1949, *Br. J. Ophthalmol.* **33**:175–182.
53. Heyndrickx, G. V., and Peeters, G., 1958, *Q. J. Exp. Physiol.* **43**:174–179.
54. Coe, J. I., 1969, *Am. J. Clin. Pathol.* **51**:741–750.
55. Bito, L. Z., 1970, *Exp. Eye Res.* **10**:102–116.
56. Salit, P. W., 1939, *Biochem. Z.* **301**:253–266.
57. Tron, E., 1926, *Albrecht von Graefes Arch. Klin. Exp. Ophthal.* **117**:677–692.
58. Sturner, W. Q., and Gantner, G. E., 1964, *Am. J. Clin. Pathol.* **42**:137–144.
59. Werner, B., 1966, *Acta Chir. Scand.* **132**:63–76.
60. Kinsey, V. E., and Reddy, D. V. N., 1965, *Invest. Ophthalmol.* **4**:104–116.
61. Kinsey, V. E., and McLean, I. W., 1970, *Invest. Ophthalmol.* **9**:769–784.
62. McLean, I., La Force, R. C., and Kinsey, V. E., 1969, *Docum. Ophthal.* **26**:171–183.
63. Bito, L. Z., Salvador, E. V., and Petrinovic, L., 1978, *Exp. Eye Res.* **26**:47–55.
64. Bito, L. Z., Davson, H., Levin, E., Murray, M., and Snider, N., 1966, *J. Neurochem.* **13**:1057–1067.
65. Palm, E., 1949, *Acta Ophthalmol.* **27**:553–562.
66. Christiansson, J., and Palm, E., 1954, *Acta Ophthalmol.* **32**:197–212.
67. Fisher, F. P., 1949, *Physiology of the Eye* (H. Davson, ed.), Blakiston, Philadelphia, p. 42.
68. Schultz, R. H., Fahning, M. L., and Graham, E. F., 1971, *J. Reprod. Fertil.* **27**:355–367.
69. Walker, A. M., 1933, *J. Biol. Chem.* **101**:269–287.
70. Naumann, H. N., 1959, *Arch. Ophthalmol.* **62**:356–363.
71. Reim, M., Cattepoel, H., Bittmann, K., and Kilp, H., 1969, *Albrecht von Graefes Arch. Klin. Exp. Ophthal.* **177**:355–368.
72. Reim, M., Lax, F., Lichte, H., and Turss, R., 1967, *Ophthalmologica* **154**:39–50.
73. Bourwieg, H., Hoffmann, K., and Riese, K., 1974, *Albrecht von Graefes Arch. Klin. Exp. Ophthalmol.* **191**:53–65.
74. Davson, H., and Duke-Elder, W. S., 1948, *J. Physiol. (Lond.)* **107**:141–152.
75. Bruun Laursen, A., 1975, *Acta Ophthalmol.* **53**:369–377.
76. Heath, H., Beck, T. C., Rutter, A. C., and Greaves, D. P., 1961, *Vision Res.* **1**:274–286.
77. Ringvold, A., 1980, *Acta Ophthalmol.* **58**:69–82.
78. Sullmann, H., 1951, *Tabulae Biologicae*, Volume 22, *Oculus*, Part 2, (F. P. Fischer, K. Steindorff, J. S. Friedenwald, J. G. Van Manen, and A. Sorsby, eds.), Dr. W. Junk, Gravenhage, pp. 1–119.
79. Stubbs, D. W., and McKernan, J. B., 1967, *Proc. Soc. Exp. Biol. Med.* **125**:1326–1328.
80. Errington, B. J., Hodgkiss, W. S., and Jayne, E. P., 1942, *Am. J. Vet. Res.* **3**:242–247.
81. Ross, K. S., and Macri, F. J., 1975, *Invest. Ophthalmol.* **14**:942–944.
82. Schonheyder, F., Ehlers, N., and Hust, B., 1976, *Ophthalmic. Res.* **8**:64–80.
83. Ehlers, N., 1981, *Acta Ophthalmol.* **59**:576–586.
84. Reddy, V. N., Thompson, M. R., and Chakrapani, B., 1977, *Exp. Eye Res.* **25**:555–562.
85. Reddy, D. V. N., Rosenberg, C., and Kinsey, V. E., 1961, *Exp. Eye Res.* **1**:175–181.
86. Bito, L. Z., Davson, H., Levin, E., Murray, M., and Snider, N., 1965, *Exp. Eye Res.* **4**:374–380.
87. Reddy, D. V. N., Chakrapani, B., and Lim, C. P., 1977, *Exp. Eye Res.* **25**:543–554.
88. Hayes, K. C., and Sturman, J. A., 1981, *Ann. Rev. Nutr.* **1**:401–425.
89. Hill, A. E., 1975, *Proc. R. Soc. Lond. [Biol.]* **190**:99–114.
90. Wurster, U., Riese, K., and Hoffman, K., 1982, *Acta Ophthalmol.* **60**:729–741.
91. Davson, H., 1972, *Physiology of the Eye*, 3rd ed., Academic Press, New York, p. 29.
92. Drinker, C. K., and Yoffey, J. M., 1941, *Lymphatics, Lymph and Lymphoid Tissue*, Harvard University Press, Cambridge, p. 152.
93. Nordmann, J., 1968, *Biologie et Chirugie du Corps Vitre* (A. Brini, A. Bronner, J.-P. Gerhard, and J. Nordmann, eds.), Masson, Paris, pp. 95–128.
94. Gershbein, L. L., 1981, *Res. Commun. Chem. Pathol. Pharmacol.* **32**:167–178.

95. Duke-Elder, S., and Davson, H., 1949, *Br. J. Ophthalmol.* **33**:21–38.
96. Glenn, W. W. L., Muus, J., and Drinker, C. K., 1943, *J. Clin. Invest.* **22**:451–460.
97. Courtice, F. C., and Morris, B., 1955, *Q. J. Exp. Physiol.* **40**:138–148.
98. Bessiere, E., Crockett, R., Le Rebeller, M. J., Maurain, C., and Grenie, D., 1973, *Albrecht Graefes Arch. Klin. Exp. Ophthal.* **187**:273–288.
99. Francois, J., Rabaey, M., and Evens, L., 1958, *Arch. Ophthalmol.* **59**:692–702.
100. Gershbein, L. L., Dan, T. C., and Shurrager, P. S., 1975, *Enzyme* **20**:165–177.
101. Kinsey, V. E., 1947, *Am. J. Ophthalmol.* **30**:1262–1266.
102. Becker, B., and Linner, E., 1952, *Acta Physiol. Scand.* **26**:79–85.
103. Becker, B., 1967, *Invest. Ophthalmol.* **6**:410–415.
104. Hockwin, O., Kietzmann, M. T., Vahar-Matiar, H., and Edelbi, A., 1978, *Ophthalmic Res.* **10**:250–258.
105. Balazs, E. A., Toth, L. Z. J., Jutheden, G. M., and Collins, B., 1965, *Exp. Eye Res.* **4**:237–48.
106. Bito, L. Z., DiBenedetto, F. E., and Stetz, D., 1982, *Exp. Eye Res.* **34**:229–237.
107. Altman, P. L., and Dittmer, D. S., 1974, *Biology Data Book*, 2nd ed., Volume 3, Federation of American Societies for Experimental Biology, Bethesda.

21

Retina

Nicolas G. Bazan and T. Sanjeeva Reddy

1. INTRODUCTION

In the past 15 years, since Richard N. Lolley wrote his inspiring overview of the retina in the first edition of this *Handbook*,[1] the biochemical techniques used in retina research have become increasingly sophisticated and progressively more widespread, yielding many advances that have been reviewed elsewhere in considerable detail.[2-16] In retinal neurochemistry, as in other areas of neurobiology, the need for a multidisciplinary approach has been recognized, and such approaches have been, in fact, applied. The current upsurge in interest in the retina is a result of two major factors: recent progress in vision science and the use of the retina as a model for the central nervous system.

In the last decade, vision scientists and ophthalmologists have adapted, modified, and developed many biochemical techniques to further our understanding of both visual cells and the neural retina in terms of normal visual processes and abnormal conditions and disorders such as retinitis pigmentosa and diabetic retinopathy. Considerable attention has been focused on underlying mechanisms, and although much of this work is still in the relatively early stages of development, a basic science foundation is being constructed on which to build future clinical applications.

In brain research, the suitability of the retina as an experimental model for the central nervous system is based on its similarity to cerebral cortex (excluding the visual cells); the two tissues are relatively indistinguishable from the standpoint of embryology, phylogenetic origin, cellular organization, ultrastructure, physiological properties, and biochemistry. One major advantage of the retinal model for the central nervous system is that its major physiological stimulus, light, can be easily used in experimentation. Also, obtaining retina for *in vitro* studies involves simpler dissecton and less trauma to the tissue compared to the preparation of brain slices.[17] Although retinal preparations require the severing of the optic nerve and anterior sectioning behind the ora

---

Nicolas G. Bazan and T. Sanjeeva Reddy • Lions Eye Research Laboratories, LSU Eye Center, Louisiana State University Medical Center School of Medicine, New Orleans, Louisiana 70112.

serrata, the result is a specimen with two undamaged surfaces, i.e., a "natural" slice of cerebral cortex.

Radiolabeled precursors, drugs, or other compounds are often administered directly to the brain by intracerebral, intracisternal, or intraventricular routes. In the retina, injection into the vitreous body is analogous to these procedures. Furthermore, because the axons of the retinal ganglion cells form the optic nerve and thus have their synaptic endings in the brain, we can use this system to follow the axoplasmic transport of substances injected into the vitreous body and taken up by the ganglionic cells as well as the influences of light stimulation on the synaptic chemistry of the ganglion cell.

The salient biochemical developments in retinal neurochemistry over the past 15 years can be summarized as follows.

The structures of the polypeptide and carbohydrate moieties of the rhodopsin molecule in the visual cells have been elucidated. In addition, the biosynthesis of rhodopsin, including related cytological events, has been studied and the specific presence of fucose has been pinpointed in cone receptors.

A light-triggered enzyme cascade has been shown to amplify the events subsequent to photon absorption, and the central role of Ca^{2+} in visual function has been documented.[18-21] Also, the presence and functional roles of minor proteins in the outer segments of photoreceptors have been investigated. We have included some of these topics in this chapter; others can be found in Chapter 22 (H. Shichi, this volume).

An active interrelationship has been demonstrated between the photoreceptor cells of the retina and the cells of the retinal pigment epithelium, with which they are in intimate contact. It has been shown that the pigment epithelium participates actively in the renewal of photosensitive membranes in the rods and cones and that there are daily rhythms of retinal metabolism. The phagocytosis of photoreceptor membranes and the role of surface cell receptors and lipofuscin granules in the pigment epithelial cells are being explained in biochemical terms. Our knowledge of the metabolism of vitamin A in the retinal pigment epithelium and its transport between photoreceptors and pigment epithelium has been greatly expanded. Retinoid-binding proteins of the retina, photoreceptors, interphotoreceptor matrix, and pigment epithelium have been identified, isolated, and purified, and their roles in the transport of retinoids and of other substances have been studied.[22]

Cyclic nucleotides have proven to be of great interest because cyclic GMP appears to play a role in the metabolism and/or function of rod-dominant retinas. Furthermore, in certain retinal degenerations, this nucleotide accumulates as a result of a reduced level of the catabolic enzyme, cyclic GMP phosphodiesterase. In addition, the relationship between cyclic nucleotide metabolism and the transduction of light has received considerable attention.[19,23] Not only are cyclic AMP and adenylate cyclase enriched in photoreceptors of cone-dominant retinas, but they also form an integral part of the dopaminergic neurotransmitter system in the neural portion of the retina.

Major advances have been made in the exploration of synaptic chemical communication between retinal cells. Techniques developed for transmitter neurobiology studies in the brain have been adapted and applied to the retina.

The cellular localization of neurotransmitters and the role of uptake systems in retinal function are being investigated. Biochemical evidence has established the identity of several neurotransmitters: the roles of acetylcholine, dopamine, GABA, and glycine have been well documented; the roles of 5-hydroxytryptamine, glutamate, and aspartate have been reasonably well documented; and those of norepinephrine, taurine, and serine less so. The physiological regulation of various neurochemical processes involving neurotransmitters (e.g., light-stimulated dopamine metabolism, acetylcholine release, and GABA uptake) has been demonstrated, and a variety of retinal neuropeptides, some of which are potent stimulators of retinal adenylate cyclase, have been identified.

Extensive studies have been carried out on the phospholipid, fatty acid, neutral lipid, and ganglioside content and composition in the retina. New information on the metabolism of membrane lipids has provided insight into membrane dynamics and opened the way to a better understanding of the regulation of membrane functions in the visual cells and in the neural retina. The main findings center on the enrichment of docosahexaenoic acid in photoreceptor membranes, the high rate of *de novo* biosynthesis of phospholipids, the effects of drugs on the synthesis of retinal membrane lipids, and the light-induced modifications of retinal lipids.

The aim of the present review is threefold—first, to update our current knowledge of the retina; second, to bring together and integrate various aspects of our knowledge to serve as a ready source of information for young scientists and others with a nascent interest in this area; and third, to provide a comprehensive bibliography covering the present body of information available in the field of retinal research today.

2. CELLULAR ORGANIZATION

Retina contains six types of neuronal cells, namely, photoreceptor cells (rods and cones), bipolar cells, horizontal cells, amacrine cells, interplexiform cells, and ganglion cells, as well as glial cells known as Müller cells. These cells and the pigment epithelium are organized in ten layers (see Fig. 1):

1. The retinal pigment epithelium (RPE), which is in intimate contact with the outer segments of the photoreceptor cells. This cell layer serves as a mediator in the exchange of various substances between the outer retina and the choroidal circulation,[13] participates in the phagocytosis of disks shed from the rods and cones, and plays a role in the catabolism of photoreceptor outer segment components.[13-15]
2. The photoreceptor outer segment layer (POS), containing mainly the outer segments of rods and cones.
3. The outer limiting membrane (OLM), a network of Müller cell apical processes that is contiguous with the photoreceptor cell inner segments.
4. The outer nuclear layer (ONL), composed of photoreceptor cell nuclei.

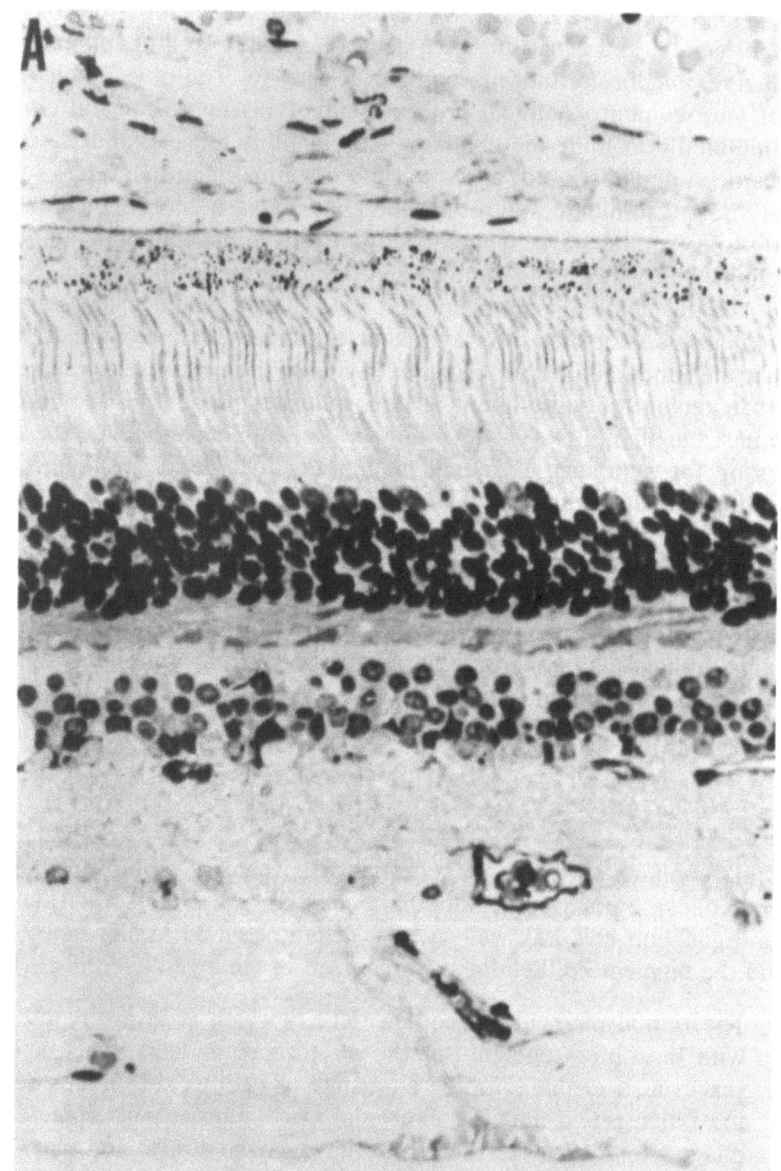

Fig. 1. A: Histological cross section of a human retina. B: Schematic diagram depicting the cellular layers and representative cell types of the retina and their spatial arrangement. a, amacrine cell; b, bipolar cell; c, cone; g, ganglion cell, h, horizontal cell; i, interplexiform cell; r, rod; cap, capillary; BM, Bruch's membrane; RPE, retinal pigment epithelium; POS, photoreceptor outer

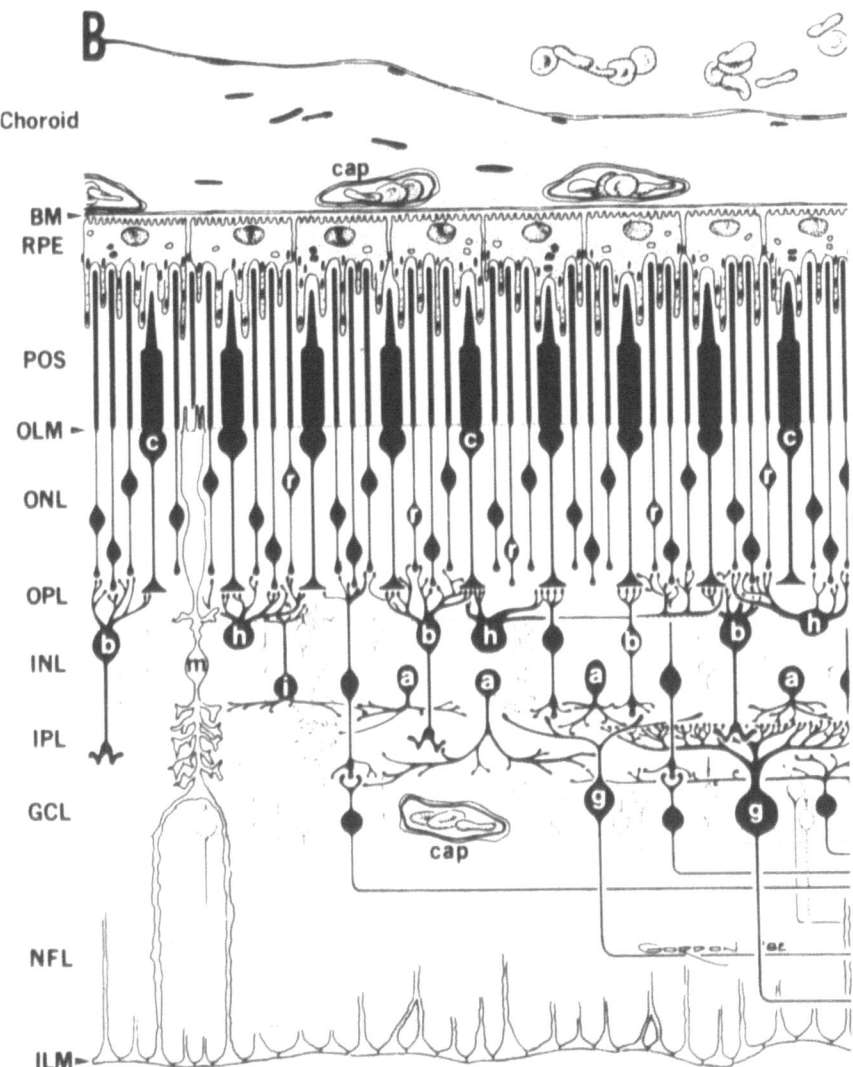

segment; OLM, outer limiting membrane; ONL, outer nuclear layer; OPL, outer plexiform layer; INL, inner nuclear layer; IPL, inner plexiform layer; GCL, ganglion cell layer; NFL, nerve fiber layer; ILM, inner limiting membrane. (Kindly provided by Drs. S. Fliesler and R. E. Anderson from ref. 8 with publisher's permission.)

5. The outer plexiform layer (OPL), comprising the synaptic region where photoreceptor, bipolar, and horizontal cells are in contact with one another.
6. The inner nuclear layer (INL), where the cell bodies of the bipolar, horizontal, and amacrine cells are located, as well as the interplexiform cells, which are in contact with the outer and inner plexiform layers.
7. The inner plexiform layer (IPL), the other synaptic region of the retina, where synaptic contacts between bipolar, amacrine, and ganglion cells are established.
8. The ganglion cells, from which extend the axons that form the optic nerve.
9. The nerve fiber layer (NFL), containing the collective axonal processes of ganglion cells that ultimately exit the eye and form the optic nerve.
10. The inner limiting membrane (ILM), made up of the network of Müller cell basal processes forming the retinal–vitreal interface. In contrast to the retinal neurons, the Müller cells extend almost all the way through the retina from the vitreous interface to the photoreceptor cell layer.

The nutrient and blood supplies to the retina are maintained by two independent circulatory systems.[24] The choroidal system is located between the sclera and the RPE and consists of an arborized network of fenestrated capillaries. The RPE and choroid are separated by a thin endothelial lining and Bruch's membrane (BM) (see Fig. 1). A second blood supply enters the retina through the optic disk in most animal species. This system differs from the choroidal supply in that it does not contain fenestrated capillaries and is therefore restricted to passive exchange of small molecules, ions, and gases.[25]

3. PHOTORECEPTOR CELLS

There are two distinct types of photoreceptor cells in the retina, rods and cones. As the names imply, the rod cells are thin and elongated, and the cone cells are shorter and in some species have a conical outer segment. Functionally, the rods are sensitive to low levels of light and cannot withstand high light intensities. Cones, however, are insensitive to low levels of light and sensitive to high intensities and thus are specialized for color vision. The proportion of rods to cones varies among species; with very few exceptions, the number of rods is much higher than the number of cones. In the rat and other nocturnal animals, the number of cones is very low. In the human retina, the rods outnumber the cones by 100 million to 6 million.[26] Large numbers of cones are found in lizards, whose retinas are devoid of rods,[27] chickens, whose cone:rod radio is 6:1,[28] and goldfish, whose cone:rod ratio varies from 6:1[29] to 15:1.[30]

The typical rod cell can be subdivided into several regions: the rod outer segment (ROS), the connecting cilium (CC), the rod inner segment (RIS), the cell body (CB), the inner fiber (IF), and the synaptic pedicle (SP). The outer segment is located at the apex of the cell and extends into the pigment epithelial layer that forms the outer margin of the retina. The ROS contains a rigid, ordered stalk composed of 500–2000 flattened membranous sacs (disks) enclosed by the plasma membrane of the cell. Most of the disks in the outer segments float freely; i.e., they are not in physical contact with each other or the plasma membrane[31–35] except for some that are localized at the base of the ROS. The human rod outer segment is about 1–2 µm in diameter and 20–40 µm in length and contains approximately 600 to 1000 disks,[36] whereas the frog ROS is about 5–7 µm × 35–50 µm and contains about 1700 to 2000 disks.[37] The ROS is connected to the inner segment by a stalk called the connecting cilium (CC), which joins the two cellular compartments. The RIS contains mitochondria, smooth and rough endoplasmic reticulum, and Golgi bodies. The cell body (CB), which contains a large nucleus, is attached to the inner segment. The inner fiber or the axonal process connects the cell body with the synaptic pedicle (SP) of the cell.

Cones have a structure generally similar to that of rods. However, the outer and inner segments and cell body are usually broader, and the synaptic pedicle is larger than that in the rod. Unlike the situation in ROS, the disks in the cone outer segments are open to the extracellular space because they are formed by infoldings of the outer cell membrane and are not free-floating entities. Also, rods and cones have different pigments; the rods have rhodopsin, whereas cone outer segments contain at least three other pigments that mediate color vision.

4. CHEMICAL COMPOSITION AND METABOLISM

Our current understanding of the chemical composition and metabolic pathways of carbohydrates, lipids, nucleic acids, and proteins in the retina has been derived from a number of different approaches. Work has been done with a variety of retinal preparations, including whole retina (e.g., *in vivo* experiments, following *in vitro* incubation, or direct extraction of the tissue following excision), subcellular fractions (e.g., rod outer segments, synaptosomes, microsomes), isolated cells, or cross-section slices. The complexity of the preparations is often a concern in the experimental design. The microsomal fraction, for instance, is a composite obtained from the endoplasmic reticulum of various retinal cells containing fragments of other cellular membranes as well. Isolated cell preparations have proven useful in the study of the electrical and biochemical properties of retinal cells. Microdissected, lyophilized, and frozen cross-sections of the retinal layers have provided valuable information. In addition, there is increasing evidence that each of the distinct layers of the retina can be divided into numerous subpopulations of cells. For example, the INL is made up of at least eighteen different types of cells, whose identities are derived from functional subdivisions of the horizontal, bipolar, and amacrine cells. The IPL consists of about 12 sublayers separated on the basis of physiological criteria, and the photoreceptor layer in amphibians contains at least four subpopulations of receptors.

In retinal research today, there is a definite trend toward gaining insight into biochemical events at the cellular and subcellular levels and correlating these findings with the retinal circuitry to enhance our knowledge of the molecular basis of information flow through the retina. To accomplish some of these aims, the cellular localization of biochemical sequences has been demonstrated by light and electron microscopic autoradiography.

4.1. Nucleic Acids

The nucleic acids have been used as quantitative indices of cell number (DNA) and cell volume (RNA).[38,39] A unique feature of the retina is its high content of DNA compared to RNA. The values for DNA and RNA vary in different species, e.g., for mice 11 and 3.5,[40] rat 10 and 2.5, and rabbit 8 and 2.25[41] mg/g wet tissue DNA and RNA, respectively. The DNA/RNA ratio of the retina, which is between 3 and 4, is much higher than that of other tissues such as gray matter (0.66), white matter (1.25),[42] and liver (0.2).[43] This high ratio of DNA/RNA in retina may reflect the densely packed photoreceptor cell populations, which have a small cytoplasmic volume.[40,44] A large ratio (15) of DNA/RNA has been reported in photoreceptor cells of the rabbit[44]; however, this was not confirmed in studies by Lolley,[40] and it was suggested that the high ratio found in the earlier work may have been a result of leakage of RNA from the cells during dissection. Lolley reported a slightly higher DNA/RNA ratio in the photoreceptor layer compared to the inner layers of the mouse retina.[40]

Bok has studied the properties of RNA in the nucleus and inner segments of rods in the rat and the frog.[45] Using [^{3}H]cytidine, he demonstrated that RNA is continuously renewed in the mature visual cells of the retinas of rats and frogs and that the synthesis of RNA takes place only in the nucleus.[45–47] More recently, Dawson and LaVail[48] used [^{3}H]thymidine autoradiography to demonstrate that peak cone cell genesis in the mouse retina occurs between the 13th and 14th days of embryonic development and ceases by the 16th day. In contrast, rod cell genesis reaches a peak on the eighth postnatal day and slows by the 15th day.[48] Although the DNA, once formed, does not undergo renewal, Young[49] reported the incorporation of tritiated thymidine into the DNA of nuclei in irradiated cells exposed to UV light. He suggested that specific repair mechanisms may compensate for the nonrenewal of the DNA.

During embryonic development, the amount of DNA and RNA in retina increases rapidly from birth to the fifth day of postnatal life in mice.[40] In rats, the increase occurs from birth to 2 weeks,[50] and the adult level is attained by the 20th day of postnatal life. This observation is consistent with rapid cell division and maturation in retina during this period.[51]

4.2. Proteins

Proteins form one of the major components of the retina. The relative amounts of retinal protein (40–50 mg/g wet tissue) are similar in various animals, such as cattle, rats, frogs, and toads.[50,52] Also, the amount of protein in the rat retina shows no appreciable changes during development.[50] Studies of retinal protein metabolism have been concerned mainly with following the time course of metabolic processes by means of labeled amino acids. Most of these studies have been restricted to the photoreceptor cells and the visual protein, rhodopsin, which forms about 80% of the photoreceptor membranes.

Labeled amino acids have been used to demonstrate active protein synthesis in the whole retina both *in vitro* and *in vivo* by autoradiographic[53] and biochemical[54] techniques. The autoradiographic studies carried out by Steinman and Ames[55] using [^{3}H]leucine *in vitro* showed high rates of protein synthesis in the inner segments of the photoreceptor cells, perikarya of ganglion cells, and cells of the inner nuclear layer, low rates of synthesis in receptor cell bodies, plexiform layers, and Müller cells, and no synthesis in receptor cell outer segments. Similar *in vivo* studies have been reported in frogs, rodents, and primates.[27,53,56]

Ames and his associates[57-60] used [^{14}C]- or [^{3}H]leucine to perform a more detailed study of protein turnover in the retina. Total protein synthesis occurred at a rate of 103 nmol/g protein per h. Both rate of synthesis and turnover appeared to be related to molecular weight. Proteins with molecular weights of 33–43,000 daltons were synthesized more rapidly (2 nmol/g per h); those above and below these weights were synthesized more slowly. Mean protein turnover was 0.52%/h, slower for smaller proteins (0.1% at about 10,000 daltons) and faster for larger proteins (1.4% at about 140,000 daltons). Turnover was little affected by continuous light or flashing light of increasing intensity.[60] Similarly, diurnal light–dark cycles (12L–12D) did not cause measurable changes[61]; however, major differences in protein synthesis throughout the retina were observed in animals kept in continuous darkness for several days.

As mentioned above, a great deal of work has been carried out on the synthesis and renewal of proteins in photoreceptor cells.[62-64] Virtually all new protein molecules are synthesized in the myoid region of the photoreceptor cells, where rough and smooth endoplasmic reticulum are located.[65] A few proteins enter the nucleus and synaptic body, and a few are transported to mitochondria, but the majority are transported to the outer segment via the membrane stalks through the connecting cilium.[65]

Opsin, which forms more than 80% of the total protein in the rod cell, is a glycoprotein that contains mannose and N-acetylglucosamine moieties in bovine retina.[66-69] The peptide moiety of opsin is phosphorylated to form the glycophosphoprotein rhodopsin, which is then sequentially glycosylated first on the ribosomes and then in the Golgi bodies of the myoid region.[70] Although the metabolism and characterization of this rod visual pigment has been studied extensively, there are few data on the visual pigments of cone cells. In the chicken, the cone visual pigment, iodopsin, appears to be a glycoprotein.[71] More recently it has been reported that [^{3}H]fucose is incorporated in large amounts into red- and blue-sensitive cones and in low amounts into green-sensitive cones.[72] The labeled molecule has been isolated and found to be an integral membrane protein with a molecular weight of 38,000. In the goldfish retina, fucosylated protein occurs in cone outer segments but not in rod outer segments.[72] The presence of rhodopsin[73-75] and other minor high-molecular-weight proteins[76,77] has been reported in rod outer segment plasma membrane.

4.3. Carbohydrates

The general carbohydrate metabolism in the retina follows a pattern similar to that of other tissues such as liver and brain. Most of the work carried out in the late 1960s has been summarized in detail by Lolley in Volume 1 of the first edition of this series.[1] Therefore, only selected features of the carbohydrate metabolism are discussed here.

Glucose is the major source of energy in the adult retina. However, in young animals, the retina can also use fatty acids for energy purposes.[78] There is a rapid transport of glucose from blood into the retina.[79,80] The retina has been shown to have the highest capacity for aerobic glycolysis of any tissue *in vitro*; oxygen is used more rapidly than in any other tissue, and a large quantity of lactic acid is produced.[1,81,82] Retinal glycolysis is so active that the citric acid cycle cannot keep pace, and the rate of respiration is twice that of brain cortex. Aerobic lactic acid production is sixfold, and anaerobic production threefold that of brain cortex.[83] However, retina does not accumulate lactic acid *in vivo*, suggesting that the supply of oxygen to the tissue is adequate.[78]

4.3.1. Glycolysis

In retina, as in other tissues, the Embden–Meyerhof pathway converts glucose to pyruvate and is controlled by phosphofructokinase and hexokinase enzymes.[84] The pyruvic acid is metabolized into carbon dioxide and water in the presence of oxygen. In the adult retina, lactic acid accumulates aerobically and anaerobically. The mechanism by which lactic acid is produced under aerobic conditions is not clear. The enzymes, subtrates, and cofactors of glycolysis have been measured in retinas from different species such as rat, rabbit, monkey and frog.[1,85–89] Studies of retinal development in rats show enzyme changes in the early postnatal period. The activities of hexokinase, phosphofructokinase, and glyceraldehyde phosphate dehydrogenase increase with age between 4 and 20 days after birth, and the profile of lactate dehydrogenase isoenzymes also changes,[87–90] but the amount of glucose 6-phosphate dehydrogenase and the activity of total lactate dehydrogenase remain the same.[87,91] The isolated retina has been known to exhibit both Pasteur and Crabtree effects,[92] the latter decreasing with age in the developing rabbit retina.[93] The processes of glycolysis and the Krebs cycle seem to have been compartmentalized within different layers of the retina. Thus, the glycolytic enzymes and substrates are concentrated in the inner layers of the retina, where there are only small amounts of Krebs cycle enzymes.[85,89]

4.3.2. Respiration

Immature retina shows a low rate of glucose oxidation, which increases sixfold between 8 and 20 days after birth, after which it stabilizes at the adult level.[51,93,94] The main pathway for glucose oxidation in the retina is the citric acid or Krebs cycle, which takes place in mitochondria. The number of mitochondria per unit area in retina seems to be lower than that in cerebral cortex, but the mitochondria present are capable of carrying out glucose oxidation to produce ATP. The Krebs cycle intermediates, pyruvate, α-ketoglutarate, succinate, and glutamate, turn over rapidly in preparations of isolated retina and retinal mitochondria.[95-97] Using labeled glucose, Cohen and Noell found that more CO_2 is produced from C_3 and C_4 of glucose than from C_6. About 20% of the labeled glucose at the C_6 position enters into glutamate production, with a small amount going toward aspartate.[93] Glutamate serves as an immediate carbon reservoir for Krebs cycle intermediates.[98] In retina, the enzymes needed for the functioning of the Krebs cycle are localized mainly in the inner segments of the visual cells.[1,88]

4.3.3 Hexose Monophosphate Shunt

Much less glucose enters the hexose monophosphate shunt compared to the Krebs cycle in the retina. However, when electron acceptors are added to the incubation medium under anaerobic conditions, a higher proportion of glucose can be oxidized via the shunt.[93] The enzymes involved in the hexose monophosphate shunt are measurable in both immature and mature retina[54,93,99] and are localized mainly in photoreceptor and bipolar cell layers.[88,89] The decrease in shunt capacity in the adult animal results not from the reduction in enzymatic capability but from a restricted rate of oxidation of $NADPH_2$ in the tissue.[99,100]

4.4. Lipids

Lipids constitute about 20 to 30 mg/g wet weight of retinal tissue in various species[101,102] and about 20% on a dry-weight basis. The highest lipid concentration (34% by dry wt.) is found in the nerve fiber layer, and the lowest concentration (12%) in the outer nuclear layer, with an intermediate value (22-25%) for the outer retinal layer or photoreceptor layer.[103] When lipids classes were studied by more elaborate techniques, it was found that phospholipids and highly unsaturated fatty acyl chains are concentrated in the photoreceptors.

Among the lipid classes, phospholipids make up about 75% of the total lipids, followed by cholesterol (10-12%),[101,102] neutral glycerides and free fatty acids (5-7%),[104-107] and gangliosides (2%).[108,109] There does not seem to be much variation in the lipid composition among different species.[8] Phosphatidylcholine (40-50%) and phosphatidylethanolamine (30-35%) are the major phospholipids, followed by phosphatidylserine (5-10%), phosphatidylinositol (3-6%), and sphingomyelin (2-8%) in lesser amounts (Table I). Minor phospholipid classes include phosphatidic acid, cardiolipin, and lysophospholipids (Table I). However, a new method of isolation has demonstrated that phosphatidic acid amounts to about 0.5% of the total phospholipids[110] and is very active metabolically compared to other phospholipids.[111-113] Plasmalogens form 30-35% of the ethanolamine phosphoglycerides and 2-3% of choline phosphoglycerides.[114,115]

Analysis of the phospholipid composition of rat retinal subcellular fractions revealed that phosphatidylcholine (40-55%) and phosphatidylethanolamine

Table I
Phospholipid Composition of the Retina (% Total Lipid Phosphorus)

Class	Toad[a]	Rat[b]	Bovine		Human[e]
Phosphatidylcholine	50.1	48.2	43.2[c]	54.4[d]	47.8
Phosphatidylethanolamine	30.9	32.5	34.1	32.3	31.7
Phosphatidylserine	9.8	10.5	10.0	5.7	8.6
Phosphatidylinositol	3.5	4.0	5.6	3.3	4.4
Sphingomyelin	2.5	3.7	2.1	4.3	4.3
Phosphatidic acid	0.6	0.54	—	—	—
Diphosphatidylglycerol (Cardiolipin)	1.9	2.8	—	—	—
Lysophosphatidylcholine	0.3	—	0.2	—	0.3

[a] Bazan and Bazan.[157]
[b] Careaga and Bazan[113]
[c] Anderson et al.[119]
[d] Dorman et al.[115]
[e] Anderson.[133]

(20–40%) form the major fractions, followed by phosphatidylserine (4–10%), phosphatidylinositol (3–5%), cardiolipin (3–8%), and sphingomyelin (4%).[113] Phosphatidic acid amounts to 0.5–1.5% in different subcellular fractions from various species, with an enrichment in ROS and microsomes.[113,116] The phospholipid composition is similar in retinal subcellular fractions from different species such as rats,[110] and cattle[117–119] (Table II). The phospholipids are distributed nearly symmetrically over the two faces of the disk membrane in the photoreceptor cells, with slightly more phosphatidylcholine and phosphatidylethanolamine at the outer surface[120–122] Preferential localization of phosphatidylethanolamine on the outer surface of the rod outer segment membranes has also been reported.[32,123,124]

Although the retina is, in a way, an extension of the brain and is comparable to the gray matter in several respects, there are striking differences in the glycolipid concentration and composition of these two tissues. In adult chicken retina, the amounts of cerebrosides and sulfatides (about 590 and 295 nmol/g wet wt., respectively) are 15 and 8 times lower than the values for adult chicken brain.[125,126] Similarly the amounts of gangliosides in retina are five and ten times lower than in whole brain and gray matter, respectively. Although there are a number of reports describing ganglioside distribution patterns in various species (Table III), the results are not consistent because of differences in the experimental techniques used for separation and quantitation.[8]

In general, however, the gangliosides in retina differ appreciably from those in brain. In mammals, GD_3 is the major retinal ganglioside, whereas GD_{1a} is predominant in brain; GD_{1a}, GD_{1b}, and GT_{1b} are also present in moderate amounts in the mammalian retina.[125,127] During the development of the chick retina, the ganglioside content increases at three distinct stages: between 8 and 11 prenatal days, between the 16th prenatal day and hatching, and between the 18th postnatal day and adulthood.[125,128] Most of the changes in the ganglioside pattern occur before hatching. It was reported that the concentration and com-

Table II
Phospholipid Composition of the Retinal Subcellular Fractions (% Total Lipid Phosphorus)

Class	ROS			Nuclei (P$_1$)[d]			Mitochondria (P$_2$)[d]			Microsomes		
	Rat[a]	Toad[b]	Cattle[c]	Rat	Toad	Cattle	Rat	Toad	Cattle	Rat	Toad	Cattle
PC	40.3	42.7	35.7	52.9	46.4	57.9	49.3	49.1	46.4	54.6	54.4	54.8
PE	37.9	39.8	45.0	33.6	38.7	23.5	30.0	39.9	31.7	20.4	31.8	22.1
PS	10.5	14.2	15.8	3.8	11.6	6.9	9.1	11.6	9.0	6.8	7.9	7.2
PI	2.6	1.9	1.5	5.6	1.9	5.8	4.0	2.7	4.2	8.1	4.8	9.6
SPH	3.9	—	0.8	3.6	—	4.1	3.8	—	5.2	4.0	—	4.7
PA	0.5	1.4	—	0.8	1.4	—	0.3	1.7	—	1.2	1.1	—
DPG	2.7	—	—	—	—	—	7.8	—	—	2.7	—	—
OR	—	—	—	—	—	0.2	—	—	0.2	—	—	0.3
SF	—	—	0.8	—	—	1.3	—	—	3.4	—	—	1.5

[a] Careaga and Bazan.[113]
[b] Bazan et al.[52]
[c] Anderson et al.[117]
[d] The P$_1$ and P$_2$ fractions of Bazan et al.[52] and Careaga and Bazan[113] correspond to the crude nuclear and mitochondrial fractions. PC, phosphatidylcholine; PE, phosphatidylethanolamine; PS, phosphatidylserine; PI, phosphatidylinositol; SPH, sphingomyelin; PA, phosphatidic acid; DPG, diphosphoglycerides; OR, origin; SF, solvent front.

Table III
Ganglioside Composition of the Retina

Ganglioside	Rat[a]	Rabbit[a]	Cattle[b]
Total (μg NANA/g wet wt)	146	174	175
Individual gangliosides (% total)			
GM_3	6.1	3.8	2.8
GM_2	0	0	0.6
GM_1	2.7	7.3	2.1
GD_3	36.5	41.8	37.3
GD_{1a}[c]	11.8	16.5	13.2
GD_{1b}	17.7	8.2	15.0
GT_b	15.3	13.9	22.3
GQ_1	9.9	0.9	2.4

[a] From Dreyfus et al.[125]
[b] From Urban et al.[127]
[c] Migrates just below GD_{1a}.

position of various gangliosides in retinas from RCS rats exhibiting a hereditary retinal degeneration were not different from those of normal rats. This suggests that photoreceptor cells contain relatively small amounts of gangliosides compared to whole retina.[129]

One of the unique features of the retina is the very high content of polyenoic fatty acids esterified in the glycerolipids.[130,131] More than 50% of the fatty acids in the retinal lipids are unsaturated, and of these, more than 60% are polyunsaturated fatty acids derived from essential fatty acids.[132,133] The fatty acid composition of retinal phospholipids from different mammalian species seems to be similar.[105,106,134] On a relative weight basis, the fatty acid composition of phosphatidylcholine is 43% palmitic acid, 19% stearic acid, 23% oleic acid, 3% arachidonic acid, and 9% docosahexaenoic acid. In contrast, phosphatidylethanolamine and phosphatidylserine contain relatively high levels of stearic acid (29% and 41%, respectively) and docosahexaenoic acid (27% and 23%, respectively). Phosphatidylinositol contains large amounts of stearic acid (60%) and arachidonic acid (32%) but no docosahexaenoic acid. Sphingomyelin has 23% palmitic acid and 45% stearic acid but lacks polyunsaturated fatty acids.[134]

However, the fatty acid composition of lipids may vary among different species. Phospholipids from the toad retina contain large amounts of docosahexaenoic acid (40%), followed by palmitic and stearic acids (about 15% each).[106] In contrast, cattle and rabbit retinas contain equal amounts of these three fatty acids (about 25% each). Similarly, the fatty acid composition of neutral lipids, such as diglycerides and triglycerides, varies appreciably in rabbit, cattle, and toad retinas,[130,135] and the fatty acid composition of phosphatidic acid from toad retina differs from that of cattle.[130] Toad retina contains larger amounts of docosahexaenoic acid (33.3%), stearic acid (16.7%), and arachidonic acid (14.7%), whereas cattle retinas contain more stearic acid (26.3%) followed by palmitic acid (20%) and docosahexaenoic acid (14.9%). Thus, amphibian retinas seem to contain higher amounts of polyunsaturated fatty acids than do mammalian retinas.

Among the subcellular fractions, ROSs differ from other organelles in that they contain large amounts of polyunsaturated fatty acids[116] (Table IV). Similarly, large amounts of polyunsaturated fatty acids have been found in synaptic plasma membrane among the brain subcellular organelles[136–138] (Table IV). The fatty acid compositions of various phospholipids of ROS from different species do not differ appreciably,[118] and microsomes isolated from bovine retina also have a high content of polyunsaturated fatty acids.[139]

Recent studies have described the different molecular species of phospholipids from bovine ROS and microsomes.[140–142] In ROS, the supraenes make up 51% of the phosphatidylserine, 31% of the phosphatidycholine, 21% of the phosphatidylethanolamine, and 9% of the phosphatidylinositol. Hexaenes form 36% of the phosphatidylserine, 67% of the phosphatidylethanolamine, and 12% of the phosphatidylinositol, whereas tetraenes form 50% of the total phosphatidylinositol. A new supraenoic molecular species (didocosahexaenoyl) was found in several phospholipids. The disaturates and monoenes compose 14% and 6%, respectively, of the phosphatidylcholine. The phospholipids of microsomes, however, contain less of the docosahexaenoate-containing species and more of the saturated to tetraenoic species. Hexaenes form 50% and 42% of phosphatidylethanolamine and phosphatidylserine, respectively; tetraenes make up 71% of the phosphatidylinositol, and monoenes form 35% of the phosphatidylcholine. Supraenes form less than 15% of all the phosphoglycerides.

4.4.1. Glycerolipid and Fatty Acid Metabolism

Most of our current understanding of the metabolism of glycerolipids and fatty acids in the retina comes from studies of the incorporation of radioactive precursors and from a few reports on enzymes involved in these pathways.[143–148] In general, these studies have provided evidence for the operation of pathways similar to those reported in other tissues such as liver and brain.[149,150] In many of these studies, [2-^{3}H]glycerol was used as the labeled precursor because there is essentially no recycling of the label; any radiolabeled glycerol that is channeled toward glycolysis loses the tritium to cellular water at the dehydrogenase step.

In 1976, it was found that the retina displays a very high rate of *de novo* biosynthesis of glycerolipids both *in vivo* and *in vitro*.[12,111,116,151–156] When radiolabeled glycerol was incubated for a short period with whole retina from toad or cattle, most of the label was incorporated into phosphatidic acid, phosphatidylinositol, diglycerides, and triglycerides. The route of incorporation appeared to be through glycerol kinase and subsequent acylation. When [^{14}C]glycerol was used as a precursor, little of the radioactivity appeared in carbon dioxide.[151,155] Further studies showed active synthesis of phosphatidylethanolamine and, after a lag period, of phosphatidylcholine.[156] *In vivo* labeling resembled the profile seen *in vitro*.[156]

[^{14}C]Glycerol incorporation into toad retinal lipids was found to be temperature dependent,[157] with incorporation increasing linearly up to 23°C. No significant changes occurred above this point. Incorporation was higher in media with larger amounts of divalent cations (e.g., Ames–Hastings me-

Table IV
Fatty Acid Composition of Whole Retina, Rod Outer Segment, Microsomes, and Synaptic Plasma Membrane Phosphatidylcholine, Phosphatidylethanolamine, and Phosphatidylserine from Rat

Fatty acid	Whole retina[a]			ROS[b]			Retinal microsomes[c]			SPM[d]		PS[f]	Brain microsomes[e]		
	PC	PE	PS	PC	PE	PS	PC	PE	PS	PC	PE		PC	PE	PS
14:0[g]	0.3	—	—	0.9	0.2	—	2.0	—	—	—	—	—	—	—	—
16:0	36.5	8.6	2.2	29.5	5.9	2.5	45.1	14.7	5.1	56.5	11.3	3.2	28.8	11.9	1.5
16:1	—	—	0.5	1.0	0.3	—	—	—	0.7	0.2	0.5	—	0.4	—	—
18:0	27.9	38.7	37.6	11.7	29.3	29.1	11.5	36.3	44.1	11.4	17.7	44.8	25.0	30.1	42.0
18:1	13.8	1.8	2.2	8.3	3.2	1.9	21.9	7.7	8.7	23.2	9.4	14.0	28.2	18.7	31.8
18:2	0.5	0.1	0.2	0.8	0.6	—	0.8	0.6	0.9	1.0	0.64	—	0.3	0.3	0.6
18:3	—	—	—	—	—	—	—	—	—	0.02	0.17	—	—	—	—
20:2,n−6	—	—	—	0.8	0.8	0.4	—	0.3	—	0.25	—	—	—	0.6	0.3
20:3,n−9	0.8	—	—	—	—	—	0.2	0.5	0.10	—	—	—	—	—	3.5
20:4,n−6	3.3	7.6	2.4	3.4	2.9	1.9	0.9	14.0	14.0	4.2	14.0	2.5	5.6	10.4	3.4
22:4,n−6	0.2	1.1	3.4	—	0.2	—	0.7	3.4	2.7	0.18	3.7	—	—	5.8	—
22:5,n−6	—	—	—	3.2	1.1	—	0.8	3.8	2.3	0.08	1.3	—	—	—	—
22:5,n−3	—	—	0.8	—	—	—	0.8	1.4	1.0	—	0.05	—	—	2.2	—
22:6,n−3	0.2	42.2	50.0	38.4	54.8	61.8	5.3	19.0	18.3	1.5	15.6	29.7	9.0	14.5	13.4
Others	—	—	0.5	1.4	—	—	9.5	—	1.4	2.1	16.6	5.8	2.7	5.4	3.6

[a] Anderson and Maude. 1972,[139]
[b] Anderson et al.[142]
[c] Anderson et al.[117]
[d] Foot et al.[140]
[e] James et al.[138]
[f] Butler and Abood. 1982,[141]
[g] ROS, rod outer segment; SPM, synaptic plasma membrane; FA, fatty acid; PC, phosphatidylcholine; PE, phosphatidylethanolamine; PS, phosphatidylserine; 14:0, tetradecanoic acid; 16:0, palmitic acid; 16:1, palmitoleic acid; 18:0, stearic acid; 18:1, oleic acid; 18:2, linoleic acid; 18:3, linolenic acid; 20:2,n−6, eicosadienoic acid; 20:3,n−9, eicosatrienoic acid; 20:4,n−6, eicosatetraenoic acid; 22:4,n−6, docosatetraenoic acid; 22:5,n−6, docosapentaenoic acid; 22:5,n−3, docosapentaenoic acid; 22:6,n−3, docosahexaenoic acid.

dium).[157,158] Incubating bovine retinas under basal aerobic conditions in Ames ionic medium showed a remarkable chanelling of the label from glycerol toward triacylglycerols. Under anoxia, the incorporation of glycerol into retinal lipids was totally abolished.[155]

The *de novo* biosynthesis of glycerolipids also occurred in subcellular fractions of the bovine retina when the entire retina was exposed to [2-^{3}H]glycerol for a short time.[116,159] Early in the incubation period, most of the radioactivity in lipids was found in microsomes, specifically in phosphatidic acid. However, thereafter, labeling was rapid in lipids from other fractions such as the nuclear and mitochondrial–synaptosomal fractions and ROS. The soluble fraction contained small amounts of phospholipids with high specific activities.

In rat retinas, intravitreally injected [2-^{3}H]glycerol produced rapid labeling of lipids,[113] notably in phosphatidic acid and diglycerides at early time periods. The highest rate of incorporation was attained by phosphatidylcholine, followed by phosphatidylinositol. Among the subcellular fractions, most of the label was found in the phosphatidylcholine and phosphatidylinositol fractions of microsomes. Diglycerides and phosphatidylethanolamine incorporated 10 and 20%, respectively, of the labeled glycerol.[113]

Recently, studies of the incorporation of [2-^{3}H]glycerol into various molecular species[135,160] revealed that relatively more radioactivity is taken up by polyenoic molecular species in the glycerolipids of the toad retina compared to bovine retinal glycerolipids. This is in agreement with the relatively higher proportion and higher degree of unsaturation of phosphatidic acid and diglycerides in the toad retina.[104,105] The percentage distribution of radioactivity was high among phosphatidylethanolamine, phosphatidylcholine, and phosphatidylinositol. However, the specific activity of phosphatidylinositol was much higher, with the largest proportions of the radioactivity in the penta-, hexa-, and supraene fractions. In general, the incorporation of [^{3}H]glycerol into docosahexaenoate-containing molecular species of glycerolipids (hexaenes and dipolyunsaturates) was very high, supporting the proposal that docosahexaenoic acid is introduced during *de novo* synthesis.[105,135,139,152,160]

Among the other lipid precursors studied in terms of *in vitro* incorporation, [^{14}C]palmitic acid showed the highest specific activity in diglycerides, followed by phosphatidic acid, at all incubation times.[151] Similarly, phosphatidylcholine and phosphatidylinositol actively incorporated the labeled fatty acid. In contrast, phosphatidylethanolamine and phosphatidylserine were the less labeled polar lipids.[151,154]

Incubation of the retina *in vitro* with [^{3}H]serine yielded a 50-fold higher incorporation of this precursor compared to labeled glycerol.[159] When labeled serine and glycerol were both incubated with retina, the highest ratio of [^{3}H]serine/[^{3}H]glycerol was observed in phosphatidylserine, suggesting that active uptake of serine into phosphatidylserine is effected by a base-exchange reaction.[159] In fact, the presence of active base-exchange reactions in the retina has recently been reported.[161] Similarly, the labeling of phosphatidylserine and phosphatidylethanolamine during short-term incubations of retinal homogenates with labeled serine suggests the presence of phosphatidylserine decarboxylation in the retina.[159] Schmidt[162,163] reported an active incorporation of

[³H]inositol into phosphatidylinositol of whole retina, inner retinal layers, and photoreceptor cell layers. In the latter, incorporation was highest and exhibited a light-dependent increase.

The incorporation of [³H]glycerol and [³H]inositol into polyphosphoinositides [1-(3-sn-phosphatidyl)-L-myo-inositol-4-phosphate and 1-(3-sn-phosphatidyl)-L-myo-inositol-4,5-biphosphate, also called diphosphoinositide and triphosphoinositide, respectively] has also been reported recently,[160] as has the incorporation of ^{32}P into phosphoinositides in octopus retina.[164] In the same study, decreased incorporation of ^{32}P into diphosphoinositide and triphosphoinositide and increased incorporation into phosphatidic acid were seen in the presence of isobutylmethylxanthine (IBMX), a phosphodiesterase inhibitor. Based on the results of these studies, a general pathway for retinal phospholipid biosynthesis has been suggested[149] (Fig. 2). The enzymes needed for some of the steps shown in this figure have already been shown to exist in retina.

Swartz and Mitchell[143] demonstrated the presence of phosphocholine transferase, which catalyzes the formation of phosphatidylcholine from CDP-choline in the presence of diacylglycerol. They found that this enzyme is active in microsomes and less so in mitochondria but is absent in the cytosolic and rod outer segment fractions.[143,145] Later, in detailed kinetic studies of retinal phosphocholine transferase and phosphoethanolamine transferase in the developing chick retina, Dreyfus et al.[145] showed that the activity of the latter enzyme peaks by the 18th day of embryonic life and decreases thereafter. Phosphocholine transferase also showed a peak in specific activity at the time of hatching. The increases in these enzyme specific activities have been correlated with the period of development of rod outer segments in chick retina. The apparent K_m values for CDP-choline and CDP-ethanolamine do not differ much during embryonic life but increase thereafter, whereas the apparent K_m of phosphocholine transferase does not change during ontogenesis.

With the use of [1-^{14}C]-labeled palmitate, oleate, and linoleate, active acyltransferase reactions have been shown to occur in bovine retina in the presence of lysophospholipid acceptors. These reactions are ATP and CoA dependent and appear to transfer the fatty acid to lysophosphatidylcholine and lysophosphatidylethanolamine in homogenates and other subcellular fractions, but their activities in the ROS fraction were not consistent.[144] More recently, the presence of choline kinase in rabbit retinal homogenates[146,147] and cytosol[148] has been demonstrated. The fatty acid activation–acylation reactions for polyunsaturated fatty acids, such as arachidonate and docosahexaenoate, were shown to be very active in rat, frog, and bovine retina, mainly in the microsomal fraction, followed by the mitochondria. These enzymes are not significantly active in ROS, which contain large quantities of highly unsaturated fatty acyl chains in phospholipids.[165,166]

Except for the work of Swartz and Mitchell[167] on phospholipases in bovine retina and studies of the changes in retinal free fatty acids,[106,141,168] mainly arachidonic acid, as an index of phospholipase A activity, little has been done on the catabolism of glycerolipids in the retina. The existing studies have reported the presence of an acid and alkaline phospholipase A_1 and A_2 in the bovine retina and pigment epithelium. Acidic phospholipases are present in the

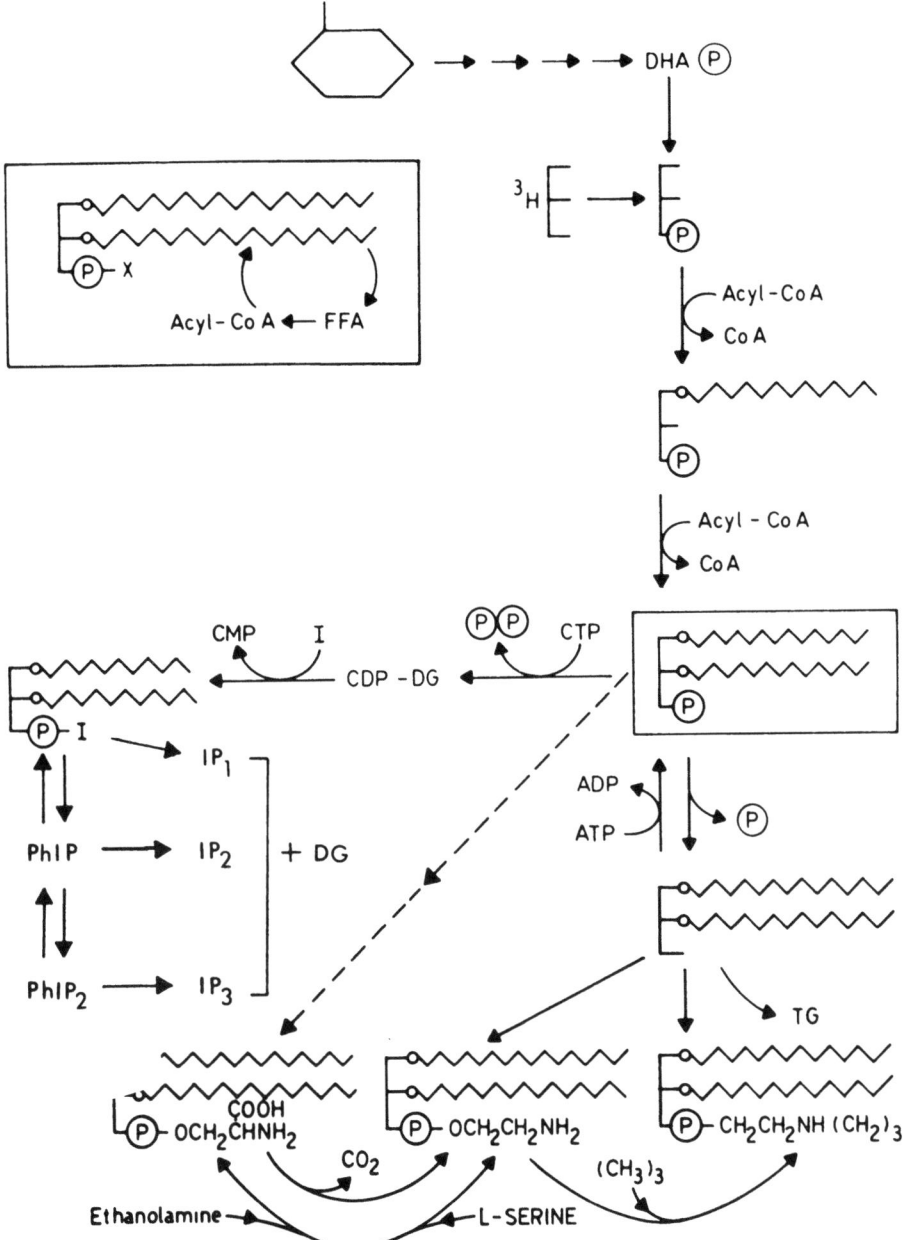

Fig. 2. Pathways of phospholipid biosynthesis in the retina. Inset: turnover of acyl groups. The compound in a box along the pathway depicts phosphatidic acid. P, phosphate; I, inositol; TG, triglyceride; CDP-DG, cytidine diphosphate diglyceride; PhIP, phosphatidylinositol 4-phosphate; PhIP$_2$, phosphatidylinositol 4,5 bisphosphate; IP$_1$, IP$_2$ and IP$_3$ are inositol 1-, 1,4- and 1,4,5-phosphate, respectively; DG, diacylglycerol. From top to bottom on right: DHAP, dihydroxyacetone phosphate; glycerol-3-phosphatidylcholine. Bottom left: phosphatidylethanolamine and phosphatidylserine. Arrows on dashed line indicate possible metabolic link (Modified from Bazan et al.[149] with publishers permission).

lysosomal fraction, whereas the alkaline phospholipases A_1 and A_2 are found in microsomes and mitochondria, respectively. No significant amounts of phospholipase activity are found in light-adapted rod outer segments. The substrate specificity and cofactor requirements of these enzymes vary widely within the subcellular fractions.

Dog retinal homogenates are shown to incorporate labeled malonyl-CoA rapidly into saturated (80%) and polyenoic fatty acids (20%), suggesting the *de novo* biosynthesis of fatty acids in retina.[169] Similarly, radiolabeled acetate was found to be incorporated into various fatty acids in the rat retina.[170] More recently, the formation of docosahexaenoic acid (22:6,*n*-3) and docosapentaenoic acid (22:5,*n*-3) from [1-^{14}C]eicosapentaenoic acid (20:5,*n*-3) *in vivo* has been demonstrated to occur 3 min after the intravitreal injection of the radiolabeled fatty acid into the rat.[171] Although 20:5 does not accumulate in retinal lipids as judged by fatty acid composition studies, these experiments show that radiolabeled 20:5 can trace the pathway leading to 22:6 synthesis in the retina. Moreover, as expected, 30 min after intravitreal injection, there was still some unconverted 20:5. Further studies have shown that in longer postinjection periods, essentially all the labeled 20:5 is converted to 22:6 (N. G. Bazan, unpublished data). These results suggest the presence of fatty acid elongation and desaturation enzymes in the retina. Further work is needed on the enzymes involved in the metabolism of fatty acids in the retina.

Dudley and Anderson[172] have reported that bovine retina contains phosphatidylcholine transfer protein, which, however, shows only 50% of the liver enzyme activity per unit protein. Both retina and liver exchange proteins are more active in transferring the phospholipid from liposomes to ROS than from liposomes to bovine heart mitochondrial membranes. It is possible that other phospholipid-specific transfer proteins may exist in the retina and that they may be involved in the molecular renewal of phospholipids in rod outer segments. In fact, newly synthesized phospholipids appear in different subcellular fractions after toad retinas are incubated with radiolabeled glycerol.[52]

Evidence for the existence of base-exchange reactions in the retina has been published.[161,173] These reactions, which are localized in the microsomal fraction, are active over a broad pH range (7.0–8.5) and require calcium for maximum activity. In rabbit retinal microsomes, labeled serine and ethanolamine are incorporated into phospholipids.[173] Similar exchange activities among serine, ethanolamine, and choline in bovine retinal microsomes have also been reported.[161] Although the rates of exchange reactions are similar in both studies, the K_m values seem to be different. The overall contribution of these base-exchange reactions to the net synthesis of phospholipids in the retina is not known.

4.4.2. Effects of Cationic Amphiphilic Drugs on Glycerolipid Metabolism

Cationic amphiphilic drugs such as propranolol and phentolamine were found to redirect the *de novo* biosynthesis of glycerolipids in retina.[12] Since these drugs exert identical metabolic effects on the lipid synthesis in retina, it was concluded that their mechanism of action was not related to their adrenergic antagonistic properties.[151] The entire molecule of the drug seems to be necessary to elicit the observed changes in glycerolipid metabolism in retina, since propranololglycol, a major CNS metabolite of propranolol[174] that is devoid of the isopropylamine group, does not mimic the lipid effect of propranolol.[12,175]

Both propranolol and phentolamine enhance the biosynthesis of phosphatidic acid, phosphatidylserine, and phosphatidylinositol in retina. In contrast, the synthesis of phosphatidylcholine, phosphatidylethanolamine, and triglycerides is inhibited.[151,153,175] The synthesis of diacylglycerols from labeled glycerol shows biphasic changes—inhibition early in the incubation period followed by later stimulation. The propranolol-induced drug effect on the lipid metabolism in whole toad and bovine retinas is localized mainly in microsomes and in the soluble fractions during the first few minutes of incubation and spreads to other fractions thereafter.[52,176] During short-term incubation with propranolol, there is an increased content of phosphatidic acid in the endoplasmic reticulum (estimated in terms of lipid phosphorus), increased uptake of [2-^{3}H]glycerol, and increased amounts of fatty acyl chains.[159,176]

Recently, it has been shown that the synthesis of polyphosphoinositides (DPI and TPI) labeled by [^{3}H]glycerol and [^{3}H]inositol *in vitro* is stimulated in the presence of propranolol.[160] Also, propranolol stimulates the incorporation of [2-^{3}H]glycerol into monoenes and saturates of phosphatidylinositol and was found to inhibit incorporation of tritiated glycol into similar molecular species of phosphatidylcholine, suggesting a common metabolic origin.[160] Even though no specific molecular species of diglyceride is inhibited during the first 10 min of propranolol incubation, the tetraene molecular species of diglycerides are stimulated at 30 min of incubation time. Recently, it has been shown that although a complete reversal is not achieved, propranolol-induced lipid biosynthetic changes are reversible and that incorporation profiles approach control values 60 min after incubation in drug-free medium.[177]

On the basis of the above studies, it has been suggested that phosphatidic acid phosphohydrolase is inhibited by cationic amphiphilic drugs[12,151,176] and also that these drugs may form complexes with phospholipid substrates of lipases. Alternatively, the drugs may interact with proteins of the endoplasmic reticulum, some of which may be enzymes active in glycerolipid metabolism,[12] or the drugs may have a nonspecific effect that alters membrane functions, as has been shown for local anesthetics.[178] Recent studies carried out in our laboratory on docosahexaenoyl- and arachidonyl-CoA synthetases from rat brain microsomes preincubated with propranolol showed decreased enzymatic activity. Inhibition increased linearly with increasing concentrations of propranolol; at higher levels of drug, the enzymes were inhibited completely. Similar results were obtained with certain local anesthetics such as tetracaine, but the potency of these drugs seemed to be lower than that of propranolol.[178a]

4.4.3. Sterol Biosynthesis

Studies on the biosynthesis and metabolism of sterols in the retina have been rather limited. The conversion of radiolabeled mevalonic acid to sterols and sterol precursors has been examined *in vitro* in isolated bovine retinas as well as in cell-free homogenates.[168,179-181] In these studies, the labeled products were identified by extensive chemical and chromatographic analyses. Squalene and lanosterol are the major labeled nonsaponifiable components, whereas cholesterol represents less than 1% of the label incorporated into nonsaponifiable lipids. The saponifiable fraction contains a substantial percentage of the total incorporated radioactivity, most of which corresponds to components having the chromatographic properties of 15- and 20-carbon isoprenoid acids.

In vivo, the metabolic fate of intraocularly injected [^{14}C]-labeled mevalonic acid was followed in the calf.[180] The chromatographic properties of the labeled nonsaponifiable products isolated from retinas were primarily those of squalene and, to a lesser extent, C_{30} monohydroxysterols (e.g., lanosterol); very little of the incorporated label behaved chromatographically like C_{27} monohydroxysterols (e.g., cholesterol). Therefore, the results obtained from *in vitro* and *in vivo* experiments are consistent and suggest that the adult bovine retina has a very limited capacity for *de novo* sterol biosynthesis.

Cholesterol accounts for at least 99% of the total sterols of the bovine retina,[101] and the steady-state levels of isoprenoid precursors of sterols in the bovine retina are negligible.[180] On the basis of these findings and the results of the metabolic studies described above, it has been proposed[180] that the adult retina obtains most of its sterol from exogenous sources such as the blood (e.g., circulating lipoproteins) and perhaps the brain (e.g., retrograde transport via the ganglion cell projections from the optic tectum). Consistent with this proposal, a preliminary experiment (S. J. Fliesler and L. D. Andrews, unpublished results) has demonstrated that [^{3}H]-labeled cholesterol injected systemically into frogs results in the labeling of all layers of the retina.

4.4.4. Ganglioside Metabolism

Dreyfus and his associates have carried out a detailed study on the *in vivo* and *in vitro* metabolism of gangliosides in the retina.[125,182,183] They injected labeled N-acetyl-D-mannosamine intraocularly into both eyes of 9-day-old chicks and followed the incorporation into gangliosides. Specific activity (cpm/mg NANA) reached a maximum by 1 to 5 days after injection and decreased thereafter.[125,184] The individual gangliosides showed a similar picture, with the highest specific activity in disialogangliosides, specifically GD_3, at 24 h after injection followed by polysialogangliosides, GT_1 and GQ_1; GM_1 showed the lowest specific activity, suggesting that different pools of sialic acid are involved in the synthesis of gangliosides. Similar results were obtained in rabbit retina,[185] and GD_3 and GM_3 gangliosides achieved the highest labeling in chick embryonic retina cultured in the presence of [^{3}H]glucosamine.[186,187]

The *in vitro* synthesis of gangliosides has been studied in developing chick retinal homogenates.[125,188,189] The specific activity of UDP-glucose:ceramide glucosyltransferase (C Glu T) was highest in 10-day-old embryonic retina and decreased during development, reaching the adult value by 14 days of posthatching life.[125,188] Similarly, the UDP-galactose:glucosylceramide galactosyltransferase (C Glu Gal T) specific activity was maximal at 11 days of embryonic life and decreased thereafter. Adult levels represented only 20% of the maximum activity.[125,189] Different sialyltransferases (ST) were studied during ontogenesis in the chick retina by means of endogenous and exogenous substrates.[182,183] The highest specific activity was found in CMP-NANA:GM_1 ST (ST_3), followed by CMP-NANA:GM_3 ST (ST_2), GMP-NANA:GD_{1b} ST (ST_5), CMP-NANA:GD_{1a} ST (ST_4), and CMP-NANA:Lac-cer (ST_1) with exogeneous substrates.[182,183] However with endogenous substrates, the activities of sialyltransferases were highest for ST_1, followed by ST_3 and ST_2.[183] All the sialyltransferases showed a decrease in activity during development; the highest activities occurred on the eighth day of embryonic life.[183,187] It is interesting to note that all of the changes in ganglioside patterns and synthetic enzyme activities were observed prior to the time when the retina becomes functional.[183]

Among the ganglioside-catabolizing enzymes, only neuraminidase has been studied in retina.[125,183,190,191] This enzyme was found to be particulate bound; no soluble neuraminidase has been detected.[125,190] It acts on both gangliosides and glycoproteins of endogenous origin and appears to be selective for ganglioside NANA (90% of the NANA released) compared to glycoprotein NANA (10%).[182] GD_{1a} is the main ganglioside attached to this enzyme.[191] In contrast to the ganglioside-synthesizing enzymes discussed above, neuraminidase showed low activity at 8 days of embryonic life, and the highest activity was reached at the time of hatching. Activity decreased slightly thereafter but remained high even during adulthood.[125,190,191] Recently, neuraminidase has been found in the rod outer segments from bovine retina.[192]

4.4.5. Prostaglandins

Arachidonic acid, a precursor of prostaglandins, is enriched in the phospholipids of the mammalian retina, notably in phosphatidylinositol and phosphatidylethanolamine.[105,135] It is highly active metabolically,[193] and is released rapidly under various experimental conditions.[194,195] However, studies on the retinal prostaglandins are very scarce. The formation of prostaglandins by rabbit retinal microsomes and an inhibition of formation in the presence of indomethacin have been reported.[196] Recently, it has been shown that PGE_2, $PGF_{2\alpha}$, PGD_2, and TxB_2 are formed from labeled arachidonic acid by incubated rabbit retinal microsomes.[197] In vitro studies involving bovine retina incubated with labeled arachidonic acid showed the production of cyclooxygenase (CO) and lipoxygenase (LO) reaction products.[198,198a] The CO products were $PGD_2, F_{2\alpha}$, E_2, 6-keto-$PGF_{1\alpha}$, and TxB_2, and the LO products were HETE-type derivatives. Indomethacin and acetylsalicylic acid greatly inhibited the production of CO products. Nordihydroguaiaretic acid markedly inhibited both CO and LO reactions.

The role of prostaglandins in the retina is not known. When PGE_2, PGE_1, and arachidonic acid are injected intravitreally, the b-wave amplitude of the ERG shows a dose-dependent effect that, at least in the case of arachidonic acid, can be inhibited by peripheral administration of indomethacin.[199] Also, the intravitreally injected lipid hydroperoxides have been shown to inhibit a-, b-, and c-waves of the rabbit ERG.[200]

4.4.6. Dolichols

Dolichols make up a rather large and complex subclass of long-chain isoprenoids.[201–204] In mammalian tissues, dolichols are unique in at least three respects: (1) they are the largest lipids known, containing 16–23 isoprene residues per molecule; (2) with the exception of two internal *trans* olefinic bonds, the configuration of the olefinic bonds is *cis*; (3) the β-isoprene residue (containing the C–O bond) is saturated instead of allylic.

The absolute amounts and chemical forms in which dolichols exist vary widely as a function of the tissues and organisms from which they are isolated. In most of the mammalian tissues that have been examined, dolichols exist primarily in the free alcohol form, although several tissues contain significant levels of fatty acid esters.[205–212] A positive correlation between the amounts of free dolichols and lipofuscin ("aging pigment") with increasing age has been reported in whole mouse brain and in human gray and white matter.[213] In addition to free alcohol (Dol) and fatty acid ester (Dol-FA) forms, dolichols may exist as the monophosphate ester (Dol-P), the pyrophosphate ester (Dol-PP), and various mono-, di-, tri-, and oligosaccharide phosphate esters (primarily containing N-acetylglucosamine, mannose, and glucose). The steady-state levels of these alternate forms of dolichols are usually extremely low in mammalian tissues. However, the "biologically active" form of dolichols is thought to be the monophosphate ester, a key intermediate in the glycosylation of the so-called asparagine-linked glycoproteins.[201,202,204,214]

Figure 3 outlines the metabolic relationships among the various forms of dolichols and their involvement in glycoprotein synthesis (i.e., the so-called "dolichol phosphate cycle"). It is paradoxical that the dolichol phosphate cycle is localized to the rough endoplasmic reticulum of the cell, whereas Dol-P (in rat liver) is enriched in the nuclear and Golgi body fractions,[215] and its synthesis is highest (on a specific activity basis) in the mitochondrial fraction.[216] Other than their role in protein glycosylation, the biological functions of dolichols are not known.

Knowledge of retinal chemistry and metabolism of dolichols is relatively limited; particularly lacking is quantitative information with regard to the amounts, chemical forms, biosynthesis, turnover, or subcellular location. Kean and co-workers[217] have documented the existence of the dolichol phosphate cycle in the retina, demonstrating the presence of several requisite enzymatic activities and (indirectly) endogenous intermediates. Cell-free preparations of retinas from a variety of animal species exhibit the capacity to transfer labeled mannose from GDP-[^{14}C]Man to endogenous acceptors, resulting in the formation of [^{14}C]-labeled Dol-P-Man, Dol-PP-oligosaccharides, and glycoproteins. Compared to a wide range of isoprenyl phosphates, Dol-P selectively stimulates the mannosyl transferase reaction and exhibits saturation kinetics [K_m = 7.3 μM for Dol-P, 1.4 μM for GDP-Man). Synthetic Dol-P-[^{14}C]Man incubated with retinal cell-free preparations results in the formation of labeled

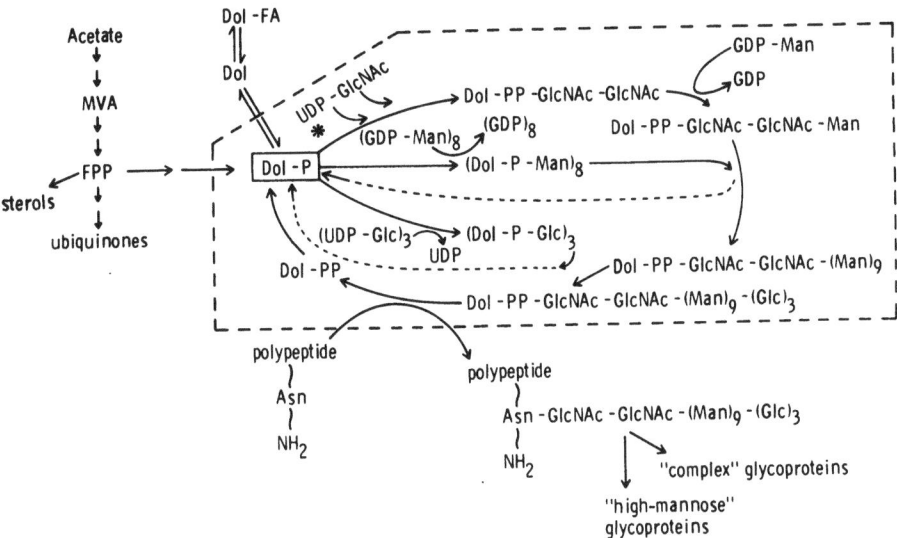

Fig. 3. A proposed pathway for the dolichol phosphate cycle. MVA, mevalonic acid; FPP, farensyl pyrophosphate; Dol, dolichol; Dol-P, dolichol phosphate; Dol-PP, dolichol pyrophosphate; Dol-FA, dolichol ester; Glc, glucose; GlcNAc, N-acetylglucosamine; Man, mannose. Asterisk represents the site of inhibition by tunicamycin. (Kindly prepared by Dr. S. Fliesler.)

Dol-PP-oligosaccharides and glycoproteins; furthermore, labeled Dol-PP-oligosaccharides isolated from the incubation mixture are capable of transferring label to protein on subsequent incubation. Curiously, the retinal subcellular fraction with the highest mannosyl transferase activity is the $750 \times g$ pellet (i.e., nuclei and cell debris), whereas microsomes exhibit relatively moderate enzymatic activity. Incubation of embryonic chick retinal cell-free homogenates with UDP-[^{3}H]GlcNAc produces three early intermediates of "core oligosaccharide" biosynthesis: GlcNAc-PP-lipid, GlcNAc-GLcNAc-PP-lipid, and Man-GlcNAc-GlcNAc-PP-lipid. More recently, it has been shown that incubating intact bovine retinas with [^{3}H]Man or [^{3}H]Glc results in the formation of several lipid-oligosaccharides, with the carbohydrate moieties having the following structures[218]: $(Glc)_3(Man)_9(GlcNAc)_2$, $Glc(Man)_9(GlcNAc)_2$, $(Man)_{8-9}(GlcNAc)_2$, and $(Man)_{4-7}(GlcNAc)_2$.

Although, taken together, these results strongly support the existence of the dolichol phosphate cycle in the retina, they do not address the question of the possible involvement of this system in the biosynthesis of the visual pigment, rhodopsin. Rhodopsin contains two relatively short, asparagine-linked oligosaccharide chains composed of mannose and N-acetylglucosamine.[219–221] The results of an early study[222] indicated that little label was transferred from GDP-[^{14}C]Man to rhodopsin in cell-free preparations.

However, more recent results obtained with intact bovine retinas incubated *in vitro*[223] have shown that tunicamycin can inhibit the incorporation of [^{3}H]Man or [^{3}H]GlcN into rhodopsin by 96% and 70%, respectively, with only moderate inhibition of [^{35}S]methionine incorporation. Tunicamycin is an antibiotic that selectively blocks the first step in the dolichol phosphate cycle (Fig. 3), namely, the condensation of UDP-GlcNAc with Dol-P to form Dol-PP-GlcNAc.[224–226] These results implicate the dolichol phosphate cycle in the biosynthesis of rhodopsin. Furthermore, Fliesler and co-workers[227,228] have shown recently that tunicamycin can selectively inhibit the incorporation of [^{3}H]Man in the photoreceptor layer of human and amphibian retinas under conditions in which the incorporation of [^{3}H]leucine is not affected. These results suggest that the dolichol phosphate cycle may be preferentially localized in the photoreceptor cells, thereby subserving the biosynthesis of rhodopsin and possibly other glycoproteins.

During the course of his studies on sterol biosynthesis in the bovine retina, Fliesler[181] examined the nonsaponifiable lipids for incorporation of radioactivity and noted that [^{3}H]mevalonic acid is not converted to products having the chromatographic properties of dolichols. He therefore speculated that the *de novo* synthesis and turnover of dolichols in the retina must be extremely slow. In view of the details of the dolichol phosphate cycle (Fig. 3) and the potential for interconversion of the various chemical forms of dolichols (e.g., conversion of dolichol fatty acid esters to free dolichol, followed by phosphorylation to Dol-P), it is reasonable to conclude that a given pool of Dol-P molecules can be recycled almost indefinitely without the need for further *de novo* synthesis. Clearly, this area of retinal metabolism warrants further investigation.

5. NEUROTRANSMITTER METABOLISM

The transfer of visual information from retina to brain involves a series of integrative and propagative events that are initiated by light-activation of the photoreceptors, integrated within the retinal circuitry, and transmitted by ganglion cells to the midbrain for relay to the cortex. The mode of transmission between retinal neurons appears to be chemical in nature.[229] As shown in Fig. 1, the two well-defined synaptic regions of the retina are the outer plexiform layer, containing the synapses between photoreceptor, bipolar, and horizontal cells, and the inner plexiform layer, containing the synapses between bipolar, amacrine, and ganglion cells. It is likely that the recently discovered interplexiform cells send information centrifugally from the inner to the outer plexiform layer.[230]

To date, there are at least 15 different chemical substances (Table V) that have been suspected of functioning as neurotransmitters within the retina. However, these transmitters do not account for all of the cell types within the retina that are suspected of using chemical mediators.

It is difficult, because of technical problems, to establish beyond doubt that a substance acts as a transmitter at a specific synapse. However, recently, a variety of techniques have been developed to study the specific criteria that must be met to establish a given substance as a transmitter.[232,233] The criteria are: (1) the substance must be synthesized and stored in the presynaptic terminal; (2) it must be released on appropriate stimulation of the presynaptic neuron; (3) the appropriate response in the postsynaptic neuron must be seen

Table V
Putative Neurotransmitters in the Retina[a]

Candidates	Localization method	Animal species	Neurotransmitter function
Acetylcholine		All species	Evidence excellent
Monoamines			
Dopamine		All species	Evidence excellent
Norepinephrine		All species	Evidence weak
Serotonin		Mammals	Evidence good
		Nonmammalian species	Evidence very good
		Invertebrates (e.g. horseshoe crab, *Limulus polyphemus*	Possible
Histamine		Mammals (rabbit)	No evidence
Epinephrine		All species	Evidence very weak
Amino acids			
Glycine		All species	Evidence good
γ-Aminobutyrate (GABA)		All species	Evidence good
Taurine		All species	Evidence contradictory
Glutamate		All species	Evidence good
Aspartate		All species	Evidence good
Proline		All species	Evidence weak
Undecapeptide			
Substance P	Immunohisto-fluorescence	Mammals: rat, rabbit, guinea pig, monkey, cow	Subclass of amacrine cells
		Avians: chick, pigeon	
		Amphibians: bullfrog,[b] mudpuppy	
		Fish: carp	Spontaneous and light-induced activity of ganglion cells
Tridecapeptide			
Neurotensin	Immunohisto-fluorescence	Several species (not specified)	Subclass of amacrine cells
Tetradecapeptide			
Somatostatin (SRIF)	Immunohisto-fluorescence	Mammals: rat, rabbit, guinea pig, cow, monkey, man	Subclass of amacrine cells
		Avians: chick[b]	
		Amphibians: bullfrog,[b] frog	
		Fish: goldfish	
Octacosapeptide			
VIP (vasoactive intestinal peptide)	Immunohisto-fluorescenmce	Variety of mammalian species, rat	Sub class of amacrine cells
Nonacosapeptide			
Glucagon	Immunohisto-flurescence	Variety of nonmammalian species (avians, amphibians, reptiles)	Subclass of amacrine cells

[a] Modified from Schorderet and Magistretti.[231]
[b] Highest concentration compared to that of other species.

on exogenous addition of the substance; (4) a mechanism for the rapid removal of the substance must exist in the synaptic cleft in order to terminate the transmitter action; and (5) the pharmacological specificity of the tested transmitter must be identical to that of the endogenous transmitter.

Apart from these, the peptide neurotransmitters must fulfill some additional criteria. The chemically synthesized peptide and endogenous biological peptide neurotransmitter (1) must elicit the same response as the transmitter, (2) must provide antibodies that interact with the endogenous transmitter, and (3) must be easily demonstrable, i.e., localized in the cell by means of immunologic techniques.[234] The methods used to satisfy the above criteria come from various disciplines such as biochemistry, morphology, electrophysiology, immunochemistry, and pharmacology.[229,235]

The logic behind the first criterion, storage and synthesis, is that if the substance in question is absent in retina (e.g., octopamine), then the transmitter function of that substance can be ruled out.[236] The presence of a substance, however, is not complete proof of neurotransmitter function. For example, the amino acids have many functions other than their role as transmitters. A variety of histochemical techniques have been developed to demonstrate the presence of suspected neurotransmitters within specific neuronal populations of retina. For example, treatment of the cells with formaldehyde or glyoxylic acid produces fluorescent derivatives of dopamine, norepinephrine, and serotonin, which can be viewed under a fluorescence microscope.[237-240] Antibodies produced against peptide transmitters have been used to demonstrate the presence of a neurotransmitter within a specific population of cells.[241] Documentation of the existence of synthesizing enzymes and precursors has also helped in establishing many neurotransmitters such as acetylcholine and dopamine.

Demonstration of the second criterion, release of the suspected transmitter on appropriate stimulus, is difficult to achieve. However, the retina is more amenable than brain to this type of study, since light is easily applied as the appropriate stimulus. The unique features of neurotransmitter release have been demonstrated in isolated retinal synaptosomes in the presence of depolarizing agents such as potassium, veratridine, and A23187.[242-244] However, it is difficult to judge whether this type of experiment is particularly meaningful.

The third criterion, postsynaptic response to exogenously added substance, is usually tested by electrophysiological methods. The intact retina can be exposed to different chemical compounds, and the electrical response can be recorded in the form of an ERG.[245] Alternatively and more directly, intracellular recordings from postsynaptic neurons can be studied. Recordings can be made in the presence of different amounts of the transmitter and also with the pharmacological agonists or antagonists.

Recently, a number of biochemical assays have been developed to measure neurotransmitter receptor binding, which can provide a measure of postsynaptic function.[246] Redburn and associates[247-252] demonstrated the presence of receptors for GABA and dopamine in different subcelluar fractions of bovine retina. The results of these receptor binding studies have been correlated with the electrophysiological analysis of receptor function.

The last criterion, removal of the transmitter from the synaptic cleft, is studied mainly by biochemical and autoradiographic techniques. The mechanism of removal is primarily by reuptake for most of the transmitters except for acetylcholine and the neuropeptides.[253] Reuptake is achieved by carrier-mediated transport and requires binding of the neurotransmitter to a recognition site, which is specific for a given transmitter.[253]

The transmitter agents described in Table V have met all or most of the criteria as functional transmitters in some species of retina.[252,254-256] The concentrations of various neurotransmitters from different species are given in Table VI. One of the interesting things to note is that most of the transmitter substances have been localized in amacrine cells. With a few exceptions, very little is known about the neurotransmitters of horizontal, photoreceptor, bipolar, ganglion, and interplexiform cells.

5.1. Acetylcholine

In retina, the cholinergic system has been studied extensively and appears to be relatively similar in different species, such as goldfish, rat, rabbit, and cat.[257] Acetylcholine serves as a neurotransmitter in certain amacrine cells of the rabbit retina[146,258-261] and most likely also in chicken retina.[262,263] Acetyl[^{3}H]choline is synthesized from [^{3}H]choline in rabbit retina and is rapidly released by light stimulation. This release is calcium dependent and is effectively blocked by GABA.[259,264] Autoradiographic studies appeared to localize acetyl[^{3}H]choline in the amacrine cells and among the cells in the ganglion cell layer.[146] Subsequently, Hayden et al.[260] showed that the latter cell bodies were displaced amacrine cells. Cholinergic amacrine cells, whether diplaced or in the usual position, have a unique morphology and have been classified as "starburst amacrine cells."[265,266] In chicken retina, the choline acetyltransferase activity in the IPL,[267,268] the high-affinity uptake system for choline,[262,263] and the potassium-induced, calcium-dependent release of acetyl[^{3}H]choline support the notion that acetylcholine functions as a neurotransmitter in a population of amacrine cells.

Although some components of the cholinergic system have been localized in the OPL of retinas in some species, it is doubtful that visual cells are cholinergic. Acetylcholinesterase is present near photoreceptor terminals in amphibians[269-271] and in the outer plexiform layer of mammals.[257,272,273] However, acetylcholinesterase is not a reliable marker for cholinergic neurons because its distribution is not restricted to cholinergic terminals. Turtle cones synthesize acetylcholine,[274] but the rate of synthesis is lower than that of classical cholinergic neurons.[275] Gerschenfeld and Piccolino[276,277] used muscarinic blockers to show an inhibition in signal transfer from cones to horizontal cells in turtle retina, suggesting the possibility of acetylcholine release from cones. However, high concentrations of the blockers were used, and various cholinergic agonists did not show consistent effects. Recently, Lam et al.[278] reported the presence of acetylcholine, its precursors, and the enzymes involved in its metabolism in the photoreceptor cells of turtle retina (Table VII). They suggested that acetylcholine is a likely neurotransmitter candidate for cephalopod photoreceptors.

Table VI
Concentrations of Various Putative Neurotransmitters in Retina from Various Species

Species	Glutamate[a]	Aspartate[a]	GABA[a]	Taurine[b]	Serotonin[c]	Dopamine[d]	Norepinephrine	SIL[e]	CIL[f]
Human									3
Cattle	4.2	0.4	0.4		39	86[g]	62[g]	11.4	7.0
Pig					5.9	1.28			
Guinea pig					6.0	0.79			
Chicken	4.4	1.5	1.9	14.9	67	1.97			
Rabbit	3.7	1.2	1.4	36.2	36	2.95			
Rat	4.0	0.8	2.3	32.5		3.59			
Mouse									
Frog	1.7	0.3	1.4		290	6.16		2.9	23
Goldfish	2.8	0.3	1.1			5.42		4.0	
Marine fish									
Lizard					96				
Cat						0.97			
Hamster						2.42			

[a] Values are from Voaden et al.[497] and expressed as μmol/g wet weight.
[b] Values taken from Gupta et al.[338] and expressed as μmol/100 mg protein.
[c] Values obtained from Osborne[351] and expressed as ng/g wet weight.
[d] Values taken from Da Prada[346] and expressed as ng/mg protein
[e] Values obtained from Yamada et al.[380] and expressed as pmol/mg protein. SIL, somatostatinlike immunoreactivity.
[f] Values obtained from Osborne et al.[354] and expressed as pmol/g wet weight. CIL, cholecystokininlike immunoreactivity.
[g] Values obtained from Nesselhut and Osborne[343] and expressed as ng/g wet weight.

Table VII
Acetylcholine, Choline Acetyltransferase, and Acetylcholinesterase in Isolated Photoreceptors[278]

	Photoreceptors	Photoreceptor-rich fraction	Sedimented retinal cells	Retina
Acetylcholine[a]	1.96 ± 0.36	1.12 ± 0.34	0.74 ± 0.05	1.05 ± 0.10
Choline[b]	—	0.36 ± 0.12	—	0.44 ± 0.08
Choline acetyltransferase	4.0 ± 0.70	—	18.3 ± 2.6	15.9 ± 2.40
Acetylcholinesterase	—	2773 ± 209	7022 ± 795	8549 ± 641

[a] Values are means ± standard deviation of μmol/g protein for 3–4 experiments.
[b] Values are means ± standard deviation of μmol product/h per protein for 3–4 experiments.

The possibility of cholinergic bipolar cells has been based on autoradiographic demonstration of accumulation of [^{3}H]choline, the precursor of acetyl[^{3}H]choline. However, there is no accumulation of radioactivity in the OPL, where the dendrites of the bipolar cells are located. Also, bipolar cells of rabbit do not accumulate [^{3}H]choline.[146] Acetylcholinesterase has not been found in the bipolar cells of retinas from various species studied.[279]

5.2. GABA

The highest concentrations of GABA and its synthesizing enzymes have been demonstrated in the amacrine cells of the rabbit retina.[280–282] [^{3}H]GABA is accumulated by certain amacrine cells[283] and to a considerable extent by glial cells. This glial uptake disguises the neuronal uptake in most mammalian species. Therefore, attempts have been made to use other GABA neuron markers such as [^{3}H]aminocyclohexane carboxylic acid,[284–286] [^{3}H]mucimol,[286–290] and [^{3}H]isoguvacine.[283] These markers are accumulated in the cells that are thought to use GABA as a neurotransmitter, whereas glial cell labeling is insignificant. However, [^{3}H]GABA can be used successfully for labeling GABA neurons in birds, amphibians, and fish, where there is little or no glial labeling.[282,291–293] Retinas from a variety of species have been shown to contain GABA-metabolizing enzymes, demonstrated by immunohistochemical methods,[294–298] and GABA receptors, demonstrated by biochemical binding techniques[247,251,299,300]; this suggests that GABA neurons are common enough to play a significant neurotransmitter role in retina.

The H_1 horizontal cells of the teleost fish accumulate [^{3}H]GABA[301] and contain glutamate decarboxylase.[294,302] Similarly, in channel catfish, all cone horizontal cells are of the H_1 type,[303] and all accumulate [^{3}H]GABA.[304] This evidence, together with electrophysiological studies of GABA-mediated reactions, shows that GABA is likely to be a neurotransmitter of H_1 cone horizontal cells in goldfish and channel fish.[305,306] Studies carried out recently suggest that GABA may also act as a neurotransmitter in analogous cells in frogs, chickens, and pigeons.[292,307,308]

5.3. Other Amino Acid Neurotransmitters

Aspartate and glutamate are thought to be photoreceptor cell neurotransmitters because these molecules hyperpolarize the horizontal cells, which are the most likely site of action of such neurotransmitters.[309-313] Significant amounts of aspartate and glutamate are present in photoreceptor cells in a number of species (Table VI).[314,315] Furthermore, high-affinity uptake of glutamate and aspartate occurs in photoreceptors,[316,317] and [^{3}H]-d-aspartate is taken up into cones of rabbit, guinea pig,[318] and goldfish retinas.[319] Also, aspartate amino transferase (AAT) has been localized in guinea pig cones by immunohistochemical methods,[320] and the release of aspartate is reduced by light stimulation in rats and rabbits.[264,321] All of these studies suggest that glutamate, aspartate, or both serve as neurotransmitters in photoreceptors and perhaps specifically in cones in the rabbit and guinea pig. In humans, [^{3}H]-d-aspartate is accumulated by rods rather than cones.[318] Similarly, rods but not cones accumulate [^{3}H]-L-glutamate[322,323] in primates. Thus, there are separate systems for retinal transport of glutamate and aspartate in different species.

A significant amount of glycine is present in the retina, and a very active glycine uptake system exists in a number of species, e.g., goldfish, frog, chicken, pigeon, guinea pig, rabbit, cat, and man.[283,293,308,324-326] Accumulation of [^{3}H]glycine in AII amacrine cells, which link bipolar to ganglion cells, has been reported.[326,327] Release of glycine by light and potassium depolarization has been shown in a number of species.[323,328-334] Recently, strychnine was shown to bind glycine receptors in bovine retina.[335] There is some evidence that glycine may act as a neurotransmitter in the bipolar cells of cat retina[326] and interplexiform cells of frog, goldfish, and rabbit retina.[319,325,336] Thus, the presence of glycine in retina and its release by nerve stimulation, as well as the presence of receptor and high-affinity uptake systems, suggest that glycine is a likely retinal neurotransmitter. Recently, it has been reported that a selective uptake of labeled glycine, GABA, and D-aspartate was seen when autoradiographic and immunofluorescence techniques were applied to rat retinal cell cultures.[336a]

The presence of taurine in retina from various vertebrate species (Table VI) and the location of its synthesizing enzymes in different parts of the retina suggest its putative role as a neurotransmitter.[257,337-341] However, the exceedingly high levels of taurine (100 mM) found in retinal cells and its proposed role as a protective counterion for sodium make it one of the less attractive transmitter candidates.

5.4. Dopamine

The dopamine neurotransmitter system is the best understood catecholamine system in retina.[9,229] However in bovine retina, norepinephrine (62 ± 4 ng/g wet wt.) and dopamine (86 ± 3 ng/g wet wt.) were demonstrated to occur in equivalent amounts by means of a sensitive high-pressure liquid chromatographic and electrochemical detection method.[342,343] Osborne[342] demonstrated the synthesis, uptake, and release of norepinephrine, its possible association with processes in the inner plexiform layer, and its possible transmitter role in retina. Although previous studies suggest that the retina lacks epinephrine,[229] this compound has been reported recently in the bovine retina.[229a] Also, Osborne (unpublished data) recently found that dopamine β-hydroxylase immunoreactivity is associated with specific ganglion cells in the bovine retina.

In most of the species studied, dopamine is localized in the amacrine cells of the inner plexiform layer.[290,344,345] In some species, such as teleost fish and a few New World monkeys, the interplexiform cells appear to be dopaminergic. The presence of dopamine-synthesizing enzymes in retina,[346] its release by light stimulation and potassium depolarization,[347-351] and the presence of appropriate receptors[246,250,350] strongly suggest that dopamine is one of the well-established retinal neurotransmitters. However, the physiological role of this transmitter is not yet completely understood.

5.5. Serotonin

The identification of serotonin as a neurotransmitter in mammalian retina is a controversial issue in spite of its now widely accepted transmitter role in nonmammalian species.[9,351] On one hand, Osborne[351-357] argues that serotonin (5-hydroxytryptamine) meets the major criteria for a transmitter candidate in mammalian retina: (1) 5-hydroxytryptamine is present in low but consistent quantities in a variety of mammalian retinas, as has been shown by high-pressure liquid chromatography[351,355]; (2) the catabolic enzyme of 5-hydroxytryptamine, monoamine oxidase, and the major neuronal metabolite of 5-hydroxytryptamine, 5-hydroxyindoleacetic acid, are present.[358,359]; (3) radioactive tryptophan is converted into serotonin *in vitro* by bovine retina[353]; (4) there is active uptake of serotonin by retinal neurons[244,353,360,361]; (5) potassium induces calcium-dependent release of serotonin from retinas previously loaded with serotonin[244,353,355,360]; (6) serotonin receptors are present in retina.[353]

On the other hand, Ehinger and co-workers argue that the extremely low levels of 5-hydroxytryptamine in mammalian retina are not consistent with a transmitter function (Table VI). Endogenous levels of 5-hydroxytryptamine are too low to be observed by histofluorescence or immunofluorescence.[354,355,357] Likewise, tryptophan hydroxylase activity is very low.[362] Because specific amacrine cells in mammalian retinas accumulate exogenous 5-hydroxytryptamine (Fig. 4),[361-386] Ehinger has suggested that these cells should be called "serotonin-accumulating cells," because it is likely that the true transmitter of these cells is a serotonin-related substance and not serotonin.

5.6. Neuropeptides

Although there are more than 20 peptides that are thought to act as neurotransmitters or neuromodulators in brain,[369] only a few (TRH, enkephalins, somatostatin, neurotensin, substance P, VIP, glucagon, and cholecystokinin) have been demonstrated in the retina by means of immunohistochemical techniques (Table V).[354,370-379] The amounts of various neuroactive peptide trans-

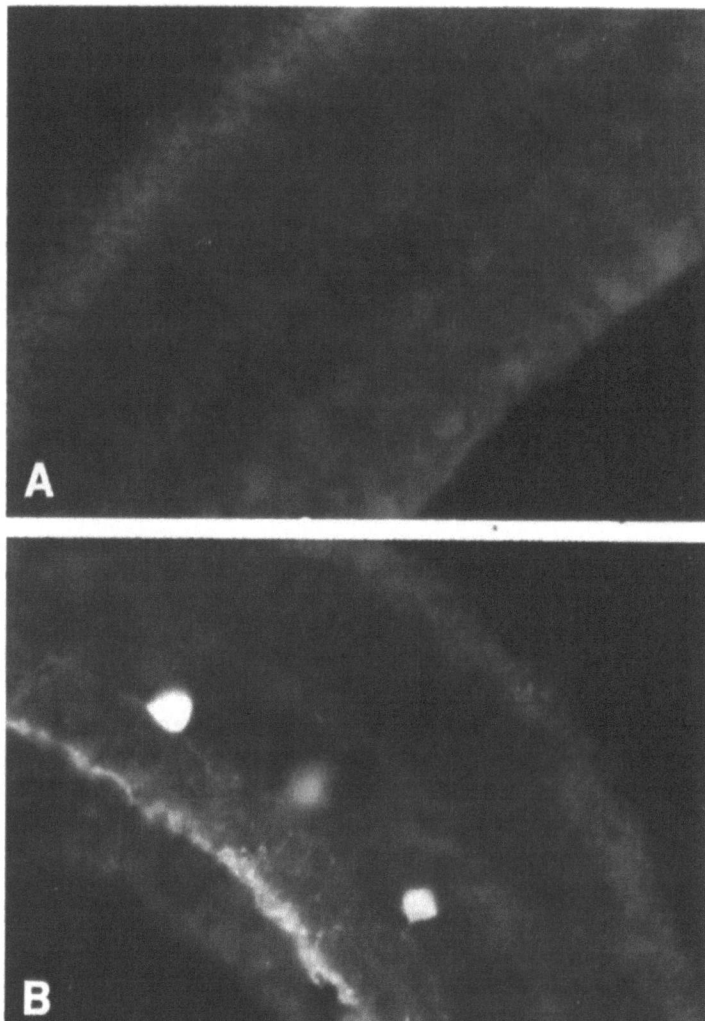

Fig. 4. The serotonin-accumulating cells in the rabbit retina. A, Rabbit retinas were processed immunohistochemically to localize serotonin. No immunoreactivity can be observed. B, Rabbit retinas were first incubated at 37°C for 30 min in physiological solution containing 10^{-7}M serotonin. Thereafter, the retinas were processed immunohistochemically to localize serotonin. Notice that the serotonin immunoreactivity is associated with some amacrine cell bodies and processes. (These data were kindly provided by Dr. N.N. Osborne.)

mitters estimated to occur in retinas from different species are shown in Table VI.

Cholecystokinin has been shown to be associated with amacrine cells of the inner inner plexiform layer.[354] Studies by Yamada et al.[380-382] have demonstrated the variations in the amounts of somatostatin and cholecystokinin in frog, goldfish, and bovine retina and their localization in the amacrine cells of the inner plexiform layer. Recently, further studies on the *in vitro* biosynthesis of somatostatin in frog retina have been reported.[383]

6. CYCLIC NUCLEOTIDE METABOLISM

6.1. Cyclic Nucleotides

The two major cyclic nucleotides, 3',5'-cyclic adenosine monophosphate (AMP) and 3',5'-cyclic guanosine monophosphate (GMP) have been demonstrated in the retina. In the mouse retina, the amount of cyclic GMP (11.5 ± 0.2 pmol/retina) is 4.5 times that of cyclic AMP (2.5 ± 0.1 pmol/retina).[384] Within the retina, these cyclic nucleotides are unequally distributed and are localized in specific cell types. Cyclic GMP is highly concentrated in the visual cells of the rod-dominant retina, where the levels of cyclic AMP are low. In the cone-dominant retina of the ground squirrel, the concentration of cyclic AMP is higher than that of cyclic GMP.[385] During postnatal life, increasing cyclic GMP levels are associated with morphological differentiation of visual cells and with the growth of rod outer segments.[386] Ninety percent of the cyclic GMP in the light-adapted rabbit retina is associated with photoreceptor cells. In photoreceptor cells, 75% of the total cyclic GMP is present in the rod outer segments.[387] The concentration of cyclic AMP in the rod outer segments is about 1/5 to 1/25 the concentration of cyclic GMP.[388,389]

6.2. Guanylate and Adenylate Cyclases

The activity of guanylate cyclase in retina (100–8000 pmol/min per mg protein) is severalfold higher than that in brain (110–120).[390-393] Most of the activity (90%) is localized in rod outer segments in bovine retina.[394] Guanylate cyclase from rod outer segments has a K_m of 0.27–0.4 mM for GTP, with a V_{max} value of 30 nmol/min per mg protein,[395,396] and requires Mn^{2+} for optimal activity and is inhibited by Ca^{2+}.[397] Guanylate cyclase activity in retina is ten times that of adenylate cyclase.[384]

Adenylate cyclase has been reported in retina from several species.[391,398-401] Among the different layers of the mouse retina, adenylate cyclase activity was found to be higher in bipolar plus ganglion cell layers (1141 pmol/min per mg protein) than in the photoreceptor cell layer.[35,402] The activity in the inner plexiform layer is three-to fourfold that of other layers.[402] It has been suggested that ROS adenylate cyclase, which is activated in darkness and inactivated in light, is involved in the sodium permeability of photoreceptor cells in the frog retina.[398]

6.3. Dopamine-Sensitive Adenylate Cyclase

Dopamine-stimulated adenylate cyclase is found only in regions of the central nervous system where dopaminergic nerve terminals are known to occur.[403] The retinas of chick, mouse, rat, rabbit, cat, calf, and monkey have been shown to contain this enzyme, which is localized in the inner layers of the retina.[404-408] In the carp retina, dopamine-sensitive adenylate cyclase seems to be the major enzyme.[349,409] The basal enzyme is localized mainly in the inner plexiform and photoreceptor cell inner segment layers of both rod-dominant (rabbit) and cone-dominant (ground squirrel) retinas.[410] Dopamine stimulates the adenylate cyclase activity differentially in various layers of the retina from rabbit and ground squirrel. Similarly, uneven distribution of the enzyme in the mouse retina and its differential regulation in inner and outer retina have been reported.[411]

Dopamine-stimulated adenylate cyclase is located in the plasma membranes of the neurons, where its activity is controlled by neurotransmitters at exoreceptor sites and intracellularly by GTP, Ca^{2+}, and macromolecular regulators as well as by the availability of ATP within the cell.[412] Ouabain (0.01 mM) and potassium ions (10–100 mM) increase cyclic AMP levels of intact retinas.[406] The effect of ouabain stimulation may be a result of inhibition of intact ATPase activity, i.e., a way of preserving ATP for use by adenylate cyclase.[404] Dowling and Watling[349] have reported that increases in cyclic AMP levels stimulated by depolarizing agents such as K^+ or veratridine are caused by the release of endogenous stores of dopamine. However, the effect of potassium or ouabain seems to be more complex. The dopaminergic adenylate cyclase system is extremely sensitive to low-intensity illumination. For example, dissection of mouse retinas under red light results in a reduction of cyclic AMP levels without affecting cyclic GMP levels.[410] However, hand dissection under infrared illumination does not appear to alter either of the cyclic nucleotides.

The dopamine analogues show a high degree of structural specificity. Thus, epinine, an N-methyl derivative of dopamine, was found to be equipotent with dopamine.[413] On the other hand, the γ-methyl analogues of dopamine are considerably less potent than the parent compound. Compounds that lack the catecholhydroxyl groups of the dopamine and substances in which the side chain contains one or three carbon atoms instead of the usual two are without activity. Thus, these results suggest that the side chain in the fully extended form is necessary for the action of dopamine. The other catecholamines, such as epinephrine, norepinephrine, and N-methyldopamine, are about one-tenth as potent as dopamine in the activation of adenylate cyclase from calf and rat retinas.[414,415]

The dopamine adenylate cyclase system displays interspecies variations in its reactions to agonists. Thus, retinal adenylate cyclase is insensitive to isoproterenol in calf retinas but is stimulated in Cebus and rhesus monkey retina.[414] In the mouse, dopamine-sensitive adenylate cyclase increases during the early postnatal life as the retina matures functionally.[402]

The neuroleptic drugs, e.g., haloperidol, chlorpromazine, and fluphenazine, which are widely used as dopamine antagonists, show essentially no effect on the basal activity of this enzyme.[406] Antipsychotic drugs such as pimozide, pimoxide, clozapine, loxapine, clothiapine and (+)-butaclamol antagonize the dopamine-sensitive adenylate cyclase system in retina.[416] Phentolamine, an α-adrenergic blocking agent, inhibits the retinal dopamine-sensitive adenylate cyclase system at a concentration 50 to 100 times that needed to effect blockade.[404] Lithium also selectively inhibits the dopamine-sensitive adenylate cyclase system in retina and striatum.[417] *In vitro*, the lowest effective dose for the inhibition of dopamine-stimulated adenylate cyclase from rabbit retina is 2 mM.[415] This is estimated to be equivalent to the highest therapeutic level of lithium achieved in human plasma during the treatment of acute mania.

6.4. Phosphodiesterase

An active cyclic nucleotide phosphodiesterase (PDE) in the retina is known to hydrolyze both cyclic AMP and cyclic GMP. The activity of this enzyme is two to six times higher than that reported for brain. In the retina, the cyclic GMP PDE specific enzyme is localized mainly in rod outer segments.[418–422] Enzyme purified from frog rod outer segments shows an approximate molecular weight of 240,000, with a K_m of 70 μM and 3 mM for cyclic GMP and cyclic AMP, respectively.[423]

In homogenates, the rate of hydrolysis of cyclic GMP is three times that of cyclic AMP.[424] The activity of phosphodiesterase in rod outer segments from rabbit and ground squirrel retina with cyclic GMP (148–318 nmol/min per mg protein) is 15 to 32 times that with cyclic AMP.[422] Recently, Lolley and his associates have reported two different cyclic GMP phosphodiesterases, one with a high K_m (300 μM) localized in the photoreceptor cell layer and another with a low K_m (30 μM) localized in the bipolar plus ganglion cell layer.[425,426]

ATP-dependent, light-stimulated phosphodiesterase has been demonstrated in rod outer segments of frog[427] and bovine[421] retinas. Another light-activated enzyme associated with rod outer segments is GTPase.[428–431] It has been proposed that these two enzymes may be involved in the regulation of cyclic GMP, which may be a transmitter in visual excitation in rod cells.[432] The role of these enzymes in visual excitation may be separated into two stages—first, the activation of several hundred phosphodiesterases and, second, the hydrolysis of nearly a thousand cyclic GMPs per second by each activated enzyme molecule.[433] Activation of phosphodiesterase requires GTP in addition to light; light activates GTPase in rod outer segment membranes. Thus, the hydrolysis of cyclic GMP in rod outer segments is associated with the light-activated, phosphodiesterase–GTPase system. Recent studies by Miller and his associates[432a,b] suggest that normal levels of intracellular cyclic GMP may be an important determinant of the latency and other response characteristics of vertebrate rod photoreceptors. The hydrolysis of cyclic GMP by light-activated phosphodiesterase is required for the normal response to illumination, the hyperpolarization of the ROS membrane.

7. RETINOID BINDING PROTEINS IN RETINA

The major compounds of the retinoid group are retinol, retinal, and retinoic acid. They are characterized structurally by a β-ionine ring and an isoprenoid side chain, and they evoke the same physiological responses as the parent compound, vitamin A. Apart from their role in vision, retinoids have been implicated in a variety of cellular functions such as rat liver lysosomal stability,[434] complement-dependent immune damage in liposomes,[435] permeability of egg lecithin liposomes,[437] microviscosity of erythrocyte membranes,[437] and increased release of lysomal degradative enzymes from retina and pigment epithelium.[438,439] The discovery by Bashor et al.[440] of a specific, soluble, cellular retinol-binding protein (CRBP) in several rat tissues shed new light on the possible mechanism of action of retinoids in cellular functions and the specific binding of retinoids in cellular proteins derived from normal and abnormal cells. Excellent reviews are available for further reading on this subject.[22,441–443]

Wiggert and Chader[444] reported the presence of retinol-binding protein in the retina and pigment epithelium–choroid unit of chick embryos. Since chick embryo retina is avascular, contamination by serum RBP was ruled out. Further support for the separate nature of cellular retinol-binding protein compared to serum RBP has been derived from fluorescence, sucrose-gradient analysis, and immunologic studies.[445] The CRBP has a molecular weight of about 14–20,000, similar to CRBP from various other tissues, and sediments at 2 S on sucrose gradients. Highly purified CRBP from retina has a molecular weight of about 16,600,[446,447] with an amino acid composition similar to that of other tissue CRBPs.[22,446] Its binding capacity is very specific for retinol and virtually nonexistent for retinoic acid.[444] The presence of another, lower affinity binding species with a sedimentation constant of 5 S has been reported in pigment epithelium–choroid complex, brain, and liver cytsosols.[448] Its lower affinity and lack of specificity is consistent with the possibility that this unknown species is albumin or a related tissue protein and that it plays little role in the physiological control of retinoid action.

The presence of specific cellular retinoic acid binding protein (CRABP) with a similar molecular weight, sedimentation constant, and amino acid composition has been observed in bovine retinal cytosol.[447,449–451] This protein binds retinoic acid selectively and is fivefold more concentrated than CRBP in bovine retina. Large amounts of this protein are also present in pigment epithelium–choroid preparations,[452] where the protein is localized primarily in the choroid, rather than the pigment epithelium cells.[453,454] Saari and his associates also have observed the presence of an 11-cis retinal binding protein specific to retina and different from CRBP or CRABP.[455–457] When this protein was isolated and characterized,[456] it was found to be more than twice the size of CRBP and CRABP, with an estimated molecular weight of 50,000. It occurs in very high concentrations and differs from other binding proteins in its specificity for 11-cis retinal and 11-cis retinol.[457]

Another retinal-binding protein with a sedimentation coefficient of 7 S is found in retinal supernatant fractions from rhesus monkey, pig, and human retinas.[458] It also appears to bind endogenous retinol.[458] A similar protein but with different characteristics has been found in bovine brain supernatant.[452] The presence of this 7 S binding protein in crude outer-segment-enriched fraction[452] and its absence in fetal retina[452] and degenerating retina[459] suggest an association with rod outer segments. However, the absence of this protein in preparations of purified ROS membranes suggests that it may be a loosely bound interphotoreceptor retinol-binding protein (IRBP), that may be lost during the isolation procedure.[452]

Recently, Adler and her associates[461,463] reported that interphotoreceptor matrix (IPM) contains mucopolysaccharides and proteins. The proteins are mainly glycoproteins and contain an average 4% carbohydrate. The IRBP association with rod outer segment has been confirmed by light–dark experiments in which a fourfold increase in 7 S CRBP was observed in light-adapted frogs.[460] This effect is specific for retinol but not for retinoic acid and also specific to retina but not to brain. Recent studies from Chader's laboratory on the characterization of this 7 S binding protein from bovine retina suggest that it is a glycoprotein with a native molecular weight of about 248,000 and a subunit molecular weight of 146,000. This protein demonstrates a differential light/dark binding to [^{3}H]-retinol in rabbit retinas (adapted *in vivo*) and has been called the interphotoreceptor retinol-binding protein.[460a] A glycoprotein of this molecular weight has also been reported in IPM.[462] Similarly, Bridges and associates[464–466] reported a mannose-containing glycoprotein in bovine IPM; this carrier of endogenous retinol showed a molecular weight of 250,000 by gel-filtration chromatography and 140,000 by SDS polyacrylamide gel electrophoresis. Further studies are needed to understand the role of these retinoid-binding proteins in the function of retinoids in retina.

8. RENEWAL OF VISUAL CELLS

The unique feature of the visual cells (rods and cones) is the continuous renewal of their outer segment disk membranes.[14,15] This renewal occurs throughout the life of the animal, with new membrane added at the base of the outer segment and older membrane shed in discrete packets at the tip. The discarded material is engulfed and digested by the pigment epithelium.[14,467] The development of autoradiography has made it possible to understand the mechanism of renewal of membrane constituents such as proteins and lipids.[468] In the late 1960s, Young used autoradiographic techniques to show that exogenously supplied radioactive amino acids are incorporated into the inner segments of the rods and eventually migrate into the base of the rod outer segments to form a discrete "band" that moves in the apical direction. After reaching the apex, the moving "band" is shed and then ingested by the pigment epithelial cells.[56,469–472]

[^{3}H]Glycerol injected intravitreally into frogs is concentrated in the myoid region of the inner segments after 1 h; with increasing time, large amounts of label are seen diffusely distributed in the outer segments.[473] However, injected radioactive fatty acids or choline produce diffuse labeling of the outer segment membranes within a few minutes of administration.[474,475] Similar *in vitro* studies involving the isolated retina incubated with [^{3}H]choline or [^{3}H]glycerol confirmed the diffuse distribution of these lipid precursors in the rod outer segments.[146,476] Recently, Anderson *et al.*[477–479] demonstrated a rapid rise followed by an equally rapid exponential decline in the specific radioactivity of individual phospholipids in isolated rod outer segments of frogs injected with [^{3}H]glycerol. Thus, both autoradiographic and biochemical studies suggest that once the lipids are synthesized in inner segments and become part of the basal disks of rod outer segments, they diffuse freely between the disks and the plasma membrane. Anderson and co-workers used labeled glycerol to show that the half-life of phosphatidylserine was 23 days in frog rod outer segments whereas that of phosphatidylcholine and phosphatidylethanolamine was 18–19 days. They also used [^{3}H]amino acids to determine that the life-span of integral proteins in ROS is about 39 days. Thus, they suggested that the rod outer segment lipids are renewed by a molecular replacement mechanism mediated via phospholipid exchange proteins[170] in addition to the membrane replacement renewal mechanism shown for rod outer segment proteins.[14]

Cones, when provided with labeled amino acids, showed labeled proteins scattered diffusely among the outer segments.[469] It was thought that mature rods renew rod outer segments by membrane replacement whereas cones replace the molecules of the existing membranes.[472,480] However, it has recently been reported that the shedding of membranes by rods in rats[481,482] and frogs[483] is a synchronized process. When the animals were exposed to 12L:12D light:dark cycles, the rod outer segment disk membranes were shed shortly after the lights were turned on. Recently, it has been reported that the process of renewal in cones seems to be similar to that in rods except for time of day. Thus, when goldfish, lizards, and chickens were exposed to cyclic 12L:12D for a few days, cone shedding seemed to occur shortly after the lights went off; by the end of the dark period, the shed disk packets were ingested completely by the pigment epithelium.[27,29,484] Inasmuch as the rods mediate vision at night and cones predominate during the day,[15] the diurnal pattern of rod and cone membrane shedding appears to fit the alternation of visual function very well.

9. EFFECT OF LIGHT ON RETINA

It is well established that substantial damage to the retina occurs from visible light at intensity levels well below thresholds for thermal burns.[485,486] It has also been suggested that retinal damage may be caused by common light sources such as room lighting, phototherapy techniques, ophthalmoscopes, and fundus cameras.[486] When albino rats are exposed to visible, constant light of high intensity for less than 1 h or moderate intensity for up to 2 days, depending on the experimental conditions, the visual cells and pigment epithelium are irreversibly affected and destroyed.[487,488] Vacuoles form at the tips of the photoreceptors at 3 h exposure, followed by swelling in the photoreceptor outer segments at 24 h; by 3–5 days, the damaged outer segments are isolated from the inner segments and become pear-shaped bodies filled with tubular material. In the final stages there is cellular damage and complete adhesion of pigment epithelium and Müller cells.[486] The affected rats also show abnormal ERG patterns, predominantly a reduction in the a-wave.[489,490] Recently, it has been suggested that hormones such as prolactin and estrogen (estradiol) are involved in the susceptibility of photoreceptors to light damage.[491-493] In the retinal dystrophic RCS rats, light accelerates retinal degeneration; the damage occurs at a much lower threshold compared to normal rats.[494] *In vitro* studies on rhodopsin suggest that one of the mechanisms of light damage of the visual cells may be related to photooxidation of the unsaturated fatty acids in the rod outer segments, which prevents normal regeneration of the visual pigment.[495,496]

In normal mouse and rat retinas, there is a light-induced reduction of cyclic GMP.[383] Similarly, decreased utilization of glucose in the light-stimulated retina has been observed.[497] In contrast, glutamate, aspartate, and GABA increase in concentration in light-adapted animals, but show decreased turnover on light stimulation.[497] Constant illumination (100–125 foot candles) causes a significant reduction in the levels of docosahexaenoic acid and an increase in lipid hydroperoxides.[498] These biochemical changes are accompanied by photoreceptor degeneration, suggesting that light-induced retinal degeneration may be a result of peroxidation of polyunsaturated fatty acids. Light stimulation produces an increased turnover of phosphatidylinositol in isolated rat retinas.[499] Furthermore, the enhanced incorporation of [^{3}H]inositol into phosphatidylinositol has been noted in the frog retina,[500] as has the incorporation of [^{3}H]inositol and ^{32}Pi into phosphatidylinositol in light-stimulated frog retinas.[501]

Recently, Schmidt[162,163,502] reported that rat retinas incubated in the light show rates of hydrolysis of existing phosphatidylinositol and incorporation of labeled glycerol, glucose, and orthophosphate into phosphatidylinositol that are two- to threefold higher than the rates of retinas incubated in the dark. When toad retinas are exposed to light flashes, the labeling of polar lipids by [^{14}C]glycerol increases, with the largest increments in phosphatidylinositol and phosphatidic acid.[157] This finding was confirmed and extended by Schmidt.[502] Moreover, it was suggested that light stimulation of the biosynthesis of phosphatidylinositol and phosphatidic acid occurs not in the photoreceptors but rather in the neural part of the retina and is related to normal neural activity.[157] It is of interest that this light effect was seen in different ionic media and that both phosphatidylinositol and phosphatidic acid biosynthesis were dependent on Ca^{2+}, as shown by addition of this cation and by ionophore studies.[157] Also, light enhances phosphorylation of phosphatidylinositol to triphosphoinositide[502] and synthesis of phosphatidylinositol within the photoreceptor cell layer. It appears that light stimulation affects different pathways of phosphatidylinositol metabolism in different retinal cells.

10. AXOPLASMIC TRANSPORT

The retina serves as a unique model for studies of axoplasmic transport because the neuronal perikarya of the ganglion cells are widely separated from their nerve endings, which are located in the brain. The axonal transport of precursors injected into the eye can be traced to the superior colliculi and lateral geniculate body or to the optic tectum in chickens and goldfish.[503-505] When

[³H]-N-acetylmannosamine was injected intraocularly, axonal transport of labeled gangliosides from retina to optic tectum was observed in the chicken.[187,503,506] Similarly, when phospholipid precursors are injected intravitreally, they are taken up by ganglion cells and incorporated into phospholipids, some of which are then transported through the axons of the optic nerve or tract to nerve endings in the lateral geniculate body or superior colliculi.[507] Maximum incorporation occurs about 5 days after injection, followed by a decline.

The turnover of phospholipids in retina labeled with intraocularly injected [³H]glycerol is very similar, with an apparent half-life of 7 days. In contrast, other axonally transported labeled phospholipids show half-life values of 10 days for choline and inositol phosphoglycerides and 13 days for serine and diacylethanolamine phosphoglycerides. Optic tract phospholipid half-life values range from 11 to 18 days. This slower turnover in optic transport may in part reflect the transfer of some lipids into a more metabolically stable lipid pool in the myelin sheath.[508] Recently, Alberghina et al.[509] reported rapid axonal transport of a relatively small amount of mono- and polyphosphoinositides. The small accumulation of labeled phosphoinositides in the lateral geniculate body and superior colliculi of the rabbit optic system even several days after labeled inositol injection differs from the larger accumulation of other phospholipids labeled with [³H]glycerol or [1-¹⁴C]palmitate, suggesting possible use of inositol phospholipids during axoplamic transport by a *myo*-inositol exchange reaction or phosphodiesterase reaction.[509,510]

11. NUTRITIONAL STUDIES

11.1. Essential Fatty Acid Deficiency

Hands et al.[511] reported a significant decrease in docosahexaenoic acid content in retinal lipids of EFA-deficient rats and concluded that this deficiency results in impaired visual activity in dim light. They also noticed a complete depletion of vitamin A in these rats. However, Futterman et al.[512,513] reported no significant changes in the unsaturated fatty acid content of retinal lipids, although there was a significant decrease in other organs. They attributed the results of the earlier study to the lack of vitamin A rather than the EFA deficiency, and this was later confirmed by Anderson and Maude[131] in rod outer segment membranes. However, Tinoco et al.[514,515] reported depletion of docosahexaenoic acid in retinal lipids of rats fed a lenolenic-acid-deficient diet. Thus, the relationship between EFA deficiency and the polyunsaturated fatty acid content of retinal lipids seems to be rather controversial and needs further exploration to verify the role of dietary fatty acids in the maintenance of important lipid constituents of the photoreceptor membranes.

11.2. Vitamin E Deficiency

Vitamin E is thought to play a role in membrane stabilization by preventing lipid peroxidation.[516] Vitamin E deficiency is known to result in degeneration of photoreceptor outer segment membranes[517,518] and accumulation of secondary lysosomes in pigment epithelium.[519,520] Since photoreceptor membranes contain large amounts of polyunsaturated fatty acids, peroxidation of these molecules may be the mechanism by which vitamin E deficiency leads to the degeneration of photoreceptor outer segments.[521] Also, in vitamin E deficiency, the lysosomal membranes rupture, releasing lysosomal enzymes that can cause the degeneration of disk membranes.[520] In addition, lipofuscin granules accumulate in the pigment epithelium of vitamin-E-deficient animals.[522,602]

The degeneration of photoreceptor cell membranes in vitamin-E-deficient monkeys[517] and dogs[523] has been documented. In a recent report on the rehabilitation of vitamin-E-deficient rats, Amemiya[520] reported that administration of vitamin E reverses the effects of mild deficiency on photoreceptor outer segment and retinal pigment epithelium, although losses caused by severe deficiency, i.e., photoreceptors with broken nuclei, could not be repaired.[520]

11.3. Vitamin A Deficiency

One of the major problems in global nutrition is vitamin A deficiency, which leads to a disease syndrome involving xerophthalmia and keratomalacia. This condition involves several ocular manifestations, including conjunctival and corneal xerosis, Bitot's spots, corneal ulceration, corneal scarring, night blindness, and retinal degeneration.[524,525] In the 1960s, the World Health Organization sponsored a global survey in which it was estimated that approximately 100,000 people are currently blind as a result of xerophthalmia.[524] Although vitamin A deficiency is known to result in the loss of visual function, very little is known about the specific physiology and biochemistry of this disorder, except that it usually occurs in conjunction with generalized protein-calorie malnutrition.

In one study,[526,527] weanling rats subjected to vitamin A deficiency supplemented with retinoic acid showed complete loss of rod outer segments and also the loss of about half of the inner segments and visual cortex at 6 months of age. At 10 months, the visual cells were absent except for one irregular row of visual cell nuclei. Similarly, rhodopsin levels declined to 5–10% of normal after 2 months, and the amount of opsin was reduced by 50%. The disappearance of the ERG preceded the loss of the photoreceptor cells.[526,527] Rod degeneration is followed by cone degeneration in vitamin-A-deficient adult rats.[528] Rats subjected to 6 months of vitamin A deficiency and rehabilitated with vitamin A for 16 days show a regeneration of normal-sized outer segments,[526] but the outer nuclear layer does not recover normal thickness. Animals maintained in the light are more susceptible to vitamin A deficiency than animals kept in the dark.[488,489,529] Recently, the relationship between vitamin A and E deficiencies has been explored. It was found that photoreceptor loss is accelerated in vitamin-A-deficient rats that are also deficient in vitamin E.[522,530] However, little is known about the biochemical changes that occur in retina during vitamin A deficiency, and this topic warrants further investigation.

11.4. Taurine Deficiency

Taurine is an essential amino acid that is highly concentrated in the retinas of all species so far studied, including man.[531,532,341] Retinal degeneration caused by taurine deficiency[533–535] is first seen as a small hyperreflective area in the center of the area centralis, with an enlarged lesion in more advanced stages.[534] Taurine-deficient cats show a reduction in ERG amplitudes that correlates well with the decrease in retinal taurine concentration prior to the death of photoreceptor cells.[536,537] Rehabilitation of the taurine-deficient animals at an early stage with the amino acid itself, but not with its precursors (e.g., methionine or cysteine), restores normal ERG patterns and preserves normal visual function.[538] Although taurine has been reported to affect ionic fluxes,[539,540] the exact mechanism by which taurine deficiency leads to the death of photoreceptor cells is not known.

11.5. Zinc Deficiency

Zinc, an essential trace element, is present in the highest concentrations in the ocular tissues of virtually all animals, including man.[541–543] Zinc deficiency is known to be involved in vision disorders, including optic neuropathy, decreased visual activity, loss of color discrimination, retinal detachment, and retinal degeneration.[544–547]

In weanling rats, this deficiency is associated with an accumulation of irregular osmiophilic inclusion bodies of presumed lipid nature in the cytoplasm of the retinal pigment epithelium.[548,549] These osmiophilic bodies differ from phagosomes, lipofuscin granules, and microperoxisomes and are associated with mitochondria and smooth endoplasmic reticulum.[549] The mechanisms that produce these osmiophilic inclusion bodies are known also to cause decreased alcohol dehydrogenase activity and decreased conversion of retinol to retinal in the soluble fraction of the rat retina.[550] However, there is no evidence that zinc is an integral part of retinol oxidoreductase, an enzyme involved in vision, in the rod outer segment membranes.[551] The formation of all cis-retinol from all trans-retinal in rod outer segments does not seem to be affected in zinc-deficient rats.[549] Although it is known that there is an interaction between zinc and vitamin A metabolism,[552–555] current results indicate that zinc has some effect on the transport of retinol to the esterification site or on the esterification process itself.[549] Further studies are needed in this area to clarify the role of zinc in vitamin A metabolism.

12. EFFECT OF ISCHEMIA

When the circulation of blood to the cell is blocked, cell functions are altered in a number of ways, and, finally, irreversible damage results. When rabbit retinas are incubated at 37°C in the absence of O_2 and glucose, the cells become swollen after 30 min, although the extracellular volume is unchanged. Cell permeability, measured by the ratio of mannitol space to inulin space, increases significantly after 60 min of O_2 and glucose deprivation.[556] Neuronal cells are irreversibly damaged within 40 min of deprivation; however, Müller cells, the principal glial cells of the retina, are not irreversibly altered even after 90 min.[557] A significant increase in the isoenzymes of lactate dehydrogenase has been reported in rabbit retinas subjected to ischemia for 2 days by means of ligation of the optic nerve.[558] Similarly, under pressure ischemia, the activities of ATPase, aldolase, and malate dehydrogenase decrease, phosphorylase activity disappears completely,[559] and the amount of ATP[560] and glycogen[561,562] also decreases. At 37°C under anaerobic conditions *in vitro*, free fatty acids, mainly docosahexaenoic acid, are released.[195] This is consistent with the release of free fatty acids by the ischemic brain reported previously.[563]

13. DIABETIC RETINOPATHY

Diabetes mellitus is characterized by abnormalities in carbohydrate and lipid metabolism and leads to retinopathy.[564-567] It has been shown that maximal intraocular pressure decreases as retinopathy progresses in diabetic patients.[568] Deposition of lipids in endothelial cells and pericytes of the retinal capillaries and their surroundings in streptozotocin-diabetic rats and human diabetics suggests a close relationship between abnormal lipid metabolism and the development of microangiopathy.[567]

A low content of polyenoic fatty acids, such as arachidonic and docosahexaenoic acid, is found in lipids from alloxan-treated diabetic rat retinas.[569] This decrease is probably a result of depressed polyenoic fatty acid synthesis, as has been shown in retinal homogenates from diabetic rats.[512] An increased turnover of phosphatidylinositol-4,5-biphosphate in the sciatic nerve of diabetic rats has been described,[570] as well as increased lipid peroxidation in alloxan-treated diabetic rat retinas.[571] Retinal protein synthesis and axonal transport have also been studied in animal models of experimental diabetes,[572] and a reduced sensitivity of adenylate cyclase to dopamine and norepinephrine in diabetic retinal homogenates has been reported.[573]

14. RETINAL DEGENERATION

Retinitis pigmentosa is a family of retinal degenerative diseases with special characteristics and a distinctive time course ultimately leading to severely impaired retinal function and blindness. Although much information is available about the morphology[574,575] and electrophysiology[576,577] of human retinitis pigmentosa, very little is known about the biochemistry or cellular physiology of this condition. The multiple etiologies of retinitis pigmentosa and the lack of clinical specimens have made it difficult to study the precise mechanisms underlying this disease process. Several experimental animal models of retinal degeneration have been developed, including the C3H mouse, RCS rat, and Irish setter dog. In all of these models, the final stage of the disease is similar: visual cell death and blindness. However, the time of onset of the disease and the morphological changes differ considerably. For example, visual cell death occurs early in life in C3H mice and Irish setter dogs and is a result of improper development of retinal cells.[578] In contrast, in the RCS rat, disease onset occurs somewhat later in life and is due primarily to improper functioning of pigment epithelium cells that leads to retinal cell death.[583] Excellent recent reviews of this subject are available for further reading.[579–584]

In C3H mice[585] and Irish setter dogs,[586] there is complete degeneration of visual cells during early postnatal life. In rats with retinal degeneration (RCS, Hunter, or Campbell type), the visual cells grow normally, but photoreceptor renewal is interrupted by a disturbance in the pigment epithelium cells involved in phagocytosis.[587] In normal animals, the shed disk membranes are phagocytized by pigment epithelium. In dystrophic rats, the synthesis and shedding of rod outer segment membranes are normal, but the inadequate rate of ingestion of disk membrane material by pigment epithelium results in an accumulation of membranous material called "debris" between rod outer segments and pigment epithelium.[583] This interferes with the normal metabolism of retina. Biochemical studies carried out on these models of retinal degeneration have described several abnormalities, which are summarized below.

14.1. Carbohydrate Metabolism

The rate of anaerobic glycolysis in the dystrophic rats decreases significantly between 16 and 25 days of age.[82,588,589] Aerobic glycolysis is not affected until 21 days of age, but the decrease in lactic acid, CO_2, and amino acid production is accompanied by a decrease in respiratory rate and glucose utilization after 21 days of age.[590] Lolley[591] reported a decreased utilization of glucose in *rd* mice after the 15th postnatal day, although there were no differences in the amounts of glucose, glycogen, lactate, ATP, and phosphocreatine within the photoreceptor layers of normal and dystrophic mice before 10 days of postnatal age. Since the changes in energy metabolism of dystrophic mice or rats are secondary to retinal degeneration, it is unlikely that glucose metabolism is involved in the primary etiology of the disease.[582]

14.2. Protein and Nucleic Acid Metabolism

The content of DNA and RNA increases with age in both normal and dystrophic rats until 2 weeks of age[50] and until 5 days after birth in mice.[40,50] The content of DNA and RNA remains constant in normal adult animals but decreases significantly in dystrophic animals after disease onset, because of cell death.[40,50] The ratio of RNA/DNA rises significantly in degenerating mouse retinas. The total protein content of retina in Campbell rats does not seem to change significantly until 7–8 weeks of age but thereafter falls to approximately 40% of the control value.[50] *In vitro* [1-^{14}C]glycine uptake into retinal proteins decreases in dystrophic rats compared to normals.[592] The differences are significant before 8 days of age, lessen with increasing age, and vanish at 21 days, suggesting a depression in the synthesis and turnover of protein between 6 and

8 days.[592] Also, the activity of the hexose monophosphate shunt pathway is increased in the dystrophic retina compared to normal retina.[87,593,594]

14.3. Lipid Peroxides

Lipids, especially unsaturated lipids, are highly susceptible to oxidation on exposure to irradiation, oxygen,[595,596] light,[597,598] and iron.[599-601] The resulting products, free radicals and peroxides, are deleterious to the cell. Since retinal membranes, particularly the photoreceptor membranes, are rich in polyunsaturated fatty acids (Table IV), they are uniquely susceptible to lipid peroxidation.[602,603] Hiramitsu and his associates[599,604,605] reported an increased accumulation of lipid peroxides in rabbit retinas exposed *in vivo* to high concentrations of oxygen, X-irradiation, or surgical implantation of an iron nail into the vitreous cavity. Chick embryos exposed to oxygen[606] and dark-adapted frogs exposed to room light for 30 min show abnormally high amounts of retinal peroxides. Frogs injected intravitreally with $FeSO_4$ show an increase in retinal lipid peroxides with a decrease in docosohexaenoic acid.[600] Similar findings were reported in the case of light-induced retinal degeneration[607,608] and vitamin E and selenium deficiencies.[609-611]

Armstrong and his associates[612] explored the role of lipid peroxides in a dog model of retinal degeneration known as canine ceroid lipofuscinosis (CCL), which is similar to human ceroid lipofuscinosis. These English setter dogs showed clinical, pathological, and biochemical changes that resemble those of the human disorder.[612] In both cases there is an accumulation of lipofuscin and ceroid pigments, which form granular, fingerprint, and/or curvilinear bodies in the viscera, nervous system, retina, and pigment epithelium. Lipid peroxides are present in the accumulated ceroid pigment in retinas with CCL[613] as well as various particulate peroxidases in the retina and pigment epithelium.[614,615] Superoxide dismutase is more active in rod outer segments than in other membranes,[615,616] and the loss of retinal pigment epithelial melanin is inversely related to increased deposition of ceroid pigment.[617] Also, in cell culture, characteristic ceroid pigment accumulates in retinal pigment epithelium cells from CCL dogs.[618]

Intravitreal injection of lipid hydroperoxides of pure polyunsaturated fatty acids such as linoleic acid, linolenic acid, arachidonic acid, and docosahexaenoic acid into the eyes of adult rabbits produced a marked decrease in the a-, b-, and c-waves of the ERG within 4 days; by 12 days, essentially no activity was seen.[619] Peroxides of phospholipids, malonaldehyde, H_2O_2, and sodium iodate were less cytotoxic, and vitamin A and related compounds had no effect on the ERG.[619] These observations suggest that purified lipid hydroperoxides alter retinal function irreversibly.

Similarly, experimental conditions involving accumulation of lipid peroxides have been shown to affect retinal physiology. A reduction in the ERG with the accumulation of lipid peroxides has been demonstrated in X-irradiated or iron-nail-implanted animals.[604,605] In CCL dogs, a reduction in b-wave amplitude was associated with increased lipid peroxides and decreased peroxidases in retina and retinal pigment epithelium.[612,619] When retinal homogenates from affected dogs were injected intravitreally into rabbit eyes, the ERG was completely abolished.[612] The ultrastructural abnormalities of the retinal pigment epithelium have been correlated with the ocular electrophysiological properties in CCL dogs. At the final stages of disease, ERG potentials are drastically reduced, suggesting cessation of neuroretinal function.[620]

All of these results suggest that the accumulation of lipid peroxides in the retina or pigment epithelial cells causes drastic changes in the cell function, leading finally to cell death. However, the exact mechanism by which these deleterious effects are produced is not properly understood.

14.4. Cyclic Nucleotide Metabolism

Among the different aspects of retinal degeneration, cyclic nucleotide metabolism probably is the best studied.[579] The concentration of cyclic GMP increase severalfold in visual cells of *rd* mice[402] and Irish setter dogs,[586] whereas cyclic AMP levels are not markedly affected.[579,621] Cyclic GMP synthesis appears to be normal; the increase in cyclic GMP in retinal degeneration is the result of a reduction in the cyclic nucleotide phosphodiesterase that breaks down cyclic GMP in the visual cells. In the RCS rat, cyclic GMP levels are lower than in normal rats[386,425] only after the accumulation of debris in the retina. In addition, the kinetics of phosphodiesterase activity are altered by the accumulated debris in both normal and dystrophic retinas *in vitro*.[386,425] Since the cyclic nucleotides are known to be involved in the morphological and functional maintenance of pigment epithelial cells,[622] the abnormal functioning of these cells in the RCS rat has been attributed to low levels of cyclic GMP.

Cyclic AMP phosphodiesterase activity is lower in retinas from Campbell rats before the onset of retinal degeneration than in controls.[623] Also, Hunter rats show a loss of retinal phosphodiesterase activity with visual cell degeneration.[623]

14.5. Gyrate Atrophy

Gyrate atrophy of the retina and choroid is one of the hereditary retinal degenerative diseases associated with an abnormal recessive gene. Seen mainly in young people, gyrate atrophy is manifested in terms of night blindness, restriction of visual field, and myopia. The disease is characterized by the peripheral degeneration of the retina with the development of scalloped margins. It is an inborn error of ornithine aminotransferase resulting in an increased level of ornithine in blood and urine.[624] Simell and Takki[625] documented the abnormal enzyme level as the putative cause of this disorder; their work has since been confirmed by others.[626,627] The increased levels of ornithine result from a deficiency of ornithine aminotransferase, a mitochondrial matrix enzyme that catalyzes the conversion of L-ornithine to glutamic γ-semialdehyde.[627–631] Ornithine aminotransferase is deficient in cultured skin fibroblasts[627,632–636] and lymphocytes[637] from gyrate atrophy patients. Experimental treatment of gyrate atrophy with high doses of pyridoxine[627,638,639] and lysine[640,641] has yielded equivocal results. Since the heterozygous carriers have one-half of the normal ornithine aminotransferase activity, measurement of this enzyme in amniotic fluid cells can be used to diagnose this disorder prenatally.[624,636,641]

15. RETINAL REGENERATION

In higher vertebrates, including man, the central nervous system cannot regenerate after injury, but this is not true for lower vertebrates. Species capable of visual regeneration can be used to examine two types of phenomena: the "missing link" in higher vertebrates that prevents regeneration and the development of the molecular mechanism of visual transduction from retina to brain in regenerating systems.[642] Over the last 8 years, Agranoff and his associates have contributed a great deal of information on the molecular biochemical mechanisms of retinal regeneration. Murray and Grafstein[643] showed that goldfish ganglion cells are capable of regenerating cut axons, a process that is accompanied by an increased incorporation of amino acids into proteins. Later, Heacock and Agranoff[644] reported that synthesis of protein, possibly of the tubulin class, increases on the side of the retina opposite the previously crushed optic nerve. Maximum label is observed 15 days after the optic nerve is crushed; normal values resume by 45 days. Furthermore, explants of denervated retina from older tadpoles show pure neuritic growth within 1–3 days *in vitro* compared to mixed cell growth in normal explants 1 week following explantation.[645] Similarly, explants of denervated retina from adult goldfish show more rapid neuritic outgrowth after optic nerve crush than explants from adult tadpoles.[646]

The functional and biochemical aspects of regneration have been studied.[647,648] In goldfish, the regeneration of the optic nerve, as determined by autoradiographic studies, can be correlated with behavioral recovery, i.e., autonomic response to illumination, ability to localize food, and an optomotor response, in that order. The rate of regeneration is markedly enhanced by increasing temperature above ambient levels (20°C).[646] Also, retinal explants grown in polycation substratum have a marked clockwise orientation of their neurites.[649]

The neuritic outgrowth from explant cultures of adult goldfish retinas arises from ganglion cells; the neurites are of axonal origin and are not derived from dendritic processes. The pattern and orientation of neuritic growth mimics that of optic nerve fibers *in vivo*.[650] Also, a single ganglion cell from a retina that underwent optic nerve crush 10–14 days before shows enhanced neuritic outgrowth with the characteristic clockwise pattern seen in retinal explants.[651] Feldman *et al.*[652] have demonstrated that the new membranes formed are added in the region of the growing tip of the neurites.

Retinal tubulin mRNA increases 4 days after nerve crush,[653] correlating well with the increased incorporation of labeled methionine into tubulin after optic nerve crush. Furthermore, the radioactivity from labeled uridine accumulates in uridine nucleotides and nucleotide sugars of postcrushed retinas[654] as a result of increased enzyme activity (e.g., uridine kinase, uridine monophosphate kinase, and uridine diphosphate kinase) in cell-free retinal preparations in goldfish.[655] The activity shows an initial increase 2 to 3 days following nerve crush, reaches a peak at 4 days, and gradually returns to basal levels by day 21. Also in the goldfish, polyamines such as putrescine are transported axonally by a regenerating optic nerve but not by the intact nerve,[656] suggesting a role for these molecules in the regeneration process. Kohsaka et al.[657] reported an increase in the activity of ornithine decarboxylase, an enzyme involved in the synthesis of putrescine, in goldfish retinas after optic nerve crush.[657] Explants of denervated retinas grown in cultures containing nerve growth factor and fetal calf serum have shown similar results.[658] Further studies revealed that various factors other than fetal calf serum and nerve growth factor can induce neuritic outgrowth from regenerating retinal explants.[659,660] Antibodies specific to gangliosides block the neuritic outgrowth from regenerative retinal explants of goldfish, suggesting that gangliosides function as receptors on the cell surface and interact with the trophic substances to induce the outgrowth.[661] Thus, from the above studies it can be concluded that the exact molecular biochemical mechanism involved in retinal regeneration is not clear and must await further research in this area.

16. CONCLUSIONS AND PERSPECTIVES

The amount of work done in the last 15 years on the investigation of the structural, metabolic, and functional aspects of retina is enormous. This chapter attempts to highlight only the most salient aspects of retinal neurochemistry, admittedly with a bias toward areas most familiar to us. We have not covered the biochemistry of photoreceptor membranes and the visual process in depth since another chapter of this book is devoted to these subjects (H. Shichi). Current neurochemical information about retinal degeneration is not described at length because a forthcoming volume of this series will contain a chapter on this subject. Among the subjects included here are a description of the structural components of retinal membranes, such as lipids and proteins, the role of lipid peroxides in retinal damage, the mechanism of photoreceptor renewal, the metabolism of cyclic GMP in normal phototransduction and in retinal degeneration, some of the factors involved in retinal regeneration, new knowledge of promising candidates for identification as neurotransmitters in the retina, and the study of daily rhythms of retinal metabolism.

Salient advances in recent years are many and various. The highly fluid environment of the photoreceptor membranes has been described and attributed to an enrichment in polyenoic fatty acids, especially docosahexaenoic acid, in various phospholipids. It has also been shown that retina has a high capacity for the synthesis of lipids. However, studies of lipid-metabolizing enzymes in retina are very scarce, and future research should focus on the regulatory mechanisms of lipid metabolism.

Extensive studies have been carried out on rhodopsin and its glycoprotein nature. Also, attempts have been made to understand the turnover of various proteins in retina. The mechanism of renewal of photoreceptor outer segment membrane has been explored, although the precise mechanism of renewal of different molecules is not completely understood. The available data suggest that the molecular replacement mechanism operates to renew soluble proteins and lipids, and the membrane replacement mechanism renews integral proteins.

A circadian rhythm of photoreceptor membrane shedding has been identified, demonstrating an elegant correlation of function and physiology; cones used in daytime vision are shed at night, and rods used at night are shed during the day.

The content and metabolism of cyclic nucleotides, i.e., cyclic AMP and cyclic GMP, have been studied, as have their distribution in different layers of retina and their role(s), especially of cyclic GMP, in retinal degeneration. Cyclic GMP has been found to be localized mainly in photoreceptor cells, whereas cyclic AMP is enriched in the inner layers of retina; cyclic GMP levels increase in retinal degeneration as a result of decreased activity of cyclic GMP phosphodiesterase. Dopamine-sensitive adenylate cyclase has also been studied extensively.

Various animal models have been developed to explain the pathophysiology of retinitis pigmentosa. Although cyclic GMP has been proposed as a factor in the etiology of retinal degeneration, the precise role is not clear and needs future attention. New approaches are needed to define molecular defects in structural membrane components, and ways should be found to halt membrane destruction. The role of immunity and the possibility of applying genetic engineering techniques to treat hereditary forms of retinitis pigmentosa are areas of future research.

The biochemical mechanism of light-induced retinal damage is unknown. Rats exposed to continuous light for only a few days demonstrate retinal degeneration and loss of retinal photoreceptors, and even ordinary levels of radiation affect the retina, not only as a result of the thermal effect. The progression of retinal degenerative diseases, including senile macular degeneration, is thought to be retarded when the retina is protected from excessive light exposure.

Light-induced damage to photoreceptors may be a useful experimental model to understand the chemical pathogenesis of retinal degenerations. Peroxidation of polyenoic fatty acids of phospholipids in photoreceptor membranes may play a central role in this disorder. Of all animal membranes, photoreceptor membranes have the highest content of the most unsaturated acyl chain of phospholipids, docosahexaenoic acid (22:6, n-3). Most of our knowledge of this fatty acid relates to its distribution in lipids; little is known about its metabolism or function.

A number of chemical substances have been added to the list of putative retinal neurotransmitters. However, few of these candidates have been proven conclusively to participate as neurotransmitters and future studies need to be directed toward the specific characterization, interaction, and roles of the various transmitters in the retina. It is of particular interest to define the role neurotransmitters and other mediators play in the chemical signaling between retinal cells (i.e. the information flow through retinal synapses). Also, the involvement of membrane components, such as phosphoproteins and phospholipids (e.g., phosphatidylinositides, arachidonic acid, prostaglandins, lipoxygenase-reaction products), in the transfer of information between retinal cells in normal and pathological conditions needs to be clarified.

The capacity of whole retina and ganglion cells to regenerate following injury has been studied in lower vertebrates with the use of tissue culture techniques. The exact factor or factors involved in regeneration and the involvement of specific molecules in this process are areas of exploration for the future. Since DNA is known to be a stable component, its renewal through the repair of broken pieces has been suggested. Future studies should focus in detail on these aspects to reveal whether cellular DNA damage could reflect alterations in the metabolism by way of its repair. In several of these studies it will be important to use isolated retinal cells and to culture them in chemically defined media. An increased understanding of growth factors and modulators of retinal cell differentiation will provide valuable insight into neuronal cell death and axonal determination of synaptic contacts.

The study of prostaglandins in retina is still in its infancy, and knowledge of the metabolism of prostaglandins is scarce. Future research will shed light on the metabolism and role of prostaglandins in normal and abnormal retinas. Although the role of dolichols in the activation of monosaccharides in glycoprotein synthesis is well known, the accumulation of these compounds in brain with increasing age may imply other functions in the nervous system, including the retina.

Although we know that experimental retinal autoimmunity and immunoreactivity can be induced in congenic strain 13 guinea pigs by retinal rod outer segments and purified rhodopsin,[662] the study of neuroimmunologic aspects of the retina is just emerging. Barnstable has produced seven monoclonal antibodies against a membrane preparation from rat retina and has succeeded in recognizing different cell types by this procedure.[663]

Another area in which basic biochemical knowledge is lacking is ocular neovascularization, particularly as it affects the retina, where it is a major cause of blindness. Little biochemical information is available about retinal capillaries and why they proliferate and invade the vitreous chamber. Other questions to be answered involve the possibility of a metabolic relationship between the retina and the vitreous, the role played by retinal pigment epithelial cells, and new blood vessel formation in retrolental fibroplasia, diabetic retinopathy, branch or central vein occlusion, sickle cell disease, and other disorders characterized by occlusion of retinal vessels.[664,665]

In diabetic retinopathy, microangiopathy leads to hypoxia; however, the biochemical bases of endothelial swelling and neuronal retinal changes are unknown. In brain at the onset of ischemia, a rapid accumulation of free fatty acids, notably arachidonic acid, takes place.[563] Anoxia also triggers the release of free fatty acids in the retina and promotes the formation of endogenous fatty acid methyl esters.[666] Whether these changes resulting from membrane phospholipid hydrolysis are merely epiphenomena or are actually involved in the pathogenesis of retinal disease remains to be established.

ACKNOWLEDGMENTS. We thank Drs. Robert E. Anderson, Gerald J. Chader, Steven Fliesler, Richard N. Lolley, Neville Osborne, and Dianna Redburn for reading this manuscript and providing valuable suggestions. Drs. Anderson, Fliesler, and Osborne kindly provided some of the figures. We are indebted to Drs. Elaine R. Berman, Dean Bok, David G. Cogan, Berndt Ehinger, Hermann Kühn, Paul J. O'Brien, H. W. Reading, Michael Schorderet, Han Wassenaar, and Richard Young for providing thoughtful comments about the salient advances in retinal neurochemistry since 1970. We also want to thank the above-mentioned colleagues and Drs. Bernard W. Agranoff, Adelbert Ames III, Donald Armstrong, Scott Basinger, C. David Bridges, Ranwell Caputto, Boris Drujan, Debora Farber, Joe Hollyfield, Richard Masland, M. Michel-Villaz, Herminia Pasantes Morales, David Papermaster, Hitoshi Schichi, Robert T. Sorbi, and Mary Voaden for their kindness in sending reprints and preprints of their work. The support of National Institutes of Health (EY04428, National Eye Institute, to N.G.B.), Louisiana Lions Foundation and the Ernest C. and Yvette C. Villere Chair for Retinal Degeneration are gratefully acknowledged. Dr. Bazan is a Research to Prevent Blindness, Inc. William and Mary Greve International Research Scholar, 1983.

REFERENCES

1. Lolley, R. N., 1969, In, *Handbook of Neurochemistry*, Volume 1 (A. Lajtha, ed.), Plenum Press, New York, pp. 473–504.
2. Bazan, N. G., and Lolley, R. N. (eds.), 1980, *Neurochemistry of the Retina*, Pergamon Press, New York.
3. Hayes, B. P., 1982, *Progress in Retinal Research*, Volume 1 (N. N. Osborne and G. J. Chader, eds.), Pergamon Press, New York. pp. 197–226.
4. Hollyfield, J. G., (ed.), 1982, *The Structure of the Eye*, Elsevier Biomedical, New York.
5. Rodieck, R. W., (ed.), 1973, *The Vertebrate Retina: Principles of Structure and Function*, W. H. Freeman, San Francisco.
6. Straatsma, B. R., Hall, M. O., Allen, R. A., and Crescitelli, F. (eds.), 1969, *The Retina: Morphology, Function and Clinical Characteristics*, University of California Press, Los Angeles.
7. Besharse, J. C., 1982, *Progress in Retinal Research*, Volume 1 (N. N. Osborne and G. J. Chader, eds.), Pergamon Press, New York, pp. 81–124.
8. Fliesler, S. J., and Anderson, R. E., 1983, *Prog. Lipid Res.* 22:79–131.
9. Ehinger, B., 1982, *Retina* 2:305–321.
10. Hargrave, P. A., 1982, *Progress in Retinal Research*, Volume 1 (N. N. Osborne and G. J. Chader, eds.), Pergamon Press, New York, pp. 1–52.
11. Schichi, H., 1984, *Handbook of Neurochemistry, Volume 8*, 2nd ed. (A. Lajtha, ed.), Plenum Press, New York, pp. 577–602.
12. Bazan, N. G., 1982, *Vision Res.* 22:1539–1548.
13. Zinn, K. M., and Marmor, M. F. (eds.), 1979, *Retinal Pigment Epithelium*, Harvard University Press, Cambridge.
14. Young, R. W., 1976, *Invest. Ophthalmol.* 15:700–725.
15. Young, R. W., 1978, *Vision Res.* 18:573–578.
16. Clayton, R. M., Haywood, J., Reading, H. W., and Write, A. (eds), 1982, *Problems of Normal and Genetically Abnormal Retinas*, Academic Press, London.
17. Ames, A. III, and Nesbett, F. B., 1981, *J. Neurochem.* 37:867–877.
18. Hagins, W., 1972, *Annu. Rev. Biophys. Bioeng.* 1:131–158.
19. Bownds, M. D., 1981, *Curr. Top. Membr. Transport* 15:203–214.

20. Bownds, M. D., 1981, *Trends Neurosci.* **4:**214–217.
21. Hermolin, J., 1981, *Cold Spring Harbor Conf. Cell Prolif.* **8:**1347–1360.
22. Chader, G. J., 1982, *Cell Biology of the Eye* (D. McDevitt, ed.), Academic Press, New York, pp. 377–433.
23. Bitensky, M., Miki, N., Keirns, J., Kerns, M., Barbaran, J., Freeman, J., Wheeler, M., Lany, J., and Marcus, F., 1975, *Adv. Cyclic Nucleotide Res.* **5:**213–240.
24. Francois, J., and Neetens, A., 1962, *The Eye* (H. Davidson, ed.), Academic Press, New York, pp. 369–416.
25. Cunha-Vaz, J. G., Shakib, M., and Ashton, N., 1966, *Br. J. Ophthalmol.* **50:**441–453.
26. Willis, W. D., Jr., and Grossman, R. G. (eds.), 1981, *Medical Neurobiology*, 3d ed., C. V. Mosby, St. Louis, pp. 273–346.
27. Young, R. W., 1977, *J. Ultrastruct. Res.* **161:**172–185.
28. Morris, V. B., 1970, *J. Comp. Neurol.* **140:**359–397.
29. O'Day, W. T., and Young, R. W., 1978, *J. Cell Biol.* **76:**593–604.
30. Stell, W. K., and Harosi, F. I., 1976, *Vision Res.* **16:**647–657.
31. Cohen, A. L., 1963, *Biol. Rev.* **38:**427–459.
32. Crain, R. C., Marinetti, G. V., and O'Brien, D. F., 1978, *Biochemistry* **17:**4186–4192.
33. Dowling, J., 1967, *Molecular Organization and Biological Function* (J. M. Allen, ed.), Harper & Row, New York, pp. 186–210.
34. Laties, A. M., and Liebman, P. A., 1970, *Science* **168:**1475–1477.
35. Laties, A. M., Bok, D., and Liebman, P. A., 1976, *Exp. Eye Res.* **23:**139–148.
36. Hogan, M. J., Alvarado, J. A., and Weddell, J. G., 1971, *Histology of the Human Eye*, W. B. Saunders, Philadelphia.
37. Rosenkranz, J., 1977, *Int. Rev. Cytol.* **50:**25–158.
38. Nass, S., and Nass, M. M. K., 1963, *J. Cell Biol.* **19:**613–629.
39. Hess, H. H., and Thalheimer, C., 1965, *J. Neurochem.* **12:**193–204.
40. Lolley, R. N., 1973, *J. Neurochem.* **20:**175–182.
41. Tsukado, Y., Oyemura, K., and Matsutani, T., 1967, *Adv. Neurol. Sci.* **10:**210–218.
42. Hess, H. H., and Thalheimer, C., 1971, *J. Neurochem.* **18:**1281–1290.
43. Ehrlich, G., and Dische, Z., 1950, *Proc. Soc. Exp. Biol. Med.* **74:**40–42.
44. Koenig, E., 1967, *J. Cell Biol.* **34:**265–274.
45. Bok, D., 1970, *Invest. Ophthalmol.* **9:**516–523.
46. Bok, D., 1966, *Anat. Rec.* **154:**320.
47. Bok, D., 1967, *Anat. Res.* **157:**217.
48. Dawson, L. D., and LaVail, M. M., 1979, *J. Comp. Neurol.* **188:**263–272.
49. Young, R. W., 1980, *Neurochemistry* **1:**123–142.
50. Yates, C. M., Dewar, A. J., Wilson, H., Winterburn, K., and Reading, H. W., 1974, *Exp. Eye Res.* **18:**119–133.
51. Sidman, R. L., 1961, *The Structure of the Eye* (K. G. Smelser and J. Jones, eds.), Academic Press, New York, pp. 487–506.
52. Bazan, H. E. P., Careaga, M. M., and Bazan, N. G., 1981, *Biochim. Biophys. Acta* **666:**63–71.
53. Droz, B., 1963, *Anat. Rec.* **145:**157–167.
54. Reading, H. W., 1965, *Biochemistry of the Retina* (C. N. Graymore and J. Jones, eds.), Academic Press, New York, pp. 73–82.
55. Steinman, L., and Ames, A. III, 1974, *Tissue Cell* **6:**137–152.
56. Young, R. W., 1967, *J. Cell Biol.* **33:**61–72.
57. Ames, A. III, and Parks, J. M., 1976, *J. Neurochem.* **27:**1017–1025.
58. Parks, J. M., Ames, A. III, and Nesbett, F. B., 1976, *J. Neurochem.* **27:**987–997.
59. Ames, A. III, Parks, J. M., and Nesbett, F. B., 1980, *J. Neurochem.* **35:**131–142.
60. Ames, A. III, Parks, J. M., and Nesbett, F. B., 1980, *J. Neurochem.* **35:**143–148.
61. Hollyfield, J. G., and Anderson, R. E., 1982, *Invest. Ophthalmol. Vis. Sci.* **23:**631–639.
62. Hall, M. O., Bok, D., and Bacharach, A. D. E., 1969, *J. Mol. Biol.* **45:**397–406.
63. Heitzman, H., 1972, *Nature (New Biol.)* **235:**114.
64. Robinson, W. E., Walker, G. A., and Bownds, D., 1972, *Nature (New Biol.)* **235:**112–114.
65. Young, R. W., and Droz, B., 1968, *J. Cell Biol.* **39:**169–184.

66. Heller, J., and Lawrence, M. A., 1970, *Biochemistry* **9**:864-869.
67. Planter, J. J., and Kean, E. L., 1976, *J. Biol. Chem.* **251**:1548-1552.
68. Liang, C. J., Yamashita, K., Muellenberg, C. G., Shichi, H., and Kobata, A., 1979, *J. Biol. Chem.* **254**:6414-6418.
69. Fukuda, M. N., Papermaster, D. P., and Hargrave, P. A., 1979, *J. Biol. Chem.* **254**:8201-8207.
70. Bok, D., Basinger, S. F., and Hall, M. O., 1974, *Exp. Eye Res.* **18**:225-240.
71. Fager, L. Y., and Fager, R. S., 1981, *Vision Res.* **21**:581-586.
72. Bunt, A. H., and Saari, J. C., 1982, *J. Cell Biol.* **92**:269-276.
73. Dewey, M. M., Davis, P. K., Blasie, J. K., and Barr, L., 1969, *J. Mol. Biol.* **39**:395-405.
74. Basinger, S., Bok, D., and Hall, M., 1976, *J. Cell Biol.* **69**:29-42.
75. Jan, L. Y., and Revel, J. P., 1974, *J. Cell Biol.* **62**:257-273.
76. Papermaster, D. S., Schneider, B. G., Zorn, M. A., and Krachenbuhl, J. P., 1978, *J. Cell Biol.* **77**:196-200.
77. Clark, V. M., and Hall, M. O., 1982, *Exp. Eye Res.* **34**:847-859.
78. Reading, H. W., 1972, *Modern Ophthalmology*, Volume 1, 2nd ed. (A. Soreby, ed.), Butterworths, London, pp. 347-382.
79. Matschinsky, F. M., Passonneau, J. V., and Lowry, O. H., 1965, *J. Histochem. Cytochem.* **13**:707.
80. Cunha-Vaz, J. G., 1966, *Br. J. Ophthalmol.* **50**:505-516.
81. Warburg, O., 1927, *Biochem. Z.* **184**:484-488.
82. Brotherton, J., 1962, *Exp. Eye Res.* **1**:234-245.
83. Adler, F. H. (ed.), 1965, *Physiology of the Eye*, 3rd ed., C. V. Mosby, St. Louis, pp. 558-583.
84. Lowry, O. H., Passonneau, J. V., Hasselberger, F. X., and Schultz, D. W., 1964, *J. Biol. Chem.* **239**:18-30.
85. Matshinsky, F. M., and McDougal, D. B., 1968, *Prog. Clin. Chem. Methods* **3**:71-86.
86. Ames, A. III, 1965, *Biochemistry of the Retina* (C. N. Graymore, ed.), Academic Press, London, pp. 22-30.
87. Bonavita, V., 1965, *Biochemistry of the Retina* (C. N. Graymore, ed.), Academic Press, London, pp. 5-13.
88. Lowry, O. H., Roberts, N. R., and Lewis, C., 1956, *J. Biol. Chem.* **220**:879-892.
89. Lowry, O. H., Roberts, N. R., Schulz, D. W., Clow, J. E., and Clark, J. R., 1961, *J. Biol. Chem.* **236**:2813-2820.
90. Graymore, C., 1964, *Nature* **201**:615-616.
91. Bonavita, V., Guarneri, R., and Ponte, F., 1967, *Vision Res.* **7**:51-58.
92. Cohen, L. H., and Noell, W. K., 1965, *Biochemistry of the Retina* (C. N. Graymore, ed.), Academic Press, London, pp. 36-50.
93. Cohen, L. H., and Noell, W. K., 1960, *J. Neurochem.* **5**:253-276.
94. Volokhov, A. A., and Shilyagina, N. N., 1966, *Fed. Proc.* **25**:T221-T226.
95. Papa, S., Lofrumento, N. E., Secchi, A. G., and Quagliariello, E., 1967, *Biochim. Biophys. Acta* **131**:288-294.
96. Futterman, S., and Kinoshita, J. H., 1959, *J. Biol. Chem.* **234**:723-726.
97. Joanny, P., Corriol. J., and Ture, A., 1965, *J. Physiol. (Lond.)* **57**:635-636.
98. Lowry, O. H., 1964, *Morphological and Biochemical Correlates of Neural Activity* (M. M. Cohen and R. S. Snider, eds.), Harper & Row, New York, pp. 178-191.
99. Graymore, C. N., 1965, *Biochemistry of the Retina* (C. N. Graymore, ed.), Academic Press, London, pp. 83-90.
100. Cohen, C. H., and Noell, W. K., 1965, *Biochemistry of the Retina* (C. N. Graymore, ed.), Academic Press, London, pp. 36-50.
101. Fliesler, S. J., and Schroepfer, G. J., Jr., 1982, *Biochim. Biophys. Acta* **711**:138-148.
102. Johnston, D., and Hudson, R. A., 1974, *Biochim. Biophys. Acta* **369**:269-277.
103. Lowry, O. H., Roberts, N. R., and Lewis, C., 1956, *J. Biol. Chem.* **220**:879-892.
104. Aveldano, M. I., and Bazan, N. G., 1972, *Biochim. Biophys. Res. Commun.* **48**:689-694.
105. Aveldano, M. I., and Bazan, N. G., 1973, *Biochim. Biophys. Acta* **296**:1-9.
106. Aveldano, M. I., and Bazan, N. G., 1974, *J. Neurochem.* **23**:1127-1135.

107. Aveldano, M. I., and Bazan, N. G., 1977, *Adv. Exp. Med. Biol.* **83**:397–404.
108. Edel-Harth, S., Dreyfus, H., Bosch, P., Rebel, G., Urban, P. F., and Mandel, P., 1973, *FEBS Lett.* **35**:284–288.
109. Holm, M., Mansson, J. E., Vanier, M. T., and Svennerholm, L., 1972, *Biochim. Biophys. Acta* **280**:356–364.
110. Rodriquez de Turco, E. B., and Bazan, N. G., 1977, *J. Chromatogr.* **137**:194–197.
111. Bazan, N. G., 1982, *Handbook of Neurochemistry*, Volume 3 (A. Lajtha, ed.), Plenum Press, New York, pp. 17–35.
112. Bazan, N. G., 1982, *Phospholipids in the Nervous System*, Volume 1 (L. A. Horrocks, G. B. Ansell, and G. Porcellati, eds.), Raven Press, New York, pp. 49–62.
113. Careaga, M. M., and Bazan, H. E. P., 1981, *Neurochem. Res.* **6**:1163–1172.
114. Broekheuyse, R. M., 1968, *Biochim. Biophys. Acta* **152**:307–315.
115. Dorman, R. V., Dreyfus, H., Freysz, L., and Horrocks, L. A., 1977, *Biochim. Biophys. Acta* **486**:55–59.
116. Bazan, N. G., di Fazio de Escalante, M. S., Careaga, M. M., Bazan, H. E. P., and Giusto, N. M., 1982, *Biochim. Biophys. Acta* **712**:702–706.
117. Anderson, R. E., Maude, M. B., and Zimmerman, W., 1975, *Vision Res.* **15**:1087–1990.
118. Anderson, R. E., and Andrews, L. D., 1982, *Visual Cells in Evolution* (J. A. Westfall, ed.), Raven Press, New York, pp. 1–22.
119. Anderson, R. E., Feldman, L. S., and Feldman, G. L., 1970, *Biochim. Biophys. Acta* **202**:367–373.
120. Drenthe, E. H. S., Bonting, S. L., and Daeman, F. J. M., 1980, *Biochim. Biophys. Acta* **603**:117–129.
121. Drenthe, E. H. S., Klompmakers, A. A., Bonting, S. L., and Daemen, F. J. M., 1980, *Biochim. Biophys. Acta* **603**:130–141.
122. Bonting, S. L., Drenthe, E., and Daemen, F. J. M., 1980, *Neurochemistry* **1**:3–16.
123. Rambach, R. A., Nemes, P. P., and Dratz, E. A., 1974, *Exp. Eye Res.* **18**:1–12.
124. Smith, H. G., Fager, R. S., and Litman, B. J., 1977, *Biochemistry* **16**:1399–1405.
125. Dreyfus, H., Urban, P. F., Harth, S., Preti, A., and Mandel, P., 1976, *Ganglioside Function: Biochemical and Pharmacological Implications* (G. Porcellati, B. Ceccarelli, and G. Tettamanti, eds.), Plenum Press, New York, pp. 163–188.
126. Dreyfus, H., Pieringer, J. A., Farooqi, A. A., Harth, S., Rebel, G., and Sarlieve, L. L., 1978, *J. Neurochem.* **30**:167–174.
127. Urban, P. F., Harth, S., Freysz, L., and Dreyfus, H., 1980, *Structure and Function of Gangliosides* (L. Svennerholm, H. Dreyfus, and P. F. Urban, eds.), Plenum Press, New York, pp. 149–157.
128. Dreyfus, H., Urban, P. F., Harth, S. E., and Mandel, P., 1975, *J. Neurochem.* **25**:245–250.
129. Seyfried, T. N., Yu, R. K., Miyazawa, N., and Lai-Y. L., 1982, *J. Neurochem.* **39**:277–279.
130. Aveldano de Caldironi, M. I., Giusto, N. M., and Bazan, N. G., 1981, *Prog. Lipid Res.* **20**:49–57.
131. Anderson, R. E., and Maude, M. B., 1972, *Arch. Biochem. Biophys.* **151**:270–276.
132. Bartley, W., Van Heymirgen, R., Nolton, B. M., and Renshaw, A., 1962, *Biochem. J.* **85**:332–335.
133. Hunds, A. R., and Bartley, W., 1963, *Biochem. J.* **87**:263–265.
134. Anderson, R. E., 1970, *Exp. Eye Res.* **10**:339–344.
135. Aveldano de Caldironi, M. I., and Bazan, N. G., 1980, *Neurochem. Int.* **1**:381–392.
136. James, O. A., MacDonald, G., and Thompson, W., 1979, *J. Neurochem.* **33**:1061–1066.
137. Foot, M., Cruz, T. F., and Clandinin, M. T., 1982, *Biochem. J.* **208**:631–640.
138. Butler, M., and Abood, L. G., 1982, *J. Membr. Biol.* **66**:1–7.
139. Giusto, N. M., and Bazan, N. G., 1979, *Biochem. Biophys. Res. Commun.* **91**:791–794.
140. Aveldano, M. I., and Bazan, N. G., 1983, *J. Lipid Res.* **24**:620–627.
141. Bazan, N. G., Aveldano de Caldironi, M. I., Giusto, N. M., and Rodriquez de Turco, E. B., 1981, *Prog. Lipid Res.* **20**:307–313.
142. Anderson, R. E., Benolken, R. M., Jackson, M. B., and Maude, M. B., 1977, *Adv. Exp. Med. Biol.* **83**:547–559.
143. Swartz, J. G., and Mitchell, J. E., 1970, *J. Lipid Res.* **11**:544–550.

144. Swartz, J. G., and Mitchell, J. E., 1974, *Biochemistry* **13**:5053–5059.
145. Dreyfus, H., Harth, S., Urban, P. F., Mandel, P., and Freysz, L., 1978, *J. Neurochem.* **31**:1157–1162.
146. Masland, R. H., and Mills, J. W., 1979, *J. Cell Biol.* **83**:159–178.
147. Masland, R. H., and Mills, J. W., 1980, *Proc. Natl. Acad. Sci. U.S.A.* **77**:1671–1675.
148. Pu, G. A. W., and Anderson, R. E., 1983, *Invest. Ophthalmol. Vis. Sci.* **24**:288–293.
149. Bazan, H. E. P., Marcheselli, V. L., Careaga, M. M., and Bazan, N. G., 1981, *New Trends in Nutrition, Lipid Research and Cardiovascular Diseases* (N. G. Bazan, R. Paoletti, and J. M. Lacono, eds.), Alan R. Liss, New York, pp. 101–110.
150. Broekhuyse, R. M., and Daemen, F. J. M., 1977, *Lipid Metabolism in Mammals*, Volume 2 (F. Snyder, ed.), Plenum Press, New York, pp. 145–188.
151. Bazan, N. G., Ilincheta de Boschero, M. G., Giusto, N. M., and Bazan, H. E. P., 1976, *Adv. Exp. Med. Biol.* **72**:138–148.
152. Bazan, N. G., and Giusto, N. M., 1980, *Control of Membrane Fluidity* (M. Kates and A. Kuksis, eds.), Humana Press, Clifton, New Jersey, pp. 223–236.
153. Bazan, N. G., Ilincheta de Boschero, M. G., and Giusto, N. M., 1977, *Adv. Exp. Med. Biol.* **83**:377–388.
154. Bazan, N. G., Aveldano, M. I., Bazan, H. E. P., and Giusto, N. M., 1976, *Lipids*, Volume 1 (R. Paoletti, G. Porcellati, and G. Jacini, eds.), Raven Press, New York, pp. 89–97.
155. Giusto, N. M., and Bazan, N. G., 1979, *Exp. Eye Res.* **29**:155–168.
156. Bazan, H. E. P., and Bazan, N. G., 1976, *J. Neurochem.* **27**:1051–1057.
157. Bazan, H. E. P., and Bazan, N. G., 1977, *Adv. Exp. Med. Biol.* **83**:489–495.
158. Giusto, M. N., and Bazan, N. G., 1977, *Adv. Exp. Med. Biol.* **83**:481–488.
159. Ilincheta de Boschero, M. G., Giusto, N. M., and Bazan, N. G., 1980, *Neurochemistry* **1**:17–28.
160. Aveldano, M. I., Pasquare de Garcia, S. J., and Bazan, N. G., 1983, *J. Lipid Res.* **24**:628–638.
161. Anderson, R. E., and Kelleher, P. A., 1981, *Exp. Eye Res.* **32**:729–736.
162. Schmidt, S. Y., 1982, *Invest. Ophthalmol. Vis. Sci. [Suppl.]* **22**:66.
163. Schmidt, S. Y., 1983, *J. Neurochem.* **40**:1630–1638.
164. Yoshioka, T., Inoue, H., Takagi, M., Hayashi, F., and Amakawa, T., 1983, *Biochim. Biophys. Acta* **755**:50–55.
165. Reddy, T. S., and Bazan, N. G., 1983, *Invest. Ophthalmol. Vis. Sci. [Suppl.]* **24**:162.
166. Bazan, H. E. P., and Bazan, N. G., 1983, *Invest. Ophthalmol. Vis. Sci. [Suppl.]* **24**:163.
167. Swartz, J. G., and Mitchell, J. E., 1973, *Biochemistry* **12**:5273–5278.
168. Fliesler, S. J., and Schroepfer, G. J., Jr., 1983, *J. Biol. Chem.* **258**:5062–5070.
169. Futterman, S., Rollins, M. H., and Vacano, E., 1968, *Biochim. Biophys. Acta* **164**:433–434.
170. Dudley, P. A., 1976, Ph.D. thesis, Baylor College of Medicine, Houston.
171. Bazan, H. E. P., Careaga, M. M., Sprecher, H., and Bazan, N. G., 1982, *Biochim. Biophys. Acta* **712**:123–128.
172. Dudley, P. A., and Anderson, R. E., 1978, *FEBS Lett.* **95**:57–60.
173. Mizuno, A., 1976, *Biochem. J.* **80**:45–47.
174. Saelens, D. A., Walle, T., Privitera, P. J., Knapp, D. R., and Gabbney, T., 1974, *J. Pharmacol. Exp. Ther.* **188**:86.
175. Ilincheta de Boschero, M. G., and Bazan, N. G., 1982, *Biochem. Pharmacol.* **31**:1049–1055.
176. Giusto, N. M., Ilincheta de Boschero, M. G., and Bazan, N. G., 1983, *J. Neurochem.* **40**:563–568.
177. Ilincheta de Boschero, M. G., and Bazan, N. G., 1983, *J. Neurochem.* **40**:260–266.
178. Singer, M. A., 1977, *Biochem. Pharmacol.* **26**:51.
178a. Reddy, T. S., and Bazan, N. G., 1983, *Soc. Neurosci. Abstr.* **9**:350.
179. Fliesler, S. J., 1980, *Fed. Proc.* **39**:1577.
180. Fliesler, S. J., 1981, *Invest. Ophthalmol. Vis. Sci. [Suppl.]* **20**(3):4.
181. Fliesler, S. J., 1979, Ph.D. thesis, Rice University, Houston.
182. Dreyfus, H., Harth, S., Urban, P. F., and Mandel, P., 1978, *Enzymes of Lipid Metabolism* (S. Gatt, L. Freysz, and P. Mandel, eds.), Plenum Press, New York, pp. 655–665.

183. Dreyfus, H., Harth, S., Yusupi, A. N., Urban, P. F., and Mandel, P., 1980, *Structure and Function of Gangliosides* (L. Svennerholm, H. Dreyfus, and P. F. Urban, eds.), Plenum Press, New York, pp. 227–237.
184. Dreyfus, H., Harth, S., Urban, P. F., and Mandel, P., 1976, *Vision Res.* **16:**1365–1369.
185. Holm, M., and Morrison, J. E., 1974, *FEBS Lett.* **46:**200–202.
186. Caputto, B. L., Maccioni, A. H. R., and Caputto, R., 1975, *Nature* **257:**492–493.
187. Caputto, R., Maccioni, A. H. R., Maccioni, H. J. F., Caputto, B. L., and Landa, C. A., 1980, *Neurochemistry* **1:**45–57.
188. Dreyfus, H., Urban, P. F., Harth, S. E., Neskovic, N. M., and Mandel, P., 1975, *Lipids* **10**542–544.
189. Dreyfus, H., Harth, S. E., Urban, P. F., Neskovic, N., and Mandel, P., 1977, *Exp. Eye Res.* **25:**1–7.
190. Dreyfus, H., Harth, S., Urban, P. F., and Mandel, P., 1976, *Life Sci.* **18:**1057–1064.
191. Preti, A., Fiorilli, A., Dreyfus, H., Harth, S., Urban, P. F., and Mandel, P., 1978, *Exp. Eye Res.* **26:**621–628.
192. Dreyfus, H., Preti, A., Harth, S., Urban, P. F., and Mandel, P., 1983, *J. Neurochem.* **40:**184–188.
193. Bazan, H. E. P., and Bazan, N. G., 1976, *Life Sci.* **17:**1671–1678.
194. Aveldano, M. I., and Bazan, N. G., 1974, *FEBS Lett.* **40:**200–202.
195. Bazan, N. G., Aveldano, M. I., and Rodriguez, E., 1982, *Prog. Lipid Res.* **20:**523–529.
196. Bhattacherjee, P., and Eakins, K., 1974, *Br. J. Pharmacol.* **50:**227–230.
197. Kass, J., and Holmberg, N., 1979, *Invest. Ophthalmol. Vis. Sci.* **18:**166–171.
198. Birkle, D., and Bazan, N. G., 1983, *Invest. Ophthalmol. Vis. Sci. [Suppl.]* **24:**163.
198a. Birkle, D., and Bazan, N. G., 1984, *Prostaglandins* **27:**203–216.
199. Siminoff, R., and Bito, L., 1982, *Curr. Eye Res.* **1:**635–642.
200. Armstrong, D., 1982, *Ceroid-lipofusciniosis (Batter's Disease)* (D. Armstrong, N. Koppang, and J. A. Rider, eds.), Elsevier, Amsterdam, pp. 247–270.
201. Hemming, F. W., 1974, *MTP International Review of Science, Biochemistry of Lipids*, Volume 4 (T. W. Goodwin, ed.), Butterworths, London, pp. 39–97.
202. Lennarz, W. J., 1975, *Science* **188:**986–991.
203. Beytia, E. D., and Porter, J. W., 1976, *Annu. Rev. Biochem.* **45:**113–142.
204. Parodi, A. J., and Leloir, L. F., 1979, *Biochim. Biophys. Acta* **559:**1–37.
205. Burgos, J., and Hemming, F. W., 1962, *Biochem. J.* **82:**454–456.
206. Butterworth, P. H. W., and Hemming, F. W., 1968, *Arch. Biochem. Biophys.* **128:**503–508.
207. Carroll, K. K., Valim, A., and Woods, M. C., 1973, *Lipids* **8:**246–248.
208. Tavares, I. A., Johnson, N. J., and Hemming, F. W., 1977, *Biochem. Soc. Trans. (Lond.)* **5:**1771–1773.
209. Keller, R. K., and Adair, W. L., Jr., 1977, *Biochim. Biophys. Acta* **489:**330–336.
210. Rupar, C. A., and Carroll, K. K., 1978, *Lipids* **13:**291–193.
211. Wenstrom J. C., and Hamilton, D. W., 1980, *Biol. Reprod.* **23:**1054–1059.
212. Tavares, I. A., Coolbear, T., and Hemming, F. W., 1981, *Arch. Biochem. Biophys.* **207:**427–436.
213. Pullarkat, R. K., and Reha, H., 1982, *J. Biol. Chem.* **257:**5991–5993.
214. Waechter, C. J., and Lennarz, W. J., 1976, *Annu. Rev. Biochem.* **45:**95–112.
215. Dallner, G., Behrens, N. H., Parodi, A. J., and Leloir, L. F., 1972, *FEBS Lett.* **24:**315–317.
216. Daleo, G. R., Hopp, E., Romero, P. A., and Pont-Lezica, R., 1977, *FEBS Lett.* **81:**411–414.
217. Kean, E. J., 1980, *Neurochemistry* **1:**103–112.
218. Plantner, J. J., and Kean, E. J., 1981, *Invest. Ophthalmol. Vis. Sci. [Suppl.]* **20(3):**4.
219. Plantner, J. J., and Kean, E. J., 1976, *J. Biol. Chem.* **251:**1548–1552.
220. Hargrave, P. A., 1977, *Biochim. Biophys. Acta* **492:**83–94.
221. Fukuda, M. N., Papermaster, D. S., and Hargrave, P. A., 1979, *J. Biol. Chem.* **254:**8201–8207.
222. Kean, J. J., and Plantner, E. J., 1976, *Exp. Eye Res.* **23:**89–104.
223. Plantner, J. J., Poncz, L., and Kena, E. J., 1980, *Arch. Biochem. Biophys.* **201:**527–532.
224. Struck, D. K., and Lennarz, W. J., 1977, *J. Biol. Chem.* **252:**1007–1013.
225. Tkacz, J. S., and Lampen, J. O., 1975, *Biochem. Biophys. Res. Commun.* **65:**248–257.

226. Lehle, L., and Tanner, W., 1976, *FEBS Lett.* **71**:167–170.
227. Tabor, G. A., Fliesler, S. J., and Hollyfield, J. G., 1982, *Invest. Ophthalmol. Vis. Sci. [Suppl.]* **22**:279.
228. Fliesler, S. J., Rayborn, M. E., and Hollyfield, J. G., 1983, *Invest. Ophthalmol. Vis. Sci. [Suppl.]* **24**:279.
229. Redburn, D. A., and Hollyfield, J., 1982, *Biochemistry of the Eye* (R. E. Anderson, ed.), American Academy of Ophthalmology, San Francisco, pp. 196–226.
229a. Osborne, N. N., and Nesselhut, T., 1983, *Neurosci. Lett.* **39**:33–36.
230. Dowling, J. E., Ehinger, B., and Hedden, W. L., 1976, *Invest. Ophthalmol.* **15**:916–926.
231. Schorderet, M., and Magistretti, P. J., 1982, *Progress in Nonmammalian Brain Research*, (L. Bolis and G. Nistico, eds.), CRC Press, Boca Raton, pp. 185–211.
232. Osborne, N. N., 1981, *Neurochem. Int.* **3**:3–16.
233. Werman, R., 1966, *Comp. Biochem. Physiol.* **18**:745–766.
234. Granier, H., and Brownstein, M. J., 1981, *Basic Neurochemistry* (G. J. Siegel, R. W. Albers, B. W. Agranoff, and R. Katzman, eds.), Little Brown, Boston, pp. 269–296.
235. Douglas, W. W., 1968, *Br. J. Pharmacol.* **34**:451–474.
236. Drujan, B. D., Borges, J. M., and Alvarez, N., 1965, *Life Sci.* **4**:473–471.
237. Frank, B., Hillarp, N. A., Thume, G., and Topr, A., 1962, *J. Histochem. Cytochem.* **10**:348–354.
238. Lindall, O., Bjorklund, A., and Svensson, L., 1974, *Histochemistry* **39**:197–227.
239. Bjorklund, A., Falck, K. B., and Onman, C., 1972, *Methods of Investigative and Diagnostic Endocrinology* (J. E. Rall and J. Kopin, eds.), North-Holland, Amsterdam, pp. 318–368.
240. Falck, B., and Owman, C., 1965, *Acta Univ. Lund.* **7**:1–23.
241. Kuljis, R. O., and Karten, H., 1983, *J. Comp. Neurol.* **217**:239–251.
242. Redburn, D. A., and Cotman, C. W., 1976, *J. Neurochem.* **26**:297–303.
243. Redburn, D. A., 1977, *Exp. Eye Res.* **25**:265–275.
244. Thomas, T. N., and Redburn, D. A., 1980, *Vision Res.* **20**:1–8.
245. Moorhead, L. C., Redburn, D. A., Merrit, J., and Garcia, C., 1979, *Am. J. Ophthalmol.* **88**:239–245.
246. Yamamura, H. I., Enna, S. J., and Kuhar, M. J., 1978, *Neurotransmitter Receptor Binding*, Raven Press, New York, pp. 1–195.
247. Redburn, D. A., Kyles, C. B., and Ferkany, J., 1979, *Exp. Eye Res.* **28**:525–532.
248. Redburn, D. A., Clement-Cormier, Y. C., and Lam, D. M. K., 1980, *Life Sci.* **27**:23–31.
249. Redburn, D., Clement-Cormier, Y., and Lam, D. M. K., 1980, *Neurochemistry* **1**:167–181.
250. Redburn, D. A., 1980, *Adv. Biochem. Psychopharmacol.* **27**:79–89.
251. Redburn, D. A., and Mitchell, C. K., 1980, *Brain Res. Bull.* **5**:189–193.
252. Redburn, D. A., and Mitchell, C. K., 1981, *J. Neurosci. Res.* **6**:487–495.
253. Krnjevic, K., 1974, *Physiol. Rev.* **54**:448–540.
254. Bonting, S. L. (ed.), 1976, *Transmitters in the Visual Process*, Pergamon Press, Oxford.
255. Graham, L. T., 1974, *The Eye*, Volume 6 (H. Davson and L. T. Graham, eds.), Academic Press, New York, pp. 283–333.
256. Watling, K. J., 1981, *Trends Pharmacol. Sci.* **9**:244–247.
257. Neal, M. J., 1976, *Transmitters in the Visual Process* (S. L. Bonting, eds.), Pergamon Press, New York, pp. 128–143.
258. Masland, R. H., and Ames, A. III., 1976, *J. Neurophysiol.* **39**:1210–1219.
259. Masland, R. H., and Livingstone, C., 1976, *J. Neurophysiol.* **39**:1210–1219.
260. Hayden, S. A., Mills, J. W., and Masland, R. M., 1980, *Science* **210**:435–437.
261. Masland, R. H., 1982, *Retina* **2**:282–287.
262. Neal, M. J., and Gilroy, J., 1975, *Brain Res.* **93**:548–551.
263. Baughman, R. W., and Bader, C. R., 1977, *Brain Res.* **138**:469–485.
264. Neal, M. J., and Massey, S. C., 1980, *Neurochemistry* **1**:191–208.
265. Farniglietti, E. V., and Siegfried, E. C., 1980, *Invest. Ophthalmol. Vis. Sci. [Suppl.]* **19**:70.
266. Vaney, D. I., Peichl, L., and Boycott, B. B., 1981, *J. Comp. Neurol.* **199**:373–391.
267. Ross, D., Cohen, A. I., and McDougal, D. B., Jr., 1975, *Invest. Ophthalmol.* **14**:756.
268. Ross, C. D., and McDougal, D. B., 1976, *J. Neurochem.* **26**:521–526.
269. Dickson, D. H., Flumerfelt, B. A., Hollenberg, M. J., and Gwyn, D. G., 1971, *Brain Res.* **35**:299–303.

270. Drujan, B. D., Diaz-Borges, J. M., and Brzin, M., 1979, *Can. J. Biochem.* **57**:43–48.
271. Miyata, M., and Tshikawa, S., 1978, *Jpn. J. Ophthalmol.* **20**:478–496.
272. Reale, E., Luciano, L., and Spitznas, M., 1971, *J. Histochem. Cytochem.* **19**:85–96.
273. Spira, A. W., 1974, *J. Histochem. Cytochem.* **22**:868–880.
274. Lam, D. M. K., 1972, *Proc. Natl. Acad. Sci. U.S.A.* **69**:1987–1992.
275. Lam, D. M. K., 1976, *Cold Spring Harbor Symp. Quant. Biol.* **40**:571–579.
276. Gerschenfeld, H. M., and Piccolino, M., 1977, *Nature* **268**:257–259.
277. Gerchenfeld, H. M., and Piccolino, M., 1979, *The Neurosciences, 4th Study Program* (F. O. Schmitt and F. G. Worden, eds.), MIT Press, Cambridge, pp. 213–226.
278. Lam, D. K., Frederick, J. M., Holleyfield, J. G., Sarthy, P. V., and Mare, R. E., 1982, *Visual Cells in Evolution* (J. A. Westfall, ed.), Raven Press, New York, pp. 65–80.
279. Nichols, C. W., and Koelle, G. B., 1968, *J. Comp. Neurol.* **133**:1–16.
280. Kuriyama, K., Sisken, B., Haber, B., and Roberte, E., 1968, *Brain Res.* **9**:165–168.
281. Graham, L. T., Jr., 1972, *Brain Res.* **36**:476–479.
282. Voaden, M. J., 1976, *Transmitters in the Visual Process* (S. L. Bonting, ed.), Pergamon Press, New York, pp. 108–125.
283. Ehinger, B., and Falck, B., 1971, *Brain Res.* **33**:157–172.
284. Neal, M. J., Cunningham, J. R., and Marshall, J., 1979, *Brain Res.* **176**:285–296.
285. Cunningham, J., Marshall, J., and Neal, M. J., 1981, *Exp. Eye Res.* **32**:445–450.
286. Agardh, E., and Ehinger, B., 1982, *J. Neural Transm.* **54**:1–18.
287. Yazulla, S., 1981, *J. Comp. Neurol.* **200**:83–93.
288. Yazulla, S., and Brecha, N., 1980, *Invest. Ophthalmol. Vis. Sci.* **19**:1415–1426.
289. Yazulla, S., and Brecha, N., 1981, *Proc. Natl. Acad. Sci. U.S.A.* **78**:643–647.
290. Pourcho, R., 1981, *Brain Res.* **215**:187–199.
291. Marc, R. E., Stell, W. K., Bok, D., and Lam, D. M. K., 1978, *J. Comp. Neurol.* **182**:221–246.
292. Hollyfield, J. G., Rayborn, M. E., Sarthy, P. V., and Lam, D. M. K., 1979, *J. Comp. Neurol.* **188**:587–598.
293. Bonaventure, N., Klioland, N., and Rousel, G., 1980, *Pfluegers Arch.* **385**:51–64.
294. Lam, D. M. K., Su, Y. Y. T., and Swain, L., 1979, *Nature* **278**:565–567.
295. Brandon, C., Lam, D. M. K., and Wu, J. Y., 1979, *Proc. Natl. Acad. Sci. U.S.A.* **76**:3557–3561.
269. Brandon, C., Lam, D. M. K., Su, Y. Y. T., and Wu, J. Y., 1980, *Brain Res. Bull. [Suppl.]* **5**:21–29.
297. Vaughn, J. E., Famiglietti, E. V., and Barber, R. P., 1981, *J. Comp. Neurol.* **197**:113–127.
298. Famiglietti, E., and Vaughn, J. E., 1981, *J. Comp. Neurol.* **197**:129–139.
299. Redburn, D. A., and Mitchell, C. K., 1981, *Life Sci.* **28**:541–549.
300. Osborne, N. N., 1981, *Br. J. Pharmacol.* **71**:259–264.
301. Lam, D. M. K., Fung, S. C., and Kong, Y. C., 1981, *J. Neurosci.* **1**:1117–1132.
302. Lam, D. M. K., 1975, *Nature* **254**:245–346.
303. Naka, K. I., 1977, *J. Neurophysiol.* **40**:26–43.
304. Lam, D. M. K., Lasater, E. H., and Naka, K. I., 1978, *Proc. Natl. Acad. Sci. U.S.A.* **75**:6310–6313.
305. Murakami, M., Shimoda, Y., and Nakatani, K. 1979, *Sensory Proc.* **2**:334–338.
306. Wu, S. M., and Dowling, J. E., 1980, *Brain Res.* **199**:401–414.
307. Marshall, J., and Voaden, M., 1974, *Invest. Ophthalmol.* **13**:602–607.
308. Voaden, M. J., Marshall, J., and Murani, N., 1974, *Brain Res.* **67**:115–132.
309. Wu, S. M., and Dowling, J. E., 1978, *Proc. Natl. Acad. Sci. U.S.A.* **75**:5205–5209.
310. Nagishi, K., and Drujan, B. D., 1979, *J. Neurosci. Res.* **4**:351–363.
311. Kondo, H., and Toyoda, J. L., 1980, *Brain Res.* **199**:240–243.
312. Ishida, A. T., and Fain, G. L., 1981, *Proc. Natl. Acad. Sci. U.S.A.* **78**:5890–5894.
313. Redburn, D. A., 1981, *Glutamate as a Neurotransmitter* (G. Dichiaza and G. L. Gessa, eds.), Raven Press, New York, pp. 79–89.
314. Berger, S. J., McDaniel, M. L., Carler, J. G., and Lowry, O. H., 1977, *J. Neurochem.* **28**:159–163.
315. Voaden, M. J., 1978, *Amino Acids as Chemical Transmitters* (F. Fonnum, ed.), Plenum Press, New York, pp. 257–274.

316. Thomas, T. N., 1978, *J. Neurochem.* **31**:63–68.
317. Redburn, D. A., and Massey, S. C., 1981, *Amino Acid Neurotransmitters* (F. V. DeFeudis and P. Mandel, eds.), Raven Press, New York, pp. 169–174.
318. Ehinger, B., 1978, *Albrecht von Graefes Arch. Klin. Exp. Ophthalmol.* **217**:1–7.
319. Marc, R. E., and Lam, D. M. K., 1981, *J. Neurosci.* **1**:152–165.
320. Altschuler, R. A., Mosinger, J. L., and Wenthold, R. J., 1981, *Soc. Neurosci. Abstr.* **7**:916.
321. Neal, M. J., Collins, G. G., and Massey, S. C., 1979, *Neurosci. Lett.* **14**:241–245.
322. Bruun, A., and Ehinger, B., 1979, *Exp. Eye Res.* **19**:435–447.
323. Lam, S. M. K., and Hollyfield, J. G., 1980, *Exp. Eye Res.* **31**:729–732.
324. Marshall, J., and Voaden, M., 1975, *Vision Res.* **15**:459–461.
325. Marshall, J., and Voaden, M., 1976, *Exp. Eye Res.* **22**:189–191.
326. Pourcho, R. G., 1980, *Brain Res.* **198**:333–346.
327. Kolb, H., 1979, *J. Neurocytol.* **8**:295–329.
328. Ehinger, B., and Lindberg, B., 1974, *Nature* **251**:727–728.
329. Ehinger, B., and Lindberg, B., 1976, *Brain Res.* **113**:535–549.
330. Lopez-Colome, A. M., Salceda, R., and Pasantes-Morales, H., 1978, *Neurochem. Res.* **3**:437–441.
331. Chin, C. A., and Lam, D. M. K., 1980, *J. Physiol.* **308**:185–195.
332. Kong, Y. C., Fung, S. C., and Lam, D. M. K., 1980, *J. Comp. Neurol.* **193**:1127–1135.
333. Coull, B. M., and Cutler, W. P., 1978, *Invest. Ophthalmol. Vis. Sci.* **17**:682–684.
334. Coull, B. M., Owens, D. K., and Cutler, R. W., 1981, *Brain Res.* **210**:301–309.
335. Borbe, H. O., Muller, W. E., and Woller, U., 1981, *Brain Res.* **205**:131–139.
336. Ehinger, B., 1981, *Exp. Eye Res.* **33**:381–391.
336a. 8eale, R., and Osborne, N. N., 1983, *Develop. Brain Res.* **7**:107–120.
337. Mandel, P., Pasantes-Morales, H., and Urban, P. F., 1976, *Transmitters in the Visual Process* (S. L. Bonting, ed.), Pergamon Press, Oxford, pp. 89–106.
338. Gupta, K., Mathur, R. L., and Agarwal, L. P., 1981, *Exp. Eye Res.* **32**:793–796.
339. Mathur, R. L., Klethi, J., Ledig, M., and Mandel, P., 1976, *Life Sci.* **18**:75–80.
340. Austin, L., Recasens, M., Mathur, R. L., and Mandel, P., 1978, *Neurosci. Lett.* **9**:51–59.
341. Voaden, M. J., Lake, N., Marshall, J., and Morjaria, B., 1977, *Exp. Eye Res.* **25**:249–257.
342. Osborne, N. N., 1981, *Exp. Eye Res.* **33**:371–380.
343. Nesselhut, T., and Osborne, N. N., 1982, *Neurosci. Lett.* **28**:41–45.
344. Dowling, J. E., and Ehinger, B., 1978, *J. Comp. Neurol.* **180**:203–220.
345. Holmgren, I., 1982, *Invest. Ophthalmol. Vis. Sci.* **22**:8–24.
346. daPrada, M., 1977, *Adv. Biochem. Psychopharmacol.* **16**:311–319.
347. Kramer, S. G., 1971, *Invest. Ophthalmol.* **10**:438–452.
348. Bauer, B., Ehinger, B., and Aberg. L., 1980, *Albrecht von Graefes Arch. Klin. Exp. Ophthalmol.* **215**:71–78.
349. Dowling, J. E., and Watling, K. J., 1981, *J. Neurochem.* **36**:569–579.
350. Schaeffer, J. M., 1980, *Exp. Eye Res.* **30**:431–437.
351. Osborne, N. N., 1982, *Biology of Serotonergic Transmission* (N. N. Osborne, ed.), John Wiley & Sons, New York, pp. 401–430.
352. Osborne, N. N., 1980, *Neurosci. Lett.* **16**:167–170.
353. Osborne, N. N., 1980, *Exp. Eye Res.* **31**:31–39.
354. Osborne, N. N., Nicholas, D. A., Cuello, A. L., and Dockray, G. J., 1981, *Neurosci. Lett.* **26**:31–35.
355. Osborne, N. N., Nesselhut, T., Nicolas, D. A., and Cuello, A. C., 1981, *Neurochem. Int.* **3**:171–176.
356. Osborne, N. N., Nesselhut, T., Nicolas D. A., Patel, S., and Cuello, A. L., 1982, *J. Neurochem.* **39**:1519–1528.
357. Osborne, N. N., 1981, *Cell Mol. Neurol.* **1**:167–174.
358. Smith, M. D., 1973, *Comp. Gen. Pharmacol.* **4**:175–178.
359. Suzuki, O., Naguchi, E., and Yagi, K., 1977, *Brain Res.* **135**:303–313.
360. Thomas, T. N., and Redburn, D. A., 1979, *Exp. Eye Res.* **28**:55–61.
361. Ehinger, B., and Floren, I., 1978, *Exp. Eye Res.* **26**:1–11.
362. Floren, I., and Hansson, C. A., 1980, *Invest. Ophthalmol. Vis. Sci.* **19**:117–125.

363. Dowling, J. E., and Ehinger, B., 1975, *Science* **188**:270–273.
364. Dowling, J. E., Ehinger, B., and Floren, I., 1980, *J. Comp. Neurol.* **192**:665–685.
365. Ehinger, B., and Holmgren, I., 1979, *Cell Tissue Res.* **197**:175–194.
366. Ehinger, B., and Floren, I., 1976, *Cell Tissue Res.* **175**:37–48.
367. Ehinger, B., and Floren, I., 1980, *Neurochemistry* **1**:209–229.
368. Floren, I., 1979, *J. Neural Transm.* **46**:1–15.
369. Snyder, S. H., 1980, *Science* **209**:976–983.
370. Brecha, N., Karten, H. J., and Laverack, C., 1979, *Proc. Natl. Acad. Sci. U.S.A.* **76**:3010–3014.
371. Brecha, K., Karten, H. J., and Schenker, C., 1981, *Neuroscience* **6**:1329–1340.
372. Buckerfield, M., Oliver, J., Chubb, I. W., and Morgan, I. G., 1981, *Neuroscience* **6**:689–695.
373. Eriksen, E. F., and Larsson, L. I., 1981, *Peptides* **2**:153–157.
374. Eskay, R. L., Furness, J. F., and Long, R. I., 1981, *Science* **212**:1049–1050.
375. Fukuda, H., Kuriyama, Y., and Shiosaka, S., 1981, *Neurosci. Lett.* **23**:239–242.
376. Jackson, I. M., Bolaffi, J. L., and Guillemin, R., 1980, *Gen. Comp. Endocrinol.* **42**:505–508.
377. Karten, H. J., and Brecha, N., 1980, *Nature* **283**:87–88.
378. Lasater, E. M., and Dowling, J. E., 1982, *Proc. Natl. Acad. Sci. U.S.A.* **79**:936–940.
379. Tornquist, K., Loren, I., Hakanson, R., and Sundler, F., 1980, *Exp. Eye Res.* **33**:55–64.
380. Yamada, T., Marshak, D., Bosinger, S., Walsh, J., Morley, J., and Stell, W., 1980, *Proc. Natl. Acad. Sci. U.S.A.* **77**:1691–1695.
381. Yamada, T., Brecha, N., Rosenquist, G., and Basinger, S., 1981, *Peptides* **2**:93–97.
382. Yamada, T., and Basinger, S., 1982, *J. Neurochem.* **39**:1539–1546.
383. Unger, W. G., Butler, J. M., Cole, D. F., Bloom, S. R., and McGregor, G. P., 1981, *Exp. Eye Res.* **32**:797–801.
384. Ferrendelli, J. A., DeVries, G. W., Cohen, A. I., and Lowry, O. H., 1980, *Neurochemistry* **1**:311–326.
385. Farber, D. B., Chase, D. G., and Lolley, R. N., 1980, *Neurochemistry* **1**:327–336.
386. Lolley, R. N., and Farber, D. B., 1976, *Exp. Eye Res.* **22**:447–486.
387. Ferrendelli, J. A., Rubin, E. H., Orr, H. T., Kinschberg, D. A., and Lowry, O. H., 1977, *Anal. Biochem.* **78**:252–259.
388. Fletcher, R. T., and Chader, G. J., 1976, *Biochem. Biophys. Res. Commun.* **70**:1297–1302.
389. Frandson, E. K., and Krishna, G., 1976, *Life Sci.* **18**:529–541.
390. Goridis, C., Virmaux, N., Urban, P. F., and Mandel, P., 1973, *FEBS Lett.* **30**:163–166.
391. Pannbalker, R. G., 1973, *Science* **182**:1138–1140.
392. Pannbalker, R. G., 1973, *Invest. Ophthalmol.* **13**:535–538.
393. Bensinger, R. E., Fletcher, R. T., and Chader, G. T., 1974, *Science* **183**:86–87.
394. Goridis, C., Virmaux, N., Weller, M., and Urban, P., 1976, *Transmitters in the Visual Process* (S. L. Bonting, ed.), Pergamon Press, New York, pp. 27–58.
395. Bensinger, R. E., Fletcher, R. T., and Chader, G. J., 1974, *Science* **183**:86–87.
396. Goridis, C., and Weller, M., 1975, *Adv. Biochem. Psychopharmacol.* **15**:391–412.
397. Krishnan, N., Fletcher, R. I., Chader, G. J., and Krishna, G., 1978, *Biochim. Biophys. Acta* **523**:506–515.
398. Bitensky, M. W., Gorman, R. E., and Miller, W. H., 1971, *Proc. Natl. Acad. Sci. U.S.A.* **68**:561–562.
399. Bitensky, M. W., Miller, W. H., Gorman, R. E., Neufeld, A. H., and Robinson, R., 1972, *Adv. Cyclic Nucleotides Res.* **1**:317–335.
400. Hendriks, K. T. H., DePont, J. J., Daemen, F. J., and Bonting, S. L., 1973, *Biochim. Biophys. Acta* **330**:156–166.
401. Manthorpe, M., and McConnell, D. G., 1974, *J. Biol. Chem.* **249**:4608–4613.
402. Lolley, R. N., Schmidt, S. Y., and Farber, D. B., 1974, *J. Neurochem.* **22**:701–707.
403. Iverson, L. L., 1975, *Science* **188**:1084–1089.
404. Brown, J. H., and Makman, M. H., 1972, *Proc. Natl. Acad. Sci. U.S.A.* **69**:539–543.
405. Bucher, M. B., and Schorderet, M., 1975, *Arch. Pharmacol.* **288**:103–104.
406. Makman, M. H., Brown, J. H., and Mishra, R. K., 1975, *Adv. Cyclic Nucleotide Res.* **5**:661–679.
407. Tunnicliff, G., 1977, *Union Med. Can.* **106**:472–474.

408. Lolley, R. N., Brown, B. M., and Farber, D. B., 1977, *Biochem. Biophys. Res. Commun.* **78**:572–578.
409. Watling, K. J., and Dowling, J. E., 1981, *J. Neurochem.* **36**:559–568.
410. DeVries, G. W., Campau, K. M., and Ferrendelli, J. A., 1982, *J. Neurochem.* **38**:759–765.
411. Ferrendelli, J. A., Campau, K. M., and DeVries, G. W., 1982, *J. Neurochem.* **38**:753–758.
412. Daly, L. W., 1977, *Int. Rev. Neurobiol.* **20**:105–168.
413. Kuman, N., and Jain, P. C., 1977, *Prog. Drug Res.* **21**:409–465.
414. Makman, M. H., Mishra, R. K., and Brown, J. H., 1975, *Adv. Neurol.* **9**:213–222.
415. Schorderet, M., 1978, *Gerontology [Suppl.]* **24**:86–93.
416. Wassenaar, J. S., and Korf, J., 1976, *Transmitters in the Visual Process* (L. Bonting, ed.), Pergamon, Oxford, pp. 199–218.
417. Schorderet, M., 1977, *Biochem. Pharmacol.* **26**:167–170.
418. Robb, R. M., 1974, *Trans. Am. Ophthalmol. Soc.* **72**:650–669.
419. Robb, R. M., 1978, *Invest. Ophthalmol. Vis. Sci.* **17**:476–480.
420. Goridis, C., and Virmaux, N., 1974, *Nature* **248**:57–58.
421. Chader, G. J., Herz, L. R. and Fletcher, R. T., 1974, *Biochim. Biophys. Acta* **347**:491–493.
422. Carter-Dawson, L. D., LaVail, M. M., and Sidman, R. L., 1978, *Invest. Ophthalmol. Vis. Sci.* **17**:489–498.
423. Miki, N., Baraban, J. M., Keirns, J. J., Boyce, J. J., and Bitensky, M. W., 1975, *J. Biol. Chem.* **250**:6320–6327.
424. Chader, G. J., Fletcher, R. T., O'Brien, P. J., and Krishna, G., 1976, *Biochemistry* **15**:1615–1620.
425. Lolley, R. N., and Farber, D. B., 1975, *Exp. Eye Res.* **20**:585–597.
426. Farber, D. B., and Lolley, R. N., 1976, *J. Cyclic Nucleotide Res.* **2**:139–148.
427. Miki, N., Keirns, J. J., Marcus, F. R., Freeman, J., and Bitensky, M. W., 1973, *Proc. Natl. Acad. Sci.* **70**:3820–3824.
428. Wheeler, G., and Bitensky, M. W., 1977, *Proc. Natl. Acad. Sci. U.S.A.* **74**:4238–4242.
429. Bitensky, M. W., Wheeler, G. L., Aloni, B., Vetwry, S., and Matuo, Y., 1978, *Adv. Cyclic Nucleotide Res.* **9**:553–572.
430. Yee, R., and Liebman, P. A., 1978, *J. Biol. Chem.* **253**:8902–8909.
431. Kawamura, S., and Bownds, M. D., 1981, *J. Gen. Physiol.* **77**:571–591.
432. Fung, B. K., and Stryer, L., 1980, *Proc. Natl. Acad. Sci. U.S.A.* **77**:2500–2504.
432a. Miller, W. H., and Nicol, G. D., 1979, *Nature* **280**:64–66.
432b. Miller, W. H., 1983, *Adv. Cyclic Nucleotide Res.* **15**:495–511.
433. Liebman, P. A., and Pugh, E. N., Jr., 1979, *Vision Res.* **19**:375–380.
434. Fell, H., Dingle, J., and Webb, M., 1962, *Biochem. J.* **83**:63–69.
435. Conrad, D., Alving, C., and Wirtz, G., 1974, *Biochim. Biophys. Acta* **332**:36–46.
436. Stillwell, W., and Ricketts, M., 1980, *Biochem. Biophys. Res. Commun.* **97**:148–153.
437. Meeks, R., Zaharevitz, D., and Chen, R., 1981, *Arch. Biochem. Biophys.* **207**:141–147.
438. Vento, R., and Cacioppo, R., 1973, *Exp. Eye Res.* **15**:43–52.
439. Dewar, A., Barron, G., and Reading, H., 1975, *Exp. Eye Res.* **20**:63–72.
440. Bashor, M., Toft, D., and Chytil, F., 1973, *Proc. Natl. Acad. Sci. U.S.A.* **70**:3484–3487.
441. Chytil, F., and Ong, D. E., 1978, *Vitam. Horm.* **36**:1–32.
442. Chytil, F., and Ong, D. E., 1978, *Receptors and Hormone Action* (B. W. O'Malley and L. Birmbaumer, eds.), Academic Press, New York, pp. 573–591.
443. Chytil, F., and Ong, D. E., 1983, *Adv. Nutr. Res.* **5**:13–29.
444. Wiggert, B., and Chader, G., 1975, *Exp. Eye Res.* **21**:143–151.
445. Abe, T., Wiggert, B., Bergsma, D., and Chader, G., 1977, *Biochim. Biophys. Acta* **498**:355–365.
446. Lion, G., Fong, S. L., and Bridges, D., 1981, *J. Biol. Chem.* **256**:3153–3155.
447. Saari, J., Futterman, S., and Bredberg, L., 1978, *J. Biol. Chem.* **253**:6432–6436.
448. Wiggert, B., Bergsma, D., and Chader, G., 1976, *Exp. Eye Res.* **22**:411–418.
449. Futterman, S., Saari, J., and Swanson, D., 1976, *Exp. Eye Res.* **22**:419–424.
450. Saari, J., and Futterman, S., 1976, *Biochim. Biophys. Acta* **444**:789–793.
451. Wiggert, B., Bergsman, D., Helsma, R., and Chader, G., 1978a, *Biochem. J.* **169**:87–94.
452. Wiggert, B., Mizukawa, A., Kuwabara, T., and Chader, G., 1978, *J. Neurochem.* **30**:653–659.

453. Saari, J., Bunt, A., Futterman, S., and Berman, E., 1977, *Invest. Ophthalmol. Vis. Sci.* **16:**797–806.
454. Wiggert, B., Masterson, E., Israel, P., and Chader, G., 1979, *Invest. Ophthalmol. Vis. Sci.* **18:**306–310.
455. Futterman, S., Saari, J., and Blair, S., 1977, *J. Biol. Chem.* **252:**3267–3271.
456. Stubbs, G., Saari, J., and Futterman, S., 1979, *J. Biol. Chem.* **254:**8529–8533.
457. Futterman, S., and Saari, J., 1977, *Invest. Ophthalmol. Vis. Sci.* **16:**768–771.
458. Wiggert, B., Bergsma, D., Lewis, M., and Chader, G., 1977, *J. Neurochem.* **29:**947–954.
459. Bergsma, D., Wiggert, B., Funahashi, M., Kuwabasa, T., and Chader, G., 1977, *Nature* **265:**66–67.
460. Wiggert, B., Derr, J., Fitzpatrick, M., and Chader, G., 1979, *Biochim. Biophys. Acta.* **582:**115–121.
460a. Lai, Y. L., Wiggert, B., Liu, Y. P., and Chader, G. J., 1982, *Nature* **298:**848–849.
461. Adler, A., and Severin, K., 1981, *Exp. Eye Res.* **32:**755–769.
462. Adler, A., and Klucznik, K., 1982, *Exp. Eye Res.* **34:**423–434.
463. Adler, A., and Klucznik, K., 1982, *Curr. Eye Res.* **1:**579–589.
464. Liou, G. I., Fong, S. L., and Bridges, C. D. M., 1982, *Fed. Proc.* **41:**683.
465. Liou, G. I., Bridges, C. D. M., and Fong, S. L., 1982, *Invest. Ophthalmol. Vis. Sci. [Suppl.]* **22:**65.
466. Liou, G. I., Bridges, C. D. M., Fong, S. L., Alvarez, R. A., and Fernandez, F. G., 1982, *Vision Res.* **22:**1457–1467.
467. Hollyfield, J. G., and Basinger, S. F., 1978, *Nature* **274:**794–796.
468. Leblond, C. P., 1965, *Am. J. Anat.* **116:**1–28.
469. Young, R. W., 1969, *Invest. Ophthalmol.* **8:**222–231.
470. Young, R. W., and Bok, D., 1969, *J. Cell Biol.* **42:**392–403.
471. Young, R. W., 1971, *Vision Res.* **11:**1–5.
472. Young, R. W., 1971, *J. Cell Biol.* **49:**303–318.
473. Bibb, C., and Young, R. W., 1974, *J. Cell Biol.* **62:**378–389.
474. Young, R. W., 1973, *Ann. Ophthalmol.* **5:**843–854.
475. Bibb, C., and Young, R. W., 1974, *J. Cell Biol.* **61:**327–343.
476. Mercurio, A. M., and Holtzman, E., 1982, *J. Neurocytol.* **11:**295–322.
477. Anderson, R. E., Maude, M. B., Kelleher, P. A., Maida, T. M., and Basinger, S. F., 1980, *Biochim. Biophys. Acta* **420:**212–226.
478. Anderson, R. E., Kelleher, P. A., and Maude, M. B., 1980, *Biochim. Biophys. Acta* **620:**227–235.
479. Anderson, R. E., Maude, M. B., and Kelleher, P. A., 1980, *Biochim. Biophys. Acta* **620:**236–246.
480. Young, R. W., 1974, *Exp. Eye Res.* **18:**215–223.
481. LaVail, M. M., 1976, *Science* **194:**1071–1074.
482. LaVail, M. M., 1976, *Exp. Eye Res.* **23:**277–280.
483. Basinger, S., Hoffman, R., and Matthews, M., 1976, *Science* **194:**1074–1076.
484. Young, R. W., 1978, *Invest. Ophthalmol. Vis. Sci.* **17:**105–117.
485. Ham, W. T., Mueller, H. A., and Shiney, D. H., 1976, *Nature* **260:**153–155.
486. Lanum, J., 1978, *Surv. Ophthalmol.* **22:**221–249.
487. Noell, W. K., Walker, V. S., Kong, B. S., and Berman, S., 1966, *Invest. Ophthalmol.* **5:**450–473.
488. Noell, W. K., 1980, *The Effects of Constant Light on Visual Processes* (T. P. Williams and B. N. Baker, eds.), Plenum Press, New York, pp. 3–28.
489. Noell, W. K., and Albrecht, R., 1971, *Science* **172:**76–80.
490. Noell, W. K., Delmelle, M. C., and Albrecht, R., 1971, *Science* **172:**72–75.
491. O'Steen, W. K., 1977, *Exp. Eye Res.* **25:**361–369.
492. O'Steen, W. K., 1979, *Photochem. Photobiol.* **29:**745–753.
493. O'Steen, W. K., 1980, *The Effects of Constant Light on Visual Process* (B. N. Baker and T. P. Williams, eds.), Plenum Press, New York, pp. 29–50.
494. Kaitz, M., and Auerbach, E., 1980, *The Effects of Constant Light on Visual Process* (B. N. Baker and T. P. Williams, eds.), Plenum Press, New York, pp. 179–194.

495. Crouch, R., Priest, D., and Duke, E., 1978, *Exp. Eye Res.* **27**:503–509.
496. Crouch, R. K., 1980, *The Effects of Constant Light on Visual Processes* (B. N. Baker and T. P. Williams, eds.), Plenum Press, New York, pp. 309–318.
497. Voaden, M. D. J., Morjaria, B., and Oraedir, A. C. I., 1980, *Neurochemistry* **1**:151–165.
498. Wiegand, R. D., Giusto, N. M., Rapp, L. M., and Anderson, R. E., 1983, *Invest. Ophthalmol. Vis. Sci.* **24**:1433–1435.
499. Abdel-Latif, A. A., Akhtar, R. A., and Hawthorne, J. N., 1977, *Biochem. J.* **162**:61–73.
500. Anderson, R. E., and Hollyfield, J. G., 1981, *Biochim. Biophys. Acta* **665**:619–622.
501. Anderson, R. E., Maude, M. B., Verner, G. E., Rayborn, M. E., and Hollyfield, J. G., 1982, *Invest. Ophthalmol. Vis. Sci. [Suppl.]* **22**:66.
502. Schimdt, S. Y., 1983, *J. Biol. Chem.* **258**:6863–6868.
503. Ledeen, R. W., Skrivanek, J. A., Tirri, L. J., Margolis, R. K., and Margolis, R. V., 1976, *Adv. Exp. Med. Biol.* **71**:83–103.
504. Landa, C. A., Maccioni, H. J. F., and Caputto, R., 1979, *J. Neurochem.* **33**:825–838.
505. Grafstein, B., Miller, J. A., Ledeen, R. W., Haley, J., and Specht, S. C., 1975, *Exp. Neurol.* **46**:261–281.
506. Maccioni, H. J. F., Landa, C. A., Arce, A., and Caputto, R., 1977, *Adv. Exp. Med. Biol.* **83**:267–282.
507. Toews, A. D., Goodrum, J. F., and Morell, P., 1979, *J. Neurochem.* **32**:1165–1173.
508. Toews, A. D., and Morell, T., 1981, *J. Neurochem.* **37**:1316–1323.
509. Alberghina, M., Karlsson, J. O., and Giuffrida, A. M., 1982, *J. Neurochem.* **39**:223–227.
510. Alberghina, M., Viola, M., Moro, F., and Giuffrida, A. M., 1981, *Neurochem. Res.* **6**:635–649.
511. Hands, A. R., Sutherland, N. S., and Bartley, W., 1965, *Biochem. J.* **94**:279–283.
512. Futterman, S., and Kupfer, C., 1968, *Invest. Ophthalmol.* **7**:105–108.
513. Futterman, S., Downer, S. L., and Hendrickson, A., 1971, *Invest. Ophthalmol.* **10**:151–154.
514. Tinoco, J., Miljanich, P., and Medwadowski, B., 1977, *Biochim. Biophys. Acta* **486**:575–578.
515. Tinoco, J., Endemann, G., Hincenbergs, I., Medwadowski, B., Miljanich, P., and Williams, M. A., 1979, *Lipids* **14**:166–173.
516. Lucy, J. A., 1978, *Tocopherol, Oxygen and Biomembranes* (C. DeDuve and O. Hayaeshi, eds.), Elsevier, Amsterdam, pp. 109.
517. Hayes, K. C., 1974, *Invest. Ophthalmol. Vis. Sci.* **13**:499.
518. Robison, W. G., Jr., Kuwabara, T., and Bieri, J. G., 1979, *Invest. Ophthalmol. Vis. Sci.* **18**:683.
519. Amemiya, T., 1981, *Albrecht von Graefes Arch. Klin. Exp. Ophthalmol.* **216**:103.
520. Amemiya, T., 1981, *Int. J. Vitam. Nutr. Res.* **51**:114–118.
521. Farnsworth, C. C., and Dratz, E. A., 1976, *Biochim. Biophys. Acta* **443**:556.
522. Robison, W. G., Kuwabara, T., and Bieri, J. G., 1982, *Retina* **2**:263–281.
523. Riis, R. C., Sheffy, B. E., Loew, E., Kern, T. J., and Smith, J. S., 1981, *Am. J. Vet. Res.* **42**:74–86.
524. Oomen, H. A. P. C., McLaren, D. S., and Escapini, H. A., 1964, *Trop. Geogr. Med.* **16**:271–315.
525. McLaren, D. S., 1980, *Nutritional Ophthalmology*, Academic Press, London.
526. Dowling, J. E., and Gibbons, I. R., 1961, *The Structure of the Eye* (G. Smelser, ed.), Academic Press, New York, pp. 85–99.
527. Dowling, J. E., and Wald, G., 1960, *Proc. Natl. Acad. Sci. U.S.A.* **46**:587–608.
528. Carter-Dawson, L., Kuwabara, T., and Bieri, J. G., 1979, *Invest. Ophthalmol. Vis. Sci.* **18**:437–446.
529. Berson, E. L., 1973, *Invest. Ophthalmol.* **18**:437–446.
530. Robison, W. G., Kuwabara, T., and Bieri, J. G., 1980, *Invest. Ophthalmol. Vis. Sci.* **19**:1030–1037.
531. Pasantes-Morales, H., Klethi, J., Ledig, M., and Mandel, P., 1972, *Brain Res.* **41**:494–497.
532. Orr, H. T., Cohen, A. I., and Lowry, O. H., 1976, *J. Neurochem.* **26**:609–611.
533. Hayes, K. C., Carey, R. E., and Schmidt, S. Y., 1975, *Science* **188**:949–951.
534. Schmidt, S. Y., Berson, E. L., and Hayes, K. C., 1976, *Invest. Ophthalmol. Vis. Sci.* **15**:47–52.

535. Berson, E. L., Hayes, K. C., Rabin, A. R., Schmidt, S. Y., and Watson, G., 1976, *Invest. Ophthalmol. Vis. Sci.* **15**:53–58.
536. Schmidt, S. Y., Berson, E. L., Watson, G., and Huang, C., 1977, *Invest. Ophthalmol. Vis. Sci.* **16**:673–678.
537. Hayes, K. C., Rabin, A. R., and Berson, E. L., 1975, *Am. J. Pathol.* **78**:505–516.
538. Schmidt, S. Y., Berson, E. L., and Hayes, K. C., 1976, *Trans. Am. Acad. Ophthalmol. Otolaryngol.* **81**:687–693.
539. Miller, S. S., and Steinberg, R. H., 1979, *J. Gen Physiol.* **74**:237–259.
540. Pasantes-Morales, H., Ademe, R. M., and Quesada, O., 1981, *J. Neurosci. Res.* **6**:337–348.
541. Galin, M. A., Nano, H. D., and Hall, T., 1962, *Invest. Ophthalmol.* **1**:142–148.
542. Eckhert, C. D., 1979, *Fed. Proc.* **38**:872.
543. Ujiie, M., 1979, *Acta Soc. Ophthalmol. Jpn.* **83**:2149.
544. Place, V. A., and Thomas, J. P., 1963, *Annu. Rev. Respir. Dis.* **87**:901.
545. Saraux, H., 1975, *Ann. Occulist (Paris)* **208**:29.
546. Figueroa, R., Weiss, H., and Smith, J. C., 1971, *Annu. Rev. Respir. Dis.* **104**:592.
547. Sorsby, A., and Harding, R., 1962, *Vision Res.* **2**:149–162.
548. Leure-dupree, A. E., 1981, *Invest. Ophthalmol. Vis. Sci.* **21**:1–9.
549. Leure-dupree, A. E., and Bridges, C. D. B., 1982, *Retina* **2**:294–302.
550. Huber, A., and Gershoff, S. N., 1975, *J. Nutr.* **105**:1486–1490.
551. Futterman, S., 1963, *J. Biol. Chem.* **238**:1145–1150.
552. Smith, J. E., Muto, Y., Milch, P. O., and Goodman, W. S., 1973, *J. Biol. Chem.* **248**:1544–1549.
553. Smith, J. C., Jr., Brown, E. D., McDaniel, E. G., and Chan, W., 1976, *J. Nutr.* **106**:569–574.
554. Smith, J. E., Brown, E. D., and Smith, J. C., Jr., 1974, *J. Lab. Clin. Med.* **84**:692–697.
555. Brown, E. D., Chan, W., and Smith, J. C., Jr., 1976, *J. Nutr.* **106**:563–568.
556. Parks, J. M., Shay, J., and Ames, A. III, 1976, *Arch. Neurol.* **33**:709–714.
557. Bours, J., Vassileva, P. I., and Hockwin, O., 1977, *Ophthalmol. Res.* **9**:201–204.
558. Shay, J., and Ames, A. III, 1976, *Arch. Neurol.* **33**:715–721.
559. Kaskel, K., Valenzuela H., Hockwin, O., Muntau, M., and Schedkler, C. M., 1976, *Albrecht Von Graefes Archiv. Klin. Exp. Ophthalmol. Res.* **200**:71–78.
560. Oberhoff, P., and Hockwin, O., 1969, *Albrecht von Graefes Arch. Klin. Exp. Ophthalmol. Res.* **178**:329–332.
561. Kaskel, D., Hockwin, O., Metzler, U., and Schedtler, C. M., 1973, *Ophthalmol. Res.* **5**:177–185.
562. Wassilewa, P., Hockwin, O., and Korte, I., 1976, *Albrecht von Graefes Arch. Klin. Exp. Ophthalmol. Res.* **199**:115–120.
563. Bazan, N. G., 1970, *Biochim. Biophys. Acta* **218**:1–10.
564. Chester, E. M., and Barber, B. H., 1967, *Arch. Int. Med.* **120**:397–419.
565. Fukada, M., Tamura, T., Abe, T., and Kashara, T., 1976, *Acta Soc. Ophthalmol. Jpn.* **80**:1514–1520.
566. Kojima, K., Tamura, T., Abe, T., and Kashara, T., 1976, *Acta Soc. Ophthalmol. Jpn.* **80**:1514–1520.
567. Kojima, K., Suzuki, M., Torii, F., Mizuno, K., Harada, T., and Ichikawa, H., 1981, *Jpn. J. Ophthalmol.* **25**:341–353.
568. Sinclair, S. H., Grunwald, J. E., Riva, C. E., Braunstein, S. N., Nichols, C. W., and Schwartz, S. S., 1982, *Ophthalmolology* **89**:748–750.
569. Futterman, S., Sturtevant, R., and Kupfer, C., 1969, *Invest. Ophthalmol.* **8**:542–544.
570. Bell, M. E., Peterson, R. G., and Eichberg, J., 1982, *J. Neurochem.* **39**:192–200.
571. Murata, T., Nishida, T., Eto, S., and Mukai, N., 1981, *Metab. Pediatr. Ophthalmol.* **5**:83–87.
572. Chihara, E., 1981, *J. Neurochem.* **37**:247–250.
573. Palmer, G. C., Wilson, G. L., and Chronister, R. B., 1983, *Life Sci.* **32**:365–374.
574. Kolb, H., and Gouras, P., 1974, *Invest. Ophthalmol.* **13**:487–498.
575. Szamier, B., and Berson, E., 1977, *Invest. Ophthalmol. Vis. Sci.* **16**:947–962.
576. Berson, E., Gouras, P., and Gunkel, R., 1968, *Arch. Ophthalmol.* **80**:273–279.

577. Ripps, H., Brin, K., and Weale, R., 1978, *Invest. Ophthalmol. Vis. Sci.* **17**:735–745.
578. Aguirre, G., Farber, D., Lolley, R., Fletcher, R. T., and Chader, G. J., 1978, *Science* **201**:1133–1134.
579. Lolley, R. N., 1980, *Curr. Top. Eye Res.* **2**:67–118.
580. Chader, G. J., Liu, Y. P., Fletcher, R. T., Aguirrie, G., Anderson, R. S., and T'So, M., 1981, *Curr. Top. Membr. Transport* **15**:133–156.
581. Dewar, A. J., and Reading, H. W., 1975, *Int. J. Biochem.* **6**:615–641.
582. Reading, H. W., and Dewar, A. J., 1976, *Biochemistry and Neurological Diseases* (A. N. Davison, ed.), Blackwells, Oxford, pp. 52–112.
583. Bok, D., and Hall, M. O., 1971, *J. Cell Biol.* **49**:664–682.
584. Chader, G. J., 1982, *Problems of Normal and Genetically Abnormal Retinas* (R. M. Clayton, J. Haywood, H. W. Reading, and A. Wright, eds.), Academic Press, London, pp. 85–98.
585. Noell, W. K., 1958, *Arch. Ophthalmol.* **60**:702–733.
586. Aguirre, G., and Rubin, L., 1975, *J. Am. Vet. Med. Assoc.* **166**:157–164.
587. Mullen, R. J., and LaVail, M. M., 1976, *Science* **192**:799–801.
588. Graymore, C. N., Tomsley, K., and Kerly, M., 1959, *Biochem. J.* **72**:459–461.
589. Walters, P. T., 1959, *Br. J. Ophthalmol.* **43**:686–696.
590. Reading, H. W., and Sorsby, A., 1962, *Vision Res.* **2**:315–325.
591. Lolley, R. N., 1972, *J. Neurochem.* **19**:175–185.
592. Reading, H. W., and Sorsby, A., 1964, *Vision Res.* **4**:209–220.
593. Reading, H. W., 1964, *Nature* **203**:491–492.
594. Yates, C. M., Reading, H. W., Bitensky, L., and Chayen, J., 1976, *Exp. Eye Res.* **23**:403–408.
595. Hiramitsu, T., Majiana, Y., Hasegaroa Y., and Hirata, K., 1974, *Acta Soc. Ophthalmol. Jpn.* **78**:819–825.
596. Yagi, K., and Ohishi, N., 1976, *Biochemical and Medical Aspects of Active Oxygen* (D. Hayaishi and K. Asada, eds.), University Park Press, Baltimore, pp. 299–307.
597. Kuwabara, T., and Gorn, R. A., 1968, *Arch. Ophthalmol.* **79**:69–78.
598. Kagan, V., Shvedova, A., Novikov, K., and Kozlov, Y., 1973, *Biochim. Biophys. Acta* **300**:76–79.
599. Hiramitsu, T., Hasegawa, Y., Hirata, K., Nishigaki, D., and Yagi, K., 1976, *Experientia* **32**:622–623.
600. Rapp, L. M., Wiegand, R. D., and Anderson, R. E., 1982, *Problems of Normal and Genetically Abnormal Retinas* (R. M. Clayton, J. Haywood, H. W. Reading, and A. Wright, eds.), Academic Press, London, pp. 109–119.
601. Kagan, V. E., Arkhipenko, Y. V., Belousova, L. V., Tyurin, V. A., Shvedova, A. A., Shukolyukov, S. A., and Kozlov, Y. P., 1981, *Vision Res.* **21**:1029–1034.
602. Mead, J., 1976, *Free Radicals in Biology*, Volume 1, Academic Press, New York, pp. 51–68.
603. Katz, M., Stone, W., and Dratz, E., 1978, *Invest. Ophthalmol. Vis. Sci.* **17**:1049–1058.
604. Hiramitsu, T., Majima, Y., Hasegawa, Y., and Hirata, K., 1978, *Acta Soc. Ophthalmol. Jpn.* **9**:819–825.
605. Hiramitsu, T., Majima, Y., Hadegawa, Y., and Hirata, K., 1979, *Acta Soc. Ophthalmol. Jpn.* **10**:1468–1473.
606. Yagi, K., Matsuoka, S., Onkawa, H., Ohishi, N., Takenchi, Y., and Sakai, L., 1980, *J. Mol. Biol.* **137**:315–348.
607. Wiegand, R. D., Giusto, N. M., and Anderson, R. E., 1982, *Problems of Normal and Genetically Abnormal Retinas* (R. M. Clayton, J. Haywood, H. W. Reading, and A. Wright, eds.), Academic Press, London, pp. 121–128.
608. Organisciak, D., Lake, J., and Wang, H., 1981, *Invest. Ophthalmol. Vis. Sci. [Suppl.]* **20**:166.
609. Farnsworth, C., Stone, W., and Dratz, E., 1979, *Biochim. Biophys. Acta* **552**:281293.
610. Joel, C., Briggs, S., Gall, D., Hannan, J., Kahlow, M., Stein, M., Tarver, A., and Yip, A., 1981, *Invest. Ophthalmol. Vis. Sci. [Suppl.]* **20**:166.
611. Joel, C. D., Briggs, S. J., Kahlow, M. A., Sievert, T. J., and Tsang, G. K. W., 1980, *Trans. Am. Soc. Neurochem.* **11**:212.
612. Armstrong, D., Neville, H., Siakotos, A., Wilson, D., Wehling, C., and Koppang, N., 1980, *Neurochemistry* **1**:405–426.

613. Armstrong, D., and Koppang, N., 1982, *Ceroid-lipofuscinosis (Batter's Disease)* (D. Armstrong, N. Koppang, and J. A. Rider, eds.), Elsevier, Amsterdam, pp. 159–165.
614. Armstrong, D., Connole, E., Feeney, L., and Berman, E. R., 1978, *J. Neurochem.* **31**:761–769.
615. Armstrong, D., Santangelo, G., and Connole, E., 1981, *Curr. Eye Res.* **1**:225–242.
616. Hall, M. O., and Hall, D. O., 1975, *Biochem. Biophys. Res. Commun.* **67**:1199–1204.
617. Armstrong, D., Hiramitsu, T., Gutteridge, J., and Nilsson, S. E., 1982, *Exp. Eye Res.* **35**:157–171.
618. O'Donnell, J. J., Armstrong, D., Koppang, N., and Wood, I., 1984, *Invest. Ophthalmol. Vis. Sci.* (in press).
619. Armstrong, D., Koppang, N., and Nilsson, S. E., 1982, *Eur. Neurol.* **21**:147–156.
620. Nilsson, S. E. G., Armstrong, D., Koppang, N., Persson, P., and Milde, K., 1983, *Invest. Ophthalmol. Vis. Sci.* **24**:77–84.
621. Farber, D. B., and Lolley, R. N., 1977, *J. Neurochem.* **28**:1089–1095.
622. Feeney, L., and Mixon, R. N., 1976, *Exp. Eye Res.* **22**:533–548.
623. Dewar, A. J., Barron, G., and Reading, H. W., 1977, *Toxicol. Appl. Pharmcol.* **42**:65–74.
624. O'Donnell, J. J., 1980, *Neurochemistry* **1**:459–463.
625. Simell, O., and Takki, K., 1973, *Lancet* **1**:1031–1033.
626. McCulloch, C., and Marliss, E. B., 1975, *Am. J. Ophthalmol.* **80**:1047–1057.
627. Berson, E. L., Schmidt, S. Y., and Rabin, A. R., 1976, *Br. J. Ophthalmol.* **60**:142–147.
628. Takki, K., 1974, *Br. J. Ophthalmol.* **58**:907–916.
629. Arshinoff, S., McCulloch, J. C., Parker, A. J., Phillips, M. J., and Marliss, E. M., 1977, *Clin. Res.* **25**:321a.
630. Kaiser-Kupfer, M. I., Valle, D., and Del Valle, L. A., 1978, *Am. J. Ophthalmol.* **85**:200–204.
631. Hayasaka, S., Matsuzawa, T., Shiono, T., Mizuno, K., and Ishigwro, I., 1982, *Exp. Eye Res.* **34**:635–638.
632. Kenneway, N. G., Weleber, R., and Buist, N. R. M., 1977, *Am. J. Hum. Genet.* **29**:61.
633. Trijbels, J. M. F., Sengers, R. C. A., Bakkeren, J. M., Dekott, A. F. M., and Dentman, F. A., 1977, *Clin. Chim. Acta* **79**:371–377.
634. O'Donnell, J. J., Sidman, R., and Martin, S., 1977, *Biochem. Biophys. Res. Commun.* **79**:396–399.
635. O'Donnell, J. J., Sidman, R., and Martin, S., 1978, *Anal. Biochem.* **90**:41–46.
636. O'Donnell, J. J., Sidman, R., and Martin, S., 1978, *Science* **200**:200–201.
637. Valle, D., Kupfer, K. M., and Del Valle, L., 1977, *Proc. Natl. Acad. Sci. U.S.A.* **74**:5159–5161.
638. Seashore, M. R., Duramt, J. L., and Rosenberg, L. E., 1972, *Pediatr. Res.* **6**:187–196.
639. Shih, V. E., Berson, E. L., and Mandel, P., 1978, *Am. J. Hum. Genet.* **30**:174–179.
640. Giordano, C., DeSanto, N. G., Pluvio, M., Santinel, R., and Stoppolo, G., 1978, *Nephrology* **22**:131–136.
641. Yatsiv, S., Statter, M., and Merin, S., 1979, *J. Lab. Clin. Med.* **93**:749–757.
642. Agranoff, B. W., Feldman, E. L., Heacock, A. M., and Schwartz, M., 1980, *Neurochemistry* **1**:487–500.
643. Murray, M., and Grafstein, B., 1969, *Exp. Neurol.* **23**:544–560.
644. Heacock, A. M., and Agranoff, B. W., 1976, **73**:828–832.
645. Agranoff, B. W., Field, P., and Gaze, R. M., 1976, *Brain Res.* **113**:225–234.
646. Landreth, G. E., and Agranoff, B. W., 1976, *Brain Res.* **118**:299–303.
647. Springer, A. D., Heacock, A. M., Schmidt, J. T., and Agranoff B. W., 1977, *Brain Res.* **128**:417–427.
648. Springer, A. D., and Agranoff, B. W., 1977, *Brain Res.* **128**:405–415.
649. Heacock, A. M., and Agranoff, B. W., 1977, *Science* **197**:64–66.
650. Johns, P. R., Heacock, A. M., and Agranoff, B. W., 1978, *Brain Res.* **142**:531–537.
651. Schwartz, M., and Agranoff, B. W., 1981, *Brain Res.* **206**:331–343.
652. Feldman, E. L., Axelrod, D., Schwartz, M., Heacock, A. M., and Agranoff, B. W., 1981, *J. Neurobiol.* **12**:591–598.
653. Burrell, H. R., Heacock, A. M., Water, R. D., and Agranoff, B. W., 1979, *Brain Res.* **168**:628–632.

654. Dokas, L. A., Kohsaka, S., Burrell, H. R., and Agranoff, B. W., 1981, *J. Neurochem.* **36**:1160–1165.
655. Kohsaka, S., Schwartz, M., and Agranoff, B. W., 1981, *Dev. Brain Res.* **1**:391–401.
656. Ingoglia, N.A., Sturman, J. A., and Eisner, R. A., 1977, *Brain Res.* **130**:433–445.
657. Kohsaka, S., Dokas, L. A., and Agranoff, B. W., 1981, *J. Neurochem.* **36**:1166–1174.
658. Schwartz, M., Kohsaka, S., and Agranoff, B. W., 1981, *Dev. Brain Res.* **1**:403–413.
659. Schwartz, M., Mizrachi, Y., and Kimbi, Y., 1982, *Dev. Brain Res.* **3**:21–28.
660. Schwartz, M., Sela, B. A., and Eshlas, N., 1982, *Dev. Brain Res.* **3**:1192–1195.
661. Spirman, N., Sela, B. A., and Schwartz, M., 1982, *J. Neurochem.* **38**:874–877.
662. Meyers-Elliott, R. H., Jacobs, D. R., and Gammon, R. A., 1982, *J. Neuroimmunol.* **4**:25–34.
663. Barnstable, C. J., 1980, *Nature* **286**:231–235.
664. Reddan, J. R., 1982, *Cell Biology of the Eye* (D. S. McDevitt, ed.), Academic Press, New York, pp. 299–375.
665. Brownlee, M., and Cerami, A., 1981, *Annu. Rev. Biochem.* **50**:385–432.
666. Giusto, N. M., and Bazan, N G., 1983, *Neurochem. Pathol.* **1**:17–41.

22

Vision

Hitoshi Shichi

1. INTRODUCTION: ANATOMY AND PHYSIOLOGY OF THE RETINA

The vertebrate eye resembles the photographic camera, equipped with a cornea (lens cover), iris, lens, and retina (film) (Fig. 1). However, this apparent similarity extends no further. Whereas photographic film is no more than a gelatin coat containing a photosensitive compound (silver halide), the retina consists of several layers of cells, which have specific functions in phototransduction and neural processing. Electrophysiological responses generated by photon absorption by the photoreceptors are processed or encoded by retinal neurons and transmitted to the brain. It is interesting to note that light has to pass through the layers of neural cells before it reaches the photoreceptor cells. Therefore, the retinal neurons are virtually colorless and transparent. Each photoreceptor cell is composed of the metabolically active inner segment, which contains the nucleus, mitochondria, endoplasmic reticulum, and other subcellular organelles, and the outer segment, which contains the visual pigment in its highly membranous structure. The two segments are connected by a cilium and share the continuous plasma membrane. The distal end (i.e., opposite to the outer segment) of the photoreceptor cell forms synapses with horizontal and bipolar cells. The bipolar cells, in turn, synapse with ganglion cells and amacrine cells (Fig. 1).

Based on the morphology of the outer segments, the photoreceptors are classified as either rod cells, which are responsible for dim light (black-and-white, scotopic) vision, or cone cells, which are essential for daylight (color, photopic) vision. There are three types of cone cells, each type containing one of three pigments that absorb blue, green, or red light. These pigments vary in the relative amounts and the absorption maxima in different species. Human cone pigments have absorption maxima at 445, 535, and 570 nm.[1] The outer segment of the rod cell contains rhodopsin (λ_{max} at about 500 nm). It is estimated that the human eye contains 120 million rod cells, mainly in the peripheral

---

Hitoshi Shichi • Institute of Biological Sciences, Oakland University, Rochester, Michigan 48063.

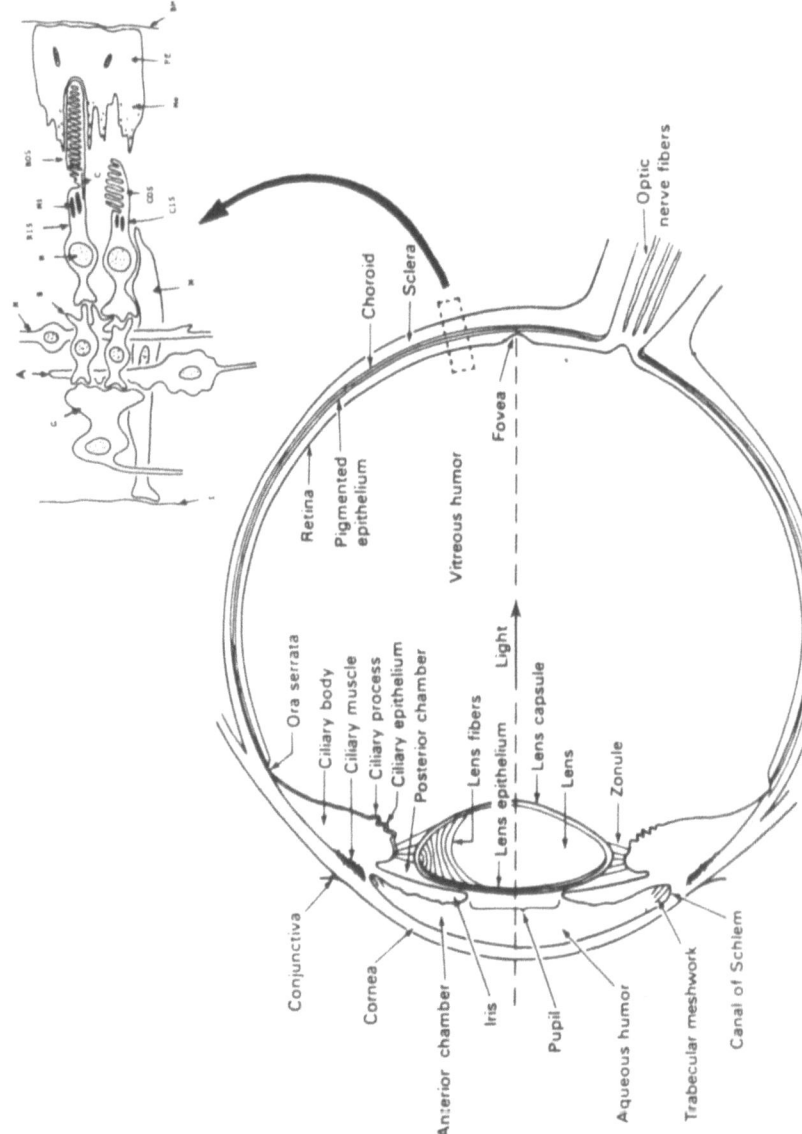

Fig. 1. Cross section of vertebrate eye. A, amacrine cell; B, bipolar cell; BM, Bruch's membrane; C, cilium; CIS, cone inner segment; COS, cone outer segment; G, ganglion cell; H, horizontal cell; I, inner limiting membranes; M, Müller cell; Me, melanin granule; Mi, mitochondrion; N, nucleus; PE, retinal pigmented epithelium; RIS, rod inner segment; ROS, rod outer segment.

region of the retina, and 6.5 million cone cells, primarily in the central (foveal) region.[2]

The visual pigments of both cone cells and rod cells contain 11-*cis*-retinal covalently bound to the opsin protein. Photoisomerization of the chromophore to the all-*trans* form is the initial photochemical event. The vertebrate photoreceptor shows a high permeability toward the Na^+ ion, and light-elicited changes in membrane potential are attributed to changes in Na^+ concentrations between inside and outside of the cell membrane.[3,4] Sodium enters the outer-segment membrane and is pumped out by the sodium pump presumably located in the inner-segment plasma membrane. Light reduces the Na^+ permeability of the outer-segment membrane but has no effect on the Na^+ pump of the inner segment. This results in a decrease in Na^+ concentration within the cell and makes the membrane potential (inside negative) more negative or hyperpolarized. Nerve cells generally increase Na^+ permeability in response to a stimulus and are depolarized or excited.[5] Therefore, it may be said that the photoreceptor cells are in an "excited state" in the dark and are deexcited or "inhibited" by light. The deexcitation of the outer-segment membrane by light propagates to the synaptic terminal of the photoreceptor cell and is believed to suppress the release of neurotransmitter that occurs in the dark.

The visual signals representing light intensity and wavelength are thus transmitted from the photoreceptor cells to the secondary neurons (e.g., bipolar cell, horizontal cell). The signals are then conveyed by the ganglion cell axons to the visual cortex of the brain via the geniculocalcarine radiations and decoded by different cerebral neurons.[6,7] Not much is known of the biochemical aspects of neural transmission in the retina, although the number of neurotransmitters reported to be present in the tissue is increasing. Therefore, this chapter deals primarily with the structural and functional aspects of photoreceptors and the molecular events that are initiated by photon absorption by the visual pigments and result in the generation of photo-induced membrane potentials.

2. PHOTORECEPTORS

2.1. Assembly and Breakdown

There are two major types of visual photoreceptors: the ciliary type and the rhabdomeric type. Vertebrate photoreceptors are of the ciliary type, whereas the rhabdomeric or villous type of photoreceptors are exemplified by invertebrate photoreceptors. The suggestion that the rhabdomeric type originated from the ciliary type remains controversial.[8] In the vertebrate photoreceptor, the growth of the cilium in early developmental stages dictates the formation of the outer segment.

It has been suggested that following the attachment of the centriole (basal body) to the plasma membrane of the inner segment, the plasma membrane starts to invaginate as the microtubules of the cilium begin to develop from the basal body.[9] The cilium, as it grows, directs the formation of disks. According to a recent hypothesis,[10] the disk surfaces (without rims) are first formed by

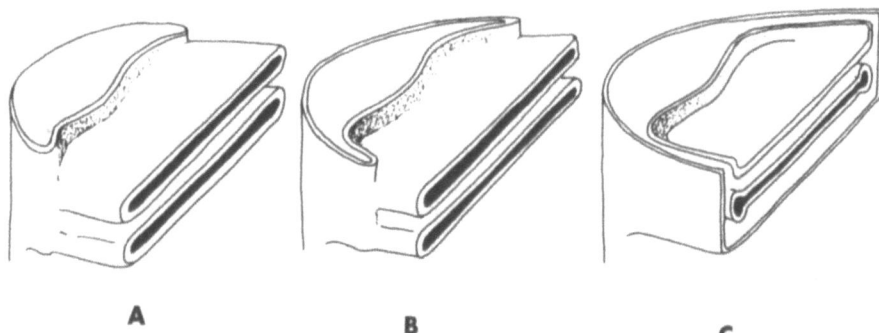

Fig. 2. Formation of the rim region of rod disk. In A, the plasma membrane begins to evaginate to form the rim. The rim progresses along the periphery of evaginated membranes (B). Rim formation extends beyond the plane of longitudinal section (C), and the rim of the detached disk is seen in the cross section. Note that the disk is formed from the adjacent membranes of two evaginations.

an evagination of the ciliary plasma membrane (plasma membrane), and a region of ciliary plasma membrane between adjacent disk surface evaginations grows around the circumferences of adjacent evaginations and seals their edges to form the complete disk (Fig. 2). Recent electron microscopic observations[11,12] on photoreceptor membranes seem to be in favor of this hypothesis. Filamentous structures are detected outside the rim region of the disk and are considered to support the stacking of disks at regular intervals.[11] The filamentous structures, probably cytoskeletons, may control the outpouching of ciliary plasma membrane. Although the newly assembled disks are thus derived from the ciliary plasma membrane, the two membranes are somewhat different in their surface appearances.[12] Processing of the plasma membrane must occur during the formation of disks. The disks thus formed in the rod outer segment are detached from the plasma membrane.

If the newly formed disks are labeled by an injection of radioactive amino acid into the animal, migration of a radioactive band from the base toward the apex of the outer segment is observed by autoradiography.[13] The disks that have reached the apical region are shed as packets from the tip of the outer segment and are phagocytized by the adjacent cells of pigmented epithelium. Therefore, the disks undergo continuous turnover. Shedding is influenced by light. Disk shedding in the rat eye follows a circadian rhythm; it is minimal in the dark and occurs vigorously soon after the onset of light.[14] Shedding of frog rod disks is also activated by light but shows a diurnal (not circadian) pattern.[15] The rod outer segment of frog maintained on a light/dark cycle shows periodic birefringence bands.[16] The period and width of bands are correlated with the rate of disk displacement. Therefore, the periodic variations in birefringence may represent disk packets shed at the tip of the outer segment.

The mechanism of shedding is not well understood yet. Binding of the sugar moiety of glycoproteins to specific receptors (lectins) is an important mechanism of cellular recognition.[17] It was suggested that sugars of cell surface proteins at the tip of the outer segment may induce shedding and phagocytosis by the pigmented epithelium.[18] This suggestion remains to be confirmed experimentally. Glycoproteins in the rod outer segment membrane bind various plant lectins such as concanavalin A,[18–22] wheat germ agglutinin, and soybean agglutinin. In RCS rats with retinal dystrophy, the outer segments are shed but not phagocytized by the pigmented epithelium, and membrane debris accumulates in the subretinal space. This debris and the outer segments of the normal retina have different affinities for plant lectins.[21] Concanavalin A receptors are present on the external surface of the rod outer segment and on the internal surface of the disk.[19,23] The location of receptors supports the mechanism of disk assembly,[10,13] involving the extension and inversion of the plasma membrane of outer segment. Phagocytosis begins with enveloping the shed membrane mass by the microvilli of the pigmented epithelium, a process aided by microfilaments.[24] The phagocytized membranes are digested by lysosomal enzymes of the epithelial cells.

Cone outer segments are also phagocytized by the pigmented epithelium[25] and are assumed, like rod disks, to be under constant turnover.[26] However, cone disks remain continuous with the plasma membrane. Therefore, radioactive amino acids incorporated in newly assembled cone disks are quickly distributed to older disks and do not form a distinct band at the base of outer segment.[13] Cone outer segments are also different from rod outer segments in the composition of cell surface oligosaccharides. In the goldfish retina, cone glycoproteins contain fucose, whereas rod glycoproteins do not.[27]

Both rod and cone outer segments possess remarkably high capabilities to assemble membranes. For example, from the rate of migration of radiolabeled disks from the base to the apex of the outer segment, the turnover rate of a frog rod is estimated to be about 8.5 weeks.[13] Assuming that a frog eye has 3 million rods (1800 disks per rod) and that a disk is 6×10^{-6} meter in diameter, the synthetic capacity of a single frog eye is calculated to be 2.5 cm^2 of disk membrane per hour. In spite of the capabilities of the rods and cones to assemble membranes, neither rod outer segment nor cone outer segment is capable of synthesizing membrane components. Components such as protein, carbohydrate, and lipid are synthesized in the inner segment and transported to the site of membrane assembly. Once incorporated into the rod disk, the integral protein rhodopsin remains membrane bound throughout the lifetime of the disk. In contrast, phospholipids, fatty acids, and vitamin A that are incorporated into the newly assembled membranes are exchangeable with their respective counterparts in the inner segment and in the older disks.[13]

2.2. Structure of Disk Membranes

Early electron microscopic studies indicated that the disk membrane is symmetrical in cross section with the hydrocarbon region of the lipids in the center zone of the membrane and that dense particles, presumably rhodopsin, are present in the lipid bilayer membrane. The bilayer membrane can be cleaved or fractured along the hydrophobic plane, and the face of the cytoplasmic half (P face) and the face of the extracellular half (E face) are distinguished. In the disk, the extracellular half corresponds to the intradiscal half. Freeze-fracture of disk membranes indicates that the E face is smooth but that the P face has a rough appearance and contains particles.[28] If the particles are rhodopsin molecules, the bulk of rhodopsin molecular mass must be present in the cytoplasmic half of the disk membrane. X-ray diffraction[29] and neutron diffraction studies support the conclusion.[30,31]

This does not rule out the possibility that rhodopsin spans the membrane, because the smooth E face may have pits that are not resolvable by current freeze-fracture replica techniques. Immunocytochemical staining indicates that rhodopsin reacts with antirhodopsin serum on both sides of the disk membrane.[32,33] According to lectin binding studies,[19,23] the oligosaccharide moiety of rhodopsin is exposed on the intradiscal surface only. This unilateral localization of the oligosaccharide moiety suggests that all rhodopsin molecules are embedded in the membrane with a specific orientation and do not distribute as two populations on both the P and E sides. Recent work[34,35] indicates that the phosphorylation sites of rhodopsin are clustered in the carboxyl terminus and exposed on the cytoplasmic surface of the disk membrane. These results support the conclusion that rhodopsin is a transmembrane protein. In a transmembrane model suggested by X-ray diffraction and neutron diffraction data,[30,31] a substantial portion of the molecule is exposed on the cytoplasmic side of membrane. By energy-transfer measurements between the light-absorbing probes placed apart on the surface of rhodopsin,[36] the size of the pigment molecule was estimated to be at least 7.5 nm. This length is sufficient for the molecule to span the disk membrane of about 4 nm in thickness.

Newly formed disks in the basal region show certain structural features that are absent in older disks and in the outer segment plasm membrane. The P face of newly formed disks contains particles in most areas but also has particle-free smooth areas.[12] Since such smooth areas are not observed in the plasma membrane, the formation of the disks does not occur by simple invagination or evagination of the plasma membrane; it must involve a more complicated membrane-processing mechanism.

Freeze-fracture replicas of frog outer segment reveal the presence of paracrystalline inclusions of both cylindrical and irregular geometries that are 0.18 to 1 μm in diameter.[37] The outer segments are neither fixed nor exposed to denaturing agents, and the inclusions do not seem to be artifacts of preparative procedures. The inclusions are deficient in particles and seem to consist of nonbilayer lipid. The physiological significance of these rare structures remains to be determined. The margin or rim of the disk has different characteristics from other regions of the disk. The rim region resists swelling in hypotonic medium.[38] A high-molecular-weight protein (incisin) is found only in the rim region by immunocytochemical techniques.[39] The high-molecular-weight protein may serve as an anchor for filamentous substances[11] that link the adjacent disks and the disks to the plasma membrane.

2.3. Properties of Disk Membranes

Rod outer segment membranes (>95% disk membranes, <5% plasma membrane) consist of 60% protein and 40% lipid. The phospholipids comprise

Table I
Lipid Composition of Vertebrate Rod Outer Segments (Relative Mole Percent)[40]

	PC[a]	PE	PS	PI	SL	NL	C	Others
Human	32.1	37.1	12.0	—	1.8	12.9	9.8	—
Rat	27.5	38.1	12.8	0.9	0.7	12.0	7.8	11.2
Frog	38.1	29.1	10.8	1.9	1.6	15.2	8.0	—

[a] PC, phosphatidylcholine; PE, phosphatidylethanolamine; PS, phosphatidylserine; PI, phosphatidylinositol; SL, sphingolipid; NL, neutral lipid, C, cholesterol.

about 90% of the total lipids.[40] The major neutral lipid is cholesterol. 1,2-Diglycerides and free fatty acids are minor constituents. In vertebrate photoreceptors, about 70% of the phospholipid is accounted for by phosphatidylethanolamine and phosphatidylcholine (Table I) The most abundant polyunsaturated fatty acid is docosahexaenoic acid (22 carbons and six unsaturated bonds), which is linked exclusively to the middle carbon (2-position) of the glycerol moiety of phospholipids (Table II). Although the 1-position of the glycerol moiety is predominantly occupied by saturated fatty acids, some polyunsaturates (e.g., diunsaturates) are also found in this position.

Polyunsaturated acyl chains of phospholipids readily undergo cis–trans isomerization at physiological temperature and disturb the ordered arrangement of lipid molecules in the bilayer. This increases membrane fluidity. Because of the high content of unsaturated fatty acids, rod membranes are highly fluid under physiological conditions and permit free movement of rhodopsin molecules within the membrane. Thus, rhodopsin can rotate with the rotational axis perpendicular to the longer axis of the disk and diffuse in a direction parallel to the longer axis of the disk. The retinal chromophore of rhodopsin is oriented in a plane almost parallel to the longer axis of the disk.[41,42] If the chromophores

Table II
Fatty Acid Composition of Phospholipids from Vertebrate Rod Outer Segments (Relative Mole Percent)[40]

		14:0 + 15:0 + 16:0 + 17:0 + 18:0	18:1 + 18:2 + 18:3	20:2 + 20:3 + 20:4 + 20:5	22:4	22:5	22:6	Others
PC[a]	Human	54.1	16.0	7.4	0.4	1.5	19.5	1.1
	Rat	44.1	9.1	4.2	—	3.2	38.4	1.0
	Frog	63.8	9.5	1.7	0.4	1.1	23.6	—
PE	Human	47.9	6.5	6.2	1.2	2.8	34.2	1.2
	Rat	35.7	3.8	3.8	0.2	1.1	54.8	0.5
	Frog	22.3	6.8	7.0	7.0	5.4	51.1	0.4
PS	Human	31.9	6.3	4.9	6.8	4.9	34.1	11.1
	Rat	31.6	2.6	2.7	—	—	61.8	1.3
	Frog	28.5	4.0	2.2	6.8	7.6	46.4	4.5

[a] PC, phosphatidylcholine; PE, phosphatidylethanolamine; PS, phosphatidylserine. The number of carbon atoms and the number of double bonds define the type of fatty acid. For example, 18:3 indicates linolenic acid, which has 18 carbon atoms and three double bonds.

do not rotate in the plane, the disk will demonstrate dichroic properties when irradiated by linearly polarized light in a direction parallel to the disk plane. This is because one of the components lineally polarized parallel to the disk plane is prefentially absorbed. Dichroism was not observed unless the photoreceptor was fixed by glutaraldehyde.[43] By means of rapid-recording spectroscopy, however, a transient dichroism was detected.[44] These results indicate that rhodopsin has rotational freedom in the disk membrane. Rhodopsin is a glycoprotein, and there is a high energy barrier for the carbohydrate moiety to go through the hydrophobic core of membrane. Therefore, the pigment molecule may not tumble from one side of the membrane to the other.

If a small area of the disk membrane is irradiated by a flash of intense light, the absorption of the irradiated area quickly decreases as a result of the bleaching of rhodopsin and recovers as unbleached rhodopsin molecules diffuse into the bleached area from the surrounding unirradiated regions. The diffusion constant of rhodopsin determined from the time course of recovery is 5×10^{-9} cm^2 s^{-1} at 22°C.[45,46] From the diffusion constant, the viscosity of the disk membrane was estimated to be about 2 poise (approximately equal to the fluidity of olive oil). Since the time course of recovery in the irradiated area is identical to the time course of absorption loss in the surrounding area, both unbleached and bleached rhodopsin molecules have the same diffusion constant.[45,46]

Cone pigments also have rotational and diffusional freedom in the membrane. Their diffusion constants are very similar to those of rhodopsin.[47] Whereas the adjacent rod disks are physically separated, cone disk membranes are continuous and permit diffusion of cone pigments throughout the cone membrane system. This explains the difference in autoradiographic data on protein renewal between rods and cones.[13] After an injection of radioactive amino acids, a distinct radioactive band is formed in rod outer segments but not in cone outer segments.

As described in Section 4, the photic bleaching of rhodopsin initiates a series of enzymic reactions. The highly fluid nature of rod membranes that allows the rotational and translational freedom of rhodopsin may be important for the pigment to interact with and activate various enzymes.

3. RHODOPSIN

3.1. Chemical Properties of the Opsin Protein

Two-dimensional crystals of rhodopsin were prepared by treatment of frog disk membranes with polysorbate 80 and subjected to optical diffraction analysis.[48] Projection maps of negatively strained crystals suggested that the pigment molecules exist as dimers 2.0–2.5 nm in width and 7.0–8.0 nm in length. This is in good agreement with the size of the molecule estimated from energy-transfer measurements[36] and supports a model in which the elongated pigment molecule spans the disk membrane. Proteolytic digestion of the disk by thermolysin releases the carboxyl terminal peptide, which is exposed on the cy-

```
                CHO                                              CHO
AcMet-Asn-Gly-Thr-Glu-Gly-Pro-Asn-Phe-Tyr-Val-Pro-Phe-Ser-Asn-Lys-Thr-Gly-Val-Val-

Arg-Ser-Pro-Phe-Glu-Ala-Pro-Gln-Tyr-Tyr-Leu-Ala-Glu-Pro-Trp-Gln-Phe-Ser-Met-Leu-

Ala-Ala-Tyr-Met-Phe-Leu-Leu-Ile-Met-Leu-Gly-Phe-Pro-Ile-Asn-Phe-Leu-Thr-Leu-Tyr-

Val-Thr-Val-Gln-His-Lys-Lys-Leu-Arg-Thr-Pro-Leu-Asn-Tyr-Ile-Leu-Leu-Asn-Leu-Ala-

Val-Ala-Asp-Leu-Phe-Met-Val-Phe-Gly-Gly-Phe-Thr-Thr-Thr-Leu-Tyr-Thr-Ser-Leu-His-

Gly-Tyr-Phe-Val-Phe-Gly-Pro-Thr-Gly-Cys-Asn-Leu-Glu-Gly-Phe-Phe-Ala-Thr-Leu-Gly-

Gly-Glu-Ile-Ala-Leu-Trp-Ser-Leu-Val-Val-Leu-Ala-Ile-Glu-Arg-Tyr-Val-Val-Val-Cys-

Lys-Pro-Met-Ser-Asn-Phe-Arg-Phe-Gly-Glu-Asn-His-Ala-Ile-Met-Gly-Val-Ala-Phe-Thr-

Trp-Val-Met-Ala-Leu-Ala-Cys-Ala-Ala-Pro-Pro-Leu-Val-Gly-Trp-Ser-Arg-Tyr-Ile-Pro-

Glu-Gly-Met-Gln-Cys-Ser-Cys-Gly-Ile-Asp-Tyr-Tyr-Thr-Pro-His-Glu-Glu-Thr-Asn-Asn-

Glu-Ser-Phe-Val-Ile-Tyr-Met-Phe-Val-Val-His-Phe-Ile-Ile-Pro-Leu-Ile-Val-Ile-Phe-

Phe-Cys-Tyr-Gly-Gln-Leu-Val-Phe-Thr-Val-Lys-Glu-Ala-Ala-Ala-Gln-Gln-Gln-Glu-Ser-

Ala-Thr-Thr-Gln-Lys-Ala-Glu-Lys-Glu-Val-Thr-Arg-Met-Val-Ile-Ile-Met-Val-Ile-Ala-

Phe-Leu-Ile-Cys-Trp-Leu-Pro-Tyr-Ala-Gly-Val-Ala-Phe-Tyr-Ile-Phe-Thr-His-Gln-Gly-

Ser-Asp-Phe-Gly-Pro-Ile-Phe-Met-Thr-Ile-Pro-Ala-Phe-Phe-Ala-Lys-Thr-Ser-Ala-Val-

Tyr-Asn-Pro-Val-Ile-Tyr-Ile-Met-Met-Asn-Lys-Gln-Phe-Arg-Asn-Cys-Met-Val-Thr-Thr-

Leu-Cys-Cys-Gly-Lys-Asn-Pro-Leu-Gly-Asp-Asp-Glu-Ala-Ser-Thr-Thr-Val-Ser-Lys-Thr-

Glu-Thr-Ser-Gln-Val-Ala-Pro-AlaCOOH
```

Fig. 3. Amino acid sequence of bovine rhodopsin. N-acetylmethionine (AcMet) and alanine (Ala) are N-terminal and C-terminal residues, respectively. CHO represents carbohydrate.

toplasmic surface of the membrane.[49] Other proteases such as trypsin,[50] papain,[51,52] and subtilisin[53] make incisions in the molecule without bleaching the pigment. On solubilization of the treated membrane, several peptide fragments are obtained. Since the oligosaccharide-bearing amino-terminal peptide, which protrudes on the intradiscal (internal) surface of the disk, is not released by proteolytic digestion of disrupted disk membranes, the pigment molecule must have limited exposure on this side of membrane. The blocked amino-terminal residue is N-acetylmethionine.[54] Various properties of rhodopsin have been discussed in terms of the reactivities of amino acid residues in the amino and carboxyl terminal sequences.[55] The complete primary structure of bovine rhodopsin elucidated recently[55a] is shown in Fig. 3.

The known sequence of rhodopsin shows several features worthy of note. Proteins synthesized on the endoplasmic reticulum and transported to the plasma membrane (for secretion from the cell or for incorporation as integral membrane proteins) generally contain extra amino-terminal sequences (about 15–29 residues) of hydrophobic nature.[56] The extra sequence or "signal" sequence is considered to facilitate the translocation of the newly synthesized polypeptide across the endoplasmic reticulum membrane and is removed before

$$\text{GlcNAc}\beta 1 \rightarrow 2\text{Man}\alpha 1 \underset{3}{\overset{\text{Man}\alpha 1 \searrow 6}{\nearrow}} \text{Man}\beta 1 \rightarrow 4\text{GlcNAc}\beta 1 \rightarrow 4\text{GlcNAc}$$

Fig. 4. The structure of the oligosaccharide moiety of bovine rhodopsin.

the nascent polypeptide folds into a matured protein. However, the opsin polypeptide synthesized on the polysomes does not contain a signal sequence.[57] More recent work shows that a number of proteins (e.g., ovalbumin, mitochondrial proteins) are synthesized without signal sequences.[58] In these proteins, signals may remain uncleaved or possibly be present in the internal region of polypeptide rather than in the amino-terminal region.

The retinal chromophore of rhodopsin is attached to $Lys^{53'}$ from the carboxylterminus.[59–61] The retinylidene lysine-containing sequence is Lys-Thr-Ser-Ala in bovine rhodopsin and Lys-Ser-Ser-Ser in sheep rhodopsin.[62] The 15 residues in the carboxyl terminal peptide of opsin contain three serine and four threonine residues (see Fig. 3). Phosphorylation of the opsin protein by ATP occurs primarily in this region.[55] In the highly phosphorylated species, about 5 mol of phosphate were incorporated per mole of rhodopsin.[63,64]

The two oligosaccharides are bound to Asp^2 and $Asp^{15,53}$ in the amino terminal peptide.[55] Glycosyl transferases that transfer oligosaccharides to the asparagine residue of protein are specific for asparagines in the sequence containing Asn-X-Thr or Asn-X-Ser (X variable).[65] The sequences Asn^2-Gly^3-Thr^4 and Asn^{15}-Lys^{16}-Thr^{17} of the opsin satisfy the requirement. The structures of the two oligossacharide moieties are probably the same and were determined to be GlcNAcβ1→2Manα1→3(Manα1→6)Manβ1→4GlcNAcβ1→4GlcNAc (Fig. 4).[66,67]

3.2. Physical Properties of the Opsin Protein

There is general agreement that rhodopsin has a molecular weight in the range of 36,000–40,000.[55] Rhodopsin has 50–60% helical structure.[68,69] Infrared linear dichroism measurements on rod outer segment membranes suggest that the helical portion of rhodopsin is oriented at right angles to the plane of the disk membrane.[70] A 60% helical content suggests that about 190 residues of rhodopsin are in α-helical conformation. Since one helical turn takes 3.6 residues and is 0.54 nm in length, the total length of the helical coil in the rhodopsin molecule is calculated to be 29 nm. The apolar core of the lipid bilayer is about 4 nm thick. Therefore, rhodopsin could have as many as seven helical columns, each of which consists of seven to eight turns and spans the membrane. The molecular model of rhodopsin is similar to the one proposed for bacteriorhodopsin, a retinal-containing protein of *Halobacteria*.[71] The helical columns of bacteriorhodopsin are believed to span the membrane.[72] The helical content of membrane-bound rhodopsin remains virtually unchanged on bleaching.[68] However, after solubilization with detergent, the bleached pigment molecule undergoes a conformational change that involves a loss of about one-third of the helical structure. Although the conformational change is not of physiolog-

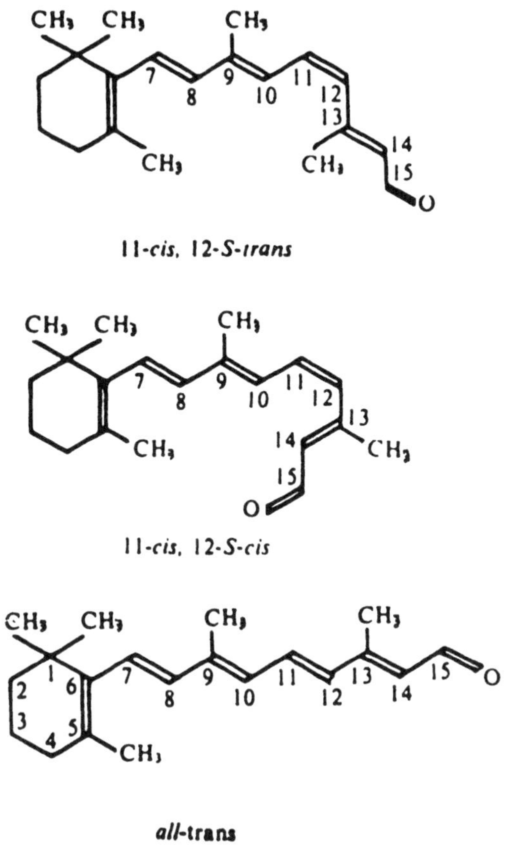

11-*cis*, 12-*S*-*trans*

11-*cis*, 12-*S*-*cis*

all-trans

Fig. 5. Structure of 11-*cis*-retinals and all-*trans*-retinal.

ical significance, it suggests that the highly helical conformation of rhodopsin within the membrane is somewhat constrained.

3.3. Chromophore and Photochemistry

The chromophore of rhodopsin is 11-*cis*-retinal which is linked with its aldehyde group to the ε-amino group of a lysine residue through Schiff base. Energy transfer studies suggest that the retinal chromophore is located nearly in the center of the hydrophobic core of the disk membrane.[73] Two stereoisomers of 11-*cis*-retinal are possible with respect to the orientation of the 13–14 double bond. If the 11–12 double bond and the 13–14 double bond are in trans configuration, the isomer is designated 11-*cis*, 12-*S*-*cis*-retinal (Fig. 5). If they are in cis configuration, the isomer is called 11-*cis*,12-*S*-*trans*-retinal. The retinal in visual pigments is believed to be in the 11-*cis*,12-*S*-*trans* confirguration on the basis of studies on visual pigment analogues. For example, 11-*cis*,14-methylretinal exists only in the 12-*S*-trans configuration because of steric hindrance between the methyl group at carbon 14 and the hydrogen atom at carbon

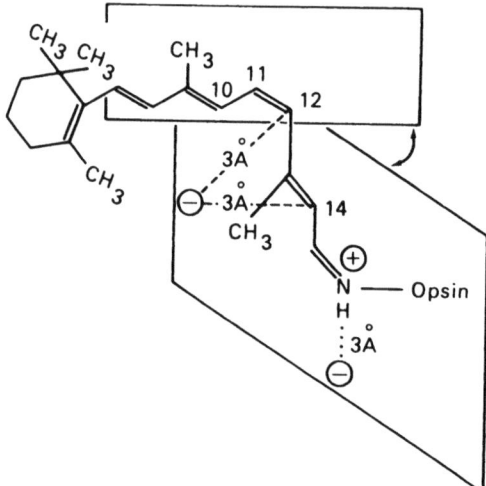

Fig. 6. Proposed model for the chromophore of rhodopsin.

10. A photobleachable visual pigment analogue can be synthesized when the opsin is mixed with the chemically synthesized 11-*cis*,14-methylretinal.[74] However, 11-*cis*-retinal in several crystalline forms exists in the 11-cis,12-S-cis confirguration.[75] Therefore, the general assumption that the chromophore of rhodopsin is 11-*cis*,12-*S-trans*-retinal should be taken with caution until other possible configurations such as a distorted 12-S-cis are entirely ruled out.

Rhodopsin has two absorption bands in the visible wavelength region: a major (α) band at 498 nm (ϵ_m = 41,000) and minor (β) band at 340 nm (ϵ_m = 10,000). Protonated Schiff base complexes between retinal and aliphatic amino acids show absorption maxima around 440 nm. Therefore, an additional 60-nm shift toward the longer wavelength side that occurs in rhodopsin is probably caused by interaction of the chromophore with the opsin. Theoretical calculations indicate that placing one negative charge near the Schiff base nitrogen and the other in a region about 0.3 nm away both from C_{12} and C_{14} would cause the required red shift.[76] The proposed model is shown in Fig. 6.

The absorption of a photon by rhodopsin results in the isomerization of the chromophore to the all-trans form (see Fig. 5). This reaction occurs before lumirhodopsin is formed. After the photoisomerization, rhodopsin undergoes a series of spectrally distinct changes (Fig. 7) that can be identified by ordinary spectroscopy at different temperatures.[77] For example, bathorhodopsin (λ_{max} = 548 nm) is detected as a stable intermediate when rhodopsin is irradiated by 437-nm light at −195°C. Warming to −140°C causes the conversion of bathorhodopsin to lumirhodopsin (λ_{max} = 497 nm), which in turn decays to metarhodopsin I (λ_{max} = 478 nm) at −40°C. If rhodopsin is irradiated by 540 nm light at −268°C, hypsorhodopsin (λ_{max} = 430 nm) is formed, which on subsequent irradiation with 430 nm light converts to bathorhodopsin.[78] The presence of two forms of hypsorhodopsin has been suggested.[79] Two distinct forms of bathorhodopsin (λ_{max} = 538 and 555 nm)[80] seem to derive from the

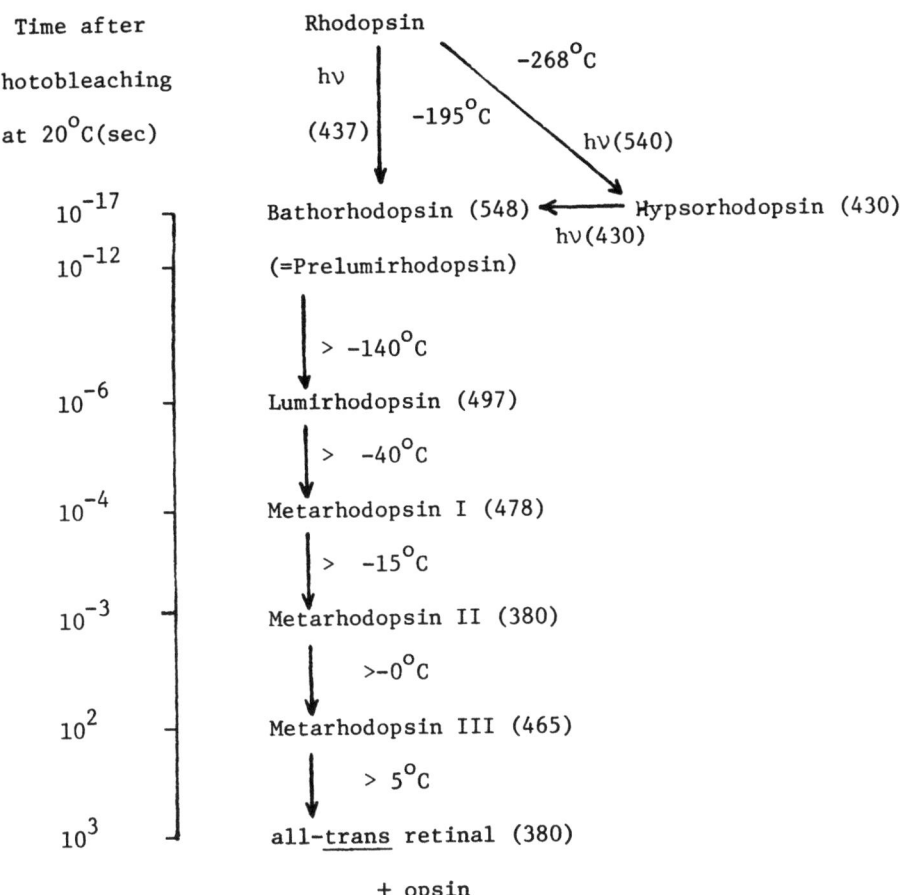

Fig. 7. Intermediates formed following photic bleaching of vertebrate rhodopsin. The absorption maximum of each intermediate is shown by the number of nanometers in parentheses.

two forms of hypsorhodopsin.[79] In spite of these findings, the question of whether hypsorhodopsin is a normal precursor to bathorhodopsin or a side product that decays to bathorhodopsin is still unsettled. Picosecond spectroscopy shows that irradiated bovine rhodopsin produces bathorhodopsin directly,[81] whereas octopus rhodopsin, upon illumination, forms bathorhodopsin before hypsorhodopsin accumulates in detectable amounts.[82] On the other hand, hypsorhodopsin formed from squid rhodopsin was concluded to be a precursor to bathorhodopsin by picosecond spectral measurements.[83]

At physiological temperature, bathorhodopsin is formed in less than 6 ps following irradiation.[84] It was then questioned whether cis–trans isomerization of the chromophore could take place in this short time period. According to recent molecular dynamics calculations, the photochemical isomerization could occur with high quantum efficiency ($\phi_{calc} \simeq 0.6$) in approximately 2 ps.[85,86] Overwhelming experimental evidence supports the conclusion that the chromophore of bathorhodopsin has a distorted all-trans conformation.[86] The energy level of ground state bathorhodopsin is about 35 kcal/mol higher than that of rhodopsin.[87] The potential energy difference is larger than the bathochromic shift (498 nm → 543 nm; *ca.* 5 kcal/mol) of the chromophore. Therefore, a considerable fraction of the energy difference must be stored by forcing the chromophore into a strained configuration. Thus, bathorhodopsin formation is the major endergonic step, and subsequent steps involving the formation of lumirhodopsin and metarhodopsins are energetically relaxing steps. A quantum efficiency (ϕ_{exp}) of about 0.6 for the overall process of rhodopsin photolysis[77] supports the conclusion that bathorhodopsin formation is the only photochemical reaction.

The time periods required for the formation of other intermediates (lumirhodopsin and metarhodopsins) are shown on the left-hand scale in Fig. 7. Decay of metarhodopsin I to metarhodopsin II (lifetime $\simeq 3 \times 10^{-4}$ s) is of particular importance because the reaction is believed to be related to the photocurrent generation (latency $\simeq 10^{-3}$ s) by the rod. The metarhodopsin I to metarhodopsin II transition is accompanied by changes in the opsin structure that are characterized by proton uptake and the appearance of SH groups.[77] The opsin conformational change, however, is not to such an extent that it can be detected by circular dichroism measurements in the far UV region. A birefringence loss was observed in rod outer segments exposed to light.[88] The chromophore of rhodopsin is optically active, and this property is retained to the level of metarhodopsin II irrespective of the isomerization of the retinal chain.[89] Metarhodopsin III is optically inactive.[90] This suggests that in going from metarhodopsin II to metarhodopsin III, retinal may be transferred from its original binding site to either a different site or a different molecule. The chick cone pigment iodopsin also decays through thermal intermediates in the light.[7] The rate of iodopsin regeneration with 11-*cis*-retinal is 500-fold faster than the rate of rhodopsin regeneration.[91]

Rhodopsin is synthesized in the inner segment of photoreceptor cell and incorporated in the disk membrane of the outer segment.[13] Once incorporated, the visual pigment serves as the photoreceptor many times until the disk membrane with which the pigment is associated is phagocytized and decomposed by the pigmented epithelium. Photobleached pigments are regenerated by exchanging all-*trans*-retinal with 11-*cis*-retinal. Wald and associates established that the biochemical basis of rhodopsin regeneration is a combination of the opsin protein with the 11-cis isomer of retinal.[92] How is 11-*cis*-retinal formed from all-*trans*-retinal? Because rhodopsin regeneration takes place in the dark, the isomerization of retinal was believed to be an enzymatic process. Nevertheless, all attempts so far have been unsuccessful in identifying a "retinal isomerase" that would catalyze the isomerization of all-*trans*-retinal to 11-*cis*-retinal in the dark. However, the failure to identify such an enzyme does not rule out the possible presence of an enzymic isomerization mechanism. In fact, the isomerization of all-*trans*-retinal to 9-*cis*-retinal (but not to 11-*cis*-retinal) in the dark is catalyzed by dihydroflavins and dithiols.[93] On the other hand, all-*trans*-retinal is photochemically isomerized to 11-*cis*-retinal in the presence of urea-treated rod outer segments.[94] Since enzyme proteins were denatured by urea treatment, the photochemical isomerization was not catalyzed by enzyme. The active agent in the rod membrane was found to be phosphatidylethanolamine. Phosphatidylethanolamine forms a protonated Schiff base with retinal and, on irradiation, effectively isomerizes the all-trans isomer to the 11-cis form. Rhodopsin is regenerated in the dark by transfer of the 11-cis isomer from the phospholipid complex to opsin. If the reaction occurs *in vivo*, this may explain rhodopsin regeneration *in situ* in the outer segment. Although the mechanism of phospholipid catalysis is unknown, orbital calculations on protonated retinylidine phosphatidylethanolamine suggest that the 11-cis isomer is the most probable product of the photoisomerization of the all-trans isomer.[95]

Another regeneration mechanism involves a shuttle of retinal (or retinol) between the photoreceptor and the pigmented epithelium. When albino rats are irradiated by intense light, all-*trans*-retinal released in the photoreceptor is reduced to retinol and stored in the pigmented epithelium.[96] If the animal is subsequently kept in darkness, the stored retinol is returned to the retina to be utilized for pigment regeneration. This shuttle of vitamin A that occurs during light and dark adaptations is called the visual cycle. A retinol-binding protein distinct from the serum retinol-binding protein is found in the retina.[97,98] It remains to be seen if the protein is involved in the transport of retinol in the visual cycle. It is not known if retinal is isomerized to the 11-cis form in the pigmented epithelium or in the retina. A membrane-associated retinol dehydrogenase in rod outer segments was shown to catalyze the conversion of all-*trans*-retinol to its retinal with $NADP^+$ as the electron acceptor.[99] 11-*cis*-Retinol did not serve as substrate.

The results suggest that 11-*cis*-retinal required for rhodopsin regeneration is formed in the outer segment by isomerizing all-*trans*-retinal. Consistent with this possibility is the absence of 11-*cis*-retinol dehydrogenase in rat retina.[100] Therefore, it is retinol (not retinal) that is transported from the pigmented epithelium to the photoreceptor. It should be noted that rhodopsin regeneration occurs without the pigmented epithelium,[101] and the vitamin A shuttle between the pigmented epithelium and the retina may not be essential for rhodopsin regeneration. This does not rule out the necessity of the shuttle; it may be an important mechanism for protecting the photoreceptor from unbound retinal and retinol, which would accumulate under intense light irradiation. Free retinal and retinol are known to have detrimental effects on biomembranes.

Visual pigment analogues are synthesized with retinal analogues that have either one (or more) of 7-cis, 9-cis, and 11-cis structures. 13-*cis*-Retinal analogues do not react with the opsin protein. Various retinals that are known to synthesize photosensitive pigments are summarized in Table III. To form pigments, the conjugated side chain (C_7–C_{15}) of retinal analogues must be of the same length as that of retinal. The binding of 11-*cis*-retinal to the opsin is inhibited by β-ionone[109] and by compounds that mimic the 9-cis and 11-cis structures of retinal.[110] These results suggest the presence of a binding site in the opsin for β-ionone and the C_{15} moiety. Thus, the retinal is positioned in the chromophore-binding cavity of the opsin both by the β-ionone binding site and the lysine residue that forms a Schiff base with the aldehyde group.

Table III
Visual Pigment Analogues Formed with Various Retinal Isomers

Isomers	Pigment formed with bovine opsin		Ref.
	λ_{max} in alcohol	λ_{max}	
11-*cis*-Retinal	380	498 (rhodopsin)	
9-*cis*-Retinal	380	485 (isorhodopsin)	
5,6-Dihydro-9-*cis*-retinal	—	465	102
5,6-Dihydro-11-*cis*-retinal	—	465	102
3-Dehydro-9-*cis*-retinal	—	500	102
3-Dehydro-11-*cis*-retinal	—	517	102
5,6-Epoxy-3-dehydro-9-*cis*-retinal	—	465	102
5,6-Epoxy-3-dehydro-11-*cis*-retinal	—	465	102
13-Desmethyl-9-*cis*-retinal	370	486	102
13-Desmethyl-11-*cis*-retinal	378	495	102
9-Desmethyl-9-*cis*-retinal	366	453	102
9-Desmethyl-11-*cis*-retinal	370	461	102
9,13-Desmethyl-9-*cis*-retinal	361	458	102
9,13-Desmethyl-11-*cis*-retinal	366	483	102
7-*cis*-3-Dehydroretinal	—	464	103
9,13-Di-*cis*-11-dehydroretinal	355	430	103
7,13-Di-*cis*-11-dehydroretinal	355	430	103
7-*cis*-11-Dehydroretinal	360	440	103
14-Methyl-11-*cis*-retinal	350	497	104
14-Methyl-9-*cis*-retinal	378	484	104
13-Desmethyl-9-*cis*-retinal	372	487	104
9-*cis*-Allenic retinal I	—	456	105
9-*cis*-Allenic retinal II	—	482	105
9,13-Di-*cis*-retinal	357	481	106
11,12-Dihydroretinal	236	315	107
9-β-Cycloheptatrienylidene-11-*cis*-retinal	375	490	108

4. VISUAL TRANSDUCTION

Invertebrate photoreceptor cells increase Na^+ permeability in the light and are depolarized. In contrast, the Na^+ permeability of vertebrate photoreceptors (both rods and cones) is decreased by light, and the plasma membrane of vertebrate cells is hypolarized. In other words, sodium permeability dictates membrane potentials in these photoreceptor systems.[111] To explain how the absorption of light by photoreceptor pigments modulates Na^+ permeability, a substance (i.e., transmitter) was postulated to be produced in response to light and to block Na^+ channels in the membrane.[6] In the rod outer segment, rhodopsin is predominantly localized in the disks, which are physically discontinuous from the plasma membrane where Na^+ conductance changes occur in response to light. Therefore, it is logical to assume that the disk releases a transmitter that travels in the cytoplasm and binds to the Na^+ channel of the plasma membrane.

The Ca^{2+} ion was proposed as the putative transmitter substance[4] on the ground that increased Ca^{2+} concentration in the medium mimics the effect of light by hyperpolarizing the photoreceptors in the dark. The membrane potential of photoreceptors is usually low (about -30 mV) in the dark, and the light-induced hyperpolarization may reach as high as 30 mV.[6] For Ca^{2+} ion to be the transmitter that causes the membrane hyperpolarization, more than 25 but fewer than 1000 Ca^{2+} ions per photolyzed rhodopsin have to be released by the disk to close a sufficient number of Na^+ channels.[112] Several other criteria also need to be satisfied.[112] However, available experimental evidence at present does not consistently support the light-dependent release of sufficient amounts of Ca^{2+} ions.[113] If Ca^{2+} is the transmitter, there must be a mechanism to sequester the released Ca^{2+} ion in the disk. Evidence is lacking for such a mechanism in the disk.[114]

While the Ca^{2+} hypothesis was losing its initial glow, 3',5'-cyclic guanosine monophosphate (cyclic GMP) was proposed as an alternative candidate for the transmitter. This hypothesis is based on several supportive findings. (1) Nearly 500 molecules of cyclic GMP phosphodiesterase are activated by photolysis of a single rhodopsin molecule.[115] This leads to the disappearance of 5×10^4 molecules of cyclic GMP with a half-life of 125 msec,[113] although the rate of cyclic GMP decrease is still controversial.[116,117] (2) Cyclic GMP concentration returns to the original level during dark adaptation. (3) Injection of cyclic GMP into the rod outer segment induces membrane depolarization, and response latency to light stimulus is increased as if the excess cyclic GMP had to be hydrolyzed in order to produce the light response.[118] If cyclic GMP is assumed to maintain plasma-membrane Na^+ channels open in the dark, the channels would close on hydrolysis of cyclic GMP by light-activated phosphodiesterase. The cyclic GMP hypothesis is summarized in Fig. 8. Following the absorption of light, rhodopsin decays to an intermediate designated here "active rhodopsin (R*)." R* then associates with a guanine-nucleotide-binding protein (G-protein) and catalyzes exchange of bound GDP with GTP. The GTP·G protein thus formed activates cyclic GMP phosphodiesterase (PDE) by removing the inhibitor (I) from the enzyme. The PDE then hydrolyzes cyclic GMP and allows the Na^+ channel to close. These reaction steps and the components involved are discussed below in some detail.

The active form of rhodopsin is probably metarhodopsin II[119,120] because the opsin undergoes appreciable conformational change at this level[77] and may expose a molecular domain for interaction with G-protein. Recent evidence indicates that the susceptibility of photolyzed rhodopsin to proteolysis is enhanced at the metarhodopsin II level.[121] G-protein has a molecular weight of 39,000–40,000[122–124] and hydrolyzes bound GTP to GDP. G-protein is associated with other proteins (subunits), which may regulate its GTPase activity. In the rod outer segment, the content of G-protein is approximately 1 molecule per 400 rhodopsin molecules.[123,124] Photolysis of one rhodopsin molecule results in the binding of about 500 molecules of a nonhydrolyzable GTP analogue to photoreceptor membranes.[122] Since one photolyzed rhodopsin molecule activates nearly 500 molecules of phosphodiesterase,[115] it turns out that one G-protein molecule will activate one phosphodiesterase molecule. Thus, the activation of G-protein by active rhodopsin was proposed to be the amplification reaction in the transduction process.[122]

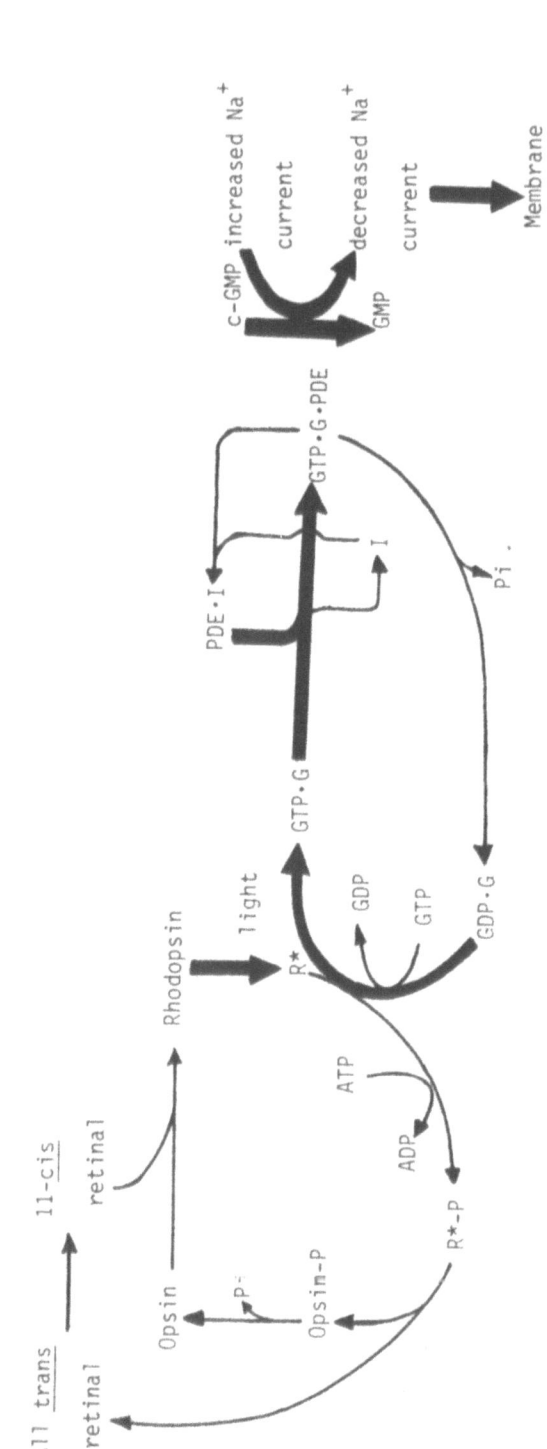

Fig. 8. Cyclic GMP hypothesis for phototransduction in vertebrate photoreceptor. Active rhodopsin (R*) produced by photolysis of rhodopsin activates G-protein by catalyzing the exchange of bound GDP for GTP; GTP·G in turn activates phosphodiesterase (PDE) by removing inhibitor I. Hydrolysis of cyclic GMP by PDE decreases Na⁺ inward current and causes membrane hyperpolarization. The main sequence of events is indicated by thick arrows.

Light-scattering changes that occur on mixing G-protein with photolyzed disk membranes suggest a 1 : 1 complex formation between G-protein and active rhodopsin.[125] Therefore, if the activation of G-protein is an amplification process of physiological significance, active rhodopsin has to contact some 500 G-protein molecules within a few milliseconds. Partial digestion of rod membranes indicates that removal of 12 amino acid residues from the carboxyl terminus of the opsin has no effect on the capability of active rhodopsin to associate with G-protein.[126] Since GTP·G-protein formed as a result of rhodopsin photolysis activates cyclic-GMP phosphodiesterase, G-protein is regarded as a mediator that transmits the photosignal received by rhodopsin to the phosphodiesterase. G-protein was therefore designated "transducin."[122]

Phosphodiesterase inhibitors were isolated from rod outer segment.[127,128] Frog phosphodiesterase has a molecular weight of 240,000,[129] whereas bovine phosphodiesterase has a molecular weight of 170,000 but associates with an inhibitor protein of molecular weight 80,000.[130] Since frog enzyme can be activated by tryptic digestion, it is possible that the purified frog enzyme was closely associated with an inhibitor protein. The activation of cyclic GMP phosphodiesterase is terminated by inactivation of GTP·G protein and reassociation of active phosphodiesterase with the inhibitor. G-protein is inactivated by hydrolysis of bound GTP to GDP. Another mechanism for the regulation of this light-activation system may be the inactivation of "active rhodopsin" by phosphorylation.[119,131] The phosphorylation reaction[132-134] is catalyzed by rhodopsin kinase,[135] which is a rhodopsin-specific protein kinase. The phosphorylation sites of the opsin protein are clustered in the carboxyl terminal region.[35] If rhodopsin phosphorylation is a regulatory mechanism, the phosphorylation site involved must be located in a region not too close to the carboxyl terminus because the site should not be included in twelve amino acid residues from the carboxyl terminus.[126]

If cyclic GMP is indeed involved in maintaining the plasma membrane Na^+ channels open in the dark, what is the mechanism of cyclic GMP action? Why do the channels close when cyclic GMP is decomposed? Are the effects of cyclic GMP and Ca^{2+} related? Because we know too little at present about the photoreceptor system to answer these questions, we shall turn to other receptor systems in which cyclic GMP is implicated.

Evidence indicates that cyclic GMP acts as the second messenger in the negative inotropic action of acetylcholine on the heart.[136] The effect of acetylcholine appears to result from a decrease in inward Ca^{2+} current. Thus, acetylcholine action mediated by increased cyclic GMP level may be related to a decrease of the inward Ca^{2+} current. Cyclic GMP and Ca^{2+} flux are similarly related in the brain.[137] From these findings, it is speculated that cyclic GMP may modulate Ca^{2+} flux in the photoreceptor cell.

Cyclic GMP at physiological concentrations decreases Ca^{2+} binding to vertebrate photoreceptor membranes.[138] In vertebrate photoreceptors, a competition in binding to the Na^+ channels between Na^+ and Ca^{2+} ions was provided as a mechanism to explain the opening and closing of the channels.[139] It is interesting to postulate that cyclic GMP bound to the Na^+ channel increases the affinity of the channel for Na^+ ion and that hydrolysis of cyclic

GMP results in a shift in the affinity of the channel toward Ca^{2+} ion. Cyclic GMP may affect the channel directly. Alternatively, the cyclic nucleotide may exert its effect through protein kinases, as in many tissues.[140] Rod outer segments contain a protein kinase that is stimulated by cyclic GMP and which phosphorylates a cytosolic protein.[141,142] It is not known whether the phosphorylated cytosolic protein plays a role in modulating the Na^+ channels.

What about a relationship between cyclic GMP and Ca^{2+}? The cyclic GMP level in the outer segment is increased by lowering extracellular Ca^{2+} concentration,[143] probably because Ca^{2+} inhibits guanylate cyclase activity.[144] Calmodulin, identified in rod outer segments,[145] may be involved when Ca^{2+} exerts its effect on enzymic reactions. However, there is no evidence at present that such a Ca^{2+} effect on enzymes plays a major role in visual transduction.

Electrophysiological studies on the effects of Ca^{2+} and cyclic GMP do not yield a consistent picture; changes in rod responses produced by these factors do not seem to be explained by a common mechanism.[146] Thus, a link between Ca^{2+} and cyclic GMP is still missing. Injection of a nonhydrolyzable GTP analogue into dark-adapted invertebrate photoreceptors (*Limulus*)[147] or exposure of partially dark-adapted vertebrate rods (toad) to a GTP analogue[148] results in the generation of electrical responses in the dark. Although the activation of phosphodiesterase by the GTP analogue and subsequent decrease in cyclic GMP level offer a reasonable explanation for the phenomena, this may not be the only reaction affected by GTP. GTP binding proteins involved in signal transduction in various receptors are known to affect diversified enzymic reactions.[149]

5. NEUROTRANSMITTERS IN THE RETINA

In order for a substance to be accepted as a neurotransmitter, it has to satisfy several criteria.[150,151] (1) The substance must be released in response to physiological stimulation. (2) When applied to the postsynaptic neuron in question, the substance must evoke the same physiological effect as that evoked by the neurotransmitter released by the presynaptic neuron. (3) The release of transmitter must be Ca^{2+} dependent. (4) The substance must be synthesized and stored in the neuron. In many cases, it is difficult to meet these criteria because of the technical difficulties involved in the isolation of specific retinal neurons and the maintenance of their viability for physiological and biochemical studies. Another major problem inherent in the study of retinal neurotransmitters is the species variation in transmitters.[150] Nevertheless, significant progress has been made in recent years in the area of retinal neurotransmitters. Several reviews are available.[150–154]

Before putative neurotransmitters are discussed, it may be helpful to see what electrophysiological changes are evoked in retinal neurons following light absorption. The cells in the outer proximal layer are generally hyperpolarized, whereas the neurons in the inner proximal layer are depolarized (Table IV).[155] There are two types of bipolar cells, those that are depolarized and those that are hyperpolarized. On-center ganglion cells are hyperpolarized, but off-center

Table IV
Response of Vertebrate Retinal Cells to Light Stimulus[6,155]

Cell type	Membrane potential change
Photoreceptors	Hyperpolarization
Horizontal cells	Hyperpolarization
Bipolar cells	Hyperpolarization, depolarization
Amacrine cells	Depolarization
Ganglion cells	Hyperpolarization, depolarization

ganglion cells are depolarized by light irradiation. The information of photoreception is transmitted from the photoreceptor to bipolars, which, in turn, transmit the information to amacrine cells and ganglion cells. Horizontal cells regulate the function of the photoreceptors.

The photosignal received by the outer segment is transmitted to the synaptic end of the photoreceptor cell. This propagation is explained by an electric cable mechanism.[4] The photosignal, i.e., a hyperpolarization of membrane, then reduces the release of neurotransmitter from the rod synapse. In the carp retina, the photoreceptor transmitter, which acts on the postsynaptic membranes of horizontal cells in the dark, is believed to be aspartate, because α-aminoadipic acid (aspartate antagonist) blocks the effect of the endogenous transmitter.[156] Externally added L-aspartate was more effective than L-glutamate in depolarizing the horizontal cells in the dark; L-asparagine, glycine, and γ-aminobutyric acid (GABA) were without effect. In the turtle retina, acetylcholine may be a photoreceptor transmitter between cones and horizontal cells.[157] Although taurine is abundant in the photoreceptor cell and is released from irradiated chick retina,[158] there is insufficient evidence to support the compound as a transmitter. Photoreceptor stimulation of bipolar cells is counteracted or inhibited by horizontal cells that release inhibitory transmitters. GABA was suggested to be an inhibitory transmitter of goldfish horizontal cells.[159] Isolated horizontal cells are capable of synthesizing GABA from glutamic acid. In the goldfish retina, there are four types of horizontal cells, which receive signal input predominantly from cone and rod photoreceptors. One type of horizontal cells is GABAergic.[160]

Dopaminergic amacrine cells have been identified in the retina of most species.[161] These amacrine cells can synthesize dopamine and release it on light stimulation.[162] The transmitter is probably involved in the synapse between amacrine cells and horizontal cells. When the carp retina is treated with trypsin and centrifuged in a Ficoll gradient, horizontal cells are collected in a band free from other retinal neurons.[163] The isolated cells possess dopamine receptors and, when stimulated by dopamine, accumulate cyclic AMP within the cell in a graded fashion. This cyclic AMP accumulation is blocked by dopamine antagonists. Thus, the dopamine receptors of the cells are associated with the adenylate cyclase system.

Indoleamine-accumulating neurons identified in the retina of many species are a subclass of amacrine cells.[164] It is unlikely that 5-hydroxytryptamine is the transmitter of the indoleamine-accumulating neurons, because 5-hydroxytryptamine level in the retina is too low for it to be a transmitter and because the activity of tryptophan hydroxylase, the enzyme involved in 5-hydroxytryptamine synthesis, is not very high in the retina. The transmitter of these cells is an unidentified indole.[164]

Both amacrine and ganglion cells are considered to possess receptors for acetylcholine, and acetylcholine mimics amacrine stimulation of ganglion cells.[154] This suggests that acetylcholine may function as an excitatory transmitter of amacrine cells. Autoradiographic experiments on [^{3}H]acetylcholine-labeled rabbit retina demonstrate that amacrine cells displaced in the ganglion layer are able to synthesize acetylcholine.[150]

Glycine-accumulating cells distinct from domaminergic cells have been tentatively identified as amacrine cells by autoradiography and physiological studies.[165] However, the level of glycine in the retina of several species was previously shown to remain unchanged between light- and dark-adapted tissues.[166]

Amacrine cells draw particular attention in recent years because several neuropeptides were found to be associated with these cells.[167] The neuropeptides so far identified in the amacrine cells by immunohistochemical techniques include [Leu^5]enkephalin, substance P, somatostatin, neurotensin, vasoactive intestinal peptide, glucagon, β-endorphin, α-melanocyte-stimulating hormone, thyrotropin-releasing hormone, and pancreatic peptide.[150] There is no firm evidence that these peptides are neurotransmitters. It is possible that the peptides are modulators if not transmitters and affect the action of true transmitters.

The vertebrate retina is an outgrowth of the forebrain and presents many advantages over the whole brain in its anatomic simplicity. In fact, the isolated retina can serve as an experimental model of the central nervous system.[168] Information on retinal neurotransmitters and their function is still scarce. However, this is one of the most rapidly advancing areas of vision research, and many exciting developments are expected in the future.

REFERENCES

1. Marks, W. B., Dobelle, W. H., and MacNichol, E. F., 1964, *Science* **143**:1181–1183.
2. Pirenne, M. H., 1967, *Vision and the Eye*, 2nd ed., Chapman and Hall, London.
3. Tomita, T., 1970, *Q. Rev. Biophys.* **3**:179–222.
4. Hagins, W. A., 1972, *Annu. Rev. Biophys. Bioeng.* **1**:131–158.
5. Stevens, C. F., 1966, *Neurophysiology: A Primer*, John Wiley & Sons, New York.
6. Cervetto, L., and Fuortes, M. G. F., 1978, *Annu. Rev. Biophys. Bioeng.* **7**:229–251.
7. Jung, R. (ed.), 1973, *Handbook of Sensory Physiology*, Vol. 7/3, Part A, *Central Processing of Visual Information*; Part B, *Visual Centers in the Brain*, Springer-Verlag, Berlin, Heidelberg, New York.
8. Westfall, J. A. (ed.), 1982, *Visual Cells in Evolution*, Raven Press, New York, pp. 91–154.
9. Greiner, J. V., 1981, *Exp. Eye Res.* **33**:433–446.
10. Steinberg, R. H., Fisher, S. K., and Anderson, D. H., 1980, *J. Comp. Neurol.* **190**:501–518.
11. Usukura, J., and Yamada, E., 1981, *Biomed. Res.* **2**:177–193.
12. Andrews, L. D., and Cohen, A. I., 1981, *Exp. Eye Res.* **33**:1–10.
13. Young, R. W., 1976, *Invest. Ophthalmol.* **15**:700–725.
14. LaVail, M. M., 1976, *Science* **194**:1071–1074.

15. Basinger, S., Hoffman, R., and Matthes, M., 1976, *Science* **194**:1074–1076.
16. Kaplan, M. W., 1981, *Invest. Ophthalmol. Vis. Sci.* **21**:395–402.
17. Barondes, S. H., 1981, *Annu. Rev. Biochem.* **50**:207–231.
18. O'Brien, P. J., 1976, *Exp. Eye Res.* **23**:127–137.
19. Adams, A. J., Tanaka, M., and Shichi, H., 1978, *Exp. Eye Res.* **27**:595–605.
20. Nir, I., and Hall, M. O., 1979, *Exp. Eye Res.* **29**:181–194.
21. McLaughlin, B. J., and Wood, J. G., 1980, *Invest. Ophthalmol. Vis. Sci.* **19**:728–742.
22. Bridges, C. D. B., 1981, *Invest. Ophthalmol. Vis. Sci.* **20**:8–16.
23. Röhlich, P., 1976, *Nature* **263**:789–791.
24. Burnside, M. B., 1976, *Exp. Eye Res.* **23**:257–275.
25. Hogan, M. J., Wood, I., and Steinberg, R. H., 1974, *Nature* **252**:305–307.
26. Young, R. W., 1978, *Invest Ophthalmol. Vis. Sci.* **17**:105–116.
27. Bunt, A. H., and Saari, J. C., 1982, *J. Cell Biol.* **92**:269–276.
28. Röhlich, P., 1981, *Acta Histochem.* [Suppl.] **23**:123–136.
29. Chabre, M., 1975, *Biochim. Biophys. Acta* **382**:322–335.
30. Saibil H., Chabre, M., and Worcester, D., 1976, *Nature* **262**:266–270.
31. Yeager, M., Schoenborn, B., Engelman, D., Moore, P., and Stryer, L., 1980, *J. Mol. Biol.* **137**:315–348.
32. Jan, L. Y., and Revel, J. P., 1974, *J. Cell Biol.* **62**:257–273.
33. Blaustein, D. I., and Dewey, M. M., 1979, *J. Histochem. Cytochem.* **27**:788–793.
34. Adams, A. J., Somers, R. L., and Shichi, H., 1979, *Photochem. Photobiol.* **29**:687–692.
35. Hargrave, P. A., Fong, S.-L., McDowell, J. H., Mas, M. T., Curtis, D. R., Wang, J. K., Juszczak, E., and Smith, D. P., 1980, *Neurochemistry of the Retina* (N. G. Bazan and R. N. Lolley, ed.), Pergamon Press, Oxford, pp. 231–244.
36. Wu, C. W., and Stryer, L., 1972, *Proc. Natl. Acad. Sci. U.S.A.* **69**:1104–1108.
37. Corless, J. M., and Costello, M. J., 1981, *Exp. Eye Res.* **32**:217–228.
38. Falk, G., and Fatt, P., 1969, *J. Ultrastruct. Res.* **28**:41–60.
39. Papermaster, D. S., Schneider, B. G., Zorn, M. A., and Kraehenbuhl, J. P., 1978, *J. Cell Biol.* **78**:415–425.
40. Anderson, R. E., and Andrews, L. D., 1982, *Visual Cells in Evolution* (J. A. Westfall, ed.), Raven Press, New York, pp. 1–22.
41. Liebman, P. A., 1962, *Biophys. J.* **2**:161–178.
42. Harosi, F. I., and Malerba, F. E., 1975, *Vision Res.* **15**:379–388.
43. Brown, P. K., 1972, *Nature* [New Biol.] **236**:35–38.
44. Cone, R. A., 1972, *Nature* [New Biol.] **236**:39–43.
45. Poo, M., and Cone, R. A., 1974, *Nature* **247**:438–441.
46. Liebman, P. A., Weiner, H. L., and Drzymala, R. E., 1982, *Methods in Enzymology* (L. Packer, ed.), Academic Press, New York, pp. 660–668.
47. Liebman, P. A., 1975, *Photoreceptor Optics* (A. W. Snyder and R. Menzel, ed.), Springer-Verlag, Berlin, Heidelberg, New York, pp. 199–214.
48. Corless, J. M., McCaslin, D. R., and Scott, B. L., 1982, *Proc. Natl. Acad. Sci. U.S.A.* **79**:1116–1120.
49. Hargrave, P. A., and Fong, S.-L., 1977, *J. Supramol. Struct.* **6**:559–570.
50. Molday, R. S., and Molday, L. L., 1979, *J. Biol. Chem.* **254**:4653–4660.
51. Fung, B. K.-K., and Hubbell, W. L., 1978, *Biochemistry* **17**:4396–4402.
52. Albert, A. D., and Litman, B. J., 1978, *Biochemistry* **17**:3893–3900.
53. Dratz E. A., Miljanich, G. P., Nemes, P. P., Gaw, J. E., and Schwartz, S., 1979, *Photochem. Photobiol.* **29**:661–670.
54. Tsunasawa, S., Narita, K., and Shichi, H., 1980, *Biochim. Biophys. Acta* **624**:218–225.
55. Hargrave, P., 1982, *J. Prog. Retinal Res.* **1**:1–51.
55a. Ovchinnikov, Y. A., Abdulaev, N. G., Feigina, M. Y., Artamonov, I. D., Zolotarev, A. S., Kostina, M. B., Bogachuk, A. S., Miroshnikov, A. I., Martinov, V. I., and Kudelin, A. B., 1982, *Bioorg. Khim.* **8**:1011–1014.
56. Blobel, G., Walter, P., Chang, C. N., Goldman, B. M., Erickson, A. H., and Lingappa, V. R., 1979, *Symp. Soc. Exp. Biol.* **33**:9–39.
57. Schechter, I., Burnstein, Y., Zemell, R., Ziv, E., Kantor, F., and Papermaster, D. S., 1979, *Proc. Natl. Acad. Sci. U.S.A.* **76**:2654–2658.

58. Kreil, G., 1981, *Annu. Rev. Biochem.* **50**:317–348.
59. Pellicone, C., Bouillon, P., and Virmaux, N., 1980, *C.R. Acad. Sci.* [D] (Paris) **290**:567–569.
60. Wang, J. K., McDowell, J. H., and Hargrave, P. A., 1980, *Biochemistry* **19**:5111–5117.
61. Mullen, E., and Akhtar, M., 1981, *FEBS Lett.* **132**:261–264.
62. Findlay, J. B. C., Brett, M., and Pappin, D. J. C., 1981, *Nature* **293**:314–316.
63. Kühn, H., and McDowell, J. H., 1977, *Biophys. Struct. Mech.* **3**:199–203.
64. Shichi, H., and Somers, R. L., 1978, *J. Biol. Chem.* **253**:7040–7046.
65. Wagh, P. V., and Bahl, O. P., 1981, *CRC Crit. Rev. Biochem.* **10**:307–377.
66. Liang C.-J., Yamashita, K., Muellenberg, C. G., Shichi, H., and Kobata, A., 1979, *J. Biol. Chem.* **254**:6416–6418.
67. Fukuda, M. N., Papermaster, D. P., and Hargrave, P. A., 1979, *J. Biol. Chem.* **254**:8201–8207.
68. Shichi, H., Lewis, M. S., Irreverre, F., and Stone, A. L., 1969, *J. Biol. Chem.* **244**:529–536.
69. Stubbs, G., Smith, H. G., and Litman, B. J., 1976, *Biochim. Biophys. Acta* **426**:46–56.
70. Michel-Villaz, M., Saibil, H. R., and Chabre, M., 1979, *Proc. Natl. Acad Sci. U.S.A.* **76**:4405–4408.
71. Engelman, D. M., and Zaccai, G., 1980, *Proc. Natl. Acad. Sci. U.S.A.* **77**:5894–5898.
72. Henderson, R., 1977, *Annu. Rev. Biophys. Bioeng.* **6**:87–109.
73. Thomas, D. D., and Streyer, L., 1982, *J. Mol. Biol.* **154**:145–157.
74. Chan, W. K., Nakanishi, K., Ebrey, T. G., and Honig, B., 1974, *J. Am. Chem. Sco.* **96**:3642–3644.
75. Drikos, G., Rüppel, Dietrich, H., and Sperling, W., 1981, *FEBS Lett.* **131**:23–27.
76. Honig, B., Dinur, U., Nakanishi, K., Balogh-Nair, V., Garvinowicz, M. A., Arnaboldi, M., and Motto, M. G., 1979, *J. Am. Chem. Soc.* **10**:7084–7086.
77. Knowles, A., and Dartnall, H. J. A., 1977, *The Eye*, Volume 2B (H. Dawson, ed.), Academic Press, New York.
78. Yoshizawa, T., 1972, *Handbook of Sensory Physiology* (H. J. A. Dartnall, ed.), Springer-Verlag, Heidelberg, pp. 149–179.
79. Sasaki, N., and Yoshizawa, T., 1981, *Photobiochem. Photobiophys.* **2**:365–371.
80. Sasaki, N., Tokunaga, F., and Yoshizawa, T., 1980, *Photochem. Photobiol.* **32**:433–441.
81. Sundstrom, V., Rentzepis, P. M., Peters, K., and Applebury, M. L., 1977, *Nature* **276**:645–646.
82. Tsuda, M., Tokunaga, F., Ebrey, T., Yue, K., Marque, T., and Eisenstein, L., 1980, *Nature* **287**:461–462.
83. Shichida, Y., Kobayashi, T., Ohtani, H., Yoshizawa, T., and Nagakura, S., 1978, *Phochem. Photobiol.* **27**:335–341.
84. Busch, G. E., Applebury, M. L., Lamola, A., and Rentzepis, P. M., 1972, *Proc. Natl. Acad. Sci. U.S.A.* **69**:2802–2806.
85. Birge, R. R., and Hubbard, L. M., 1980, *J. Am. Chem. Soc.* **102**:2195–2202.
86. Birge, R. R., 1981, *Annu. Rev. Biophys. Bioeng.* **10**:315–354.
87. Cooper, A., 1979, *Nature* **282**:531–533.
88. Liebman, P. A., Jagger, W. S., Kaplan, M. W., and Bargoot, F. G., 1974, *Nature* **251**:31–36.
89. Waggoner, A. S., and Stryer, L., 1971, *Biochemistry* **10**:3250–3254.
90. Yoshizawa, T., and Horiuchi, S., 1973, *Biochemistry and Physiology of Visual Pigments* (H. Langer, ed.), Springer-Verlag, Berlin, Heidelberg, New York, pp. 69–80.
91. Wald, G., Brown, P. K., and Smith, P. H., 1955, *J. Gen. Physiol.* **38**:623–681.
92. Wald, G., and Brown, P. K., 1956, *Nature* **177**:174–176.
93. Futterman, S., and Rollins, M. H., 1973, *J. Biol. Chem.* **248**:7773–7779.
94. Shichi, H., and Somers, R. L., 1974, *J. Biol. Chem.* **249**:6570–6577.
95. Rabinovitch, B., 1979, *Photochem. Photobiol.* **29**:567–574.
96. Dowling, J. E., 1960, *Nature* **188**:114–118.
97. Wiggert, B. O., Bergsma, D. R., and Chadler, G. J., 1976, *Exp. Eye Res.* **22**:411–418.
98. Saari, J. C., and Futterman, S., 1976, *Exp. Eye Res.* **22**:425–533.
99. Daemen, F. J. M., Rotmans, J. P., and Bonting, S. L., 1974, *Exp. Eye Res.* **18**:97–103.
100. Yoshikami, S., and Noell, G. N., 1978, *Science* **200**:1393–1395.

101. Cone, R. A., and Brown, P. K., 1969, *Nature* **221**:818–820.
102. Kropf, A., Whittenberger, B. P., Goff, S. P., and Waggoner, A. S., 1973, *Exp. Eye Res.* **17**:591–606.
103. Matsumoto, H., Asato, A. E., Denny, M., and Liu, R. S. H., 1979, *Photochem. Photobiol.* **29**:695–698.
104. Ebrey, T., Govindjee, R., Honig, B., Pollock, E., Chan, W., Crouch, R., Yudd, A., and Nakanishi, K., 1975, *Biochemistry* **14**:3933–3941.
105. Nakanishi, K., Yudd, A. P., Crouch, R. K., Olson, G. L., Cheung, H. C., Govindjee, R., Ebrey, T. G., and Patel, D. J., 1976, *J. Am. Chem. Soc.* **98**:236–238.
106. Crouch, R., Purvin, V., Nakanishi, K., and Ebrey, T., 1975, *Proc. Natl. Acad. Sci. U.S.A.* **72**:1538–1542.
107. Nakanishi K., Balogh-Nair, M. A., Gawinowicz, M. A., Arnaboldi, M., Motto, M., and Honig, B., 1979, *Photochem. Photobiol.* **29**:657–660.
108. Akita, H., Tanis, S. P., Adams, M., Balogh-Nair, V., and Nakanishi, K., 1980, *J. Am. Chem. Soc.* **102**:6370–6372.
109. Matsumoto, H., and Yoshizawa, T., 1975, *Nature* **258**:523–526.
110. Towner, P., Gaertner, W., Walckhoff, B., Oesterhelt, D., and Hopf, H., 1981, *Eur. J. Biochem.* **117**:353–359.
111. Fuortes, M. G. (ed.), 1972, *Handbook of Sensory Physiology*, Volume VII/2, Springer-Verlag, Berlin, Heidelberg, New York.
112. Brown, J. E., 1979, *Membrane Transduction Mechanisms*, Volume 33 (R. A. Cone and J. E. Dowling, eds.), Raven Press, New York, pp. 117–121.
113. Hubbell, W. L., and Bownds, M. D., 1979, *Annu. Rev. Neurosci.* **2**:17–34.
114. Szuts, E. Z., 1981, *Molecular Mechanisms of Photoreceptor Transduction* (W. H. Miller, ed.), Academic Press, New York, pp. 291–305.
115. Liebman, P. A., 1981, *Molecular Mechanisms of Photoreceptor Transduction* (W. H. Miller, ed.), Academic Press, New York, pp. 157–170.
116. Ebrey, T. G., Kilbride, P., Hurley, J. B., Calhoon, R., and Tsuda, M., 1981, *Molecular Mechanisms of Photoreceptor Transduction* (W. H. Miller, ed.), Academic Press, New York, pp. 121–132.
117. Govardovskii, V. I., and Berman, A. L., 1981, *Biophys. Struct. Mech.* **7**:125–130.
118. Miller, W. H., and Nicol, G. D., 1981, *Molecular Mechanisms of Photoreceptor Transduction* (W. H. Miller, ed.), Academic Press, New York, pp. 417–437.
119. Liebman, P. A., 1981, *New Directions in Ophthalmic Research* (M. L. Sears, ed.), Yale University Press, New Haven pp. 207–222.
120. Fukuda, Y., and Yoshizawa, T., 1981, *Biochim. Biophys. Acta* **675**:195–200.
121. Kühn, H., Mommertz, O., and Hargrave, P. A., 1982, *Biochim. Biophys. Acta* **679**:95–100.
122. Stryer, L., Hurley, J. B., and Fung, B. K.-K., 1981, *Molecular Mechanisms of Photoreceptor Transduction* (W. H. Miller, ed.), Academic Press, New York, pp. 93–108.
123. Shinozawa, T., and Bitensky, M. W., 1981, *Biochemistry* **20**:7068–7074.
124. Shichi, H., 1981, *Molecular Mechanisms of Photoreceptor Transduction* (W. H. Miller, ed.), Academic Press, New York, pp. 273–289.
125. Kühn, H., Bennett, N., Michel-Villaz, M., and Chabre, M., 1981, *Proc. Natl. Acad. Sci. U.S.A.* **78**:6873–6877.
126. Kühn, H., and Hargrave, P. A., 1981, *Biochemistry* **20**:2410–2417.
127. Dumler, I., and Etingof, R., 1976, *Biochim. Biophys. Acta* **429**:474–484.
128. Hurley, J. B., Barry, B., and Ebrey, T. G., 1981, *Biochim. Biophys. Acta* **675**:359–365.
129. Miki, N., Baraban, J., Keirns, J., Boyce, J., and Bitensky, M., 1975, *J. Biol. Chem.* **250**:6320–6327.
130. Baehr, W., Devlin, M., and Applebury, M., 1979, *J. Biol. Chem.* **254**:11669–11677.
131. Liebman, P. A., and Pugh, E. N., 1979, *Vision Res.* **19**:375–380.
132. Kühn, H., and Dreyer, W. J., 1972, *FEBS Lett.* **20**:1–6.
133. Bownds, D., Dawes, J., Miller, J., and Stahlman, M., 1972, *Nature [New Biol.]* **237**:125–127.
134. Frank, R. N., Cavanagh, H. D., and Kenyon, K. R., 1973, *J. Biol. Chem.* **248**:596–609.
135. Shichi, H., and Somers, R. L., 1978, *J. Biol. Chem.* **253**:7040–7046.

136. Kupfermann, I., 1980, *Annu. Rev. Physiol.* **42**:629–641.
137. Palmer, G. N., 1981, *Life Sci.* **28**:2785–2798.
138. Cavaggioni, A., and Sorbi, R. T., 1981, *Proc. Natl. Acad. Sci. U.S.A.* **78**:3964–3968.
139. Stieve, H., and Bruns, M., 1978, *Z. Naturforsch.* [*Biosci.*] **33C**:574–579.
140. Goldberg, N. D., and Haddox, M. K., 1977, *Annu. Rev. Biochem.* **46**:823–896.
141. Farber, D. B., Brown, B. M., and Lolley, R. N., 1979, *Biochemistry* **18**:370–378.
142. Polans, A. S., Hermolin, J., and Bownds, M. D., 1979, *J. Gen. Physiol.* **74**:595–613.
143. Cohen, A. I., 1981, *Molecular Mechanisms of Photoreceptor Transduction* (W. H. Miller, ed.), Academic Press, New York, pp. 215–229.
144. Fleischman, D., and Denisevich, M., 1979, *Biochemistry* **18**:5060–5066.
145. Kohnken, R. E., Chafouleas, J. G., Eadie, D. M., Means, A. R., and McConnell, E. G., 1981, *J. Biol. Chem.* **256**:12517–12522.
146. Brown, J. E., and Waloga, G., 1981, *Molecular Mechanisms of Photoreceptor Transduction* (W. E. Miller, ed.), Academic Press, New York, pp. 369–380.
147. Fein, A., and Corson, D. W., 1981, *Science* **212**:555–557.
148. Clack, J. W., Oakley, B., and Pepperberg, D. R., 1982, *Proc. Natl. Acad. Sci. U.S.A.* **79**:2690–2694.
149. Rodbell, M., 1980, *Nature* **284**:17–22.
150. Watling, K. J., 1981, *Trends Pharmacol. Sci* **2**:244–247.
151. Neal, M., 1976, *Gen. Pharmacol.* **7**:321–333.
152. Bonting, S. L. (ed.), *Transmitters in the Visual Process*, Pergamon Press, Oxford, 1976.
153. Graham, L. T., 1974, *The Eye* (H. Davson and L. T. Graham, ed.), Academic Press, New York, London, pp. 283–342.
154. Starr, M. S., 1977, *Essays in Neurochemistry and Neuropharmacology* (M. B. Youdin, W. Lovenberg, D. F. Sharman, and J. R. Lagnado, ed.), John Wiley & Sons, New York, pp. 152–174.
155. Dowling, J. E., 1970, *Invest. Ophthalmol.* **9**:655–680.
156. Wu, S. M., and Dowling, J. E., 1978, *Proc. Natl. Acad. Sci. U.S.A.* **75**:5205–5209.
157. Gerschenfeld, H. M., and Piccolino, M., 1977, *Nature* **268**:257–259.
158. Pasantes-Morales, H., Urban, P. E., Klethi, J., and Mandel, P., 1973, *Brain Res.* **51**:375–378.
159. Lam, D. M. K., 1976, *Cold Spring Harbor Symp. Quant. Biol.* **40**:571–579.
160. Lam, D. M. K., Su, Y. Y. T., Chin, C. A., Brandon, C., Wu, J.-Y., Marc, R. E., and Lasater, E. M., 1980, *Brain Res. Bull.* **5**(Suppl. 2):137–140.
161. Ehinger, B., 1976, *Transmitters in the Visual Process* (S. L. Bonting, ed.), Pergamon Press, Oxford, pp. 145–163.
162. Krammer, S. G., 1976, *Transmitters in the Visual Process* (S. L. Bonting, ed.), Pergamon Press, Oxford, pp. 165–198.
163. Buskirk, R., and Dowling, J. E., 1981, *Proc. Natl. Acad. Sci. U.S.A.* **78**:7825–7829.
164. Ehinger, B., and Florén, I., 1980, *Neurochemistry of the Retina* (N. G. Bazan and R. N. Lolley, ed.), Pergamon Press, Oxford, pp. 209–229.
165. Lam, D. M. K., Mark, R. E., Sarthy, P. V., Chin, C. A., Su, Y. Y. T., Brandon, C., and Wu, J.-Y., 1980, *Neurochemistry of the Retina* (N. G. Bazan and R. N. Lolley, ed.), Pergamon Press, Oxford, pp. 183–190.
166. Voaden, M. J., 1976, *Transmitters in the Visual Process* (S. L. Bonting, ed.), Pergamon Press, Oxford, pp. 107–125.
167. Stell, W. K., Marshak, D., Tadataka, Y., Brecha, N., and Karten, H., 1981, *Trends Neurosci.* **3**:292–295.
168. Ames, A., and Newbett, F. B., 1981, *J. Neurochem.* **37**:867–877.

23

Cell Cultures

L. Hertz, B. H. J. Juurlink, and S. Szuchet

1. INTRODUCTION

This chapter focuses on one specific development in the field of cell and tissue culture that has come to fruition during the last 10–15 years, i.e., the establishment of cultures of individual cell types from suspensions of dissociated cells. These cultures, which can be prepared so that they contain little or no contamination with other cell types, are in the following called monotypic cultures.* They have become of great importance in cellular neurobiology, as is illustrated by some examples. Three cell types, the astrocyte, the neuron, and the oligodendrocyte are discussed separately (Sections 3–5). Before this discussion, however, in this and the following section, we try to sketch a somewhat broader picture by defining what cell and tissue culture is (and what it is not) and by discussing commonly used principles and procedures underlying the establishment of cultures that are highly enriched in one specific cell type.

Two approaches have been made to study cells and tissues from the nervous system *in vitro*. One of these is the conventional use of brain slices, cell homogenates, isolated cells, or subcellular structures (e.g., synaptosomes), preparations that are familiar to virtually all neurochemists. After their isolation, such preparations are often maintained over a span of hours but not days. These methods have yielded a wealth of information (see chapter by McIlwain in this *Handbook*), but they are not cell or tissue culture methods. Cell and tissue culture, which is the other *in vitro* technique, requires by definition that the tissues or cells be kept *in vitro* for at least 24 h,[1,2] which enables them to recover from the dissection trauma and to start a life of their own. Often, the cells or tissues are kept much longer, i.e., weeks, months, or even years.

---

* This terminology does not necessarily imply that all monotypic cultures show characteristics of only one cell type, since certain cell lines display both neuronal and glial features.

L. Hertz • Department of Pharmacology, University of Saskatchewan, Saskatoon, Saskatchewan S7N 0W0, Canada.   *B. H. J. Juurlink* • Department of Anatomy, University of Saskatchewan, Saskatoon, Saskatchewan S7N 0W0, Canada.   *S. Szuchet* • Department of Neurology, University of Chicago, Chicago, Illinois 60637.

The purposes of using cell or tissue culture are manifold. This methodology was introduced more than three-quarters of a century ago when Harrison[3] observed that nerve cells in explants of isolated tadpole neural tissue were able to extend processes during incubation *in vitro* and thus conclusively demonstrated that the nerve fiber develops as an extension from the neuron, a question that had not been completely resolved at that time. Harrison, moreover, suggested that such a method could be applied to "the study of the influences which act upon the growing nerve," a vision that has, indeed, come true. Tissue culture techniques have been essential not only to demonstrate the existence of trophic factors and their role in cellular interactions but also to elucidate the possible mechanism of action for at least one of these factors. Another reason that it is of major importance to recognize the effects of external factors in tissue and cell culture work is the obvious one that a possible lack of these factors under ordinary culturing conditions conceivably might have a detrimental effect on the cultured cells or tissues. With ordinary cell and tissue culture techniques, including the use of serum (or specific hormones and other nutritional factors) in the medium, this seems in many cases not to be a serious impediment, but the functional normality of any cells grown in cultures has to be ascertained as well as possible.

For a long time, the predominant culture type used in the neurosciences was explant cultures, i.e., cultures that have been obtained by transferring small pieces of tissue from the animal to the tissue culture dish. A remarkable amount of such work was carried out, mainly in Europe, during the first decades of this century (for a review see Von Mihalik[4]). In North America, tissue culture studies of the nervous system were pioneered by Margaret R. Murray, who in 1965 comprehensively reviewed this area and carried its history up to date.[5] She also contributed a chapter to the first edition of this *Handbook*.[6] These two reviews have maintained their timeliness, and the only thing we wish to reiterate regarding explant cultures is that explants under appropriate conditions can develop biochemically in close parallelism with the neurochemical development in the living animal.[7]

2. PRINCIPLES USED TO OBTAIN MONOTYPIC CULTURES

2.1. Cell Lines

Two basic approaches have been used to obtain monotypic cell cultures: (1) *cell lines* have been obtained from individual cells or cultures of one cell type (often from tumors) and propagated, or (2) conditions for the preparation of *primary cultures* (i.e., cultures obtained directly from the animal) have been chosen to select for a single cell type. Cell lines are, by definition, cultures that by subculturing (passage from dish to dish) have been serially transplanted in an uninterrupted sequence of generations,[1,2] and they can often be maintained for a very long time. They can be diploid cell lines,[1,2] which consist of diploid nontumorigenic cells. Cells subcultured (passaged) from a primary culture fall in this category. Diploid cell lines have a limited life-span, as is the case for normal human glial cells, which start to senesce after 15–30 population doublings.[8,9] Diploid cell lines have only been used to a limited extent in neurochemical work (e.g., refs.10,11).

The so-called established cell lines[1,2] consist of transformed cells that have been obtained either from a spontaneous, a chemically induced, or a virally induced tumor or from cultures of originally normal cells that have transformed spontaneously (e.g., the NN glial cell line) or as a result of exposure to viral or chemical carcinogens. These cells have been widely used. They are generally nondiploid and often, but not always, tumorigenic.[2] Neurochemically, they are in general considerably less differentiated than cells in primary cultures. Established cell lines can apparently be subcultured forever, and the mere ability of a cell line to be subcultured without limits does indicate that the cell line has transformed, if for no known reason then spontaneously. Such a spontaneous transformation occurs readily in rodent cells but seldom if at all in cells from chicks or humans.[9] Clones, i.e., populations derived from one single cell by mitosis[1,2] using different techniques,[12] can in principle be obtained from primary cultures, diploid cell lines, or established cell lines. Often, the term clonal cell line is used for an established cell line that has been cloned.

The apparent advantage of cell lines is that one cell or group of cells can proliferate and be subcultured through many generations such that large numbers of presumably identical cells can be used for investigation. Several qualifications, have, however, to be made. The first and perhaps most important is that the price for the apparent immortality of established cell lines is alterations in the genome, which are correlated with a loss or quantitative diminution of at least some of the differentiated functions of the cells. Often, as a result of its regression to a less differentiated, more pluripotent state, the cell may in addition exhibit features that are not characteristic for the cell type in question *in vivo*. Thus, the glial-specific protein GFA has been demonstrated in a presumably neuronal cell line,[13] and the neuronal specific enolase has been found in a glioma cell line.[14] In addition, even cloned established cell lines do not have a stable phenotype and may exhibit spontaneous changes of morphological and/or functional characteristics over time.[2,15] The only way to at least partly correct for this phenomenon is for the investigator to maintain a supply of early generations (early passage) in a frozen state. This is done by some but not all investigators. Another risk of long-time culturing is the possibility for cross contamination with other cells utilized in the laboratory.

Compounding these difficulties is the fact that the morphological and functional characteristics of established cell lines generally change immediately after subculturing and then recover gradually.[16,17] This is probably a reflection of the role cell-to-cell interactions play in phenotypic expression. This phenomenon is of considerable biological interest, and the neurochemist who is interested in using certain cell lines as models for their *in vivo* counterparts will obviously have to select cultures at a certain stage of confluency. This phenomenon occurs not only in established cell lines but also in diploid cell lines, and an example of this (Fig. 1) is the pronounced and long-lasting reduction of the S-100 protein after subculturing of primary cultures of astrocytes, particularly at low cell densities that was observed by Labourdette and Mandel.[18]

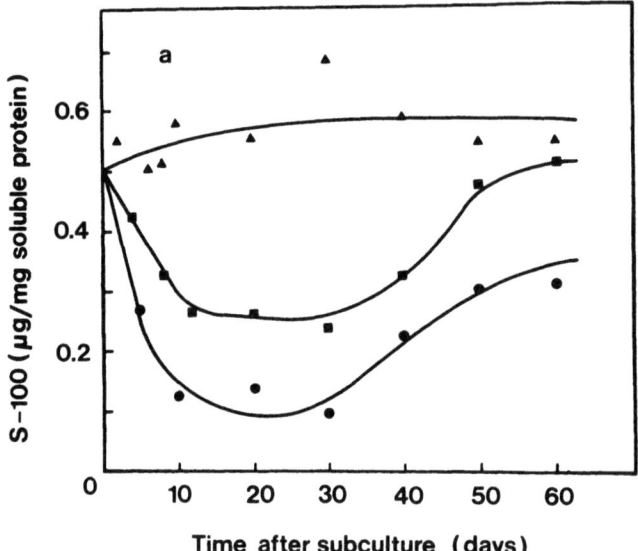

Fig. 1. Content of S-100 in cultures obtained by subculturing primary cultures of rat astrocytes, shown as a function of time after subculturing with the cells seeded at one-half (▲), one-fifth (■) or one-tenth (●) of their original density. (From ref. 18, with permission.)

2.2. Primary Cultures

2.2.1. General Characteristics

Cells taken directly from the organism (and kept *in vitro* for more than 24 h) are considered to be primary cultures.[1,2] Primary cultures of individual cell types are currently widely used in neurochemical investigations and have apparent advantages over established cell lines by being normal diploid cells [established for primary cultures of mouse astrocytes by B. H. J. Juurlink, L. Hertz, and H. C. Wang (unpublished experiments) and for human fetal neural cells by Icard *et al.*[10]] and by more closely reflecting the metabolism and functions of the cells found *in vivo*.[19–21] In general, they are obtained after an initial mechanical and/or enzymatic dissociation of the tissue. Different procedures have been used, often in combination, to enable the establishment of monotypic cultures from such mixed cell suspensions. These are discussed below.

2.2.2. Ontogenetic Considerations of Importance for the Establishment of Monotypic Cultures

In general, to establish primary neuronal or astrocytic cultures from the mammalian CNS, the starting material must be obtained from immature tissue at a developmental stage favoring culturing of the cell type in question, i.e., still proliferating and/or early postmitotic cells. An understanding of the strategies used to obtain monotypic neural cell cultures will therefore be facilitated

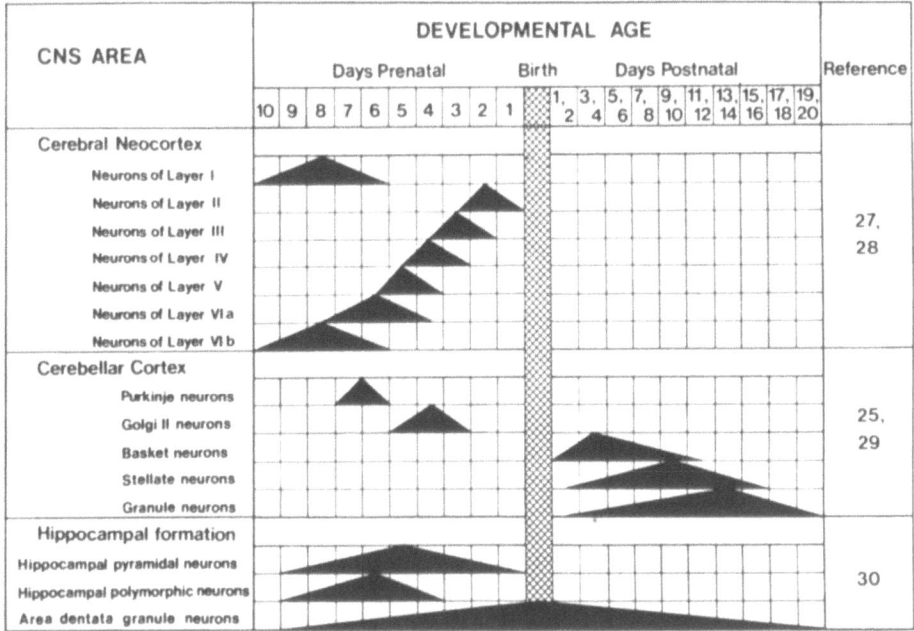

Fig. 2. Graph depicting the onset, peak, and completion of neuronal production (solid triangles) in three brain regions. Except for the hippocampal formation, which represents mouse data, all data were obtained from studies on the rat. Since the mouse and rat have the same rate of development[26] and are at comparable stages at birth, the information is applicable to both species. Note that the postnatal axis is on a compressed scale compared to the prenatal axis. For further details see ref. 26 or the references cited in the figure.

by a brief consideration of some aspects of nervous system development. More detail can be obtained from an excellent review by Jacobson.[22]

The vertebrate central nervous system can first be recognized as a thickened placode of cells referred to as the neural plate, which soon rounds up to form the neural tube. The cells composing the neural tube are pseudostratified neuroepithelial cells known as ventricular cells. These cells ultimately give rise to the various neuronal and macroglial (astrocytic and oligodendroglial) cell populations of the mature CNS. In general, if one examines the development of any given region of the CNS, there appears to be an overall temporal sequence of specific cell production. On the whole, the large long-axoned (Golgi type I) neurons are the earliest formed cells, followed by the major portion of the smaller shorter-axoned (Golgi type II) interneurons (Fig. 2). In the rat or mouse, neuronogenesis is essentially completed at birth with the exception of the formation of the interneurons of the olfactory bulbs,[23] the dentate gyrus,[24] and the cerebellar cortex.[25] The major period of gliogenesis occurs after the main period of neuronogenesis, and the majority of astrocytes are formed before the main period of oligodendrogliogenesis.

This overall temporal pattern of cell production appears to be true of all CNS regions studied, but the exact time periods involved can vary somewhat from one CNS region to another within any given species (Fig. 2). Similarly,

Table I
Comparable Stages of Development

Mouse[a]	Rat[a]	Chick[b]
9.0	10.5	1.2
9.5	11.0	1.5
10.0	11.5	1.8
10.5	12.0	2.2
11.0	12.5	2.5
11.5	13.0	2.8
12.0	13.5	3.3
12.5	14.0	4.0
13.0	14.5	4.8
13.5	15.0	5.5
14.0	15.5	6.3
14.5	16.0	7.0
15.0	16.5	8.0
15.5	17.0	9.0
16.0	17.5	10.0

[a] Day after copulation considered as day 1 of gestation.
[b] Number of days incubated.

although the temporal patterns in principle are identical in different species, the time period that embryos are at a comparable stage can vary considerably from species to species.[26] The ontogenetic development in mice and rats is relatively similar with the exception that in the mouse comparable developmental stages occur 1.5 days earlier than in the rat. Both are, however, quite different from the chick (Table I).

A consideration of the development of the rat cerebral neopallium* (Fig. 2) will illustrate some of the events and time intervals involved in CNS development. In this tissue, the Golgi I neurons of layers I and VIB are generated mainly from the 12th to 16th days of gestation (10–6 days prenatally), whereas the neurons from layers VIA to II are generated in an inside-out temporal sequence from about the 14th day of gestation to birth. Synapses can be identified in layers I and VIB as early as on the 16th to 17th days of gestation[28] and begin to develop in the intervening layers by the third day post-natum. Although synaptogenesis is initiated during the fetal stages of development, the main period of synaptogenesis and, hence, the development of the dendritic tree and neuronal maturation occur during the first 3 weeks post-natum.[31] A few astroglial cells appear as early as the 13th day of gestation within the rat neopallium,[32] but the majority of astroglial cells are generated during the late fetal stages and the first week after birth, with most being formed by the end of the third week post-natum.[33,34] Hence, the generation of astroglial cells occurs mainly concomitant with the time period of neuronal maturation, although astrocytic proliferation continues into adulthood.[35,36] Both morphological studies[37] and myelin basic protein immunocytochemistry or radioim-

---

* The neopallium comprises the major part of the cerebral cortex plus underlying white matter.

munoassay[38,39] have shown that oligodendroglial cells are absent in the neonatal rat cerebral cortex. It seems, however, that oligodendroglial cell precursors are proliferating before this stage. By the end of the first week post-natum, oligodendroglial cells can be recognized in cerebral structures such as the indusium griseum,[40] and relatively large numbers are present within the corpus callosum and cortex by 3 weeks post-natum. The major period of formation and maturation of cerebral oligodendroglial cells in the rat appears to be the time period between 3 weeks and 3 months post-natum.[33,41]

It is in accordance with this developmental pattern that preparations established from dissociated fetal CNS tissue result in mixed neuronal–glial cultures. Such cultures have been established from rat cerebral hemispheres as early as the 11th day of gestation,[42] and the production of astrocytes, ependymal cells, and oligodendrocytes in these cultures occurs on a similar time schedule as *in vivo*. Cultures from comparable (Table I) developmental stages (36–48 h of incubation) have also been established using chick cerebral hemispheres,[43] and if cell seeding density is sufficiently high to allow aggregation and intimate contact of neuronal precursor cells, neuronogenesis can occur in culture.[44,45]

The early attempts to culture dispersed dissociated neurons demonstrated that postmitotic neurons obtained from immature CNS survived the dissociation and culture procedures better than neurons obtained from more mature CNS.[46] This has been examined in some detail by Banker and Cowan,[47] who demonstrated that rat hippocampal neurons that were 1–2 days postmitotic survived the culturing procedures better than older neurons. One reason for this seems readily apparent in that maturation of the neurons leads to the elaboration of axonal and dendritic processes so that older neurons are much more susceptible to damage during dissociation. A consideration of the time periods of neuronogenesis in the rat would suggest that the optimal times in development for obtaining cerebral cortical neuronal cultures would be from the 15th to 20th day of gestation and for cerebellar interneurons (basket, stellate, or granule cells) from the second to the 15th day post-natum (Fig. 2). Unless astrocytic growth is curtailed, both of these cultures will have a mixed neuronal–glial composition.

In the peripheral nervous system (PNS) as well, neuronogenesis is completed before the main period of gliogenesis.[48] An apparently clear-cut difference from CNS neurons is, however, that peripheral neurons can be obtained in dissociated cultures from adult mouse dorsal root ganglia[49] or from adult human trigeminal and superior cervical ganglia.[50] This may reflect the less complicated morphological structure of these neurons and consequent less severe injury during the cell separation.

The strategies for the establishment of astroglial cultures have recently been reviewed.[51] In short, it is based on (1) selecting the time in development when there is a large population of proliferating astroblasts and when neuronogenesis is not only essentially completed but the neurons are also at least a couple of days postmitotic; (2) using rather vigorous cell dissociation procedures, which tend to destroy postmitotic neurons; and (3) choosing culture conditions that allow the preferential attachment and proliferation of astroglial cells. These three principles permit astrocytic cultures of 95% purity to be obtained (Section 3.1.1). For rodent cerebral neopallium, the optimal time in development for obtaining monotypic astroglial cultures is at birth. In chick, the optimal time period in development for obtaining cerebral astroglial cultures has been determined to be the 14th or 15th day of incubation (i.e., at a comparable stage of development as in the mouse or rat, as estimated from an extrapolation of Table I). At this stage, the neurons, which are already several days postmitotic, will not attach, or if they do attach they will rapidly degenerate.[46] It is disputed whether or not a significant number of oligodendrocyte precursors attach and proliferate, but under conditions ordinarily used for production of astrocytic cultures from mouse or rat brains, the number of oligodendrocytes is insignificant.[39,51,52]

It has, up to now, not been possible to prepare primary cultures highly enriched in oligodendroglial cells by simply selecting a cell suspension obtained from CNS tissue at a time when large numbers of specifically oligodendroglial cells are being formed. This is perhaps a reflection of the late ontogenetic stage at which oligodendrocytes are formed and the large amount of cellular debris that will be present as a result of the disruption of mature neurons and may interfere with cellular attachment to the substratum. Recently, however, a number of laboratories have observed that planting dissociated cells from perinatal rodent cerebral or cerebellar hemispheres in high density in culture results in preparations that contain both astrocytes and varying numbers of cells that exhibit oligodendroglial characteristics.[53-59] This is consistent with the presence of oligodendroglial cell precursors at this age. In such cultures, the putative oligodendroglial cells become situated on top of an astroglial monolayer. McCarthy and De Vellis[55] have taken advantage of this cellular stratification by shaking off the loosely attached oligodendroglial cells from the firmly attached astroglial cells and subculturing the highly enriched (95%) oligodendroglial cells. Although the latter cultures are generally regarded as primary cultures, this is not strictly correct since they, by definition, represent the first stage of a diploid cell line and can also be characterized as secondary cultures. However, provided cell density is kept reasonably high, the transfer of the cells may have little functional consequence (see Fig. 1).

2.2.3. Cell Separation Based on Physical Parameters

As an alternative approach to obtain monotypic cultures, brain cell suspensions have been separated into the constituent cell types by gradient centrifugation techniques. Relatively homogeneous populations of neurons, astrocytes, and oligodendrocytes can be obtained in this way.[60] This method has become of great importance for isolation of oligodendrocytic cells, which subsequently have been cultured. Cerebellar astrocytes and interneurons have also been cultured after separation by gradient centrifugation, but this has few if any advantages over the preparation of cultures directly from a cell suspension.[61,62] Cell separation by gradient centrifugation is based on the fact that different cell types have different densities, which are correlated, e.g., with cell size and the ratio between nucleus and cytoplasm.[60] Because of the differences in cell densities among different cell types, specific cell types will be located at different levels of a density gradient. Such a gradient can be established in many different ways.[60] Originally, sucrose was used, but it was an improvement when Rose[63] replaced sucrose with the higher-molecular-weight and thus osmotically less active Ficoll. The next significant move was the introduction of enzymes to loosen tissue connectivity prior to mechanical disruption. Trypsin was used by Norton and Poduslo[64] and was applied to the separation of oligodendrocytes by Poduslo and Norton.[65] As a result of these and other advances in cell separation techniques (reviewed in refs. 60,63,65–69), there has been an outburst of procedures for isolation of oligodendrocytes over the last few years, all yielding viable cells capable of surviving in culture for months.[70–74]

2.2.4. Cell Separation Based on Affinity Methods

Affinity methods are of considerable potential value to separate not only different cell types but also different cell classes, e.g., neurons carrying different receptors. Once the individual cells have reacted with a ligand that selectively recognizes a surface constituent, the subsequent problem is the separation of these cells from others without affinity for the ligand in question.[75] A recent elegant approach to obtain such a cell separation is the use of automatic cell sorters that recognize specific antigens on the cell surface because they have been brought to react with fluorescent or magnetic antibody complexes.[60] Such a method has apparently been successfully employed to isolate oligodendrocytes for subsequent cultures (Section 5.2.2).

A much more commonly used example of cell separation by affinity methods is differential attachment, obtained by coating of the tissue culture plastics with materials to which certain cell types adhere much more strongly than others. The classical example is polylysine, a polycation that interacts electrostatically with the negative charges of cell membranes.[76] This leads to an enhanced attachment of neurons (Section 4.2.1b) and oligodendrocytes (Section 5.3.2) but not of astrocytes. This compound is therefore widely used to prepare both oligodendrocytic and neuronal enriched cultures.[77–80] Originally, poly-L-lysine was used, but currently most investigators use the nonmetabolizable poly-D-lysine. Since neuronal cultures are prepared from very immature animals, and oligodendrocytic cultures from an isolated cell population, there will be little or no risk of cross contamination between the two cell types.

Differential attachment has been used very effectively to separate the different cell types of the PNS, which consists of essentially three cell types: neurons, glial cells (Schwann cells), and fibroblasts. One of the more common procedures used to obtain relatively pure peripheral neuronal cultures takes advantage of a more rapid attachment of nonneuronal cells than of neuronal cells[81–84] to collagen-coated Petri dishes. The former are, however, not homotypic, since they consist of both Schwann cells and fibroblasts, and in order to obtain homotypic Schwann cell cultures, the fibroblasts must be killed selectively (Section 2.2.5).

2.2.5. Differential Cytotoxicity

A commonly used example of cell separation by differential cytotoxicity is the sensitivity of dividing but not of nondividing cells to drugs interfering with DNA synthesis because they are either DNA base analogues or folic acid antagonists. Many of these drugs, e.g., fluorodeoxyuridine (FdU) and aminopterin, also affect RNA synthesis and may therefore also be toxic to nondividing cells. Cytosine arabinoside inhibits DNA synthesis rather specifically and may be a better drug to kill dividing cells with little or no effect on nondividing cells.[85] Both cytosine arabinoside and FdU have been used to curtail astrocytic growth in neuronal cultures from cerebral hemispheres (Section 4.2.1), which, in spite of the use of polylysine-coated dishes, generally cannot be prepared completely without astrocytes. It is important that these drugs be used at an early stage, i.e., while the astrocytes are still rapidly dividing.

A number of very elegant procedures using cytotoxic drugs have been developed to obtain highly purified Schwann cell cultures. These procedures generally take advantage of the high proliferative activity of fibroblasts and mitotic quiescence of Schwann cells in the absence of axons. Wood and Bunge[86] and Wood[87] explanted undissociated rat dorsal root ganglia into culture and eliminated the fibroblasts with two pulses of cytosine arabinoside. During subsequent culturing in the absence of the cytotoxic agent, there was axonal outgrowth and Schwann cell proliferation, yielding cultures that were composed of a central neuronal aggregate and a peripheral region consisting of a carpet of Schwann cells. Finally, excising the central neuronal aggregate left a culture of highly purified Schwann cells that could be serially passaged. A similar approach was used to prepare homotypic Schwann cell cultures from dissociated peripheral nerves.[88] Alternatively, the fibroblasts can be eliminated by complement-dependent lysis using antibodies against the Thy-1 antigen, which is present on the fibroblast plasmalemma but not on the Schwann cell plasmalemma.[88]

2.2.6. Use of Defined Media to Obtain Monotypic Cultures

In view of the biochemical differences that exist between different cell types in the CNS or the PNS, it may be possible to induce selective survival or destruction of individual cell types by the use of specific hormone-supplemented but serum-free media. This principle has still not been used to a large extent, but Bottenstein et al. have been able to induce selective survival of neurons in chick embryo sensory ganglionic cultures by use of a defined serum-free medium,[89] and Yavin and Yavin have used an even simpler medium to grow chick cerebral neurons.[90] Also, a medium suitable for culturing of a glioma cell line leads to astrocytic growth in primary cultures from the cerebral hemispheres of newborn mice,[91] whereas a medium suitable for a neuroblastoma cell line promotes selective survival of neurons.[91,92] Along similar lines, withdrawal of serum (from a defined medium suitable for Schwann cells) has been used to deplete fibroblasts from Schwann cell cultures, and omission of nerve growth factor has been employed to prevent the survival of peripheral neurons.[93]

2.2.7. Cell Identification

Unequivocal identification of cultured cells, including possible contaminating cell types, is a long-recognized problem in cell culture. Recently, the recognition of cell-specific antigens[13,94–97] and the availability of antisera directed against them has facilitated the task of identifying cultured cells. It should, however, be recognized that certain antigens are only transiently present (e.g., M1 in astrocytes[97]) and that others (e.g., GFA, neuronal enolase) become abundant only at a relatively late developmental stage.[98,99] Examples of the use of antigenic markers in the identification of cultured cells and in the establishment of cell contamination in astrocytic, neuronal, and oligodendrocytic cultures are given in the following.

3. ASTROCYTES

3.1. Primary Cultures

3.1.1. Procedure

The monotypic primary culture that is easiest to understand and probably also simplest to prepare is that of rodent or chick astrocytes. Shein[100] and Varon and Raiborn[101] had already successfully obtained cultures highly enriched in astrocytes when Booher and Sensenbrenner,[46] by varying the age of the brain from which the cultures were prepared, were the first to utilize the developmental characteristics of the different cell types (Section 2.2.2) to achieve an easy separation between cultures consisting almost exclusively of astrocytes (or astroblasts) and others highly enriched in neurons.

During the following decade, the establishment of astrocytic cultures has become a routine procedure in a large number of laboratories (for references see 51). Some differences are found in the ways in which they are prepared, e.g., with respect to the exact age of the animal used to produce the cultures (where Cummins and Glover[102] use 3- to 4-day-old rats and most other investigators use newborn rats or mice), the methods employed during the preparation of the cultures (e.g., use of enzymes; sieving of the cell suspension; vortexing), or the type of medium (and serum) used. Many of these differences are minor and probably of little importance. However, others are apparently not, since a considerable variation exists between different laboratories regarding the degree of contamination with other cells and even regarding the character of contaminating cells (Table II). Although the concordance between different laboratories generally is reasonably good with respect to biochemical and biophysical characteristics of the cultures, there are also some pronounced differences, most notably with respect to the intensity of potassium uptake, where Hertz and co-workers[103–105] and Kimelberg et al.[106,107] have reached

Table II
Cellular Composition of Primary Rodent Astroglial Cultures

Culture age (days)	GFA[a]	Glu-S[b]	GalCer[c]	MBP[d]	Neurons[e]	FBN[f]	Macrophages[g]	References
3 weeks[h]	95	95	---	---	0	<1	3	51
3 weeks[h]	60–70	---	---	0	0	---	10	39, 132, 135
3–5 days	80	---	1	---	8	---	3	94
3 weeks	79	---	<1	---	---	---	---	131
3 weeks	74	---	---	---	---	24	---	129
2 weeks	95	---	---	---	---	---	---	53
2–3 weeks[h]	90–95%	---	---	---	---	---	---	109
3 weeks	96%	---	---	---	---	---	---	128

[a] Glial fibrillary acidic protein, an astrocyte marker.
[b] Glutamine synthetase, an astrocyte marker.
[c] Galactocerebroside, an oligodendrocyte marker.
[d] Myelin basic protein, an oligodendrocyte marker.
[e] Morphological criteria,[82] neuron-specific enolase,[132] tetanus toxin binding.[94]
[f] Fibronectin marker of nonneural tissue, including fibroblasts, meningeal, and endothelial cells.
[g] Based on phagocytic properties[51,94] or macrophage acidic protein.[135]
[h] Grown in the presence of dBcAMP.

very different conclusions. We feel that it is beyond the scope of the present review to discuss these differences in any detail but since we are aware of no method yielding more homogeneous (Fig. 3; Table II) or better differentiated cultures of astrocytes than the one we have adapted,[51] largely using the general procedure described by Booher and Sensenbrenner,[46] we describe this method in some detail. Some of the other variants of the method may, however, yield equally good results.

The procedures used in the preparation of the cultures aim at enhancing the preferential survival and proliferation of astroglial cell precursors by (1) selecting the appropriate starting material, (2) selecting procedures for enriching the starting material in astrocyte precursor cells, and (3) using culture conditions that allow preferential survival and proliferation of astrocyte precursor cells.

An appropriate source of tissue is the neopallium of the cerebral hemispheres of newborn mice, i.e., the portion dorsal and lateral to the lateral ventricles. This preparation consists of the subventricular zone, containing the actively proliferating astrocytic precursor cells, and the overlying cortex with some still-dividing astroblasts,[108] whereas dividing neuroblasts are no longer present. In contrast to most other investigators, we remove not only the olfactory bulbs but also the basal ganglia and the hippocampal formations in order not to include neurons from these structures, which mature later than most other neurons (Section 2.2.2) and therefore might be able to survive in the cultures, although this may not be a major point. After removal of the meninges, the tissue is cut with a scalpel into 1-mm cubes. No trypsinization is employed with our procedure, although some other investigators do use trypsin (e.g., ref. 53). Thereafter, the cells are dissociated in one way or another. We use vortexing, which results in the destruction of a large proportion (about 90%) of

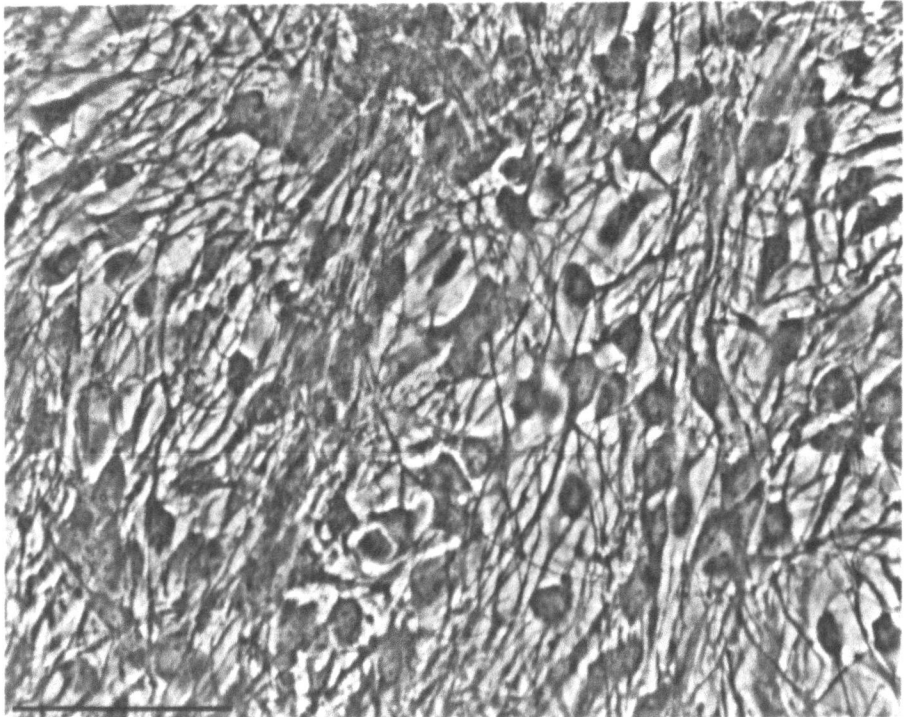

Fig. 3. Phase-contrast micrograph of living 4-week-old primary culture of mouse astrocytes prepared as described in ref. 51. Bar represents 100 μm. Note the presence of a few large flat cells, i.e., immature astrocytes.

the ramified neurons with a selective survival of the small and more immature proliferative cells (mainly astroblasts).

Subsequently, the cell suspension must be filtered. We use two sievings through nylon mesh (Nitex®), the first with a pore size of 80 μm and the next with a pore size of 10 μm. This procedure filters out blood vessels and cellular aggregates, leaving a cell suspension enriched in astrocyte precursor cells, which is seeded in uncoated plastic tissue culture dishes or flasks (Falcon or Nunc; some other tissue culture plastics have not functioned as well), which do not favor neuronal attachment. Only relatively few cells in the suspension attach,[108,109] but it is unknown whether this indicates that many astrocytes do not attach or that most cells in the suspension are nonastrocytic. However, the attached astrocytes proliferate vividly, and after 8–14 days the cultures are confluent. Provided the cell separation procedure was appropriately carried out, three or four 60-mm-dish cultures can be obtained from one mouse brain. If whole brain hemispheres are used, the number is even larger.

Cultures from rats are prepared in essentially the same manner, since the developmental patterns in mice and in rats are so similar (Table I). It can, however, not be assumed *a priori* that cultures from other animals or other brain regions, developing at a different pace, can be prepared from tissue at the same age. Cultures of chick astrocytes are generally set up from 15-day-old embryos (Section 2.2.2). The rule that immature brain has to be used as the source for astrocytic cultures is not absolute, since cultures of glial-like cells have been obtained from autopsy material by Pontén and co-workers,[8,9] and since cultures of astrocytes, verified by their GFA content (Section 3.1.2), can be obtained from adult rats after intracerebral injection of kainic acid[110] or certain surgical procedures.[111] This may well reflect the continued gliogenesis *in vivo* into adulthood (Section 2.2.2).

The growth medium we use is a modified Eagle's MEM[51] with either horse or fetal bovine serum, which is aerated with a 5% CO_2/95% air atmosphere (90% humidity). The modifications of the original MEM include doubling of the concentrations of all amino acids except glutamine (2 mM) and quadrupling of the concentrations of all vitamins. This medium is not commercially available but is very similar to Dulbecco's medium; its use in our laboratory is more by tradition than by trial and error, although we have observed that lowering or raising the glutamine concentration may affect the cultures adversely[112] and that all essential amino acids, although maybe not cystine,[113] are present in sufficient or more than sufficient amounts. Some other authors have used different media, e.g., Eagle's basal medium. The differences in composition between the two media are not pronounced, but they do have a distinct effect on the morphological differentiation of the cells.[114] The serum we use is almost always from adult horses. This gives the advantage that the same two or three horses can be used continuously, which virtually abolishes difficulties arising from the well-known batch-to-batch variation in commercially available sera.[115] The only exception to this is our chick astrocyte cultures, which, according to our limited experience, do not grow well in horse serum. This is in agreement with observations by Skaper *et al.*[92]

The culture medium is changed after 3 days (when nonattached cells are removed) and subsequently twice a week. Under these conditions, only three cell populations adhere to the plastic substratum, i.e., phagocytic cells, oligodendrocytic precursors, and astrocyte precursor cells, whereas neurons do not adhere (or rapidly die if they adhere[46]). With regular growth medium changes, the phagocytic cells undergo little proliferation, whereas astrocyte precursor cells proliferate extensively. With the culturing conditions employed, oligodendrocytic cell precursors and cells will remain insignificant. Others have, however, reported the presence of much larger amounts of oligodendrocytes (Section 3.1.2).

After culturing for 2 weeks, the cells form a confluent monolayer of flat epithelial cobblestonelike cells[51] directly adhering to the plastic substratum. Many authors use such cultures for biochemical experiments, but we generally prefer to treat them for at least 1–2 weeks with 0.25 mM dibutyryl cyclic AMP (dBcAMP), which causes a distinct morphological differentiation with pronounced extension of cell processes (Fig. 3) and some biochemical alterations[116–119] and seems to further reduce the contamination with phagocytic cells. It is unresolved whether one ought to use this compound, which in higher concentrations is cytotoxic. It does cause a modest increase of intracellular cyclic AMP,[120] which could mimic the physiological effects of certain transmitters (e.g., norepinephrine or serotonin) that seem to play a role during differentiation of the CNS *in vivo*. In support of this concept, Narumi et al.[121] reported norepinephrine to cause a similar morphological differentiation in primary cultures of astrocytes, but this compound has shown much less of an effect in our studies. This may be related to our use of different media (modified MEM *vs.* Eagle's basal medium), especially since the morphological and biochemical effects of dBcAMP tend to be fast but transient when cells are grown in Eagle's basal medium[114,122] and slow but permanent when cells are grown in the modified MEM.[117-119] The reason for this difference is not understood, but it obviously carries a message for the understanding of cell differentiation, as do experiments by Sensenbrenner et al.[59] and by Lim et al.[123,124] on brain extracts or factors that have a differentiating effect on astrocytes. The only experiments in which we do not use dBcAMP-treated cultures (but nevertheless at least 3-week-old cultures) are those on the effect of β-adrenergic agonists on cyclic AMP accumulation, which becomes drastically reduced after exposure to dBcAMP.[120,125] This might seem unexpected, since dBcAMP does not act at the receptor site, but a corresponding effect has been observed in C-6 cells.[126] Whether or not the cells have been treated with dBcAMP, they can be maintained in culture for at least one-half year.

3.1.2. Cell Identity and Cell Markers

Immature astrocytes may be difficult to identify since the morphology of the very immature glial cells bears little resemblance to the mature cells. However, astrocytes *in situ* can unequivocally be identified by their ultrastructural morphology[127] or by macromolecular markers (see below), and a number of studies by several different groups of investigators have repeatedly demonstrated that almost all the flat epithelial-like cells that send out processes during exposure to dBcAMP are astrocytic in nature and that these cells constitute 70–95% of the total cell population (Table II).

In our hands,[51] astrocytes, recognized by GFA or glutamine synthetase staining (see below), constitute 95% of the cells in the cultures (Table II), and approximately the same percentage has been reported by several other authors.[53,109,128] In contrast to Stieg et al.,[129] we have less than 1% fibronectin-positive cells (unpublished data), i.e., probably leptomeningeal cells, endothelial cells, or fibroblasts.[130] We do have a slight contamination with macrophages (Table II), but with regular growth medium changes, these cells constitute less than 5% of the total cells in the cultures.[51] As already mentioned, they seem to show a further decline during the continued culturing with dBcAMP. A third population of cells in the cultures is comprised of small bipolar or multipolar cells that do not adhere to the plastic substratum but are situated on top of the flat cells. These cells, which have been reported to be oligodendrocytes (Section 2.2.2), constitute only about 1% of the total cell population in our cultures[51] and in those employed by Chiu et al.,[131] Hansson et al.,[39] and Kimelberg et al.[52] They are more numerous in cultures seeded at higher densities.[53,54] The odd neuron is very rarely encountered, e.g., not a single neuron among 35,000 cells counted for our data in Table II. We have no indications in our cultures of the quite marked (15–20%) contamination with endothelial cells reported by Hansson et al.[132] in corresponding cultures and by Garthwaite and Balázs[62] in cultures of cerebellar astrocytes.

Certain antigenic markers with a well-established glial localization *in vivo* have become of considerable importance to establish the "astrocyticity" of cultured cells. A classical example of this is the use of the antigenic marker GFA (glial fibrillary acidic protein), which in the mature CNS is found exclusively in astrocytes.[133] This protein is abundant in primary cultures of astrocytes,[39,134,135] and its content increases after exposure to dBcAMP or brain extracts.[39,116,136] Immunohistochemically, it can be used for identification of astrocytes in primary cultures (Table II). S-100 is an antigenic marker that *in vivo* is present in both astrocytes and oligodendrocytes[137] and probably to a minor extent in neurons.[138,139] This marker is also present in astrocytes in primary cultures,[18,128] and from Fig. 1 it could be seen how slowly it returns to its previous level if one attempts to subculture these cells. Such a procedure can therefore not be recommended unless one is specifically interested in studying factors that regulate gene expression. Other astrocytic antigenic markers are the C1 and the M1 antigens described by Schachner et al.[97] Both of these antigens are present in primary cultures of astrocytes from cerebellum, although the latter only in a subpopulation of the GFA-positive cells, which could mean that the expression of this antigen in many astrocytes *in vivo* is transient. This may obviously be of importance for developmental studies. Some antigenic markers are present in but not confined to astrocytes. These include vimentin,[131,140] which is also found in fibroblasts,[97] and actin.[131,141]

Other antigenic markers of relevance for astrocytic cell cultures are enzymes with known function such as glutamine synthetase and carbonic anhydrase. Glutamine synthetase activity seems *in vivo* to be confined to astrocytes,[142] and immunohistochemical staining has shown that it is present in an overwhelming majority of cells in primary cultures of astrocytes.[143,144] Furthermore, as can be seen from Fig. 4, the activity of this enzyme develops in parallel *in vivo* and in cultured astrocytes.[116] Since astrocytes account for about 30% of the cell volume in human brain cortex but less in the mouse,[145] one could have expected to observe a higher activity in cultured astrocytes than in the brain *in vivo*. The failure to do so is partly a result of the culturing conditions, since addition of glucocorticoids to the culturing medium for 24 h increases the activity two- to threefold.[143,144] This is one of the few cases in which a normal tissue culture medium with serum has been found to be deficient.

The *in vivo* localization of carbonic anhydrase (CAII) to oligodendrocytes only[146] or to oligodendrocytes plus astrocytes[147] is disputed. Carbonic anhydrase activity is undoubtedly present in primary cultures of astrocytes.[122] However, Sensenbrenner et al.[136] have reported that staining for CAII in chick astrocytic cultures with some oligodendrocytic contamination is confined to the latter cells. We have, in contrast, claimed that the carbonic anhydrase activity in mouse cultures containing few, if any oligodendrocytes is so high, especially after prolonged culturing with dBcAMP, that astrocytes must have carbonic anhydrase activity.[118] Recently, this question was resolved by Ki-

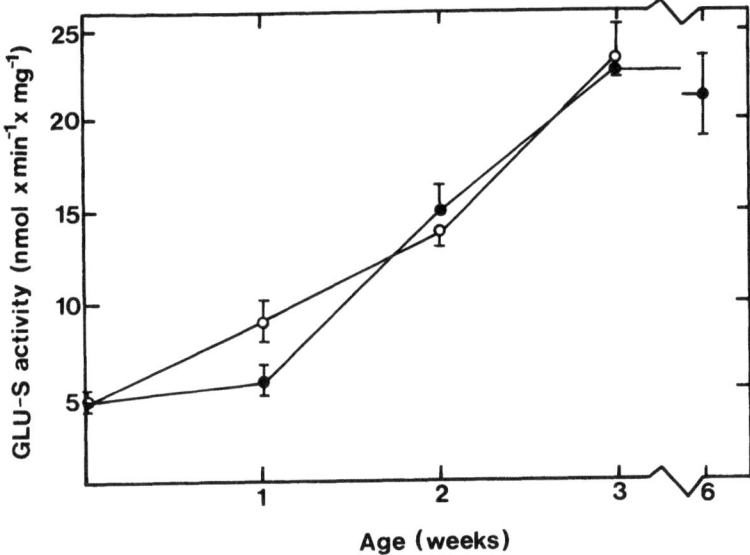

Fig. 4. Development of glutamine synthetase activity in primary cultures of astrocytes prepared from neonatal mouse brains as a function of age in culture (○) compared to the *in vivo* activities of the enzyme in brain (●) at comparable ages. (From ref. 116, with permission.)

melberg et al.,[52] who, by simultaneous immunocytochemical labeling for both GFA and carbonic anhydrase (Fig. 5), have shown unequivocally that astrocytes in primary cultures stain for CAII.

Negative markers, i.e., compounds or activities that are present in other cell types but presumably absent in astrocytes, are of at least equal importance as positive markers. Glutamate decarboxylase (GAD) is an enzyme that *in vivo* is confined to neurons (or to certain types of neurons). In our hands, primary cultures of astrocytes have no measurable activity of this enzyme[148] (J.-Y. Wu, A. Schousboe, and L. Hertz, unpublished experiments), and they produce only negligible (0.01 nmol/min per mg protein) amounts of GABA.[149] However, Bardakjian et al.[150] have reported a somewhat higher activity (0.2 nmol/min per mg). The neuronal markers D2 and D3[13] are absent from primary cultures of astrocytes. The virtual absence of fibronectin in our cultures has already been mentioned. Sulfatide synthesis and 2',3'-cyclic nucleotide 3'-phosphohydrolase (CNPase) activity, presumably oligodendrocytic functions, are also virtually absent (S. Monsma, S. M. Yim, L. Hertz and S. Szuchet, unpublished experiments). Very low CNPase activities and amounts of myelin basic protein were also reported by Hansson et al.[39] and Kimelberg et al.[52]

3.1.3. Neurochemical Use of Primary Cultures of Astrocytes

A confluent culture of astrocytes in a 60-mm dish contains a monolayer of cells attached to the dish and amounting to about 0.5 mg protein. Such a culture can be used for almost anything that can be done with a brain slice or

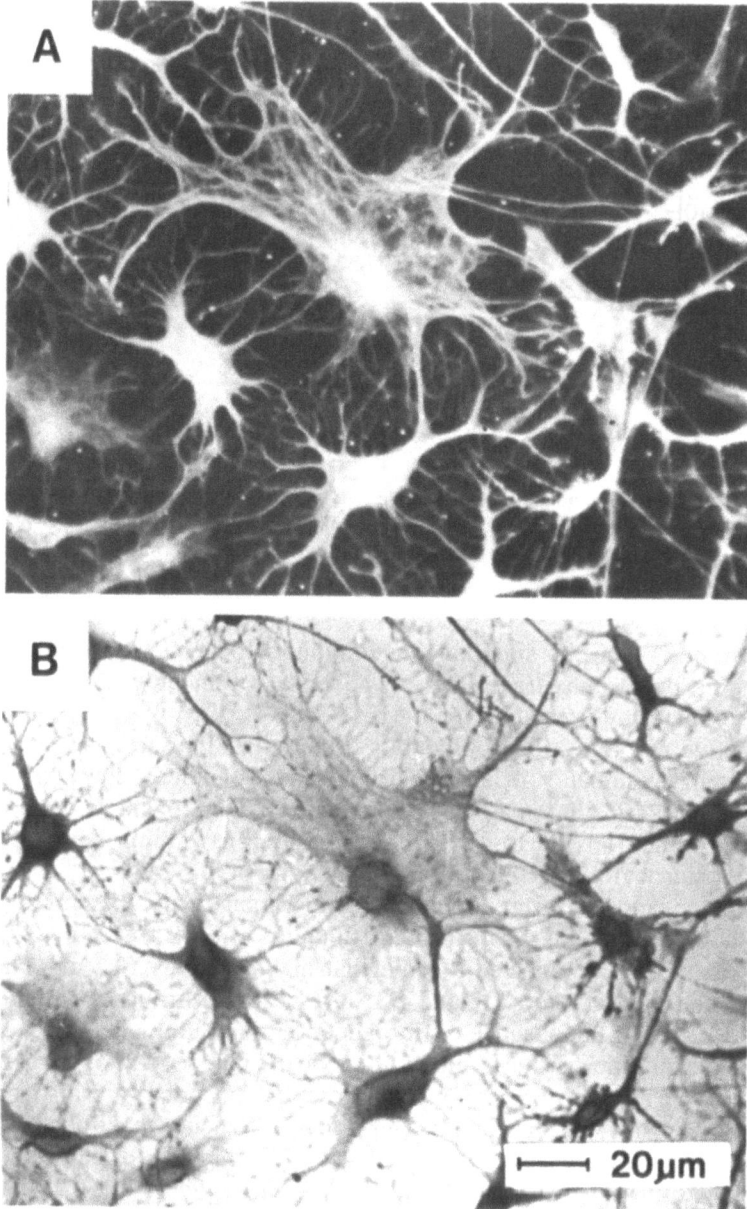

Fig. 5. Simultaneous immunocytochemical labeling for GFA protein (A) and carbonic anhydrase (B) in primary cultures of rat astrocytes. Note that the same cells stain for both markers. (From ref. 52, with permission.)

brain homogenate. Many enzyme activities, e.g., glutamine synthetase activity, can be determined in individual cultures; others, e.g., pyruvate carboxylase activity, require that a few cultures be pooled. Uptake characteristics can be much better determined than in brain slices because of the thinness of the preparation and the lack of problems with extracellular diffusion. Binding studies can be carried out both with the intact monolayer of cells (e.g., ref.151) and after homogenization of the cells[152] This is a distinct advantage over the situation with whole brain, but it also has the danger that cellular uptake could be included in the apparent binding to intact cells.

Since primary cultures of astrocytes generally are obtained from immature tissues and have developed in isolation from other cell types, the information obtained will have to be compared with that obtained using other preparations, e.g., astrocytes obtained by gradient centrifugation, which may have different sources of error but are not immature, and also with available data for whole brain. Such comparisons have, in general, demonstrated that primary cultures of astrocytes have characteristics that closely resemble those that can be expected for astrocytes *in vivo*,[19,20,153] and quantitatively the characteristics in these cultures appear reasonable when compared to the *in vivo* situation. This has already been mentioned in the case of glutamine synthetase activity (Fig. 4). Areas in which primary cultures of astrocytes have been of prime importance for the understanding of astrocytic functions include energy metabolism, uptake and metabolism of amino acid transmitters, uptake of potassium, and binding of certain transmitters and drugs. These topics are all discussed by Hertz in this *Handbook*[154] and are not repeated here, although the activities by astrocytes in these areas are likely to be of major importance in brain function.

It should, however, be noted that the uptake rate for potassium is much less in cultures of chick astrocytes than in corresponding cultures from the mouse.[21] We should also like to mention that primary cultures of astrocytes have been instrumental for the demonstration that glutamate is to a very large extent accumulated into astrocytes[61,155-157] and in these cells is not only converted to glutamine but also metabolized to CO_2.[158] This oxidative degradation is so intense that glutamate must be a major metabolic fuel for astrocytes in primary cultures. Before it is finally accepted that this is also the case for astrocytes *in vivo*, it should be taken into consideration that the culture medium contains a higher glutamine concentration (2 mM) that the extracellular glutamine concentration in the brain (probably 0.5 mM) and that this difference conceivably might have enhanced the metabolic degradation of glutamine via glutamate. It is, however, against this possibility that the activity of the glutaminase (the enzyme converting glutamine to glutamate) is unaltered when the glutamine concentration of the medium is reduced tenfold.[159]

3.2. Cell Lines

In principle, a primary culture of astrocytes can be converted to a diploid cell line if it is subcultured. This procedure has been used only rarely to study astrocytic characteristics, and it has, in our opinion, no advantages over the primary cultures themselves. An established cell line has been created by Cummins et al.[160] by treatment of primary cultures of rat astrocytes with herpes virus and was used for the very legitimate purpose of comparing possible differences in glycogen metabolism between normal and transformed cells.

The NN hamster glial cell line was derived from glial cells cultured *in vitro* and subsequently *in vivo*.[161] It has undergone spontaneous transformation, indicated not only by its immortality but also by the fact that 40% of the cells are tetraploid or octoploid.[17] This cell line, which has been cloned, has been used for a fair number of neurochemical studies.[17] Potassium-induced stimulation of oxygen uptake, an apparently astrocytic phenomenon, is absent.[162] It is occasionally stated that the cells in the NN cell line are more normal than those in cell lines derived from tumors or virally or chemically transformed cells, but the biological and biochemical characteristics of the cell line described above do not support this point of view.

By far the best known and most universally used glial cell line is the C-6 cell line. The widespread use of this cell line in neurochemical research is the reason that it is better characterized than any other astrocytic cell line, and for this reason the C-6 cells may in many cases be preferable to other cell lines. However, the possibilities for spontaneous alterations (see below) should always be kept in mind. The C-6 cells were obtained by Benda and co-workers from one of 11 brain tumors induced by intravenous injection of N-nitrosomethylurea into rats and subsequently plated in culture[163] and later cloned.[164] Some of the cells in the initial cultures were multipolar cells resembling astrocytes, but the tumor itself had the appearance of a mixed glioma that also contained small cells with a round nucleus, likely to be oligodendrocytes.[165] It is in keeping with this that C-6 cells show certain oligodendrocytic characteristics. Thus, their glycerolphosphate-3-dehydrogenase (GPDH) activity is enhanced by exposure to hydrocortisone, an oligodendrocytic characteristic.[166] In addition, C-6 cells possess a considerable activity of CNPase,[167] an enzyme located in oligodendrocytes and myelin (Section 5.4.2b). This enzyme is, as already mentioned, virtually absent from most preparations of astrocytes in primary cultures.

The C-6 cell was originally selected as a glial cell line on account of its high content of S-100, which is found in about the same amount as in the brain *in vivo*.[163-165] However, even in a single dish, C-6 cells sometimes display differences in expression of S-100.[168] Furthermore, the presence of this marker in both astrocytes and oligodendrocytes should be kept in mind, especially on account of the mixed morphology of the tumor from which the C-6 line was originally obtained and the other evidence that this cell line displays some oligodendrocytic features. The presence of the astrocyte-specific GFA protein is much more unpredictable, and it may be present in some C-6 preparations but not in others.[165]

An elegant example of the pluripotential nature of the C-6 cells and of the mutability of their phenotypic expression in relation to passage time and to the confluency state of the culture is given by the investigations of Parker et al.[169] These investigators found that the glutamine synthetase activity of confluent cultures derived from the 82nd passage is more than fourfold higher (and comparable to that in brain) than that in cultures derived from the 22nd passage.

This finding has been confirmed by Browning and Nicklas, who reported that exposure to dBcAMP further increases the glutamine synthetase activity.[170] The reverse development was noted for CNPase activity, with early-passage cultures having a much higher activity at confluency that the later-passage cultures.[169] Both enzyme activities were lower in cells shortly after subculturing than at confluency.

With respect to a possible neurochemical overlap with neuronal characteristics, several authors have observed very low GAD activities in C-6 cells, but activities of 0.2–0.5 nmol/min per mg were reported by others.[171] These values are still low compared to that in primary cultures of neurons (Section 4.1). The probably neuronal-specific proteins D2 and D3 are found in nonnegligible amounts in C-6 cells.[13]

In addition to the initial clone of C-6 established by Benda et al.,[163,164] numerous subclones have been developed.[17] They have been used for a variety of neurochemical studies.[17] Historically, it is of importance that GABA uptake into cultured glial cells was first demonstrated using these cells,[172,173] but it should also be realized that quantitatively the magnitude of this uptake was small. The same is true for potassium uptake,[174,175] which is about 1/100 of that observed by ourselves in primary cultures of astrocytes.[103,105] β-Adrenergic receptors on astrocytes were first shown using the C-6 cells[176,177] and have only subsequently been demonstrated on astrocytes in primary cultures.[128,152,178,179] Metabolically, this cell line may be more generally astrocytic than the NN cells, since a modest enhancement of respiration has been observed in response to an elevated potassium concentration (E. T. Browning, personal communication).

From Pontén's laboratory, several diploid and established cell lines have been introduced, i.e., the diploid NG (normal glia) obtained from normal adult brain and the nondiploid MG (malignant glia) obtained from tumors (see, e.g., refs. 9,17). One of these lines, the MG138, which contains both S100 and GFA,[180] has been used for neurochemical work.[11] Another cell line that is currently used is the LRM55 cell line, which was cloned from an ethylnitrosourea-induced mixed glioma from the spinal cord of a Fisher rat[181] and has a high-affinity transport system for glutamate[182] and another for taurine, β-alanine, and GABA.[182]

Many other glial and probably mainly astrocytic cell lines exist.[17] One can be mentioned as an example of the pluripotential nature of established cell lines. This is the anaplastic glioma F98 cell line, which generally is considered to be glial in nature but responds to nerve growth factor and exhibits neuron-specific enolase, i.e., possesses neuronal characteristics.[14]

With present-day availability of primary cultures of astrocytes and the often questionable expression of astrocytic phenotype by astrocytic cell lines, it seems difficult to defend the continued use of established cell lines as specific astrocytic models. What remains a legitimate reason for the continued use of astrocytic cell lines is a more biologically oriented study of processes regulating gene expression, e.g., the enhancement of glutamine synthetase activity by dBcAMP and the correlation between this phenomenon and other cellular events.[169,170] Other studies for which the cell lines seem suitable are those exploring the mechanisms of processes that have been found to be operating in a cell line, e.g., uptake of glutamate and other compounds in the LRM55 line.[181,182] For such investigations, the availability of large amounts of identical cells (provided that no spontaneous alterations in phenotypic gene expression occur) is of value.

4. NEURONS

4.1. Neuronal–Glial Cocultures

Cultures of dissociated neurons were first obtained from spinal cord by Cavanaugh[183] and from spinal ganglia by Nakai.[184] These cultures contained, in addition to neurons, other cell types. For many purposes it is, however, of no consequence if a neuronal culture is not monotypic but contains other cellular components. Thus, elegant electrophysiological studies have been done on mixed neuronal–glial cultures derived from spinal cord or dorsal root ganglia[185,186] and hippocampus.[187] Similarly, mixed neuronal–glial cocultures from the CNS have been examined for neurotransmitter uptake using autoradiographic techniques.[157,188]

It has been traditional to prepare neuronal–glial cocultures by seeding brain cell suspensions obtained from animals at appropriate developmental ages, e.g., cerebral hemispheres of 15-day-old rat or mouse embryos or 7-day-old chick embryos, onto a preformed "feeder" layer of astrocytes[43,46,189,190] or fibroblasts.[191] In addition to postmitotic neurons, the newly seeded cell suspension contains proliferating astroblasts and oligodendroglial precursor cells (Section 2.2.2), which probably attach and proliferate. Therefore, the feeder layer alone can not be used as an adequate nonneuronal control for neurochemical work unless proliferation of these newly seeded cells is prevented by cytotoxic drugs[190] (Section 2.2.5). Neuronal–glial cocultures can also be obtained from cell suspensions seeded on polylysine- or collagen-coated surfaces, since the astroblasts in the cell suspension, if no cytotoxic agent is present, will form the feeder layer.[42,46] One difference between these techniques is that the neurons remain dispersed for a longer time if seeded onto a feeder layer,[192] whereas they generally reaggregate faster in the absence of such a layer.

The developmental stages of animals used for the preparation of neuronal–glial cocultures is critical (Section 2.2.2). It may be a reflection of this that Bonkowski and Dryden,[193] using newborn mouse brains to prepare mixed neuronal–glial cocultures, obtained cells of somewhat peculiar morphology and reported much lower membrane potentials than those seen by Dichter[191] in cerebral neurons and by Peacock[187] in hippocampal neurons.

4.2. Monotypic Primary Cultures of Neurons

4.2.1. Procedure

4.2.1a. Use of Substratum and Cytotoxic Agents. Almost regardless of the type of monotypic neuronal culture one attempts to establish, the cell suspension used (e.g., from the rat cerebral cortex at 15–20 days of gestation or from the cerebellum at 2–15 days post-natum) contains, in addition to the postmitotic immature neurons large numbers of proliferating astrocytes and oligodendrocytic precursor cells (Section 2.2.2). In order to obtain neuronal cultures, a relatively simple approach is to (1) select a suitable substratum (Section 2.2.4) favoring attachement of the neurons and a suitable cell density for the culturing of the dissociated cells and (2) subsequently eliminate the proliferating—mainly glial (and in the PNS also fibroblastic)—elements by aid of agents showing differential cytotoxicity (Section 2.2.5). Therefore, the cell suspension must be cultured on a substratum other than a glial or fibroblastic monolayer to which the neurons will firmly attach.

Basically, two approaches have been used, i.e., collagen coating and polylysine coating (Section 2.2.4). In collagen-treated culture dishes, at least 25% of the dissociated neurons from chick sympathetic ganglia attach to the collagen substrate, and about 14% of them develop neurites.[81] A higher percentage of dissociated neurons (60% or more) attach to polylysine-coated surfaces.[47,78,194] The adhesion of the cells is independent of cell density[47] but is dependent on temperature and energy supply.[77] After 30 min of incubation in a polylysine-coated dish, detachment of the attached cells requires 5–10 min incubation with 0.125% trypsin,[77] and some cells are already firmly attached to polylysine within 10 min. The process of adhesion is essentially completed within 0.5–2 h,[47,77] and under our experimental conditions, most neurons attach within 15 min. Because the adhesion is rapid, no significant reaggregation of the neurons occurs.[47] Pretreatment with relatively high concentrations of polylysine has a toxic effect on both attached neurons and nonneuronal cells.[194] At a lower concentration, polylysine has no effect on neuronal survival, but it may under certain conditions entirely prevent an increase in nonneuronal cells.[47,79,194] This is not the case with the culturing conditions used by us (unpublished) or by Sensenbrenner et al.[195] Under these conditions, a cell suspension from 15-day-old mouse or 5- to 12-day-old chick embryos planted into a dish pretreated with D-polylysine will, after 2 weeks of culturing, be overgrown with astrocytes if no cytotoxic drug has been added to curtail astrocytic growth.

The neuronal survival in culture is dependent on the cell density,[47,196,197] and in order to enhance long-term neuronal survival, a high cell density has to be employed. Neuritic outgrowth also increases with a higher cell density in the cultures.[47] The effects of cell density on neuronal survival and neuritic outgrowth suggest that cellular interactions and a "conditioning" of the culture medium are important for culturing neurons.[196,198,199]

It is generally, but not always,[47,79,194] necessary to eliminate the proliferating nonneuronal cells, which usually are eliminated from monotypic neuronal cultures by incorporating cytotoxic drugs such as aminopterin, fluorodeoxyuridine (FdU), or cytosine arabinoside into the culturing medium. Aminopterin, a folic acid antagonist, has apparently been used the least, but it seems to have no obvious toxic effect on neurons even after continuous application for 10 days.[185,200] Also, FdU has been reported to be relatively harmless to nondividing cells.[87] According to Dambergs et al.[201] it is, however, more toxic to nondividing cells than is cytosine arabinoside, which is in accordance with an inhibitory effect on RNA synthesis also (Section 2.2.5). Continuous treatment with FdU for 4 days has been used to prevent the multiplication and overgrowth of nonneuronal cells in cultures of neurons from rat cerebral hemispheres or chick sensory ganglia.[202,203]

Cytosine arabinoside is a pyrimidine antagonist that differs from cytosine deoxyriboside in containing arabinose rather than deoxyribose. It is lethal to proliferating cells.[204] Postmitotic neurons *in vivo* are affected only slightly, if at all, by cytosine arabinoside,[87,204] which, in contrast to the two other drugs, has no effect on RNA metabolism. It has been used successfully in culture systems to inhibit the rapid proliferation of nonneuronal cells.[80,85,87,119,191,197,205] The time for the addition of cytosine arabinoside is critical; if the addition is too early,[197] the entire culture may suffer; if it is too late, the astrocytes will already have formed a monolayer of nondividing cells.

4.2.1b. Cerebral Cortical Neurons. As we have done for the primary cultures of astrocytes, we describe our own methodology in some detail. Again, methods used by others may be equally effective, although we do feel that the early (between day 3 and day 4) exposure to cytosine arabinoside makes the glial contamination of these cultures less than that in most other preparations of mammalian cerebral cortical neurons. The method we use[80] is a slight modification of those described by Yavin and Yavin,[77] Sensenbrenner,[43] Dichter,[191] Yu,[189] Larsson *et al.*,[206] and Yu and Hertz.[207] The cerebral hemispheres are removed from 15-day-old mouse embryos and freed of the meninges. Subsequently, the two hemispheres are split, and the olfactory bulbs, basal ganglia, and hippocampal formations are removed. This leaves the neopallium, i.e., the portion dorsal and lateral to the lateral ventricles, which is cut into small cubes and gently dissociated, i.e., trypsinized for 2 min using 0.2% trypsin in Puck's solution[80] at room temperature. We do not use collagenase, DNase, RNase, and hyaluronidase, as do some investigators.[91]

After inhibition of the trypsin by addition of modified Eagle's minimum essential medium (same medium as used for the astrocytes) containing 20% horse serum, the tissue is triturated with a fire-polished pipette. The resulting cell suspension is centrifuged for 2 min at 900 × g. The pellet is resuspended in serum-free medium with an increased glucose concentration (30 mM instead of 7.5 mM) and filtered through a Nitex ® mesh with a pore size of 80 μm. A cell suspension corresponding to 1–1½ brain per dish is seeded into 60-mm Falcon plastic tissue culture dishes that have been coated by exposure overnight to 12.5 μg/ml of D-polylysine in water. After 15 min of incubation at 37°C, unattached cells (mostly nonneuronal cells) are removed together with the medium, which is replaced with similar, fresh modified MEM with 5% horse serum. The cultures are incubated at 37°C in a 95%/5% (v/v) mixture of atmospheric air and carbon dioxide (90% humidity). After 3 days of culturing, the cultures are exposed to 40 μM cytosine arabinoside for 24 h to curtail astrocytic growth. Thereafter, they are refed fresh modified MEM with serum but without the cytotoxic agent and are used for biochemical studies between the ages of 7 and 17 days.

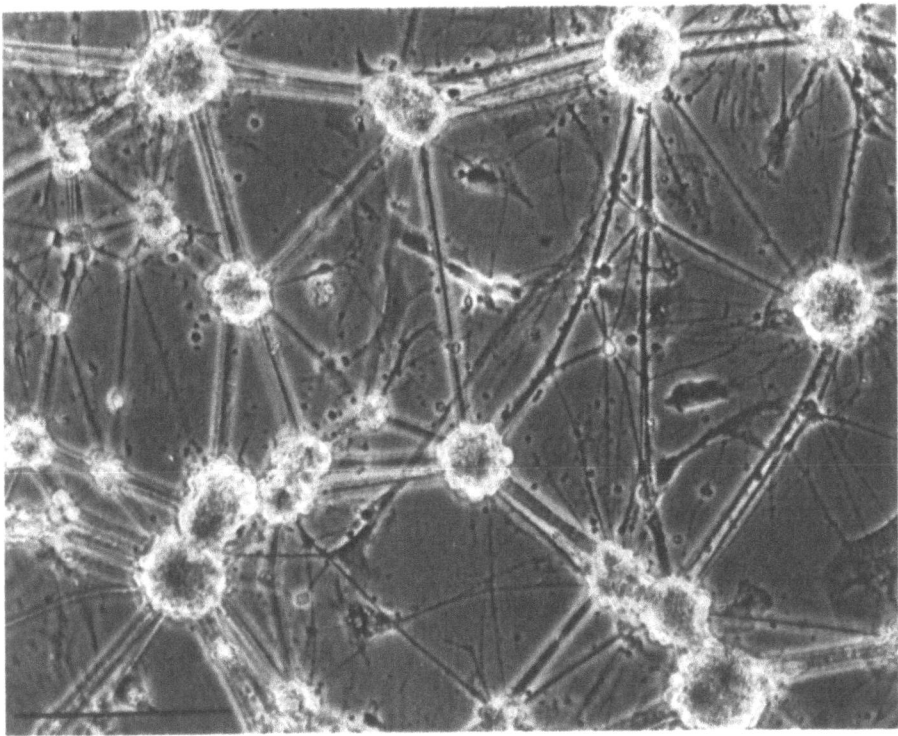

Fig. 6. Phase-contrast micrograph of living 12-day-old culture of mouse cerebral cortical neurons prepared as described in ref. 80. Bar represents 200 μm.

Initially, the cells are evenly distributed in the culture, but with time they gather in clumps as has been described by other authors.[47,195] This can be seen from Fig. 6, which also shows that the contamination with astrocytes is relatively slight. Because of the high glucose concentration in the medium, no further change of medium is necessary. This seems to enhance the viability of the cells, but, nevertheless, they can not be maintained longer than 17–21 days. Such degeneration of neurons in relatively pure cultures seem invariably to occur after at most 2–3 weeks.[47,79,208] It may perhaps be a reflection of the well-known phenomenon that the survival of many neurons (particularly Golgi type I) is dependent on their receiving the appropriate synaptic inputs and/or on the presence of the appropriate target cells (see ref. 22; Chapter 7). It seems, however, to be against this concept that neurons survive much longer if seeded on a glial feeder layer.[186,209]

Very pure cultures of chick cerebral neurons have been prepared by Pettmann et al.[79] using polylysine coated dishes but no cytotoxic agent. They found that optimal chick neuronal cultures are obtained from 8- to 11-day-old chick embryos, i.e., developmental stages comparable to those in rodents that give optimal neuronal cultures (Table I). The absence of astrocytes in such cultures seems to depend on the detailed experimental conditions, since the same laboratory has previously observed a considerable astrocytic contamination in cultures from 8- to 12-day-old chick embryos grown under relatively similar conditions.[195] Skaper et al.[92] and Brunner et al.[91] have developed a different procedure for obtaining highly enriched neuronal cultures from rodent or chick cerebral hemispheres with little glial contamination. They observed that serum-free Eagle's MEM containing the supplements described by Bottenstein et al.[89] for culturing of neuroblastoma cells allowed the survival of postmitotic neurons but did not allow the proliferation of the glial elements. The ability of neurons, but not astrocytes, to grow in the absence of serum has also been reported by Yavin and Yavin.[90]

4.2.1c. Cerebellar Neurons. The culture of cerebellar granule cells was pioneered by Lasher and Zagon[210] and Messer.[197] These cells, which are in all likelihood glutamatergic,[211,212] develop at a relatively late ontogenetic stage (Fig. 2). Therefore, cerebella from 7- to 8-day-old mice are generally used to prepare these cultures.[61,197,205,207] The principles for the preparation are otherwise the same as for cerebral cortical neurons.[207] It has been reported that neuronal viability is enhanced in the presence of an increased potassium concentration,[210] which is used by most investigators.[61,197] According to A. Schousboe (personal communication), they also grow better with 0.8 than with 2.0 mM glutamine in the medium. Monotypic cerebellar granule cells can be maintained in culture for a longer time (21–30 days) than cerebral cortical neurons.[189,197]

4.2.1d. Peripheral Neurons. Within the framework of this chapter, we can not cover in any detail peripheral neuronal (sensory and autonomic) cultures. As indicated in Section 2.2.4, highly purified peripheral neuronal cultures can be obtained using differential adhesion techniques and/or cytotoxic agents with selective toxicity for dividing cells. In contrast to CNS neurons, highly purified peripheral neurons require the presence of nerve growth factor in the culture medium for survival.[213] There is also good evidence that their survival may be enhanced by elevated (20–40 mM depending on species) potassium concentration.[214] No elevated potassium levels were, however, employed by Kim et al.[50]

4.3. Cell Identity and Cell Markers

4.3.1. Neurons versus Nonneurons

Typical neurons in mixed cultures or in highly enriched primary cultures of neurons are, in general, easy to recognize. A neuron by definition is postmitotic, and after final mitosis, the neurons rapidly develop unique morphological and biochemical properties with which they can be readily identified. One can use neurophysiological criteria, e.g., action potentials[187,191] or morphological features such as the presence of synapses,[191,197,215,216] Nissl substance,[217] or enzymes related to transmitter synthesis.[217–219] Another possibility is to assay for the presence of unique neuronal macromolecular markers such as neuron-specific enolase,[220,221] synaptin, D_2, and D_3.[13,62,189] Also, tetanus toxin binding to neuron-specific gangliosides has been widely used.[61,79,222] It has already been mentioned that the presence of some of these markers depends not only on the identity of a cell but also on its maturation (Section 2.2.7). Their absence in a specific cell, therefore, does not necessarily identify that cell. This is the case with the neuronal-specific enolase. Two forms of enolase are present in the mature brain, a nonneuronal enolase, which is present in all cell types other than neurons, and a neuronal enolase.[13,221] Immature neurons only contain nonneuronal enolase, but as the neurons mature, the nonneuronal enolase is replaced by the neuron-specific enolase.[99,221]

As in the case of astrocytes, negative markers are as important as positive markers. Relatively little work has been done in this area, but Yu[189] and Currie[61] both found small amounts of GFA (corresponding to the morphological observation of a slight astrocytic contamination) in cerebral and cerebellar neurons kept in culture for about 1 week. The latter author estimated the contamination with astrocytes to be 7.6% and that with oligodendrocytes, recognized by staining with antigalactocerebroside, to be 0.2%.

4.3.2. Neuronal Classes

Determination of the class of neurons is in general difficult. At the level of the individual cell, it is possible to use electrophysiological characteristics and autoradiographically measured uptake, or lack of uptake, of a specific transmitter candidate. It is often used as a criterion for the predominance of the presumably glutamatergic granule cells in cerebellar cultures that they do not accumulate GABA.[197,223,224] These cells constitute 86% of the cells in cultures from 7-day-old rat cerebella,[61] with GABA-accumulating neurons accounting for another 6%. The latter neurons are probably GABAergic, i.e., stellate or basket cells. These cells develop earlier than the granule cells (Fig. 2), and, hence, one finds a higher proportion of GABAergic neurons (16%) in cultures from 2-day-old rat cerebella.[223] The glutamatergic character of the granule cells may seem at variance with the fact that autoradiographic studies have failed to show an accumulation of glutamate into cerebellar granule cells in culture.[225,226] In spite of this, such cells accumulate glutamate but very little GABA by a high-affinity uptake process from a medium containing micromolar amounts of this amino acid.[205,207] In the hippocampus, the situation seems more clear-cut, since some neurons accumulate GABA but not glutamate, others glutamate but not GABA.[157] In agreement with the early development of Purkinje cells (Fig. 2), few if any Purkinje cells are present in cultures of cerebellar neurons obtained from 7-day-old rats,[61,62,212] but they can be obtained in mixed neuronal–glial cultures of cerebellum from late fetal rats.[227]

Most of the neurons in cerebral cortical neuronal cultures seem to be GABAergic. This is indicated not only by electrophysiological characteristics,[228] a high-affinity GABA uptake,[80,207,209,216,229,230] and a potassium-induced, calcium-dependent GABA release (Fig. 7, top)[80,209,230] but also by GAD activity,[80,190,209] which may become higher than it is in the brain cortex *in vivo*.[80,209] This can be seen from Fig. 8, which also shows that the GAD activity shows a steep increase during culturing.

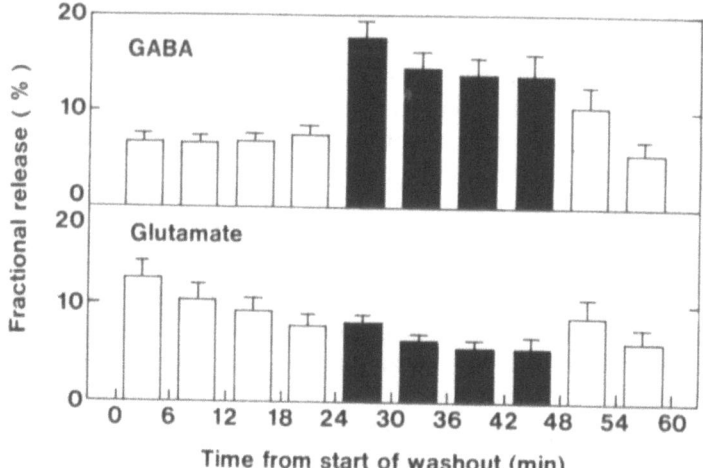

Fig. 7. Fractional release of GABA (top) or glutamate (bottom) from 17-day-old cultures of mouse neurons during incubation in normal medium (open bars) or during exposure to excess (55 mM) potassium (solid bars). The cultures had been preloaded with radioactive GABA or glutamate, and the fractional release is the activity released during each consecutive 6-min period as a percentage of the radioactivity remaining in the culture at the start of that period. Thus, the increased fractional release of GABA in response to an elevated potassium concentration reflects a potassium-induced release of GABA. (From ref. 80, with permission.)

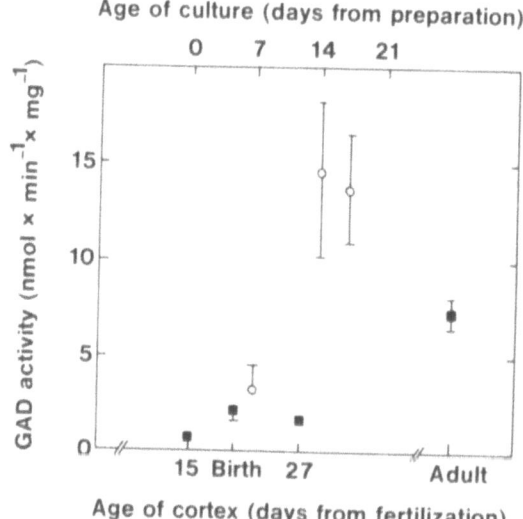

Fig. 8. Development of glutamate decarboxylase (GAD) activity in primary cultures of cerebral cortical neurons prepared from the brains of 15-day-old mouse embryos as a function of age in culture (○) compared to the *in vivo* activities of the enzyme (●) in the brain (neopallium) at comparable ages. (From ref. 80, with permission.)

Such a developmental increase in GAD activity in neuronal cultures that are almost devoid of glial cells shows that manifestation of the phenotypic expression of GAD activity does not require neuronal–glial interactions. This is also true in the case of tyrosine hydroxylase, since a similar although less pronounced increase in the activity of this enzyme has been observed in highly purified chick cerebral cultures, but the absolute activities of this enzyme are much lower than the GAD activities.[79] In the case of the PNS, neuronal interactions with nonneuronal cells do play a role in determining the identity of the neuron. Thus, Mains and Patterson[231] demonstrated that autonomic neurons obtained from newborn rat superior cervical ganglia, when grown in the absence of nonneuronal cells, synthesize and store large amounts of norepinephrine and only small amounts of acetylcholine. However, when these cells are cultured in the presence of cardiac or skeletal muscle cells[232] or in the presence of conditioned medium from such nonneuronal cultures,[233,234] the acetylcholine production rises between 40 (conditioned medium) and 1000 (cocultures) times, and a concomitant reduction occurs in norepinephrine synthesis. That this is a case of the environment dictating neurotransmitter synthesis and not of selective survival of cholinergic or noradrenergic neurons can be seen from the fact that similar transmitter switches occur in microcultures that each contain a single neuron.[235]

The demonstration of a potassium-induced release of GABA from a cerebral neuronal preparation does not mean that all neurons in the culture are GABAergic, and Snodgrass et al. have demonstrated a slight potassium-induced release of glutamate from their cultures, which were prepared from whole cerebral hemispheres.[209] This is in contrast to our own finding (Fig. 7, bottom) that excess potassium does not enhance glutamate release in cultures prepared exclusively from neopallium.[80] This difference might obviously reflect the more narrow range of surviving neurons in the latter case and fits the concept that most neurons in the cerebral cortex are probably GABAergic.[228]

4.4. Neurochemical Use of Primary Cultures of Neurons

What has been said about the kind of experiments that can be carried out using primary cultures of astrocytes also applies to neurons. These cultures are generally fairly densely seeded (to enhance survival), and each 60-mm dish of cerebral cortical neurons prepared according to our procedure contains approximately 1 mg protein.

In contrast to astrocytes, which can be kept in culture for many months and are known to undergo a pronounced functional differentiation, mouse cerebral neuronal cultures can only survive in isolation for 2–3 weeks (Section 4.2.2b). Since such cultures are obtained from 15-day-old embryos, and the mouse gestation period is 20 days, this corresponds to a postnatal age of 9–16 days, i.e., an animal with a still immature brain. It therefore becomes imperative to possess information about the degree of functional maturation of the cultures.

The finding that GAD activity increases during the culturing period to levels 2–3 times those in the cortex (Fig. 8) is reassuring in light of the fact that neurons account for about 40% of the brain cortex.[145] It does, however, give reason for some concern that the GAD activity reaches its maximum levels at a time when the *in vivo* activity is still increasing (Fig. 8). In contrast, cortical neuronal cultures with a larger content of astrocytes (as a result of later addition of cytosine arabinoside) continue to show an increase in GAD activity for a longer time,[209] but the maximum activity (10 nmol/min per mg protein) seems to be slightly lower than that shown in Fig. 8 (probably reflecting the "dilution" with astrocytes). Thus, it seems that GABAergic development may occur at an accelerated rate in the virtual absence of astrocytes. Another indication of an extensive GABAergic development in the highly enriched cerebral cortical cultures is that the GABA content, measured in 17-day-old cultures, reaches a level of 75 nmol/mg protein (compared to 15–20 nmol/mg in 7-day-old cultures).[80] This is reasonably comparable with the value of 50–150 nmol/mg* in individual GABAergic neurons *in vivo*,[236,237] especially since the exact proportion of GABAergic neurons in the cultures is unknown. The previously mentioned potassium-induced increase in GABA release (Fig. 7) also appears during the second week of culturing.[80,230] Similar maturational phenomena have been observed by Yavin and Yavin[238] with respect to phospholipid content in cultured cortical neurons. For the use of monotypic cerebral cortical neuronal cultures as good models of their *in vivo* counterparts, it therefore appears essential that they be kept in culture for a minimum of 12–14 days.

At present, cultures of cerebral cortical neurons are mainly used to study uptake, content, release, metabolism, and formation of GABA, glutamate, and glutamine, i.e., questions connected with metabolic compartmentation.[80,206,207,209,216,229,239] In addition, these cultures have been used in studies of differences in calcium transport[240] and in potassium[104,105,241] transport between neurons and astrocytes. Some of these studies have been carried out in cultures younger than 12–14 days and may therefore not have given optimal results. The lack of any major nonneuronal contamination and the normality of the cells are obviously of major importance in such studies.

The ontogenic development of cerebellar granule cells in cultures has apparently not been studied in detail, but a potassium-induced release of glutamate, which presumably is the granule cell transmitter, has been observed by Drejer *et al.*[205] and Levi *et al*[242] in 7- to 12-day-old cultures. This is at about the same time that a pronounced potassium-induced GABA release occurs from the inhibitory interneurons in these cultures, but an inhibition of this release by tetrodotoxin does not become apparent until later.[226]

Currently, much of the work done with purified peripheral neuronal cultures is directed toward defining the trophic factors involved in the survival and differentiation of neurons and examining the cell-to-cell interactions involved in myelination (see Section 6). It has already been mentioned that peripheral sympathetic neurons require nerve growth factor for their survival and trophic factors from nonneuronal sources for the acquisition of cholinergic characteristics (Section 4.2.2). This area has recently been reviewed by Black and Patterson.[243] Similar studies are being performed on the regulation of syn-

---

* Estimated on the assumption of 20% solids, one-half of which is protein.

thesis of substance P and somatostatin in sensory neurons by nonneuronal cells.[244,245]

4.5. Neuronal Cell Lines

The most widely used neuronal cell lines have been derived from neuroblastoma tumors. The earliest report of growing such tumors *in vivo* is by Murray and Stout,[246] who reported that explants of human neuroblastoma tumors on plasma clots resulted in a rapid outgrowth of axons and suggested that such *in vitro* methodologies could be used as diagnostic tools. Little further work was done until Goldstein and Pinkel,[247] who again used human explant cultures, and demonstrated that the cells could proliferate for long periods in culture (up to 8 months) and still retain the capacity to differentiate into mature-looking ganglion cells. It was not until the late 1960s and early 1970s that human neuroblastoma cells were grown as dissociated cultures, and a number of clonal cell lines were established.[248–250] The majority of these and other more recently established human neuroblastoma cell lines are pseudodiploid.[251]

The neuronal cell line used most commonly in neurochemical research is the C-1300 neuroblastoma line, which is derived from a tumor that arose spontaneously in an albino A/J male mouse in 1940 and was identified as a sympathetic neuroblastomal tumor.[252] The tumor was maintained for almost 30 years by serial passage through mice before it was adapted to culture.[253–255] Since then, a number of clonal cell lines have been established, all of which vary considerably with respect to acetylcholinesterase, choline acetyltransferase, and tyrosine hydroxylase activities as well as with respect to modal chromosome number (varying from 59 to 207, compared to a diploid chromosome number of 40 in the mouse) and morphology.[253,256–258] Electrophysiological[259] as well as morphological[255] characteristics early established that the C-1300 neuroblastoma cell lines consist of a heterogeneous population of cells with characteristics that range from immature to mature autonomic neurons, depending on clonal cell line and culture conditions. Cell lines have also been established from rat adrenal medullary pheochromocytoma cells (e.g., the PC12 cell line) that display many of the properties associated with pheochromocytoma cells as well as their nonneoplastic counterparts, the adrenal medullary chromaffin cells.[260,261]

A number of established CNS neuronal cells lines are available. Some of these were derived from chemically induced tumors in rat CNS[262,263] or by infection of primary cultures of rat CNS with a temperature-sensitive Rous sarcoma virus.[264] Others resulted from spontaneous transformations in primary CNS cultures of rat or mouse,[265,266] two species in which cultured cells are known to be relatively prone to spontaneous transformation.[9]

Established neuronal lines offer the advantage of having large numbers of more or less identical cells in culture for long periods of time. This may be especially attractive in the case of neurons, since currently one cannot maintain homotypic primary neuronal cell cultures of the CNS for more than a month. The question is how closely the established neuronal cell lines reflect the behavior of their counterparts *in vivo*. Like established glial cell lines, neuronal cell lines are characterized by great biological diversity both in phenotypic expression and in genetic content. As an example, certain C-1300 neuroblastoma clones respond to applied dopamine by membrane hyperpolarization,[267] whereas other clones respond by depolarization.[268] Neuronal cell lines that have been virally transformed by the aid of temperature-sensitive Rous sarcoma virus[264] may, however, hold great promise if the large numbers of transformed cells that can be grown at permissive temperatures are able to differentiate into normal neurons when grown at nonpermissive temperatures.

There is an extensive body of literature associated with the neurochemistry, neurophysiology, and differentiation of neuronal cell lines, particularly with respect to the C-1300 neuroblastoma. Thus, a considerable amount of work has been done on C-1300 neuroblastoma clones with respect to types of receptors present (e.g., acetylcholine, serotonin, dopamine), electrophysiological properties, and synaptic development (reviewed in refs. 269–271). Neuroblastoma cells have been used extensively as neuronal models for studying the biochemical properties and regulation of receptors for neurotransmitters and neuromodulators,[272,273] and it has been shown histochemically that the same cells may contain both catecholamines and acetylcholinesterase.[274] There is also a large amount of literature dealing with morphological and biochemical differentiation of neuroblastoma cells.[272,274-276] The PC12 pheochromocytoma cell line has been used extensively,[261] e.g., in examining the mechanism of action of nerve growth factor.[277]

Less work has been done on established CNS neuronal cell lines. However, the presence and concentration of putative neurotransmitters as well as the presence of neuronal-specific antigens have been examined.[262,263] In view of the considerable amount of work that has been carried out on cerebellar granule cells in primary cultures, it seems of special interest that the cerebellar neuronal ε_c^t cell line may represent a granule cell.[266,278] This line has been used to study a coupled transport of glutamate and sodium.[278] As in the case of the astrocytic cell lines, such detailed investigation of phenomena displayed by a given cell line is probably one of the most meritorious uses of such cells.

5. OLIGODENDROCYTES

5.1. Strategies to Obtain Monotypic Cultures of Oligodendrocytes

As has already been mentioned, three approaches have been used to isolate pure populations of oligodendrocytes: one of them depends on selection of a tissue enriched in oligodendrocytes and subsequent fractionation of cells based on physical parameters such as size and density (Section 2.2.3); the second uses antigenic specificity to obtain purified fractions (Section 2.2.4); and the third starts with a mixed astrocytic–oligodendrocytic culture and selectively separates oligodendrocytes because of layering and adherence properties. In what follows, we briefly describe prototype procedures and provide a critical evaluation of each. In addition, a few oligodendrocytic cell lines are discussed.

5.2. Procedures for Cell Isolation

5.2.1. Use of Density Gradient Centrifugation

Attempts to isolate oligodendrocytes span a period of over 20 years. Korey[279] made the first try at isolating a "neuroglial" fraction. Although his preparation was crude at best, he made his mark by introducing a technical approach that has perpetuated to this day. Three key elements from Korey's method have been passed on: (1) the selection of an enriched tissue, e.g., white matter for separation of oligodendroglial cells; (2) dispersion of tissue with the aid of size-controlled sieves; and (3) density gradient centrifugation of crude cell suspensions. Rose[63] exchanged sucrose for Ficoll, but no attempt was made to distinguish different glial cell types until Fewster and co-workers[280,281] adopted Rose's technique for the isolation of oligodendrocytes from bovine and rat white matter. Another important technical advance was the introduction of enzymes to loosen tissue connectivity prior to mechanical disruption.[65]

The middle 1970s denote a turning point in the history of studies on oligodendrocytes, as we see a shift from purified fractions to pure preparations to viable cells. The first report that oligodendrocytes can be maintained *in vitro* is that of Fewster and Blackstone[282]; soon others emulated them. Poduslo and McKhann[283] kept oligodendrocytes in suspension culture for 2 days and showed that they had a metabolism compatible with their counterparts *in situ*. Similar findings followed.[284–286] The next leap forward was the achievement of long-term primary cultures of oligodendrocytes. These have now been established from ovine brains,[71,72,287,288] feline and porcine brains,[289] bovine brains,[73] and rat brains.[74]

We present a detailed account of a recently modified version of the method of Szuchet *et al.*[71,72] because we have first-hand experience with it. This modification was introduced in order to (1) simplify the procedure, (2) shorten it wherever possible, and (3) scale it up so that twice the number of brains can be processed in the same amount of time. The entire operation is carried out at 4°C in a laminar-flow hood with the exception of the trypsinization and centrifugation steps. However, the centrifuge is right next to the hood, and the water bath is in the same room. The room is kept under ultraviolet light when not in use. A refrigerated centrifuge with a rate controller and a swinging bucket rotor is essential. We use a Sorvall RC-5 refrigerated centrifuge and the HS-4 and GSA rotors. The machine has been modified by the addition of a high resolution slow speed control, which permits one to set and control any speed from 500 to 3000 rpm within less than 10 rpm. Speed is monitored with a frequency counter. Recently, an integrator has been added to the circuitry that allows one to maintain the total work constant; i.e., $\int_{t_1}^{t_2} \psi^2 dt$ = constant. It also permits one to vary speed and time and still obtain the same resolution. Although this bit of sophistication is very convenient, it is not essential, and the lack of it should not deter anybody from using this technique. Provided the line voltage is fairly constant, and provided the centrifuge has a rate controller, a good resolution of bands can be obtained without the modification.

Table III presents a flow chart of the procedure for isolation of oligodendrocytes as currently practiced in our laboratory.[290] Briefly, brains from 4- to

Table III
Procedure for Isolation of Oligodendrocytes[290]

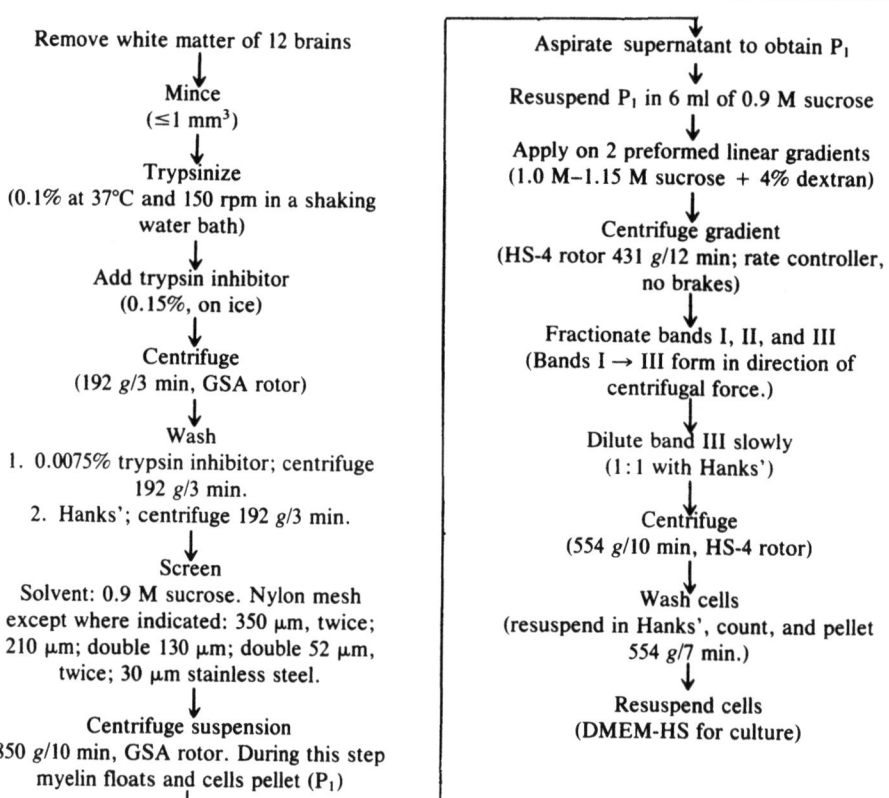

6-month-old lambs are obtained at a slaughterhouse immediately after the death of the animals; they are immersed in chilled Hanks' solution,* placed into a bucket of ice, and brought to the laboratory within 1 h of death. The centra ovales and corpora callosa are dissected free of gray matter and placed in four preweighed beakers containing 80 ml of cold Hanks' each. We have found that it is more efficient to have four people doing this step. Two of them cut and preclean the white matter; the others finish the cleaning. This takes 20–30 min. From this point, a single person continues with the entire preparation.

After all pieces have been cleaned, net weight is determined. This usually amounts to 20–22 g per beaker. The pieces are transferred to a cold stainless-steel plate, minced finely and put into four 500-ml Erlenmeyer flasks; 200 ml

---

* Hanks' balanced salt solution (Hanks') without Ca^{2+}, Mg^{2+}, bicarbonate, and phenol red is used as solvent either at full (H) or at half strength (H/2). All solutions are sterilized by filtration through 0.22-μm filters and cooled at 4°C with the exception of the trypsin solution, which is kept at room temperature.

of the trypsin solution† is added to each, and the flasks are positioned in a water bath at 37°C for 2.5 min/g white matter. The subdivision of white matter is done for two reasons: first, not to extend the time of trypsinization beyond 1 h; second, to insure a large surface area of contact between tissue and trypsin solution. The flasks are cooled for 2 min; 60 ml of trypsin inhibitor‡ is added; the contents are then mixed, distributed into six 300-ml polycarbonate bottles, and centrifuged. The softened tissue is pooled into two bottles, washed with 0.008% trypsin inhibitor and next with Hanks', with centrifugation after each washing.

Final disruption of tissue is accomplished by screening. The base of a 50-ml plastic syringe is sawed off and replaced by a 7-cm^2 piece of 350 μm nylon screen, which is secured with a 2.5-cm-wide Tygon® sleeve and further fastened with a #20 stainless-steel tubing clamp (see ref. 71). The tissue is suspended in 50 ml of 0.9 M sucrose* in H/2, poured into the syringe, pushed (gently) through the screen with the plunger, and collected in a beaker, care being taken to recover material that adheres to the outer surface of the screen. The procedure is repeated using another syringe and adding more solvent. A two-piece polypropylene funnel is mounted with a 210-μm screen, and the crude suspension is passed through it by spreading the material with fingertips (surgical gloves are worn), squirting solvent (i.e., 0.9 M sucrose in H/2) with a wash bottle, and slightly tapping the funnel. No vacuum is used. Sequentially, the suspension is passed through a series of screens (Table III). A total of 800 ml of solvent is used during the screening. The crude suspension is centrifuged. During this step, myelin floats to the top of the tube while the cells pellet (P_1). Myelin and supernatant are aspirated off, leaving only P_1. It is critical that the suspension be totally removed, since failure to do so leads to streaming during gradient centrifugation. The P_1 is suspended in 5–6 ml of 0.9 M sucrose in H/2 and applied on a linear gradient made of 75 ml of 1.0 M sucrose and 60 ml of 1.15 M sucrose containing 4% (w/v) dextran 70. Two such gradients are used in the preparation. They are layered simultaneously with the aid of a two-channel peristaltic pump and two home-made glass gradient makers (two inverted Erlenmyer flasks, bottoms cut off, are joined at the narrow end with a glass capillary tubing and a Teflon® stopcock). Gradients are prepared in the cold room within a small hood, and the timing is such that the gradients are completed at the same time as the cell suspension is ready for application. The latter is layered carefully on top of the gradients and centrifuged at 431 g for 12 min. The rate controller in conjunction with the frequency counter and integrator are used to start and stop the machine. No brakes are employed.

Three bands separate on this gradient (I, II, and III). Band I contains predominantly red blood cells and is discarded. We have previously shown

---

† Solution 0.1% trypsin and 0.02% EDTA dissolved in Hanks' containing 2% (w/v) Ficoll and adjusted to pH 6.00 ± 0.05. This solution is made 10 × concentrated and stored frozen.
‡ Trypsin inhibitor in Hanks'. The concentration of this solution is varied according to the manufacturer's specification of activity so as to inactivate all trypsin used. We use a 1.5- to two-fold excess on a molar basis.
* Solutions of 0.9 M and 1.0 M sucrose in H/2, and 1.15 M sucrose containing 4% (w/v) dextran 70 in H/2 are made daily.

that cells from bands II and III contain oligodendrocytes.[72] Because only work pertaining to cells isolated from band III is presented here, what follows is limited to these cells. Band III cells are carefully removed with a Pasteur pipette, diluted one-to-one by slow addition of Hanks', and centrifuged. The cells are washed once with Hanks' and pelleted again. Further handling of cells depends on experimental requirements. The method as described above yields on the average a total of 2×10^8 cells (bands III, only), i.e., approximately 2.5×10^6 cells per gram of wet white matter.

Other authors have developed somewhat different methods based on the use of Percoll for the density gradient step. Gebicke-Härter et al.[289] use 8- to 12-week-old cat or 6-month-old pig white matter as their starting material. Brains are perfused *in situ* with a solution containing 10 mM potassium phosphate, 5% glucose, 5% fructose, 0.1% collagenase, 0.1% hyaluronidase, and 1.5 mM $CaCl_2$ for 20 min. The same solution is used throughout the cell isolation. After dissection of the white matter, the tissue is further disintegrated by passing it through a series of screens. The suspension is then mixed with a solution of Percoll and centrifuged for 30 min at 10,000 g in an SW30 rotor. Oligodendrocytes are collected close to the bottom of the tube. Yield is not reported. Lisak et al.[73] use an approach very similar to that of Gebicke-Härter et al.[289] They start with 2- to 5-month-old calf brain white matter, which is trypsinized prior to disruption through screens. Oligodendrocytes are also isolated on a Percoll gradient. Yield is $15-45 \times 10^6$ cells per 40 g of white matter, which is equivalent to $0.4-1.2 \times 10^6$ cells per gram tissue.

The procedures of Gebicke-Härter et al.[289] Lisak et al.,[73] and Szuchet et al.[70,71] are comparable in terms of amount of labor involved and cost of operation, with perhaps the latter one being the most complex. They are also comparable in the degree of purity of the initial preparation. Szuchet et al.[70,71] have a better yield per gram of tissue (about twice) than Lisak et al.[73] Percoll[73,289] could have an advantage over sucrose because of its low osmolarity and might, therefore, be less harmful to cells. It can, however, be surmised from the experience of Szuchet et al.[70,71] that within the limits of sucrose concentration used, oligodendrocytes are able to recover well from hypertonicity. On the other hand, it appears that sucrose is probably the best medium to separate oligodendrocytes from myelin precisely because of its high tonicity. By separating out myelin at an early step, Szuchet et al.[71] are able to use a shallow linear gradient that permits the separation of subsets of oligodendrocytes in a highly reproducible fashion.

5.2.2. Procedures Based on Immunoaffinity

Although the general applicability of this approach has yet to be tested, and hard data on long-term culture of oligodendrocytes isolated by this method have yet to appear, we have included this approach here because (1) of the promise it holds and because (2) this is the direction where emphasis will be placed in future developments for cell fractionation. Meier and Schachner[291] and Meier et al.[292] have shown that it is possible to isolate oligodendrocytes with high selectivity (91 ± 4%) from dissociated mouse cerebella (from 7- to

8-day-old animals) using Magnetogel® beads coated with a monoclonal antibody against oligodendrocytes (Section 2.2.4). Highly purified cells survive in culture for 1 week, but in cultures of lesser degree of purity, oligodendrocytes can be maintained for longer times. Joseph et al.[293] have used an antigalactocerebroside serum to separate oligodendrocytes from a similar source. Only 5–8% of the total amount of cells are retained by the beads. They have maintained these cells for up to 6 weeks and claim 95% purity. It is too early to assess the impact of this approach.

5.2.3. Procedures Based on Cell Stratification

This procedure was initially developed by McCarthy and De Vellis[55] and has since been adopted and modified by others (e.g., ref. 294). The method stems from an observation that planting of dissociated cells from perinatal rodents at high density results in cultures consisting of an astrocytic monolayer on top of which oligodendrocytes rest (Section 2.2.2). McCarthy and de Vellis[55] devised a way to separate the two populations of cells to give independent astrocytic and oligodendrocytic cultures. Briefly, their procedure consists of maintaining the mixed culture for 10 days. At day 10, the culture flask is placed on a rotary shaker at 250 rpm at 37°C for 12 h. Suspended cells are removed and filtered through Nitex® (33 and 17 µm). The filtrate is centrifuged at 40 g for 5 min, and pelleted cells are resuspended in culture medium and plated at 3×10^4 cells/cm^2. In this way, oligodendrocyte cultures of 98% purity can be obtained. Repeating the procedure once increases the purity of the cultures to better than 99%. The authors state that approximately $1-1.2 \times 10^7$ viable cells can be obtained. However, they fail to say the amount of tissue required to achieve this.

McCarthy and De Vellis' method offers an elegant way of preparing highly purified fractions of oligodendrocytes and astrocytes simultaneously. Moreover, oligodendrocytes isolated this way are virgin cells from the standpoint of myelination and thus are distinct from the cells isolated by the other procedures. However, this technique too has its drawbacks: first, it can only be used at certain times during development; second, purity of cultures does not last more than a couple of days, as the cultures are overgrown by astrocytes. Bhat et al.[294] have addressed the latter issue by introducing a lysis step based on an observation by Snyder et al.[295] that astrocytes are lysed when incubated in a balanced salt solution at pH 7.2. and claim that in this manner they can preserve the purity of their cultures.

5.3. Establishment of Oligodendrocyte Cultures from Cells Isolated by Gradient Centrifugation

5.3.1. Procedure

There are three prerequisites for the establishment of oligodendrocyte cultures: first, to have viable oligodendrocytes; second, to find a substratum to which the cells will attach; and third, to concoct a nutritional medium that

Table IV
Procedure for Oligodendrocyte Culture[290]

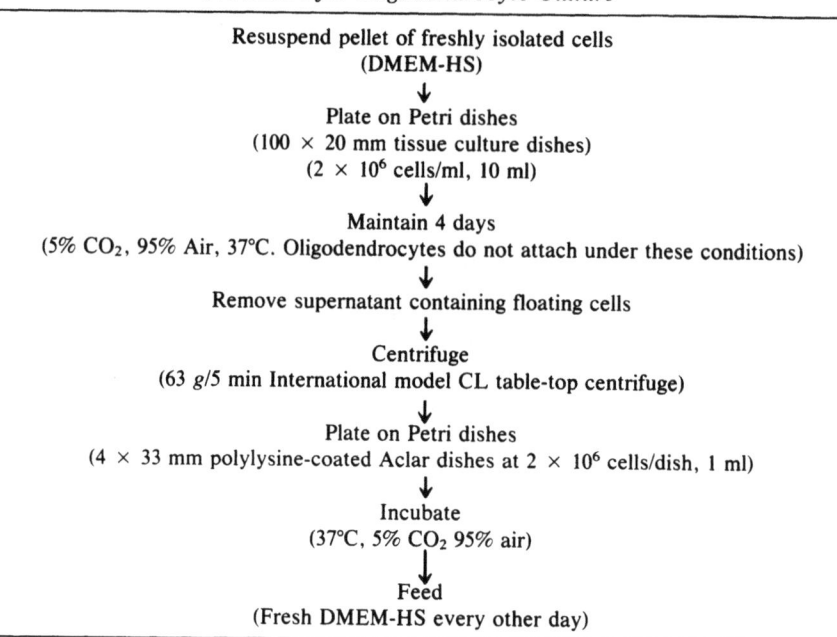

would support the growth of oligodendrocytes but at the same time be inhibitory to the growth and proliferation of other cell types.

There are as many "recipes" for culturing oligodendrocytes as there are methods for their isolation. The time is not yet ripe for a comparison of the virtues or demerits of these because most systems have not been put to a test above and beyond cell identification. For this reason, we only describe the procedure used in our laboratory and refer the reader to the appropriate references to seek out the others.

After examining a variety of media and supplements we opted for Dulbecco's modified Eagle's medium supplemented with 20% horse serum (DMEM-HS) plus 2 mM glutamine, 30 μg/ml of amphotericin B, and 5 μg/ml of gentamycin. The antibiotics can be discontinued after the first week *in vitro*. The technique is outlined in the flow chart (Table IV). A pellet of freshly isolated cells from band III (see ref. 72) is washed once by gentle resuspension in DMEM followed by centrifugation at 63 g for 5 min; the new pellet is resuspended in DMEM-HS and plated at a concentration of 2×10^6 cells/ml. Cultures are sustained in an incubator at 37°C in an atmosphere of 95% air/5% CO_2 at 90% humidity. Oligodendrocytes grow well in the above medium; they are metabolically very active and appear highly differentiated. On the other hand, potential contaminants such as fibroblasts and astrocytes do not seem to prosper. This might well be because of conditions employed during the isolation procedure, since the medium is very similar to the one used for astrocytic cultures (Section 3.1.1).

Oligodendrocytes exhibit selectivity in their interaction with the substratum, and the behavior of these cells in culture is the resultant of an interplay of cell–substratum and cell–cell forces. Thus, if cells are plated on a nonattractive surface, cell–cell interaction predominates, and the cells do not attach. Instead, they form cell aggregates (floating clusters). If, on the other hand, the cell–substratum interaction is strong, the cells attach and spread out in a monolayer, retaining close proximity to each other. We have used this property to isolate a highly differentiated subpopulation of oligodendrocytes from less differentiated cells[289]: initially, oligodendrocytes are seeded on plastic tissue culture dishes in which 60–80% of the cells will not attach but will form floating clusters. These cells are referred to as $B_{3,f}$ oligodendrocytes to distinguish them from the cells that do attach and also express oligodendrocytic functions, though to a lesser extent.[290,296,297] The cells are left to float for 96 h, at which time the medium containing the floating clusters is removed and centrifuged (Table IV). Subsequently, the supernatant is removed by simply inverting the tube, and the pellet is resuspended by gently tapping the tube in the small volume of medium that drains from the walls of the tube. It is important, at this stage, to make sure that the pellet is completely disintegrated and that there are no visible clumps. Then DMEM-HS is added in sufficient amount to give the desired cell concentration, and cells are plated on polylysine-coated plastic Petri dishes (35 mm or 100 mm) or home-made Aclar (Allied Chemical Corp. Pottsville, PA) Petri dishes (25 mm). Under these conditions, the cells attach, usually within 24 h, and extend processes (see below). Feeding schedules are twice weekly for cultures kept in large dishes (35 mm and 100 mm) and every other day for those in the small dishes.

5.3.2. Importance of Substratum

The influence of the substratum was already touched on above. Oligodendrocytes appear to have restrictive requirements in their interaction with the substratum; they also have a strong tendency to self-associate. *In vivo*, an equilibrium must be reached between these opposing forces that results in an arraying of cells in rows with the formation of tight junctions between neighboring cells.[298] A similar balance of forces seems to be operative *in vitro*.[290] Thus, when substrate requirements are fulfilled, oligodendrocytes attach and align not too differently from their counterparts *in situ*; they also form tight junctions.[299,300] But if the surface does not suit the needs of the cells, they choose to interact among themselves by forming aggregates. Junctional complexes are, however, also formed between the latter.[299-301]

Szuchet and Yim[290] state that oligodendrocytes do best on a surface that carries a positive charge, such as can be provided by polylysine (Section 2.2.4). The orientation of the charges is of importance. For instance, these authors find that polylysine-coated glass is not as effective a surface as polylysine-coated Aclar®; they speculate that this may be because polylysine attaches to the glass via electrostatic forces, creating a hydrophobic surface, whereas it must interact hydrophobically with Aclar®, leaving the charges free. Meier and Schachner[291] do not seem to have experienced similar difficulties, since they maintain oligodendrocytes on polylysine-coated glass. However, they do report considerable cell death. The importance of the nature of the substratum was also noticed by Gebicke-Härter et al.[289] They compared polylysine, fetuin, and mucin and concluded that the latter, at a concentration of 5 mg/100 ml, was the best for attachment but did not lead to outgrowth of processes. Gebicke-Härter et al.[289] therefore chose fetuin over polylysine as the surface for oligodendrocyte growth because they found the latter to induce cell aggregation. The tendency of oligodendrocytes to "clump" has also been reported by Bhat et al.[294] It is not, however, seen in little-differentiated or dedifferentiated oligodendrocytes (S. Szuchet, unpublished observations).

5.3.3. Cell Survival

Perhaps the closest approach to a quantitative study on oligodendrocyte survival was done by Meier and Schachner,[291] who found that in cultures of high purity (>90% oligodendrocytes) obtained by immunoaffinity methods, the cell density, defined arbitrarily as the number of cells in the field of vision, dropped drastically so that after 1 week, barely any cell was left. On the other hand, in cultures in which oligodendrocytes represented approximately 70% of the cell population, the cells were faring well, and after a week, most oligodendrocytes were still viable. A plating efficiency of 10% was given by Gebicke-Härter et al.[289] and of 40–70% by Lisak et al.[73] This point is not addressed by others.[55,294] We concur with Gebicke-Härter[289] that cell survival is a function of many variables, some of which (e.g., the precise age of the animals obtained at an abattoir) escape our control.

We have not performed a quantitative study of cell survival and hence can only give a qualitative appreciation. Currently, we plate cells at a density of $4.1 \times 10^5/cm^2$. This is double the density we have used in the past[72]; it is an order of magnitude higher than that used by Meier and Schachner[291] and others. It is our observation that oligodendrocytes survive better when plated at high rather than low concentrations. This might explain the previously mentioned observation by several different authors that densely seeded dissociated cultures from the perinatal rodent brain contain a substantial amount of oligodendrocytes, whereas sparsely seeded cultures are virtually devoid of this cell type (Section 2.2.2). It is also our experience that oligodendrocytes fare better in highly purified cultures than in less pure ones. In this, we disagree with Meier and Schachner.[291] Under normal circumstances, cell death in our cultures does not exceed 20%.

After 4 to 5 weeks in culture, $B_{3,f}$ oligodendrocytes start to aggregate.[290] This occurs relatively slowly, and it may take a week or more until all processes are retracted and only isolated clumps are left. No explanation can be offered at present for this observation. We suspect that it might have to do with the accumulation of substance(s) inhibiting continued cell attachment. The phenomenon is unlikely to originate from cell senescence because mere trypsinization and replating "revive" the cells so that they are able to go on for another 5 weeks or so.

5.3.4. Cell Identity and Cell Markers

Most methods currently in use for isolating oligodendrocytes[55,65,71–73,289,295,302] yield preparations of high degree of purity. The crux of the matter is to be able to retain or, even better, improve the pureness of the cells as they are grown *in vitro*.

In most recent publications, cell identification is largely based on enumeration of cells by immunocytochemical staining rather than by morphological and ultrastructural criteria. No doubt this is the most powerful tool now available, but the need of caution in characterizing antisera and in interpreting results is great.

Both positive markers that tell what the cells are and negative markers to know what the cells are not have been used. Among the positive markers, GalCer (galactocerebroside) has long been accepted as a membrane component that specifically identifies oligodendrocytes.[303] Accordingly, most data reported make use of anti-GalCer sera to identify both freshly isolated cells and cells that have been maintained in culture. Two other positive markers have been used, anti-MBP (myelin basic protein)[38] and antibodies raised against oligodendrocytic membrane antigens.[97] In Table V we have compiled the available data. Where no numbers were given by the authors, we have translated their qualitative statements into pluses. Table V shows two important things. First, independent of the method used for cell isolation, better than 95% of the freshly isolated cells are GalCer positive. However, neither GalCer nor MBP are good markers for freshly isolated cells derived from white matter because of the danger that these components might arise from myelin and are simply adsorbed on the surface of the cells. For example, Szuchet et al.[72] found that freshly isolated cells stained positively with anti-MBP independently of whether they were fixed (MBP is an intracellular antigen and should not stain unfixed cells). They explain their results as being caused by remnants of myelin or adsorption of the protein on the surface of the cells. Second, where stated, it is seen that beyond 1 week, the cultures no longer retain their initial purity. An exception to this is the data given by Szuchet and Yim[290]; their cultures are 98–99% GalCer and antioligodendrocytic serum positive after 15–25 days in culture (Fig. 9d).

Among negative markers, GFA that specifically detects astrocytes is most commonly used, although others, e.g., fibronectin, which identifies fibroblasts and certain other cells,[130] and neuron-specific enolase (Section 4.2.1), have also been used (Table V). It can be seen that at least in one case astrocytes are responsible for overtaking the cultures.

The importance of using more than one antiserum to identify the growing cells stems from recent reports[304,305] that claim that anti-GalCer and antisulfatide may possibly stain cells other than oligodendrocytes. This is against a large volume of data that point to the contrary. Although these observations must await further judgment from other laboratories, they do point to the need and desire to use more than one criterion for classifying cells.

Table V
Immunocytochemical Staining of Freshly Isolated and Cultured Oligodendrocytes

Culture age (days)	GalCer[a]			MBP[b]			OLG[c]			S-100[d] F/I	GFA[e] F/I	NSE[f] F/I	FBN[g] S	IgG-FITC[h] S	Reference
	S[i]	F/I[j]	P[k]	F/I	S	F/I	P								
1	>95	>95	---	>95	---	---	---			2	<1	<1	---	---	290
15	98–99	---	98–99	---	0	98–99	0			0	<0.5	0.01	0	<1	
25	98–99	---	98–99	---	0	98–99	0			0	<0.5	---	0	<1	
1	97–99	---	---	50–60	---	---	---			---	<1	---	<1	---	73
7	90–95	---	---	30–40	---	---	---			---	5–10	---	<1	---	
14	50–70	---	---	30–40	---	---	---			---	30–50	---	<1	---	
21	30–50	---	---	30–40	---	---	---			---	50–70	---	<1	---	
1	---	+++	+++	+	---	---	---			---	0	---	---	---	294
20	---	---	---	+++	---	---	---			---	---	---	---	---	
1	---	90–95	---	---	---	---	---			---	<1	---	---	---	289
14	---	---	---	---	---	---	---			---	0	---	---	---	
1	---	---	---	---	91	---	---			---	+	---	---	---	291
3	---	---	---	---	95	---	---			---	+	---	---	---	
7	---	---	---	---	97	---	---			---	+	---	---	---	

[a] Galactocerebroside, an oligodendrocytic marker.
[b] Myelin basic protein, an oligodendrocytic marker.
[c] Oligodendrocytic antibodies, oligodendrocytic markers.
[d] Glial (oligodendrocytic and astrocytic) marker.
[e] Glial fibrillary acidic protein, an astrocytic marker.
[f] Neuron-specific enolase.
[g] Fibronectin, a marker of nonneural tissue, including fibroblasts, meningeal, and endothelial cells.
[h] Rabbit IgG fluorescein conjugate, an Fc receptor (macrophage) marker.
[i] Surface staining of unfixed cells.
[j] Fixed cells and/or intracellular staining.
[k] Staining of processes.

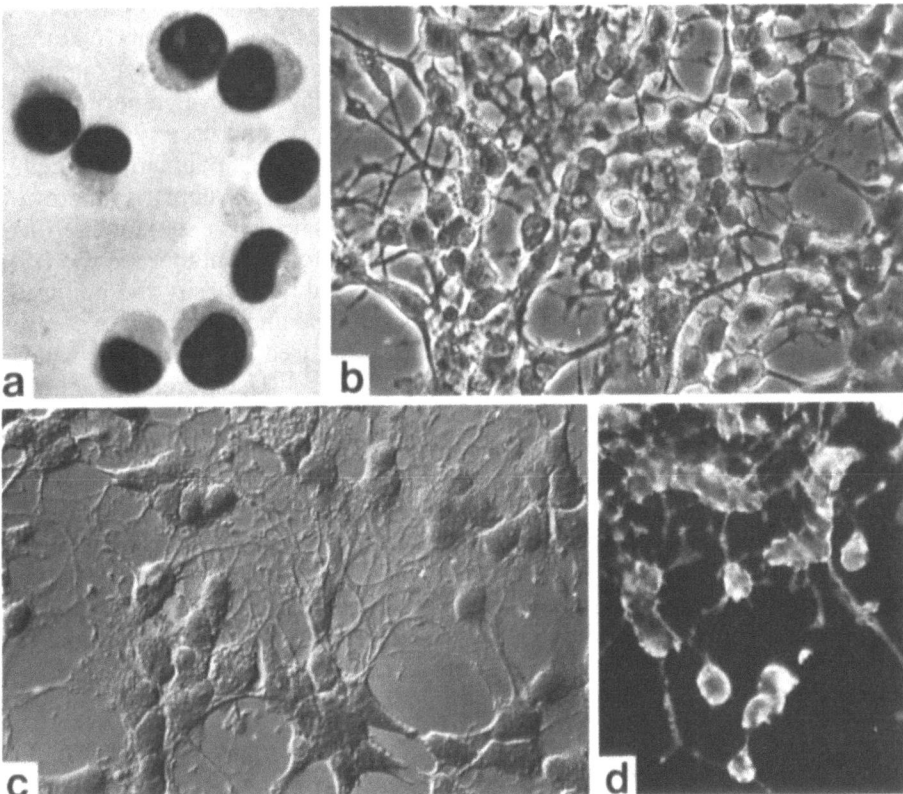

Fig. 9. Light microscopic photographs of cultured oligodendrocytes. a: Floating oligodendrocytes after 4 days *in vitro*. Cells are stained with 2% orceine in lactic–acetic acid. b: Phase-contrast micrograph of live oligodendrocytes after 20 days in culture showing alignment of cells in rows and an extensive network of processes. Note cell–cell contacts; most of these are tight junctions. c: Fixed oligodendrocytes after 20 days *in vitro* as viewed by Nomarsky optics. Note intricate mesh of fine processes not visualized by phase contrast. d: Cultured oligodendrocytes after 15 days *in vitro*; surface stained by indirect immunofluorescence with a monoclonal antibody against galactocerebroside (generous gift of Dr. B. Ranscht). More than 98% of the cells show bright fluorescence.

5.4. Properties of Cultured Oligodendrocytes

5.4.1. Morphology

Unattached oligodendrocytes do not differ in appearance from freshly isolated cells (cf. Fig. 1 in ref. 72) except, perhaps, that they appear more robust than the latter, having had a chance to recover from the trauma of isolation (Fig. 9a). While in the floating state, the cells do not extend processes, nor do they synthesize redundant membranes. Process extensions occur only after the cells have attached to the substratum. The variability of process morphology has been described.[289,290] It is evident that with time in culture oligodendrocytes build an intricate network of processes (Figs. 9a and c) and align in rows (Fig. 9b).

The ultrastructure of cultured oligodendrocytes[301, 306] exhibits features common to oligodendrocytes *in situ*.[307] Recently, Massa *et al*[299,300] reexamined the ultrastructure of $B_{3,f}$ oligodendrocytes after 1 day and over a period of 4 weeks using transmission and freeze–fracture electron microscopy. They describe the cells, viewed by thin sections, as a homogeneous population of cells of medium to high electron density, rich in organelles, particularly microtubules, that have a well developed Golgi apparatus. The cells extend numerous processes, cylindrical and flattened, which adhere to one another, to the cell body, and to the dish (Fig. 9). No attempt was made to classify the cultured cells according to the subgroups described by Leblond and co-workers.[33,41] On the other hand, Gonatas *et al*[308] characterized cultured rat oligodendrocytes as belonging to the medium-type oligodendrocytes.[33,41] Using freeze-fracture, Massa *et al*.[299,300] report that oligodendrocytes in 1-day suspension cultures show no cell junctions, although tight junctional arrays are formed *de novo* in culture and become increasingly numerous with time. No gap junctions were found by these authors. These results are in contrast with the observations of Gonatas *et al.*,[308] who describe pentalaminar and heptalaminar profiles consistent with tight and gap junctions between plasma membranes of processes and between perikarya and processes of rat oligodendrocytes. In trying to account for this discrepancy, Massa *et al*.[300] state that "culture conditions and/or species differences may modify junction formation by purified oligodendrocytes in culture."

5.4.2. Biochemical Properties

5.4.2a. Lipid Synthesis. In this review, we have placed emphasis on long-term cultures and therefore did not present earlier work dealing with short-term cultures, which has been discussed in a recent review.[288] It should, however, be stated that those early studies[283–285,309,310] were important in establishing the fact that isolated oligodendrocytes are metabolically active.

Mack and Szuchet[311,312] were the first to follow the incorporation of [^{3}H]galactose and $^{35}SO_4^{2-}$ into myelin-typical glycolipids, galactocerebrosides, and sulfatides by cultured oligodendrocytes as a function of time in culture for 70 days. They also examined the incorporation of galactose into gangliosides.[313] Their work showed that oligodendrocytes preferentially synthesize components of myelin and that there is a significant increase in the levels of incorporation as a function of time in culture, with a maximum at approximately 20 days. Szuchet *et al*.[297] examined the overall lipid metabolism of $B_{3,f}$ cells as a function of time in culture for up to 35 days using four precursors: $^{35}SO_4^{2-}$, [^{3}H]galactose, [^{14}C]acetate, and [^{3}H]glycerol (Fig. 10). Not surprisingly, they found that $^{35}SO_4^{2-}$ and galactose were metabolized into glycolipids, whereas [^{14}C]acetate and [^{3}H]glycerol were incorporated preferentially into neutral lipids and phospholipids, respectively. The age dependence of each of these compounds was different. Thus, the uptake of [^{3}H]galactose into glycolipids was almost constant for the first 20 days but increased moderately thereafter. In contrast, the intake of $^{35}SO_4^{2-}$ into sulfatide followed a pattern that at 35 days was still on the increase. A fourfold increase in the incorporation [^{3}H]glycerol

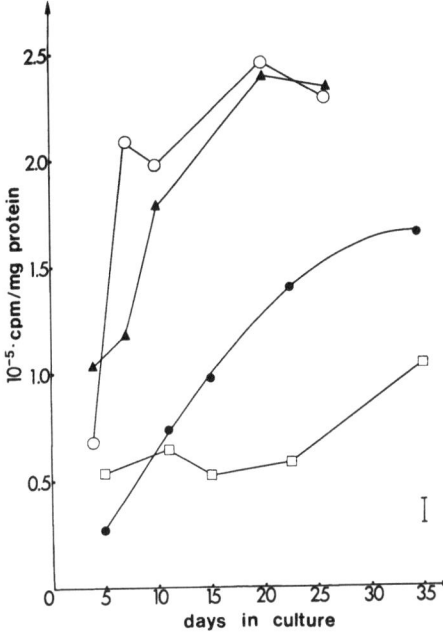

Fig. 10. Overall lipid metabolism of cultured oligodendrocytes as a function of time in culture. At selected times, cultures were exposed for a total of 72 h to either 15 μCi/ml of $H_2^{35}SO_4$ (782 mCi/mmol, carrier-free) plus 1 μCi/ml of [^{3}H]galactose (11.5 Ci/mmol) or to 1.5 μCi/ml of [2-^{3}H]glycerol (500 mCi/mmol) plus 2 μCi/ml of [1-^{14}C]acetate (56.7 mCi/mmol). Thereafter, the cultures were harvested and washed three times with Hank's BSS, centrifuging after each wash. The final wet pellet was processed to separate the various lipids as described by Szuchet et al.[297] ○, incorporation of [^{3}H]glycerol into total phospholipids; ▲, [^{14}C]acetate into neutral lipids; ●, $^{35}SO_4^{2-}$ into sulfatide; □, [^{3}H]galactose into glycolipids.

into phospholipids was observed from day 4 to day 7. Compared to this, the increase in the uptake of [^{14}C]acetate into neutral lipids (essentially cholesterol) during the same span was modest at first, but by 10 days it reached levels comparable to those of [^{3}H]glycerol. From then on, these two precursors ran a parallel course that peaked at day 20 and started on a downward trend thereafter.

The correlation of these observations with morphological events is both important and interesting. The measurements taken at 4 days correspond to cells that have never attached and have not sent out processes. Thus, the burst in the synthesis of phospholipids followed by cholesterol coincides with a period of extensive membrane accumulation required for cell attachment and process extension. Differentiation of these membranes through the addition of specific glycolipids proceeds at a relatively slower pace (Fig. 10). This sequence of events fits remarkably well with what takes place in vivo.[314]

Phospholipids, neutral lipids, and glycolipids account for almost all lipids present in oligodendrocytes and myelin.[65,302,314] Their relative distribution is, however, different: myelin is enriched in cholesterol and glycolipids at the expense of phospholipids. Within the latter, the predominance of classes is shifted. For instance, phosphatidylcholine, the major component of oligodendrocytes, becomes second to phosphatidylethanolamine within myelin.[314-319] This modulation is a slow process that occurs during development and transforms what was once an extension of oligodendrocyte plasma membrane into an independent entity, myelin. When Fig. 11 is analyzed in the light of the above, it becomes evident that oligodendrocytes isolated post-myelination and maintained in culture are able to carry out functions associated with myeli-

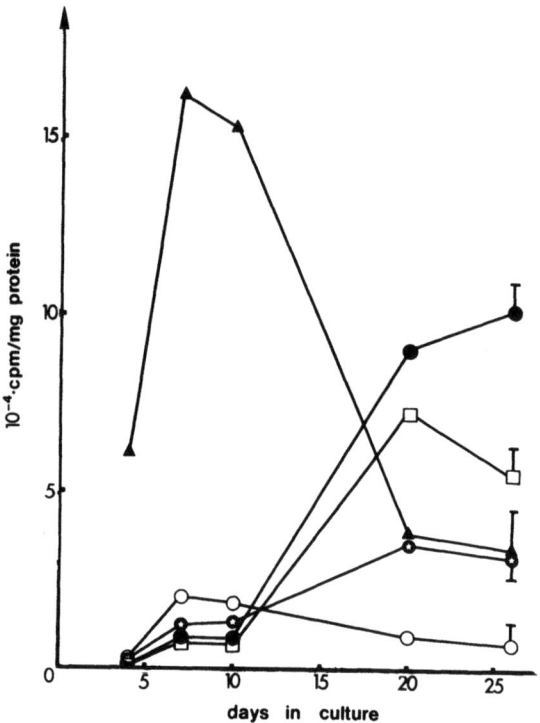

Fig. 11. Incorporation of [^{3}H]glycerol into phospholipids by cultured oligodendrocytes. Experimental details as given in Fig. 10. ▲, phosphatidylcholine; ●, phosphatidylserine plus phosphatidylethanolamine; □, phosphatidylinositol; ⊙, lysolecithin; ○, sphingomyelin. Bars are standard deviations. (From ref. 297, with permission.)

nogenesis. Examination of Figs. 10 and 11 shows that after 15 days *in vitro* (about 10 days after attachment), oligodendrocytes undergo a change from a metabolism suitable for the synthesis of their own membranes to one geared towards myelinogenesis. Moreover, the fact that the relative proportions of each of the phospholipids fall within expected levels[320] is yet another indication that oligodendrocytes remain highly differentiated in the *in vitro* microenvironment provided and that the absence of other cell types or specific humoral factors does not lead to metabolic aberrations.

5.4.2b. Enzymes Specific to Oligodendrocytes and Myelin. Cyclic nucleotide phosphodiesterase constitutes the major component of Wolfgram myelin proteins.[321,322] It appears to be localized in oligodendrocyte plasma membrane, polysomes, and myelin[323,324] but has also been found in "transitional" myelin membranes.[325] The function of CNPase and its natural substrate are unknown. *In vitro*, CNPase hydrolyses 2′,3′-cyclic AMP, a property that is used for its quantitative determination. However, because CNPase has a developmental pattern not unlike those of CST (sulfogalactosyltransferase) and CGalT (UDP-galactose:galactosyltransferase), and because its maximum ac-

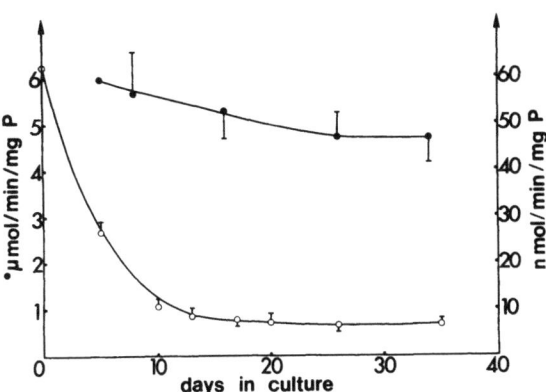

Fig. 12. Activities of CNPase and GPDH in cultured oligodendrocytes as a function of time in culture. ●, CNPase; ○, GPDH. Bars are SEM ($n = 10$). (From ref. 290, with permission.)

tivity coincides with myelination, this enzyme is an accepted marker for myelination.

Szuchet and Yim[290] followed the specific activity of CNPase in oligodendrocyte cultures as a function of time (Fig. 12). They observed a very high activity (6 μmol/min per mg protein) for floating $B_{3,f}$ (5 days *in vitro*), which matches or exceeds that reported for explant cultures,[326,327] maybe reflecting the content of other cell types in the latter. A similar value (7.1 μmol/min per mg) was observed by Bhat *et al.*[294] in 20-day-old oligodendrocyte cultures derived from mixed astrocytic–oligodendrocytic cultures prepared from newborn rat brain, i.e., premyelinating oligodendrocytes (Section 5.2.3). A much lower value (0.44 μmol/min per mg) was, however, reported by McCarthy and De Vellis[55] for cultures younger than those of Bhat *et al.*[294] After the $B_{3,f}$ oligodendrocytes attach, there is an almost linear decline in CNPase activity with a shallow slope. At 26 days *in vitro*, an apparent "steady state" is reached at 78% of the initial value (Fig. 12). The reason for this decay in activity is not known, but it is probably not related to the absence of other cell types, because it has been reported that in explant cultures in which all cell types are present, the mere fact of placing the system *in vitro* results in a 50% drop in activity.[326] *In vivo* and in explant cultures, the specific activity of this enzyme increases at first, goes through a maximum, undergoes a slight decrease, and then stays constant. Thus, this is one instance where the *in vitro* system follows a different path, but it should be kept in mind that the oligodendrocytes used by Szuchet and Yim[290] were obtained from 4- to 6-month-old lambs.

Another enzyme that has been primarily associated with oligodendrocytes is GPDH. This enzyme catalyzes the conversion of dihydroxyacetone phosphate to glycerol phosphate in the presence of nicotinamide adenine dinucleotide (NAD). Although this is a ubiquitous enzyme in rat brain, induction by glucocorticoids appears to be cell and tissue specific.[166,328] Thus, De Vellis and his co-workers found this enzyme to be a normal component of oligodendrocyte cytosol and its induction by hydrocortisone to be a unique feature of rat oligodendrocytes, and on this basis GPDH induction is advocated as a specific marker for oligodendrocytes.[55,329,330] The validity of these observations for other species has yet to be proven.[331]

In freshly isolated ovine oligodendrocytes, the specific activity of GPDH is very high (62 nmol/min per mg protein), but it drops exponentially with time in culture, and by 17 days *in vitro* it is only 10% of the value at isolation. Initial attempts to induce its activity with hydrocortisone proved unsuccessful in our cultures, but stimulation from 43 to 690 nmol/min per mg has been reported by McCarthy and De Vellis[55] in their rat brain cultures. Recently Cammer *et al.*[332] compared the activity of GPDH in rat forebrain homogenates with that found in oligodendrocytes isolated from the same brain. Interestingly, though perhaps coincidentally, the specific activity they report in oligodendrocytes isolated from 120-day-old animals is equal to the "steady-state" value reported by Szuchet and Yim[290] (Fig. 12).

5.4.2c. Immunocytochemical Detection of Myelin Proteins. One of the motives for striving to obtain oligodendrocyte cultures of a high degree of purity is because such systems, in addition to permitting a characterization of the cells, offer the opportunity to address fundamental questions pertaining to oligodendrocyte–neuron–astrocyte interactions and their role in the synthesis of myelin components and ultimately in the assembly of myelin itself (see also Section 6). Ideal targets for testing hypotheses concerning the involvement of the various cell types are proteins such as myelin basic protein (MBP), proteolipid protein (PLP), and myelin-associated glycoprotein (MAG) that have been shown by indirect immunofluorescence techniques to be present in oligodendrocytes at the time of active myelination but have not been detected in the mature cell.[38,333-335] The finding of these proteins in monotypic cultures of oligodendrocytes can be taken as evidence that their syntheses do not require the presence of other cells. That this is the case of MBP has been shown convincingly in several laboratories.[72,73,294] A similar observation of MAG has now been made in oligodendrocytes after 20 days in culture.[290]

5.5. Cell Lines

The partial oligodendrocytic characteristics of the C-6 cell line have already been described (Section 3.2). More purely oligodendrocytic cell lines are the 33B and 21A lines developed by Fields *et al.*[336] from ethylnitrosourea-treated rats and the G26 series obtained by Sundarraj *et al.*[337] from a transplantable mouse glioma. The latter cell lines have been studied in some detail. They do express oligodendrocytic characteristics, e.g., sulfate incorporation into sulfogalactosylceramide,[338] but many of these characteristics are expressed to only a limited extent.[17,69] Some of the lines in this series do, however, also contain the astrocytic-specific GFA[339] or are positive for the Na^+ ionophore, a neuronal characteristic.[340] A recent approach is the production of cell hybrids between C-6 glioma cells and normal oligodendroglia cells obtained by gradient centrifugation.[341] Some of these hybrids express a wide range of oligodendrocytic characteristics,[69,341] but their stability after subculturing remains to be established (Section 6).

6. CONCLUDING REMARKS

Cell and tissue culture was initiated by Harrison[3] to answer one specific question, i.e., what is the origin of the nerve fiber? This ingenious endeavor was followed by a period in which the major efforts were to demonstrate that tissues can be grown *in vitro*. The recent developments in monotypic cell cultures rest on these efforts but also, as we hope we have made clear, on a vast basis of knowledge about the development and general biology of cells not only *in vitro* but also *in vivo*. The whole area of cellular neurobiology has benefited greatly from this development, as we are now beginning to obtain solid knowledge about what each of the three basic cell types, neurons, astrocytes, and oligodendrocytes, can do in isolation. One remarkable finding is the extent to which each of the cell types is able to develop or maintain its differentiated characteristics in the absence of other cell types. This, however, does not mean that cell-to-cell interactions are of no importance, since the cells used to establish monotypic cultures have already undergone a considerable development *in situ* while interacting with other cells.

Cell interactions are necessary for certain aspects of neuronal and glial differentiation, and tissue culture offers a unique possibility to study these interactions by omitting and/or recombining specific populations of cells. These aspects include the cytoarchitectonic development[342,343] and certain features of enzyme induction.[221,344,345] A fascinating area, which may be of considerable potential importance for understanding dysmyelination or demyelination, involves studies of interactions between neurons and myelinating cells from the peripheral[48] or the central[346] nervous system. In such studies, Wood *et al.*[346] observed that coculturing oligodendrocytes with dorsal root ganglion neurons induced cell division (recognized by [^{3}H]thymidine incorporation) in some of the oligodendrocytes as well as myelin formation.

Such experiments show very clearly that breaking the whole into its parts and then attempting to put the pieces back together, one at a time, is a good strategy for a complex system such as the CNS, and it is from such an approach that our understanding of the CNS will ultimately emerge. Another example of neuronal–glial interactions includes those that may occur between cells of different types exposed to a compound accumulated into either cell type. K_m and V_{max} values do not alone define the uptake rates to be expected in the CNS *in vivo*, since the extent to which the extracellular space is bounded by neurons or glial cells is uncertain. In addition to an uptake into either cell type, there is the fascinating possibility of direct transfer from one cell type to another. That such a transfer can, indeed, occur has been elegantly concluded by Latzkovits *et al.*[241] from the complex uptake curve for ^{86}Rb, a potassium analogue, into mixed neuronal–astrocytic chick brain cultures compared to the simple exponential uptake into monotypic astrocytic cultures (Fig. 13).

An example of a quite different interaction between the genomes of different cell types is cell hybridization, i.e., cell fusion. These cell hybrids are characterized by chromosomal loss, which often occurs selectively from one of the parental chromosomal complements,[271,347] before they end up with a

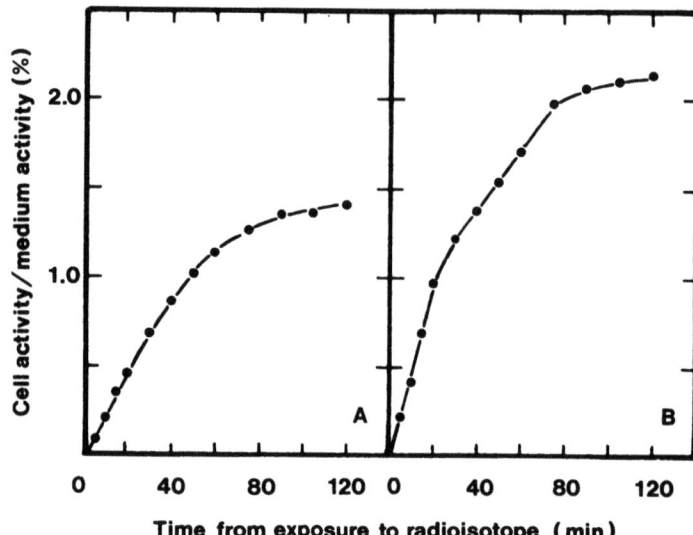

Fig. 13. Uptake of ^{86}Rb, a potassium analogue, into monotypic primary cultures of chick astrocytes (A) or corresponding mixed neuronal–astrocytic cultures (B). The complex uptake curve in the mixed cultures compared to an exponential uptake in the monotypic cultures indicates that in addition to an uptake of ^{86}Rb from the medium into either cell type, there also appears to be a direct transfer of ^{86}Rb between the two cell types. (From ref. 241, with permission.)

stable chromosomal complement derived from the two parental cells. Neuronal/nonneuronal cell hybrids have been used to study certain neurochemical aspects such as the biochemical properties and regulation of neurotransmitter receptors.[348] Their principal interest appears to be in the area of genetics. Like other cell lines, they may also be useful for studies of the biological mechanisms they do display. As such, they cannot be regarded as reflecting functional interactions between neurons and glial cells, which are characterized by interactions between autonomous cells.

ACKNOWLEDGMENTS. The authors' research has been supported by grants from the National Multiple Sclerosis Society (to S.S.) and from the Medical Research Council of Canada (to B.H.J.J. and L.H.). Dr. A. C. H. Yu is cordially thanked for permission to use passages from his M.Sc. thesis.

REFERENCES

1. Fedoroff, S., 1966, *Exp. Cell Res.* **46**:648–672.
2. Fedoroff, S., 1977, *Cell, Tissue and Organ Cultures in Neurobiology*, (S. Fedoroff and L. Hertz, eds.), Academic Press, New York, pp. 265–286.
3. Harrison, R. G., 1907, *Proc. Soc. Exp. Biol. Med.* **4**:140–143.
4. Von Mihálik, P., 1935, *Arch. Exp. Zellforsch.* **17**:119–176.
5. Murray, M. R., 1965, *Cells and Tissues in Culture, Methods, Biology and Physiology*, (E. N. Willmer, ed.), Academic Press, New York, London, pp. 373–455.

6. Murray, M. R., 1971, *Handbook of Neurochemistry*, Volume 5 (A. Lajtha, ed.), Plenum Press, New York, pp. 373–438.
7. Lehrer, G. M., Bornstein, M. B., Weiss, C., Furman, M., and Lichtman, C., 1970, *Exp. Neurol.* **27**:410–425.
8. Pontén, J., and McIntyre, E. H., 1968, *Acta Pathol. Microbiol. Scand.* **74**:465–486.
9. Pontén, J., and Westermark, B., 1980, *Advances in Cellular Neurobiology*, Volume 1 (S. Fedoroff and L. Hertz, eds.), Academic Press, New York, pp. 209–227.
10. Icard, C., Liepkalns, V. A., Yates, A. J., Singh, N. P., Stephens, R. E., and Hart, R. W., 1981, *J. Neuropathol. Exp. Neurol.* **40**:512–525.
11. Walum, E., Westermark, B., and Pontén, J., 1981, *Brain Res.* **212**:215–218.
12. Ham, R. G., 1972, *Methods in Cell Biology* (D. M. Prescott, ed.), Academic Press, New York, pp. 37–73.
13. Bock, E., 1977, *Cell, Tissue and Organ Cultures in Neurobiology* (S. Fedoroff and L. Hertz, eds.), Academic Press, New York, London, pp. 407–422.
14. Viñores, S. A., Marangos, P. J., and Ko, L., 1982, *Dev. Brain Res.* **5**:23–28.
15. Varon, S., and Saier, M., 1975, *Exp. Neurol.* **48**:135–162.
16. Rosenberg, R. N., Vandeventer, L., De Francesco, L., and Friedkin, M. E., 1971, *Proc. Natl. Acad. Sci. U.S.A.* **68**:1436–1440.
17. Pfeiffer, S. E., Betchart, B., Cook, J., Mancini, P., and Morris, R., 1977, *Cell, Tissue and Organ Cultures in Neurobiology* (S. Fedoroff and L. Hertz, eds.), Academic Press, New York, pp. 287–346.
18. Labourdette, G., and Mandel, P., 1978, *Biochem. Biophys. Res. Commun.* **85**:1307–1313.
19. Schousboe, A., 1977, *Cell, Tissue and Organ Cultures in Neurobiology* (S. Fedoroff and L. Hertz, eds.), Academic Press, New York, pp. 441–446.
20. Hertz, L., 1979, *Prog. Neurobiol.* **13**:277–323.
21. Hertz, L., and Chaban, G., 1982, *Neuroscience Approached through Cell Culture*, Volume 1 (S. E. Pfeiffer, ed.), CRC Press, Boca Raton, pp. 157–174.
22. Jacobson, M., 1978, *Developmental Neurobiology*, Plenum Press, New York.
23. Hinds, J. W., 1966, *J. Comp. Neurol.* **134**:287–304.
24. Schlessinger, A. R., Cowan, W. M., and Gottlieb, D. I., 1975, *J. Comp. Neurol.* **159**:149–167.
25. Altman, J., 1972, *Developmental Neurobiology* (W. A. Himwich, ed.), Charles C. Thomas, Springfield, Illinois, pp. 197–237.
26. Butler, H., and Juurlink, B. H. J., 1984, *An Atlas for Staging Mammalian and Chick Embryos*, CRC Press, Boca Raton (in press).
27. Brückner, G., Mareš, V., and Biesold, D., 1976, *J. Comp. Neurol.* **166**:245–256.
28. Raedler, E., and Raedler, A., 1978, *Anat. Embryol.* **154**:253–266.
29. Das, G. D., and Nornes, H. O., 1972, *Z. Anat. Entwickl. Gesch.* **138**:155–165.
30. Angevine, J. B., 1965, *Exp. Neurol.* [*Suppl.*] **2**:1–39.
31. Juraska, J. M., and Fifkova, E., 1979, *J. Comp. Neurol.* **183**:257–268.
32. Rickmann, M., and Wolff, J. R., 1976, *Exp. Brain Res.* [*Suppl.*] **1**:239–243.
33. Ling, E. A., and Leblond, C. P., 1973, *J. Comp. Neurol.* **149**:73–82.
34. Ichikawa, M., and Hirata, Y., 1982, *Dev. Brain Res.* **4**:369–377.
35. Smart, I., and Leblond, C. P., 1961, *J. Comp. Neurol.* **116**:349–367.
36. Latov, N., Nilaver, G., Zimmerman, E. A., Johnson, W. G., Silverman, A. J., Defendi, R., and Cote, L. D., 1979, *Dev. Biol.* **72**:382–384.
37. Sturrock, R. R., 1978, *J. Anat.* **122**:521–537.
38. Sternberger, N. H., Itoyama, Y., Kies, M. W., and Webster, H. deF., 1978, *J. Neurocytol.* **7**:251–263.
39. Hansson, E., Rönnbäck, L., Lowenthal, A., Noppe, M., Alling, C., Karlsson, B., and Sellstrom, A., 1982, *Brain Res.* **231**:173–183.
40. Sturrock, R. R., 1978, *J. Anat.* **125**:433–445.
41. Imamoto, K., Patterson, J. A., and Leblond, C. P., 1978, *J. Comp. Neurol.* **180**:115–138.
42. Abney, E. R., Bartlett, P. P., and Raff, M. C., 1981, *Dev. Biol.* **83**:301–310.
43. Sensenbrenner, M., 1977, *Cell, Tissue and Organ Cultures in Neurobiology* (S. Fedoroff and L. Hertz, eds.), Academic Press, New York, pp. 191–213.

44. Barakat, I., and Sensenbrenner, M., 1981, *Dev. Brain Res.* **1**:355–368.
45. Barakat, I., Sensenbrenner, M., and Vincendon, G., 1982, *Neurochem. Res.* **7**:287–300.
46. Booher, J., and Sensenbrenner, M., 1972, *Neurobiology* **2**:97–105.
47. Banker, G. A., and Cowan, W. M., 1977, *Brain Res.* **126**:397–425.
48. Bunge, R. P., Bunge, M. B., and Cochran, M., 1978, *Neurology (Minneap.)* **28**:59–67.
49. Scott, B. S., 1977, *J. Neurobiol.* **8**:417–427.
50. Kim, S. U., Warren, K. G., and Kalia, M., 1979, *Neurosci. Lett.* **11**:137–141.
51. Hertz, L., Juurlink, B. H. J., Fosmark, H., and Schousboe, A., 1982, *Neuroscience Approached Through Cell Culture*, Volume 1 (S. E. Pfeiffer, ed.), CRC Press, Baca Raton, pp. 175–186.
52. Kimelberg, H. K., Stieg, P. E., and Mazurkiewicz, J. E., 1982, *J. Neurochem.* **39**:734–742.
53. Manthorpe, M., Adler, R., and Varon, S., 1979, *J. Neurocytol.* **8**:605–621.
54. Labourdette, G., Roussel, G., and Nussbaum, J. L., 1980, *Neurosci. Lett.* **18**:203–209.
55. McCarthy, K. D., and De Vellis, J., 1980, *J. Cell Biol.* **85**:890–902.
56. Mirsky, R., Winter, J., Abney, E. R., Pruss, R. M., Gavrilovic, J., and Raff, M. C., 1980, *J. Cell Biol.* **84**:483–494.
57. Bologa-Sandru, L., Siegrist, H. P., Z'Graggen, A., Hofmann, K., Wiesmann, U., Dahl, D., and Herschkowitz, N., 1981, *Brain Res.* **210**:217–219.
58. Berg, G., and Schachner, M., 1982, *Neurosci. Lett.* **28**:75–80.
59. Sensenbrenner, M., Barakat, I., Delaunoy, J. P., Labourdette, G., and Pettmann, B., 1982 *Neuroscience Approached through Cell Culture*, Volume 1 (S. E. Pfeiffer, ed.), CRC Press, Boca Raton, pp. 87–105.
60. Henn, F. A., 1980, *Advances in Cellular Neurobiology*, Volume 1 (S. Fedoroff and L. Hertz, eds.), Academic Press, New York, pp. 373–403.
61. Currie, D. N., 1980, *Tissue Culture in Neurobiology* (E. Giacobini, A. Vernadakis, and A. Shahar, eds.), Raven Press, New York, pp. 75–87.
62. Garthwaite, J., and Balázs, R., 1981, *Advances in Cellular Neurobiology*, Volume 2 (S. Fedoroff, and L. Hertz, eds.), Academic Press, New York, pp. 461–489.
63. Rose, S. P. R., 1967, *Biochem. J.* **102**:33–34.
64. Norton, W. T., and Poduslo, S., 1970, *Science* **167**:1144–1146.
65. Poduslo, S. E., and Norton, W. T., 1972, *J. Neurochem.* **19**:727–736.
66. Sellinger, O. Z., and Azcurra, J. M., 1974, *Res. Meth. Neurochem.* **2**:3–37.
67. Nagata, Y., Mikoshiba, K., and Tsukada, Y., 1974, *J. Neurochem.* **22**:493–503.
68. Poduslo, S. E., and Norton, W. T., 1975, *Methods in Enzymology*, Volume 35 (S. P. Colowick, and N. Kaplan, eds.), Academic Press, New York, pp. 561–579.
69. Norton, W. H., 1983, *Advances in Cellular Neurobiology*, Volume 4 (S. Fedoroff, and L. Hertz, eds.), Academic Press, New York, London, pp. 3–55.
70. Szuchet, S., Arnason, B. G. W., and Polak, P. E., 1978, *Biophys. J.* **21**:51.
71. Szuchet, S., Arnason, B. G. W., and Polak, P. E., 1980, *J. Neurosci. Methods* **3**:7–19.
72. Szuchet, S., Stefansson, K., Wollmann, R. L., Dawson, G., and Arnason, B. G. W., 1980, *Brain Res.* **200**:151–164.
73. Lisak, R. P., Pleasure, D. E., Silberberg, D. H., Manning, M. C., and Saida, T., 1981, *Brain Res.* **223**:107–122.
74. Hirayama, M., Silberberg, D. H., Lisak, R. P., and Pleasure, D., 1983, *J. Neuropathol. Exp. Neurol.* **42**:16–28.
75. Varon, S., and Manthorpe, M., 1980, *Advances in Cellular Neurobiology*, Volume 1 (S. Fedoroff and L. Hertz, eds.), Academic Press, New York, pp. 405–442.
76. Katchalsky, A., Danon, D., and Nevo, A., 1953, *Biochim. Biophys. Acta* **33**:120–138.
77. Yavin, E., and Yavin, Z., 1974, *J. Cell Biol.* **62**:540–546.
78. Letourneau, P. C., 1975, *Dev. Biol.* **44**:77–91.
79. Pettmann, B., Louis, J. C., and Sensenbrenner, M., 1979, *Nature* **281**:378–380.
80. Yu, A. C. H., Hertz, E., and Hertz, L., 1984, *J. Neurochem.* **42**:951–960.
81. Varon, S., and Raiborn, C., 1972, *J. Neurocytol.* **1**:211–221.
82. Varon, S., Raiborn, C., and Tyszka, E., 1973, *Brain Res.* **54**:51–63.
83. McCarthy, K. D., and Partlow, L. M., 1976, *Brain Res.* **114**:391–414.
84. Barde, Y. A., Edgar, D., and Thoenen, H., 1982, *Neuroscience Approached through Cell Culture*, Volume 1 (S. E. Pfeiffer, ed.), CRC Press, Boca Raton, pp. 83–86.

Cell Cultures

85. Sotelo, J., Gibbs, C. J., Gajdusek, D. C., Toh, B. H., and Wurth, M., 1980, *Proc. Natl. Acad. Sci. U.S.A.* **77**:653–657.
86. Wood, P. M., and Bunge, R. P., 1975, *Nature* **256**:662–664.
87. Wood, P. M., 1976, *Brain Res.* **115**:361–375.
88. Brockes, J. P., Fryxell, K. J., and Lemke, G. E., 1981, *J. Exp. Biol.* **95**:215–230.
89. Bottenstein, J. E., Skaper, S. D., Varon, S., and Sato, G. H., 1980, *Exp. Cell Res.* **125**:183–190.
90. Yavin, Z., and Yavin, E., 1980, *Dev. Biol.* **75**:454–459.
91. Brunner, G., Lang, K., Wolfe, R. A., McClure, D. B., and Sato, G. H., 1981, *Dev. Brain Res.* **2**:563–575.
92. Skaper, S. D., Adler, R., and Varon, S., 1979, *Dev. Neurosci.* **2**:233–237.
93. Varon, S., and Manthorpe, M., 1982, *Advances in Cellular Neurobiology*, Volume 3 (S. Fedoroff and L. Hertz, eds.), Academic Press, New York, pp. 36–95.
94. Raff, M. C., Fields, K. L., Hakomori, S. I., Mirsky, R., Pruss, R. M., and Winter, J., 1979, *Brain Res.* **174**:283–308.
95. Varon, S., 1978, *Dynamic Properties of Glia Cells* (E. Schoffeniels, E. Franck, L. Hertz, and D. B. Tower, eds.), Pergamon Press, Oxford, pp. 93–103.
96. Eng, L. F., and Bigbee, J. W., 1978, *Adv. Neurochem.* **3**:43–98.
97. Schachner, M., Sommer, I., Lagenaur, C., Schnitzer, J., and Berg, G., 1983, *Neuroscience Approached Through Cell Culture*, Volume 2 (S. E. Pfeiffer, ed.), CRC Press, Boca Raton, pp. 115–139.
98. Bignami, A., Dahl, D., and Rueger, D. C., 1980, *Advances in Cellular Neurobiology*, Volume 1 (S. Fedoroff and L. Hertz, eds.), Academic Press, New York, and London, pp. 285–310.
99. Schmechel, D. E., Brightman, M. W., and Marangos, P. J., 1980, *Brain Res.* **190**:195–214.
100. Shein, H. M., 1965, *Exp. Cell Res.* **40**:554–569.
101. Varon, S., and Raiborn, C., 1969, *Brain Res.* **12**:180–199.
102. Cummins, C. J., and Glover, R. A., 1978, *J. Anat.* **125**:117–125.
103. Hertz, L., 1978, *Brain Res.* **145**:202–208.
104. Walz, W., and Hertz, L., 1982, *J. Neurochem.* **39**:70–77.
105. Walz, W., and Hertz, L., 1983, *Brain Res.* **277**:321–328.
106. Kimelberg, H. K., Narumi, S., Biddlecome, S., and Bourke, R. S., 1978, *Dynamic Properties of Glia Cells*, (E. Schoffeniels, G. Franck, L. Hertz, and D. B. Tower, eds.), Pergamon Press, Oxford, pp. 347–357.
107. Kimelberg, H. K., Bowman, C., Biddlecome, S., and Bourke, R. S., 1979, *Brain Res.* **177**:533–550.
108. Juurlink, B. H. J., Fedoroff, S. Hall, C., and Nathaniel, E. J. H., 1981, *J. Comp. Neurol.* **200**:375–391.
109. Olson, J. E., and Holtzman, D., 1980, *J. Neurosci. Res.* **5**:497–500.
110. Singh, V. K., and Van Alstyne, D., 1978, *Brain Res.* **155**:418–421.
111. Lindsay, R. M., Barber, P. C., Sherwood, M. R., Zimmer, J., and Raisman, G., 1982, *Brain Res.* **243**:329–343.
112. Schousboe, A., Fosmark, H., and Hertz, L., 1975, *J. Neurochem.* **25**:909–911.
113. Yu, A. C. H., Farmer, M. J., and Hertz, L., 1983, *Abstracts 9th Meeting International Society Neurochemistry, J. Neurochem.* **41**(Suppl. 566).
114. Moonen, G., Cam, Y., Sensenbrenner, M., and Mandel, P., 1975, *Cell Tissue Res.* **163**:365–372.
115. Boone, C. W. N., Mantel, N., Caruso, T. D., Karzan, E., and Stevenson, R. E., 1972, *In Vitro* **7**:174–189.
116. Hertz, L., Bock, E., and Schousboe, A., 1978, *Dev. Neurosci.* **1**:226–238.
117. White, F. P., and Hertz, L., 1981, *Neurochem. Res.* **6**:353–364.
118. Schousboe, A., Nissen, C., Bock, E., Sapirstein, V. S., Juurlink, B. H. J., and Hertz, L., 1980, *Tissue Culture in Neurobiology* (E. Giacobini, A. Vernadakis, and A. Shahar, eds.), Raven Press, New York, pp. 397–409.
119. Yu, P. H., and Hertz, L., 1982, *J. Neurochem.* **39**:1493–1495.
120. Hertz, L., 1981, *Glial and Neuronal Cell Biology* (S. Fedoroff, ed.), Alan R. Liss, New York, pp. 45–58.

121. Narumi, S., Kimelberg, H. K., and Bourke, R. S., 1978, *J. Neurochem.* **31**:1479–1490.
122. Kimelberg, H. K., Narumi, S., and Bourke, R. S., 1978, *Brain Res.* **153**:55–77.
123. Lim, R., Mitsunobu, K., and Li, W. K. P., 1975, *Exp. Cell Res.* **79**:243–246.
124. Lim, R., Turriff, D. E., Troy, S. S., and Kato, T., 1977, *Cell, Tissue and Organ Cultures in Neurobiology* (S. Fedoroff and L. Hertz, eds.), Academic Press, New York, pp. 223–235.
125. Ciesielski-Treska, J., and Ulrich, G., 1980, *Abstr. Int. Soc. Dev. Neurosci.*, p. 285.
126. Terasaki, W. L., Brooker, G., De Vellis, J., Inglish, D., Hsu, C. Y., and Moylan, R. D., 1978, *Adv. Cyclic Nucleotide Res.* **9**:33–52.
127. Privat, A., 1978, *Dynamic Properties of Glia Cells* (E. Schoffeniels, G. Franck, L. Hertz, and D. B. Tower, eds.), Pergamon Press, New York, pp. 55–64.
128. Ebersolt, C., Perez, M., and Bockaert, J., 1981, *Brain Res.* **213**:139–150.
129. Stieg, P. E., Kimelberg, H. K., Mazurkiewicz, J. E., and Banker, G. A., 1980, *Brain Res.* **199**:493–500.
130. Schachner, M., Schoonmaker, G., and Hynes, R. O., 1978, *Brain Res.* **158**:149–158.
131. Chiu, F. C., Norton, W. T., and Fields, K. L., 1981, *J. Neurochem.* **37**:147–155.
132. Hansson, E., Sellström, Å., Persson, L. I., and Rönnbäck, L., 1980, *Brain Res.* **188**:233–246.
133. Bignami, A., Eng, L. F., Dahl, D., and Uyeda, C. T., 1972, *Brain Res.* **43**:429–435.
134. Bock, E., Jörgensen, O. S., Dittmann, L., and Eng, L. F., 1975, *J. Neurochem.* **25**:867–870.
135. Hansson, E., 1982, *Primary Astroglial Cultures: Aspects of Morphology, Biochemistry and Transmitter Metabolism*, D.Sc. Thesis, University of Gotenburg, Sweden.
136. Sensenbrenner, M., Labourdette, G., Delaunoy, J. P., Pettmann, B., Devellliers, G., Moonen, G., and Bock, E., 1980, *Tissue Culture in Neurobiology* (E. Giacobini, A. Vernadakis, and A. Shahar, eds.), Raven Press, New York, pp. 385–395.
137. Ludwin, S. K., Kosek, J. C., and Eng, L. F., 1976, *J. Comp. Neurol.* **165**:197–208.
138. Hydén, H., and McEwen, B. S., 1966, *Proc. Natl. Acad. Sci. U.S.A.* **55**:354–358.
139. Haglid, K. G., Hansson, H. A., and Rönnbäck, L., 1977, *Brain Res.* **123**:331–345.
140. Fedoroff, S., White, R. V. Subrahmanyan, L., and Kalnins, V. I., 1981, *Glial and Neuronal Cell Biology* (S. Fedoroff, ed.), Alan R. Liss, New York, pp. 1–19.
141. Ciesielski-Treska, J., Guerold, B., and Aunis, D., 1982, *Neuroscience* **7**:509–522.
142. Norenberg, M. D., and Martinez-Hernandez, A., 1979, *Brain Res.* **161**:303–310.
143. Juurlink, B. H. J., Schousboe, A., Jörgensen, O. S., and Hertz, L., 1981, *J. Neurochem.* **36**:136–142.
144. Hallermayer, K., Harmening, C., and Hamprecht, B., 1981, *J. Neurochem.* **37**:43–52.
145. Pope, A., 1978, *Dynamic Properties of Glia Cells* (E. Schoffeniels, G. Franck, L. Hertz, and D. B. Tower, eds.), Pergamon Press, Oxford, pp. 13–20.
146. Ghandour, M. S., Langley, O. K., Vincendon, G., and Gombos, G., 1979, *J. Histochem. Cytochem.* **27**:1634–1637.
147. Roussel, G., Delaunoy, J. P., Nussbaum, J. L., and Mandel, P., 1979, *Brain Res.* **160**:47–55.
148. Schousboe, A., Hertz, L., and Svenneby, G., 1977, *Neurochem. Res.* **2**:217–229.
149. Wu, P. H., Durden, D. A., and Hertz, L., 1979, *J. Neurochem.* **32**:379–390.
150. Bardakjian, J., Tardy, M., Pimoule, C., and Gonnard, P., 1979, *Neurochem. Res.* **4**:517–527.
151. Hertz, L., Baldwin, F., and Schousboe, A., 1979, *Can. J. Physiol. Pharmacol.* **57**:223–226.
152. Hertz, L., and Richardson, J. S., 1983, *J. Neurosci. Res.* **9**:173–182.
153. Hertz, L., and Schousboe, A., 1980, *Brain Res. Bull.* **5**(Suppl. 2):403–409.
154. Hertz, L., 1982, *Handbook of Neurochemistry*, Volume 1, 2nd ed. (A. Lajtha, ed.), Plenum Press, New York, pp. 319–355.
155. Schousboe, A., Svenneby, G., and Hertz, L., 1977, *J. Neurochem.* **29**:999–1005.
156. Hertz, L., Schousboe, A., Boechler, N., Mukerji, S., and Fedoroff, S., 1978, *Neurochem. Res.* **3**:1–14.
157. Walker, C. R., and Peacock, J. H., 1982, *Dev. Brain Res.* **2**:541–555.
158. Yu, A. C. H., Schousboe, A., and Hertz, L., 1982, *J. Neurochem.* **39**:954–966.
159. Kvamme, E., Svenneby, G., Hertz, L., and Schousboe, A., 1982, *Neurochem. Res.* **7**:761–770.
160. Cummins, C. J., Lust, W. D., and Passonneau, J. V., 1983, *J. Neurochem.* **40**:128–136.

161. Shein, S. H., Britva, A., Hess, H. H., and Selkoe, D. J., 1970, *Brain Res.* **19**:497–501.
162. Ciesielski-Treska, J., Roth-Schechter, B., Beya, D., Mandel, P., Nissen, C., and Hertz, L., 1976, *J. Neurochem.* **26**:197–199.
163. Benda, P., Lightbody, J., Sato, G., Levine, L., and Sweet, W., 1968, *Science* **161**:370–371.
164. Benda, P., Someda, K., Messer, J., and Sweet, W. H., 1971, *J. Neurosurg.* **34**:310–323.
165. Benda, P., 1978, *Dynamic Properties of Glia Cells* (E. Schoffeniels, G. Franck, L. Hertz, and D. B. Tower, eds.), Pergamon Press, Oxford, pp. 66–81.
166. De Vellis, J., McGinnis, J. F., Breen, G. A. M., Leveille, P., Bennett, K., and McCarthy, K., 1977, *Cell, Tissue and Organ Cultures in Neurobiology* (S. Fedoroff, and L. Hertz, eds), Academic Press, New York, London, pp. 485–511.
167. Zanetta, J. P., Benda, P., Gombos, G., and Morgan, I. G., 1972, *J. Neurochem.* **19**:881–883.
168. Liao, C. L., Eng, L. F., Herman, M. M., and Bensch, K., 1978, *J. Neurochem.* **30**:1181–1186.
169. Parker, K. K., Norenberg, M. D., and Vernadakis, A., 1980, *Science* **208**:179–181.
170. Browning, E. T., and Nicklas, W. J., 1982, *J. Neurochem.* **39**:336–341.
171. Schousboe, A., 1982, *Neuroscience Approached through Cell Culture*, Volume 1 (S. E. Pfeiffer, ed.), CRC Press, Boca Raton, pp. 106–141.
172. Hutchison, H. T., Werrbach, K., Vance, C., and Haber, B., 1974, *Brain Res.* **66**:265–274.
173. Schrier, B. K., and Thompson, E. J., 1974, *J. Biol. Chem.* **249**:1769–1780.
174. Kimelberg, H. K., 1974, *J. Neurochem.* **22**:971–976.
175. Kukes, G., De Vellis, J., and Elul, K., 1974, *Brain Res.* **104**:93–105.
176. Gilman, A. G., and Nirenberg, M., 1971, *Proc. Natl. Acad. Sci. U.S.A.* **68**:2165–2168.
177. McCarthy, K. D., and De Vellis, J., 1978, *J. Cyclic Nucleotide Res.* **4**:15–26.
178. McCarthy, K. D., and Harden, T. K., 1981, *J. Pharmacol. Exp. Ther.* **216**:183–191.
179. Maderspach, K., and Fajszi, C., 1983, *Dev. Brain Res.* **6**:251–257.
180. Edström, A., Haglid, K. G., Kanje, M., Rönnbäck, L., and Walum, E., 1974, *Exp. Cell Res.* **83**:426–429.
181. Martin, D. L., and Shain, W., 1979, *J. Biol. Chem.* **254**:7076–7084.
182. Waniewski, R. A., and Martin, D. L., 1983, *Brain Res.* **268**:390–394.
183. Cavanaugh, M. W., 1955, *Exp. Cell Res.* **9**:42–48.
184. Nakai, J., 1956, *Am. J. Anat.* **99**:81–99.
185. Peacock, J. H., Nelson, P. G., and Goldstone, M. W., 1973, *Dev. Biol.* **30**:137–152.
186. Ransom, B. R., and Barker, J. L., 1981, *Advances in Cellular Neurobiology*, Volume 2 (S. Fedoroff, and L. Hertz, eds.), Academic Press, New York, London, pp. 83–114.
187. Peacock, J. H., 1979, *Brain Res.* **169**:247–260.
188. Burry, R. W., and Lasher, R. S., 1978, *Brain Res.* **151**:19–29.
189. Yu, A. C. H., 1980, *Uptake of Glutamine and Glutamate by Cultured Neurons*, M.Sc. Thesis, University of Saskatchewan, Saskatoon.
190. Hauser, K., and Bernasconi, R., 1980, *Tissue Culture in Neurobiology* (E. Giacobini, A. Vernadakis, and A. Shahar, eds.), Raven Press, New York, pp. 205–219.
191. Dichter, M. A., 1978, *Brain Res.* **149**:279–293.
192. Athias, P. M., Sensenbrenner, M., and Mandel, P., 1974, *Differentiation* **2**:99–106.
193. Bonkowski, L., and Dryden, W. F., 1976, *Brain Res.* **107**:69–84.
194. Varon, S., 1979, *Neurochem. Res.* **4**:155–173.
195. Sensenbrenner, M., Maderspach, K., Latzkovits, L., and Jaros, G. G., 1978, *Dev. Neurosci.* **1**:90–101.
196. Bornstein, M. B., and Modek, P. G., 1972, *Brain Res.* **37**:287–293.
197. Messer, A., 1977, *Brain Res.* **130**:1–12.
198. Ludueña, M. A., 1973, *Dev. Biol.* **33**:268–284.
199. Kozak, L. P., Dahl, D., and Bignami, A., 1978, *Brain Res.* **150**:631–637.
200. Nelson, P. G., and Peacock, J. H., 1973, *Brain Res.* **61**:163–174.
201. Dambergs, R., Leah, J., and Kidson, C., 1978, *Exp. Neurol.* **59**:296–303.
202. Godfrey, E., Nelson, P., Schrier, B., Breuer, A. C., and Ransom, B. R., 1975, *Brain Res.* **90**:1–21.
203. Rieske, E., and Kreutzberg, G. W., 1977, *Neurosci. Lett.* **5**:153–157.
204. Aguayo, A., Roamine, J., and Bray, G., 1975, *J. Neurocytol.* **4**:633–674.

205. Drejer, J., Larsson, O. M., and Schousboe, A., 1982, *Exp. Brain Res.* **47:**259–269.
206. Larsson, O. M., Thorbek, P., Krogsgaard-Larsen, P., and Schousboe, A., 1981, *J. Neurochem.* **37:**1509–1516.
207. Yu, A. C. H., and Hertz, L., 1982, *J. Neurosci. Res.* **7:**23–35.
208. Hendrickson, C. M., and Lin, S., 1977, *Fed. Proc.* **36:**515.
209. Snodgrass, S. R., White, W. F., Biales, B., and Dichter, M., 1980, *Brain Res.* **190:**123–138.
210. Lasher, R. S., and Zagon, I. S., 1972, *Brain Res.* **41:**482–488.
211. Young, A. B., Oster-Granite, M. L., Herndon, R. M., and Snyder, S. H., 1974, *Brain Res.* **73:**1–13.
212. Messer, A., 1980, *Advances in Cellular Neurobiology*, Volume 1, (S. Fedoroff and L. Hertz, eds.), Academic Press, New York, London, pp. 179–207.
213. Varon, S., and Adler, R., 1981, *Advances in Cellular Neurobiology*, Volume 2 (S. Fedoroff and L. Hertz, eds.), Academic Press, New York, pp. 115–163.
214. Scott, B. S., and Fisher, K. C., 1970, *Exp. Neurol.* **27:**16–22.
215. Peacock, J. H., Rush, D. F., Mathers, L. H., 1979, *Brain Res.* **169:**231–246.
216. Borg, J., Ramaharobandro, N., Mark, J., and Mandel, P., 1980, *J. Neurochem.* **34:**1113–1122.
217. Sensenbrenner, M., Booher, J., and Mandel, P., 1973, *Experientia* **29:**699–701.
218. Mandel, P., Ciesielski-Treska, J., and Sensenbrenner, M., 1976, *Molecular and Functional Neurobiology* (W. H. Grispen, ed.), Elsevier, Amsterdam, pp. 111–155.
219. Vernadakis, A., and Arnold, E. B., 1980, *Advances in Cellular Neurobiology*, Volume 1 (S. Fedoroff and L. Hertz, eds.), Academic Press, New York, pp. 229–283.
220. Schengrund, C. L., and Marangos, P. J., 1980, *J. Neurosci. Res.* **5:**305–311.
221. Ledig, M., Tholey, G., and Mandel, P., 1982, *Dev. Brain Res.* **4:**451–454.
222. Mirsky, R., Wendon, L. M. B., Black, P., Stolkin, C., and Bray, D., 1978, *Brain Res.* **148:**251–259.
223. Lasher, R. S., 1974, *Brain Res.* **69:**235–254.
224. Currie, D. N., and Dutton, G. R., 1980, *Brain Res.* **199:**473–481.
225. Gordon, R. D., Wilkin, G. P., Hunt, A., Patel, A. J., and Balázs, R., 1981, *Trans. Am. Soc. Neurochem.* **12:**258.
226. Pearce, B. R., and Dutton, G. R., 1982, *Neuroscience Approached through Cell Culture*, Volume 1 (S. E. Pfeiffer, ed.), CRC Press, Boca Raton, pp. 143–156.
227. Moonen, G., Neale, E. A., MacDonald, R. L. Gibbs, W., and Nelson, P. G., 1982, *Dev. Brain Res.* **5:**59–73.
228. Dichter, M. A., 1980, *Brain Res.* **190:**111–121.
229. Balcar, W. J., Mark, J., Borg, J., and Mandel, P., 1979, *Neurochem. Res.* **4:**339–354.
230. Larsson, O. M., Drejer, J., Hertz, L., and Schousboe, A., 1983, *J. Neurosci. Res.* **9:**291–302.
231. Mains, R. E., and Patterson, P. H., 1973, *J. Cell Biol.* **59:**329–345.
232. Patterson, P. H., and Chun, L. L. Y., 1974, *Proc. Natl. Acad. Sci. U.S.A.* **71:**3607–3610.
233. Patterson, P. H., and Chun, L. L. Y., 1977, *Dev. Biol.* **56:**263–280.
234. Patterson, P. H., and Chun, L. L. Y., 1977, *Dev. Biol.* **60:**473–481.
235. Reichardt, L. F., and Patterson, P. H., 1977, *Nature* **270:**147–151.
236. Otsuka, M., Obata, K., Miyata, Y., and Tanaka, Y., 1971, *J. Neurochem.* **18:**287–295.
237. Okada, Y., 1982, *Problems in GABA Research: from Brain to Bacteria*, (Y. Okada and E. Roberts, eds.), Excerpta Medica, Amsterdam, pp. 23–29.
238. Yavin, E., and Yavin, Z., 1980, *Tissue Culture in Neurobiology*, (E. Giacobini, A. Vernadakis, and A. Shahar, eds.), Raven Press, New York, pp. 277–289.
239. Ramaharobandro, N., Borg, J., Mandel, P., and Mark, J., 1982, *Brain Res.* **244:**113–121.
240. Barnes, E. M., and Mandel, P., 1981, *J. Neurochem.* **36:**82–85.
241. Latzkovits, L., Sensenbrenner, M., and Mandel, P., 1974, *J. Neurochem.* **23:**193–200.
242. Levi, G., Gallo, V., Ciotti, M. T., Coletti, A., and Aloisi, F., 1982, *Trans. Am. Soc. Neurochem.* **13:**250.
243. Black, I. B., and Patterson, P. H., 1980, *Curr. Topics. Dev. Biol.* **15:**27–40.
244. Mudge, A. W., Leeman, S. E., and Fischbach, G. D., 1979, *Proc. Natl. Acad. Sci. U.S.A.* **76:**526–530.

245. Mudge, A. W., 1981, *J. Exp. Biol.* **95**:195–203.
246. Murray, M. R., and Stout, A. P., 1947, *Am. J. Pathol.* **23**:429–441.
247. Goldstein, M. N., 1968, *J. Pediatr. Surg.* **3**:166–169.
248. Goldstein, M. N., and Pinkel, D., 1958, *J. Natl. Cancer Inst.* **20**:675–689.
249. Tumilowicz, J. J., Nichols, W. W., Cholon, J. J., and Green, A. E., 1970, *Cancer Res.* **30**:2110–2118.
250. Biedler, J. N., Helson, L., and Spengler, B. A., 1973, *Cancer Res.* **33**:2643–2652.
251. Biedler, J. N., Meyers, M. B., and Spengler, B., 1983, *Advances in Cellular Neurobiology*, Volume 4 (S. Fedoroff and L. Hertz, eds.), Academic Press, New York, pp. 268–307.
252. Dunham, L. J., 1953, *J. Natl. Cancer Inst.* **13**:1299–1377.
253. Augusti-Tocco, G., and Sato, G., 1969, *Proc. Natl. Acad. Sci. U.S.A.* **64**:311–315.
254. Klebe, R. J., and Ruddle, F. H., 1969, *J. Cell Biol.* **43**:69a.
255. Schubert, D., Humphreys, S., Baroni, C., and Cohn, M., 1969, *Proc. Natl. Acad. Sci. U.S.A.* **64**:316–323.
256. Seeds, N. W., Gilman, A. G., Amano, T., and Nirenberg, M. W., 1970, *Proc. Natl. Acad. Sci. U.S.A.* **66**:160–167.
257. Amano, T., Richelson, E., and Nirenberg, M., 1972, *Proc. Natl. Acad. Sci. U.S.A.* **69**:258–263.
258. Prasad, K. N., Mandel, B., Waymire, J. C., Lees, G. J., Vernadakis, A., and Weiner, N., 1973, *Nature [New Biol.]* **241**:117–119.
259. Nelson, P. G., Ruffner, W., and Nirenberg, M., 1969, *Proc. Natl. Acad. Sci. U.S.A.* **64**:1004–1010.
260. Greene, L. A., and Tischler, A. S., 1976, *Proc. Natl. Acad. Sci. U.S.A.* **73**:2424–2428.
261. Greene, L. A., and Tischler, A. S., 1983, *Advances in Cellular Neurobiology*, Volume 4 (S. Fedoroff and L. Hertz, eds.), Academic Press, New York, pp. 373–414.
262. Schubert, D., Heinemann, S., Carlisle, W., Tarikas, H., Kimes, B., Patrick, J., Steinbach, J. H., Culp, W., and Brandt, B. L., 1974, *Nature* **249**:224–227.
263. Schubert, D., Carlisle, W., and Look, C., 1975, *Nature* **254**:341–343.
264. Giotta, G. J., and Cohn, M., 1982, *Neuroscience Approached through Cell Culture*, Volume 1 (S. E. Pfeiffer, ed.), CRC Press, Boca Raton, pp. 203–225.
265. Bulloch, K., Stallcup, W. B., and Cohn, M., 1976, *Brain Res.* **135**:25–36.
266. Bulloch, K., Stallcup, W. B., and Cohn, M., 1978, *Life Sci.* **22**:495–504.
267. Peacock, J. H., and Nelson, P. G., 1973, *J. Neurobiol.* **4**:363–374.
268. Kato E., and Narahashi, T., 1982, *J. Physiol. (Lond.)* **333**:213–226.
269. Giller, E. L., Breakefield, X. O., Christian, C. N., Neale, E. A., and Nelson P. G., 1975, *Golgi Centennial Symposium*, (M. Santini, ed.), Raven Press, New York, pp. 603–623.
270. Nelson, P. G., 1975, *Physiol. Rev.* **55**:1–61.
271. Nelson, P. G., 1977, *Cell, Tissue and Organ Cultures in Neurobiology*, (S. Fedoroff and L. Hertz, eds.), Academic Press, New York, pp. 347–365.
272. Prasad, K. N., 1980, *Regulation of Differentiation in Mammalian Nerve Cells*, Plenum Press, New York.
273. Cohen, N. M., Schmidt, D. M., McGlennen, R. C., and Klein, W. L., 1983, *J. Neurochem.* **40**:547–554.
274. Mandel, P., Ciesielski-Treska, J., Hermetet, J. C., Hertz, L., Nissen, C., Tholey, G., and Warter, F., 1973, *Central Nervous System—Studies on Metabolic Regulation and Function* (E. Genazzani and H. Herken, eds.), Springer-Verlag, Berlin, pp. 223–230.
275. Denis-Donini, S., and Augusti-Tocco, G., 1980, *Curr. Top. Dev. Biol.* **16**:323–348.
276. De Laat, S. W., and Van Der Saag, P. T., 1982, *Int. Rev. Cytol.* **74**:1–54.
277. Yankner, B. A., and Shooter, E. M., 1982, *Annu. Rev. Biochem.* **51**:845–868.
278. Stallcup, W. B., Bulloch, K., and Baetge, E. E., 1979, *J. Neurochem.* **32**:57–65.
279. Korey, S. R., 1957, *Metabolism of the Nervous System* (D. Richter, ed.), Pergamon Press, Oxford, New York, pp. 87–90.
280. Fewster, M. E., Scheibel, A. B., and Mead, J. F., 1967, *Brain Res.* **6**:401–408.
281. Fewster, M. E., Blackstone, S. C., and Ihrig, T. J., 1973, *Brain Res.* **63**:263–271.
282. Fewster, M. E., and Blackstone, S. C., 1975, *Neurobiology* **5**:316–328.
283. Poduslo, S. E. and McKhann, G. M., 1977, *Neurosci. Lett.* **5**:159–163.

284. Poduslo, S. E., Miller, K., and McKhann, G. M., 1978, *J. Biol. Chem.* **253:**1592–1597.
285. Pleasure, D., Abramsky, O., Silberberg, D., Quinn, B., Parkis, J., and Saida, T., 1977, *Brain Res.* **134:**377–382.
286. Pleasure, D., Hardy, M., Johnson, G., Lisak, R., and Silberberg, D., 1981, *J. Neurochem.* **37:**452–460.
287. Szuchet, S., Stefansson, K., Dawson, G., and Arnason, B. G. W., 1979, *Trans. Am. Soc. Neurochem.* **10:**82.
288. Szuchet, S., and Stefansson, K., 1980, *Advances in Cellular Neurobiology*, Volume 1 (S. Fedoroff and L. Hertz, eds.), Academic Press, New York, London, pp. 313–346.
289. Gebicke-Härter, P. J., Althans, D., Schwartz, P., and Neuhoff, V., 1981, *Dev. Brain Res.* **1:**497–518.
290. Szuchet, S., and Yim, S. H., 1984, *J. Cell Biol. J. Neurosci. Res.* **11:**131–144.
291. Meier, D., and Schachner, M., 1982, *J. Neurosci. Res.* **7:**135–145.
292. Meier, D., Langenauer, C., and Schacher, M., 1982, *J. Neurosci. Res.* **7:** 119–134.
293. Joseph, K. C., Silberberg, D. H., Kronick, P. L., and Cambell, G., 1982, *Trans. Am. Soc. Neurochem.* **13:**268.
294. Bhat, S., Barbarese, E., and Pfeiffer, S. E., 1981, *Proc. Natl. Acad. Sci. U.S.A.* **78:**1283–1287.
295. Snyder, D. S., Raine, C. S., Farooq, M., and Norton, W. T., 1980, *J. Neurochem.* **34:**1614–1621.
296. Yim, S. H., and Szuchet, S., 1981, *Trans. Am. Soc. Neurochem.* **11:**224.
297. Szuchet, S., Yim, S. H., and Monsma, S., 1983, *Proc. Natl. Acad. Sci. U.S.A.* **80:**7019–7023.
298. Massa, P. T., and Mugnaini, E., 1982, *Neuroscience* **7:**523–538.
299. Massa, P. T., Szuchet, S., and Mugnaini, E., 1983, *Biophys, J.* **41:**69.
300. Massa, P. T. Szuchet, S., and Mugnaini, E., 1983, *J. Neurosci.* (in press).
301. Wollmann, R. L., Szuchet, S., Barlow, J., and Jerkovic, M., 1981, *J. Neurosci. Res.* **6:**757–769.
302. Farooq, M., Crammer, W., Snyder, D. S., Raine, C. S., and Norton, W. T., 1981, *J. Neurochem.* **36:**431–440.
303. Raff, M. C., Mirsky, R., Fields, K. L., Lisak, R. P., Dorfman, S. H., Silberberg, S., Greoson, H. A., Leibowitz, S., and Kennedy, M. C., 1978, *Nature* **274:**813–816.
304. Zalc, B., Duponey, M. P., Hause, J. J., and Baumann, N. A., 1981, *Brain. Res.* **211:**34–54.
305. Uchida, T., Takahashi, K., Yamaguchi, H., and Jagai, Y., 1981, *Jpn. J. Exp. Med.* **51:**29–35.
306. Wollman, R. L., and Szuchet, S., 1981, *Soc. Neurosci. Abstr.* **7:**422.
307. Peters, A., Palay, S. L., and Webster, H. deF., 1976, *Fine Structure of the Nervous System*, W. B. Saunders, Philadelphia, pp. 232–263.
308. Gonatas, N. K., Hirayama, M., Stieber, A., and Silberberg, D. H., 1982, *J. Neurocytol.* **11:**997–1008.
309. Poduslo, S. E., McFarland, H. F., and McKhann, G. M., 1977, *Science* **197:**727–736.
310. Poduslo, S. E., 1978, *Myelination and Demyelination* (J. Palo, ed.), Plenum Press, New York, pp. 71–94.
311. Mack, S. R., and Szuchet, S., 1980, *Trans. Am. Soc. Neurochem.* **11:**104.
312. Mack, S. R., and Szuchet, S., 1981, *Brain Res.* **214:**180–185.
313. Mack, S. R., Szuchet, S., and Dawson, G., 1981, *J. Neurosci. Res.* **6:**361–367.
314. Norton, W. T., 1977, *Myelin* (P. Morell, ed.), Plenum Press, New York, pp. 161–199.
315. Horrocks, L. A., Mechler, R. J., and Collins, R. L., 1966, *Variation in Chemical Composition of the Nervous System* (T. V. Ansell, ed.), p. 46.
316. Horrocks, L. A., 1968, *J. Neurochem.* **15:**483–488.
317. Eng, L. F., and Noble, E. P., 1968, *Lipids* **3:**157–162.
318. Dalal, K. B., and Roboz-Einstein, E., 1969, *Brain Res.* **16:**441–451.
319. Norton, W. T., and Poduslo, S. E., 1973, *J. Neurochem.* **21:**759–773.
320. Mandel, P., Nussbaum, J. L., Neskovic, N. M., Sarlieve, L. L., and Kurihara, T., 1972, *Advances in Enzyme Regulation* (G. Weber, ed.), Pergamon Press, Oxford, pp. 203–243.
321. Sprinkle, T. J., Zaruba, M. E., and McKhann, G. M., 1978, *J. Neurochem.* **31:**309–314.

322. Drummond, R. J., and Dean, G., 1980, *J. Neurochem.* **35:**1155-1165.
323. Roussel, G., Delaunoy, J. P., Nussbaum, J. L., and Mandel, P., 1977, *Neuroscience* **2:**307-313.
324. Roussel, G., Delaunoy, J. P., Mandel, P., and Nussbaum, J. L., 1978, *J. Neurocytol.* **7:**155-163.
325. Waehneldt, T. V., 1975, *Biochem. J.* **151:**435-443.
326. Sheppard, J. R., Brus, D., and Weliner, J. M., 1978, *J. Neurobiol.* **9:**309-314.
327. Sarlieve, L. L., Suba, R., Campbell, G. M., and Pieringier, R. A., 1980, *Brain Res.* **184:**79-90.
328. De Vellis, J., and Inglish, D., 1973, *Prog. Brain Res.* **40:**321-330.
329. Weingarten, D., and De Vellis, J., 1980, *Biochem. Biophys. Res. Commun.* **93:**1297-1304.
330. Hirsch, H. E., Branco, C. E., and Parks, M. E., 1980, *J. Neurochem.* **34:**760-762.
331. Varon, S., and Somjen, G. G., 1979, *Neurosci. Res. Prog. Bull.* **17:**36-40.
332. Cammer, W., Snyder, D. S., Zimmerman, T. R., Farooq, M., and Norton, W. T., 1982, *J. Neurochem.* **38:**360-367.
333. Sternberger, N., Quarles, R. H., Itoyama, Y., and Webster, H. D., 1979, *Proc. Natl. Acad. Sci. U.S.A.* **76:**1510-1514.
334. Hartman, B. K., Agrawal, H. C., Kalmbach, S., and Shearer, W. T., 1979, *J. Comp. Neurol.* **188:**273-290.
335. Hartman, B. K., Agrawal, H. C., Agrawal, D., and Kalmbach, T., 1982, *Proc. Natl. Acad. Sci. U.S.A.* **79:**4217-4220.
336. Fields, K. L., Gosling, C., Megson, M., and Stern, P. L., 1975, *Proc. Natl. Acad. Sci. U.S.A.* **72:**1296-1300.
337. Sundarraj, N., Schachner, M., and Pfeiffer, S. E., 1975, *Proc. Natl. Acad. Sci. U.S.A.* **72:**1927-1931.
338. Dawson, G., and Kernes, S. M., 1979, *J. Biol. Chem.* **254:**163-167.
339. Bignami, A., and Stoolmiller, A. C., 1979, *Brain Res.* **163:**353-357.
340. West, G. J., Uki, J., Stahn, R., and Herschman, H. R., 1977, *Brain Res.* **130:**387-392.
341. McMorris, F. A. Miller, S. L., Pleasure, D., and Abramsky, O., 1981, *Exp. Cell Res.* **133:**395-404.
342. Crain, S. M., 1977, *Cell, Tissue and Organ Cultures in Neurobiology* (S. Fedoroff and L. Hertz, eds.), Academic Press, New York, pp. 147-190.
343. Garber, B. B., 1977, *Cell, Tissue and Organ Cultures in Neurobiology* (S. Fedoroff and L. Hertz, eds.), Academic Press, New York, pp. 515-537.
344. Linser, P., Saad, A. D., Soh, B. M., and Moscona, A. A., 1982, *Embryonic Development, Part B: Cellular Aspects* (M. M. Burger and R. Weber, eds.), Alan R. Liss, New York, pp. 445-458.
345. Seeds, N. W., 1983, In: *Advances in Cellular Neurobiology*, Volume 4, (S. Fedoroff and L. Hertz, eds.), Academic Press, New York, London, pp. 57-79.
346. Wood, P. Szuchet, S., Williams, A. K., and Bunge, R. P., 1983, *Trans. Am. Soc. Neurochem.* **14:**212.
347. Lewin, B., 1980, *Gene Expression*, John Wiley & Sons, New York, pp. 242-280.
348. Kahn, D. J., Mitrius, J. C., and U'Prichard, D. C., 1982, *Mol. Pharmacol.* **21:**17-26.

Index

Acetylcholine, ACTH, 394
Acetylcholine, retina, 527
Acetylcholinesterase, gonadal steroids, 43
ACTH
  ACh, 394
  aging, 391
    catecholamines, 393
  conditioned avoidance behavior, 378
  cyclic AMP, 394
  excessive grooming, 381
  GABA, 394
  learning impairment, 359
  lipid phosphorylation, 399
  membrane fluidity, 401
  phosphoprotein B-50, 398
  B-50 protein kinase, 398
  protein phosphorylation, 395
  serotonin (5–HT), 394
  sexual behavior, 388
  social behavior, 389
  synaptic plasma membrane, 397, 401
Adenylate cyclase, retina, 533
Adrenal cortex
  adrenocortical hormones, 47
  adrenal steroids: see Adrenal steroids
Adrenal steroids
  biotransformation, 51
  control of behavior, 78
  enzymes, effect on, 74
  interaction with nerve cell, 53
  morphology, 74
  neuroendocrinology, 76
  neuropeptides, effects on, 73
  neurotransmitters, 69
  proteins, effects on, 73
  receptors
    activation, 51
    aging, 67
    binding, 59
    corticosterone, cellular localization, 54
    dexamethasone, cellular localization, 54
    heterogeneity, 60

Adrenal steroids (*cont.*)
  receptors (*cont.*)
    mineralocorticoids, cellular localization, 54
    nuclear uptake in slices, 58
    ontogeny, 67
    regulation, 64
    steroid-receptor complex, translocation, 52
    transformation to DNA binding state, 51
Adrenocortical hormones
  function, 48
  uptake by cells, 49
  see also: Adrenal steroids
Aging, ACTH, 391
Aging, adrenal steroid receptors, 67
Amino acid systems, interaction with peptidergic systems, 264
Amino acids
  intraocular fluids, 486
  neurotransmitters, retina, 530
  plasma composition
    brain precursor concentration, 309
    hormonal influences, 313
    sequence, β-lipotropin, 138
    sleep neurochemistry, 450
  see also: individual amino acids
Analgesia, cystokinin, 121
Angiotensin
  central action, 231
  distribution, 230
  peripheral action, 231
Antibodies, behavior, 349
Ascorbic acid, intraocular fluids, 484
Astrocytes, cultures: *see* Cell cultures, astrocytes
Axolemma, NSG-axolemma fusion, 298
Axonal transport, 297
  gonadal steroids, 40
  retina, 539
  ventromedial nucleus, 40
Axons, branching, 283

663

Behavior
    antibodies, 349
    conditioned avoidance, 378
    drugs, 352
    gonadal steroids, 29
    kindling, 348
    imprinting, 348
    microtubule binding agents, 349
    molluscan preparations, 346
    neuropharmacological approaches, 352
    non-invasive studies, 351
    peptides, 350
    protein phosphorylation, 350
    protein synthesis, 350
    sexual, ACTH, 388
    social, ACTH, 389
    steroid control, 78
Bicarbonate, intraocular fluids, 475
Blood-ocular barrier systems, transport functions, 474
Bombesin
    central effects, 227
    distribution, 226
    effects on GI tract, 227
    endocrine effects, 227
    peripheral effects, 227
Brain lesion, learning impairment, 359
Brain slices
    electrical response, 331
    hippocampus, 333
        CA1 region, 334
        CA3 region, 336
        dentate gyrus, 334
    histological examination, 331
    lateral geniculate, 337
    neocortex, 332
    neurotransmitter synthesis, 330
    optic tract, 337
    piriform lobe
        lateral olfactory tract, 332
        olfactory cortex, 332
    superior colliculus, 337
    superfusion, 329

Calcium
    conductance, 286
    dependence, 286, 298
    intraocular fluids, 478
Calmodulin
    dopamine receptor regulation
        desensitized, 211
        supersensitive, 211
    membrane proteins, phosphorylation, 206
        drug effects, 209
    myelin phosphorylation, 209
    neuronal function, 201

Calmodulin (cont.)
    postsynaptic densities, phosphorylation, 208
    protein phosphorylation, 207
    receptor regulation, 210
    tryptophan metabolism, 202
    tubulin phosphorylation, 208
    tyrosine metabolism, 205
Carbohydrates
    metabolism, 508
        retinal degeneration, 543
    retinal composition, 505
Cardiovascular effects, TRH, 187
Catecholamines, ACTH, 393
Catecholamine synthesis, tyrosine availability, 316
Cell cultures
    astrocytes, 605
        lines, 613
        markers, 609
        monotypic, 596
            cell identification, 602
            cell separation, 602
            cytotoxicity, 664
            media, 604
            primary cultures, 598
    neurons, 616
        cell lines, 625
        cerebellar interneurons, 620
        cortical, 618
        -glial cocultures, 616
        markers, 620
        monotypic cultures, 616
    oligodendrocytes, 626
        cell survival, 634
        density gradient centrifugation, 627
        enzymes, 640
        immunoaffinity, 630
        lipid synthesis, 638
        markers, 635
        monotypic culture, 626
        morphology, 637
        myelin proteins, 642
Cells
    identification, 605
    neuroendocrine, neurosecretion: see Neurosecretion
    photoreceptor, retina, 504
    separation, 602
    see also: Cell cultures
Cerebrum, neurons, cultures, 620
Chloride, intraocular fluids, 475
Cholecystokinin
    analgesia, 121
    CNS depression, 122
    converting enzymes, 118
    distribution in CNS, 116

Cholecystokinin (cont.)
  hormone release, 123
  hyperglycemia, 121
  hypothermia, 121
  monoamines, 127
  neurons, 120
  receptors, 119
  rotational syndrome, 122
  satiety, 124
Choline, cholinergic systems and peptidergic systems, 263
Clones, cDNA, nucleotide sequencing for β-lipotropin precursor, 147
Colchicine, gonadal steroids, 40
Compartments, intraocular fluids, 470
Cortex, neuronal cultures, 618
Cortex, pain transmission, 429
Corticosterone, cellular localization, 54
Corticosterone, nuclear localization, 58
Cultures
  cells: see Cell cultures
  monotypic: see Cell cultures, monotypic
  primary: see Cell cultures, primary
Cyclic nucleotide metabolism
  adenylate and guanylate cyclases, 533
  retinal degeneration, 546
  phosphodiesterase, 535
Cyclic nucleotide phosphodiesterase, learning impairment, 366
Cytotoxicity, cell cultures, 664

Degeneration, retinal, 543
Dentate gyrus, brain slices, 334
Depression, CNS, cholecystokinin, 122
Dexamethasone, cellular localization, 54
Diabetic retinopathy, 543
Diencephalon, pain transmission, 428
Differentiation, neurons, thyroid hormones, 8
DNA
  binding, steroid receptors, 51
  cloned cDNA sequencing, β-lipotropin precursor, 147
  learning impairment, 361, 365
Dolichols, retina, 522
Dopamine
  dopaminergic systems and peptidergic systems, 260
  neurotensin interaction, 105
  receptors, calmodulin, 211
  retina, 530
Dorsal horn systems, pain transmission, 419
Drugs
  behavior, 352
  calmodulin-dependent phosphorylation, membrane proteins, 209
  glycerol metabolism, retina, 518

Egg-laying hormone, neurosecretion, 291, 300
Electrophysiology, neurotensin, 96
Endocrine effects, TRH, 185
Energy metabolism, sleep neurochemistry, 442
Enzymes
  intraocular fluids, 489
  myelin, oligodendrocyte culture, 640
  steroid effects, 74

Fatty acid deficiency, retina, 540
Fatty acid metabolism, retina, 513
Funicular pathways, pain transmission, 423

GABA, retina, 529
Ganglioside metabolism, retina, 520
Gastrin
  central effects, 225
  endocrine effects, 226
  G.I. tract effects, 225
  peripheral effects, 225
Gastrointestinal effects, TRH, 187
Gastrointestinal polypeptides, 217
  see also: individual polypeptides
Genomic mechanisms, gonadal steroids, 32
Glucose, intraocular fluids, 481
Glycerol metabolism, drug effects, 518
Glycerolipids, metabolism, retina, 513
Glycolysis, retina, 508
Glycopeptides, precursors, 291
Golgi, neurosecretion, 285
Gonadal steroids
  acetylcholinesterase, 43
  activational effects, 30
  antisomycin, 32
  axonal transport, 40
  colchicine, 40
  genomic mechanisms, 32
  hypothalamic nuclei, 35
  memory formation, 34, 43
  neural gene expression, 32, 41
  organizational actions, 40
  proteins
    identification, 35
    induction, 35
    steroid modification, 35
    synthesis, 32
  receptors
    estrogen, 35, 41
    5-HT, 40
    muscarinic cholinergic, 38
    progestin, 41
    progesterone, 30
    translocation to nucleus, 42
  RNA synthesis, 32
  sexual behavior, 29

Gonadal steroids (*cont.*)
  synaptic properties of cells, 40, 43
  target cells, 34
  testosterone, 30
  ventromedial nucleus, 35
Grooming, excessive, ACTH, 381
Guanylate cyclase, retina, 508

Hormones
  adrenocortical: *see* Adrenocortical hormones
  plasma amino acid pattern, 313
  release, cholecystokinin, 123
  thyrotropin-releasing: *see* Thyrotropin-releasing hormone
Hyperglycemia, cholecystokinin, 121
Hypothalamus
  nuclei, gonadal steroids, 35
  peptidergic systems, 249
  peptides, 249
    enkephalins, 253
    hypophysiotropic hormones, 251
    oxytocin, 250
    pro-opiocortin-related, 252
    vasopressin, 250
Hypothermia, cholecystokinin, 121

Immunoaffinity, oligodendrocyte isolation, 630
Imprinting, behavior, 349
  learning impairment, 358, 361
Insulin
  binding in CNS, 235
  distribution, 235
  receptors in brain, 235
Intraocular fluids, 469
  blood-ocular barrier systems, transport function, 474
  compartments, 470
  composition
    amino acids, 486
    ascorbic acid, 484
    bicarbonate, 475
    calcium, 478
    chloride, 475
    enzymes, 489
    glucose, 481
    lactic acid, 481
    magnesium, 478
    osmolality, 488
    phosphate, 481
    potassium, 478
    proteins, 489
    sodium, 475
  concentration gradients, 473

Intraocular fluids (*cont.*)
  retina
    homeostasis of microenvironment, 491
    physiological solutions for research, 492
    transport, function of blood-ocular barrier, 474
Ischemia, retina, 542

Kinases, B-50 protein, ACTH, 398
Kindling, behavior, 348

Lactic acid, intraocular fluids, 481
Lateral geniculate, slices, 337
Lateral olfactory tract, slices, 332
Learning impairment
  ACTH, 359
  brain lesion, 359
  cyclic nucleotide phosphodiesterase, 366
  DNA content of brain, 365
  DNA synthesis inhibition, 361
  imprinting, 358, 361
  memory substances, 359
  monoamines, 366
  protein synthesis, 357
    inhibition, 359
  RNA synthesis, 357
    inhibition, 359
Light, effect on retina, 538
Lipids
  peroxides, retinal degeneration, 545
  phosphorylation, ACTH, 399
  retina, 508
  synthesis, oligodendrocyte cultures, 638
β-Lipotropin
  amino acid sequence, 138
  biosynthesis, 145
  conformation, 142
  intermediates, 147
  peripheral occurrence, 151
  pituitary, 145, 151
  precursor processing, proteinases, 150
  properties, 142
  sequencing, cloned cDNA and intermediates, 147
  structure, 138

Magnesium, intraocular fluids, 478
Media, monotypic cell cultures, 604
Medulla, pain transmission, 426
Membrane proteins, calmodulin stimulation of phosphorylation, 206
  drug effects, 209
Membranes
  fluidity, ACTH, 401
  neurosecretion, electrophysiology, 286

Index

Membranes (*cont.*)
  NSG-axolemma fusion, 298
  recycling, 285
  see also: Neurosecretion
Memory, 343
  formation, gonadal steroids, 34, 43
  learning impairment, 359
  see also: Behavior
Mesencephalon, pain transmission, 427
Metabolism
  energy, sleep neurochemistry, 442
  intermediary, sleep neurochemistry, 442
  neurotensin, 95
  TRH, 175
Microtubules, binding agents, behavior, 349
Modulation, TRH, 175
Modulatory systems, pain transmission, 432
Molluscan preparation, behavior, 346
Monoamines
  cholecystokinin, 127
  learning impairment, 366
  synthesis control
    precursor availability, 309
    tryptophan availability and serotonin synthesis, 315
    tryosine availability and catecholamine synthesis, 316
Motilin
  action in CNS, 227
  distribution, 226
  effect on electrical activity of brain, 227
  GI tract effects, 227
  growth hormone effects, 227
  peripheral effects, 227
Motor activity, neurotensin effect, 98
Muscle tone, neurotensin effect, 98
Myelin, phosphorylation, calmodulin, 209
Myelin proteins, oligodendrocyte cultures, 642
Myelination, thyroid hormones, 12

Neocortex, slices, 332
Nerve cells, steroid interaction, 53
Neuroendocrinology, steroid control, 76
Neurons
  calmodulin regulation of function, 201
  cholecystokinin, 120
  cultures: *see* Cell cultures, neurons
  gelatinosa neurons, pain transmission, 420
  lamina V neurons, pain transmission, 421
  neurosecretion: *see* Neurosecretion
Neuropeptides, retina, 531
Neuropeptides, steroid effects, 73
Neurophysin, neurosecretion, 289
Neuropils, development, thyroid hormones, 10
Neurons, differentiation, thyroid hormones, 8

Neurosecretion
  axonal transport, 297
  Ca^{2+} dependence, 286, 298
  egg-laying hormone, 291, 300
  electrophysiology, 286
    action potentials, 286
    Ca^{2+} conductance, 286
    K^+ channels, 287
    membranes, 287
  granules, neurosecretory (NSG), 283, 298
  membranes
    NSG-axolemma fusion, 298
    recycling, 285
  morphology, 283
    Aplysia, 284
    axonal branching, 283
    bag cells, 284
    Golgi, 285
    neuroendocrine cells, 283
    neurohypophyseal neurons, 284
    neurosecretory granules (NSG), 284
    peptidergic neurons, 283
    perikaryal regions, 285
    RNA, 285
    synaptic role of peptides, 283
    transmitter-secretory cells, 285
    transmitter synthesis, 285
    vesicles, 285
  protein synthesis, 285
  RNA, 285, 288, 300
  synthesis of neurosecretory products, 288
    Aplysia, 288
    enzymatic cleavage (converting enzymes), 294
    glycopeptide precursors, 291
    neurophysin, 289
    oxytocin, 289
    peptide precursors, 288
    pre-prohormones, 288
    proinsulin, 288
    ribosomes, 288
    RNA, 288
    vasopressin, 289
  transmitters, 285
Neurosecretory granules (NSG)
  -axolemma fusion, 298
  neurosecretion: *see* Neurosecretion
Neurotensin
  distribution, 93
  dopamine interaction, 105
  electrophysiological studies, 96
  metabolism in CNS, 95
  motor activity, 98
  muscular tone, 98
  neurotransmitters, 104

Neurotransmitters
  brain slices, 330
  development, thyroid hormones, 12
  monoamines, synthesis control, 309
  neurotensin: see Neurotensin
  retina, 524, 588
  steroid effects, 69
Nigrostriatal system, interaction with peptidergic systems, 261
Noradrenergic systems, interaction with peptidergic systems, 262
Nuclei, uptake of steroids, slices, 58
Nucleic acids
  retinal composition, 506
  retinal degeneration, 544
  sleep neurochemistry, 456
Nutrition, retina, 540

Olfactory cortex, slices, 332
Oligodendrocytes, cultures: see Cell cultures, oligodendrocytes
Ontogeny, adrenal steroid receptors, 67
Optic tract, slices, 337
Osmolality, intraocular fluids, 488
Oxytocin, neurosecretion, 289

Pain transmission
  cortex, 429
  diencephalon, 428
  dorsal horn systems, 419
  dorsofunicular pathways, 424
  funicular pathways, 423
  gelatinosa neurons, 420
  intersegmental links, 424
  "lamina V" neurons, 421
  marginal cells, 420
  medulla, 426
  mesencephalon, 427
  modulatory systems, 432
    spinal cord, 432
  primary afferents, 414
  rostral projection system, 430
  supraspinal systems, 425
Pancreatic polypeptide
  CNS effects, 234
  distribution, 232
  GI tract effects, 234
  peripheral effects, 234
  respiration effects, 234
Peptidergic systems, 243
  interactions with other systems, 260
    aminergic, 260
    amino acid, 264
    cholinergic, 263
    dopaminergic, 260
    nigrostriatal, 261

Peptidergic systems (cont.)
  interactions with other systems (cont.)
    non-monoaminergic, 263
    noradrenergic systems, 262
    peptide-peptide, 264
    serotonergic, 263
    tuberofundibular, 260
  localization, 247
    enkephalin, 253
    hypophysiotropic hormones, 251
    hypothalamic origin, 249
    oxytocin, 250
    peripheral nerve, 257
    pituitary origin, 249
    pro-opiocortin-related peptides, 252
    spinal cord, 257
    vasopressin, 250
  neuroregulation, 258
  receptors, 266
  RNA
    content, 271
    labeling, 271
    protein synthesis, 272
  second messengers, 266
  trophic functions, 270
Peptides
  behavior, 350
  hypothalamic: see Hypothalamus, peptides
  as neuroregulators, 258
  neurosecretion: see Neurosecretion
  peptidergic systems: see Peptidergic systems
  -peptide interaction, 264
  peripheral nerve, 257
  pituitary, 247
  spinal cord, 257
  trophic functions, 270
Peripheral nerve, peptidergic systems, 257
Phosphate, intraocular fluids, 481
Phosphodiesterase, retina, 535
Phosphoprotein B-50, ACTH, 398
Phosphorylation, membrane proteins, calmodulin stimulation, 206
Photoreceptor cells, retina, 504
Piriform lobe, slices, 332
Pituitary
  hypophysiotropic hormones in hypothalamus, 251
  β-lipotropin biosynthesis, 145
  β-lipotropin occurrence, 151
  peptidergic systems, 247
  peptides, 247
  prolactin regulation, 161
Polypeptides, gastrointestinal, 217
Postsynaptic densities, phosphorylation, calmodulin stimulation, 208

Index

Potassium, intraocular fluids, 478
Potassium channels, 287
Prohormones, neurosecretion: see Neurosecretion
Proinsulin, neurosecretion, 288
Prolactin
  degradation, 164
  pituitary regulation, 161
  -producing cells, 162
  regulation, 161
    production, 164
    release, 164
  release, 164
  secretion
    man, 165
    rats, 166
    regulation, 167
  storage, 163
  synthesis, 159, 163
Prostaglandins, retina, 521
Proteinases, β-lipotropin precursor processing, 150
Proteins
  gonadal steroids: see Gonadal steroids, proteins
  intraocular fluids, 489
  metabolism, retinal degeneration, 506
  phosphorylation
    ACTH, 395
    behavior, 350
    calmodulin stimulation, 207
    retinal composition, 506
    steroid effects, 73
  synthesis
    behavior, 350
    learning impairment, 357
    thyroid hormones, 19
Pyroglutamate aminopeptidase, TRH metabolism, 181

Receptors
  adrenal steroids, 54
  calmodulin regulation, 210
    dopamine, 211
  cholecystokinin, 119
  gonadal steroids: see Gonadal steroids, receptors
  peptidergic systems, 266
  thyroid hormones, 17
Regeneration, retina, 547
Respiration, carbohydrates, retina, 508
Respiratory effects, TRH, 188
Retina
  anatomy, 569
  axoplasmic transport, 539

Retina (cont.)
  carbohydrates
    composition and metabolism, 508
    glycolysis, 508
    respiration, 508
    retinal degeneration, 543
  chromophores, 579
  cyclic neucleotides
    adenylate and guanylate cyclases, 533
    phosphodiesterase, 535
    retinal degeneration, 546
  degeneration, 543
  diabetic retinopathy, 543
  homeostasis of microenvironment
    avian, 491
    mammalian, 491
  ischemia, 542
  light effect, 538
  lipids
    composition and metabolism, 508
    dolichols, 522
    gangliosides, 520
    glycerol, drug effects on, 518
    glycerolipids and fatty acids, 513
    prostaglandins, 521
    retinal degeneration, 545
    sterols, 519
  neurotransmitters, 524, 588
    acetylcholine, 527
    amino acids, 530
    dopamine, 530
    GABA, 529
    neuropeptides, 531
    serotonin, 531
  nucleic acids
    composition, 506
    retinal degeneration, 544
  nutritional deficiencies
    essential fatty acids, 540
    taurine, 542
    vitamin A, 540
    vitamin E, 540
    zinc, 542
  photoreceptors
    assembly, 571
    breakdown, 571
    cells, 504
    disk membranes, 574
  physiological solutions for research, 492
  physiology, 569
  proteins
    composition and metabolism, 506
    retinal degeneration, 544
  regeneration, 547
  rhodopsin, 576
  visual cell renewal, 537

Rhodopsin: see Retina, photoreceptors
Ribonucleic acid (RNA)
    neurosecretion: see Neurosecretion
    synthesis
        gonadal steroids, 32
        learning impairment, 357

Second messengers, peptidergic systems, 266
Secretin
    distribution, 220
    effect on endocrine system, 220
    immunoreactivity, 220
    transmitter modulation, 220
Sequencing
    biosynthetic intermediates of β-lipotropin, 147
    cloned cDNA for β-lipoprotein precursor, 147
Serotonergic systems, interaction with peptidergic systems, 263
Serotonin
    ACTH, 394
    retina, 531
    synthesis, tryptophan availability, 315
Sleep
    amino acids, 450
    deprivation, 442, 453, 456
    energy metabolism, 442
    intermediary metabolism, 442
    nucleic acids, 456
    proteins, 450
    sleep-inducing factors, 459
Sodium, intraocular fluids, 475
Spinal cord
    pain transmission, 432
        modulatory systems, 432
    peptides, 257
    thyrotropin-releasing hormone (TRH), 188
Steroids, adrenal: see Adrenal steroids
Sterols, biosynthesis, retina, 519
Superfusion, brain slices, 329
Superior colliculus, brain slices, 337
Supraspinal systems, pain transmission, 425
Synapses
    gonadal steroids, 40, 43
    plasma membrane, ACTH, 397, 401
    synaptogenesis, thyroid hormones, 12
Synthesis, monoamines, precursor availability, 309
Synthesis, prolactin, 159, 163

Taurine, deficiency, retina, 542
Target cells, gonadal steroids, 34
Temperature, body, neurotensin effect, 97
Testosterone, gonadal steroids, 30

Thyroid hormones
    -binding proteins, 17
    cell acquisition, 9
    cell division, 9
    cell migration, 10
    development of the neuropil, 10
    mechanism of action, 14
        metabolism, 14
    myelination, 12
    neuronal differentiation, 8
    protein synthesis, 19
    receptors, 17
    secretion, 4
    synaptogenesis, 12
    synthesis, 3
    thyroxine conversion to triiodothyronine, 16
    transmitter development, 12
    transport, 15
    uptake, 14
Thyrotropin-releasing hormone (TRH)
    behavioral effects, 184
    cardiovascular effects, 187
    distribution, 175
        brain, 188
        spinal cord, 188
    endocrine effects, 185
    gastrointestinal effects, 187
    metabolism, 175
    modulation, 175
    pyroglutamate aminopeptidase, 181
    receptor, 191
    respiratory effects, 188
    synaptosomal localization, 190
    TRH-amidase, 181
    uptake and release, 190
Transmitters: see Neurotransmitters
Transport
    axonal: see Axonal transport
    blood-ocular barrier, 474
    thyroid hormones, 15
Trophic functions, peptidergic systems, 270
Tryptophan
    availability, and monoamine synthesis, 315
        serotonin synthesis, 315
    metabolism, calmodulin, 202
    sources, 309
    uptake by brain, 312
Tyrosine
    availability, and monoamine synthesis, 316
        catecholamine synthesis, 316
    calmodulin, 205
    source, 309
    uptake by brain, 312
Tubero-infundibular system, interaction with peptidergic systems, 260

Tubulin phosphorylation, calmodulin
    stimulation, 208

Uptake
    cellular, adrenocortical hormones, 49
    TRH, 190
    tryptophan monoamine precursor
        availability, 312
    tyrosine monoamine precursor availability,
        312

Vasoactive intestinal polypeptide (VIP)
    cAMP effects, 222
    central effects, 222
    distribution, 221

Vasoactive intestinal polypeptide (cont.)
    endocrine effects, 222
    GI tract effects, 221
    neurotransmitter, 221
    peripheral action, 221
    receptor binding, 222
    salivary action, 221
    urogenital effects, 222
Vasopressin, neurosecretion, 289
Vesicles, neurosecretion; see Neurosecretion
Vision: see Retina
Visual cells, renewal, retina, 537
Vitamin A deficiency, retina, 540
Vitamin E deficiency, retina, 540

Zinc deficiency, retina, 542

If you have any concerns about our products,
you can contact us on
ProductSafety@springernature.com

In case Publisher is established outside the EU,
the EU authorized representative is:
**Springer Nature Customer Service Center GmbH
Europaplatz 3, 69115 Heidelberg, Germany**

Printed by Libri Plureos GmbH
in Hamburg, Germany